LIBRARY
Lyndon State College
Lyndonville, VT 05851

D1605040

Who Gave You the Epsilon?

and Other Tales of Mathematical History

Copyright © 2009 by
The Mathematical Association of America

ISBN 978-0-88385-569-0

Library of Congress number 2008935288

Printed in the United States of America

Current printing (last digit):
10 9 8 7 6 5 4 3 2 1

Who Gave you the Epsilon?

and Other Tales of Mathematical History

Edited by

Marlow Anderson
Colorado College

Victor Katz
University of the District of Columbia

Robin Wilson
Open University

Published and Distributed by
The Mathematical Association of America

SPECTRUM SERIES

The Spectrum Series of the Mathematical Association of America was so named to reflect its purpose: to publish a broad range of books including biographies, accessible expositions of old or new mathematical ideas, reprints and revisions of excellent out-of-print books, popular works, and other monographs of high interest that will appeal to a broad range of readers, including students and teachers of mathematics, mathematical amateurs, and researchers.

Coordinating Council on Publications
Paul Zorn, *Chair*

Spectrum Editorial Board
Gerald L. Alexanderson, *Editor*

William Dunham	Edward W. Packel
Richard K. Guy	Kenneth A. Ross
Michael A. Jones	Marvin Schaefer
Keith M. Kendig	Sanford Segal
Jeffrey L. Nunemacher	Franklin F. Sheehan

777 Mathematical Conversation Starters, by John de Pillis

99 Points of Intersection: Examples—Pictures—Proofs, by Hans Walser. Translated from the original German by Peter Hilton and Jean Pedersen.

All the Math That's Fit to Print, by Keith Devlin

Calculus Gems: Brief Lives and Memorable Mathematics, by George F. Simmons

Carl Friedrich Gauss: Titan of Science, by G. Waldo Dunnington, with additional material by Jeremy Gray and Fritz-Egbert Dohse

The Changing Space of Geometry, edited by Chris Pritchard

Circles: A Mathematical View, by Dan Pedoe

Complex Numbers and Geometry, by Liang-shin Hahn

Cryptology, by Albrecht Beutelspacher

The Early Mathematics of Leonhard Euler, by C. Edward Sandifer

The Edge of the Universe: Celebrating 10 Years of Math Horizons, edited by Deanna Haunsperger and Stephen Kennedy

Euler and Modern Science, edited by N. N. Bogolyubov, G. K. Mikhailov, and A. P. Yushkevich. Translated from Russian by Robert Burns.

Euler at 300: An Appreciation, edited by Robert E. Bradley, Lawrence A. D'Antonio, and C. Edward Sandifer

Five Hundred Mathematical Challenges, Edward J. Barbeau, Murray S. Klamkin, and William O. J. Moser

The Genius of Euler: Reflections on his Life and Work, edited by William Dunham

The Golden Section, by Hans Walser. Translated from the original German by Peter Hilton, with the assistance of Jean Pedersen.

The Harmony of the World: 75 Years of Mathematics Magazine, edited by Gerald L. Alexanderson with the assistance of Peter Ross

How Euler Did It, by C. Edward Sandifer

Is Mathematics Inevitable? A Miscellany, edited by Underwood Dudley

I Want to Be a Mathematician, by Paul R. Halmos

Journey into Geometries, by Marta Sved

JULIA: a life in mathematics, by Constance Reid

The Lighter Side of Mathematics: Proceedings of the Eugène Strens Memorial Conference on Recreational Mathematics & Its History, edited by Richard K. Guy and Robert E. Woodrow

Lure of the Integers, by Joe Roberts

Magic Numbers of the Professor, by Owen O'Shea and Underwood Dudley

Magic Tricks, Card Shuffling, and Dynamic Computer Memories: The Mathematics of the Perfect Shuffle, by S. Brent Morris

Martin Gardner's Mathematical Games: The entire collection of his Scientific American columns

The Math Chat Book, by Frank Morgan

Mathematical Adventures for Students and Amateurs, edited by David Hayes and Tatiana Shubin. With the assistance of Gerald L. Alexanderson and Peter Ross.

Mathematical Apocrypha, by Steven G. Krantz

Mathematical Apocrypha Redux, by Steven G. Krantz

Mathematical Carnival, by Martin Gardner

Mathematical Circles Vol I: In Mathematical Circles Quadrants I, II, III, IV, by Howard W. Eves

Mathematical Circles Vol II: Mathematical Circles Revisited and Mathematical Circles Squared, by Howard W. Eves

Mathematical Circles Vol III: Mathematical Circles Adieu and Return to Mathematical Circles, by Howard W. Eves

Mathematical Circus, by Martin Gardner

Mathematical Cranks, by Underwood Dudley

Mathematical Evolutions, edited by Abe Shenitzer and John Stillwell

Mathematical Fallacies, Flaws, and Flimflam, by Edward J. Barbeau

Mathematical Magic Show, by Martin Gardner

Mathematical Reminiscences, by Howard Eves

Mathematical Treks: From Surreal Numbers to Magic Circles, by Ivars Peterson

Mathematics: Queen and Servant of Science, by E.T. Bell

Memorabilia Mathematica, by Robert Edouard Moritz

Musings of the Masters: An Anthology of Mathematical Reflections, edited by Raymond G. Ayoub

New Mathematical Diversions, by Martin Gardner

Non-Euclidean Geometry, by H. S. M. Coxeter

Numerical Methods That Work, by Forman Acton

Numerology or What Pythagoras Wrought, by Underwood Dudley

Out of the Mouths of Mathematicians, by Rosemary Schmalz

Penrose Tiles to Trapdoor Ciphers ... and the Return of Dr. Matrix, by Martin Gardner

Polyominoes, by George Martin

Power Play, by Edward J. Barbeau

Proof and Other Dilemmas: Mathematics and Philosophy, edited by Bonnie Gold and Roger Simons

The Random Walks of George Pólya, by Gerald L. Alexanderson

Remarkable Mathematicians, from Euler to von Neumann, Ioan James

The Search for E.T. Bell, also known as John Taine, by Constance Reid

Shaping Space, edited by Marjorie Senechal and George Fleck

Sherlock Holmes in Babylon and Other Tales of Mathematical History, edited by Marlow Anderson, Victor Katz, and Robin Wilson

Student Research Projects in Calculus, by Marcus Cohen, Arthur Knoebel, Edward D. Gaughan, Douglas S. Kurtz, and David Pengelley

Symmetry, by Hans Walser. Translated from the original German by Peter Hilton, with the assistance of Jean Pedersen.

The Trisectors, by Underwood Dudley

Twenty Years Before the Blackboard, by Michael Stueben with Diane Sandford

Who Gave You the Epsilon? and Other Tales of Mathematical History, edited by Marlow Anderson, Victor Katz, and Robin Wilson

The Words of Mathematics, by Steven Schwartzman

MAA Service Center
P.O. Box 91112
Washington, DC 20090-1112
800-331-1622   FAX 301-206-9789

Introduction

For over one hundred years, the Mathematical Association of America has been publishing high-quality articles on the history of mathematics, some written by distinguished historians and mathematicians such as J. L. Coolidge, B. L. van der Waerden, Hermann Weyl and G. H. Hardy. Many well-known historians of the present day also contribute to the MAA's journals, such as Ivor Grattan-Guinness, Judith Grabiner, Israel Kleiner and Karen Parshall.

Some years ago, we decided that it would be useful to reprint a selection of these papers and to set them in the context of modern historical research, so that current mathematicians can continue to enjoy them and so that newer articles can be easily compared with older ones. The result was our MAA volume *Sherlock Holmes in Babylon*, which took the story from earliest times up to the time of Euler in the eighteenth century. The current volume is a sequel to our earlier one, and continues with topics from the nineteenth and twentieth centuries. We hope that you will enjoy this second collection.

A careful reading of some of the older papers shows that although modern research has introduced some new information or has fostered some new interpretations, in large measure they are neither dated or obsolete. Nevertheless, we have sometimes decided to include two or more papers on a single topic, written years apart, to show the progress in the history of mathematics.

We wish to thank Don Albers, Director of Publications at the MAA, and Gerald Alexanderson, former chair of the publications committee of the MAA, for their support for the history of mathematics at the MAA in general, and for this project in particular. We also want to thank Beverly Ruedi for her technical expertise in preparing this volume for publication.

Contents

Introduction vii

Analysis
Foreword 3
Who Gave You the Epsilon? Cauchy and the Origins of Rigorous Calculus, *Judith V. Grabiner* 5
Evolution of the Function Concept: A Brief Survey, *Israel Kleiner* 14
S. Kovalevsky: A Mathematical Lesson, *Karen D. Rappaport* 27
Highlights in the History of Spectral Theory, *L. A. Steen* 36
Alan Turing and the Central Limit Theorem, *S. L. Zabell* 52
Why did George Green Write his Essay of 1828 on Electricity and Magnetism?,
   I. Grattan-Guinness 61
Connectivity and Smoke-Rings: Green's Second Identity in its First Fifty Years,
   Thomas Archibald 69
The History of Stokes' Theorem, *Victor J. Katz* 78
The Mathematical Collaboration of M. L. Cartwright and J. E. Littlewood,
   Shawnee L. McMurran and James J. Tattersall 88
Dr. David Harold Blackwell, African American Pioneer, *Nkechi Agwu, Luella Smith and
   Aissatou Barry* 98
Afterword 109

Geometry, Topology and Foundations
Foreword 113
Gauss and the Non-Euclidean Geometry, *George Bruce Halsted* 115
History of the Parallel Postulate, *Florence P. Lewis* 120
The Rise and Fall of Projective Geometry, *J. L. Coolidge* 125
Notes on the History of Geometrical Ideas, *Dan Pedoe* 133
A note on the history of the Cantor set and Cantor function, *Julian F. Fleron* 137
Evolution of the Topological Concept of "Connected", *R. L. Wilder* 142
A Brief, Subjective History of Homology and Homotopy Theory in this Century, *Peter Hilton* ... 148
The Origins of Modern Axiomatics: Pasch to Peano, *H. C. Kennedy* 157
C. S. Peirce's Philosophy of Infinite Sets, *Joseph W. Dauben* 161
On the Development of Logics between the two World Wars, *I. Grattan-Guinness* 172
Dedekind's Theorem: $\sqrt{2} \times \sqrt{3} = \sqrt{6}$, *David Fowler* 185
Afterword 192

Algebra and Number Theory

Foreword 197
Hamilton's Discovery of Quaternions, *B. L. van der Waerden* 200
Hamilton, Rodrigues, and the Quaternion Scandal, *Simon L. Altmann* 206
Building an International Reputation: The Case of J. J. Sylvester (1814–1897),
    Karen Hunger Parshall and Eugene Seneta 220
The Foundation Period in the History of Group Theory, *Josephine E. Burns* 230
The Evolution of Group Theory: A Brief Survey, *Israel Kleiner* 237
The Search for Finite Simple Groups, *Joseph A. Gallian* 254
Genius and Biographers: The Fictionalization of Evariste Galois, *Tony Rothman* 271
Hermann Grassmann and the Creation of Linear Algebra, *Desmond Fearnley-Sander* 291
The Roots of Commutative Algebra in Algebraic Number Theory, *Israel Kleiner* 299
Eisenstein's Misunderstood Geometric Proof of the Quadratic Reciprocity Theorem,
    Reinhard C. Laubenbacher and David J. Pengelley 309
Waring's Problem, *Charles Small* 313
A History of the Prime Number Theorem, *L. J. Goldstein* 318
A Hundred Years of Prime Numbers, *Paul T. Bateman and Harold G. Diamond* 328
The Indian Mathematician Ramanujan, *G. H. Hardy* 337
Emmy Noether, *Clark H. Kimberling* 349
"A Marvelous Proof," *Fernando Q. Gouvêa* 360
Afterword 375

Surveys

Foreword 381
The International Congress of Mathematicians, *George Bruce Halsted* 383
A Popular Account of some New Fields of Thought in Mathematics, *G. A. Miller* 385
A Half-century of Mathematics, *Hermann Weyl* 391
Mathematics at the Turn of the Millennium, *Philip A. Griffiths* 411
Afterword 423

Index   427
About the Editors   431

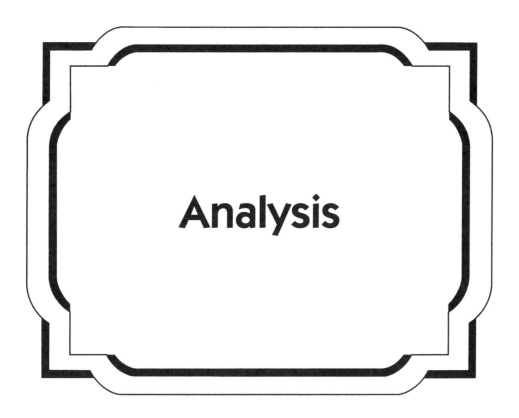

Foreword

In this chapter, we look at some critical ideas in the history of analysis in the nineteenth and twentieth centuries and also consider some of the people involved in their creation.

Certainly, the most important idea in the rigorous study of calculus is the idea of a *limit*. Although Newton and Leibniz both understood this idea heuristically and used it in their discussions of fluxions and differentials, respectively, it was not until the nineteenth century that our modern treatment in terms of epsilons and deltas was created, mostly by Augustin-Louis Cauchy. Judy Grabiner explores this creation in her article, showing how Cauchy took the ideas of approximation worked out by d'Alembert, Lagrange, and others in the eighteenth century and turned them into the foundations of the rigorous definitions of the central concepts of calculus, including limit, continuity, and convergence.

Today, *functions* are the basic objects studied in calculus, although they were not so for Newton and Leibniz. Israel Kleiner surveys the history of this concept, showing how some of the vague ideas of the creators of calculus were turned into precise definitions by Euler and others. He continues by discussing how the notion of function developed under the influence of the development of Fourier series and then was further affected by the invention of various pathological functions in the second half of the nineteenth century. Kleiner concludes with a treatment of the various extensions of the notion of function in the twentieth century, including L_2 functions and distributions.

The set of L_2 functions was one of the first examples of a *Hilbert space*. Lynn Steen discusses the history of this concept, concentrating on the notion of the spectrum of a linear operator. Beginning with the development of the principal axis theorem in two and three dimensions in connection with quadratic forms, Steen follows the thread of "diagonalization" through such concepts as infinite systems of linear equations and the theory of integral equations into the work of Hilbert himself. He then continues with the work of numerous mathematicians in the twentieth century, showing in particular how spectral theory became central to the development of quantum mechanics in the 1920s.

Three articles in this chapter deal with potential theory, whose mathematical foundations begin with George Green's essay on electricity and magnetism of 1828. Ivor Grattan-Guinness looks at the circumstances surrounding the writing of Green's essay, discussing his motivation, his sources, and the influence the essay had. Thomas Archibald looks at the history of one particular result from this essay, now called Green's second identity, and traces its importance in Riemann's and later Thomson's conceptions of multiple-connectivity, in Helmholtz's work in fluid dynamics, and in Tait's reworking of the results through the use of quaternions. Finally, Katz looks at the history of the various results that were unified in the twentieth century into what is now called the generalized

Stokes's theorem. These include the three theorems that are studied in any multi-variable calculus course: Green's theorem, the divergence theorem, and the original Stokes's theorem.

Each of the other articles in this chapter each deal specifically with the work of a particular mathematician who contributed greatly to analysis in the nineteenth and twentieth century. Karen Rappaport gives us details on the life and work of Sonya Kovalevsky, from studying calculus from the wallpaper in her nursery to her important work in partial differential equations and the motion of rigid bodies. S. L. Zabell looks at some relatively unknown work by Alan Turing on the central limit theorem, accomplished before his crucial work on the decision problem convinced him to work on problems in logic and led ultimately to his accomplishments at Bletchley Park during the Second World War. Shawnee McMurran and James Tattersall look at the mathematical work of Dame Mary Cartwright, particularly her collaborative work with James Littlewood dealing with the solution of the van der Pol differential equation.

Finally, Nkechi Agwu, Luell Smith, and Aissatou Barry present the biography of David Blackwell, the first prominent African-American mathematician, best known for his work in statistics. The authors show how he overcame the prejudices of his time not only to achieve a Ph.D. in mathematics but also to become a full professor of statistics at the University of California, Berkeley and later Chair of the department.

Who Gave You the Epsilon? Cauchy and the Origins of Rigorous Calculus

JUDITH V. GRABINER

American Mathematical Monthly **90** (1983), 185–194

Student: The car has a speed of 50 miles an hour. What does that mean?

Teacher: Given any $\varepsilon > 0$, there exists a δ such that if $|t_2 - t_l| < \delta$, then $\left| \frac{s_2 - s_1}{t_2 - t_1} - 50 \right| < \varepsilon$.

Student: How in the world did anybody ever think of such an answer?

Perhaps this exchange will remind us that the rigorous basis for the calculus is not at all intuitive—in fact, quite the contrary. The calculus is a subject dealing with speeds and distances, with tangents and areas—not inequalities. When Newton and Leibniz invented the calculus in the late seventeenth century, they did not use delta-epsilon proofs. It took a hundred and fifty years to develop them. This means that it was probably very hard, and it is no wonder that a modern student finds the rigorous basis of the calculus difficult. How, then, did the calculus get a rigorous basis in terms of the algebra of inequalities?

Delta-epsilon proofs are first found in the works of Augustin-Louis Cauchy (1789–1867). This is not always recognized, since Cauchy gave a purely verbal definition of limit, which at first glance does not resemble modern definitions ([6], p. 19):

> When the successively attributed values of the same variable indefinitely approach a fixed value, so that finally they differ from it by as little as desired, the last is called the *limit* of all the others.

Cauchy also gave a purely verbal definition of the derivative of $f(x)$ as the limit, when it exists, of the quotient of differences $(f(x + h) - f(x))/h$ when h goes to zero, a statement much like those that had already been made by Newton, Leibniz, d'Alembert, Maclaurin, and Euler. But what is significant is that Cauchy translated such verbal statements into the precise language of inequalities when he needed them in his proofs. For instance, for the derivative Cauchy ([8], p. 44) makes the following statement:

> Let δ, ε be two very small numbers; the first is chosen so that for all numerical [i.e., absolute] values of h less than δ, and for any value of x included [in the interval of definition], the ratio $(f(x + h) - f(x))/h$ will always be greater than $f'(x) - \varepsilon$ and less than $f'(x) + \varepsilon$. ($*$)

This one example will be enough to indicate how Cauchy did the calculus, because the question to be answered in the present paper is not, "how is a rigorous delta-epsilon proof constructed?" As Cauchy's intellectual heirs we all know this. The central question is, how and why was Cauchy able to put the calculus on a rigorous basis, when his predecessors were not?

The answers to this historical question cannot be found by reflecting on the logical relations between the concepts, but by looking in detail at the past and seeing how the existing state of affairs in fact developed from that past. Thus we will examine the mathematical situation in the seventeenth and eighteenth centuries — the background against which we can appreciate Cauchy's innovation. We will describe the powerful techniques of the calculus of this earlier period and the relatively unimpressive views put forth to justify them. We will then discuss how a sense of urgency about rigorizing analysis gradually developed in the eighteenth century. Most important, we will explain the development of the mathematical techniques necessary for the new rigor from the work of men like Euler, d'Alembert, Poisson, and especially Lagrange. Finally, we will show how these mathematical results, though often developed for purposes far removed from establishing foundations for the calculus, were used by Cauchy in constructing his new rigorous analysis.

The practice of analysis: from Newton to Euler

In the late seventeenth century, Newton and Leibniz, almost simultaneously, independently invented the calculus. This invention involved three things. First, they invented the general concepts of differential quotient and integral (these are Leibniz's terms; Newton called the concepts "fluxion" and "fluent"). Second, they devised a notation for these concepts which made the calculus an algorithm: the methods not only worked, but were easy to use. Their notations had great heuristic power, and we still use Leibniz's dy/dx and $\int y dx$, and Newton's $\dot{x}$, today. Third, both men realized that the basic processes of finding tangents and areas, that is, differentiating and integrating, are mutually inverse—what we now call the Fundamental Theorem of Calculus.

Once the calculus had been invented, mathematicians possessed an extremely powerful set of methods for solving problems in geometry, in physics, and in pure analysis. But what was the nature of the basic concepts? For Leibniz, the differential quotient was a ratio of infinitesimal differences, and the integral was a sum of infinitesimals. For Newton, the derivative, or fluxion, was described as a rate of change; the integral, or fluent, was its inverse. In fact, throughout the eighteenth century, the integral was generally thought of as the inverse of the differential. One might imagine asking Leibniz exactly what an infinitesimal was, or Newton what a rate of change might be. Newton's answer, the best of the eighteenth century, is instructive. Consider a ratio of finite quantities (in modern notation, $(f(x+h) - f(x))/h$ as h goes to zero). The ratio eventually becomes what Newton called an "ultimate ratio". Ultimate ratios are ([26], p. 39)

> limits to which the ratios of quantities decreasing without limit do always converge, and to which they approach nearer than by any given difference, but never go beyond, nor ever reach until the quantities vanish.

Except for "reaching" the limit when the quantities vanish, we can translate Newton's words into our algebraic language. Newton himself, however, did not do this, nor did most of his followers in the eighteenth century. Moreover, "never go beyond" does not allow a variable to oscillate about its limit. Thus, though Newton's is an intuitively pleasing picture, as it stands it was not and could not be used for proofs about limits. The definition sounds good, but it was not understood or applied in algebraic terms.

But most eighteenth-century mathematicians would object, "Why worry about foundations?" In the eighteenth century, the calculus, intuitively understood and algorithmically executed, was applied to a wide range of problems. For instance, the partial differential equation for vibrating strings was solved; the equations of motion for the solar system were solved; the Laplace transform and the calculus of variations and the gamma function were invented and applied; all of mechanics was worked out in the language of the calculus. These were great achievements on the part of eighteenth-century mathematicians. Who would be greatly concerned about foundations when such important problems could be successfully treated by the calculus? Results were what counted.

This point will be better appreciated by looking at an example which illustrates both the "uncritical" approach to concepts of the eighteenth century and the immense power of eighteenth-century techniques, from the work of the great master of such techniques: Leonhard Euler. The problem is to find the sum of the series

$$\frac{1}{1} + \frac{1}{4} + \frac{1}{9} + \cdots + \frac{1}{k^2} + \cdots.$$

It clearly has a finite sum since it is bounded above by the series

$$1 + \frac{1}{1 \cdot 2} + \frac{1}{2 \cdot 3} + \frac{1}{3 \cdot 4} + \cdots + \frac{1}{(k-1) \cdot k} + \cdots,$$

whose sum was known to be 2; Johann Bernoulli had found this sum by treating

$$1 + \frac{1}{1 \cdot 2} + \frac{1}{2 \cdot 3} + \frac{1}{3 \cdot 4} + \cdots$$

as the difference between the series $\frac{1}{1} + \frac{1}{2} + \frac{1}{3} + \cdots$ and the series $\frac{1}{2} + \frac{1}{3} + \frac{1}{4} + \cdots$, and observing that this difference telescopes ([2], IV, 8; cited in [28], p. 321).

Euler's summation of $\sum 1/k^2$ makes use of a lemma from the theory of equations: given a polynomial equation whose constant term is 1, the coefficient of the linear term is the sum of the reciprocals of the roots with the signs changed. This result was both discovered and demonstrated by considering the equation $(x-a)(x-b) = 0$, having roots a and b. Multiplying and then dividing out ab, we obtain

$$\frac{1}{ab}x^2 - \left(\frac{1}{a} + \frac{1}{b}\right)x + 1 = 0;$$

the result is now obvious, as is the extension to equations of higher degree.

Euler's solution then considers the equation $\sin x = 0$. Expanding this as an infinite series, Euler obtained

$$x - \frac{x^3}{3!} + \frac{x^5}{5!} - \cdots = 0.$$

Dividing by x yields

$$1 - \frac{x^2}{3!} + \frac{x^4}{5!} - \cdots = 0.$$

Finally, substituting $x^2 = u$ produces

$$1 - \frac{u}{3!} + \frac{u^2}{5!} - \cdots = 0.$$

But Euler thought that power series could be manipulated just like polynomials. Thus, we now have a polynomial equation in u, whose constant term is 1. Applying the lemma to it, the coefficient of the linear term with the sign changed is $1/3! = 1/6$. The roots of the equation in u are the roots of $\sin x = 0$ with the substitution $u = x^2$, namely $\pi^2, 4\pi^2, 9\pi^2, \ldots$. Thus the lemma implies

$$\frac{1}{6} = \frac{1}{\pi^2} + \frac{1}{4\pi^2} + \frac{1}{9\pi^2} + \cdots.$$

Multiplying by π^2 yields the sum of the original series ([4], p. 487; [13], (2) Vol. 14, pp. 73–86):

$$\frac{1}{1} + \frac{1}{4} + \frac{1}{9} + \cdots + \frac{1}{k^2} + \cdots = \frac{\pi^2}{6}.$$

Though it is easy to criticize eighteenth-century arguments like this for their lack of rigor, it is also unfair. Foundations, precise specifications of the conditions under which such manipulations with infinities or infinitesimals were admissible, were not very important to men like Euler, because without such specifications they made important new discoveries, whose results in cases like this could readily be verified. When the foundations of the calculus were discussed in the eighteenth century, they were treated as secondary. Discussions of foundations appeared in the introductions to books, in popularizations, and in philosophical writings, and were not—as they are now and have been since Cauchy's time—the subject of articles in research-oriented journals.

Thus, where we once had one question to answer, we now have two. The first remains, where do Cauchy's rigorous techniques come from? Second, one must now ask, why rigorize the calculus in the first place? If few mathematicians were very interested in foundations in the eighteenth century ([16], Chapter 6), then when, and why, were attitudes changed?

Of course, to establish rigor, it is necessary—though not sufficient—to think rigor is significant. But more important, to establish rigor, it is necessary (though also not sufficient) to have a set of techniques in existence which are suitable for that purpose. In particular, if the calculus is to be made rigorous by being reduced to the algebra of inequalities, one must have both the algebra of inequalities, and facts about the concepts of the calculus that can be expressed in terms of the algebra of inequalities.

In the early nineteenth century, three conditions held for the first time: Rigor was considered important; there was a well-developed algebra of inequalities; and, certain properties were known about the basic concepts of analysis—limits, convergence, continuity, derivatives, integrals—properties which could be expressed in the language of inequalities if desired. Cauchy, followed by Riemann and Weierstrass, gave the calculus a rigorous basis, using the already-existing algebra of inequalities, and built a logically-connected structure of theorems about the concepts of the calculus. It is our task to explain how these three conditions—the developed algebra of inequalities, the importance of rigor, the appropriate properties of the concepts of the calculus—came to be.

The algebra of inequalities

Today, the algebra of inequalities is studied in calculus courses because of its use as a basis for the calculus, but why should it have been studied in the eighteenth century when this application was unknown? In the eighteenth century, inequalities were important in the study of a major class of results: approximations.

For example, consider the equation $(x+1)^\mu = a$, for μ not an integer. Usually a cannot be found exactly, but it can be approximated by an infinite series. In general, given some number n of terms of such an approximating series, eighteenth-century mathematicians sought to compute an upper bound on the error in the approximation—that is, the difference between the sum of the series and the nth partial sum.

This computation was a problem in the algebra of inequalities. Jean d'Alembert [10] solved it for the important case of the binomial series; given the number of terms of the series n, and assuming implicitly that the series converges to its sum, he could find the bounds on the error—that is, on the remainder of the series after the nth term—by bounding the series above and below with convergent geometric progressions. Similarly, Joseph-Louis Lagrange [24] invented a new approximation method using continued fractions and, by extremely intricate inequality calculations, gave necessary and sufficient conditions for a given iteration of the approximation to be closer to the result than the previous iteration. Lagrange also derived the Lagrange remainder of the Taylor series [23], using an inequality which bounded the remainder above and below by the maximum and minimum values of the nth derivative and then applying the intermediate-value theorem for continuous functions. Thus, through such eighteenth-century work ([16], pp. 56–68; [14], Chapters 2–4), there was by the end of the eighteenth century a developed algebra of inequalities, and people used to working with it. Given an n, these people are used to finding an error—that is, an epsilon.

Changing attitudes toward rigor

Mathematicians were much more interested in finding rigorous foundations for the calculus in 1800 than they had been a hundred years before. There are many reasons for this: no one enough by itself, but apparently sufficient when acting together. Of course one might think that eighteenth-century mathematicians were always making errors because of the lack of an explicitly-formulated rigorous foundation. But this did not occur. They were usually right, and for two reasons. One is that if one deals with real variables, functions of one variable, series which are power series, and functions arising from physical problems, errors will not occur too often. A second reason is that mathematicians like Euler and Laplace had a deep insight into the basic properties of the concepts of the calculus, and were able to choose fruitful methods and evade pitfalls. The only "error" they committed was to use methods that shocked mathematicians of later ages who had grown up with the rigor of the nineteenth century.

What then were the reasons for the deepened interest in rigor? One set of reasons was philosophical. In 1734, the British philosopher Bishop Berkeley ([1], Section 35) had attacked the calculus on the ground that it was not rigorous. In *The Analyst, or a Discourse Addressed to an Infidel Mathematician*, he said that mathematicians had no business attacking the unreasonableness of religion, given the way they themselves reasoned. He ridiculed fluxions— "velocities of evanescent increments"—calling the evanescent increments "ghosts of departed quantities". Even more to the point, he correctly criticized a number of specific arguments from the writings of his mathematical contemporaries. For instance, he attacked the process of finding the fluxion (our derivative) by reviewing the steps of the process: if we consider $y = x^2$, taking the ratio of the differences $((x+h)^2 - x^2)/h$, then simplifying to $2x + h$, then letting h vanish, we obtain $2x$. But is h zero? If it is, we cannot meaningfully divide by it; if it is not zero, we have no right to throw it away. As Berkeley put it ([1], Section 15), the quantity we have called h "might have signified either an increment or nothing. But then, which of these soever you make it signify, you must argue consistently with such its signification".

Since an adequate response to Berkeley's objections would have involved recognizing that an equation involving limits is a shorthand expression for a sequence of inequalities—a subtle and difficult idea—no eighteenth-century analyst gave a fully adequate answer to Berkeley. However, many tried. Maclaurin, d'Alembert, Lagrange, Lazare Carnot, and possibly Euler, all knew about Berkeley's work, and all wrote something about foundations. So Berkeley did call attention to the question. However, except for Maclaurin, no leading mathematician spent much time on the question because of Berkeley's work, and even Maclaurin's influence lay in other fields.

Another factor contributing to the new interest in rigor was that there was a limit to the number of results that could be reached by eighteenth-century methods. Near the end of the century, some leading mathematicians had begun to feel that this limit was at hand. D'Alembert and Lagrange indicate this in their correspondence, with Lagrange calling higher mathematics "decadent" ([2], Vol. 13, p. 229). The philosopher Diderot went so far as to claim that the mathematicians of the eighteenth century had "erected the pillars of Hercules" beyond which it was impossible to go ([11], pp. 180–181). Thus, there was a perceived need to consolidate the gains of the past century.

Another factor was Lagrange, who became increasingly interested in foundations, and, through his activities, interested other mathematicians. In the eighteenth century, scientific academies offered prizes for solving major outstanding problems. In 1784, Lagrange and his colleagues posed the problem of the foundations of the calculus as the Berlin Academy's prize problem. Nobody solved it to Lagrange's satisfaction, but two of the entries in the competition were later expanded into full-length books, the first on the Continent, on foundations: Simon L'Huilier's *Exposition Elémentaire des Principes des Calculs Supérieurs*, Berlin, 1787, and Lazare Carnot's *Réflexions sur la Métaphysique du Calcul Infinitésimal*, Paris, 1797. Thus Lagrange clearly helped revive interest in the problem.

Lagrange's interest stemmed in part from his respect for the power and generality of algebra; he wanted to gain for the calculus the certainty he believed algebra to possess. But there was another factor increasing interest in foundations, not only for Lagrange, but for many other mathematicians by the end of the eighteenth century: the need to teach. Teaching forces one's attention to basic questions. Yet before the mid-eighteenth century, mathematicians had often made their living by being attached to royal courts. But royal courts declined; the number of mathematicians increased; and mathematics began to look useful. First in military schools and later on at the Ecole Polytechnique in Paris, another line of work became available: teaching mathematics to students of science and engineering. The Ecole Polytechnique was founded by the French revolutionary government to train scientists, who, the government believed, might prove useful to a modern state. And it was as a lecturer in analysis at the Ecole Polytechnique that Lagrange wrote his two major works on the calculus which treated foundations; similarly, it was 40 years earlier, teaching the calculus at the Military Academy at Turin, that Lagrange had first set out to work on the problem of foundations. Because teaching forces one to ask basic questions about the nature of the most important concepts, the change in the economic circumstances of mathematicians—the need to teach—provided a catalyst for the crystallization of the foundations of the calculus out of the historical and mathematical background. In fact, even well into the nineteenth century, much of foundations was born in the teaching situation: Weierstrass's foundations come from his lectures at Berlin; Dedekind first thought of the problem of continuity while teaching at Zurich; Dini and Landau turned to foundations while teaching analysis; and, most important for our present purposes, so did Cauchy. Cauchy's foundations of analysis appear in the books based on his lectures at the Ecole Polytechnique; his book of 1821 was the first example of the great French tradition of *Cours d'Analyse*.

The concepts of the calculus

Arising from algebra, the algebra of inequalities was now there for the calculus to be reduced to; the desire to make the calculus rigorous had arisen through consolidation, through philosophy, through teaching, through Lagrange. Now let us turn to the mathematical substance of eighteenth-century analysis, to see what was known about the concepts of the calculus before Cauchy, and what he had to work out for himself, in order to define, and prove theorems about, limit, convergence, continuity, derivatives, and integrals.

First, consider the concept of limit. As we have already pointed out, since Newton the limit had been thought of as a bound which could be approached closer and closer, though not surpassed. By 1800, with the work of L'Huilier and Lacroix on alternating series, the restriction that the limit be one-sided had been abandoned. Cauchy systematically translated this refined limit-concept into the algebra of inequalities, and used it in proofs once it had been so translated; thus he gave reality to the oft-repeated eighteenth-century statement that the calculus could be based on limits.

For example, consider the concept of convergence. Maclaurin had said already that the sum of a series was the limit of the partial sums. For Cauchy, this meant something precise. It meant that, given an ε, one could find n such that, for more than n terms, the sum of the infinite series is within ε of the nth partial sum. That is the reverse of the error-estimating procedure that d'Alembert had used. From his definition of a series having a sum, Cauchy could prove that a geometric progression with radius less in absolute value than 1 converged to its usual sum. As we have said, d'Alembert had shown that the binomial series for, say, $(1+x)^{p/q}$ could be bounded above and below by convergent geometric progressions. Cauchy assumed that if a series of positive terms is bounded above, term-by-term, by a convergent geometric progression, then it converges; he then used such comparisons to prove a number of tests for convergence: the root test, the ratio test, the logarithm test. The treatment is quite elegant ([6], (2), Vol. 3, pp. 114–138). Taking a technique used a few times by men like d'Alembert and Lagrange on an *ad hoc* basis in approximations, and using the definition of the sum of a series based on the limit concept, Cauchy created the first rigorous theory of convergence.

Let us now turn to the concept of continuity. Cauchy ([6], p. 43) gave essentially the modern definition of continuous function, saying that the function $f(x)$ is continuous on a given interval if for each x in that interval "the numerical [i.e., absolute] value of the difference $f(x + a) - f(x)$ decreases indefinitely with a". He used this definition in proving the intermediate-value theorem for continuous functions ([6], pp. 378–380; English translation in [16], pp. 167–168). The proof proceeds by examining a function $f(x)$ on an interval, say $[b, c]$, where $f(b)$ is negative, $f(c)$ positive, and dividing the interval $[b, c]$ into m parts of width $h = (c - b)/m$. Cauchy considered the sign of the function at the points

$$f(b), f(b + h), \ldots, f(b + (m - 1)h), f(c);$$

unless one of the values of f is 0, there are two values of x differing by h such that f is negative at one, positive at the other. Repeating this process for new intervals of width

$$(c - b)/m, (c - b)/m^2, \ldots,$$

gives an increasing sequence of values of x: $b, b_1, b_2, \ldots$ for which f is negative, and a decreasing sequence of values of x: $c, c_1, c_2, \ldots$ for which f is positive, and such that the difference between b_k and c_k goes to 0. Cauchy asserted that these two sequences must have a common limit a. He then argued that since $f(x)$ is continuous, the sequence of negative values $f(b_k)$ and of positive values $f(c_k)$ both converge toward the common limit $f(a)$, which must therefore be 0.

Cauchy's proof involves an already existing technique, which Lagrange had applied in approximating real roots of polynomial equations. If a polynomial was negative for one value of the variable, positive for another, there was a root in between, and the difference between those two values of the variable bounded the error made in taking either as an approximation to the root ([20], p. 87; [21]). Thus again we have the algebra of inequalities providing a technique which Cauchy transformed from a tool of approximation to a tool of rigor.

It is worth remarking at this point that Cauchy, in his treatment both of convergence and of continuity, implicitly assumed various forms of the completeness property for the real numbers. For instance, he treated as obvious that a series of positive terms, bounded above by a convergent geometric progression, converges: also, his proof of the intermediate-value theorem assumes that a bounded monotone sequence has a limit.

While Cauchy was the first systematically to exploit inequality proof techniques to prove theorems in analysis, he did not identify all the implicit assumptions about the real numbers that such inequality techniques involve. Similarly, as the reader may have already noticed, Cauchy's definition of continuous function does not distinguish between what we now call point-wise and uniform continuity; also, in treating series of functions, Cauchy did not distinguish between point-wise and uniform convergence. The verbal formulations like "for all" that are involved in choosing deltas did not distinguish between "for any epsilon and for all x" and "for any x, given any epsilon" (Grattan-Guinness [18] puts it well: "Uniform convergence was tucked away in the word "always," with no reference to the variable at all."). Nor was it at all clear in the 1820s how much depended on this distinction, since proofs about continuity and convergence were in themselves so novel. We shall see the same confusion between uniform and point-wise convergence as we turn now to Cauchy's theory of the derivative.

Again we begin with an approximation. Lagrange gave the following inequality about the derivative:

$$(**) \qquad f(x + h) = f(x) + hf'(x) + hV,$$

where V goes to 0 with h. He interpreted this to mean that, given any D, one can find h sufficiently small so that V is between $-D$ and $+D$ ([21], p. 87; [23], p. 77). Clearly this is equivalent to statement $(*)$ above, Cauchy's delta-epsilon characterization of the derivative. But how did Lagrange obtain this result? The answer is surprising; for Lagrange, formula $(**)$ was a consequence of Taylor's theorem. Lagrange believed that any function (that is, any analytic expression, whether finite or infinite, involving the variable) had a unique power-series expansion (except possibly at a finite number of isolated points). This is because he believed that there was an "algebra of infinite series", an algebra exemplified by work of Euler such as the example we gave above. And Lagrange said that the way to make the calculus rigorous was to reduce it to algebra. Although there is no "algebra" of infinite series that gives power-series expansions without any consideration of convergence and limits, this assumption led Lagrange to define $f'(x)$ without reference to limits, as the coefficient of the linear term in h in the Taylor series expansion for $f(x + h)$. Following

Euler, Lagrange then said that, for any power series in h, one could take h sufficiently small so that any given term of the series exceeded the sum of all the rest of the terms following it; this approximation, said Lagrange, is assumed in applications of the calculus to geometry and mechanics ([23], p. 29; [21], p. 101; [12]). Applying this approximation to the linear term in the Taylor series produces (∗∗), which I call the Lagrange property of the derivative. (Like Cauchy's (∗), the inequality-translation Lagrange gives for (∗∗) assumes that, given any D, one finds h sufficiently small so $|V| \leq D$ with no mention whatever of x.)

Not only did Lagrange state property (∗∗) and the associated inequalities, he used them as a basis for a number of proofs about derivatives: for instance, to prove that a function with positive derivative on an interval is increasing there, to prove the mean-value theorem for derivatives, and to obtain the Lagrange remainder for the Taylor series. (Details may be found in the works cited in [16] and [17].) Lagrange also applied his results to characterize the properties of maxima and minima, and orders of contact between curves.

With a few modifications, Lagrange's proofs are valid—provided that property (∗∗) can be justified. Cauchy borrowed and simplified what are in effect Lagrange's inequality proofs about derivatives, with a few improvements, basing them on his own (∗). But Cauchy made these proofs legitimate because Cauchy defined the derivative precisely to satisfy the relevant inequalities. Once again, the key properties come from an approximation. For Lagrange, the derivative was *exactly*—no epsilons needed—the coefficient of the linear term in the Taylor series; formula (∗∗), and the corresponding inequality that $f(x+h) - f(x)$ lies between $h(f'(x) \pm D)$, were approximations. Cauchy ([16], Chapter 5 [17]) brought Lagrange's inequality properties and proofs together with a definition of derivative devised to make those techniques rigorously founded.

The last of the concepts we shall consider, the integral, followed an analogous development. In the eighteenth century, the integral was usually thought of as the inverse of the differential. But sometimes the inverse could not be computed exactly, so men like Euler remarked that the integral could be approximated as closely as one liked by a sum. Of course, the geometric picture of an area being approximated by rectangles, or the Leibnizian definition of the integral as a sum, suggests this immediately. But what is important for our purposes is that much work was done on approximating the values of definite integrals in the eighteenth century, including considerations of how small the subintervals used in the sums should be when the function oscillates to a greater or lesser extent. For instance, Euler [23] treated sums of the form

$$\sum_{k=0}^{n} f(x_k)(x_{k+1} - x_k)$$

as approximations to the integral $\int_{x_0}^{x_n} f(x)dx$.

In 1820, S.-D. Poisson, who was interested in complex integration and therefore more concerned than most people about the existence and behavior of integrals, asked the following question. If the integral F is defined as the anti-derivative of f, and if $b - a = nh$, can it be proved that $F(b) - F(a) = \int_a^b f(x)dx$ is the limit of the sum

$$S = hf(a) + hf(a+h) + \cdots + hf(a + (n-1)h)$$

as h gets small? (S is an approximating sum of the eighteenth-century sort.) Poisson called this result "the fundamental proposition of the theory of definite integrals." He proved it by using another inequality-result: the Taylor series with remainder. First, he wrote $F(b) - F(a)$ as the telescoping sum

$$(∗∗∗) \quad F(a+h) - F(a) + F(a+2h) - F(a+h)$$
$$+ \cdots + F(b) - F(a + (n-1)h).$$

Then, for each of the terms of the form

$$F(a + kh) - F(a + (k-1)h),$$

Taylor's series with remainder gives, since by definition $F' = f$,

$$F(a + kh) - F(a + (k-1)h)$$
$$= hf(a + (k-1)h) + R_k h^{1+w},$$

where $w > 0$, for some R_k. Thus the telescoping sum (2) becomes

$$hf(a) + hf(a+h) + \cdots + hf(a + (n-1)h)$$
$$+ (R_1 + \cdots + R_n)h^{1+w}.$$

So $F(b) - F(a)$ and the sum S differ by $(R_1 + \cdots + R_n)h^{1+w}$. Letting R be the maximum value for the R_k,

$$(R_1 + \cdots + R_n)h^{1+w} \leq n \cdot R(h^{1+w})$$
$$= R \cdot nh \cdot h^w = R(b-a)h^w.$$

Therefore, if h is taken sufficiently small, $F(b) - F(a)$ differs from S by less than any given quantity [27].

Poisson's was the first attempt to prove the equivalence of the antiderivative and limit-of-sums conceptions of the integral. However, besides the implicit assumptions of the existence of antiderivatives and bounded first derivatives for f on the given interval, the proof assumes that the subintervals on which the sum is taken are all equal. Should the result not hold for unequal divisions also? Poisson thought so, and justified it by saying ([27], pp. 329–330),

> if the integral is represented by the area of a curve, this area will be the same, if we divide the difference ... into an infinite number of equal parts, or an infinite number of unequal parts following any law.

This, however, is an assertion, not a proof. And Cauchy saw that a proof was needed.

Cauchy ([6], (2) Vol. 3, p. iii) did not like formalistic arguments in supposedly rigorous subjects, saying that most algebraic formulas hold "only under certain conditions, and for certain values of the quantities they contain." In particular, one could not assume that what worked for finite expressions automatically worked for infinite ones. Thus, Cauchy showed that the sum of the series $\frac{1}{1} + \frac{1}{4} + \frac{1}{9} + \cdots$ was $\pi^2/6$ by actually calculating the difference between the nth partial sum and $\pi^2/6$ and showing that it was arbitrarily small ([6], (2) Vol. 3, pp. 456–457).

Similarly, just because there was an operation called taking a derivative did not mean that the inverse of that operation always produced a result. The existence of the definite integral had to be proved. And how does one prove existence in the 1820s? One constructs the mathematical object in question by using an eighteenth-century approximation that converges to it. Cauchy defined the integral as the limit of Euler-style sums $\sum f(x_k)(x_{k+1} - x_k)$ for $x_{k+1} - x_k$ sufficiently small. Assuming explicitly that $f(x)$ was continuous on the given interval (and implicitly that it was uniformly continuous), Cauchy was able to show that all sums of that form approach a fixed value, called by definition the integral of the function on that interval. This is an extremely hard proof ([5], pp. 122–125; [16], pp. 171–175 in English translation). Finally, borrowing from Lagrange the mean-value theorem for integrals, Cauchy proved the Fundamental Theorem of Calculus ([5], pp. 151–152).

Conclusion

Here are all the pieces of the puzzle we originally set out to solve. Algebraic approximations produced the algebra of inequalities; eighteenth-century approximations in the calculus produced the useful properties of the concepts of analysis: d'Alembert's error-bounds for series, Lagrange's inequalities about derivatives, Euler's approximations to integrals. There was a new interest in foundations. All that was needed was a sufficiently great genius to build the new foundation.

Two men came close. In 1816, Carl Friedrich Gauss gave a rigorous treatment of the convergence of the hypergeometric series, using the technique of comparing a series with convergent geometric progressions; however, Gauss did not give a general foundation for all of analysis. Bernhard Bolzano, whose work was little known until the 1860s, echoing Lagrange's call to reduce the calculus to algebra, gave in 1817 a definition of continuous function like Cauchy's and then proved—by a different technique from Cauchy's—the intermediate-value theorem [3]. But it was Cauchy who gave rigorous definitions and proofs for all the basic concepts; it was he who realized the far-reaching power of the inequality-based limit concept; and it was he who gave us—except for a few implicit assumptions about uniformity and about completeness—the modern rigorous approach to calculus.

Mathematicians are used to taking the rigorous foundations of the calculus as a completed whole. What I have tried to do as a historian is to reveal what went into making up that great achievement. This needs to be done, because completed wholes by their nature do not reveal the separate strands that go into weaving them—especially when the strands have been considerably transformed. In Cauchy's work, though, one trace indeed was left of the origin of rigorous calculus in approximations—the letter epsilon. The ε corresponds to the initial letter in the word 'erreur' (or "error"), and Cauchy in fact used ε for "error" in some of his work on probability [9]. It is both amusing and historically appropriate that the ε, once used to designate the 'error' in approximations, has become transformed into the characteristic symbol of precision and rigor in the calculus. As Cauchy transformed the algebra of inequalities from a tool of approximation to a tool of rigor, so he transformed the calculus from a powerful method of generating results to the rigorous subject we know today.

References

1. George Berkeley, *The Analyst, or a Discourse Addressed to an Infidel Mathematician*, 1736.

2. Johann Bernoulli, *Opera Omnia*, 1742.

3. B. Bolzano, *Rein Analytischer Beweis des Lehrsatzes dass Zwischen je Zwey Werthen, die ein Entgegengesetzes Resultat gewaehren, Wenigstens eine Reele Wurzel*

der Gleichung Liege, Prague, 1817; English version, S. B. Russ, A translation of Bolzano's paper on the intermediate value theorem, *Hist. Math.* **7** (1980), 156–185.

4. C. Boyer, *History of Mathematics*, John Wiley, 1968.

5. A.-L. Cauchy, *Calcul Infinitésimal*; in [7], (2), Vol. 4.

6. A.-L. Cauchy, *Cours d'Analyse*, Paris, 1821; in [7], (2), Vol. 3.

7. A.-L. Cauchy, *Oeuvres Complètes d'Augustin Cauchy*, Gauthier-Villars, 1899.

8. A.-L. Cauchy, Résumé des leçons données à l'école royale polytechnique sur le calcul infinitésimal, 1823; in [7], (2), Vol. 4.

9. A.-L. Cauchy, Sur la plus grande erreur à craindre dans un résultat moyen, et sur le système de facteurs qui rend cette plus grande erreur un minimum, *C. R. Acad. Sci. Paris* **37**, 1853; in [7], (1), Vol. 12.

10. J. d'Alembert, Réflexions sur les suites et sur les racines imaginaires, in *Opuscules Mathématiques*, Vol. 5, Briasson, 1768, pp. 171–215.

11. D. Diderot, De l'interprétation de la nature, in *Oeuvres Philosophiques* (ed. P. Vernière), Garnier, 1961, pp. 180–181.

12. L. Euler, *Institutiones Calculi Integralis*, 3 vols, St. Petersburg, 1768–1770; in [13], (1), Vol. II.

13. L. Euler, *Opera Omnia*, Leipzig, 1912.

14. H. Goldstine, *A History of Numerical Analysis from the 16th through the 19th Century*, Springer-Verlag, 1977.

15. J. V. Grabiner, Cauchy and Bolzano: tradition and transformation in the history of mathematics, in [25].

16. J. V. Grabiner, *The Origins of Cauchy's Rigorous Calculus*, MIT Press, 1981.

17. J. V. Grabiner, The origins of Cauchy's theory of the derivative, *Hist. Math.* **5** (1978), 379–409.

18. I. Grattan-Guinness, *Development of the Foundations of Mathematical Analysis from Euler to Riemann*, MIT Press, 1970.

19. A. P. Iushkevich, O vozniknoveniya poiyatiya ob opredelennom integrale Koshi, *Trudy Instituta Istorii Estestvoznaniya, Akademia Nauk SSSR* **1** (1947), 373–411.

20. J.-L. Lagrange, *Equations Numériques*; in [22], Vol. 8.

21. J.-L. Lagrange, *Leçons sur le Calcul des Fonctions*, Paris, 1806; in [22], Vol. 10.

22. J.-L. Lagrange, *Oeuvres de Lagrange*, Gauthier-Villars, 1867–1892.

23. J.-L. Lagrange, *Théorie des Fonctions Analytiques*, 2nd ed., Paris, 1813; in [22], Vol. 9.

24. J.-L. Lagrange, *Traité de la Résolution des Équations Numériques de Tous les Degrés*, 2nd ed., Courcier, 1808; in [22], Vol. 8.

25. E. Mendelsohn, *Transformation and Tradition in the Sciences*, Cambridge, 1984.

26. Isaac Newton, *Mathematical Principles of Natural Philosophy* (transl. A. Motte, revised by Florian Cajori), 3rd ed., 1726, Univ. of California Press, 1934.

27. S. D. Poisson, Suite du mémoire sur les integrales définies, *J. de l'Ecole Polytechnique*, Cah. **18** II (1820), 295–341.

28. D. J. Struik, *A Source Book in Mathematics, 1200–1800*, Harvard, 1969.

Evolution of the Function Concept: A Brief Survey

ISRAEL KLEINER

Introduction

The evolution of the concept of function goes back 4000 years; 3700 of these consist of anticipations. The idea evolved for close to 300 years in intimate connection with problems in calculus and analysis. (A one-sentence definition of analysis as the study of properties of various classes of functions would not be far off the mark.) In fact, the concept of function is one of the distinguishing features of 'modern' as against 'classical' mathematics. W. L. Schaaf ([24], p. 500) goes a step further:

> The keynote of Western culture is the function concept, a notion not even remotely hinted at by any earlier culture. And the function concept is anything but an extension or elaboration of previous number concepts—it is rather a complete emancipation from such notions.

The evolution of the function concept can be seen as a tug-of-war between two elements, two mental images: the geometric (expressed in the form of a curve) and the algebraic (expressed as a formula—first finite and later allowing infinitely many terms, the so-called *analytic expression*). (See [7], p. 256.) Subsequently, a third element enters, namely, the logical definition of function as a correspondence (with a mental image of an input-output machine). In the wake of this development, the geometric conception of function is gradually abandoned. A new tug-of-war soon ensues (and is, in one form or another, still with us today) between this novel logical (*abstract, synthetic, postulational*) conception of function and the old algebraic (*concrete, analytic, constructive*) conception.

In this article, we will elaborate these points and try to give the reader a sense of the excitement and the challenge that some of the best mathematicians of all time confronted in trying to come to grips with the basic conception of function that we now accept as commonplace.

Precalculus developments

The notion of function in explicit form did not emerge until the beginning of the eighteenth century, although implicit manifestations of the concept date back to about 2000 B.C. The main reasons that the function concept did not emerge earlier were:

- lack of algebraic prerequisites—the coming to terms with the continuum of real numbers, and the development of symbolic notation;
- lack of motivation. Why define an abstract notion of function unless one had many examples from which to abstract?

In the course of about two hundred years (c. 1450–1650), there occurred a number of developments that were fundamental to the rise of the function concept:

- extension of the concept of number to embrace real and (to some extent) even complex numbers (Bombelli, Stifel, *et al.*);
- the creation of a symbolic algebra (Viète, Descartes, *et al.*);
- the study of motion as a central problem of science (Kepler, Galileo, *et al.*);
- the wedding of algebra and geometry (Fermat, Descartes, *et al.*).

The seventeenth century witnessed the emergence of modern mathematized science and the invention of analytic geometry. Both of these developments suggested a dynamic, continuous view of the functional relationship as against the static, discrete view held by the ancients.

In the blending of algebra and geometry, the key elements were the introduction of *variables* and the expression of the relationship between variables by means of *equations*. The latter provided a large number of examples of curves (potential functions) for study and set the final stage for the introduction of the function concept. What was lacking was the identification of the independent and dependent variables in an equation ([2], p. 348):

> Variables are not functions. The concept of function implies a unidirectional relation between an independent and a dependent variable. But in the case of variables as they occur in mathematical or physical problems, there need not be such a division of roles. And as long as no special independent role is given to one of the variables involved, the variables are not functions but simply variables

See [6], [15], [27] for details.

The calculus developed by Newton and Leibniz had not the form that students see today. In particular, it was not a calculus of *functions*. The principal objects of study in seventeenth-century calculus were (geometric) curves. (For example, the cycloid was introduced geometrically and studied extensively well before it was given as an equation.) In fact, seventeenth-century analysis originated as a collection of methods for solving problems about curves, such as finding tangents to curves, areas under curves, lengths of curves, and velocities of points moving along curves. Since the problems that gave rise to the calculus were geometric and kinematic in nature, and since Newton and Leibniz were preoccupied with exploiting the marvelous tool that they had created, time and reflection would be required before the calculus could be recast in algebraic form.

The variables associated with a curve were geometric—abscissas, ordinates, subtangents, subnormals, and the radii of curvature of a curve. In 1692, Leibniz introduced the word *function* (see [25], p. 272) to designate a geometric object associated with a curve. For example, Leibniz asserted that "a tangent is a function of a curve" ([12], p. 85).

Newton's method of fluxions applies to *fluents*, not functions. Newton calls his variables fluents—the image (as in Leibniz) is geometric, of a point flowing along a curve. Newton's major contribution to the development of the function concept was his use of power series. These were important for the subsequent development of that concept.

As increased emphasis came to be placed on the formulas and equations relating the functions associated with a curve, attention was focused on the role of the symbols appearing in the formulas and equations and thus on the relations holding among these symbols, independent of the original curve. The correspondence (1694–1698) between Leibniz and Johann Bernoulli traces how the lack of a general term to represent quantities dependent on other quantities in such formulas and equations brought about the use of the term *function* as it appears in Bernoulli's definition of 1718 (see [3], p. 9 and [27], p. 57 for details):

> One calls here Function of a variable a quantity composed in any manner whatever of this variable and of constants ([23], p. 72).

This was the first formal definition of function, although Bernoulli did not explain what "composed in any manner whatever" meant. See [3], [6], [12], [27] for details of this section.

Euler's *Introductio in Analysin Infinitorum*

In the first half of the eighteenth century, we witness a gradual separation of seventeenth-century analysis from its geometric origin and background. This process of degeometrization of analysis ([2], p. 345) saw the replacement of the concept of variable, applied to geometric objects, with the concept of function as an algebraic formula. This trend was embodied in Euler's classic *Introductio in Analysin Infinitorum* of 1748, intended as a survey of the concepts and methods of analysis and analytic geometry needed for a study of the calculus.

Euler's *Introductio* was the first work in which the concept of function plays an explicit and central role. In the preface, Euler claims that mathematical analysis is the general science of variables and their functions. He begins by defining a function as an *analytic expression* (that is, a formula):

> A function of a variable quantity is an analytical expression composed in any manner from that variable quantity and numbers or constant quantities ([23], p. 72).

Euler does not define the term *analytic expression*, but tries to give it meaning by explaining that admissible analytic expressions involve the four algebraic operations, roots, exponentials, logarithms, trigonometric functions, derivatives, and integrals. He classifies functions as being algebraic or transcendental; single-valued or multi-valued; and implicit or explicit. The *Introductio* contains one of the earliest treatments of trigonometric functions as numerical ratios (see [13]), as well as the earliest algorithmic treatment of logarithms as exponents. The entire approach is algebraic. Not a single picture or drawing appears (in Volume 1).

Expansions of functions in power series play a central role in this treatise. In fact, Euler claims that any function can be expanded in a power series: "If anyone doubts this, this doubt will be removed by the expansion of every function" ([3], p. 10). This remark was certainly in keeping with the spirit of mathematics in the eighteenth century. Hawkins ([10], p. 3) summarizes Euler's contribution to the emergence of function as an important concept:

> Although the notion of function did not originate with Euler, it was he who first gave it prominence by treating the calculus as a formal theory of functions.

As we shall see, Euler's view of functions was soon to evolve. (See [2], [3], [6], [27] for details of the above.)

The vibrating-string controversy

Of crucial importance for the subsequent evolution of the concept of the function was the vibrating-string problem:

An elastic string having fixed ends (0 and ℓ, say) is deformed into some initial shape and then released to vibrate. The problem is to determine the function that describes the shape of the string at time t.

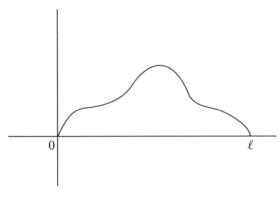

The controversy centered around the meaning of *function*. In fact, Grattan-Guinness ([9], p. 2) suggests that in the controversy over various solutions of this problem,

> the whole of eighteenth-century analysis was brought under inspection: the theory of functions, the role of algebra, the real line continuum and the convergence of series ....

To understand the debates that surrounded the vibrating-string problem, we must first mention an article of faith of eighteenth-century mathematics:

> If two analytic expressions agree on an interval, they agree everywhere.

This was not an unnatural assumption, given the type of functions (analytic expressions) considered at that time. On this view, the whole course of a curve given by an analytic expression is determined by any small part of the curve. This implicitly assumes that the independent variable in an analytic expression ranges over the whole domain of real numbers, without restriction.

In view of this, it is baffling (to us) that as early as 1744, Euler wrote to Goldbach stating that

$$\frac{\pi - x}{2} = \sum_{n=1}^{\infty} \frac{\sin nx}{n}.$$

(see [27], p. 67). Here, indeed, is an example of two analytic expressions that agree on the interval $(0, 2\pi)$, but nowhere else.

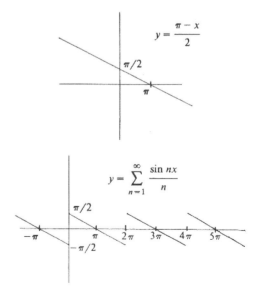

In 1747, d'Alembert solved the vibrating-string problem by showing that the motion of the string is governed by the partial differential equation

$$\frac{\partial^2 y}{\partial t^2} = a^2 \frac{\partial^2 y}{\partial x^2} \quad (a \text{ is a constant}),$$

the so-called *wave equation*. Using the boundary conditions $y(0,t) = 0$ and $y(\ell,t) = 0$, and the initial conditions

$$y(x,0) = f(x) \quad \text{and} \quad \left.\frac{\partial y}{\partial t}\right|_{t=0} = 0,$$

he solved this partial differential equation to obtain

$$y(x,t) = \frac{1}{2}[\phi(x+at) + \phi(x-at)]$$

as the most general solution of the vibrating-string problem, ϕ being an arbitrary function. It follows readily that

$$y(x,0) = f(x) = \phi(x) \text{ on } (0,\ell),$$
$$\phi(x+2\ell) = \phi(x),$$
$$\text{and } \phi(-x) = -\phi(x).$$

Thus, ϕ is determined on $(0,\ell)$ by the initial shape of the string, and is continued (by the article of faith) as an odd periodic function of period 2ℓ.

D'Alembert believed that the function $\phi(x)$ (and hence $f(x)$) must be an analytic expression—that is, it must be given by a formula. (To d'Alembert, these were the only permissible functions.) Moreover, since this analytic expression satisfies the wave equation, it must be twice differentiable.

In 1748, Euler wrote a paper on the same problem in which he agreed completely with d'Alembert concerning the solution but differed from him on its interpretation. Euler contended that d'Alembert's solution was not the most general, as the latter had claimed. Having himself solved the problem mathematically, Euler claimed his experiments showed that the solution $y(x,t) = \frac{1}{2}[\phi(x+at) + \phi(x-at)]$ gives the shapes of the string for different values of t, even when the initial shape is not given by a (single) formula. From physical considerations, Euler argued that the initial shape of the string can be given

(a) by several analytic expressions in different subintervals of $(0,\ell)$ (say, circular arcs of different radii in different parts of $(0,\ell)$

or, more generally,

(b) by a curve drawn free-hand.

But according to the article of faith prevalent at the time, neither of these two types of initial shapes could be given by a single analytic expression, since such an expression determines the shape of the entire curve by its behavior on any interval, no matter how small. Thus, d'Alembert's solution could not be the most general.

D'Alembert, who was much less interested in the vibrations of the string than in the mathematics of the problem, claimed that Euler's argument was "against all rules of analysis."

Langer ([16], p. 17) explains the differing views of Euler and d'Alembert concerning the vibrating-string problem in terms of their general approach to mathematics:

Euler's temperament was an imaginative one. He looked for guidance in large measure to practical considerations and physical intuition, and combined with a phenomenal ingenuity, an almost naive faith in the infallibility of mathematical formulas and the results of manipulations upon them. D'Alembert was a more critical mind, much less susceptible to conviction by formalisms. A personality of impeccable scientific integrity, he was never inclined to minimize shortcomings that he recognized, be they in his own work or in that of others.

Daniel Bernoulli entered the picture in 1753 by giving yet another solution of the vibrating-string problem. Bernoulli, who was essentially a physicist, based his argument on the physics of the problem and the known facts about musical vibrations (discovered earlier by Rameau *et al.*). It was generally recognized at the time that musical sounds (and, in particular, vibrations of a musical string) are composed of fundamental frequencies and their harmonic overtones. This physical evidence, and some loose mathematical reasoning, convinced Bernoulli that the solution to the vibrating-string problem must be given by

$$y(x,t) = \sum_{n=1}^{\infty} b_n \sin\frac{n\pi x}{\ell} \cos\frac{n\pi at}{\ell}.$$

This, of course, meant that an arbitrary function $f(x)$ can be represented on $(0,\ell)$ by a series of sines,

$$y(x,0) = f(x) = \sum_{n=1}^{\infty} b_n \sin\frac{n\pi x}{\ell}.$$

(Bernoulli was only interested in solving a physical problem, and did not give a definition of function. By an arbitrary function he meant an arbitrary shape of the vibrating string.)

Both Euler and d'Alembert (as well as other mathematicians of that time) found Bernoulli's solution absurd. Relying on the eighteenth-century article of faith, they argued that since $f(x)$ and the sine series $\sum_{n=1}^{\infty} b_n \sin(n\pi x/\ell)$ agree on $(0, \ell)$, they must agree everywhere. But then one arrived at the manifestly absurd conclusion that an arbitrary function $f(x)$ is odd and periodic. (Since Bernoulli's initial shape of the string was given by an analytic expression, Euler rejected Bernoulli's solution as being the most general solution.) Bernoulli retorted ([22], p. 78) that d'Alembert's and Euler's solutions constitute "beautiful mathematics but what has it to do with vibrating strings?"

The debate lasted for several more years (it was joined later by Lagrange) and then died down without being resolved. Ravetz ([22], p. 81) characterized the essence of the debate as one between d'Alembert's mathematical world, Bernoulli's physical world, and Euler's no-man's land between the two. The debate did, however, have important consequences for the evolution of the function concept. Its major effect was to extend that concept to include:

(a) Functions defined piecewise by analytic expressions in different intervals. Thus,
$$f(x) = \begin{cases} x & x \geq 0 \\ -x & x < 0 \end{cases}$$
was now, for the first time, considered to be a *bona fide* function.

(b) Functions drawn freehand and possibly not given by any combination of analytic expressions.

As Lützen [17] put it:

D'Alembert let the concept of function limit the possible initial values, while Euler let the variety of initial values extend the concept of function. We thus see that this extension of the concept of function was *forced* upon Euler by the physical problem in question.

To see how Euler's own view of functions evolved over a period of several years, compare the definition of function he gave in his 1748 *Introductio* with the following definition given in 1755, in which the term *analytic expression* does not appear ([23], pp. 72–73):

If, however, some quantities depend on others in such a way that if the latter are changed the former undergo changes themselves then the former quantities are called functions of the latter quantities. This is a very comprehensive notion and comprises in itself all the modes through which one quantity can be determined by others. If, therefore, x denotes a variable quantity then all the quantities which depend on x in any manner whatever or are determined by it are called its functions....

Euler's view of functions was reinforced later in that century by work in partial differential equations ([22], p. 86):

The work of Monge in the 1770s, giving a geometric interpretation to the integration of partial differential equations, seemed to provide a conclusive proof of the fact that functions more general than those expressed by an equation were legitimate mathematical objects....

(See [3], [4], [9], [16], [18], [19], [22], [27] for details on this section.)

Fourier and Fourier series

Fourier's work on heat conduction (submitted to the Paris Academy of Sciences in 1807, but published only in 1822 in his classic *Analytic Theory of Heat*) was a revolutionary step in the evolution of the function concept. Fourier's main result of 1822 was the following.

Theorem. *Any function $f(x)$ defined over $(-\ell, \ell)$ is representable over this interval by a series of sines and cosines,*
$$f(x) = \frac{a_0}{2} + \sum_{n=1}^{\infty} \left[a_n \cos \frac{n\pi x}{\ell} + b_n \sin \frac{n\pi x}{\ell} \right],$$
where the coefficients a_n and b_n are given by
$$a_n = \frac{1}{\ell} \int_{-\ell}^{\ell} f(t) \cos \frac{n\pi t}{\ell} dt \quad \text{and}$$
$$b_n = \frac{1}{\ell} \int_{-\ell}^{\ell} f(t) \sin \frac{n\pi t}{\ell} dt.$$

Fourier's announcement of this result met with incredulity. It upset several tenets of eighteenth-century mathematics. The result was known to Euler and Lagrange (among others), but only for certain functions. Fourier, of course, claimed that it is true for all functions, where the term *function* was given the most general contemporary interpretation ([23], p. 73):

In general, the function $f(x)$ represents a succession of values or ordinates each of which is arbitrary. An infinity of values being given to the abscissa x, there are an equal number of ordinates $f(x)$. All have actual numerical values, either positive or negative or null. We do not suppose these ordinates to be subject to a common law; they succeed each other in any manner whatever, and each of them is given as if it were a single quantity.

Fourier's proof of his theorem was loose even by the standards of the early nineteenth century. In fact, it was formalism in the spirit of the eighteenth century—"a play upon symbols in accordance with accepted rules but without much or any regard for content or significance" ([16], p. 33). To convince the reluctant mathematical community of the reasonableness of his claim, Fourier needed to show that:

(a) The coefficients of the Fourier series can be calculated for *any* $f(x)$.

(b) *Any* function $f(x)$ can be represented by its Fourier series in $(-\ell, \ell)$.

He showed this by:

(a') Interpreting the coefficients a_n and b_n in the Fourier series expansion of $f(x)$ as areas (which made sense for arbitrary functions $f(x)$, not necessarily given by analytic expressions).

(b') Calculating the a_n and b_n (for small values of n) for a great variety of functions $f(x)$, and noting the close agreement in $(-\ell, \ell)$ (but not outside that interval) between the initial segments of the resulting Fourier series and the functional values of $f(x)$.

Fourier accomplished all this using mathematical reasoning that would be clearly unacceptable to us today. However,

> It was, no doubt, partially because of his very disregard for rigor that he was able to take conceptual steps which were inherently impossible to men of more critical genius ([16], p. 33).

Fourier's work raised the analytic (algebraic) expression of a function to at least an equal footing with its geometric representation (as a curve). His work had a fundamental and far-reaching impact on subsequent developments in mathematics. (For example, it forced mathematicians to re-examine the notion of integral, and was the starting point of the researches that led Cantor to his creation of the theory of sets.) As for its impact on the evolution of the function concept, Fourier's work:

- did away with the article of faith held by eighteenth-century mathematicians. (Thus, it was now clear that two functions given by different analytic expressions can agree on an interval without necessarily agreeing outside the interval.)

- showed that Euler's concept of discontinuous was flawed. (Some of Euler's discontinuous functions were shown to be representable by a Fourier series—an analytic expression—and were thus continuous in Euler's sense.)

- gave renewed emphasis to analytic expressions.

As we shall see, all this forced a re-evaluation of the function concept. (See [3], [6], [7], [9], [16], [19] for details.)

As we have noted, the period 1720–1820 was characterized by a development and exploitation of the tools of the calculus bequeathed by the seventeenth century. These tools were employed in the solution of important 'practical' problems (e.g., the vibrating-string problem, the heat-conduction problem). These problems, in turn, clamored for attention to important 'theoretical' concepts (e.g., function, continuity, convergence). A new subject—analysis—began to take form, in which the concept of function was central. But both the subject and the concept were still in their formative stages. It was a period of formalism in analysis—formal manipulations dictated the rules of the game, with little concern for rigor. The concept of function was in a state of flux—an analytic expression (an 'arbitrary' formula), then a curve (drawn freehand), and then again an analytic expression (but this time a specific formula, namely a Fourier series). Both the subject of analysis (certainly its basic notions) and the concept of function were ripe for a re-evaluation and a reformulation. This is the next stage in our development.

Dirichlet's concept of function

Dirichlet was one of the early exponents of the critical spirit in mathematics ushered in by the nineteenth century (others were Gauss, Abel, Cauchy). He undertook a careful analysis of Fourier's work to make it mathematically respectable. The task was not simple:

> To make sense out of what he [Fourier] did took a century of effort by men of "more critical genius", and the end is not yet in sight ([4], p. 263).

Fourier's result that any function can be represented by its Fourier series was, of course, incorrect. In a fundamental paper of 1829, Dirichlet gave sufficient conditions for such representability:

Theorem. *If a function f has only finitely many discontinuities and finitely many maxima and minima in $(-\ell, \ell)$, then f may be represented by its Fourier series on $(-\ell, \ell)$.* (*The Fourier series converges pointwise to f where f is continuous, and to $[f(x+) + f(x-)]/2$ at each point x where f is discontinuous.*)

For a mathematically rigorous proof of this theorem, one needed

(a) clear notions of continuity, convergence, and the definite integral, and

(b) clear understanding of the function concept.

Cauchy contributed to the former, and Dirichlet to the latter. We first turn very briefly to Cauchy's contributions.

Cauchy was one of the first mathematicians to usher in a new spirit of rigor in analysis (see the article by Judith Grabiner earlier in this volume). In his famed *Cours d'Analyse* of 1821 and subsequent works, he rigorously defined the concepts of continuity, differentiability, and integrability of a function in terms of limits. (Bolzano had done much of this earlier, but his work went unnoticed for fifty years.)

In dealing with continuity, Cauchy addresses himself to Euler's conceptions of continuous and discontinuous. He shows that the function

$$f(x) = \begin{cases} x, & x \geq 0 \\ -x, & x < 0 \end{cases}$$

(which Euler considered discontinuous) can also be written as $f(x) = \sqrt{x^2}$, and

$$f(x) = \frac{2}{\pi} \int_0^\infty \frac{x^2}{x^2 + t^2} \, dt,$$

which means that $f(x)$ is also continuous in Euler's sense. This paradoxical situation, Cauchy claims, cannot happen when *his* definition of continuity is used.

Cauchy's conception of function is not very different from that of his predecessors:

> When the variable quantities are linked together in such a way that, when the value of one of them is given, we can infer the values of all the others, we ordinarily conceive that these various quantities are expressed by means of one of them which then takes the name of *independent variable*; and the remaining quantities, expressed by means of the independent variable, are those which one calls the *functions* of this variable ([3], p. 104).

Although Cauchy gives a rather general definition of a function, his subsequent comments suggest that he had in mind something more limited (see [10], p. 10). He classifies functions as *simple* and *mixed*. The simple functions are

$$a + x, \; a - x, \; ax, \; a/x, \; x^a, \; a^x, \; \log x, \; \sin x,$$
$$\cos x, \; \arcsin x, \; \arccos x,$$

and the mixed functions are composites of the 'simple' ones—say, $\log(\sin x)$. See [3], [6], [8], [9], [12], [14] for Cauchy's contribution.

Now let us consider Dirichlet's definition of function [19]:

> y is a function of a variable x, defined on the interval $a < x < b$, if to every value of the variable x in this interval there corresponds a definite value of the variable y. Also, it is irrelevant in what way this correspondence is established.

The novelty in Dirichlet's conception of function as an arbitrary correspondence lies not so much in the definition as in its application. Mathematicians from Euler through Fourier to Cauchy had paid lip service to the arbitrary nature of functions; but in practice they thought of functions as analytic expressions or curves. Dirichlet was the first to take seriously the notion of function as an arbitrary correspondence (but see [3], p. 201). This is made abundantly clear in his 1829 paper on Fourier series, at the end of which he gives an example of a function (the Dirichlet *function*),

$$D(x) = \begin{cases} c, & x \text{ is rational} \\ d, & x \text{ is irrational,} \end{cases}$$

that does not satisfy the hypothesis of his theorem on the representability of a function by a Fourier series (see [10], p. 15). The Dirichlet function:

- was the first explicit example of a function that was not given by an analytic expression (or by several such), nor was it a curve drawn freehand;

- was the first example of a function that is discontinuous (in our, not Euler's sense) *everywhere*;

- illustrated the concept of function as an arbitrary pairing.

Another important point is that Dirichlet, in his definition of function, was among the first to restrict explicitly the domain of the function to an interval; in the past, the independent variable was allowed to range over all real numbers. (See [3], [5], [9], [10], [15], [17], [27] for details about Dirichlet's work.)

Pathological functions

With his new example $D(x)$, Dirichlet 'let the genie escape from the bottle'. A flood of 'pathological functions', and classes of functions, followed in the succeeding half century. Certain functions were introduced to test the domain of applicability of various results (e.g., the Dirichlet function was introduced in connection with the representability of a function in a Fourier series). Certain classes of functions were introduced in order to extend various concepts or results (e.g., functions of bounded variation were introduced to test the domain of applicability of the Riemann integral).

The character of analysis began to change. Since the seventeenth century, the processes of analysis were assumed to be applicable to *all* functions, but it now turned out that they are restricted to particular *classes* of functions. In fact, the investigation of various classes of functions—such as continuous functions, semi-continuous functions, differentiable functions, functions with non-integrable derivatives, integrable functions, monotonic functions, continuous functions that are not piecewise monotonic—became a principal concern of analysis. (One example is Dini's study of continuous non-differentiable functions, for which he defined the so-called *Dini derivatives*.) Whereas mathematicians had formerly looked for order and regularity in analysis, they now took delight in discovering exceptions and irregularities. The towering personalities connected with these developments were Riemann and Weierstrass, although many others made important contributions (e.g., du Bois Reymond and Darboux).

The first major step in these developments was taken by Riemann in his *Habilitationsschrift* of 1854, which dealt with the representation of functions in Fourier series. As we recall, the coefficients of a Fourier series are given by integrals. Cauchy had developed his integral only for continuous functions, but his ideas could be extended to functions with finitely many discontinuities. Riemann extended Cauchy's concept of integral and thus enlarged the class of functions representable by Fourier series. This extension (known today as the *Riemann integral*) applies to functions of bounded variation, a much broader class of functions than Cauchy's continuous functions. Thus, a function can have infinitely many discontinuities (which can be dense in any interval) and still be Riemann integrable.

Riemann gave the following example (published in 1867) in his *Habilitationsschrift*:

$$f(x) = 1 + \frac{(x)}{1^2} + \frac{(2x)}{2^2} + \frac{(3x)}{3^2} + \cdots,$$

where for any real number α the function (α) is defined as 0 if $\alpha = \frac{1}{2} + k$ (k an integer), and α minus the nearest integer when $\alpha \neq \frac{1}{2} + k$ (k an integer). This function is discontinuous for all $x = m/2n$, where m is an integer relatively prime to $2n$ (see [6], p. 325). In contrast to Dirichlet's function $D(x)$, this one is given by an analytic expression and is Riemann-integrable.

Riemann's work may be said to mark the beginning of a theory of the mathematically discontinuous (although there are isolated examples in Fourier's and Dirichlet's works). It planted the discontinuous firmly upon the mathematical scene. The importance of this development can be inferred from the following statement of Hawkins ([10], p. 3):

> The history of integration theory after Cauchy is essentially a history of attempts to extend the integral concept to as many discontinuous functions as possible; such attempts could become meaningful only after existence of highly discontinuous functions was recognized and taken seriously.

In 1872, Weierstrass startled the mathematical community with his famous example of a continuous nowhere-differentiable function

$$f(x) = \sum_{n=1}^{\infty} b^n \cos(a^n \pi x),$$

where a is an odd integer, b a real number in $(0, 1)$, and $ab > 1 + 3\pi/2$ (see [12], p. 387). (Bolzano had given such an example in 1834, but it went unnoticed.) This example was contrary to all geometric intuition. In fact, up to about 1870, most books on the calculus 'proved' that a continuous function is differentiable except possibly at a finite number of points! (See [10], p. 43.) Even Cauchy believed that.

Weierstrass's example began the disengagement of the continuous from the differentiable in analysis. Weierstrass's work (and others' in this period) necessitated a re-examination of the foundations of analysis and led to the so-called arithmetization of analysis, in which process Weierstrass was a prime mover. As Birkhoff notes ([3], p. 71):

Weierstrass demonstrated the need for higher standards of rigor by constructing *counter-examples* to plausible and widely held notions.

Counter-examples play an important role in mathematics. They illuminate relationships, clarify concepts, and often lead to the creation of new mathematics. (An interesting case study of the role of counter-examples in mathematics can be found in the book *Proofs and Refutations* by I. Lakatos.) The impact of the developments we have been describing was, as we already noted, to change the character of analysis. A new subject was born—the theory of functions of a real variable. Hawkins ([10], p. 119) gives a vivid description of the state of affairs:

> The nascent theory of functions of a real variable grew out of the development of a more critical attitude, supported by numerous counter-examples, towards the reasoning of earlier mathematicians. Thus, for example, continuous non-differentiable functions, discontinuous series of continuous functions, and continuous functions that are not piecewise monotonic were discovered. The existence of exceptions came to be accepted and more or less expected. And the examples of non-integrable derivatives, rectifiable curves for which the classical integral formula is inapplicable, non-integrable functions that are the limit of integrable functions, Harnack-integrable derivatives for which the Fundamental Theorem II is false, and counterexamples to the classical form of Fubini's Theorem appear to have been received in this frame of mind. The idea, as Schoenflies put it in his report ... was to proceed, as in human pathology, to discover as many exceptional phenomena as possible in order to determine the laws according to which they could be classified.

It should be pointed out, however, that not everyone was pleased with these developments (at least in analysis), as the following quotations from Hermite (in 1893) and Poincaré (in 1899), respectively, attest ([15], p. 973):

> I turn away with fright and horror from this lamentable evil of functions which do not have derivatives. [Hermite]

> Logic sometimes makes monsters. For half a century we have seen a mass of bizarre functions which appear to be forced to resemble as little as possible honest functions which serve some purpose. More of continuity, or less of continuity, more derivatives, and so forth. Indeed, from the point of view of logic, these strange functions are the most general; on the other hand those which one meets without searching for them, and which follow simple laws appear as a particular case which does not amount to more than a small corner.

> In former times when one invented a new function it was for a practical purpose; today one invents them purposely to show up defects in the reasoning of our fathers and one will deduce from them only that.

> If logic were the sole guide of the teacher, it would be necessary to begin with the most general functions, that is to say with the most bizarre. It is the beginner that would have to be set grappling with this teratologic museum. [Poincaré]

The effect of the events we have been describing on the function concept can be summarized as follows. Stimulated by Dirichlet's conception of function and his example $D(x)$, the notion of function as an arbitrary correspondence is given free rein and gains general acceptance; the geometric view of function is given little consideration. (Riemann's and Weierstrass's functions could certainly not be 'drawn', nor could most of the other examples given during this period.) After Dirichlet's work, the term *function* acquired a clear meaning independent of the term *analytic expression*. During the next half century, mathematicians introduced a large number of examples of functions in the spirit of Dirichlet's broad definition, and the time was ripe for an effort to determine which functions were actually describable by means of analytic expressions, a vague term in use during the previous two centuries. (See [3], [10], [14], [15] for details of this period.)

Baire's classification scheme

The question whether every function in Dirichlet's sense is representable analytically was first posed by Dini in 1878 (see [5], p. 31). Baire had undertaken to give an answer in his doctoral thesis of 1898. The very notion of analytic representability had to be clarified, since it was used in the past in an informal way. Dini himself used it vaguely, asking

> if every function can be expressed analytically, for all values of the variable in the interval, by a

finite or infinite series of operations ("opérations du calcul") on the variable ([5], p. 32).

The starting point for Baire's scheme was the Weierstrass Approximation Theorem (published in 1885):

Every continuous function $f(x)$ on an interval $[a, b]$ is a uniform limit of polynomials on $[a, b]$.

Baire called the class of continuous functions *class* 0. Then he defined the functions of *class* 1 to be those that are not in class 0, but which are (point-wise) limits of functions of class 0. In general, the functions of class m are those functions which are not in any of the preceding classes, but are representable as limits of sequences of functions of class $m-1$. This process is continued, by transfinite induction, to all ordinals less than the first uncountable ordinal Ω. (Since the Baire functions thus constructed are closed under limits, nothing new results if this process is repeated.) This classification into Baire classes α ($\alpha < \Omega$) is called the *Baire classification*, and the functions which constitute the union of the Baire classes are called *Baire functions*.

Baire called a function *analytically representable* if it belonged to one of the Baire classes. Thus, a function is analytically representable (in Baire's sense) if it can be built up from a variable and constants by a finite or denumerable set of additions, multiplications, and passages to pointwise limits.

The collection of analytically representable functions (Baire functions) is very encompassing. For example, discontinuous functions representable by Fourier series belong to class 1. Thus, functions representable by Fourier series constitute only a part of the totality of analytically representable functions. (Recall Fourier's claim that *every* function can be represented by a Fourier series!) As another example, Baire showed that the pathological Dirichlet function $D(x)$ is of class 2, since

$$D(x) = \begin{Bmatrix} c, & x \text{ is rational} \\ d, & x \text{ is irrational} \end{Bmatrix}$$
$$= (c-d) \lim_{n \to \infty} \lim_{m \to \infty} (\cos n!\pi x)^{2m} + d.$$

Moreover, any function obtained from a variable and constants by an application of the four algebraic operations and the operations of analysis (such as differentiation, integration, expansion in series, use of transcendental functions)—the kind of function known in the past as an analytic expression—was shown to be analytically representable.

Lebesgue pursued these studies and showed (in 1905) that each of the Baire classes is non-empty, and that the Baire classes do not exhaust all functions. Thus, Lebesgue established that there are functions which are not analytically representable (in Baire's sense). This he did by actually exhibiting a function outside the Baire classification, "using a profound but extremely complex method" [19]. According to Luzin [19], "the impact of Lebesgue's discovery was just as stunning as that of Fourier in his time." (See [5], [19], [20], [21] for details.)

Not all functions in the sense of Dirichlet's conception of function as an arbitrary correspondence are analytically representable (in the sense of Baire), although it is (apparently) very difficult to produce a specific function that is not. Do such non-analytically representable functions *really* exist? This is part of our story in the next section.

Debates about the nature of mathematical objects

Function theory was characterized by some at the turn of the twentieth century as the branch of mathematics which deals with counter-examples. This view was not universally applauded, as the earlier quotations from Hermite and Poincaré indicate. In particular, Dirichlet's general conception of function began to be questioned. Objections were raised against the phrase in his definition that "it is irrelevant in what way this correspondence is established." Subsequently, the arguments for and against this point linked up with the arguments for and against the axiom of choice (explicitly formulated by Zermelo in 1904) and broadened into a debate over whether mathematicians are free to create their objects at will.

There was a famous exchange of letters in 1905 among Baire, Borel, Hadamard, and Lebesgue concerning the current logical state of mathematics (see [5], [20], [21] for details). Much of the debate was about function theory—the critical question being whether a definition of a mathematical object (say a number or a function), however given, legitimizes the existence of that object; in particular, whether Zermelo's axiom of choice is a legitimate mathematical tool for the definition or construction of functions. In this context, Dirichlet's conception of function was found to be too broad by some (e.g., Lebesgue) and devoid of meaning by others (e.g., Baire and Borel), but was acceptable to yet others (e.g., Hadamard). Baire, Borel, and Lebesgue supported the requirement of a definite 'law' of correspondence in the definition of a function. The 'law', moreover, must be reasonably explicit—that is, understood by and communicable to anyone who wants to study the function.

To illustrate the point, Borel compares the number π (whose successive digits can be unambiguously determined, and which he therefore regards as well defined) with the number obtained by carrying out the following 'thought experiment'. Suppose we lined up infinitely many people and asked each of them to name a digit at random. Borel claims that, unlike π, this number is not well defined since its digits are not related by any law. This being so, two mathematicians discussing this number will never be certain that they are talking about the *same* number. Put briefly, Borel's position is that without a definite law of formation of the digits of an infinite decimal, one cannot be certain of its identity.

Hadamard had no difficulty in accepting as legitimate the number resulting from Borel's thought experiment. By way of illustration, he alluded to the kinetic theory of gases, where one speaks of the velocities of molecules in a given volume of gas although no one knows them precisely. Hadamard felt that "the requirement of a law that determines a function ... strongly resembles the requirement of an *analytic expression* for that function, and that this is a throwback to the eighteenth century" [19].

The issues described here were part of broad debates about various ways of doing analysis—synthetic versus analytic, or idealist versus empiricist. These debates, in turn, foreshadowed subsequent 'battles' between proponents and opponents of the various philosophies of mathematics (e.g., formalism and intuitionism) dealing with the nature and meaning of mathematics. And, of course, the issue has not been resolved. (See [4], [5], [19], [20], [21] for details.)

The period 1830–1910 witnessed an immense growth in mathematics, both in scope and in depth. New mathematical fields were formed (complex analysis, algebraic number theory, non-Euclidean geometry, abstract algebra, mathematical logic), and older ones were deepened (real analysis, probability, analytic number theory, calculus of variations). Mathematicians felt free to create their systems (almost) at will, without finding it necessary to seek motivation from or applications to concrete (physical) settings. At the same time there was, throughout the nineteenth century, a reassessment of gains achieved, accompanied by a concern for the foundations of (various branches of) mathematics. These trends are reflected in the evolution of the notion of function. The concept unfolds from its modest beginnings as a formula or a geometric curve (eighteenth and early nineteenth centuries) to an arbitrary correspondence (Dirichlet). This latter idea is exploited throughout the nineteenth century by way of the construction of various 'pathological' functions. Toward the end of the century, there is a re-evaluation of past accomplishments (Baire classification, controversy relating to use of the axiom of choice), much of it in the broader context of debates about the nature and meaning of mathematics.

Recent developments

Here we briefly touch on three more recent developments relating to the function concept.

(A) L_2 *functions*. The set

$$L_2 = \{f(x) : f^2(x) \text{ is Lebesgue-integrable}\}$$

forms a *Hilbert space*—a fundamental object in functional analysis. Two functions in L_2 are considered to be the same if they agree everywhere except possibly on a set of Lebesgue measure zero. Thus, in L_2 Function Theory, one can always work with representatives in an equivalence class rather than with individual functions. These notions, as Davis and Hersh observed ([4], p. 269),

> involve a further evolution of the concept of function. For an element in L_2 is not a function, either in Euler's sense of an analytic expression, or in Dirichlet's sense of a rule or mapping associating one set of numbers with another. It is function-like in the sense that it can be subjected to certain operations normally applied to functions (adding, multiplying, integrating). But since it is regarded as unchanged if its values are altered on an arbitrary set of measure zero, it is certainly not just a rule assigning values at each point in its domain.

(B) *Generalized functions* (*Distributions*). The concept of a distribution or generalized function is a very significant and fundamental extension of the concept of function. The theory of distributions arose in the 1930s and 1940s. It was created to give mathematical meaning to the differentiation of non-differentiable functions—a process which the physicists had employed (unrigorously) for some time. Thus, Heaviside (in 1893) differentiated the function

$$f(x) = \begin{cases} 1, & x > 0 \\ 1/2, & x = 0 \\ 0, & x < 0 \end{cases}$$

to obtain the impulse function

$$\delta(x) = \begin{cases} 0, & x \neq 0 \\ \infty, & x = 0. \end{cases}$$

(In 1930, Dirac introduced $\delta(x)$ as a convenient notation in the mathematical formulation of quantum theory.)

In formal terms, a *distribution* is a continuous linear functional on a space D of infinitely differentiable functions (called *test functions*) that vanish outside some interval $[a, b]$. To any continuous (or locally integrable) function F, there corresponds a distribution $\Phi_F : D \mapsto \mathbb{C}$ given by

$$\Phi_F(x) = \int_{-\infty}^{\infty} F(t)x(t)dt.$$

However, not every distribution comes from such a function. The distribution $\delta : D \mapsto \mathbb{C}$ given by $\delta(x) = x(0)$ corresponds to the Dirac δ-function mentioned above, and does not arise from any function F in the way described above. (See [4], [18], [26].)

A basic property of distributions is that each distribution has a derivative that is again a distribution. In fact ([26], p. 338),

the enduring merit of distribution theory has been that the basic operations of analysis, differentiation and convolution, and the Fourier/Laplace transforms and their inversion, which demanded so much care in the classical framework, could now be carried out without qualms by obeying purely algebraic rules.

(C) *Category theory.* The notion of a function as a mapping between arbitrary sets gradually became dominant in the mathematics of the twentieth century. Algebra had a major impact on this development, in which the concept of a function was placed in the general framework of the concept of a mapping from one set into another. Thus, linear transformations of vector spaces (principally, $\mathbb{R}^n$ and $\mathbb{C}^n$) were dealt with throughout much of the nineteenth century. Homomorphisms of groups and automorphisms of fields were introduced in the latter part of that century. As early as 1887, Dedekind gave a fairly 'modern' definition of the term *mapping* ([23], p. 75):

By a mapping of a system S a law is understood, in accordance with which to each determinate element s of S there is associated a determinate object, which is called the image of s and is denoted by $\varphi(s)$; we say too, that $\varphi(s)$ corresponds to the element s, that $\varphi(s)$ is caused or generated by the mapping φ out of s, that s is transformed by the mapping φ into $\varphi(s)$.

Analysis, too, played a major role in this extension of the domain and range of definition of a function to arbitrary sets. (Recall that Dirichlet's definition of function was as an arbitrary correspondence between (real) *numbers*.) Thus, Euler and others in the eighteenth century treated (informally) functions of several variables. In 1887, considered the year of birth of functional analysis, Volterra defined the notion of a 'functional' which he called a "function of functions". (A *functional* is a function whose domain is a set of functions and whose range is the real or complex numbers.) In the first two decades of the twentieth century, the notions of metric space, topological space, Hilbert space, and Banach space were introduced; functions (operators, linear operators) between such spaces play a prominent role. (See [15] for details.)

In 1939, Bourbaki gave the following definition of a function ([3], p. 7):

Let E and F be two sets, which may or may not be distinct. A relation between a variable element x of E and a variable element y of F is called a *functional relation* in y if, for all $x \in E$, there exists a unique $y \in F$ which is in the given relation with x.

We give the name of *function* to the operation which in this way associates with every element $x \in E$ the element $y \in F$ which is in the given relation with x; y is said to be the value of the function at the element x, and the function is said to be *determined* by the given functional relation. Two equivalent functional relations determine the *same* function.

Bourbaki then also gave the definition of a function as a certain subset of the Cartesian product $E \times F$. This is, of course, the definition of function as a set of ordered pairs.

All of these 'modern' general definitions of function were given in terms of sets, and hence their logic must receive the same scrutiny as that of set theory.

In category theory, which arose in the late 1940s to give formal expression to certain aspects of homology theory, the concept of function assumes a fundamental role. It can be described as an 'association' from an 'object' A to another 'object' B. The objects A and B need not have any elements (that is, they need not be sets in the usual sense). In fact, the objects A and B can be entirely dispensed with. A *category* can then be defined as consisting of functions (or 'maps'), *which are taken as undefined (primitive) concepts* satisfying certain relations or axioms. In fact, in 1966 Lawvere outlined how category theory can replace set theory as a foundation for mathematics. (See [11] for details.)

In the recent developments outlined in this section, we have seen the function concept modified (L_2 functions), generalized (distributions), and finally 'generalized out of existence' (category theory). Have we come full circle?

References

1. G. Birkhoff, *A Source Book in Classical Analysis*, Harvard Univ. Press, 1973.
2. H. Bos, Mathematics and rational mechanics, *Ferment of Knowledge* (eds. G. S. Rousseau and R. Porter), Cambridge Univ. Press, 1980, pp. 327–355.
3. U. Bottazzini, *The Higher Calculus: A History of Real and Complex Analysis from Euler to Weierstrass*, Springer-Verlag, 1986.
4. P. J. Davis and R. Hersh, *The Mathematical Experience*, Birkhäuser, 1981.
5. P. Dugac, Des fonctions comme expressions analytiques aux fonctions représentables analytiquement, *Mathematical Perspectives* (ed. J. Dauben), Academic Press, 1981, pp. 13–36.
6. C. H. Edwards, *The Historical Development of the Calculus*, Springer-Verlag, 1979.
7. A. Gardiner, *Infinite Processes*, Springer-Verlag, 1982.
8. J. V. Grabiner, *The Origins of Cauchy's Rigorous Calculus*, MIT Press, 1981.
9. I. Grattan-Guinness, *The Development of the Foundations of Mathematical Analysis from Euler to Riemann*, MIT Press, 1970.
10. T. Hawkins, *Lebesgue's Theory of Integration—its Origins and Development*, Chelsea, 1975 (orig. 1970).
11. H. Herrlich and G. E. Strecker, *Category Theory*, Allyn and Bacon, 1973.
12. R. F. Iacobacci, *Augustin-Louis Cauchy and the Development of Mathematical Analysis*, Ph.D. Dissertation, New York University, 1965. (University Microfilms, no. 65-7298, 1986.)
13. V. J. Katz, Calculus of the trigonometric functions, *Hist. Math.* **14** (1987), 311–324.
14. P. Kitcher, *The Nature of Mathematical Knowledge*, Oxford Univ. Press, 1983.
15. M. Kline, *Mathematical Thought from Ancient to Modern Times*, Oxford Univ. Press, 1972.
16. R. E. Langer, Fourier series: The genesis and evolution of a theory, The First Herbert Ellsworth Slaught Memorial Paper, Math. Assoc. of America, 1947. (Suppl. to *Amer. Math. Monthly* **54** (1947), 1–86.)
17. J. Lützen, The development of the concept of function from Euler to Dirichlet (in Danish), *Nordisk. Mat. Tidskr.* **25/26** (1978), 5–32.
18. J. Lützen, Euler's vision of a generalized partial differential calculus for a generalized kind of function, *Math. Mag.* **56** (1983), 299–306.
19. N. Luzin, Function (in Russian), *The Great Soviet Encyclopedia*, **59** (c. 1940), pp. 314–334.
20. A. F. Monna, The concept of function in the nineteenth and twentieth centuries, in particular with regard to the discussion between Baire, Borel and Lebesgue, *Arch. Hist. Exact Sci.* **9** (1972/73), 57–84.
21. G. H. Moore, *Zermelo's Axiom of Choice: Its Origins, Development, and Influence*, Springer-Verlag, 1982.
22. J. R. Ravetz, Vibrating strings and arbitrary functions, *The Logic of Personal Knowledge: Essays Presented to M. Polanyi on his Seventieth Birthday*, The Free Press, 1961, pp. 71–88.
23. D. Rüthing, Some definitions of the concept of function from Joh. Bernoulli to N. Bourbaki, *Math. Intelligencer* **6:4** (1984), 72–77.
24. W. L. Schaaf, Mathematics and world history, *Math. Teacher* **23** (1930), 496–503.
25. D. J. Struik, *A Source Book in Mathematics, 1200-1800*, Harvard Univ. Press, 1969.
26. F. Treves, Review of *The Analysis of Linear Partial Differential Operators* by L. Hormander, *Bull. Amer. Math. Soc.* **10** (1984), 337–340.
27. A. P. Youschkevitch, The concept of function up to the middle of the nineteenth century, *Arch. Hist. Exact Sci.* **16** (1976), 37–85.

S. Kovalevsky: A Mathematical Lesson

KAREN D. RAPPAPORT

American Mathematical Monthly **88** (1981), 564–574

Sofya Kovalevsky was a noted writer whose works include both fiction and non-fiction. She was also a political activist and a public advocate of feminism. In addition, she was a brilliant mathematician who made significant contributions despite the enormous educational and political obstacles that she had to overcome. Somehow her many achievements have been forgotten. In those few instances where her work has not been lost it has been denigrated by such studies as Felix Klein's history of nineteenth-century mathematics. Klein dismisses Kovalevsky's work in the following manner ([3], p. 294): "Her works are done in the style of Weierstrass and so one doesn't know how much of her own ideas are in them." He finds something wrong with all her research and credits her with only one positive accomplishment, drawing Weierstrass out of his shell through their correspondence. It is time to set this record straight and to let the facts speak for themselves. (Although Kovalevskaya, the feminine version of the name Kovalevsky, is used in much of the current literature, the author will refer to Kovalevsky in this paper, as that was the name under which Sofya Kovalevsky worked and published. However, in the references the author has used the version of the name that appeared in the work cited.)

Sofya Krukovsky, known affectionately as Sonya, was born in Moscow in 1850. Her father, a Russian army officer, retired in 1858 and moved the family—Sofya, her older sister, Anyuta, and her younger brother, Fedya—to Palibino, an estate near the Lithuanian border. After settling at Palibino, the household discovered that they had not brought a sufficient amount of wallpaper with them. Rather than travel a great distance to obtain new wallpaper, they decided to use old newspapers on the wall. Since only the nursery required the paper, this was deemed an adequate solution. However, while searching the attic for newspaper, they discovered paper of a better quality. On it were the lecture notes from a calculus course taken by General Krukovsky. This is how the nursery walls came to be covered with the calculus notes that, in her later years, Sofya claimed to have studied. Sofya often repeated this anecdote and enjoyed reporting how her calculus teacher exclaimed ([17], p. 216): "You have understood them as though you knew them in advance."

Kovalevsky claimed that her interest in mathematics was aroused by her Uncle Peter, who would discuss numerous abstractions and mathematical concepts with her. When the family tutor, Joseph Malevich, read of this in Sofya's autobiographical work, *Memories of Childhood*, he was incensed. He wrote a long essay in a Russian newspaper explaining why he should receive credit for Kovalevsky's mathematical development. In response to this criticism Kovalevsky wrote the following tribute in *An Autobiographical Sketch* ([17], p. 216): "It is to Joseph Malevich that I am indebted for my first systematic study of mathematics. It happened so long ago that I no longer remember his lessons at all.... It was arithmetic that Malevich taught best... I have to confess that arithmetic held little interest for me."

Kovalevsky studied mathematics against her father's wishes. When she was thirteen, she smuggled an algebra text into her room and studied it. When she was fourteen she taught herself trigonometry in order to study a physics book written by her neighbor Professor Tyrtov—trigonometry was necessary for the optics section, and the young Sofya taught herself without tutor or text. By constructing a chord on a circle, she was able to explain the sine function and to develop the other trigonometric formulas. When Professor Tyrtov saw her work, he was struck by its similarity to the actual mathematical development. Calling her a new Pascal, Tyrtov pleaded with the General to permit Sofya to study mathematics.

After a year of exhortation, General Krukovsky relented and allowed Sofya to go to Petersburg to study calculus and other subjects.

After completing her studies in 1867, Sofya wanted to continue her education, but the Russian university system was closed to women. The only option for study was to go to Switzerland, but General Krukovsky would not allow his daughters to go abroad.

Sofya's older sister, Anyuta, felt imprisoned at Palibino and sought a way out. She found it through the radical politics of the times. This was a period of political ferment in Russia. The nihilists, feminists, and radicals were all active, and their ideas were brought to Palibino by the local priest's son on his vacation from school in Petersburg. While they scandalized the neighborhood, these ideas had great influence on Anyuta, who in turn influenced Sofya. Anyuta joined a radical group that advocated higher education for women and promoted the concept of the "fictitious husband" to enable women to obtain more freedom. A married woman did not need her father's signature for a passport, and so a fictitious husband would enable Anyuta to travel abroad for her education. Anyuta and her friend Zhanna found a 26-year-old university student, Vladimir Kovalevsky, who agreed to marry one of them. Unfortunately for Anyuta, she brought Sofya to one of their meetings. Vladimir became infatuated with Sofya and insisted on marrying her. After several secret meetings and much intrigue, General Krukovsky consented, and Vladimir and Sofya were married in September 1868.

Following their marriage, the Kovalevskys left for Petersburg to study, and to search for a husband for Anyuta. With little effort Sofya had won the freedom to pursue her education, the freedom and independence that Anyuta had been fighting so hard for. Sofya's feelings of guilt about this can be seen in her letters to Anyuta, who was still confined at Palibino. She wrote ([16], p. 232; cited in [17], p. 11): "At times a strange anguish comes over me and I feel ashamed that everything is coming to me so easily and without any struggle."

In Petersburg, Sofya received permission from the instructors to attend classes unofficially. She wrote ([16], p. 225) to her sister: "... Lectures begin tomorrow and so my real life begins at 9 A.M. ... [Vladimir] and friends will solemnly escort me by way of the backstairs so that there is hope of hiding from the administration and from curious stares." It was at Petersburg that Sofya decided to concentrate on mathematics. In a letter to Anyuta she said ([22], p. 25): "I have become convinced that one cannot learn everything and one life is barely sufficient to accomplish what I can in my chosen field."

The Kovalevskys and Anyuta, who was still unmarried but chaperoned by the young couple, left for Europe in 1869. Sofya intended to study mathematics, and Vladimir geology. Anyuta planned to pursue her revolutionary activities. Sofya and Vladimir settled in Heidelberg, but Sofya was not permitted to matriculate at the university. She appealed to both the faculty and the administration. A special committee was formed, and it was decided that each individual professor could choose whether to permit Sofya Kovalevsky to attend his lectures unofficially.

Kovalevsky was now able to attend lectures, and her outstanding mathematical ability became the talk of Heidelberg. As a firm believer in education for women, she used her reputation to assist other Russian women in their efforts to attend the university. One of these women was her friend Yulya Lermontov, who later became the first female chemist in Russia. For many years Bunsen had described Sofya as "a dangerous woman" because, according to him, Sofya had tricked him into permitting Yulya to use the previously all-male chemistry labs. In 1874, Karl Weierstrass asked Sofya for confirmation of the story because "he [Bunsen] writes fiction even if he doesn't publish it." (See [27], p. 51.)

Sofya, Vladimir, and Yulya lived and studied together in Heidelberg until the fall of 1869. When Anyuta arrived for a visit, she was quite surprised to find Sofya still living with her fictitious husband and proceeded to evict Vladimir from the apartment. A short time later, Vladimir left Heidelberg to study at Jena.

As her mathematics progressed, Kovalevsky felt the need to study with Karl Weierstrass, the most noted mathematician of the time, at the University of Berlin. She traveled to Berlin for the start of the fall semester of 1870, only to find the university closed to women. Sofya wrote ([17], p. 218): "The capital of Prussia proved to be backward. Despite all my pleadings and efforts I had no success in obtaining permission to attend the University of Berlin."

Determined to study mathematics, Kovalevsky personally presented herself to Weierstrass as an aspiring student. On the basis of recommendations from Koenigsberger at Heidelberg, Weierstrass was willing to see her. He assigned her a set of problems on hyperelliptic functions that he had just given to his class. Weierstrass was so impressed with the ability she demonstrated in her solutions that he personally requested the university to allow her to attend his classes unofficially. However, the university was intransigent in its decision.

Not wanting to waste this mathematical talent, Weierstrass offered to tutor Kovalevsky privately. The lessons, begun in the fall of 1870, lasted for four years. The working relationship lasted a lifetime. When Weierstrass made his offer, he had no idea that Kovalevsky would become the closest of his disciples and remain so until her death.

The lessons began with twice-weekly meetings, Sundays at Weierstrass's home and weekdays at Kovalevsky's. The mathematics lessons are partially documented in a series of forty-one letters from Weierstrass to Kovalevsky, spanning the period of March 1871 to August 1874. They show that the emphasis was on Weierstrass's favorite topic: Abelian functions. Kovalevsky's responses are unavailable, because Weierstrass had burned them on hearing of her death. However, some idea of the importance of this correspondence to Kovalevsky can be found in her writings ([17], p. 218): "These studies had the deepest possible influence on my entire career in mathematics. They determined finally and irrevocably the direction I was to follow in my later scientific work: all my work has been done precisely in the spirit of Weierstrass."

These letters also show the development of a close personal relationship between Kovalevsky and Weierstrass. They give a glimpse of the growing affection that Weierstrass felt for his pupil. Weierstrass was unaware of Kovalevsky's marriage arrangement and did not understand Vladimir's appearances. It was not until two years later (October 1872) that Weierstrass learned the truth. The series of letters at this time indicate that the topic of a relationship between Kovalevsky and Weierstrass was raised and the truth about her marital status was made known. Both were upset at the scene, and Weierstrass made several attempts to reassure his pupil that he would thereafter only discuss science. For a short period the correspondence remained strictly mathematical, but it did not stay that way. Weierstrass was not able to mask his concern and affection for Kovalevsky.

In October of 1872, Weierstrass had suggested several possible topics for Kovalevsky's dissertation. By 1874, she had completed three original works, any one of which Weierstrass felt would be acceptable. Now he needed to find a university that would award Kovalevsky a degree. In July 1874, the University of Göttingen awarded Sofya Kovalevsky a Ph.D., in absentia, *summa cum laude*, without either orals or defense. Needless to say, this was an unprecedented event.

The three papers presented for the degree were:

1. On the Theory of Partial Differential Equations

2. On the Reduction of a Certain Class of Abelian Integrals of the Third Rank to Elliptic Integrals

3. Supplementary Remarks and Observations on Laplace's Research on the Form of Saturn's Rings

The first of the papers, on partial differential equations, was published in Crelle's journal in 1875. This was considered a great honor, especially for a novice mathematician, since Crelle's journal was considered the most serious mathematical publication in Germany.

In the first paper, Kovalevsky had generalized a problem that had been posed by Cauchy. Cauchy had examined an existence theorem for partial differential equations, and Kovalevsky generalized Cauchy's results to systems of order r containing time derivatives of order r. The mathematician H. Poincaré said that ([23], p. 234) "Kovalevsky significantly simplified the proof and gave the theorem its definitive form." Today, this theorem on the existence and uniqueness of solutions of partial differential equations is often, but not always, known as the Cauchy-Kovalevsky Theorem. While studying the partial differential equation problem, Kovalevsky examined the heat equation. Some of her results were helpful to Weierstrass and he wrote: "So you see, dear Sonya, that your observation (which seemed so simple to you) on the distinctive property of partial differential equations... was for me the starting point for interesting and very elucidating researches." ([23], p. 235)

Shortly after Sofya received her degree from Göttingen, her friend Yulya was awarded a degree in chemistry from Göttingen. Vladimir had received his degree in paleontology two years earlier.

Although she had earned her Ph.D. and had written a highly acclaimed paper, Sofya Kovalevsky was unable to get a job. Even the efforts of her mentor, Weierstrass, were fruitless. So, in 1875, together with Vladimir, Anyuta, and Anyuta's husband, Sofya returned to the family home in Russia. Weierstrass ([27], p. 52) encouraged her to relax at home and "enjoy the pleasures of big city life. I know you won't give up your scientific work." However, Kovalevsky did not actively pursue her work that winter. She felt guilty and wrote ([17], p. 220): "I worked far less zealously than I had done in Germany and, indeed, the situation was far less propitious for scholarly work... The only thing that still gave me scholarly support was the exchange of letters and ideas with my beloved teacher Weierstrass." As important as this exchange may have been, the correspondence with Weierstrass stopped in 1875: It was resumed for a short time in 1878 but did not become regular until 1880. Weierstrass was very hurt by this neglect. However, it must be acknowledged that Kovalevsky was never a good correspondent, even while in Germany, so that it is not surprising to find this lapse in writing after her return to Russia.

In Russia, Kovalevsky again tried to find a job. The only position in mathematics available to a woman was teaching arithmetic in the lower grades of a girls' school, and since Sofya admitted ([26], p. 147; cited in [24], p. 287) that she was "unfortunately weak in the multiplication table", she could not seriously consider such a position. Therefore, Kovalevsky turned to other intellectual pursuits. She wrote fiction, theater reviews, and popular-science reports for a newspaper. She was instrumental in the organization of the Bestuzhev School for Women, but because of what were considered her radical views she was not permitted to teach there.

During this period in Russia, General Krukovsky died and left Sofya a small inheritance. Vladimir invested this money in business enterprises that eventually went bankrupt.

By 1878, Sofya had become bored with her activities and wanted to return to mathematics. She wrote to Weierstrass for advice. Weierstrass was excited by this letter, the first he had received from his pupil in three years. However, Kovalevsky's return to mathematics was delayed by the birth of a daughter, Sofya Vladimirovna, in October 1878.

On their return to Russia the Kovalevskys had assumed the obligations of a real marriage. This was done partly as an obligation to Sofya's parents and partly because of their new politics. It was their feeling to end lying relationships of all kinds, and so the marriage was finally consummated.

Kovalevsky's return to mathematics was encouraged by a scientific conference held in Petersburg in 1880. The Russian mathematician Chebyshev invited Kovalevsky to present a paper at this conference. She found her unpublished dissertation on Abelian Integrals, translated it from German to Russian in one night, and presented it to the conference. Although it had lain untouched for six years, it was well received by the mathematicians.

Following her presentation, the Swedish mathematician Gösta Mittag-Leffler, who had met Kovalevsky earlier while she was a student in Germany, offered to find her a position in his country. Kovalevsky was very appreciative of this offer and wrote, in 1881, to Mittag-Leffler ([19], p. 51): "[If I can teach] I may in this way open the universities to women, which have hitherto only been open by special favor, a favor which can be denied at any moment."

Kovalevsky's desire for a position was spurred not only by her feminism but also by her need to do mathematics. She wished ([19], p. 52) "at the same time, to be able to live for my work, surrounded by those who are occupied with the same questions."

While waiting for word from Sweden, Kovalevsky looked in Berlin for research work. While Vladimir was on a business trip, Sofya secretly visited Weierstrass. When she decided that she would return to Berlin to pursue research, Vladimir was quite angry with the decision and their marriage ended. For the next two years, Kovalevsky lived a student's life in Berlin, and the care of her daughter was shared by Sofya's friend Yulya and her brother-in-law Alexander.

While Sofya Kovalevsky was busy conducting her research on the refraction of light in crystals, Vladimir Kovalevsky again managed to get himself into financial difficulties. There was a stock scandal and Vladimir, faced with ruin, committed suicide in the spring of 1883. Sofya took solace in her mathematics and hoped to find a badly needed job in Stockholm.

Mittag-Leffler had recently been appointed the head of the mathematics department at the newly founded University of Stockholm and was able to offer Sofya a position there. However, the job had conditions. In order to prove her competence, Kovalevsky was to teach for a probationary year, with no pay and no official university affiliation. Kovalevsky agreed to this because she had no other options.

In the fall of 1883, Sofya Kovalevsky arrived in Stockholm to become a lecturer at the University of Stockholm. Her reception was mixed. Although hailed by many, others agreed with A. Strindberg ([22], p. 50), who wrote in the local paper, "A female professor is a pernicious and unpleasant phenomenon—even, one might say, a monstrosity."

In the spring of 1884, Kovalevsky lectured in German on partial differential equations. These lectures were well received, and Mittag-Leffler was able to obtain the funds for her appointment as a Professor of Higher Analysis in July 1884. Word of the appointment was sent to Sofya in Moscow, where she was spending the summer with her daughter. However, Sofya did not bring her daughter to Stockholm in the fall because she was still unsure of her position. She was publicly criticized for her child-care arrangements but chose to ignore this. She didn't bring her daughter to Sweden until 1885, when the child was eight years old.

In addition to joining the Stockholm faculty in 1884, Kovalevsky became an editor for *Acta Mathematica* and published her first paper on crystals. In 1885, she received a second appointment to the Chair of Mechanics. She also published a second paper on the propagation of light in crystals but was embarrassed when Volterra found a serious error in her work: she had used a multi-valued function as if it were a single-valued one. Since this research was performed after a three-year hiatus in her mathematical career, Kovalevsky felt that her teacher, Weierstrass, should have caught the error prior to publication. Weierstrass, quite distraught, blamed illness and overwork. (It might be added that Weierstrass was 70 years old at the time.)

While in Stockholm, Sofya lived for a time in Mittag-Leffler's home, where she met and developed a friendship with his sister, Anna Leffler. Anna Leffler, a well-known advocate of women's rights and a writer, encouraged Sofya's literary leanings. In 1887 they collaborated on a play entitled *The Struggle for Happiness*. It was based on an idea that had occurred to Sofya while she sat at the bedside of her dying sister.

After Anyuta died in the fall of 1887, Sofya felt lonely and despondent. The sisters had been close, and Sofya felt the loss deeply. However, at this time two events occurred that helped to assuage her grief. Both the announcement of a new competition for the Prix Bordin and the arrival in Stockholm of a Russian lawyer named Maxim Kovalevsky were to have profound effects on the life of Sofya Kovalevsky.

Early in 1888 the French Academy of Science announced a new competition for the Prix Bordin. Papers on the theory of the rotation of a solid body would be considered for the prize competition at the end of the summer. Gösta Mittag-Leffler encouraged Sofya to work on a paper for the competition.

While Sofya was engaged in her research on the paper, Maxim Kovalevsky arrived to give a series of lectures at Stockholm University. He had been dismissed from Moscow University for criticizing Russian constitutional law. Aside from politics, Sofya and Maxim had many interests in common, and their attraction resulted in a scandalous affair. Eventually Maxim proposed marriage, on the condition that Sofya give up her research. Even if she had wished to give up her mathematics, Sofya was too far into her work for the prize competition to stop. In order to free Sofya to do her work, Mittag-Leffler invited Maxim to his summer home in Uppsala. This was a wise move, for, as Sofya stated ([10], p. 124), "If burly Maxim had stayed longer, I do not know how I should have got on with my work."

With Maxim gone, Sofya was able to finish her work and the paper was submitted on time. Of the fifteen papers, which were submitted anonymously, one was considered so outstanding that the award was increased from 3000 francs to 5000 francs. The Prix Bordin was awarded to Sofya Kovalevsky in December of 1888. Sofya attended the awards ceremony with Maxim. Special recognition was given to her work. In his congratulatory speech, the President of the Academy of Sciences said ([22], p. 61): "Our co-members have found that her work bears witness not only to profound and broad knowledge, but to a mind of great inventiveness."

Prior to Sofya Kovalevsky's work the only solutions to the motion of a rigid body about a fixed point had been developed for the two cases where the body is symmetric. In the first case, developed by Euler, there are no external forces, and the center of mass is fixed within the body. This is the case that describes the motion of the earth. In the second case, derived by Lagrange, the fixed point and the center of gravity both lie on the axis of symmetry of the body. This case describes the motion of the top. Sofya Kovalevsky developed the first of the solvable special cases for an unsymmetrical top. In this case the center of mass is no longer on an axis in the body. She solved the problem by constructing coordinates explicitly as ultra-elliptic functions of time. Kovalevsky continued this work in two more papers on a rigid body motion. These both received awards from the Swedish Academy of Sciences in 1889. Her later works on the subject have been lost.

Kovalevsky's professorship in Sweden was due to expire in 1889. Desirous of returning to her native country, she inquired about a position in Russia. Her request was flatly denied. Russian mathematicians were indignant at this slight of Kovalevsky and decided to honor her. It was suggested that an honorary membership in the Russian Academy of Sciences would provide that recognition. However, in order to do that, the charter had to be amended to allow for female membership. In November 1889, Chebyshev sent the following telegram to Kovalevsky ([16], p. 135; cited in [25]): "Our Academy of Sciences has just now elected you a corresponding member, having just permitted this innovation for which there has been no precedent until now. I am very happy to see this fulfillment of one of my most impassioned and justified desires."

Kovalevsky sought positions throughout Europe but was again unsuccessful. She therefore had to accept the renewal of the professorship in Stockholm.

While working in Stockholm, Kovalevsky regularly commuted to France, where she visited Maxim at his villa. During these visits Maxim encouraged her literary interests. She wrote *Memories of Childhood* in Russian. It was translated into Swedish and published in 1889. In order not to shock Swedish society with its personal revelations, it was released as a novel entitled *The Raevsky Sisters*. The original was published in Russian in 1890. Kovalevsky's novel *A Nihilist Girl* was written in Swedish in 1890. A Russian version had been started, but Sofya's sudden death left it unfinished. Maxim edited the two versions and was responsible for the posthumous publication of the novel. This book was highly praised by critics in Russia and Scandinavia.

It was Kovalevsky's frequent trips to France to visit Maxim that eventually caused her death. On returning to Stockholm from a visit, early in 1891, she fell ill. The illness was misdiagnosed, and by the time it was finally found to be pneumonia it was out of control. On February 10, 1891, Sofya Kovalevsky died. Although there was widespread mourning and eulogies were given around the world, the Russian Minister of the Interior, I. N. Durnovo, was concerned that too much attention was being paid to "a woman who was, in the last analysis, a Nihilist." ([16], p. 532; cited in [25])

The mathematical world was more generous in its praise. Mittag-Leffler gave the official eulogy for the University of Stockholm. Speaking of her as a teacher he said ([20], p. 171): "We know with what inspiring zeal she explained [her] ideas... and how willingly she gave the riches of her knowledge." In his eulogy, Kronecker, of the University of Berlin, spoke [18] of Kovalevsky as "one of the rarest investigators." Karl Weierstrass, who felt her loss most deeply, having burned all of her letters, said [21] " 'People die, ideas endure': it would be enough for the eminent figure of Sofya to pass into posterity on the lone virtue of her mathematical and literary work."

In her short lifetime, Sofya Kovalevsky left a notable collection of political, literary, and mathematical works. Her contributions, completed in spite of many obstacles, certainly warrant her a place in our intellectual and mathematical history.

During her mathematical career Sofya Kovalevsky published ten papers in mathematics and mathematical physics. Three of these papers, [4], [6], [9] were written during her student days under Weierstrass (1870–1874). The articles on the refraction of light [5], [7], [8] were written years later in Berlin (1881–1883) after her return to her mathematical researches. The remaining research on rigid body motion and Bruns's Theorem [11–14] was completed during her tenure at the University of Stockholm (1883–1891). The only complete collection of Kovalevsky's works is published in Russian [15]. Portions of her work in partial differential equations and rigid body motion appear in English, and since these are Kovalevsky's most important works they will be discussed in some detail.

The proof of the first general existence theorem in partial differential equations was presented by Cauchy in 1842, in his publications on integrating differential equations with initial conditions. Here he showed the existence of analytic solutions for ordinary differential equations and certain linear partial differential equations.

The Cauchy problem for the ordinary differential equation $du/dt = f(t, u)$, with the initial conditions $u = u_0$ and $t = t_0$ states:

If $f(t, u)$ is an analytic function in a neighborhood of the point (t, u) there exists a unique solution of $du/dt = f(t, u)$ with the given initial conditions.

For systems of first-order partial differential equations of the form

$$\frac{\partial u_i}{\partial t} = F_i\left(t, x_1, \ldots, x_n; u_1, \ldots, u_n; \frac{\partial u_1}{\partial x_1}, \ldots, \frac{\partial u_m}{\partial x_m}\right)$$

with initial conditions when $t = 0$, that is,

$$u_i(0, x_1, \ldots, x_n) = w_i(x_1, \ldots, x_n)$$

for $i = 1, \ldots, m$, the Cauchy problem is to find a solution $u(x, t)$ that satisfies the initial conditions ([1], p. 318).

To solve this problem Cauchy assumed that F_i and w_i were analytic. He obtained a locally convergent power series solution by using his "method of majorants." The original function F_i is replaced by a simple analytic function whose power series expansion coefficients are non-negative and greater than or equal to the absolute value of the corresponding coefficients for F_i. The derived system is then explicitly integrated to give a solution which is the majorant for the solution to the original with $t = 0$.

In her thesis Kovalevsky generalized Cauchy's results to systems of an order r containing time derivatives, $\partial^r u_i/\partial t^r$, of order r. It is this generalization that is known as the Cauchy-Kovalevsky theorem.

Kovalevsky used majorants of the form

$$\frac{M}{1 - [(t_1 + t_2 + \cdots t_r)/\rho]},$$

where ρ and M are constants.

For the higher order system

$$\frac{\partial^{n_i} u_i}{\partial t^{n_i}}$$

$$= F_i\left(t, x_1, \ldots, x_n; u_1, \ldots, u_m; \frac{\partial^k u_j}{\partial t^{k_0} \partial x_1^{k_1} \cdots \partial x_n^{k_n}}\right)$$

with $i, j = 1, 2, \ldots, m$; $k_0 + k_1 + \cdots + k_n = k \leq n_j, k_0 < n_j$, and initial values given at some point $t = t_0$ for the u_i and their first $n_i - 1$ derivatives with respect to t, Kovalevsky proved the following. If all the functions F_i are analytic in a neighborhood of the point

$$(t, x_1, \ldots, x_n; \ldots, w_{j,k_0,k_1,\ldots,k_n})$$

and all the functions $w_j^{(k)}$ are analytic in a neighborhood of the point $(x_1, \ldots, x_n)$, then Cauchy's problem has an analytic solution in a certain neighborhood of the point $(t, x_1, \ldots, x_n)$, and it is the unique solution in the class of analytic functions. Note ([23], p. 233) that

$$w_{j,k_0,k_1,\cdots,k_n} = \frac{\partial^{k-k_0} u_j}{\partial x_1^{k_1} \cdots \partial x_n^{k_n}}.$$

The simplest form of this theorem states that any equation of the form

$$\partial u/\partial t = f(t, x, u, \partial u/\partial x),$$

where f is analytic in its arguments for values near $(t_0, x_0, u_0, \partial u_0/\partial x_0)$, possesses one and only one solution $u(t, x)$ which is analytic near (t_0, x_0).

The Cauchy-Kovalevsky Theorem has significant limitations. It is restricted to analytic functions, and convergence may fail in a region of interest. Also the computation of the coefficients of the series may be too tedious to be practical. However, it is important in that it shows that within a class of analytic solutions of analytic equations the number of arbitrary functions needed for a general solution is the same as the order of the equation and these arbitrary functions involve one less independent variable than the number occurring in the equation.

The work for which Kovalevsky was best known in her time was her research on the motion of a rigid body about a fixed point. The equations of motion of a rigid body moving about a fixed point were derived by Euler in 1750. They are as follows:

$$\frac{d\gamma}{dt} = r\gamma' - q\gamma'',$$

$$A\frac{dp}{dt} + (C - B)qr = Mg(y_0\gamma'' - z_0\gamma'),$$

$$\frac{d\gamma'}{dt} = p\gamma'' - r\gamma,$$

$$B\frac{dq}{dt} + (A - C)rp = Mg(z_0\gamma - x_0\gamma''),$$

$$\frac{d\gamma''}{dt} = q\gamma - p\gamma',$$

$$C\frac{dr}{dt} + (B - A)pq = Mg(x_0\gamma' - y_0\gamma).$$

Here A, B, C are the principal axes of the ellipsoid of the body relative to the fixed point; M is the mass; g is the acceleration due to gravity; $\gamma, \gamma', \gamma''$ are the direction cosines of the angles which the three axes make at each moment; their direction is the same as the force of the rigid body; p, q, r are the components of the angular velocity along the principal axes; x_0, y_0, z_0 are the coordinates of the center of gravity of the body considered in a system of coordinates of which the origin is the fixed point and whose direction coincides with that of the principal axes of the ellipsoid of inertia.

The problem to be solved was the integration of this system of equations so that the position of the moving body at any time could be obtained. Before 1888, the integration had been completed for only two cases. The first was for the condition $x_0 = y_0 = z_0 = 0$. This case, studied by Euler and Poisson, is the one where the center of gravity of the moving body coincides with the fixed point. This is the motion of a force-free symmetric body. There are no external forces acting on the body and the motion is about a fixed point within the body, the center of mass. If the fixed point is the center of gravity, then gravity does not influence the motion. The axes of rotation are thus fixed in the body. An example of this force-free motion is the free rotation of the earth. In the case of the earth's free rotation, the axis of rotation does not coincide with the axis of symmetry. It is very slightly tilted, although it passes through the center of mass of the earth. What then happens is that the instantaneous axis precesses slowly about the axis of symmetry.

In the second case, studied by Lagrange, $A = B, x_0 = y_0 = 0$. Here the fixed point and the center of gravity lie on the same axis. When this axis is the axis of symmetry, the motion is that of the spinning top. "A top is defined to be a material body which is symmetrical about an axis and terminates in a sharp point... at one of the axes" ([28], p. 164). This top spins about a fixed point that is not the center of gravity, but the center of gravity and the fixed point both lie on the axis of symmetry of the top. The weight of the top gives rise to a moment of force as it spins about the fixed point on the plane.

In both of these cases the rigid body was symmetrical. Sofya Kovalevsky developed the first of the soluble special cases for the heavy or unsymmetrical top. In this case, the center of mass no longer lies on the axis of the body. Instead, it is in the equatorial plane (the plane perpendicular to the axis) and passes through the fixed point. In addition, two of the principal moments of inertia are equal and double the third. The center of gravity is in the plane of the equal moments of inertia.

Euler's equations containing six unknown functions has the following first integrals:

$$Ap^2 + Bq^2 + Cr^2$$
$$-2Mg(x_0\gamma + y_0\gamma' + z_0\gamma'') = C_1$$
$$Ap\gamma + Bq\gamma' + Cr\gamma'' = C_2$$
$$\gamma^2 + \gamma'^2 + \gamma''^2 = C_3 = 1.$$

It is sufficient to find one more integral to obtain a complete solution in quadratures. This occurs because the time variable appears only in the form dt and can be eliminated, leaving only five equations.

Kovalevsky [12] derived the fourth integral for the case $A = B = 2C$, $z_0 = 0$, and showed that the only conditions necessary for a given series to be a solution to Euler's system are that the constants A, B, C, x, y, z satisfy one of four conditions:

1. $A = B = C$,
2. $x_0 = y_0 = z_0 = 0$ (Euler's case),
3. $A = B, x_0 = y_0 = 0$ (Lagrange's case),
4. $A = B = 2C, z_0 = 0$ (Kovalevsky's case [13]).

Kovalevsky obtained a solution for her case by rotating the coordinate axes in the xy-plane and choosing a unit of length so that $y_0 = 0$ and $C = 1$. With $c_0 = Mgx_0$ the Euler equations become:

$$2\frac{dp}{dt} = qr \qquad \frac{d\gamma}{dt} = r\gamma' - q\gamma''$$
$$2\frac{dg}{dt} = -pr - c_0\gamma'' \qquad \frac{d\gamma'}{dt} = p\gamma'' - r\gamma$$
$$2\frac{dr}{dt} = c_0\gamma' \qquad \frac{d\gamma''}{dt} = q\gamma - p\gamma'.$$

The three algebraic integrals are:

$$2(p^2 + q^2) + r^2 = 2c_0 x + 6l_1$$
$$2(p\gamma + q\gamma') + r\gamma'' = 2l$$
$$\gamma^2 + \gamma'^2 + \gamma''^2 = 1,$$

where l and l_1 are constants of integration.

Kovalevsky then derived a fourth integral:

$$\left[(p+qi)^2 + c_0(\gamma + i\gamma')\right]$$
$$\times \left[(p-qi)^2 + c_0(\gamma - \gamma'i)\right] = k^2,$$

where k is an arbitrary constant. Defining $x_1 = p + qi$, $x_2 = p - qi$, she made several transformations of the variables. After some algebraic manipulations she obtained the equations:

$$0 = \frac{ds_1}{\sqrt{R_1(s_1)}} + \frac{ds_2}{\sqrt{R_1(s_2)}}$$
$$dt = \frac{s_1 ds_1}{\sqrt{R_1(s_1)}} + \frac{s_2 ds_2}{\sqrt{R_1(s_2)}},$$

where $R_1(S)$ is a fifth-degree polynomial whose zeros are unique and s_1 and s_2 are polynomials in x_1 and x_2. This system results in hyperelliptic integrals which Kovalevsky solved by using theta functions.

For this highly praised research Kovalevsky was awarded the Bordin Prize in 1888 and a prize from the Swedish Academy in 1889. Historians, however, have found this work "not of sufficient interest for the theory of a top to find a place here." ([2], p. 369) Today the value of her work is seen not in the results or the originality of the method but in the interest it stimulated in the problem of the rotation of a rigid body. Several researchers have continued the work of finding new cases of special solutions. This includes Chaplygin's development of the integral for a symmetrical top turning about a fixed point, with moments of inertia $A = B = 4C$ and the Hesse-Schiff equations of motion for a top. There is also a body of work by others (for example Bobylev, Steklov, Goryachev, and V. Kovalevsky).

The remaining works by Sofya Kovalevsky are of lesser importance and will be only briefly reviewed.

In the paper on Abelian integrals, Kovalevsky showed how a certain type of Abelian integral could be expressed as an elliptic integral.

The paper on Saturn's rings is concerned with the stability of motion of liquid ring-shaped bodies. Laplace found the form of the ring to be a skewed cross-section of an ellipse. Kovalevsky, using a series expansion, proved that the rings were egg-shaped ovals symmetric about a single axis. However, the subsequent proof that Saturn's rings consist of discrete particles and not a continuous liquid made this work inapplicable.

In the articles on the refraction of light in crystals, Kovalevsky applied a method developed by Weierstrass to differentiate Lamé's partial differential equations. However, Volterra discovered a basic error in her work. She had (as had Lamé) treated a multivalued function as though it were single-valued, and the solution could not be applied to the equations in her form. However, the paper did demonstrate the previously unpublished theory of Weierstrass.

Kovalevsky's last article derived a simpler proof of Bruns's Theorem based on a property of the potential function of a homogeneous body.

References

1. G. Birkhoff (ed.), *A Source Book in Classical Analysis*, Harvard Univ. Press, 1973.

2. A. Gray, *A Treatise on Gyrostatics and Rotational Motion, Theory and Applications*, Dover, 1959 (first appeared 1918).

3. Felix Klein, *Vorlesungen über die Entwicklung der Mathematik im 19 Jahrhundert*, Vol. I, Springer-Verlag, 1926.

4. S. Kovalevsky, Zur Theorie der partiellen Differentialgleichungen, *J. Reine Angew. Math.* **80** (1875), 1–32.

5. S. Kovalevsky, Über die Brechung des Lichtes in cristallinischen Mitteln, *Acta Mathematica* **6** (1883), 249–304.

6. S. Kovalevsky, Über die Reduction einer bestimmten Klasse von Abel'scher Integrale 3-ten Ranges auf elliptische Integrale, *Acta Mathematica* **4** (1884), 393–414.

7. S. Kovalevsky, Sur la propagation de la lumière dans un milieu cristallisé, *C. R. Acad. Sci. Paris* **98** (1884), 356–357.

8. S. Kovalevsky, Om ljusets fortplantning uti ett kristalliniskt medium, *Öfversigt af Kongl. Vetenskaps-Akademiens Forhandlinger* **41** (1884), 119–121.

9. S. Kovalevsky, Züsatze und Bermerkungen zu Laplace's Untersuchung über die Gestalt des Saturnringes, *Astronomische Nachrichten* **111** (1885), 37–48.

10. S. Kovalevskaya to A. Leffler, 1888. Cited in [19], p. 124.

11. S. Kovalevsky, Sur le problème de la rotation d'un corps solide autour d'un point fixe, *Acta Mathematica* **12** (1889), 177–232.

12. S. Kovalevsky, Mémoire sur un cas particulier du problème de la rotation d'un corps pesant autour d'un point fixe, où l'intégration s'effectue à l'aide de fonctions ultraelliptiques du temps, *Mémoires Présentés par Divers Savants à l'Académie des Sciences de l'Institut National de France* **31** (1890), 1–62.

13. S. Kovalevsky, Sur une propriété du système donations différentielles qui définit la rotation d'un corps solide autour d'un point fixe, *Acta Mathematica* **14** (1890), 81–93.

14. S. Kovalevsky, Sur un théorème de M. Bruns, *Acta Mathematica* **15** (1891), 45–52.

15. S. V. Kovalevskaya, *Nauchnye raboty* (Scientific Works) (ed. P. Y. Polubarinova-Kochina), AN SSSR, Moscow, 1948.

16. S. V. Kovalevskaya, *Vospominaniya i pis'ma* (ed. S. Ya. Shtraikh), 1951.

17. Sofya Kovalevskaya, *A Russian Childhood* (transl. and ed. Beatrice Stillman), Springer-Verlag, 1978.

18. L. Kronecker, *J. Reine Angew. Math.* **108** (1891), 1–18.

19. Anna Carlotta Leffler, Duchess of Cajanello, *Sonya Kovalevsky* (transl. A. de Furuhjelm), T. Fisher Unwin, 1895.

20. G. Mittag-Leffler, Commemorative Speech as Rector of Stockholm University, 1891. Cited in [19], p. 171.

21. G. Mittag-Leffler, Weierstrass et Sonja Kowalewsky, *Acta Mathematica* **39** (1923), 170.

22. P. Y. Polubarinova-Kochina, *Sophia Vasilyevna Kovalevskaya, Her Life and Work*, Men of Russian Science (transl. P. Ludwick), Foreign Languages Press, 1957.

23. P. Y. Polubarinova-Kochina, On the Scientific Work of Sofya Kovalevskaya (transl. N. Koblitz), in [17].

24. B. Stillman, Sofya Kovalevskaya: Growing Up in the Sixties, *Russian Literature Triquarterly* **9** (1974), 287.

25. G. J. Tee, Sofya Vasil'yevna Kovalevskaya, *Math. Chronicle* **5** (1977), 132–133.

26. L. A. Vorontsova, *Sofia Kovalevskaia*, Moscow, 1957.

27. Karl Weierstrass, *Briefe von Karl Weierstrass an Sofie Kowalevskaja 1871–1891 – Pis'ma Karla Veierstrassa k Sof'ye Kovalevskoi*, Nauka, 1973.

28. E. T. Whittaker, *Analytical Dynamics of Particles and Rigid Bodies*, Cambridge Univ. Press, 1965.

Highlights in the History of Spectral Theory

L. A. STEEN

American Mathematical Monthly **80** (1973), 359–381

Not least because such different objects as atoms, operators and algebras all possess spectra, the evolution of spectral theory is one of the most informative chapters in the history of contemporary mathematics. The central thrust of the modern spectral theorem is that certain linear operators on infinite dimensional spaces can be represented in a "diagonal" form. At the beginning of the twentieth century neither this spectral theorem nor the word "spectrum" itself had entered the mathematician's repertoire. Thus, although it has deep roots in the past, the mathematical theory of spectra is a distinctly twentieth-century phenomenon.

Today every student of mathematics encounters the spectral theorem not later than his first course in functional analysis and often as early as his first course in linear algebra. Usually he studies one specimen of the spectral theorem, plucked out of historical context and imbedded in the logical context of his particular course. Although this scheme is pedagogically efficient and logically aesthetic, it does often obscure the fact that the spectral theorem was (and perhaps still is) an evolving species. Its evolution is an outstanding example of the counterpoint between pure and applied mathematics, for while the motive force in its evolution was the attempt to provide adequate mathematical theories for various physical phenomena, the forms through which it evolved are precisely those which have marked the development of modern abstract analysis.

So we offer here an austere outline of the evolution of the spectral theorem as a microcosmic example of the history of twentieth-century mathematics. To understand the significance of contemporary achievements and to recognize their continuity with the past, we begin with the principal historical roots of our subject.

1 Principal axes theorem

The only theorem available at the turn of the twentieth century which we can with hindsight recognize as a direct forerunner of the modern spectral theorem is the principal axes theorem of analytical geometry. It should not be surprising that the simplest form of this theorem is contained in the writings of the founders of analytical geometry, Pierre de Fermat (1601–1665) and René Descartes (1596–1650). For the Euclidean plane $\mathbb{R}^2$, this theorem says that a quadratic form $ax^2 + 2bxy + cy^2$ can be transformed by a rotation of the plane into the normal form $\alpha x^2 + \beta y^2$, where the principal axes of the normal form coincide with the new coordinate axes. The essential content of this theorem—that the algebraic reduction to normal form corresponds to the geometric rotation onto principal axes—is contained in Descartes' *La Géométrie* [1637], and was known at about the same time by Fermat but not published until after his death [1679]. The term "principal axes" was introduced by Leonhard Euler (1707–1783) in his investigation of the mechanics of rotating bodies [1765]; Euler also discussed (in [1748]) the reduction of quadratic forms in two and three dimensions.

The general form of the principal axes theorem asserts that any symmetric quadratic form $(Ax, x) = \sum \alpha_{ij} x_i x_j$ on $\mathbb{R}^n$ can be rewritten by means of an orthogonal transformation $T: \mathbb{R}^n \mapsto \mathbb{R}^n$ in the normal form $\sum \lambda_i x_i^2$. (*A* is *symmetric* if $\alpha_{ij} = \alpha_{ji}$, and T is *orthogonal* if it leaves invariant the Euclidean metric on $\mathbb{R}^n$.) The generalization from $\mathbb{R}^3$ to $\mathbb{R}^n$ of the algebraic part of this theorem (that a quadratic form can be written as a sum of squares) was discussed by Joseph Louis Lagrange (1736–1813) in a paper [1759] on the maxima and minima of functions of several variables. In [1827] Carl Gustav Jacob

Jacobi (1804–1851) investigated the principal axes of various quadratic surfaces, and about the same time Augustin-Louis Cauchy (1789–1867) showed in [1829] and [1830] that the coefficients λ_i of the normal form of a symmetric quadratic form must be real.

But it was not until the second half of the nineteenth century that the general form of the principal axes theorem was achieved when James Joseph Sylvester (1814–1897) and Arthur Cayley (1821–1895) used the notation of matrices to systematize the algebraic description of n-dimensional space. In [1852] Sylvester showed explicitly that the coefficients λ_i in the normal form of (Ax, x) are the roots of the characteristic polynomial $\det(\lambda I - A) = 0$; in [1858] Cayley inaugurated the calculus of matrices, in which the reduction to normal form corresponded to a diagonalization process on the matrix A. Specifically, the principal axes theorem says in the language of matrices that each symmetric real matrix A is orthogonally equivalent to a diagonal matrix D; in other words, for some orthogonal matrix T, the matrix $D = T^{-1}AT$ is in diagonal form. The diagonal entries of D are the *eigenvalues* of A—that is, the roots of the polynomial equation $\det(\lambda I - A) = 0$.

Although the new concepts of matrix theory had an immediate and profound influence on British mathematics, their impact on the continent was relatively minor. Especially in Germany bilinear forms continued well into the twentieth century to be the principal tool of analytical geometry, and in [1878] Georg Frobenius (1849–1917) published a systematic account of matrix algebra entirely in the language of bilinear forms. So by the end of the nineteenth century we can discern two versions of the principal axes theorem: the reduction to normal form of a symmetric bilinear form, and the diagonalization of a real symmetric matrix.

2 Infinite systems of linear equations

The central fact of modern spectral theory is that certain linear operators on infinite-dimensional spaces can also be presented in "diagonal" form. Thus the second historical taproot of spectral theory is the evolution of infinite-dimensional theory from finite-dimensional cases. This evolution occurred first in algebra—in the solution of systems of linear equations—and only much later in geometry. Finite systems of linear equations were solved most often throughout the eighteenth and nineteenth centuries by the method of elimination, as expounded, for instance, in [1770] and [1779] by Euler and Etienne Bézout (1730–1783). In [1750] Gabriel Cramer (1704–1752) introduced for 3×3 systems the rule which now bears his name, although he did not, of course, use the concept or notation of determinants.

Infinite systems of equations were used throughout the eighteenth and nineteenth centuries to obtain formal solutions to differential equations by the method of undetermined coefficients: if a formal power series with unknown coefficients is substituted for the unknown in a given differential equation, the task of solving the differential equation is reduced to that of determining the infinitely many unknown coefficients. (Of course, few at that time worried very much about the convergence of the power series thus obtained.) If all went well, the infinite system of equations in the unknown coefficients would exhibit a recursive pattern which made it possible to solve the infinite system by finite-dimensional tools. But for this reason precisely, these recursive techniques contributed little to the development of a general theory of infinite-dimensional systems.

Joseph Fourier (1768–1830) launched the first significant general attack on the problem of infinite systems of equations when he attempted to show [1822] that every function can be expressed as an infinite linear combination of trigonometric terms. The problem of determining the unknown coefficients in these linear combinations led him directly to the general problem of solving an infinite system of linear equations. Fourier's approach (called the *principe des réduites* by Frédéric Riesz [1913a]) was to solve the first $n \times n$ system by ordinary means and let $n \to \infty$.

Although Fourier's assertion about the expansion of "arbitrary" functions into trigonometric series stimulated intense work on the theory of integration, his method of solving infinite systems of linear equations was virtually ignored. More than fifty years passed before Theodor Kötteritzsch of Saxony reopened the investigation with a paper [1870] in which he attempted to extend Cramer's rule to infinite systems. Seven years later the American astronomer George William Hill (1838–1914) published in Cambridge, Massachusetts, a monograph [1877b] in which he successfully applied to the infinite-dimensional case the theory of determinants which had at that time only been established for finite-dimensional systems. Hill's work was first disseminated in Europe in [1886a] when G. Mittag-Leffler reprinted it in *Acta Mathematica* in the year following the appearance in France of a paper [1885a] by Paul Appell (1855–1930) in which he applied the *principe des réduites* to determine the coefficients of the power series expansion of elliptic functions.

At this point Henri Poincaré (1854–1912) entered the discussion with two papers ([1885b], [1886b]) in which he provided a rigorous definition for an infinite determinant in order to clarify the works of Hill and Appell. The work begun in Paris by Poincaré was continued in Stockholm by Helge von Koch (1870–1924) who developed between 1890 and 1910 an extensive theory of infinite determinants. Von Koch's first major papers on this subject appeared in [1891] and [1892]; his own survey of the field in [1910d] provides further references. The more recent survey [1968] by Michael Bernkopf includes a complete discussion of these fundamental papers.

3 Integral equations

The theory of infinite matrices and determinants might have led directly to an elementary spectral theorem if someone had generalized the diagonalization form of the principal axes theorem. But the road to spectral theory was not that straight: the first spectral theorem was achieved only after infinite determinants were applied to integral equations, thereby extending the theory from the countably to the uncountably infinite. The formal study of integral equations is usually traced back to [1823] and [1826] when the young Norwegian genius Niels Henrik Abel (1802–1829) used an integral equation to solve a generalized tautochrone problem concerning the shape of a wire along which a frictionless bead slides under the influence of gravity. Somewhat later Joseph Liouville (1809–1882) introduced (in [1837]) the method of iteration to solve a specific type of integral equation; in [1877a] Carl Neumann (1832–1925) extended Liouville's iterative method to a more general setting while investigating a boundary value problem for harmonic functions.

Neumann's work precipitated considerable research in integral equations, especially by Poincaré in France and in Rome by Vito Volterra (1860–1940). But it was not until 1900 that the theory of integral equations became especially relevant to the history of spectral theory, for in that year the Swedish mathematician Ivar Fredholm (1866–1927), then a docent at the University of Stockholm, applied to integral equations the theory of infinite matrices and determinants as developed by his colleague von Koch. By mimicking von Koch's technique for expanding infinite determinants, Fredholm developed in [1900] his now famous "alternative" theorem concerning the solutions ϕ of the integral equation

$$\phi(x) + \int_0^1 K(x,y)\phi(y)\,dy = \psi(x), \quad 0 \leq x \leq 1. \tag{1}$$

Just as Daniel Bernoulli (1700–1784) nearly two centuries earlier had represented the vibrating string as the limit of n oscillating particles [1732], so Fredholm considered the integral equation (1) to be the limiting case of the corresponding linear system

$$\phi(x_i) + \sum_{j=1}^n K(x_i, y_j)\phi(y_j) = \psi(x_i), \quad 1 \leq i \leq n. \tag{2}$$

Fredholm defined a "determinant" D_K for the integral equation (1) which is the continuous analog of the classical determinant of the $n \times n$ system (2) and showed—in exact analogy to the classical theory for (2)—that the integral equation (1) has a unique solution which can be expressed as the quotient of two "determinants" whenever $D_K \neq 0$; or alternatively, if $D_K = 0$, then the transposed homogeneous equation

$$\phi(x) + \int_0^1 K(y,x)\phi(y)\,dy = 0$$

has non-trivial solutions and (1) is solvable if and only if ψ is orthogonal to each of these solutions. Fredholm's major paper on this subject appeared in [1903a]; a summary of this work together with later developments is the substance of his survey article [1910e].

4 David Hilbert

Although there is very little in the papers of either von Koch or Fredholm that could be construed as a logical ancestor of the modern spectral theorem, we have discussed these developments for two particular reasons—one mathematical, the other historical. The twentieth-century evolution of infinite-dimensional spectral theory from the much simpler finite-dimensional theory is foreshadowed by the nineteenth-century development of linear equation and determinant theory, from the finite to the infinite (von Koch) to the continuous (Fredholm). But there is even a more direct connection, for when Fredholm's ideas were introduced (by Fredholm's colleague Eric Holmgren) in David Hilbert's 1900–01 seminar at Göttingen, Hilbert, in the words of Hermann Weyl [1944], "caught fire at once". For the next ten years Hilbert (1862–1943) focused his impressive mathematical talent exclusively on integral equations, and through a series of six papers published in *Göttingen Nachrichten* from 1904 to 1910 (collected and published as one volume in [1912a]) he outlined the basic definitions and theorems of spectral theory (which he named) and Hilbert space theory (which he did not name, or even define directly).

Hilbert worked primarily with the integral equation

$$\phi(x) - \lambda \int_0^1 K(x,y)\phi(y)\, dy = \psi(x) \quad (3)$$

together with the analogous finite- or infinite-dimensional matrix equation

$$\phi(x_i) - \lambda \sum_j K(x_i, y_j)\phi(y_j) = \psi(x_i). \quad (4)$$

In the process of constructing the machinery necessary to solve these equations, Hilbert defined the spectrum of the quadratic form K, distinguished the point spectrum from the continuous spectrum, and defined the concept of complete continuity which served to separate those forms that had pure point spectra from those with more complicated spectra. But most important from the viewpoint of this essay, he formulated and proved the spectral theorem—not only for completely continuous forms, but for bounded forms as well.

Hilbert's papers on integral equations contain an astonishing quantity of what we now recognize as modern analysis in classical language. Because he was primarily concerned with solving integral equations, Hilbert never applied his results specifically to matrices or operators; furthermore, because of the position of the parameter λ in equation (3), all of Hilbert's eigenvalues and spectral points are reciprocals of those in use today. And while his theorems had a most modern thrust, his basic method of proof was that of Bernoulli and Fredholm—a laborious passage to the limit from the corresponding finite case.

Beginning in 1905 with his doctoral dissertation under Hilbert, Erhard Schmidt (1876–1959) generalized and simplified Hilbert's work by introducing the suggestive language of Euclidean geometry. In [1907a], [1907b] and [1908a] Schmidt presented a definitive theory of "Hilbert's space"—what we now call ℓ^2, the space of square-summable sequences—replete with the language of norms, linearity, subspaces and orthogonal projections. (It was Schmidt who generalized to ℓ^2 the iterative algorithm for orthonormalization first introduced in [1883] by Jörgen Pederson Gram of Copenhagen.) Schmidt's conceptual simplifications were immediately incorporated by Ernst Hellinger (1883–1950) and Hermann Weyl (1885–1955) in their 1907 and 1908 dissertations under Hilbert. In [1909a] Hellinger reformulated the theory of quadratic forms in the new language of Hilbert and Schmidt, and in the same year Weyl published an extensive study of bounded forms and their spectra [1909d]. So by the end of the first decade of the twentieth century we can perceive in the writings of Hilbert and his pupils the major part of spectral theory for bounded linear transformation on ℓ^2.

5 Hilbert-Schmidt spectral theory

Recall, as did Hilbert at the beginning of his first paper on integral equations [1904a], the principal axes theorem for finite-dimensional spaces. Let $\lambda_1 \leq \lambda_2 \leq \cdots \leq \lambda_n$ be the n (real) eigenvalues of the symmetric $n \times n$ matrix K, listed according to multiplicity. Let $\phi_1, \ldots, \phi_n$ be an orthonormal collection of corresponding eigenvectors, so $K\phi_i = \lambda_i \phi_i$, for $1 \leq i \leq n$. Then the action of K is represented, with respect to the basis, by the diagonal matrix L with entries λ_i on the main diagonal. The matrix T whose rows are the vectors $\phi_1, \ldots, \phi_n$ is an orthogonal transformation which maps the new basis vectors $\phi_1, \ldots, \phi_n$ back to the original (canonical) basis vectors. Thus $L = T^{-1}KT$ is the diagonalization of K by the orthogonal transformation T. The matrix L can be written as $\sum_{i=1}^n \lambda_i P_i$, where P_i is the *projection* (i.e., the transformation which projects $\mathbb{R}^n$ onto the one-dimensional subspace spanned by ϕ_i).

Hilbert's first step in extending this theorem was to generalize the concept of eigenvalue to the case of an infinite symmetric form K. His new concept was the *spectrum* of K, denoted by $\sigma(K)$, which is the set of λ for which the transformation $\lambda I - K$ is not invertible. (Actually, Hilbert used $I - \lambda K$ while Schmidt used $\lambda I - K$.) The subset of $\sigma(K)$ consisting of those λ for which the equation $K\phi = \lambda\phi$ has non-trivial solutions is called the *point spectrum* of K; this is the strict analog of the set of eigenvalues. The complement of the point spectrum in $\sigma(K)$ is called the *continuous spectrum*. Much of

Hilbert's fourth paper [1906a] is devoted to a study of the relationships between a transformation K and its spectrum $\sigma(K)$.

One of the simplest relationships Hilbert discovered was that the spectrum of K is a bounded set whenever K is a bounded transformation—that is, whenever the set

$$S = \{\|Kx\| : \|x\| \leq 1\}$$

is bounded, where the notation $\|\ \|$, due to Schmidt, is the ℓ^2 norm. In fact, whenever K is symmetric, the least upper bound of S, called the bound (or norm) of K and denoted by $\|K\|$, is the same as the least upper bound of $\{|\lambda| : \lambda \in \sigma(K)\}$; this fact is now called the *spectral radius theorem*. The bounded linear transformations on ℓ^2 are important from another point of view, also due to Hilbert: they are precisely the continuous linear transformations, in the sense that they preserve strong convergence (i.e., $\|Kx_n - Kx\| \to 0$ whenever $\|x_n - x\| \to 0$).

Hilbert extended the principal axes theorem to symmetric bounded linear transformations; the spectra of these transformations are bounded subsets of the real axis. Those λ in the point spectrum $\rho(K)$ of K are like eigenvalues, since there exists an orthonormal collection of corresponding eigenvectors ϕ_λ satisfying $K\phi_\lambda = \lambda\phi_\lambda$. If P_λ denotes the projection onto the subspace generated by ϕ_λ, we can form the diagonal transformation $L = \sum \lambda P_\lambda$ where λ ranges over the point spectrum $\rho(K)$. The transformation L reflects accurately the action of K on the subspace generated by the eigenvectors ϕ_λ, but since this subspace will in general be strictly smaller than ℓ^2—since we have omitted the continuous spectrum—we cannot say that L and K represent the same transformation.

To express the contribution of the continuous spectrum, Hilbert set up an integral patterned after one defined in [1894] by the Dutch mathematician Thomas-Jean Stieltjes (1856–1894). In his study of continued fractions, Stieltjes was led (via the problem of moments) to the integral $\int_a^b f(x)dg(x)$ as the limit of the sum

$$\sum f(\xi_i))[g(x_i) - g(x_{i-1})]$$

(for continuous f and increasing g). By rewriting the sum $\sum \lambda_i P_{\lambda_i}$ as

$$\sum \lambda_i \left[E_{\lambda_i} - E_{\lambda_{i-1}} \right],$$

where $E_{\lambda_i} = \sum_{j=1}^{i} P_{\lambda_j}$, Hilbert constructed for the continuous spectrum $s(K)$ the Stieltjes-type integral

$$\int_{s(K)} \lambda \, dE_\lambda$$

as the limit of sums of the form $\sum \lambda_i \left[E_{\lambda_i} - E_{\lambda_{i-1}} \right]$. Then Hilbert's spectral theorem was that every symmetric bounded linear transformation on ℓ^2 can be represented (by means of an orthogonal transformation) in the "diagonal" form

$$\sum_{\rho(K)} \lambda P_\lambda + \int_{s(K)} \lambda \, dE_\lambda, \qquad (5)$$

where the summation is over the point spectrum, and the integral is over the continuous spectrum.

Hilbert completed his spectral theory by identifying a large class of transformations whose continuous spectra were empty. He called these transformations *completely continuous*, and Schmidt characterized them by the property of mapping weakly convergent sequences to strongly convergent sequences. In other words, the linear transformation K is completely continuous if

$$\|Kx_n - Kx\| \to 0$$

whenever $(y, x_n) \to (y, x)$ for all y. The completely continuous transformations are the nearest infinite-dimensional analogs to the finite-dimensional transformations, since their spectra consist entirely of eigenvalues with zero as the only possible accumulation point; furthermore, every completely continuous symmetric linear transformation K can be expressed (by an orthogonal transformation) in the diagonal form $\sum \lambda P_\lambda$ (since $s(K) = \varnothing$).

Although Hilbert originally used infinite matrices merely as convenient approximations to integral equations, he concluded his theoretical investigation by establishing a major link between these two theories, namely that of a *complete orthogonal system*. Such a system $\{\phi_n\}$, either of vectors in the sequence space ℓ^2 or of continuous functions on the interval $[0, 1]$, is characterized by the orthogonality relation $(\phi_n, \phi_m) = 0$ if $n \neq m$, together with the fact that every vector (or continuous function) ϕ can be represented by the Fourier-type series $\phi = \sum_{n=1}^{\infty} a_n \phi_n$. The matrix equation (4) can then be derived (by mathematics, rather than by analogy) from the integral equation (3) by replacing each continuous function ϕ, ψ by its Fourier expansion with respect to the complete orthogonal system $\{\phi_n\}$. This application of a complete orthogonal system enabled Hilbert to derive

Fredholm's alternative theorem for the integral equation (1) directly from the corresponding theorem for the infinite linear system (2).

To keep the record straight, we should emphasize again that Hilbert introduced spectral theory in the language of quadratic forms, whereas we have reported his work primarily in the language of linear transformations on the infinite-dimensional space ℓ^2. Barely fifty years had elapsed since Cayley in England and Hermann Grassman (1809–1877) in Germany had begun, in [1843] and [1844], the systematic study of Euclidean n-dimensional space for $n > 3$. Hilbert and Schmidt were the first to explore the totally unknown depths of an infinite-dimensional space and it was not until other such spaces were studied that the broad outlines of a theory of linear transformations became clear. The early twentieth-century development of infinite-dimensional (function) spaces is recorded in [1966].

6 The Lebesgue integral

At about the same time as Hilbert was creating his spectral theory for spaces of square summable sequences, Henri Lebesgue (1875–1941) was developing the new integral which now bears his name ([1901], [1904c]). In three brief papers in 1907 Friedrich Riesz (1880–1956) and Ernst Fischer (1875–1959) joined together the works of Hilbert and Lebesgue by showing that Hilbert's space ℓ^2 is isomorphic to the space L^2 of functions whose square is Lebesgue integrable. In a subsequent paper [1910c] (in which he introduced the more general L^p spaces), Riesz derived a spectral theory for L^2 entirely analogous to that developed for ℓ^2 by Hilbert and Schmidt.

In the year preceding the appearance of his paper on L^p spaces, Riesz proved in [1909b] his now famous representation theorem in which he solved a problem first studied in [1903b] by Jacques Hadamard (1865–1963). What Riesz showed was that every continuous linear functional on $C([a, b])$ is a Stieltjes integral $\int f \, dg$ with respect to some function g of bounded variation. Lebesgue then showed in [1910f], in direct response to Riesz's paper, that every Stieltjes integral can be interpreted as a Lebesgue integral under a proper interpretation of the heuristic formula

$$\int f(x) \, dg(x) = \int f(x) g'(x) \, dx.$$

This led Johann Radon (1887–1956) to develop (in [1913b]) integration with respect to a measure (i.e., a countably additive set function), thus encompassing the integrals of both Lebesgue and Stieltjes and providing the foundation for all modern theories of the abstract integral.

We can see from this digression that the evolution of the modern integral was closely connected to Hilbert's creation of spectral theory. Although neither theory depended logically on the other, the historical dependence of each on the other is quite clear: Hilbert used Stieltjes' integral to obtain the spectral theorem for ℓ^2, while Riesz, following Hilbert, used and thereby immortalized Lebesgue's integral by developing the spectral theory of L^2.

The second decade of spectral theory was rather uneventful. In Göttingen, Hilbert had turned his attention to the axiomatization of physics, a task which he had proposed to the International Congress of Mathematicians in 1900 as the sixth of his famous 23 problems for twentieth-century mathematics. "Physics," he said, "is much too hard for physicists" ([1970]). In the United States Eliakim Hastings Moore (1862–1932) at the University of Chicago developed a system of "general analysis" ([1908b], [1912b]), which was designed to include as special cases the work of Hilbert, Fredholm, and Riesz. But Moore's results were constrained by the fact that European investigators were not then accustomed to receiving new mathematical ideas from America. So while Moore's research had a profound effect on the development of mathematics in the United States, it did not influence significantly the direction of research on spectral theory.

Many European efforts from 1910 to 1925 were devoted to exposition and recapitulation. Riesz [1913a], Fredholm [1910c], and von Koch [1910d] published surveys of the theory of infinitely many variables and integral equations, each of which contained various forms of Hilbert's spectral theory. Hilbert's collected papers on integral equations were themselves published in book form in [1912a]. But certainly the most impressive survey work of this period was the massive *Enzyklopädie der Mathematischen Wissenschaften* which contains in volume II.3.2 a comprehensive discussion of integral equations and spectral theory by Hellinger and Otto Toeplitz (1881–1940); this survey paper was also published separately [1928a].

7 Quantum mechanics

In Göttingen in 1925–26 Werner Heisenberg (1901–1976) and Erwin Schrödinger (1887–1961) created the theory of quantum mechanics. In Heisenberg's theory the physical fact that certain atomic observations cannot be made simultaneously was interpreted mathematically to mean that the operations which represented these observations were not commutative. Since the algebra of matrices is noncommutative, Heisenberg together with Max Born and Pascual Jordan ([1925a], [1926a]) represented each physical quantity by an appropriate (finite or infinite) matrix, called a *transformation*; the set of possible values of the physical quantity was the spectrum of the transformation. (So the spectrum of the transformation which represented the energy of an atom was precisely the spectrum of the atom.) Schrödinger, in contrast, advanced a less unorthodox theory based on his partial differential wave equation. Following some initial surprise that Schrödinger's "wave mechanics" and Heisenberg's "matrix mechanics"—two theories with substantially different hypotheses—should yield the same results, Schrödinger unified the two approaches by showing, in effect, that the eigenvalues (or more generally, the spectrum) of the differential operator in Schrödinger's wave equation determine the corresponding Heisenberg matrix. Similar results were obtained simultaneously ([1925b], [1926b]) by the British physicist Paul A. M. Dirac (1902–1984). Thus interest in spectral theory once again became quite intense.

Hilbert himself was astonished that the spectra of his quadratic forms should come to be interpreted as atomic spectra. "I developed my theory of infinitely many variables from purely mathematical interests, and even called it 'spectral analysis' without any presentiment that it would later find an application to the actual spectrum of physics" [1970]. It quickly became clear, however, that Hilbert's spectral theory was the proper mathematical basis for the new mechanics. Finite and infinite matrices were interpreted as transformations on a Hilbert space (still thought of primarily as ℓ^2 or L^2) and physical quantities were represented by these transformations. The mathematical machinery of quantum mechanics became that of spectral analysis and the renewed activity precipitated the publication by Aurel Wintner (1903–1958) of the first book [1929b] devoted to spectral theory.

Hilbert's original spectral theorem applied to real quadratic forms (or infinite matrices) that were bounded and symmetric. This theorem was quickly and easily extended (by Schmidt and others) to bounded complex matrices $A = (a_{ij})$ for which $a_{ij} = \overline{a_{ji}}$; such matrices are called *Hermitian* after the French mathematician Charles Hermite (1822–1901) who introduced them (in [1855]) and proved their eigenvalues real. Both symmetric and Hermitian forms may be characterized in terms of their respective inner product by the relation $(Ax, y) = (x, Ay)$ for all x, y. Like symmetric matrices, Hermitian transformations have real spectra and, more generally, play the role of the real number line in the algebra of transformations.

Almost miraculously, it was precisely the Hermitian transformations which qualified in the new mechanics to represent a physical quantity. One reason for this is that physical quantities are measured by real numbers, so it is natural to represent them by those transformations which behave like real numbers. Perhaps a more compelling justification is that the hypothesis that the transformations of mathematical physics are Hermitian implies certain fundamental laws (or assumptions) of physics: if A is Hermitian, the wave equation $\ddot{\phi} = A\phi$ implies the conservation of energy, a fundamental law of classical mechanics, and the solutions of Schrödinger's equation $\dot{\phi} = iA\phi$ will have constant norm, which is a fundamental assumption of quantum mechanics.

Although every observable was represented in the new mechanics by a Hermitian transformation, it was not necessarily true that every such transformation represented an observable. Dirac [1930b] added the crucial hypothesis that a Hermitian transformation represents an observable if and only if its eigenvectors form a complete (orthogonal) system: his hypothesis was designed to insure that any vector (representing a quantum-mechanical state) could be expressed as a (possibly infinite) linear combination of eigenvectors of any given observable. The identification of transformations with this property is part of the Hilbert-Schmidt spectral theory, but this theory provided only a partial answer: those Hermitian transformations which are completely continuous have a complete set of eigenvalues.

This theorem did not provide a satisfactory elucidation of Dirac's hypotheses since the transformations of quantum mechanics are usually not completely continuous. Most of the important transformations in physics involve differentiation of, say, functions in L^2. The theorem on integration by parts shows that differentiation is formally symmetric, for in this case $(Af, g) = (f, Ag)$ means $\int f'g = \int fg'$. But since the derivative of a function has practically no relation to the magnitude of the function, differentiation is neither continuous nor bounded, nor even defined everywhere. In fact, if a symmetric or Hermitian transformation (like differentiation) were defined everywhere, it would have to be bounded. This rather surprising result—which says, in effect, that a candidate for the spectral theorem which fails to be bounded must fail to be everywhere defined— was demonstrated as early as [1910b] by Hellinger and Toeplitz.

Thus many of the transformations of quantum mechanics, although Hermitian, failed nevertheless to satisfy the second of Hilbert's hypotheses—namely, that they be bounded. Like differentiation, they were unbounded and defined only on a dense subset of L^2. Paul Dirac attempted to overcome the exceptional behavior of differentiation by introducing his δ-function to provide derivatives where none existed and thereby to enlarge the set of functions to which the differentiation transformation could be applied. Dirac's approach was highly successful in explaining the new quantum mechanics and led eventually to Laurent Schwartz's theory of distributions, precisely because it lacked an adequate mathematical foundation. But in 1926 Dirac's approach represented more an alternative to spectral theory than an extension of it, and it did not really help to extend Hilbert's theory to unbounded transformations.

8 John von Neumann

After Hilbert, the only major study of unbounded transformations was that published in [1923] by Torsten Carleman (1892–1949) in Sweden. In this monograph Carleman showed that many of the results of Fredholm and Hilbert still hold under a weaker type of boundedness hypothesis. But from the viewpoint of spectral theory, the major breakthrough came in 1927–29 when the 25-year-old Hungarian John von Neumann (1903–1957) revolutionized the study of spectral theory by introducing the abstract concept of a linear operator on Hilbert space. In [1927] von Neumann expressed the transformation theory of quantum mechanics in terms of operators on a Hilbert space, and explicitly recognized the need to extend from the bounded to the unbounded case the spectral theory of Hermitian operators. In [1929a] he carried out that extension.

Before von Neumann, the name "Hilbert space" had been applied principally to the space ℓ^2 of square-summable sequences (often called "Hilbert's space") or to the space L^2 of Lebesgue square-integrable functions which Riesz had proved isomorphic to ℓ^2. The essential properties of these spaces, widely recognized, were those of a vector space with an inner product which was complete and separable (i.e., which had a countable dense subset). Von Neumann's first step in his theory of linear operators was to define an (abstract) Hilbert space axiomatically as any separable complete inner product space. He then defined a general linear operator on the abstract Hilbert space H as a linear transformation defined on some subset of H. This subset, called the *domain* D_T of the operator T, is usually assumed to be a linear subspace of H which, like the domain of the differentiation operator, is dense in H. Von Neumann's linear operators thus comprehend both the matrices and quadratic forms of Hilbert's theory, and the transformations of quantum mechanics.

A linear operator is continuous if and only if it is bounded, and a bounded linear operator with a dense domain can be uniquely extended to a bounded linear operator on the whole space H. Every linear operator T with a dense domain has a unique *adjoint* operator T^* defined by the relation

$$(Tx, y) = (x, T^*y) \qquad (6)$$

for all $x \in D_T$; the domain of T^* is the set of $y \in H$ for which (6) holds for all x. An operator T is called *self-adjoint* if $T = T^*$, and symmetric if T^* is an extension of T, or equivalently, if $(Tx, y) = (x, Ty)$ whenever $x, y \in D_T$. (In von Neumann's papers, the self-adjoint operators were called *hypermaximal*.)

Every self-adjoint operator is clearly symmetric, and every symmetric operator which is everywhere defined must be self-adjoint. Thus for bounded linear operators (which either are everywhere defined or may be extended to become so) the concept of symmetric and self-adjoint coincide. The Hellinger-Toeplitz theorem, cited in Section 7 above, can be extended to von Neumann's operators and shows that any symmetric operator which is everywhere defined must be bounded. (This result is closely related to a more general theorem due to Stefan Banach (1892–1945), now commonly known as the *closed graph theorem* [1932b].) Thus in von Neumann's theory there are precisely three types of symmetric operators:

I bounded, self-adjoint and everywhere defined;

II unbounded, self-adjoint and densely but not everywhere defined;

III unbounded, not self-adjoint, and densely but not everywhere defined.

Hilbert's original theory applied to operators of type I, while von Neumann's spectral theorem encompassed those of type II as well since it applies to all self-adjoint operators. This theory, though initiated by von Neumann, was developed by Riesz [1930c] and more extensively, by Marshall H. Stone (1903–1989) at Yale University who expounded it in great detail in [1932a]. The combined (but largely independent) efforts of von Neumann and Stone for the five-year period 1927–1932 provided for spectral theory the largest collection of new methods since Hilbert's five-year effort of 1901–1906.

9 Von Neumann–Stone spectral theory

Hilbert's general spectral theorem says that every bounded symmetric linear transformation T can be written in the form

$$\sum_{\rho(T)} \lambda P_\lambda + \int_{s(T)} \lambda \, dE_\lambda.$$

By rewriting the first sum as a Stieltjes-type integral and combining it with the second integral, we may express Hilbert's spectral theorem in the concise form

$$T = \int_{\sigma(T)} \lambda \, dE_\lambda, \tag{7}$$

where the integral is over the entire (bounded) spectrum of T. The operators E_λ are projections with the following properties:

(i) if $\lambda < \mu$, the range of E_λ is contained in the range of E_μ;

(ii) if $\varepsilon > 0$, $E_{\lambda+\varepsilon} \to E_\lambda$ as $\varepsilon \to 0$;

(iii) $E_\lambda \to 0$ as $\lambda \to -\infty$;

(iv) $E_\lambda \to I$ as $\lambda \to \infty$.

Stone called such a family of operators a *resolution of the identity*; in more intuitive language, properties (i)–(iv) require that the function $\lambda \to E_\lambda$ be increasing, continuous from the right, with 0 and I as left and right limiting values.

The von Neumann-Stone extension of the spectral theorem for self-adjoint operators from the bounded to the unbounded case corresponds to the extension of (7) from bounded to unbounded spectra $\sigma(T)$. Specifically, it says that to each self-adjoint operator T there corresponds a unique resolution of the identity $\{E_\lambda\}$ such that (7) holds.

Despite the power of this theorem, many differential operators are not covered by it since they are rarely self-adjoint. For instance, to make the operator $D = d/dt$ symmetric on a dense subset Δ of the Hilbert space $L^2(0,1)$, we should select for Δ the subset consisting of those continuously differentiable functions f which satisfy $f(0) = f(1) = 0$ (in order to insure that the relation $(Df, g) = (f, Dg)$ would follow by integration by parts). But the domain Δ is too small to permit D to be self-adjoint, for every continuously differentiable L^2 function is in the domain of D^*. To make D self-adjoint we would have to enlarge its domain appropriately—thereby risking a loss of symmetry. Each symmetric operator of type III suffers from the same disease: its domain is smaller than that of its adjoint. Moreover the cure—namely, extension of the domain—is often fatal since with a larger domain the operator may fail to be symmetric.

To apply his spectral theorem to symmetric operators von Neumann had to know which types of symmetric operators admit self-adjoint extensions. He [1929a] and Wintner [1929b] identified a large class of such operators, namely, those operators T, called *semibounded*, for which there is a positive constant M satisfying either $(Tx, x) < M\|x\|$ for all $x \in D_T$, or $-M\|x\| \le (Tx, x)$ for all $x \in D_T$. The best statement of this result is due to Stone [1932a] and Kurt O. Friedrichs [1934]: every semibounded symmetric operator may be extended to a semibounded self-adjoint operator with the same bound.

Whereas the central focus of the von Neumann-Stone spectral theory (and of Hilbert's also) is on operators with real spectra, the spectral theorem does apply, at least in two cases, to operators with more general spectra. The simplest case concerns isometric operators which leave the inner product on H invariant; from this definition it follows easily that the spectrum of an isometric operator is a subset of the unit circle. An isometric operator that maps H onto H is called *unitary* and is characterized by the fact that its adjoint is its inverse (i.e., $TT^* = T^*T = I$). Unitary operators were first studied in [1909c] by Isaac Schur, following their introduction by Léon Autonne in [1902]. In [1929a] von Neumann employed the Cayley transform

$$C : T \to (T - iI)(T + iI)^{-1}$$

to map symmetric operators T into isometric operators $C(T)$; he showed that T is self-adjoint if and only if $C(T)$ is unitary. Thus the spectral theory for unitary operators follows from that for self-adjoint operators by use of a spectral integral on the unit circle instead of on the real line.

Now every bounded linear operator T can be written in the "Cartesian" form $T = A + iB$, where A and B are bounded and self-adjoint; in fact,

$$A = \frac{1}{2}(T + T^*), \quad B = \frac{1}{2i}(T - T^*).$$

Thus it would appear likely that the spectral theorem could be extended to all bounded linear operators by using this decomposition. However, the details of that extension require that $AB = BA$ (or equivalently, that $TT^* = T^*T$). So the desired extension works only for those operators which commute with their adjoints: such operators are called *normal*, after Toeplitz [1918a]. Toeplitz extended Hilbert's spectral theorem to completely continuous normal quadratic forms by showing that such a form was unitarily equivalent to a diagonal form. More generally, the spectral resolution

$$T = \int_{\sigma(T)} \lambda \, dE_\lambda$$

extends to bounded normal operators, where the integration is over the spectrum of T which is a compact subset of the complex plane contained in the disc of radius $\|T\|$. Von Neumann [1930a] and Stone [1932a] extended both the definition and spectral theory of normal operators to the unbounded case as well.

We have come a long way from the principal axes theorem, and the spectral theorems of von Neumann and Stone reflect far more analysis than geometry. The geometric content of the spectral theorem for finite-dimensional space is that the entire space can be expressed as the direct sum of subspaces on each of which the given transformation acts like simple multiplication. But this theorem fails in the infinite-dimensional cases as soon as the continuous spectrum appears. In a paper written in 1938 but not published until [1949a], von Neumann effectively resuscitated the geometrical spectral theorem by defining a direct integral of Hilbert spaces (in strict analogy with the direct sum). He then showed that the action of a self-adjoint operator on any Hilbert space could be represented as the accumulated effect of simple multiplications on certain subspaces whose direct integral was (unitarily equivalent to) the original space.

10 Gelfand-Naimark theorem

The collection of all operators on a Hilbert space forms a ring; such rings, with various topologies, were extensively investigated by von Neumann and Francis J. Murray in [1936a], [1937a] and [1940a]. During the same period 1936–1940 S. W. P. Steen published in England a series of five papers ([1936b], [1937b], [1938a], [1939], (1940b)) devoted to an axiomatic theory of operators. But the papers that offered the most significant insight into the spectral theorem were [1941a], [1941b] and [1943] published in the U.S.S.R. by Israel M. Gelfand, Mark A. Naimark and Georgii E. Silov. Gelfand and his colleagues created a theory of normed rings which not only subsumed much of the work of von Neumann, Murray and Steen on rings of operators, but also provided a beautiful general setting for the study of Fourier transforms and harmonic analysis. Related studies were carried out in the United States by Stone ([1940c], [1941c]) and Shizuo Kakutani [1941d].

Normed rings were first introduced in [1936c] by the Japanese mathematician Mitio Nagumo under the name of "linear metric rings". In [1946] Charles E. Rickart christened Gelfand's normed rings "Banach algebras" to avoid misunderstanding due to the algebraic meaning of "ring"; as a consequence Russian mathematicians now use the former name, while Americans use the latter. But regardless of its name, the properties of a *Banach algebra* are those of a complete normed algebra (over the complex field $\mathbb{C}$) satisfying the multiplicative triangle inequality $\|x\|\|y\| \leq \|xy\|$. We shall assume that each Banach algebra contains an identity element e, where $\|e\| = 1$. The set of all bounded linear operators on a Hilbert space is a Banach algebra, as is the set of all continuous complex-valued functions on a compact topological space X, with the sup norm

$$\|f\| = \sup\{|f(x)| : x \in X\}.$$

The part of Banach algebra theory germane to spectral theory is the relation between these two examples.

Gelfand's theory of commutative Banach algebras depends on three fundamental concepts: homomorphisms, maximal ideals and spectra. A *homomorphism* of a commutative Banach algebra B is a nonzero multiplicative linear functional; its kernel is a *maximal ideal* since it is contained in no larger proper ideal. Moreover, every maximal ideal I is the kernel of some homomorphism, for in this case the factor algebra B/I is the field $\mathbb{C}$ of complex numbers (according to a result announced by Stanislaw Mazur [1938b] and proved by Gelfand [1941a]) so the composite map $B \to B/I \to \mathbb{C}$ is a homomorphism of B whose kernel is I. The set M_B of homomorphisms (or equivalently, of maximal ideals) of B is given the weakest topology relative to which all of the functions $\hat{x}: h \to h(x)$ are continuous, for all $x \in B$. Then the topological space M_B is compact and Hausdorff, and each element x of B is represented in $C(M_B)$ (the Banach algebra of continuous complex valued functions on M_B) by its "Gelfand transform" $\hat{x}$.

In strict analogy with the spectral theory of operators on a Hilbert space, Gelfand defines the *spectrum* $\sigma(x)$ of an element $x \in B$ to be the set of complex numbers λ for which the element $x - \lambda e$ has no inverse. The set $\sigma(x)$ is compact, non-empty and contained in the disc of radius $\|x\|$. Furthermore, $\sigma(x)$ happens to be precisely equal to the range of the Gelfand transform $\hat{x}$: $\sigma(x) = \{h(x) \mid h \in M_B\}$. For this reason the space M_B of maximal ideals is often called the *spectrum* of the Banach algebra B. If B is the algebra generated by a single element x (such as a particular operator on H), then the spectrum of the algebra B is mapped homeomorphically by $\hat{x}$ onto the spectrum of x.

In [1943] Gelfand and Naimark showed that the commutative Banach algebra $C(M_B)$ is characterized by the presence of an involution—namely, the operation of complex conjugation $* : f \to \bar{f}$. Specifically, they showed that any commutative Banach algebra with an involution (called a B^*-algebra) is isometrically isomorphic to the algebra $C(M_B)$ for some Banach algebra B. In particular, the commutative B^*-algebra $B(T)$ generated by a given bounded normal operator T is isomorphic to the algebra $C(\sigma(T))$ of all continuous functions on $\sigma(T)$, the spectrum of $B(T)$; T is assumed normal in order that the presence of the involution $* : T \to T^*$ should not destroy the commutativity of the algebra $B(T)$.

The impact of the Gelfand-Naimark theorem on spectral theory is this: the spectral theorem for a bounded normal operator T can be inferred via the isomorphism between $B(T)$ and $C(\sigma(T))$ from a corresponding theorem concerning continuous complex-valued functions on $\sigma(T)$. The required theorem is just that every continuous function on $\sigma(T)$ (in particular, the identity function $f(\lambda) = \lambda$) can be approximated uniformly by measurable step functions of the form $\sum f(\lambda_i) \chi_{\Lambda_i}$, where χ_{Λ_i} is the characteristic function of the measurable set Λ_i. The translation of this theorem to the algebra $B(T)$ (in the special case $f(\lambda) = \lambda$) is the spectral resolution of the bounded normal operator T: $T = \int \lambda \, dE_\lambda$. In words instead of symbols, the approximation theorem says that a continuous function can be approximated by linear combinations of characteristic functions, while the spectral theorem says that bounded normal operators can be approximated by linear combinations of projections. Thus Gelfand's theory of Banach algebras revealed that the spectral theorem is in some fundamental sense equivalent to a most rudimentary fact in the theory of functions.

Gelfand's theory actually yields a spectral theorem far stronger than those which we have so far discussed. By translating the approximation theorem for an arbitrary continuous function f we obtain a spectral resolution of the form $f(T) = \int f(\lambda) \, dE_\lambda$. This formula was originally introduced by von Neumann and Stone as the basis of their "operational calculus". A related general spectral theorem, also due to von Neumann [1930a], can be inferred from the Gelfand-Naimark isomorphism: any commutative family of bounded normal operators admits a simultaneous diagonalization—that is, a single resolution of the identity which simultaneously represents all operators in the family by means of the integral $\int f(\lambda) \, dE_\lambda$ for various functions f.

11 Unfinished business

This concludes our saga of the spectral theorem. Our historical vision has been deliberately narrow, focused throughout on the evolution of just one theorem, and only rarely have we glanced at the many fascinating applications and extensions of the basic theory. For example, spectral theory for spaces without inner products can be traced back to Riesz [1918b] and T. H. Hildebrandt [1931], while the rudiments of spectral theory for differential operators are contained in the work [1908c] of George Birkhoff; in [1928b] and [1930d] Norbert Wiener developed a theory of *spectral analysis* for functions in an attempt to analyze mathematically the spectrum of white light, while twenty years later Arne Beurling [1949b] inaugurated the complementary study of *spectral synthesis*; and in [1942] Edgar R. Lorch, continuing work begun in [1913a] by F. Riesz, investigated spectral sets in the plane by means of contour integrals. Had we stopped to investigate each such offshoot our evolutionary tree (Figure 1) would have looked like a forest. Indeed, it took Nelson Dunford and Jacob Schwartz nearly 3000 pages to survey spectral theory ([1958a], [1963], [1971]). So any who are inspired to examine the fruits of spectral theory are invited to read this treatise or any of its many less ambitious companions ([1951], [1953], [1958b], [1962]). Our mission to describe the roots and main trunk of spectral theory is accomplished.

STEEN: Highlights in the History of Spectral Theory

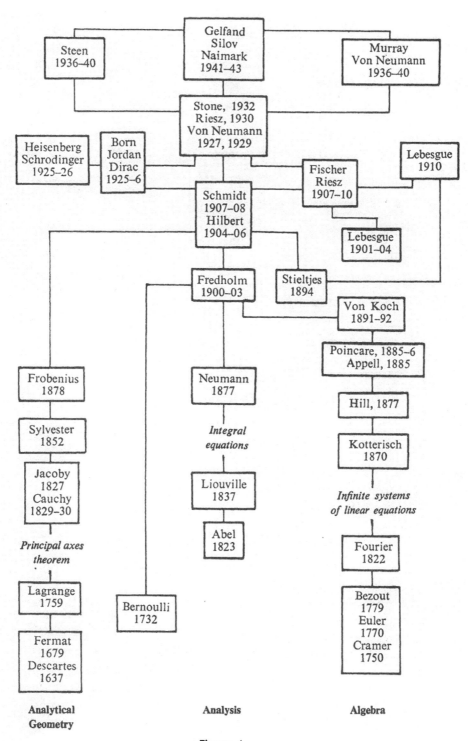

Figure 1.

References

1637 René Descartes, *La Géométrie*, Appendix I to *Discours de la Méthode*, Leiden, 1637 (*Oeuvres* VI, pp. 367–514).

1679 Pierre de Fermat, Ad locus planos et solidos isagoge, *Varia Opera Mathematica*, 1679 (*Oeuvres* I, pp. 91–110).

1732 Daniel Bernoulli, Theoremata de oscillationibus corporum filo flexii connaxorum et catenae verticaliter suspensae, *Comm. Acad. Scient. Imper. Petropolitanae* **6** (1732), 108–122.

1748 Leonhard Euler, *Introductio in Analysin Infinitorum*, 1748 (*Opera Omnia* (1) IX, pp. 379–392).

1750 Gabriel Cramer, *Introduction à l'analyse des lignes courbes algébriques*, Geneva, 1750.

1759 Joseph-Louis Lagrange, Recherches sur la méthode de maximis et minimis, *Miscellanea Taurinensia* **1** (1759), 18–42 (*Oeuvres* Vol. 1, pp. 3–20).

1765 Leonhard Euler, *Theoria Motus Corporum Solidorum Seu Rigidorum*, 1765 (*Opera Omnia* (2) III, pp. 193–214).

1770 Leonhard Euler, *Vollständige Ableitung zur Algebra*, St. Petersburg, 1770.

1779 Etienne Bézout, *Théorie Générale des Équations Algébriques*, Paris, 1779.

1822 Joseph B. J. Fourier, *Théorie Analytique de la Chaleur*, Paris, 1822.

1823 Niels Henrik Abel, Solution de quelques problèmes à l'aide d'intégrales définies, *Magazin for Naturv.* **1** (1823), 205–215 (*Oeuvres* 1, pp. 11–27).

1826 Niels Henrik Abel, Résolution d'un problème de mécanique, *J. Reine Angew. Math.* **1** (1826), 153–157 (*Oeuvres* 1, 97–101).

1827 Carl G. J. Jacobi, Über die Hauptaxen der Flächen der Zweiten Ordnung, *J. Reine Angew. Math.* **2** (1827), 227–233 (*Werke* III, 45–53).

1829 Augustin-Louis Cauchy, Sur l'équation à l'aide de laquelle on détermine les inégalités séculaires des mouvements des planètes, *Exercices de Mathématiques*, Paris, 1829 (*Oeuvres* (2) IX, pp. 174–195).

1830 Augustin-Louis Cauchy, Mémoire sur l'équation qui a pour racines les moments d'inertie principaux d'un corps solide et sur diverses équations du même genre, *Mém. Acad. Sci. Inst. France* **9** (1830), 111–113 (*Oeuvres* (1) II, pp. 79–81).

1837 Joseph Liouville, Sur le développement des fonctions... (second mémoire), *J. Math. Pures Appl.* **2** (1837), 16–35.

1843 Arthur Cayley, Chapters in the analytical geometry of n dimensions, *Cambridge Math. J.* **4** (1843), 119–127 (*Math. Papers* I, pp. 55–62).

1844 Hermann Grassmann, *Die Ausdehnungslehre*, Leipzig, 1844.

1852 James Joseph Sylvester, A demonstration of the theorem that every homogeneous quadratic polynomial is reducible by real orthogonal substitution to the form of a sum of positive and negative squares, *Phil. Mag.* **4** (1852), 138–142 (*Math. Papers* I, pp. 378–381).

1855 Charles Hermite, Remarque sur un théorème de M. Cauchy, *C. R. Acad. Sci. Paris* **41** (1855), 181–183.

1858 Arthur Cayley, A memoir on the theory of matrices, *Phil. Trans. Roy. Soc. London* **148** (1858), 17–37 (*Math. Papers* II, pp. 475–496).

1870 Theodor Kötteritzsch, Über die Auflösung eines Systems von unendlich vielen linearen Gleichungen, *Z. Math. Physik* **15** (1870), 1–15, 229–268.

1877a Carl Neumann, *Untersuchungen über des logarithmische und Newtonsche Potential*, Teubner, 1877.

1877b George William Hill, *On the Part of the Motion of the Lunar Perigee which is a Function of the Mean Motions of the Sun and Moon*, John Wilson, 1877.

1878 Georg Frobenius, Über lineare Substitutionen und bilineare Formen, *J. Reine Angew. Math.* **84** (1878), 1–63 (*Gesammelte Abhandlungen* I, 343–405).

1883 Jörgen Pederson Gram, Über die Entwickelung realer Funktionen in Reihen mittelst der Methode der kleinsten Quadrate, *J. Reine Angew. Math.* **94** (1883), 41–73.

1885a Paul Appell, Sur une méthode élémentaire pour obtenir les développements en série trigonométrique des fonctions elliptiques, *Bull. Soc. Math. France* **13** (1885), 13–18.

1885b Henri Poincaré, Remarques sur l'emploi de la méthode précédente, *Bull. Soc. Math. France* **13** (1885), 19–27 (*Oeuvres* V, pp. 85–94).

1886a George William Hill, On the part of the motion of the lunar perigee which is a function of the sun and the moon, *Acta Mathematica* **8** (1886), 1–36 (*Coll. Math. Works* I, pp. 243–270).

1886b Henri Poincaré, Sur les determinants d'ordre infini, *Bull. Soc. Math. France* **14** (1886), 77–90 (*Oeuvres* V, pp. 95–107).

1891 Helge von Koch, Sur une application des déterminants infinis à la théorie des équations différentielles linéaires, *Acta Mathematica* **15** (1891), 53–63.

1892 Helge von Koch, Sur les determinants infinis et les équations différentielles linéaires, *Acta Mathematica* **16** (1892), 217–295.

1894 Thomas-Jean Stieltjes, Recherches sur les fractions continués, *Ann. Fac. Sci. Toulouse* **8** (1894), 1–122 (*Oeuvres* II, 402–566).

1900 Ivar Fredholm, Sur une nouvelle méthode pour la resolution du problème de Dirichlet, *Öfver. Vet. Akad. Förhand, Stockholm* **57** (1900), 39–46.

1901 Henri Lebesgue, Sur une généralisation de l'intégrale définie, *C. R. Acad. Sci. Paris* **132** (1901), 1025–1028.

1902 Léon Antonne, Sur l'Hermitien, *Rend. Circ. Mat. Palermo* **16** (1902), 104–128.

1903a Ivar Fredholm, Sur une classe d'équations fonctionnelles, *Acta Mathematica* **27** (1903), 365–390.

1903b Jacques Hadamard, Sur les opérations fonctionnelles, *C. R. Acad. Sci. Paris* **136** (1903), 351–354.

1904a David Hilbert, Grundzüge einer allgemeinen Theorie der linearen Integralgleichungen, Erste Mitteilung, *Göttingen Nachrichten* (1904), 49–91.

1904b David Hilbert, Grundzüge einer allgemeinen Theorie der linearen Integralgleichungen, Zweite Mitteilung, *Göttingen Nachrichten* (1904), 213–259.

1904c Henri Lebesgue, *Leçons sur l'Intégration et la Recherche des Fonctions Primitives*, Gauthier-Villars, 1904.

1905 David Hilbert, Grundzüge einer allgemeinen Theorie der linearen Integralgleichungen, Dritte Mitteilung, *Göttingen Nachrichten* (1905), 307–338.

1906a David Hilbert, Grundzüge einer allgemeinen Theorie der linearen Integralgleichungen, Vierte Mitteilung, *Göttingen Nachrichten* (1906), 157–227.

1906b David Hilbert, Grundzüge einer allgemeinen Theorie der linearen Integralgleichungen, Fünfte Mitteilung, *Göttingen Nachrichten* (1906), 439–480.

1907a Erhard Schmidt, Zur Theorie der linearen und nichtlinearen Integralgleichungen I, *Math. Annalen* **63** (1907), 433–476.

1907b Erhard Schmidt, Zur Theorie der linearen und nichtlinearen Integralgleichungen II, *Math. Annalen* **64** (1907), 161–174.

1907c Friedrich Riesz, Sur les systémes orthogonaux de fonctions, *C. R. Acad. Sci. Paris* **144** (1907), 615–619.

1907d Friedrich Riesz, Über orthogonale Funktionensysteme, *Göttingen Nachrichten* (1907), 116–122.

1907c Ernst Fischer, Sur la convergence en moyenne, *C. R. Acad. Sci. Paris* **144** (1907), 1022–1024.

1908a Erhard Schmidt, Über die Auflösung lineare Gleichungen mit unendlich vielen Unbekannten, *Rend. Circ. Mat. Palermo* **25** (1908), 53–77.

1908b Eliakim Hastings Moore, On a form of general analysis with application to linear differential and integral equations, *Cong. Int. Math. Rome* (1908), 98–114.

1908c George D. Birkhoff, Boundary value and expansion problems of ordinary linear differential equations, *Trans. Amer. Math. Soc.* **9** (1908), 373–395 (*Coll. Papers* I, 14–36).

1909a Ernst Hellinger, Neue Begründung der Theorie quadratischer Formen von unendlichvielen Veränderlichen, *J. Reine Angew. Math.* **136** (1909), 210–271.

1909b Friedrich Riesz, Sur les opérations fonctionnelles linéaires, *C. R. Acad. Sci. Paris* **149** (1909), 974–977.

1909c Isaac Schur, Über die charakteristischen Wurzein einer linearen Substitution mit enier Anwendung auf die Theorie der Integralgleichungen, *Math. Annalen* **66** (1909), 488–510.

1909d Hermann Weyl, Über beschrinkte quadratische Formen deren Differenz vollstetig ist, *Rend. Circ. Mat. Palermo* **27** (1909), 373–392.

1910a David Hilbert, Grundzüge einer allgemeinen Theorie der linearen Integralgleichungen, Sechste Mitteilungen, *Göttingen Nachrichten* (1910), 355–419.

1910b Ernst Hellinger and Otto Toeplitz, Grundlagen für eine Theorie der unendlichen Matrizen, *Math. Annalen* **69** (1910), 289–330.

1910c Friedrich Riesz, Untersuchungen über Systeme integrierbarer Funktionen, *Math. Annalen* **69** (1910), 449–497.

1910d Helge von Koch, Sur les systèmes d'une infinité d'équations linéaires à une infinité d'inconnues, *C. R. Cong. Math., Stockholm* (1910), 43–61.

1910c Ivar Fredholm, Les équations intégrales linéaires, *C. R. Cong. Math., Stockholm* (1910), 92–100.

1910f Henri Lebesgue, Sur l'intégrale de Stieltjes et sur les opérations linéaires, *C. R. Acad. Sci. Paris* **150** (1910), 86–88.

1912a David Hilbert, *Grundzüge einer Allgemeinen Theorie der Linearen Integralgleichungen*, Teubner, 1912.

1912b Eliakim Hastings Moore, On the foundations of the theory of linear integral equations, *Bull. Amer. Math. Soc.* **18** (1912), 334–362.

1913a Frédéric Riesz, *Les Systèmes d'Équations Linéaires à une Infinité d'Inconnues*, Gauthier-Villars, 1913.

1913b Johann Radon, Theoric und Anwendungen der absolut additiven Mengenfunktionen, *Sitz. Akad. Wiss., Wien* **122** (1913), 1295–1438.

1918a Otto Toeplitz, Das algebraische Analogen zu einem Satze von Fejér, *Math. Zeitschrift* **2** (1918), 187–197.

1918b Friedrich Riesz, Über lineare Funktionalgleichungen, *Acta Mathematica* **41** (1918), 71–98.

1923 Torsten Carleman, *Sur les équations intégrales singulières à noyau réel et symétrique*, Uppsala Univ. Årsskrift, 1923.

1925a Max Born and Pascual Jordan, Zur Quantenmechanik, *Z. Physik* **34** (1925), 858–888.

1925b Paul A. M. Dirac, The fundamental equations of quantum mechanics, *Proc. Roy. Soc. London* (A) **109** (1925), 642–658.

1926a Max Born, Werner Heisenberg and Pascual Jordan, Zur Quantenmechanik II, *Z. Physik* **35** (1926), 557–615.

1926b Paul A. M. Dirac, On the theory of quantum mechanics, *Proc. Roy. Soc. London* (A) **112** (1926), 661–677.

1927 John von Neumann, Mathematische Begründung der Quantenmechanik, *Göttingen Nachrichten* (1927), 1–57 (*Coll. Works* I, pp. 151–207).

1928a Ernst Hellinger and Otto Toeplitz, *Integralgleichungen und Gleichungen mit unendlichvielen Unbekannten*, Teubner, 1928.

1928b Norbert Wiener, The spectrum of an arbitrary function, *Proc. London Math. Soc.* **27** (1928), 483–496.

1929a John von Neumann, Allgemeine Eigenwerttheorie Hermitescher Funktionaloperation, *Math. Annalen* **102** (1929), 49–131 (*Coll. Works* II, 3–85).

1929b Aurel Wintner, *Spektraltheorie der unendlichen Matrizen*, Hirzel, 1929.

1930a John von Neumann, Zur Algebra der Funktionaloperationen und Theorie der normalen Operatoren, *Math. Annalen* **102** (1930), 370–427 (*Coll. Works* II, 86–143).

1930b Paul A. M. Dirac, *The Principles of Quantum Mechanics*, Oxford, 1930.

1930c Frederich Riesz, Über die linearen Transformationen des komplexen Hilbertschen Raumes, *Acta Litt. Sci. (Szeged)* **5** (1930), 23–54.

1930b Norbert Wiener, Generalized harmonic analysis, *Acta Mathematica* **55** (1930), 117–258.

1931 T. H. Hildebrandt, Linear functional transformations in general spaces, *Bull. Amer. Math. Soc.* **37** (1931), 185–212.

1932a Marshall H. Stone, *Linear Transformations in Hilbert Space*, Amer. Math. Soc. Colloq. Publ. XV, 1932.

1932b Stefan Banach, *Théorie des Operations Linéaires*, Warsaw, 1932.

1934 Kurt O. Friedrichs, Spektraltheorie halbbeschränkter Operatoren und Anwendung auf die Spektralzerlegung von Differentialoperatoren, *Math. Annalen* **109** (1934), 465–487, 685–713.

1936a Francis J. Murray and John von Neumann, On rings of operators, I, *Ann. Math.* **37** (1936), 116–229 (*Von Neumann Coll. Works* III, 6–119).

1936b S. W. P. Steen, An introduction to the theory of operators, I, *Proc. London Math. Soc.* **41** (1936), 361–392.

1936c Mitio Nagumo, Einige analytische Untersuchungen in linearen metrischen Ringen, *Japan. J. Math.* **13** (1936), 61–80.

1937a Francis J. Murray and John von Neumann, On rings of operators, II, *Trans. Amer. Math. Soc.* **41** (1937), 208–248 (*Von Neumann Coll. Works* III, pp. 120–160).

1937b S. W. P. Steen, An introduction to the theory of operators, II, *Proc. London Math. Soc.* **43** (1937), 529–543.

1938a S. W. P. Steen, An introduction to the theory of operators, III, *Proc. London Math. Soc.* **44** (1938), 398–441.

1938b Stanislaw Mazur, Sur les anneaux linéaires, *C. R. Acad. Sci. Paris* **207** (1938), 1025–1027.

1939 S. W. P. Steen, An introduction to the theory of operators, IV, *Proc. Cambridge Phil. Soc.* **35** (1939), 562–578.

1940a John von Neumann, On rings of operators III, *Ann. Math.* **41** (1940), 94–161 (*Coll. Works* III, pp. 161–228).

1940b S. W. P. Steen, An introduction to the theory of operators V, *Proc. Cambridge Phil. Soc.* **36** (1940), 139–149.

1940c Marshall H. Stone, A general theory of spectra I, *Proc. Nat. Acad. Sci. U.S.A.* **26** (1940), 280–283.

1941a Israel M. Gelfand, Normierte Ringe, *Mat. Sbornik* **9** (1941), 3–24.

1941b Israel M. Gelfand and Georgii E. Silov, Über verschiedene Methoden der Einführung der Topologie in die Menge der Maximalen Ideale eines normierten Ringes, *Mat. Sbornik* **9** (1941), 25–38.

1941c Marshall H. Stone, A general theory of spectra II, *Proc. Nat. Acad. Sci. U.S.A.* **27** (1941), 83–87.

1941d Shiuo Kakutani, Concrete representation of abstract (L) spaces and the mean ergodic theorem, *Ann. Math.* **42** (1941), 523–537.

1942 Edgar R. Lorch, The spectrum of linear transformations, *Trans. Amer. Math. Soc.* **52** (1942), 238–248.

1943 Israel M. Gelfand and M. A. Naimark, On the imbedding of normed rings into the ring of operators in Hilbert space, *Mat. Sbornik* **12** (1943), 197–213.

1944 Hermann Weyl, David Hilbert and his mathematical work, *Bull. Amer. Math. Soc.* **50** (1944), 612–654.

1946 Charles E. Rickart, Banach algebras with an adjoint operation, *Ann. Math.* **47** (1946), 528–550.

1949a John von Neumann, On Rings of Operators, Reduction Theory, *Ann. Math.* **50** (1949), 401–485 (*Coll. Works*, III, pp. 400–484).

1949b Arne Beurling, On the spectral synthesis of bounded functions, *Acta Mathematica* **81** (1949), 225–238.

1951 Paul R. Halmos, *Introduction to Hilbert Space*, Chelsea, 1951.

1953 Richard G. Cooke, *Linear Operators*, Macmillan, 1953.

1958a Nelson Dunford and Jacob T. Schwartz, *Linear Operators I: General Theory*, Interscience, 1958.

1958b Paul R. Halmos, *Finite Dimensional Vector Spaces*, 2nd ed., Van Nostrand, 1958.

1962 Edgar R. Lorch, *Spectral Theory*, Oxford Univ. Press, 1962.

1963 Nelson Dunford and Jacob T. Schwartz, *Linear Operators II: Spectral Theory*, Interscience, 1963.

1966 Michael Bernkopf, The development of function spaces with particular reference to their origins in the integral equation theory, *Arch. Hist. Exact Sci.* **3** (1966), 1–96.

1968 Michael Bernkopf, A history of infinite matrices, *Arch. Hist. Exact Sci.* **4** (1968), 308–358.

1970 Constance Reid, *Hilbert*, Springer-Verlag, 1970.

1971 Nelson Dunford and Jacob T. Schwartz, *Linear Operators III: Spectral Operators*, Wiley, 1971.

Alan Turing and the Central Limit Theorem

S. L. ZABELL

American Mathematical Monthly **102** (1995), 483–494

Because the English mathematician Alan Mathison Turing (1912–1954) is remembered today primarily for his work in mathematical logic (Turing machines and the "Entscheidungsproblem"), machine computation, and artificial intelligence (the "Turing test"), his name is not usually thought of in connection with either probability or statistics. One of the basic tools in both of these subjects is the use of the normal or Gaussian distribution as an approximation, one basic result being the Lindeberg-Feller central limit theorem taught in first-year graduate courses in mathematical probability. No-one associates Turing with the central limit theorem, but in 1934 Turing, while still an undergraduate, rediscovered a version of Lindeberg's 1922 theorem and much of the Feller-Lévy converse to it (then unpublished). This paper discusses Turing's connection with the central limit theorem and its surprising aftermath: his use of statistical methods during World War II to break key German military codes.

1 Introduction

Turing went up to Cambridge as an undergraduate in the Fall Term of 1931, having gained a scholarship to King's College. (Ironically, King's was his second choice; he had failed to gain a scholarship to Trinity.) Two years later, during the course of his studies, Turing attended a series of lectures on the Methodology of Science, given in the autumn of 1933 by the distinguished astrophysicist Sir Arthur Stanley Eddington. One topic Eddington discussed was the tendency of experimental measurements subject to errors of observation to often have an approximately normal or Gaussian distribution. But Eddington's heuristic sketch left Turing dissatisfied, and Turing set out to derive a rigorous mathematical proof of what is today termed the central limit theorem for independent (but not necessarily identically distributed) random variables.

Turing succeeded in his objective within the short span of several months (no later than the end of February 1934). Only then did he find out that the problem had already been solved, twelve years earlier in 1922, by the Finnish mathematician Jarl Waldemar Lindeberg (1876–1932). Despite this, Turing was encouraged to submit his work, suitably amended, as a Fellowship Dissertation. (Turing was still an undergraduate at the time; students seeking to become a Fellow at a Cambridge college had to submit evidence of original work, but did not need to have a Ph.D. or its equivalent.) This revision, entitled "On the Gaussian Error Function," was completed and submitted in November 1934. On the strength of this paper Turing was elected a Fellow of King's four months later (March 16, 1935) at the age of 22; his nomination was supported by the group theorist Philip Hall and the economists John Maynard Keynes and Alfred Cecil Pigou. Later that year the paper was awarded the prestigious Smith's prize by the University (see [26]).

Turing never published his paper. Its major result had been anticipated, although, as will be seen, it contains other results that were both interesting and novel at the time. But in the interim Turing's mathematical interests had taken a very different turn. During the spring of 1935, awaiting the outcome of his application for a Fellowship at King's, Turing attended a course of lectures by the topologist M. H. A. Newman on the Foundations of Mathematics. During the International Congress of Mathematicians in 1928, David Hilbert had posed three questions: is mathematics *complete* (that is, can every statement in the language of number theory be either proved or disproved?), is it *consistent*, and is it *decidable*? (This last is the *Entscheidungsproblem*, or the "decision problem"; does there exist an *algorithm* for deciding whether or not a specific mathematical assertion does or does not have a proof.) Kurt Gödel had shown in 1931 that the answer to the first question is *no* (the so-called "first incompleteness theorem"); and that if number theory is consistent, then a proof of this fact does not exist using the methods of the first-order predicate calculus (the "second incompleteness theorem"). Newman had proved the Gödel theorems in his course, but he pointed out that the third of Hilbert's questions, the *Entscheidungsproblem*, remained open.

This challenge attracted Turing, and in short order he had arrived at a solution (in the negative), using the novel device of *Turing machines*. The drafting of the resulting paper (Turing, 1937) dominated Turing's life for a year from the spring of 1935 ([26], p. 109); and thus Turing turned from mathematical probability, never to return.

A copy of Turing's Fellowship Dissertation survives, however, in the archives of the King's College Library, and its existence raises an obvious question. Just how far did a mathematician of the calibre of Turing get in this attack on the central limit theorem, one year before he began his pioneering research into the foundations of mathematical logic? The answer to that question is the focus of this paper.

2 The central limit theorem

The earliest version of the central limit theorem (CLT) is due to Abraham de Moivre (1667–1754). If $X_1, X_2, X_3, \ldots$ is an infinite sequence of 1s and 0s recording whether a success ($X_n = 1$) or failure ($X_n = 0$) has occurred at each stage in a sequence of repeated trials, then the sum $S_n := X_1 + X_2 + \cdots + X_n$ gives the total number of successes after n trials. If the trials are independent, and the probability of a success at each trial is the same, say $P[X_n = 1] = p, P[X_n = 0] = 1 - p$, then the probability of seeing exactly k successes in n trials has a binomial distribution:

$$P[S_n = k] = \frac{n!}{k!(n-k)!} p^k (1-p)^{n-k}.$$

If n is large (for example, $10,000$), then as de Moivre noted, the direct computation of binomial probabilities "is not possible without labor nearly immense, not to say impossible"; and for this reason he turned to approximate methods (see [9]): using Stirling's approximation (including correction terms) to estimate the individual terms in the binomial distribution and then summing, de Moivre discovered the remarkable fact that

$$\lim_{n \to \infty} P\left[a \leq \frac{S_n - np}{\sqrt{np(1-p)}} \leq b \right]$$
$$= \frac{1}{\sqrt{2\pi}} \int_a^b \exp\left[-\frac{1}{2} x^2 \right] dx,$$

or $\Phi(b) - \Phi(a)$, where $\Phi(x)$ is the cumulative distribution function of the standard normal (or Gaussian) distribution:

$$\Phi(x) := \frac{1}{\sqrt{2\pi}} \int_{-\infty}^{x} \exp\left[-\frac{1}{2} t^2 \right] dt.$$

During the nineteenth and twentieth centuries this result was extended far beyond the simple coin-tossing set-up considered by de Moivre, important contributions being made by Laplace, Poisson, Chebyshev, Markov, Liapunov, von Mises, Lindeberg, Lévy, Bernstein, and Feller; see [1], [34], [29], and [36] for further historical information. Such investigations revealed that if $X_1, X_2, X_3, \ldots$ is *any* sequence of independent random variables having the same distribution, then the sum S_n satisfies the CLT provided suitable centering and scaling constants are used: the centering constant np in the binomial case is replaced by the sum of the *expectations* $E[X_n]$; the scaling constant $\sqrt{np(1-p)}$ is replaced by the square root of the sum of the *variances* $\text{Var}[X_n]$ (provided these are finite).

Indeed, it is not even necessary for the random variables X_n contributing to the sum S_n to have the same distribution, provided that no one term dominates the sum. Of course this has to be made precise. The best result is due to Lindeberg. Suppose $E[X_n] = 0, 0 < \text{Var}[X_n] < \infty, s_n^2 := \text{Var}[S_n]$, and

$$\Lambda_n(\varepsilon) := \sum_{k=1}^n E\left[\left(\frac{X_k}{s_n} \right)^2 ; \frac{|X_k|}{s_n} \geq \varepsilon \right].$$

(The notation $E[X; Y \geq \varepsilon]$ means that the expectation of X is restricted to outcomes ω such that $Y(\omega) \geq \varepsilon$.) The *Lindeberg condition* is the requirement that

$$\Lambda_n(\varepsilon) \to 0, \quad \text{for all } \varepsilon > 0; \qquad (2.1)$$

and the *Lindeberg central limit theorem* [33] states that if the sequence of random variables $X_1, X_2, \ldots$ satisfies the Lindeberg condition (2.1), then for all $a < b$,

$$\lim_{n \to \infty} P\left[a < \frac{S_n}{s_n} < b \right] = \Phi(b) - \Phi(a). \qquad (2.2)$$

Despite its technical appearance, the Lindeberg condition turns out to be a natural sufficient condition for the CLT. There are two reasons for this. First, the Lindeberg condition has a simple consequence: if $\sigma_k^2 := \text{Var}[X_k]$, then

$$\rho_n^2 := \max_{k \leq n} \left(\frac{\sigma_k^2}{s_n^2} \right) \to 0. \qquad (2.3)$$

Thus, if the sequence $X_1, X_2, X_3, \ldots$ satisfies the Lindeberg condition, the variance of an individual term X_k in the sum S_n is asymptotically negligible. Second, for such sequences the Lindeberg condition is *necessary* as well as sufficient for the CLT to hold, a beautiful fact discovered (independently) by William Feller and Paul Lévy in 1935. In short: (2.1) ⇔ (2.2) + (2.3).

If, in contrast, the Feller-Lévy condition (2.3) fails, then it turns out that convergence to the normal distribution can occur in a fashion markedly different from that of the CLT. If (2.3) does *not* hold, then there exist a number $\rho > 0$ and two sequences of positive integers $\{m_k\}$ and $\{n_k\}$, where $\{n_k\}$ is strictly increasing, such that

$$1 \leq m_k \leq n_k \quad \text{for all } k$$
$$\text{and} \quad \text{Var}\left[\frac{X_{m_k}}{s_{n_k}}\right] = \frac{\sigma_{m_k}^2}{s_{m_k}^2} \to \rho^2 > 0. \qquad (2.4)$$

Feller [13] showed that if normal convergence occurs (that is, condition (2.2) holds), but condition (2.4) also obtains, then

$$\frac{1}{\rho} \frac{X_{m_k}}{s_{n_k}} \Rightarrow N(0,1);$$

that is, there exists a subsequence X_{m_k} whose contributions to the sums S_n are non-negligible (relative to s_n) and which, properly scaled, converges to the standard normal distribution. (The symbol ⇒ denotes convergence in distribution; $N(\mu, \sigma^2)$, the normal distribution, has expectation μ and variance σ^2.)

Note. For the purposes of brevity, this summary of the contributions of Feller and Lévy simplifies a much more complex story; see [29] for a more detailed account. (Or better, consult the original papers themselves!)

3 Turing's fellowship dissertation

Turing's fellowship dissertation was written twelve years after Lindeberg's work had appeared, and shortly before the work of Feller and Lévy. There are several aspects of the paper that demonstrate Turing's insight into the basic problems surrounding the CLT. One of these is his decision, contrary to a then common textbook approach (see, e.g., [4], pp. 87–90), but crucial if the best result is to be obtained (and the approach also adopted by Lindeberg), to work at the level of distribution functions (i.e., the function $F_X(t) := P[X \leq t]$) rather than densities (the derivatives of the distribution functions). In Appendix B Turing notes:

> I have attempted to obtain some results [using densities]... but without success. The reason is clear. In order that the shape frequency functions $u_n(x)$ of $f_n(x)$ should tend to the shape frequency function $\phi(x)$ of the Gaussian error, much heavier restrictions on the functions $g_n(x)$ are required than is needed if we only require that $U_n \to \Phi$. It became clear to me... that it would be better to work in terms of distribution function throughout.

This was an important insight. Although versions of the central limit theorem do exist for densities, these ordinarily require stronger assumptions than just the Lindeberg condition (2.1); see, e.g., [14], pp. 516–517, and [35], Chapter 7. Let us now turn to the body of Turing's paper, and consider it, section by section.

Basic structure of the paper

The first seven sections of the paper (pp. 1–6) summarize notation and the basic properties of distribution functions. Section 1 summarizes the problem; Section 2 defines the distribution function F (abbreviated DF) of an "error" ε; Section 3 summarizes the basic properties of the expectation and mean square deviation (MSD) of a sum of independent errors; rigorous proofs in terms of the distribution function are given in an appendix at the end of the paper (Appendix C). Section 4 discusses the distribution function of a sum of independent errors, the *sum distribution function* (SDF), in terms of the distribution functions of each term in the sum, and derives the formula for $F \oplus G$, the convolution of two distribution functions. Section 5 then introduces the concept of the *shape function* (SF): the standardization of a distribution function F to have zero expectation and unit MSD; thus, if F has expectation μ and MSD σ^2 ($\sigma > 0$), then the shape function of F is $U(x) := F(\sigma(x - \mu))$. (Turing uses the symbols a and k^2 to denote μ and σ^2; several other minor changes in notation of this sort are made below.)

In Section 6 Turing then states the basic problem to be considered: given a sequence of errors ε_k, having distribution functions G_k, shape functions V_k, means μ_k, mean square deviations σ_k^2, sum distribution functions F_n, and shape functions U_n for each F_n, under what conditions do the shape functions $U_n(x)$ converge uniformly to $\Phi(x)$, the "SF of the Gaussian Error"? Turing then assumes for simplicity that $\mu_k = 0$ and $\sigma_k^2 < \infty$. In Section 7 ("Fundamental Property of the Gaussian Error"), he notes that the only properties of Φ that are used in deriving sufficient conditions for normal convergence are that it is an SF, and the "self-reproductive property" of Φ: that is, if $X_1 \sim N(0, \sigma_1^2)$ and $X_2 \sim N(0, \sigma_2^2)$ are independent, then $X_1 + X_2 \sim N(0, \sigma_1^2 + \sigma_2^2)$. (The notation $X \sim N(\mu, \sigma^2)$ means that the random variable X has the distribution $N(\mu, \sigma^2)$.)

The quasi-necessary conditions

It is at this point that Turing comes to the heart of the matter. In Section 8 ("The Quasi-Necessary Conditions") Turing notes

> The conditions we shall impose fall into two groups. Those of one group (the quasi-necessary conditions) involve the MSDs only. They are not actually necessary, but if they are not fulfilled U_n can only tend to Φ by a kind of accident.

The two conditions that Turing refers to as the "quasi-necessary" conditions are:

$$\sum_{k=1}^{\infty} \sigma_k^2 = \infty \quad \text{and} \quad \frac{\sigma_n^2}{s_n^2} \to 0. \qquad (3.1)$$

It is easy to see that Turing's condition (3.1) is equivalent to condition (2.3). (That (2.3) implies (3.1) is immediate. To see (3.1) implies (2.3): given $\varepsilon > 0$, choose $M \geq 1$ so that $\sigma_n^2/s_n^2 < \varepsilon$ for $n \geq M$, and $N \geq M$ so that $\sigma_n^2/s_N^2 < \varepsilon$ for $1 \leq k < M$; if $n \geq N$, then $\sigma_k^2/s_n^2 < \varepsilon$ for $1 \leq k \leq n$.)

In his Theorems 4 and 5, Turing explores the consequences of the failure of either part of condition (3.1). Turing's proof of Theorem 4 requires his

Theorem 3. *If X and Y are independent, and both X and $X + Y$ are Gaussian, then Y is Gaussian.*

This is a special case of a celebrated theorem proven shortly thereafter by Harald Cramér (1936): if X and Y are independent, and $X + Y$ is Gaussian, then both X and Y must be Gaussian. Lévy had earlier conjectured Cramér's theorem to be true (in 1928 and again in 1935), but had been unable to prove it. Cramér's proof of this result in 1936 in turn enabled Lévy to arrive at necessary and sufficient conditions for the CLT of a very general type (using centering and scaling constants other than the mean and standard deviation), and this in turn led Lévy to write his famous monograph, *Théorie de l'Addition des Variables Aléatoires* [31]; see [29], pp. 80–81, 90.

Cramér's theorem is a hard fact; his original proof appealed to Hadamard's theorem in the theory of entire functions. The special case of the theorem needed by Turing is much simpler; it is an immediate consequence of the characterization theorem for characteristic functions. To see this, let $\phi_X(t) := E[\exp(itX)]$ denote the characteristic function of a random variable X; and suppose that X and Y are independent, $X \sim N(0, \sigma^2)$, and $X + Y \sim N(0, \sigma^2 + \tau^2)$. Then

$$\exp\left(-\frac{\sigma^2 + \tau^2}{2}t^2\right) = \phi_{X+Y}(t)$$
$$= \phi_X(t)\phi_Y(t) = \exp\left(-\frac{\sigma^2}{2}t^2\right)\phi_Y(t),$$

and hence $\phi_Y(t) = \exp(-(\tau^2/2)t^2)$; thus $Y \sim N(0, \tau^2)$, because the characteristic function of a random variable uniquely determines the distribution of that variable. Turing's proof, which uses distribution functions, is not much longer.

It is an immediate consequence of Cramér's theorem that if $S_n/s_n \Rightarrow N(0, 1)$, but $\lim_{n \to \infty} S_n^2 < \infty$, then all the summands X_j must in fact be Gaussian. But Turing did not have this fact at his disposal, only his much weaker Theorem 3. His Theorem 4 (phrased in the language of random variables) thus makes the much more limited claim that if (a) $\sum \sigma_n^2 < \infty$, (b) S_n converges to a Gaussian distribution, and (c) X_0 is a random variable at once independent of the original sequence $X_1, X_2, \ldots$ and having a distribution other than Gaussian, then the sequence $S_n^* = X_0 + S_n$ *cannot* converge to the Gaussian distribution. In other words: if $\sum \sigma_n^2 < \infty$, then "the convergence ... to the Gaussian is so delicate that a single extra term in the sequence ... upsets it".

Turing's Theorem 5 in turn explores the consequences of the failure of (3.1) in the case that $\sum \sigma_n^2 = \infty$ but $\rho_n^2 := \sigma_n^2/s_n^2$ does not tend to zero as $n \to \infty$. The statement of the theorem is somewhat technical in nature, but Turing's later summary of it captures the essential phenomenon involved:

> If F_n [the distribution function of S_n] tends to Gaussian and σ_n^2/s_n^2 does not tend to zero [but $\sum \sigma_n^2 = \infty$] we can find a subsequence of G_n [the distribution function of X_n] tending to Gaussian.

Thus Turing had by some two years anticipated Feller's discovery of the subsequence phenomenon. (In Turing's typescript, symbols such as F_n are entered by hand; in the above quotation the space for F_n has by accident been left blank, but the paragraph immediately preceding this one in the typescript makes it clear that F_n is intended.)

The sufficient conditions

Turing states in his preface that he had been "informed that an almost identical proof had been given by Lindeberg." This comment refers to the *method* of proof Turing uses, not the *result* obtained. Turing's method is to smooth the distribution functions $F_n(x)$ of the sum by forming the convolution $F_n * \Phi(x/\rho)$, expand the result in a Taylor series to third order, and then let the variance ρ^2 of the convolution term tend to zero. This is similar to the method employed by Lindeberg. (There is an important difference, however: Turing does not use Lindeberg's "swapping" argument. For an attractive modern presentation of the Lindeberg method, see [2], pp. 167–170; for a discussion of the method, see Pollard's comments in [29], pp. 94–95.)

Turing does *not*, however, succeed in arriving at the Lindeberg condition (2.1) as a sufficient condition for convergence to the normal distribution; the most general sufficient condition he gives is complex in appearance (although it necessarily implies the Lindeberg condition). Turing concedes that his "form of the sufficiency conditions is too clumsy for direct application," but notes that it can be used to "derive various criteria from it, of different degrees of directness and of comprehensiveness". One of these holds if the summands X_k all have the same *shape* (that is, the shape functions $V_k(x) := p[X_k/\sigma_k \leq x]$ coincide); and thus includes the special case of identically distributed summands having a second moment. (This was no small feat, since even this special case of the more general Lindeberg result had eluded proof until the publication of Lindeberg's paper.)

One formulation of this criterion, equivalent to the one actually stated by Turing, is: there exists a function $J : \mathbb{R}^+ \to \mathbb{R}^+$ such that $\lim_{t \to \infty} J(t) = 0$, and

$$E\left[\left(\frac{X_k}{\sigma_k} - t\right)^2 ; \left|\frac{X_k}{\sigma_k}\right| \geq t\right] \leq J(t),$$
$$\text{for all } k \geq 1,\ t \geq 0. \qquad (3.2)$$

In turn one simple sufficient condition for this given by Turing is that there exists a function ϕ such that $\phi(x) > 0$ for all x, $\lim_{x \to \pm\infty} \phi(x) = \infty$ and

$$\sup_k E\left[\left(\frac{X_k}{x_k}\right)^2 \phi\left(\frac{X_k}{\sigma_k}\right)\right] < \infty. \qquad (3.3)$$

(Note that unfortunately one important special case not covered by either of these conditions is that the X_k are *uniformly bounded*: $|X_k| \leq C$, for some C and all $k \geq 1$.)

In assessing this portion of Turing's paper, it is important to keep two points in mind. First, Turing states in his preface that "since reading Lindeberg's paper I have for obvious reasons made no alterations to that part of the paper which is similar to his." The manuscript is thus necessarily incomplete; it presumably would have been further polished and refined had Turing continued to work on it; the technical sufficient conditions given represent how far Turing had gotten on the problem *prior* to seeing Lindeberg's work. Second, in 1934 the Lindeberg condition was only known to be *sufficient*, not necessary; thus even in discussing his results in other sections of the paper (where he felt free to refer to the Lindeberg result), it may not have seemed important to Turing to contrast his own particular technical sufficient conditions with those of Lindeberg; the similarity in method must have seemed far more important.

One counter-example

Turing concludes by giving a simple example of a sequence $X_1, X_2, \ldots$ that satisfies the quasi-necessary conditions (3.1), but not the CLT. For $n \geq 1$, let

$$P[X_n = \pm n] = \frac{1}{2n^2}; \quad P[x_n = 0] = 1 - \frac{1}{n^2}.$$

Then $E[X_n] = 0$, $\text{Var}[X_n] = E[X_n^2] = 1$, $s_n^2 = \text{Var}[S_n] = n \to \infty$, and $\rho_n^2 = 1/n \to 0$; thus (3.1) is satisfied. Turing then shows that if S_n/s_n converges, the limit distribution must have a discontinuity at zero, and therefore cannot be Gaussian.

It is interesting that Turing should happen to choose this particular example; although he does not note it, the sequence $\{S_n/s_n : n > 1\}$ has the property that $\text{Var}[S_n/s_n] = 1$, but $\lim_{n\to\infty} S_n(\omega)/s_n = 0$ for almost all sample paths ω. This is an easy consequence of the first Borel-Cantelli lemma: because

$$\sum_{n=1}^{\infty} P[x_n \neq 0] = \sum_{n=1}^{\infty} \frac{1}{n^2} = \zeta(2) = \frac{\pi^2}{6} < \infty,$$

it follows that $P[X_n \neq 0 \text{ infinitely often}] = 0$; thus $P[\sup_n |S_n| < \infty] = 1$ and $P[\lim_{n\to\infty} S_n/s_n = 0] = 1$.

The existence of such sequences has an interesting consequence for the CLT. Let $\{Y_n : n \geq 1\}$ be a sequence of independent random variables, jointly independent of the sequence $\{X_n : n \geq 1\}$ and such that $P[Y_n = \pm 1] = \frac{1}{2}$. Let $T_n := Y_1 + Y_2 + \cdots + Y_n$; then a trite calculation shows that $S_n + T_n$ satisfies the Feller condition (2.3), but not the Lindeberg condition (2.1). Let $t_n^2 := \text{Var}[T_n]$; then $T_n/t_n \Rightarrow N(0,1)$ and $\text{Var}[S_n + T_n] = s_n^2 + t_n^2$, hence

$$\frac{S_n + T_n}{\sqrt{\text{Var}[S_n + T_n]}} = \frac{s_n}{\sqrt{s_n^2 + t_n^2}} \left(\frac{S_n}{s_n}\right)$$
$$+ \frac{t_n}{\sqrt{s_n^2 + t_n^2}} \left(\frac{T_n}{t_n}\right)$$
$$= \left(\frac{1}{\sqrt{2}}\right)\left(\frac{S_n}{s_n}\right) + \left(\frac{1}{\sqrt{2}}\right)\left(\frac{T_n}{t_n}\right)$$
$$\Rightarrow N\left(0, \tfrac{1}{2}\right).$$

Thus the sequence $S_n + T_n$ does converge to a Gaussian distribution! This does not, however, contradict the Feller converse to the Lindeberg CLT; that result states that $S_n + T_n$, rescaled to have unit variance, cannot converge to the *standard* Gaussian $N(0,1)$.

4 Discussion

Turing's Fellowship Dissertation tells us something about Turing, something about the state of mathematical probability at Cambridge in the 1930s, and something about the general state of mathematical probability during that decade.

I. J. Good ([19], p. 34) has remarked that when Turing "attacked a problem he started from first principles, and he was hardly influenced by received opinion. This attitude gave depth and originality to his thinking, and also it helped him to choose important problems." This observation is nicely illustrated by Turing's work on the CLT. His dissertation is, viewed in context, a very impressive piece of work. Coming to the subject as an undergraduate, his knowledge of mathematical probability was apparently limited to some of the older textbooks such as "Czuber, Morgan Crofton, and others" (*Preface*, p. ii).

Despite this, Turing immediately realized the importance of working at the level of distribution functions rather than densities; developed a method of attack similar to Lindeberg's; obtained useful sufficient conditions for convergence to the normal distribution; identified the conditions necessary for true central limit behavior to occur; understood the relevance of a Cramér-type factorization theorem in the derivation of such necessary conditions; and discovered the Feller subsequence phenomenon. If one realizes that the defects of the paper, such as they are, must largely reflect the fact that Turing had ceased to work on the main body of it after being apprised of Lindeberg's work, it is clear that Turing had penetrated almost immediately to the heart of a problem whose solution had long eluded many mathematicians far better versed in the subject than he. (It is interesting to note that Lindeberg was also a relative outsider to probability theory, and only began to work in the field a few years before 1922.)

The episode also illustrates the surprisingly backward state of mathematical probability in Cambridge at the time. Turing wrote to his mother in April 1934: "I am sending some research I did last year to Czuber in Vienna [the author of several excellent German textbooks on mathematical probability], not having found anyone in Cambridge who is interested in it. I am afraid however that he may be dead, as he was writing books in 1891" ([26], p. 88). (Czuber had in fact died nearly a decade before, in 1925.)

This disinterest is particularly surprising in the case of G. H. Hardy, who was responsible for a number of important results in probabilistic number theory. But anyone who has studied the Hardy-Ramanujan proof of the distribution of prime divisors of an integer (1917), and compared it to Turán's ([27], p. 71–74), will realize at once that the even most rudimentary ideas of modern probability must have been foreign to Hardy; see also [10], [11]. Indeed, Paul Erdős believes that "had Hardy known the even least little bit of probability, with his amazing talent he would certainly have been able to prove the law of the iterated logarithm" [8]. Perhaps this reflected in part the limited English literature on the subject. In 1927, when Harald Cramér visited England and mentioned to Hardy (his friend and former teacher) that he had become interested in probability theory, Hardy replied that "there was no mathematically satisfactory book in English on this subject, and encouraged me to write one" ([7], p. 516).

Finally, Turing's thesis illustrates the transitional nature of work in mathematical probability during the decade of the 1930s, before the impact of Kolmogorov's pioneering book *Grundbegriffe der Wahrscheinlichkeitsrechnung* [28] had been felt. In his paper Turing had thought it necessary to state and prove some of the most basic properties of distribution functions and their convolutions (in Sections 3 and 4, and Appendix C of the dissertation). His comment that his Appendix C "is only given for the sake of logical completeness and it is of little consequence whether it is original or not" illustrates that such results, although "known", did not enjoy general currency at the time. (It is all too easy to overlook today the important milestone in the literature of the subject marked by the publication in 1946 of Harald Cramér's important textbook *Mathematical Methods of Statistics* [6].)

It is also interesting to note Turing's approach to the problem in terms of convolutions of distribution functions rather than sums of independent random variables. Feller had similarly avoided the use of the language of random variables in his 1935 paper, formulating the problem instead in terms of convolutions. The reason, as Le Cam ([29], p. 87) notes, was that "Feller did not think that such concepts [as random variable] belonged in a mathematical framework. This was a common attitude in the mathematical community."

Current mathematical attitudes towards probability have changed so markedly from the distrust and scepticism of earlier times that today the sheer magnitude of the shift is often unappreciated. Joseph Doob, whose own work dates back to this period, notes that "even as late as the 1930s it was not quite obvious to some probabilists, and it was certainly a matter of doubt to most nonprobabilists, that probability could be treated as a rigorous mathematical discipline. In fact it is clear from their publications that many probabilists were uneasy in their research until their problems were rephrased in what was then nonprobabilistic language" ([29], p. 93–94).

5 Epilogue: Bletchley Park

After his Fellowship dissertation Turing "always looked out for any statistical aspects of [a] problem under consideration" [3]. This trait of Turing is particularly striking in the case of his cryptanalytic work during the second world war.

Turing left England for Princeton in 1936, to work with the logician Alonzo Church; he returned in 1938, after his Fellowship at King's College had been renewed. Recruited almost immediately by GC and CS (the Government Code and Cipher School), on September 4th, 1939 (one day after the outbreak of war) Turing reported to Bletchley Park, the British cryptanalytic unit charged with breaking German codes, soon rising to a position of considerable importance. (Turing's work at Bletchley was the subject of a 1987 London play, "Breaking the Code", written by Hugh Whitemore and starring Derek Jacobi, of "I, Claudius" fame.)

The staff at Bletchley Park included many gifted people, distinguished in a number of different fields; among these were the mathematicians M. H. A. Newman, J. H. C. Whitehead, Philip Hall, Peter Hilton, Shaun Wylie, David Rees, and Gordon Welchman; the international chessmasters C. H. O'D. Alexander, P. S. Milner-Barry, and Harry Golombek; and others such as Donald Mitchie (today an important figure in artificial intelligence), Roy Jenkins (the later Chancellor of the Exchequer), and Peter Benenson (the founder of Amnesty International). Turing's chief statistical assistant in the later half of 1942 was another mathematician, I. J. Good, fresh from studies under Hardy and Besicovitch at Cambridge. (Good arrived at Bletchley on May 27, 1942, the day the *Bismarck* was sunk.) In recent years Good has written several papers ([18], [19], [21], [22]) discussing Turing's *ad hoc* development of Bayesian statistical methods at Bletchley to assist in the decrypting of German messages. (More general accounts of the work at Bletchley include [32], [38] and [25]; see also the bibliography in [21].)

The specific details of Turing's statistical contributions are too complex to go into here. (Indeed, much of this information was until recently still classified and, perhaps for this reason, Good's initial papers on the subject do not even describe the specific cryptanalytic techniques developed by Turing; they give instead only a general idea of the type of statistical methods used. But in his most recent paper on this subject [22], Jack Good does provide a detailed picture of the various cryptanalytic techniques that Turing developed at Bletchley Park.) Three of Turing's most important statistical contributions were:

1. his discovery, independently of Wald, of some form of sequential analysis;

2. his anticipation of empirical Bayes methods (later further developed in the 1950s by Good and independently by Herbert Robbins); and

3. his use of logarithms of the Bayes factor (termed by Good the "weight of evidence") in the evaluation and execution of decryption.

(For many references to the concept of weight of evidence, see, for example, [23] and the two indices of [20].) The units for the logarithms, base 10, were termed *bans* and *decibans* [18]:

> One reason for the name ban was that tens of thousand of sheets of paper were printed in the town of Banbury on which weights of evidence were entered in decibans for carrying out an important process called Banburismus. . . .

"Tens of thousands of sheets of paper . . . ." This sentence makes it clear that Turing's contributions in this area were not mere idle academic speculation, but an integral part of the process of decryption employed at Bletchley.

One episode is particularly revealing as to the importance with which the Prime Minister, Winston Churchill, viewed the cryptanalytic work at Bletchley. On October 21, 1941, frustrated by bureaucratic inertia, Turing, Welchman, Alexander, and Milner-Barry wrote a letter *directly* to Churchill (headed "Secret and Confidential; Prime Minister only") complaining that inadequate personnel had been assigned to them; immediately upon its receipt Churchill sent a memo to his principal staff officer directing him to "make sure they have all they want on extreme priority and report to me that this had been done" ([26], pp. 219–221).

Much of I. J. Good's own work in statistics during the decades immediately after the end of the war was a natural outgrowth of his cryptanalytic work during it; this includes both his 1950 book, *Probability and the Weighing of Evidence*, and his papers on the sampling of species (e.g., [15]) and the estimation of probabilities in large sparse contingency tables (much of it summarized in [16]). Some of this work was stimulated either directly (see, e.g., [17], p. 936) or indirectly (the influence being somewhat remote, however, in the case of contingency tables) by Turing's ideas:

> Turing did not publish these war-time statistical ideas because, after the war, he was too busy working on the ground floor of computer science and artificial intelligence. I was impressed by the importance of his statistical ideas, for other applications, and developed and published some of them in various places (see [21], p. 211).

References

1. W. J. Adams, *The Life and Times of the Central Limit Theorem*, Kaedmon, 1974.

2. L. Breiman, *Probability*, Addison-Wesley, 1968.

3. J. L. Britton, *The Collected Works of A. M. Turing: Pure Mathematics*, North-Holland, 1992. [Contains the two-page Preface to Turing's Fellowship Dissertation.]

4. W. Burnside, *Theory of Probability*, Cambridge Univ. Press, 1928.

5. H. Cramér, Über eine Eigenschaft der normalen Verteilungsfunktion, *Math. Zeitschrift* **41** (1936), 405–414.

6. H. Cramér, *Mathematical Methods of Statistics*, Princeton University Press, 1946.

7. H. Cramér, Half of a century of probability theory: some personal recollections, *Annals of Probability* **4** (1976), 509–546.

8. P. Diaconis, Personal communication, 1993. [The quotation is a paraphrase from memory.]

9. P. Diaconis and S. Zabell, Closed form summation for classical distributions: variations on a theme of De Moivre, *Statistical Science* **6** (1991), 284–302.

10. P. D. T. A. Elliott, *Probabilistic Number Theory I: Mean Value Theorems*, Springer-Verlag, 1979.

11. P. D. T. A. Elliott, *Probabilistic Number Theory II: Central Limit Theorem*, Springer-Verlag, 1980.

12. W. Feller, Über den zentralen Grenzwertsatz der Wahrscheinlichkeitsrechnung, *Math. Zeitschrift* **40** (1935), 521–559.

13. W. Feller, Über den zentralen Grenzwertsatz der Wahrscheinlichkeitsrechnung II, *Math. Zeitschrift* **42** (1937), 301–312.

14. W. Feller, *An Introduction to Probability Theory and its Applications*, Vol. 2, 2nd ed., Wiley, 1971.

15. I. J. Good, The population frequencies of species and the estimation of population parameters, *Biometrika* **40** (1953), 237–264.

16. I. J. Good, *The Estimation of Probabilities: An Essay on Modern Bayesian Methods*, MIT Press, 1965.

17. I. J. Good, The joint probability generating function for run-lengths in regenerative binary Markov chains, with applications, *Annals of Statistics* **1** (1973), 933–939.

18. I. J. Good, A. M. Turing's statistical work in World War II, *Biometrika* **66** (1979), 393–396.

19. I. J. Good, Pioneering work on computers at Bletchley, *A History of Computing in the Twentieth Century* (eds. N. Metropolis, J. Howlett, and G.-C. Rota), Academic Press, 1980, pp. 31–45.

20. I. J. Good, *Good Thinking*, Minnesota Univ. Press, 1983.

21. I. J. Good, Introductory remarks for the article in *Biometrika* **66** (1979), in [3], pp. 211–223.

22. I. J. Good, Enigma and Fish, in [25], pp. 149–166.

23. I. J. Good, Causal tendency, necessitivity and sufficientivity: an updated review, in *Patrick Suppes, Scientific Philosopher* (ed. P. Humphreys), Kluwer, 1993.

24. G. H. Hardy and S. Ramanujan, The normal number of prime factors of a number, *Quarterly J. Math.* **48** (1917), 76–92.

25. F. H. Hinsley and A. Stripp (eds.), *Codebreakers: The Inside Story of Bletchley Park*, Oxford Univ. Press, 1992.

26. A. Hodges, *Alan Turing: The Enigma*, Simon and Schuster, 1983.

27. M. Kac, *Statistical Independence in Probability, Analysis and Number Theory*, Math. Assoc. of America, 1959.

28. A. A. Kolmogorov, *Grundbegriffe der Wahrscheinlichkeitsrechnung*, Springer-Verlag, 1933.

29. L. Le Cam, The central limit theorem around 1935 (with discussion), *Statistical Science* **1** (1986), 78–96.

30. P. Lévy, Propriétés asymptotiques des sommes de variables indépendantes on enchaînées, *J. Math. Pures Appl.* **14** (1935), 347–402.

31. P. Lévy, *Théorie de l'Addition des Variables Aléatoires*, Gauthier-Villars, 1937.

32. R. Lewin, *Ultra Goes to War*, McGraw-Hill, 1978.

33. J. W. Lindeberg, Eine neue Herleitung des Exponentialgesetzes in der Wahrscheinlichkeitsrechnung, *Math. Zeitschrift* **15** (1922), 211–225.

34. L. E. Maistrov, *Probability Theory: A Historical Sketch*, Academic Press, 1974.

35. V. V. Petrov, *Sums of Independent Random Variables*, Springer-Verlag, 1975.

36. S. M. Stigler, *The History of Statistics*, Harvard Univ. Press, 1986.

37. A. M. Turing, *On the Gaussian Error Function*, Unpublished Fellowship Dissertation, King's College Library, Cambridge, 1934.

38. G. Welchman, *The Hut Six Story*, McGraw-Hill, 1982.

Why did George Green Write his Essay of 1828 on Electricity and Magnetism?

I. GRATTAN-GUINNESS

American Mathematical Monthly **102** (1995), 387–396

1 Honor to Green

Among the centenaries of mathematicians and scientists celebrated in 1993, perhaps the most remarkable was the bicentenary of the birth of a professional miller and part-time mathematician, one George Green (1793–1841) of Sneinton, then near Nottingham. Among other achievements, he was the creator of theorems and functions now named after him which make him a principal contributor to potential theory and to its applications in mechanics and mathematical physics.

During the week corresponding to that of his birth (which occurred on 14 July) various events took place. A three-day conference was held at the University of Nottingham, mainly on the use of his work in modern mathematics and physics. It included a visit to the mill at Sneinton which had been restored and opened as a science center in 1985. The next day a stained glass window was dedicated at the Gonville and Caius College, Cambridge, where he was resident from 1833 to 1837 as an extremely mature student, and (only) for the winter of 1839–1840 as a Fellow. Finally, on Friday 16 July a meeting was held at the Royal Society of London on his life and work and the modern importance of the latter. It was followed by a quite exceptional event: the unveiling of a plaque in his memory in the floor of the nave of Westminster Abbey, close to the tomb of Isaac Newton and to the plaques for his first publicist Lord Kelvin, Michael Faraday and Clerk Maxwell.

These events had been preceded by the publication in May of an excellent biography of Green [5]—a daunting task to write, as his life is so obscure (for example, no surviving likeness or portrait has ever been found, and his manuscripts seem to have been destroyed). It is clear, though, that in virtual isolation at Sneinton he taught himself Continental mathematics, and produced first-class research work. It was published in 1828 (his 35th year) as a 72-page *Essay on the Mathematical Analysis of Electricity and Magnetism* [11], put out at his own expense with the help of a subscription list. Largely ignored during the author's lifetime, it has since been reprinted no less than seven times and translated into German. How and why was it created?

2 Three strands in eighteenth-century mechanics

One of the most profound influences exercised by this fugitive work is that it raised both the status of potential theory in mathematics and the quality of the theorems that could be stated in it. Prior to this time three strands of thought in potential theory (as we now understand it) were active, though not necessarily with close links between them ([26], Vol. 1).

The most significant strand was the attraction of spheroids to an external point. Isaac Newton had found various special properties in the *Principia*, in his synthetic style: they were extended from the 1770s onwards by P. S. Laplace and A. M. Legendre using analytical methods, especially the Legendre functions and surface and zonal harmonics.

Another line came from Alexis Clairaut on the Continent from the 1730s (with some contributions from Colin MacLaurin in Britain soon afterwards), where properties of equipotential surfaces were studied; this work laid stress on the exact differential of a function of several variables, and assisted in the birth soon afterwards of the full partial differential calculus. With d'Alembert, some aspects of Euler's work, and especially J. L. Lagrange, variational mechanics was developed, in which force and velocity potentials were often used in the formation of differential equations.

A third strand grew out of Daniel Bernoulli's *Hydrodynamica* (1738), where considerations of 'ascensis actualis et potentialis' led to conservation of energy as a basis for (much) mechanics; his notions were to end up in the next century as kinetic and potential energy, respectively, although with substantial changes in conception in which potential theory was to play a role.

In addition, an isolated contribution came from Lagrange in 1762. While pondering ways of solving the equations for the propagation of sound in three dimensions, he formed volume integrals of the solution in each coordinate direction, integrated them by parts to create surface integrals, and then added up the resulting equations to obtain a simpler differential equation to integrate. A clever but *ad hoc* manoeuvre, it had little influence even upon its distinguished innovator; but it was closer to the way ahead pursued in the early nineteenth century than the strands just mentioned.

3 Poisson and the appearance of divergence theorems

Enter Siméon-Dénis Poisson (1781–1840), the leading supporting actor in this drama, student and then professor and graduation examiner at the *École Polytechnique*, devout follower of Laplace and Lagrange in mathematical methods and physical modeling. Poisson inaugurated mathematical electrostatics (I shall use the word 'electricity' of the time) in two papers ([16], [18]) published by the Paris Academy of Sciences, in which he analyzed arrangements studied experimentally by C. A. Coulomb 30 years previously; equilibrium on a charged spheroid, and between two spheres. The principal mathematical exercise was to modify Legendre functions and related potential theory to fit the assumptions made about the phenomena ([10], pp. 496–513).

In a short paper written soon after these two [17], Poisson rectified an important oversight of his masters when he pointed out that the differential equation governing the potential V to a body A relative to an interior point M was not Laplace's equation

$$\Delta V = 0 \text{ (with `Δ' as the Laplacian operator)}, \quad (1)$$

but

$$\Delta V = -4\pi\rho, \quad (2)$$

where ρ was the density of material at M. He might have got this insight from his recent work on electricity; another strong candidate is a paper of that year on the attraction of spheroids by Carl Friedrich Gauss, which contained a result which in vectors reads

$$\int_S d\mathbf{s}.\mathbf{r} = 0 \text{ or } -4\pi, \quad (3)$$

where $\mathbf{r} = BM$ and B is an arbitrary point in A, according as M is outside or inside the surface S of A. Neither man dealt with the case where M is *on* S, when -2π obtains in (2) ([10], pp. 418–424).

Twelve years later Poisson came to the Academy of Sciences with another pair of papers, this time analyzing magnetism ([21], [22]). Taking a magnetic body A to be composed of discrete 'magnetic elements' D, he set to zero certain surface integrals over D expressing internal equilibrium, and wrote down volume integrals to state the components of attraction of A to an external point M relative to his imposed rectangular coordinate system (x, y, z). The second part of the first paper dealt with a 'simplification of the preceding formulae'; integrating these integrals by parts with respect to (say) z led him to convert the volume integral to an integral over the surface S of A. I write his finding in the form

$$\iiint_A H_z(x, y, z)\, dx\, dy\, dz$$
$$= \iint_S H(x_S, y_S, z_S) \cos n\, dS, \quad (4)$$

where H was the function expressing the components of magnetic attraction, and n was the angle between the z-axis and the normal at the point (x_S, y_S, z_S) of S. Adding this formula to its brothers for the x- and y-directions gave him the first general divergence theorem in mathematics; imitating the notation of (4), it can be written

$$\iiint_A [F_x + G_y + H_z](x, y, z)\, dx\, dy\, dz$$
$$= \iint_S [F \cos l + G \cos m + H \cos n]$$
$$\times (x_S, y_S, z_S)\, dS. \quad (5)$$

He modified it for the case when M was inside A by the manner of his proof of (2), and found a new term involving a factor $-4\pi/3$.

Poisson knew that his result was not restricted to convex bodies (a sum of integrals of the form (4) is required as the z-axis goes in and out of A), nor to magnetism. But he saw it simply as a convenience; triple integrals are replaced by double integrals ([10], pp. 948–953). This point will be crucial for Green, as we shall see in the next section.

In a third paper, published by the Academy in the *Mémoires* [23], Poisson analyzed the process of magnetization in moving bodies. A most complicated analysis used Legendre functions, once again; but an important detail was his recollection of his equation (2) for interior points, and first presentation of the version with $-2\pi\rho$ for surface points.

Surface integrals were enjoying a springtime in French mathematics at this time. For example, Adrien Marie Ampère had been studying electromagnetism and electrodynamics (his word) since 1820; his analysis made adroit use of both surface and line integrals, the latter arising naturally in connection with the attraction caused by current-bearing wires ([10], pp. 941–961). One of his most remarkable results, published in 1826, was to show that Poisson's basic formulae for magnetism could be restated in his own preferred conception, which saw magnetism as a special case of electricity and so replaced Poisson's 'magnetic elements' with a tiny electrical solenoid (his word again).

Another source was Joseph Fourier's pioneering work on heat diffusion, created in the mid-1800s, fully published only in the early 1820s, especially in his book *Théorie Analytique de la Chaleur* (1822), and then receiving much attention from the new generation of French mathematicians. In particular, around 1826 Jean Duhamel and Russian visitor Mikhail Ostrogradsky independently sought to justify mathematically Fourier's use of trigonometric series solutions ([10], pp. 1168–1176). Let f and g be two different special solutions for diffusion in a body A and consider $I := \iiint_A fg\, dV$. Integrating by parts through A led to a divergence theorem like Poisson's (4); and applying Fourier's external surface condition showed that in fact $I = 0$. Hence f and g were orthogonal over A, like the sine and cosine functions. We can see that this does not provide the justification sought; more to the point is the use again of surface integrals and a divergence theorem.

Although these integrals were making an appearance, their presence in mathematics was still slight. Good evidence is provided by Augustin Louis Cauchy (1789–1857), former pupil of the *École Polytechnique* (when Poisson was professor) and now professor there himself, inaugurating his revision of the calculus and mathematical analysis by his famous new approach with the theory of limits at the centre, emphasis laid upon continuity of functions. (Poisson and others there protested vigorously.) Above all, the derivative and integral were defined *separately*, so that the fundamental theorem of calculus became a proper theorem for the first time ([10], pp. 707–804, including Cauchy's concurrent inauguration of complex-variable analysis). However, he never furnished a definition of either the line or the surface integral, although the required forms of definition would not have been hard to devise; they were too marginal to be worth such attention.

Then Green started thinking.

4 Green and the place of surface integrals

Possible sources for Green's essay will be appraised in the next section; here its main contents are described. Pages are cited from the printing in the edition of his works [12].

After various preliminaries, the essay contains two roughly equal parts on electricity and on magnetism, in that order. These latter analyses draw heavily on various largely known integral expressions to state external and internal potentials (the latter maybe learned from Poisson's (2)), and Legendre functions to express the potentials in analytical form. He extended various results due to Poisson, and considered some variant situations, such as when the spheres are connected by a wire [27].

The chief novelties were presented in the 'general preliminary results' stated in the opening. First was the explicit specification of 'the potential function', as he called it, and now named after him ([11], pp. 9–10):

> It only remains therefore to find a function V' which satisfies the partial differential equation, becomes equal to [a given function] $\bar{V}'$ when [the point] p is upon the surface A, vanishes when p is at an infinite distance from A, and is besides such that none of its differential coefficients shall be infinite when the point p is exterior to A.

(Note the inadequate specification of $V'(\infty)$, and that the prime does not denote differentiation.) This formulation anticipated in certain ways the 'Dirichlet principle', which was to assume such status in potential theory when its author began lecturing on the subject from 1839 in Berlin: a decade earlier he also was in Paris, but working on Fourier's heat theory, Cauchy's analysis, and number theory.

Secondly was Green's type of divergence theorem, expressed entirely within the rectangular coordinate system (x, y, z) rather than with surface differentials of Poisson's (4): for two 'continuous functions' $U(x, y, z)$ and $V(x, y, z)$,

$$\int dx\,dy\,dz\,U\delta V + \int d\sigma\,U(dV/dw)$$
$$= \int dx\,dy\,dz\,V\delta U + \int d\sigma\,V(dU/dw) \quad (6)$$

([11], p. 23). I follow his use of δ for the Laplacian operator (an unusual symbol, perhaps required by the limitations of his printer's font box), $d\sigma$ for the element of the surface, all integrals stated with only one sign $\int$ (unlike Poisson's use of multiple integral signs), and round brackets to indicate partial 'differential coefficients' (Euler's practice, and name also, both of which Green followed). He modified his result for 'singularities' in U (or V) at points G by adding in terms of the form $-4\pi U(x_G, y_G, z_G)$ to the appropriate side of the equation ([11], p. 27), like Poisson's own modification; he may also have known of Poisson's equation (2) from its reappearance in [23]. I wonder at the import of the continuity imposed upon U and V, and the reference to 'singularities'; had he also been reading Cauchy on reforming the calculus?

Green had taken up a current research interest in mathematical physics in using volume and surface integrals to analyze electricity and magnetism; and with his insights he surpassed all contemporaries. This theorem (6), while similar in mathematical form to Poisson's (5), was understood at a far deeper level as physics (and also surpassed Gauss's (3) in generality). Whereas Poisson saw only simplification in his integral, Green recognized that

> the importance of his own theorem lay in relating properties inside bodies to properties on their surfaces and *vice versa*.

He must have realized that theorems of this kind served for multiple integral calculus like the fundamental theorem of the calculus itself, hence the importance of integration by parts.

These insights doubtless led Green further to the novelty of his function V' in which conditions in a body and on its surface were imposed. Such functions were found for various cases with the help of his theorem; one of them followed it in its symmetrical form ([11], pp. 37–39), and launched what have become known as 'reciprocity relations'.

One may guess therefore, with some confidence, that

> Poisson's first two papers on magnetism were the source of inspiration for Green's research, especially the divergence theorem (5).

Up to then Green had doubtless been learning mathematical skills and theories, but he had not found a deep problem: Poisson (unintentionally) provided this, in the form of an unexceptionable but somewhat limited use of Legendre functions to analyze the distribution of magnetism, and especially in a 'simplification' which held much deeper consequences than its author had realized.

5 Sources and influences

While it is possible to guess at Green's original motivation, his training in mathematics remains unknown. Among local figures, headmaster John Toplis (1774–1857) would have been a crucial figure in forming the interests of his former pupil at Nottingham Grammar School: a deplorer of the state of British mathematics in the *Philosophical Magazine* in 1804, a translator of Lacroix there a year later, and of Book One of Laplace's *Mécanique Céleste* in a book published in Nottingham in 1814. However, in 1819 he returned to his college (Queens', Cambridge), and was *not* to be one of the subscribers. The only other likely supporter is Sir Edward Bromhead (1789–1855: so Green's senior by a mere four years), member with Charles Babbage and John Herschel of the Analytical Society at Cambridge in the mid-1810s, and a subscriber to the essay; but his letter of April 1828 acknowledging receipt of his presentation copy ([5], p. 67) shows that he had *not* been aware of its contents before it arrived.

Green's access to literature is also little understood. In his essay he cited as mathematical sources Laplace's *Mécanique Céleste*, Book 3 (1799) for Legendre functions, Fourier's *Théorie Analytique de la Chaleur*, and of course, Poisson's three papers on magnetism and the two on electricity; Boit's *Traité de Physique* (1816) was used for information on Coulomb's experiments. A passing reference ([11], p. 103) to Lagrange's follower L. F. A. Arbogast shows his familiarity with some of the current French operator techniques. A sentence in his introduction comparing Fourier with Cauchy and Poisson on methods of solving differential equations in hydrodynamics ([11], p. 8) suggests that he had read the paper [8] on precisely this matter, which had been published in a Paris journal ([10], pp. 683–686).

How did Green gain access to these works? While British texts would have been available in the local library, access there to foreign literature is much less certain, even presuming that he had funds available to buy it. The point is particularly perplexing for journals—in particular, the Paris *Mémoires* with its Poisson papers. How did Green know that those papers were published there in the first place? Although presentations to the Academy were reported in Paris journals and sometimes abroad, the news did not circulate very much, and Poisson had not given any warning in earlier papers that research in magnetism was in progress. The best chance was that some summary version was translated into a foreign language such as English—and indeed this did happen to summaries of these two papers, in the *Quarterly Journal of Science* ([19], [20]).

Each summary paper began with a virtually verbatim repeat of parts of the opening preamble of the parent paper, and then summarized some later results and features. The accounts concentrated mostly on physical and experimental aspects; mathematical procedures were only mentioned (and three formulae quoted in the second summary), although not in a manner to reveal any major novelties. In particular, the divergence theorem (5) was described only in general terms, and with reference to simplication: 'by means of certain transformations, the triple integrals which they contain are reduced to double integrals, and the equations become much more simple' ([19], p. 327). No reader of the time could have guessed that surface integrals were involved; but Green might have been alerted to watch out for the full versions of the papers.

Regarding timetable, the volume of the Paris Academy *Mémoires* containing these papers appeared right at the end of 1826 ([1], p. 473). Allowing for the usual delay for ships to deliver copies across the channel, one can guess that the spring of 1827 was in hand before Green read at least Poisson's first paper and had his inspiration. Since his essay was to appear in April 1828, this would have given him a maximum of around a year to carry out the research—not an excessive time, even for a part-time mathematician. His motivation was high, most of the required skills and familiarity with the literature were already available—and above all his ideas were fruitful, so that the fruit would grow freely and quickly.

6 Options for publication

However, in contrast to this splendid piece of research and development, Green's sales and marketing were hopeless. He cannot be blamed for his scientific isolation in Nottingham, but he was somewhat naive in resorting to the traditions of publication by public subscription. For the increase in scientific activity in Britain in recent years, together with advances in printing technology, had raised the chances and opportunities for publication, especially for an author like him with financial means available to assist with the costs of production. He could have tried Deighton's of Cambridge, who were then publishing quite a lot in mathematics [9] and in fact stocked the essay when it came out; or maybe Taylor (now Taylor and Francis), regular producers of scientific books. He could have written a paper summarizing his findings for their *Philosophical Magazine*, which was widely distributed in the scientific world: although it did not publish mathematics frequently, there were papers from time to time, and indeed there had been an exchange in there in 1826 on another aspect of potential theory (namely, properties of equipotential surfaces) between Poisson and the Scottish-born mathematician James Ivory ([10], pp. 1190–1195). In fact, if he had felt it proper so to act he could have sought advice from Ivory, the mathematician most conversant with potential theory in Britain at that time. He might also have treated his manuscript as a long paper instead of a short book, and tried to submit it to the Royal Society, or the Cambridge Philosophical Society, or the Royal Society of Edinburgh.

I have no doubt that Green never considered any of these possibilities. His essay went to his 52 supporting subscribers, most of whom could not have read a page of it ([13], pp. 45–48); and so it vanished from sight. Very rarely has it appeared even in booksellers' catalogues.

7 On Green's second period

Green's later career was somewhat less unorthodox than previously, in as much as he was resident at Gonville and Caius College, Cambridge (Bromhead's *alma mater*) from 1833 to 1837 and for some months of 1839–1840 as a Fellow. He had a small overlap in residence with someone capable of understanding his work, indeed the first mathematician to cite the essay; but this was the eccentric Robert Murphy (1806–1843), who spoilt a promising career by financial incompetence. [On Green's and Murphy's work see [6]; the reference to Green's book is in [15]. The nature of Murphy's misdemeanors has not been clear; my information comes from a letter [7] of perhaps 1835 to Babbage written by Augustus De Morgan, who was Professor in London University, where Murphy was then trying to make a living (British Library, Additional Mss. 37189, no. 241). Compare [5], pp. 112–113.]

Green published eight papers (and a supplement to one of them), mostly in the *Transactions* of the Cambridge Philosophical Society with Bromhead as communicator. However, his lack of marketing skills were again to the fore: he cited his essay only twice ([12], pp. 120, 192), and on neither occasion did he even give the reader the publication details, never mind a comment to explain its importance.

The other papers fall into two partly related groups, both showing strong French influence in both content and methodology ([3], Ch. 13). One group deals with elastic bodies, which could be construed to be physically bending objects, or else the elastic ether (and perhaps with luck, both at once). The task was to study the propagation of longitude and transverse vibrations; Green also tackled the difficult question of behavior at the interface between different substances. He sought generality by making no stipulations about the constituted properties of the substances. The principal influence seems to have been the non-molecular studies of elasticity made from 1827 onwards by Cauchy, which had been partly inspired by Fresnel's work in waval optics ([28], Ch. 5).

The other group examined the potentials of fluids, which again might cover sound and water, but also the supposed electric and magnetic fluids. Green made this analogy quite explicit in the title of the first of these papers, when referring to the 'laws of the equilibrium of fluids analogous to the electric fluid' ([12], p. 117).

For methods Green used both his function and theorem, and some of the special results from his essay. He played a little more with operator methods, and produced some solutions in terms of elliptic integrals (although he seemed to be unaware of the recently introduced elliptic functions). A paper on the motion of waves and canals took a step towards the approximating asymptotic solution method now known as 'WKB' ([24], pp. 309–314), although he limited himself to working within the linearizing models of his time. He worked with potentials to the inverse nth power; and on one occasion he required of his potential function that it be invariant under infinitely small rotations, a step that Sophus Lie was to bring (independently) to great generality and prominence 60 years later.

Somewhat separate from Green's other papers was one dealing with the motion of the 'simple' pendulum. This was a favorite topic at this time, a typical example of small-effect science; for the pendulum was required to work to a great degree of accuracy for the purposes of geodesy. Laplace and Poisson, and also F. W. Bessel and G. B. Airy, had been among its many earlier students [29].

8 Recognition

Green's marketing skills increased at least to the extent that he sent some of these papers to Carl Jacobi ([5], p. 104; the copies are in private possession), and presumably while at Cambridge he gave copies of his essay to William Hopkins, who passed either two or three copies on to the young William Thomson (1824-1907) in 1845. [One of these copies of Green's essay is now kept at Nottingham ([5], p. 105).] Then, as is famously known, the essay found its first enthusiastic reader, four years after the death of its author in 1841. Thomson introduced the name 'Green's theorem', and soon came to his 'method of images' as a result of reading the analysis in the essay of the effect on the electrical charge in a body at an interior/exterior point of a source at a given exterior/interior point ([11], pp. 50–54). He soon arranged for the essay to be reprinted in Crelle's journal, although it did not appear until 1850–1854. [The circumstances of this reprinting of the essay are strange. Firstly, Thomson asked Crelle and not his friend Joseph Liouville, who also edited a journal (still often known after Liouville) and was actively interested in potential theory. Secondly, while Crelle had agreed enthusiastically to the suggestion by 1846 ([13], pp. 41–43), he did not reprint it for several years, and then in three parts over five years.] Later he and P. G. Tait called the Dirichlet principle 'Green's problem' ([25], arts. 499–518). The name 'Green's function' for functions satisfying conditions like Green's own is due to Bernhard Riemann and Carl Neumann ([4], art. 18).

Today, Green's function and his theorem are extolled because of the roles which they continue to play in modern physics and in engineering; but it would be a misunderstanding of history to think that their importance is *due* to these applications. On the contrary, their rise occurred during the period of *classical* physics, when there appeared a mountainous production of books and papers on potential theory and its use in mathematical physics [2]; all the applications mentioned above were involved, and in due course new ones such as thermodynamics and meteorology, and also mathematical economics. All major applied mathematicians took part, along with many minor ones, and some pure mathematicians also (in particular Karl Weierstrass, who sabotaged standard methods of manipulation in 1870 with his famous counter-example to the Dirichlet principle using the inverse tangent function).

Not only Green's insights and results were used by his successors; his own work, especially the essay, were made available *four times* in the last thirty years of the nineteenth century to an extent surpassing all other literature of his own time. The edition of his works by N. Ferrers appeared in London in 1871, and was reprinted in facsimile in 1903 in (of all places) Paris. The essay itself was also reprinted in facsimile, in 1890 in Berlin, in a series of classic reprints of science; five years later it appeared in an annotated German translation by A. van Oettingen and A. Wangerin, in Wilhelm-Ostwald's famous booklet series of editions of major scientific works. Green's successors in the classical phase not only absorbed his contributions into their own heritage; they wanted to read the words of the master himself.

Their modern successors have maintained the tradition; for Ferrers's edition appeared again in 1970, and the *Essay* itself in 1993, in the university of his home town Nottingham, as part of their bicentennial celebrations of their remarkable citizen.

References

1. *Procès-Verbaux des Séances de l'Académie des Sciences Tenues Depuis la fondation* [in 1795] *Jusqu'au Mois d'Août, 1835*, Vol. 8, Hendaye (Observatoire) Academy of Sciences, 1918.

2. M. Bacharach, *Abriss der Geschichte der Potentialtheorie*, Thein, 1883.

3. H. Burkhardt, Entwicklungen nach oscillirenden functionen und integration der differential-gleichungen der mathematischen physik, *Jahresbericht der Deutschen Mathematiker-Vereinigung* **10** (1908).

4. H. Burkhardt and F. W. F. Mayer, Potentialtheorie, *Encyklopädie der Mathematischen Wissenschaften* **2A** (1900), 464–503.

5. D. M. Cannell, *George Green*, Athlone Press, 1993.

6. J. J. Crow, Integral theorems in Cambridge mathematical physics, 1830–55, in [14], pp. 112–148.

7. A. De Morgan, Letter to C. Babbage, British Library, Additional Mss. 37189, no. 241.

8. J. B. J. Fourier, Note relative aux vibrations des surfaces élastiques, *Bulletin des Sciences, par la Société Philomatique de Paris* (1818), 129–136 (*Oeuvres* Vol. 2, pp. 255–265).

9. I. Grattan-Guinness, Mathematics and mathematical physics at Cambridge, 1815–1840..., in [14], pp. 84–111.

10. I. Grattan-Guinness, *Convolutions in French Mathematics, 1800–1840. From the Calculus and Mechanics to Mathematical Analysis and Mathematical Physics*, 3 vols., Birkhäuser (Deutscher Verlag der Wissenschaften), 1990.

11. G. Green, *An Essay on the Mathematical Analysis of Electricity and Magnetism*, Nottingham, 1828. repr. in *J. Reine Angew. Math.* **39** (1850), 75–89; **44** (1852), 356–374; and **47** (1854), 161–211.

12. G. Green, *Mathematical Papers* (ed. N. M. Ferrers), Macmillan, 1871.

13. H. Green, A biography of George Green..., in *Studies and Essays in... Honor of George Sarton* (ed. A. Montagu), Schuman, 1946, pp. 545–594.

14. P. M. Harman, *Wranglers and Physicists...*, Manchester Univ. Press, 1985.

15. R. Murphy, On the inverse method of definite integrals..., *Trans. Cambridge Philosophical Society* **4** (1833), 353–408.

16. S. D. Poisson, Mémoire sur la distribution de l'électricité à la surface des corps conducteurs, *Mémoires Paris. Acad. Sci.* (1811), pt. 1, 1–92.

17. S. D. Poisson, Remarques sur l'équation qui se présente dans la théorie des attractions des sphéroïdes, *Nouveau Bulletin des Sciences, par la Société Philomatique de Paris* **3** (1812–13), 388–392.

18. S. D. Poisson, Second mémoire sur la distribution de l'électricité à la surface des corps conducteurs, *Mémoires Paris. Acad. Sci.* (1811), pt. 2, 163–274.

19. S. D. Poisson, [Translation of a summary of [21]], *Quarterly J. Science* **17** (1824), 317–324.

20. S. D. Poisson, [Translation of a summary of [22]], *Quarterly J. Science* **19** (1825), 122–131.

21. S. D. Poisson, Mémoire sur la théorie du magnétisme, *Mémoires Paris. Acad. Sci.* **5** (1821–22), 247–338.

22. S. D. Poisson, Second mémoire sur la théorie du magnétisme, *Mémoires Paris. Acad. Sci.* **5** (1821–22), 488–533.

23. S. D. Poisson, Mémoire sur la théorie du magnétisme en mouvement, *Mémoires Paris. Acad. Sci.* **6** (1823), 441–570.

24. A. I. Schlissel, The development of asymptotic solutions to ordinary differential equations, 1817–1920, *Arch. Hist. Exact Sci.* **16** (1977), 307–378.

25. W. Thomson and P. G. Tait, *Treatise on Natural Philosophy*, 2 pts., Cambridge Univ. Press, 1883.

26. I. Todhunter, *A History of the Mathematical Theories of Attraction and Figure of the Earth...*, 2 vols., Macmillan, 1873; repr. 1962, Dover.

27. G. J. Whitrow, George Green (1793–1841): a pioneer of modern mathematical physics and its methodology, *Annali dell'Istituto di Storia della Scienza di Firenze* **2** (1984), 47–68.

28. E. T. Whittaker, *History of the Theories of Aether and Electricity. The Classical Theories*, Nelson, 1951.

29. C. Wolf, *Mémoires sur le pendule...*, 2 pts., Gauthier-Villars, 1889–1891.

Connectivity and Smoke-Rings: Green's Second Identity in its First Fifty Years

THOMAS ARCHIBALD

Mathematics Magazine **62** (1989), 219–232

Introduction

James Clerk Maxwell, in his review of Thomson and Tait's *Treatise on Natural Philosophy*, noted an important innovation in the authors' approach to mathematics ([8], Vol. 2, p. 777):

> The first thing which we observe in the arrangement of the work is the prominence given to kinematics, ... and the large space devoted under this heading to what has been hitherto considered part of pure geometry. The theory of curvature of lines and surfaces, for example, has long been recognized as an important branch of geometry, but in treatises on motion it was regarded as lying as much outside of the subject as the four rules of arithmetic or the binomial theorem.
>
> The guiding idea however ... is that geometry itself is part of the science of motion, and that it treats, not of the relations between figures already existing in space, but of the process by which these figures are generated by the motion of a point or a line.

This "guiding idea," which treats geometric entities as physical objects in some sense, had been influential with mathematicians for many years. Countless mathematical problems have their origin in the investigation of the natural world. However, it also happens that the solutions of some problems may be facilitated by attributing physical properties to the mathematical objects under study. In addition, mathematical constructs usually thought of as "purely geometric" may be created by considering such mathematico-physical entities.

It is my purpose in this article to illustrate some aspects of the cross-fertilization of mathematics and physics by examining the development of Green's second identity (known to physicists as Green's theorem) and its generalizations over a fifty-year period, from 1828 to 1878. During this period, despite an increased emphasis on logical rigor in some circles, many mathematicians continued to accept physical proofs of analytic theorems as valid. Such proofs used hypothetical physical properties such as incompressibility to characterize the regions in space; this trend was found most strongly in nineteenth-century British mathematics, though it was not unknown elsewhere.

Peter Guthrie Tait

Green's second identity, well known from vector calculus, states that

$$\iiint \left(\varphi \nabla^2 \psi - \psi \nabla^2 \varphi \right) dx\,dy\,dz = \iint \left(\varphi \frac{\partial \psi}{\partial n} - \psi \frac{\partial \varphi}{\partial n} \right) da.$$

The integration on the left is performed over a region bounded by a closed surface S. The integral on the right is then a surface integral over S, and n is an outward normal to S. Finally, φ and ψ are continuously differentiable real-valued functions (scalar fields) on $\mathbb{R}^3$. This theorem first appeared in a paper by George Green published in 1828, along with a number of other lemmas which Green employed in his study of electrostatics and magnetism. Green's results remained virtually unknown, however, until William Thomson (later Lord Kelvin) obtained two copies of Green's pamphlet in 1845. Green's results subsequently became widely known, and were central to the mathematical theory of potential, one of the most important tools of mathematical physics in the following decades.

Potential theory had originated as a body of results which arose in connection with the efforts of French mathematical physicists (notably Poisson, Laplace, and Biot), to extend the methods of Newton. Laplace attempted to explain many natural phenomena as the result of forces proportional to the inverse square of the distance between the interacting objects. To achieve this, it was necessary to determine the integrals of vector forces. Laplace showed that such forces could be treated as what we now term the gradient of a scalar function, and hence was able to simplify the calculations greatly. Such a function, the gradient of which is a force, is known as a potential for that force. (We will also see velocity potentials in the course of this article, which are functions the gradient of which gives a velocity.) Like that of Laplace, Green's work was a contribution both to mathematical physics and to potential theory, since it expresses relationships between potentials and their integrals as well as applying the results to physical problems [4].

At its beginning, the idea of potential was a mathematical convenience. However, by the 1850s it had acquired physical interpretations. In particular, if a vector function has a potential, the integral of that potential along a curve depends only on the endpoints of the integration, and integrals around closed paths are zero. This expresses the fact that the vector function is an exact differential. Physically, this implies that the force described by the function is conservative, so that potential functions are closely associated with potential energy.

Green's 1828 essay

Green's paper, which was published privately in Nottingham in 1828, was called *An Essay on the Application of Mathematical Analysis to the Theories of Electricity and Magnetism* [1]. George Green (1793–1841), a miller's son, had been given access to the library of a local aristocrat interested in science. This opportunity, and Green's ability, permitted him to master basic works by Laplace, Lagrange, and Poisson. Inspired by Laplace's work on gravitation and Poisson's on electrostatics and magnetism, Green set forth to investigate electrostatics using similar hypotheses but new methods.

Of particular interest to us are Green's general mathematical theorems, presented at the beginning of the paper, which he later applied to particular electrical and magnetic calculations. Green's second identity is the key theorem in this section. It is the essential tool in solving the Laplace equation and the Poisson equation by the method which Green introduced. A detailed discussion of this method, today known as the method of *Green's functions*, would take us too far afield.

In modern notation, the identity Green proved was

$$\iiint U \nabla^2 V d^3 x + \iint U \frac{\partial V}{\partial n} d\sigma$$
$$= \iiint V \nabla^2 U d^3 x + \iint V \frac{\partial U}{\partial n} d\sigma. \quad (1)$$

Here U and V are any two functions which are continuously differentiable in the region of differentiation, and n is now the inward normal from the surface σ. Green used the symbol δ to express what we have denoted by ∇^2. Green's proof of this identity rests on applying integration by parts to the expression

$$\iiint \left\{ \frac{\partial V}{\partial x} \frac{\partial U}{\partial x} + \frac{\partial V}{\partial y} \frac{\partial U}{\partial y} + \frac{\partial V}{\partial z} \frac{\partial U}{\partial z} \right\} dx\, dy\, dz$$
$$= \iiint (\nabla V) \cdot (\nabla U). \quad (2)$$

Assuming U, V are sufficiently differentiable, we can integrate by parts in each variable. For example, let

$$u = \frac{\partial U}{\partial x} dx \quad \text{and} \quad v = \frac{\partial V}{\partial x} dx.$$

Then substitution in (2) yields

$$\iint dy\, dz \left(\int \frac{dV}{dx} \frac{dU}{dx} dx \right)$$
$$= \iint V(x_1) \frac{dU}{dx} \bigg|_{x=x_1} dy\, dz$$
$$- \iint V(x_0) \frac{dU}{dx} \bigg|_{x=x_0} dy\, dz - \iiint V \frac{\partial^2 U}{\partial x^2} dx\, dy\, dz.$$

Green then argued that, if σ is a surface element and n an inward normal, we have

$$\iint dy\,dz\left(V(x_1)\frac{dU}{dx}\bigg|_{x_1} - V(x_0)\frac{dU}{dx}\bigg|_{x_0}\right)$$
$$= -\iint d\sigma \frac{\partial x}{\partial n} V \frac{dU}{dx}.$$

Hence the partial integral becomes

$$\iint dy\,dz\left(\int \frac{\partial V}{\partial x}\frac{\partial U}{\partial x}dx\right) = -\iint d\sigma \frac{\partial x}{\partial n} V \frac{\partial U}{\partial x}$$
$$- \iiint V\frac{\partial^2 U}{\partial x^2}.$$

Consequently, the result of integration with respect to all three variables gives

$$\iiint dx\,dy\,dz\,(\nabla V)\cdot(\nabla U)$$
$$= -\iint d\sigma V\frac{\partial U}{\partial n} - \iiint V\nabla^2 U. \quad (3)$$

This is often known as Green's first identity. By symmetry, we may interchange U and V in (3) to obtain the second identity:

$$\iint d\sigma V\frac{\partial U}{\partial n} - \iiint V\nabla^2 U$$
$$= -\iint d\sigma U\frac{\partial V}{\partial n} - \iiint V\nabla^2 U.$$

Many present-day niceties in the proof of this identity were not considered by Green. A full proof involves dealing properly with the relationship between the infinitesimals and the finite, and we must use the equivalence of multiple and iterated integrals, which Green did not distinguish. The advances in rigorous analysis due to Cauchy may well have been unknown to Green at this time, since he mentions his limited access to the latest work. Instead his arguments rely on the geometry of infinitesimals. In this, his work resembles that of most of his contemporaries, even in France.

Green's work went almost entirely unnoticed for many years. None of the private subscribers who purchased the pamphlet appears to have been capable of appreciating its worth, and his results and methods remained little known [2]. Green's work might have been forgotten had it not been mentioned by the Irish electrician Robert Murphy. Murphy referred to Green as the originator of the term potential, though Murphy's own definition of potential was erroneous, indicating that he had not actually seen Green's work (see [13], pp. 1–2 and 126 for more about Murphy).

Green himself did not revive interest in his earlier work. His efforts in the interim were devoted to further research, and to an education at Cambridge. His other papers met a happier immediate reception; several were published in the *Transactions of the Cambridge Philosophical Society*, where they attracted the interest of the British scientific community. Green thus made a name for himself before his death in 1841, though his reputation was considerably enhanced by the rediscovery of the 1828 *Essay*.

William Thomson, later Lord Kelvin

Thomson rediscovers Green

It was William Thomson (later Lord Kelvin) who first drew the attention of the international scientific world to Green's results. Sometime in 1842 Thomson had read a reference by Murphy to Green's paper; his interest was piqued for several reasons. Thomson was himself then engaged in research on the theory of attraction, and published papers on the subject in 1842 and 1843. Murphy had referred to Green's use of the term *potential*, a notion which, as Thomson states, was also employed by Gauss with great success in his 1839 paper on inverse-square forces. Thomson doubtless wondered how Green, whose name he knew well, had employed the notion of potential, and was curious about the exact nature of his results.

Thomson was unable to see a copy of Green's work until January 25, 1845, shortly before he was about to embark on a trip to France following the completion of his studies at Cambridge. By chance, Thomson's tutor, Hopkins, had two copies which he had apparently never examined, and sent them with Thomson.

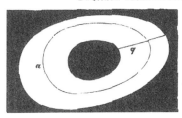

Figure 1. Riemann's illustration of a doubly connected surface.

Thomson was very impressed with the generality of Green's results, and was soon endeavoring to apply them in his own research. On his arrival in France, Thomson showed the paper to Liouville, Sturm, and Chasles, among others. Soon the Paris mathematical community was well aware of Green's work: for example, Liouville gave Green and Gauss equal credit for the introduction of the term potential in an 1847 paper ([7] p. 174). Thomson sent the other copy of Green's paper to Germany with Cayley, who delivered it to August Crelle, editor of the *Journal für die Reine und Angewandte Mathematik*. Crelle published a translation of Green's paper in three installments between 1850 and 1854, hence it became well known to interested researchers in Germany. Thus, 25 years after Green's original publication, his methods began to find their way into the scientific literature and textbooks of Europe ([12], pp. 113–119). One of the first to make use of Green's work was Bernhard Riemann (1826–1866), who was then writing his doctoral dissertation at Göttingen.

Riemann and multiply-connected regions

Riemann's principal interest in Green's work was in the method of Green's functions, which Green had used to solve boundary-value problems involving functions satisfying Laplace's equation

$$\nabla^2 \varphi = 0.$$

Here φ is to be interpreted as the potential function of the electrostatic force due to a charge density on a conductor. Riemann, however, noticed that Green's methods could be useful in the study of functions of a complex variable, since the real and complex parts of such functions must satisfy Laplace's equation. Employing this insight, Riemann developed methods that enabled him to specify a complex function by its boundary values and discontinuities. In so doing, Riemann presented the idea of *multiply-connected regions* of the plane: a region is simply connected if a cross-cut divides it in two, and has connectivity equal to the number of cuts taken to separate it. (See Fig. 1.) This notion was published in Riemann's dissertation (1851) and found wider circulation with the appearance of his paper on abelian integrals (1857) [9]. It was here that it was seen by Hermann von Helmholtz, who was attempting to employ Green's ideas in a different way.

Helmholtz and vortices

Hermann von Helmholtz's 1858 paper *On Integrals of Hydrodynamic Equations which yield Vortex Motion* was also deeply influenced by Green's work [3]. Helmholtz (1821–1894) had become interested in the solution of boundary-value problems in fluid mechanics in connection with his investigation of the physiology of the ear. (He was at that time a professor of anatomy in Bonn ([6], pp. 307–312).) Furthermore, Helmholtz saw a parallel between certain problems in hydrodynamics and problems in electromagnetic theory, a long-standing interest of his. Attempts to provide a detailed theoretical treatment of the analogy between electromagnetic theory and fluid dynamics may have been sparked by the superficial resemblance between electrical and hydrodynamical phenomena. The electric current was widely viewed in the mid-nineteenth century as the flow of one or two "electric fluids" along a conductor. The motion of this fluid produces a magnetic effect. André-Marie Ampère demonstrated in the 1820s that magnetism may be explained as the result of hypothetical microscopic electric currents in a body, and hence the existence of the electromagnetic phenomenon should mean that a current gives rise to other currents. If the original current flows in a straight line, the currents responsible for magnetic effects must be helical, forming microscopic vortices.

The researches of Helmholtz and others in this area aimed to make this rather vague picture precise. In the 1858 paper, Helmholtz examined the following question: suppose we are given a closed container filled with a frictionless incompressible fluid. How does action on the boundary of the container affect the motion inside? Helmholtz apparently saw the value of Green's theorems in such an investigation soon after reading Green's paper, but was kept from working out his ideas because of other academic obligations. However, Helmholtz had also recently read Riemann's paper of 1857, which made it clear to him that Green's theorem could only be used when the regions involved were simply connected. This is because functions with potentials—what we would now term *conservative vector fields*—may in fact be multiple-valued in multiply-connected regions.

Let us discuss how Helmholtz used Green's theorem. He began with Euler's equation of fluid dynamics, which we may write in vector notation as

$$\vec{F} = \frac{1}{\rho}\nabla \vec{p} + \left(\frac{\partial}{\partial t} + \vec{v}\cdot\nabla\right)\vec{v}, \quad (4)$$

$$\nabla\cdot\vec{v} = 0. \quad (5)$$

Here ρ is the density of the fluid, $\vec{p}$ the pressure, $\vec{v}$ the velocity. Helmholtz supposed that

$$\vec{F} = \nabla V, \quad \text{where } V \text{ is a force potential}$$

and

$$\vec{v} = \nabla\varphi, \quad \text{where } \varphi \text{ is a velocity potential.}$$

From (5) we have that φ satisfies Laplace's equation, since $\nabla\cdot\vec{v} = \nabla^2\varphi = 0$. Helmholtz then noted that this implies

$$\nabla\times\vec{v} = \nabla\times(\nabla\varphi) = 0.$$

Bernhard Riemann

If the walls of the container are rigid, this means that the component of velocity perpendicular to the boundary is zero. Hence, if n is an outward normal, $\partial\varphi/\partial n$ is equal to zero everywhere. But by Green's first identity,

$$\iiint_R (\nabla U \cdot \nabla V)\,dx^3$$
$$= \iint V\frac{\partial U}{\partial n}\,ds - \iiint_R V\nabla^2 U\,dx^3.$$

If $U = V = \varphi$ we have

$$\iiint_R (\nabla\varphi)^2\,dx^3 = \iint \varphi\frac{\partial\varphi}{\partial n}\,ds = 0,$$

remembering that $\nabla^2\varphi = 0$. Thus $\nabla\varphi = 0$, i.e., no motion is induced in the fluid. Hence, any motion of a fluid in a closed vessel (with simply-connected interior) which has a velocity potential must depend exclusively on a motion of the boundary. Helmholtz went on to show that a motion of the boundary uniquely determines such a motion in the fluid. A further important conclusion stated that vortices can only be produced by a motion which has no velocity potential. More important still, if vortices do exist initially, they are stable under the action of conservative forces. Either they must be closed tubes, or else they extend from the boundary to the boundary.

When vortices do exist one can consider the portion of fluid without vortices as a multiply-connected region, the vortices being the "holes." Thus to solve boundary value problems in such a region, one would ideally have an extension of Green's theorem to deal with such cases. Helmholtz stressed the desirability of such a generalization, and noted that Riemann's notion of connectivity could readily be extended to three dimensions for this purpose.

In Helmholtz's work the geometric entities immediately become physical. For one thing, they are three dimensional. Also the points of space become associated with the molecules of a fluid, and holes in the space correspond to vortices. In this instance, the geometric entities may be given clear physical interpretation, and physical questions (the solution of specific boundary-value problems, for example) dictate the mathematical problems which are important.

Hermann von Helmholtz

Helmholtz's research was received with greater sympathy by British mathematical physicists, especially Thomson, than by his German colleagues. This occurred in part because of the shared interests of Helmholtz and Thomson in hydrodynamic models for electromagnetic theory, an interest that arose because of their attitude toward the then-prevailing thought on electromagnetic theory in Germany. This theory, based on work by Gauss's collaborator Wilhelm Weber, explained electrical phenomena on the basis of a velocity-dependent force law. Both Helmholtz and Thomson felt that such a force could not satisfy the energy conservation principle. Helmholtz apparently felt as well that his mathematical skills were dimly regarded by his German contemporaries, because he had not been formally trained as a mathematician. Thus it was among British mathematical physicists that Helmholtz's papers were read with greatest interest and understanding.

Tait, Thomson, smoke-rings and atoms

Helmholtz's approach found an enthusiastic admirer in Peter Guthrie Tait (1837–1901), a Cambridge-educated Scot who was teaching in Belfast in 1858. Tait was attempting at that time to master William Rowan Hamilton's method of quaternions, and to demonstrate the physical usefulness of the method by obtaining significant applications. In this respect Helmholtz's work interested him, and he made an English translation for his own use ([12], p. 511).

A parenthetic note about quaternions: Nowadays, this set of objects is most likely to show up in algebra courses or proofs in algebraic number theory which can make use of its properties as a non-commutative division ring. This is quite remote from their original intended use in geometry and analysis. Hamilton invented quaternions in 1843, and introduced with them the idea of operators. Particularly important was the del or nabla operator, our ∇. For Hamilton and Tait, a quaternion described a quotient of what we would term *vectors*; such a quotient consists of a 4-tuple which describes the stretch and the three rotations which bring an arbitrary pair of vectors into coincidence ([11], pp. 37–38). Later on we shall see how Tait used this approach to obtain what he called "physical proofs" of analytic statements.

Tait moved to Edinburgh in 1860, where he began a collaboration with William Thomson, then at Glasgow. In 1866 and 1867 their collaboration was at its peak, as they prepared their *Treatise on Natural Philosophy* (which was to become the standard introductory physics text in Britain for decades). Early in 1867, Tait showed Thomson an experimental demonstration of the stability properties of vortices by means of smoke-rings, as well as Helmholtz's mathematical treatment of the problem. Thomson described this event to Helmholtz in a letter:

> Just now, however, vortex motions have displaced everything else, since a few days ago Tait showed me in Edinburgh a magnificent way of producing them. Take one side (or a lid) off a box (any old packing box will serve) and cut a large hole in the opposite side. Stop the open side AB loosely with a piece of cloth, and strike the middle of the cloth with your hand. If you leave anything smoking in the box, you will see a magnificent ring shot out by every blow.

Thomson then went on to describe what he found particularly interesting about the phenomenon and the theory.

> The absolute permanence of the rotation, and the unchangeable relation you have proved between it and the portion of the fluid once acquiring such motion in a perfect fluid, shows that if there is a perfect fluid all through space, constituting the substance of all matter, a vortex-ring would be as permanent as the solid hard atoms assumed by Lucretius and his followers (and predecessors) to account for the permanent properties of bodies ... thus if two vortex rings were once created in a perfect fluid, passing through one another like links of a chain, they could never come into collision, or break one another, they would form an indestructible atom ([12], pp. 514–515).

Figure 2. William Thomson's Knots

Thomson embarked on the mathematical theory of these apparently indestructible vortex atoms at once, and his results were read before the Royal Society of Edinburgh a little over three weeks later. His paper *On Vortex Motion*, much augmented, appeared in 1878 [14]. Here he encountered and solved the problem posed by Helmholtz of extending Green's theorem to multiply-connected regions. For in order to investigate the properties of vortex atoms, it was necessary to solve boundary value problems where complicated vortices (such as those pictured) formed part of the boundary. (See Fig. 2.)

In this paper, Thomson wrote the original version of Green's theorem thus:

$$\iiint_R \nabla\varphi \cdot \nabla\varphi' \, dV$$
$$= \iint \varphi \frac{\partial \varphi'}{\partial n} \, da - \iiint_R \varphi \nabla^2 \varphi' \, dV$$
$$= \iint \varphi' \frac{\partial \varphi}{\partial n} \, da - \iiint_R \varphi' \nabla^2 \varphi \, dV.$$

Here φ and φ' must be single-valued. Thomson then investigated what happens if φ' is multi-valued—that is to say, if we consider R to be a multiply-connected region, and then considered how to set the problem up in general. A multiply-connected space can be made simply connected by making cuts, or by inserting what Thomson calls "stopping barriers." The integral around an (almost) closed path from one side of the barrier to the other has a constant value k_i, which is the same for all such paths; such constants k_i exist for all stopping barriers and the integral in question becomes

$$\iiint_R \nabla\varphi \cdot \nabla\varphi' \, dV$$
$$= \iint \varphi \frac{\partial \varphi'}{\partial n} \, d\sigma + \sum_i k_i \iint \frac{\partial \varphi'}{\partial n} \, d\sigma'$$
$$- \iiint_R \varphi \nabla^2 \varphi' \, dV$$
$$= \iint \varphi' \frac{\partial \varphi}{\partial n} \, d\sigma + \sum_i k_i \iint \frac{\partial \varphi}{\partial n} \, d\sigma'$$
$$- \iiint_R \varphi' \nabla^2 \varphi \, dV$$

Here $d\sigma'$ represents a surface element of the barrier surface.

Thus armed, Thomson was able to examine fluid motion in multiply-connected regions, concluding that the normal component of velocity of a fluid at every point of the boundary determines the motion inside a multiply-connected region (provided we know the circulation of the fluid in each region). He then considered how best to define the order of connectivity, noting that for some of the surfaces shown the stopping barriers must be self-intersecting and difficult to distinguish. He therefore proposed a definition using what he called "irreconcilable paths"—which we would now call homotopy classes of closed paths with base point. He selected a point on the surface, and noted that the connectivity of the surface is determined if we see how many mutually irreconcilable paths can be drawn on its surface. For a simply-connected region, for example, all closed paths on the surface are homotopic. Although this affords an unambiguous definition of connectivity, it is not easily possible using this method to get the generalization of Green's theorem. For that the stopping barriers are required.

Thomson's interest in vortex atoms thus led him directly to a generalization of Green's theorem, and to the question of the proper definition of connectivity. His proof technique is in essence the same as that of Green, augmented by the stopping barriers, and it is this method that is usually taught today in courses on vector calculus.

Tait's quaternion version of Green's theorem

By the late 1860s, Tait's interest in quaternions had turned into a crusade. To the British Association in 1871, he said:

> comparing a Cartesian investigation... with the equivalent quaternion one... one can hardly help making the remark that they contrast even more strongly than the decimal notation with the binary scale or with the old Greek arithmetic, or than the well-ordered subdivision of the metrical system with the preposterous non-systems of Great Britain.

In the same address, Tait pointed out that from the quaternion point of view:

> Green's celebrated theorem is at once seen to be merely the well-known equation of continuity expressed for a heterogeneous fluid, whose density at every point is proportional to one electric potential, and its displacement or velocity proportional to and in the direction of the electric force due to another potential [5].

Let us see exactly what he means. In his 1870 paper *On Green's and Other Allied Theorems*, Tait supposed a spatial region R to be uniformly filled with points ([10], pp. 136–150). If points inside and outside the regions are displaced by a vector then we may have a net decrease or increase of the volume—that is, of the number of points—in the region R. This can be calculated in two ways:

1. We can find the total increase in density throughout R

$$\iiint_R \operatorname{div} \sigma \, dV \quad (6)$$

(in Tait's notation $\iint S \cdot \nabla \sigma \, d\varsigma$).

2. We can estimate the excess of those that pass inwards through the surface over those that pass outwards:

$$\iint \sigma \cdot \vec{n} \, da \quad (7)$$

(in Tait's notation $\iint S \cdot \sigma U \nu \, dx$).

The expressions (6) and (7) must be equal, yielding what we now call the divergence theorem from the equation of continuity. If we consider that the density—for example, of electric fluid—is given by a potential P_1, and the displacement is proportional to a force σ with potential P, then we have:

$$\nabla(PP_1) = P\nabla P_1 + P_1 \nabla P$$

and

$$\nabla^2(PP_1) = P\nabla^2 P_1 + P_1 \nabla^2 P + 2(\nabla P \cdot \nabla P_1). \quad (8)$$

But by the divergence theorem

$$\iiint_R \nabla^2(PP_1) \, dV = \iiint_R \operatorname{div}(\nabla PP_1) \, dV$$
$$= \iint_{\partial R} (\nabla PP_1) \cdot \vec{n} \, da$$
$$= \iint_{\partial R} (P\nabla P_1 + P_1 \nabla P) \cdot \vec{n} \, da.$$

Hence, from (8),

$$\iint_{\partial R} (P\nabla P_1 + P_1 \nabla P) \cdot \vec{n} \, da$$
$$= \iiint_R (P\nabla^2 P_1 + P_1 \nabla^2 P) \, dV$$
$$+ 2 \iiint_R \nabla P \cdot \nabla P_1 \, dV.$$

But the left side here, by the divergence theorem, is

$$\iiint_R (P\nabla^2 P_1 - P_1 \nabla^2 P) \, dV.$$

Combining these two yields Green's theorem in the form

$$\iiint_R (\nabla P \cdot \nabla P_1) \, dV$$
$$= -\iiint_R P_1 \nabla^2 P \, dV + \iint_{\partial R} P \frac{\partial P}{\partial n} \, da$$
$$= -\iiint_R P\nabla^2 P_1 \, dV + \iint_{\partial R} P \frac{\partial P_1}{\partial n} \, da.$$

Notice that the argument depends on treating geometric points as mobile physical entities, with continuity properties like those of a fluid.

We find Tait's views nicely summarized in his 1892 review of Poincaré's *Thermodynamique* ([5], p. 273):

> Some forty years ago, in a certain mathematical circle at Cambridge, men were wont to deplore the necessity of introducing words at all in a physico-mathematical textbook: the unattainable, though closely approachable Ideal being regarded as a world devoid of aught but formulae! But one learns something in forty years, and accordingly the surviving members of that circle now take a very different view of the matter. They have been taught alike by experience and by example to regard mathematics, so far at least as physical enquiries are concerned, as a mere auxiliary to thought... this is one of the great truths which were enforced by Faraday's splendid career.

Conclusion

Our excursion from Green to Tait has taken us from electrostatics and potential theory, via complex analysis and fluid dynamics, to homotopy classes of maps and vector analysis. While I have only touched on a few of the interesting problems associated with these developments, I hope that I have shown that physical thinking is important, not only in posing mathematical problems, but also in solving them. In particular, physical thinking may lead to the creation of certain mathematical notions, such as connectivity, which are of interest in their own right—for example in the classification of the knots described by Thomson. Tait undertook this classification problem around 1870, achieving the first basic results of knot theory.

References

1. George Green, *Mathematical Papers* (ed. N. Ferrers), Macmillan, 1871; reprint, Chelsea, 1970.
2. H. Gwynedd Green, A Biography of George Green, in *Studies and Essays in the History of Science and Learning* (ed. A. Montagu), Schumann, 1956.
3. H. Helmholtz, Über Integrale der hydrodynamischen Gleichungen, welche den Wirbelbewegungen entsprechen, *Journal für Mathematik* **55** (1858), 25–55.
4. O. D. Kellogg, *Foundations of Potential Theory*, Springer, 1929.
5. C. G. Knott, *Life and Scientific Work of P. G. Tait*, Cambridge Univ. Press, 1911.
6. Leo Königsberger, *Hermann von Helmholtz*, Vieweg, 1902.
7. J. Liouville, reprinted in [13].
8. J. Clerk Maxwell, Review of Thomson and Tait in *Nature*, Vol. xx, reprinted in *Scientific Papers*, Vol. 2.
9. G. F. B. Riemann, *Gesammelte Mathematische Werke*, Teubner, 1892.
10. P. G. Tait, *Scientific Papers*, Cambridge Univ. Press, 1911.
11. P. G. Tait and P. Kelland, *Introduction to Quaternions*, Macmillan, 1904.
12. S. P. Thompson, *Life of Lord Kelvin*, Chelsea, 1976.
13. W. Thomson, *Reprint of Papers on Electrostatics and Magnetism*, Macmillan, 1872.
14. W. Thomson, On vortex motion, *Trans. Royal Soc. Edinburgh* **25** (1869), 217–260.

The History of Stokes's Theorem

VICTOR J. KATZ

Mathematics Magazine **52** (1979), 146–156

Most current American textbooks in advanced calculus devote several sections to the theorems of Green, Gauss, and Stokes. Unfortunately, the theorems referred to were not original to these men. It is the purpose of this paper to present a detailed history of these results from their origins to their generalization and unification into what is today called the generalized Stokes' theorem.

Origins of the theorems

The three theorems in question each relate a k-dimensional integral to a $(k-1)$-dimensional integral; since the proof of each depends on the fundamental theorem of calculus, it is clear that their origins can be traced back to the late seventeenth century. Toward the end of the eighteenth century, both Lagrange and Laplace actually used the fundamental theorem and iteration to reduce k-dimensional integrals to those of one dimension less. However, the theorems as we know them today did not appear explicitly until the nineteenth century.

The first of these theorems to be stated and proved in essentially its present form was the one known today as Gauss's theorem or the *divergence theorem*. In three special cases it occurs in an 1813 paper of Gauss [8]. Gauss considers a surface (superficies) in space bounding a solid body (corpus). He denotes by PQ the exterior normal vector to the surface at a point P in an infinitesimal element of surface ds and by QX, QY, QZ the angles this vector makes with the positive x-axis, y-axis, and z-axis, respectively. Gauss then denotes by $d\Sigma$ an infinitesimal element of the yz-plane and erects a cylinder above it, this cylinder intersecting the surface in an even number of infinitesimal surface elements $ds_1, ds_2, \ldots, ds_{2n}$. For each j, $d\Sigma = \pm ds_j \cos QX_j$, where the positive sign is used when the angle is acute, the negative when the angle is obtuse. Since if the cylinder enters the surface where QX is obtuse, it will exit where QX is acute (see Fig. 1), Gauss obtains $d\Sigma = -ds_1 \cos QX_1 = ds_2 \cos QX_2 = \cdots$ and concludes by summation that "The integral $\int ds \cos QX$ extended to the entire surface of the body is 0."

He notes further that if T, U, V are rational functions of only y, z, only x, z, and only x, y, respectively, then

$$\int (T\cos QX + U\cos QY + V\cos QZ)ds = 0.$$

Gauss then approximates the volume of the body by taking cylinders of length x and cross-sectional area $d\Sigma$ and concludes in a similar way his next theorem:

The entire volume of the body is expressed by the integral $\int ds\, x(\cos QX)$ extended to the entire surface.

We will see below how these results are special cases of the divergence theorem.

In 1833 and 1839 Gauss published other special cases of this theorem, but by that time the general theorem had already been stated and proved by Michael Ostrogradsky. This Russian mathematician, who was in Paris in the late 1820s, presented a paper [15] to the Paris Academy of Sciences on February 13, 1826, entitled *Proof of a Theorem in Integral Calculus*. In this paper Ostrogradsky introduces a surface with element of surface area ε bounding a solid with element of volume ω. He denotes by α, β, γ the same angles which Gauss called QX, QY, QZ, and by p, q, r three differentiable functions of x, y, z. He states the divergence theorem in the

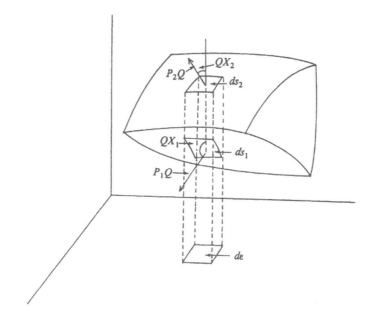

Figure 1.

form:

$$\int \left(a\frac{\partial p}{\partial x} + b\frac{\partial p}{\partial y} + c\frac{\partial p}{\partial z} \right) \omega$$
$$= \int (ap\cos\alpha + bq\cos\beta + cr\cos\gamma)\varepsilon,$$

where a, b, c are constants and where the left-hand integral is taken over a solid, the right-hand integral over the boundary surface.

We note that Gauss's results are all special cases of Ostrogradsky's theorem. In each case $a = b = c = 1$; Gauss's first result has $p = 1, q = r = 0$; his second has

$$\frac{\partial p}{\partial x} = \frac{\partial q}{\partial y} = \frac{\partial r}{\partial z} = 0;$$

and his third has $p = x$, $q = r = 0$. We also will see that Gauss's proof is a special case of that of Ostrogradsky.

Ostrogradsky proves his result by first considering $\frac{\partial p}{\partial x}\omega$. He integrates this over a 'narrow cylinder' going through the solid in the x-direction with cross-sectional area $\bar{\omega}$, using the fundamental theorem of calculus to express this integral as

$$\int \frac{\partial p}{\partial x}\omega = \int (p_1 - p_0)\bar{\omega},$$

where p_0 and p_1 are the values of p on the pieces of surface where the cylinder intersects the solid.

Since $\bar{\omega} = \varepsilon_1 \cos\alpha_1$ on one section of surface and $\bar{\omega} = -\varepsilon_0 \cos\alpha_0$ on the other (α_1 and α_0 being the appropriate angles made by the normal, ε_1 and ε_0 being the respective surface elements), we get

$$\int \frac{\partial p}{\partial x}\omega = \int p_1\varepsilon_1 \cos\alpha_1 + \int p_0\varepsilon_0 \cos\alpha_0$$
$$= \int p\cos\alpha\varepsilon,$$

where the left integral is over the cylinder and the right one is over the two pieces of surface (see Fig. 2). Adding up the integrals over all such cylinders gives one-third of the final result, the other two-thirds being done similarly. We note that this proof can easily be modified to suit modern standards, and is in fact used today, e.g., in Taylor and Mann [24].

Though the above proof applies to arbitrary differentiable functions p, q, r, we will note for future reference that Ostrogradsky uses the result only in the special case where

$$p = v\frac{\partial u}{\partial x} - u\frac{\partial v}{\partial x}, \quad q = v\frac{\partial u}{\partial y} - u\frac{\partial v}{\partial y},$$
$$r = v\frac{\partial u}{\partial z} - u\frac{\partial v}{\partial z},$$

with u and v also being differentiable functions of three variables.

Ostrogradsky presented this theorem again in a paper in Paris on August 6, 1827, and finally in

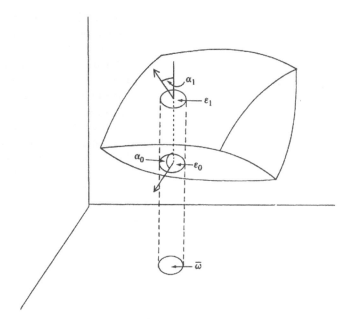

Figure 2.

St. Petersburg on November 5, 1828. The latter presentation was the only one published by Ostrogradsky, appearing in 1831 in [16]. The two earlier presentations have survived only in manuscript form, though they have been published in Russian translation.

In the meantime, the theorem and related ones appeared in publications of three other mathematicians. Simeon Denis Poisson, in a paper presented in Paris on April 14, 1828 (published in 1829), stated and proved an identical result [19]. According to Yushkevich in [28], Poisson had refereed Ostrogradsky's 1827 paper and therefore presumably learned of the result. Poisson neither claimed it as original nor cited Ostrogradsky, but it must be realized that references were not made then with the frequency that they are today.

Another French mathematician, Frederic Sarrus, published a similar result in 1828 in [21], but his notation and ideas are not nearly so clear as those of Ostrogradsky and Poisson. Finally, George Green, an English mathematician, in a private publication of the same year [9], stated and proved the following:

$$\int u\Delta v\,dx\,dy\,dz + \int u\frac{dv}{dw}d\sigma$$
$$= \int v\Delta u\,dx\,dy\,dz + \int v\frac{du}{dw}d\sigma,$$

where u, v are functions of three variables in a solid body 'of any form whatever,' Δ is the symbol for the Laplacian, and d/dw means the normal derivative; the first integrals on each side are taken over the solid and the second over the boundary surface. Green proved his theorem using the same basic ideas as did Ostrogradsky. In addition, if we use again the special case where

$$p = v\frac{\partial u}{\partial x} - u\frac{\partial v}{\partial x}, \quad q = v\frac{\partial u}{\partial y} - u\frac{\partial v}{\partial y},$$
$$r = v\frac{\partial u}{\partial z} - u\frac{\partial v}{\partial z},$$

we can conclude by a short calculation that the two theorems are equivalent. Nevertheless, Green did not so conclude; he was interested in the theorem in the form in which he gave it. It would thus be difficult to attribute the divergence theorem to him.

All of the mathematicians who stated and proved versions of this theorem were interested in it for specific physical reasons. Gauss was interested in the theory of magnetic attraction, Ostrogradsky in the theory of heat, Green in electricity and magnetism, Poisson in elastic bodies, and Sarrus in floating bodies. In nearly all cases, the theorems involved occurred in the middle of long papers and were only thought of as tools toward some physical end. In fact, for both Green and Ostrogradsky the functions u and v mentioned above were often solutions of Laplace-type equations and were used in boundary value problems.

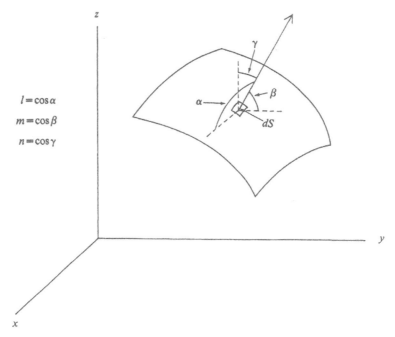

Figure 3.

The theorem generally known as *Green's theorem* is a two-dimensional result which was also not considered by Green. Of course, one can derive this theorem from Green's version by reducing it to two dimensions and making a brief calculation. But there is no evidence that Green himself ever did this.

On the other hand, since Green's theorem is crucial in the elementary theory of complex variables, it is not surprising that it first occurs, without proof, in an 1846 note of Augustin Cauchy [5], in which he proceeds to use it to prove 'Cauchy's theorem' on the integral of a complex function around a closed curve. Cauchy presents the result in the form:

$$\int \left(p \frac{dx}{ds} + q \frac{dy}{ds} \right) = \pm \iint \left(\frac{\partial p}{\partial y} - \frac{\partial q}{\partial x} \right) dx\, dy,$$

where p and q are functions of x and y, and where the sign of the second integral depends on the orientation of the curve which bounds the region over which the integral is taken. Cauchy promised a proof in his private journal *Exercises d'analyse et de physique mathématique*, but he apparently never published one.

Five years later, Bernhard Riemann presented the same theorem in his inaugural dissertation [20], this time with proof and in several related versions; again he uses the theorem in connection with the theory of complex variables. Riemann's proof is quite similar to the proof commonly in use today; essentially he uses the fundamental theorem to integrate $\partial q/\partial x$ along lines parallel to the x-axis, getting values of q where the lines cross the boundary of the region; then he integrates with respect to y to get

$$\int \left[\int \frac{\partial q}{\partial x} dx \right] dy = -\int q\, dy = -\int q \frac{dy}{ds} ds.$$

The other half of the formula is proved similarly.

The final theorem of our triad, *Stokes's theorem*, first appeared in print in 1854. George Stokes had for several years been setting the Smith's Prize Exam at Cambridge, and in the February 1854 examination, question #8 is the following [22] (see Fig. 3):

If X, Y, Z be functions of the rectangular coordinates x, y, z, dS an element of any limited surface, l, m, n the cosines of the inclinations of the normal at dS to the axes, ds an element of the boundary line, shew that

$$\iint \left\{ l \left(\frac{\partial Z}{\partial y} - \frac{\partial Y}{\partial z} \right) + m \left(\frac{\partial X}{\partial z} - \frac{\partial Z}{\partial x} \right) \right.$$
$$\left. + n \left(\frac{\partial Y}{\partial x} - \frac{\partial X}{\partial y} \right) \right\} dS$$
$$= \int \left(X \frac{dx}{ds} + Y \frac{dy}{ds} + Z \frac{dz}{ds} \right) ds$$

... the single integral being taken all around the perimeter of the surface.

It does not seem to be known if any of the students proved the theorem. However, the theorem had already appeared in a letter of William Thomson (Lord Kelvin) to Stokes on July 2, 1850, and the left-hand expression of the theorem had appeared in two earlier works of Stokes. The first published proof of the theorem seems to have been in a monograph of Hermann Hankel in 1861 [10]. Hankel gives no credit for the theorem, only a reference to Riemann with regard to Green's theorem, which theorem he calls well known and makes use of in his own proof of Stokes' result.

In his proof Hankel considers the integral $\int X\,dx + Y\,dy + Z\,dz$ over a curve bounding a surface given explicitly by $z = z(x,y)$. Then

$$dz = \frac{\partial z}{\partial x}dx + \frac{\partial z}{\partial y}dy,$$

and so the given integral becomes

$$\int \left(X + \frac{\partial z}{\partial x}Z\right)dx + \left(Y + \frac{\partial z}{\partial y}Z\right)dy.$$

By Green's theorem, this integral in turn becomes

$$\iint \left\{ \frac{\partial\left(X + \frac{\partial z}{\partial x}Z\right)}{\partial y} - \frac{\partial\left(Y + \frac{\partial z}{\partial y}Z\right)}{\partial x} \right\} dx\,dy.$$

An explicit evaluation of the derivatives then leads to the result:

$$\int (X\,dx + Y\,dy + Z\,dz)$$
$$= \iint \left\{ \left(\frac{\partial X}{\partial y} - \frac{\partial Y}{\partial x}\right) + \left(\frac{\partial Z}{\partial y} - \frac{\partial Y}{\partial z}\right)\frac{\partial z}{\partial x} \right.$$
$$\left. + \left(\frac{\partial X}{\partial z} - \frac{\partial Z}{\partial x}\right)\frac{\partial z}{\partial y} \right\} dx\,dy.$$

Since a normal vector to the surface is given by $(-\partial z/\partial x, -\partial z/\partial y, 1)$ and since the components of the unit normal vector are the cosines of the angles which that vector makes with the coordinate axes, it follows that $\partial z/\partial x = -l/n$, $\partial z/\partial y = -m/n$, and $dS = dx\,dy/n$. Hence by substitution, Hankel obtains the desired result.

Of course, this proof requires the surface to be given explicitly as $z = z(x,y)$. A somewhat different proof, without that requirement, is sketched in Thomson and Tait's *Treatise on Natural Philosophy* (1867) [25] without reference. In 1871 Clerk Maxwell wrote to Stokes asking about the history of the theorem [12]. Evidently Stokes answered him, since in Maxwell's 1873 *Treatise on Electricity and Magnetism* there appears the theorem with the reference to the Smith's Prize Exam [13]. Maxwell also states and proves the divergence theorem.

Vector forms of the theorem

All three theorems first appeared, as we have seen, in their coordinate forms. But since the theory of quaternions was being developed in the mid-nineteenth century by Hamilton and later by Tait, it was to be expected that the theorems would be translated into their quaternion forms. First we must note that Hamilton's product of two quaternions

$$\mathbf{p} = x_0 + x_1\mathbf{i} + x_2\mathbf{j} + x_3\mathbf{k} \quad \text{and}$$
$$\mathbf{q} = y_0 + y_1\mathbf{i} + y_2\mathbf{j} + y_3\mathbf{k}$$

may be written as

$$\mathbf{pq} = (x_0y_0 - x_1y_1 - x_2y_2 - x_3y_3)$$
$$+ (x_2y_3 - y_2x_3)\mathbf{i} + (x_3y_1 - y_3x_1)\mathbf{j}$$
$$+ (x_1y_2 - x_2y_1)\mathbf{k}.$$

Again we denote the scalar part by $S\nabla\boldsymbol{\sigma}$ and the vector part by $V\nabla\boldsymbol{\sigma}$.

Tait, then, in an 1870 paper [23] was able to state the divergence theorem in the form

$$\iiint S\cdot\nabla\boldsymbol{\sigma}\,d\zeta = \iint S\cdot\boldsymbol{\sigma}U\nu\,ds,$$

where $d\zeta$ is the element of volume, ds an element of surface, and $U\nu$ a unit normal vector to the surface. Furthermore, Stokes's theorem took the form

$$\int S\cdot\boldsymbol{\sigma}\,d\rho = \iint S\cdot V\nabla\boldsymbol{\sigma}U\nu\,d\zeta,$$

where $d\rho$ is an element of length of the curve bounding the surface.

Maxwell, in his treatise of three years later, repeated Tait's formulas, but also came one step closer to our current terminology. He proposed to call $S\nabla\boldsymbol{\sigma}$ the *convergence* of $\boldsymbol{\sigma}$ and $V\nabla\boldsymbol{\sigma}$ the *curl* of $\boldsymbol{\sigma}$. Of course, Maxwell's convergence is the negative of what we call the divergence. Furthermore, we note that when two quaternions $\mathbf{p}$ and $\mathbf{q}$ are pure vectors, Hamilton's $S\cdot\mathbf{pq}$ is precisely the negative of the inner product $\mathbf{p}\cdot\mathbf{q}$. (This particular idea was first developed by Gibbs and Heaviside about twenty years later.)

Putting these notions together, we get the modern vector form of the divergence theorem

$$\iiint_M (\text{div}\,\boldsymbol{\sigma})\, dV = \iint_S \boldsymbol{\sigma} \cdot \mathbf{n}\, dA,$$

where $\boldsymbol{\sigma}$ is a vector field $X\mathbf{i} + Y\mathbf{j} + Z\mathbf{k}$, dV is an element of volume, dA is an element of surface area of the surface S bounding the solid M, and $\mathbf{n}$ is the unit outward normal to this surface. Stokes's theorem then takes the form

$$\iint_S (\text{curl}\,\boldsymbol{\sigma}) \cdot \mathbf{n}\, dA = \int_\Gamma (\boldsymbol{\sigma} \cdot \mathbf{t})\, ds,$$

where ds is the element of length of the boundary curve Γ of the surface S and $\mathbf{t}$ is the unit tangent vector to Γ. Obviously, a similar result can be given for Green's theorem.

Generalization and unification

The generalization and unification of our three theorems took place in several stages. First of all, Ostrogradsky himself in an 1836 paper in Crelle [17] generalized his own theorem to the following:

$$\int_V \left(\frac{\partial P}{\partial x} + \frac{\partial Q}{\partial y} + \frac{\partial R}{\partial z} + \cdots \right) dx\, dy\, dz \ldots$$

$$= \int_S \frac{\left(P\frac{\partial L}{\partial x} + Q\frac{\partial L}{\partial y} + R\frac{\partial L}{\partial z} + \cdots \right)}{\sqrt{\left(\frac{\partial L}{\partial x}\right)^2 + \left(\frac{\partial L}{\partial y}\right)^2 + \left(\frac{\partial L}{\partial z}\right)^2 + \cdots}}\, dS.$$

Here Ostrogradsky lets $L(x, y, z, \ldots)$ be "a function of as many quantities as one wants," V be the set of values $x, y, z, \ldots$ with $L(x, y, z, \ldots) > 0$, and S be the set of values with $L(x, y, z, \ldots) = 0$. In modern terminology, if there are n values, S would be an $(n-1)$-dimensional hypersurface bounding the n-dimensional volume V.

Ostrogradsky's proof here is similar to his first one. He integrates $\partial P/\partial x$ with respect to x and after a short manipulation gets

$$\int \frac{\partial P}{\partial x} dx\, dy\, dz \ldots = \int \frac{P\frac{\partial L}{\partial x}}{\sqrt{\left(\frac{\partial L}{\partial x}\right)^2}}\, dy\, dz \ldots$$

with, of course, a similar expression for every other term. Then, putting

$$dS = \sqrt{dy^2 dz^2 \cdots + dx^2 dz^2 \cdots + dx^2 dy^2 \cdots + \cdots},$$

he shows that

$$\frac{dy\, dz \ldots}{\sqrt{\left(\frac{\partial L}{\partial x}\right)^2}} = \frac{dx\, dz \ldots}{\sqrt{\left(\frac{\partial L}{\partial y}\right)^2}} = \cdots$$

$$= \frac{dS}{\sqrt{\left(\frac{\partial L}{\partial x}\right)^2 + \left(\frac{\partial L}{\partial y}\right)^2 + \left(\frac{\partial L}{\partial z}\right)^2 + \cdots}}$$

and concludes the result by summation.

To understand how Ostrogradsky gets his expression for dS, we note that for a parametrized surface in three-space,

$$dS = \Delta du\, dv$$
$$= \sqrt{\left(\frac{\partial(y,z)}{\partial(u,v)}\right)^2 + \left(\frac{\partial(z,x)}{\partial(u,v)}\right)^2 + \left(\frac{\partial(x,y)}{\partial(u,v)}\right)^2}\, du\, dv$$

and

$$\frac{\partial(y,z)}{\partial(u,v)} du\, dv = dy\, dz, \quad \frac{\partial(z,x)}{\partial(u,v)} du\, dv = dz\, dx,$$

$$\frac{\partial(x,y)}{\partial(u,v)} du\, dv = dx\, dy.$$

Hence

$$dS = \sqrt{dy^2 dz^2 + dz^2 dx^2 + dx^2 dy^2}.$$

For more details, see [1].

If we further note that

$$\left(\sqrt{\left(\frac{\partial L}{\partial x}\right)^2 + \left(\frac{\partial L}{\partial y}\right)^2 + \left(\frac{\partial L}{\partial z}\right)^2 + \cdots} \right)^{-1}$$

$$\times \left(\frac{\partial L}{\partial x}, \frac{\partial L}{\partial y}, \frac{\partial L}{\partial z}, \ldots \right)$$

is the unit outward normal $\mathbf{n}$ to S, and if we let $\boldsymbol{\sigma}$ denote the vector function $(P, Q, R, \ldots)$, then Ostrogradsky's result becomes

$$\int (\text{div}\,\boldsymbol{\sigma}) dV = \int \boldsymbol{\sigma} \cdot \mathbf{n}\, dS,$$

a direct generalization of the original divergence theorem.

The first mathematician to include all three theorems under one general result was Vito Volterra in 1889 in [27]. Before quoting the theorem, we need to understand his terminology. An r-dimensional hyperspace in n-dimensional space is given parametrically by the n functions

$$x_i = (u_1, u_2, u_3, \ldots, u_r), \quad i = 1, \ldots, n.$$

Volterra considers the n by r matrix $J = (\partial x_i / \partial u_j)$ and denotes by $\Delta_{i_1 i_2 \ldots i_r}$ the determinant of the r by

r submatrix of J consisting of the rows numbered $i_1, i_2, \ldots, i_r$. Letting

$$\Delta = \left(\sum_{i_1 < \cdots < i_r} \Delta_{i_1, i_2, \ldots, i_r}^2 \right)^{\frac{1}{2}},$$

he calls $\Delta du_1 \, du_2 \ldots du_r$ an "element of hyperspace" and

$$\alpha_{i_1 i_2 \ldots i_r} = (\Delta_{i_1 i_2 \ldots i_r} / \Delta)$$

a direction cosine of the hyperspace. (The α's are, of course, functions of the u's.) For the case where $r = 2$ and $n = 3$ we have already calculated Δ above in our discussion of Ostrogradsky's theorem. Since the determinants $\partial(x_i, x_j)/\partial(u_1, u_2)$ are precisely the components of the normal vector to the surface, the α_{ij} are then the components of the unit normal vector, hence are the cosines of the angles which that vector makes with the appropriate coordinate axes.

We now quote Volterra's theorem, translated from the Italian:

Let $L_{i_1 i_2 \ldots i_r}$ be functions of points in a hyperspace S_n defined and continuous in all their first derivatives and such that any transposition of indices changes only the sign. Let the forms

$$M_{i_1 i_2 \ldots i_{r+1}} = \sum_{s=1}^{r+1} (-1)^{s-1} \frac{\partial L_{i_1 i_2 \ldots i_{s-1} i_{s+1} \ldots i_{r+1}}}{\partial x_{i_s}}.$$

We denote by S_r the boundary of a hyperspace S_{r+1} of $r+1$ dimensions open and immersed in S_n; by $\alpha_{i_1 i_2 \ldots i_{r+1}}$ the direction cosines of S_{r+1} and by $\beta_{i_1 i_2 \ldots i_r}$ those of S_r. The extension of the theorem of Stokes consists of the following formula:

$$\int_{S_{r+1}} \sum_i M_{i_1 i_2 \ldots i_{r+1}} \alpha_{i_1 i_2 \ldots i_{r+1}} dS_{r+1}$$
$$= \int_{S_r} \sum_i L_{i_1 i_2 \ldots i_r} \beta_{i_1 i_2 \ldots i_r} dS_r.$$

Let us check the case where $r = 1$ and $n = 3$ to see how this result generalizes Stokes's theorem. In that case we have three functions L_1, L_2, L_3 of points in three-dimensional space. The M functions are then given as follows:

$$M_{12} = \frac{\partial L_2}{\partial x_1} - \frac{\partial L_1}{\partial x_2}, \quad M_{31} = \frac{\partial L_1}{\partial x_3} - \frac{\partial L_3}{\partial x_1},$$
$$M_{23} = \frac{\partial L_3}{\partial x_2} - \frac{\partial L_2}{\partial x_3}.$$

Since $r = 1$, S_r is a curve given by three functions $x_1(u), x_2(u)$, and $x_3(u)$. So

$$\Delta = \sqrt{\left(\frac{dx_1}{du}\right)^2 + \left(\frac{dx_2}{du}\right)^2 + \left(\frac{dx_3}{du}\right)^2}$$

and $ds = \Delta du$. Then

$$\beta_i = \frac{dx_i/du}{\Delta} \text{ for } i = 1, 2, 3, \quad \text{and} \quad \beta_i ds = \frac{dx_i}{du} ds.$$

The α_{12}, α_{31} and α_{23} are the appropriate cosines as mentioned above. Hence the theorem will read:

$$\int_{S_2} \left(\frac{\partial L_3}{\partial x_2} - \frac{\partial L_2}{\partial x_3} \right) \alpha_{23} + \left(\frac{\partial L_1}{\partial x_3} - \frac{\partial L_3}{\partial x_1} \right) \alpha_{31}$$
$$+ \left(\frac{\partial L_2}{\partial x_1} - \frac{\partial L_1}{\partial x_2} \right) \alpha_{12} \, dS_2$$
$$= \int_{S_1} \left(L_1 \frac{dx_1}{du} + L_2 \frac{dx_2}{du} + L_3 \frac{dx_3}{du} \right) du,$$

the result being exactly Stokes's theorem. A similar calculation will show that the case $r = 2, n = 3$ will give the divergence theorem, the case $r = 1, n = 2$ will give Green's theorem, and the case $r = n - 1$ is precisely Ostrogradsky's own generalization.

We note further that if we replace α_{ij} by $(\partial(x_i, x_j)/\partial(u, v))/\Delta$, dS_2 by $\Delta du \, dv$, and $(\partial(x_i, x_j)/\partial(u, v)) du \, dv$ by $dx_i dx_j$, and if we set $x_1 = x, x_2 = y, x_3 = z$, we get another familiar form of Stokes's theorem:

$$\int_{S_2} M_{23} dy \, dz + M_{31} dz \, dx + M_{12} dx \, dy$$
$$= \int_{S_1} L_1 dx + L_2 dy + L_3 dz.$$

Similarly, the divergence theorem becomes

$$\int_{S_3} L_{23} dy \, dz + L_{31} dz dx + L_{12} dx \, dy$$
$$= \int_{S_2} \left(\frac{\partial L_{23}}{\partial x} + \frac{\partial L_{31}}{\partial y} + \frac{\partial L_{12}}{\partial z} \right) dx \, dy \, dz.$$

Although Volterra used his theorem in several papers in his study of differential equations, he did not give a proof of the result; he only said that it "is obtained without difficulty."

If one studies Volterra's work, it becomes clear that it would be quite useful to simplify the notation. This was done by several mathematicians around the turn of the century. Henri Poincaré, in his 1899 work *Les Méthodes Nouvelles de la Mécanique Céleste* [18], states the same generalization as Volterra, but in a much briefer form:

$$\int \sum A d\omega = \int \sum \sum_k \pm \frac{dA}{dx_k} dx_k d\omega.$$

Here the left-hand integral is taken over the $(r-1)$-dimensional boundary of an r-dimensional variety in n-space, while the right-hand integral is over the entire variety. Hence A is a function of n variables and $d\omega$ is a product of $r-1$ of the dx_is, the sum being taken over all such distinct products. Poincaré's form of the theorem is more compact than that of Volterra, in part because the direction cosines are absorbed into the expressions $d\omega$. (See [1] for more details.) Poincaré, like Volterra, in this and other works of the same period, was chiefly interested in integrability conditions of what we now call differential forms; i.e., in when a form ω is an exact differential.

The mathematician chiefly responsible for clarifying the idea of a differential form was Elie Cartan. In his fundamental paper of 1899 [2] he first defines an 'expression différentielle' as a symbolic expression given by a finite number of sums and products of the n differentials $dx_1, dx_2, \ldots, dx_n$ and certain coefficient functions of the variables $x_1, x_2, \ldots, x_n$. A differential expression of the first degree, $A_1 dx_1 + A_2 dx_2 + \cdots + A_n dx_n$, he calls an 'expression de Pfaff.'

Cartan states certain rules of calculation with these expressions. In particular, his rule for 'evaluating' a differential expression requires that the value of

$$A dx_{m_1} dx_{m_2} \cdots dx_{m_n}$$

be the product of A with the determinant $|\partial x_{m_j}/\partial \alpha_i|$, where the xs are functions of the parameters α_i. The standard rules for determinants then require that if any dx_i is repeated, the value is 0 and that any permutation of the dx_is requires a sign change if the permutation is odd. For instance, Cartan concludes that $A dx_1 dx_2 dx_3 = -A dx_2 dx_1 dx_3$, or just that $dx_1 dx_2 = -dx_2 dx_1$.

Cartan further discusses changes of variable; if $x_1, x_2, \ldots, x_n$ are functions of $y_1, y_2, \ldots, y_n$, then

$$dx_i = \frac{\partial x_i}{\partial y_1} dy_1 + \frac{\partial x_i}{\partial y_2} dy_2 + \cdots + \frac{\partial x_i}{\partial y_n} dy_n,$$
$$i = 1, 2, \ldots, n.$$

Then, for instance, in the case $n=2$, we get

$$dx_1 dx_2 = \frac{\partial(x_1, x_2)}{\partial(y_1, y_2)} dy_1 dy_2.$$

One might note, on the other hand, that if one assumes a change of variable formula of this type, then one is forced to the general rule $dx_i dx_j = -dx_j dx_i$.

Finally, Cartan defines the "derived expression" of a first-degree differential expression $\omega = A_1 dx_1 + A_2 dx_2 + \cdots + A_n dx_n$ to be the second-degree expression $\omega' = dA_1 dx_1 + dA_2 dx_2 + \cdots + dA_n dx_n$, where, of course,

$$dA_i = \sum_j \frac{\partial A_i}{\partial x_j} dx_j.$$

For the case $n=3$ one can calculate by using the above rules that if $\omega = A_1 dx + A_2 dy + A_3 dz$, then

$$\omega' = \left(\frac{\partial A_3}{\partial y} - \frac{\partial A_2}{\partial z}\right) dy dz + \left(\frac{\partial A_1}{\partial z} - \frac{\partial A_3}{\partial x}\right) dz dx$$
$$+ \left(\frac{\partial A_2}{\partial x} - \frac{\partial A_1}{\partial y}\right) dx dy.$$

Comparing this with the example we gave in discussing Volterra's work, it is clear that Volterra's M_{23}, M_{31}, and M_{12} are precisely the coefficients of Cartan's ω'.

Cartan in [2] did not discuss the relationship of his differential expressions to Stokes's theorem; nevertheless, by the early years of the twentieth century the generalized Stokes's theorem in essentially the form given by Poincaré was known and used by many authors, although proofs seem not to have been published.

By 1922, Cartan had extended his work on differential expressions in [3]. It is here that he first uses the current terminology of "exterior differential form" and "exterior derivative". He works out specifically the derivative of a 1-form (as we did above) and notes that for $n=3$ Stokes's theorem states that $\int_C \omega = \iint_S \omega'$, where C is the boundary curve of the surface S. (This is, of course, exactly Volterra's result in the same special case.) Then, defining the exterior derivative of any differential form $\omega = \sum A dx_i dx_j \ldots dx_l$ to be $\omega' = \sum dA \, dx_i dx_j \ldots dx_l$ (with dA as above), he works out the derivative of a 2-form Ω in the special case $n=3$, and shows that for a parallelepiped P with boundary S, $\iint_S \Omega = \iiint_P \Omega'$. One can easily calculate that this is the divergence theorem, and we must assume that Cartan realized its truth in more general cases. He was, however, not yet ready to state the most general result.

The "d" notation for exterior derivative was used in 1902 by Theodore DeDonder in [6], but not again until Erich Kähler reintroduced it in his 1934 book

Einführung in die Theorie der Systeme von Differentialgleichungen [11]. His notation is slightly different from ours, but in a form closer to ours it was adopted by Cartan for a course he gave in Paris in 1936–1937 (published as *Les Systèmes Différentiels Extérieurs et leurs Applications Géométriques* [4] in 1945). Here, after discussing the definitions of the differential form ω and its derivative $d\omega$, Cartan notes that all of our three theorems (which he attributes to Ostrogradsky, Cauchy-Green, and Stokes, respectively) are special cases of $\int_C \omega = \int_A d\omega$, where C is the boundary of A. To be more specific, Green's theorem is the special case where ω is a 1-form in 2-space; Stokes's theorem is the special case where ω is a 1-form in 3-space; and the divergence theorem is the special case where ω is a 2-form in 3-space. Finally, Cartan states that for any $(p+1)$-dimensional domain A with p-dimensional boundary C one could demonstrate the general Stokes's formula:

$$\int_C \omega = \int_A d\omega.$$

(For examples of the use of these theorems, see any advanced calculus text, e.g., [1] or [24]. For more information on differential forms, one can consult [7].)

Appearance in texts

A final interesting point about these theorems is their appearance in textbooks. By the 1890s all three theorems were appearing in the analysis texts of many different authors. The third of our theorems was always attributed to Stokes. The French and Russian authors tended to attribute the first theorem to Ostrogradsky, while others generally attributed it to Green or Gauss; this is still the case today. Similarly, Riemann is generally credited with the second theorem by the French, while Green is named by most others. Before Cartan's 1945 book, about the only author to attribute that result to Cauchy was H. Vogt in [26].

The generalized Stokes's theorem, first published, as we have seen, in 1945, has only been appearing in textbooks in the past twenty years, the first occurrence probably being in the 1959 volume of Nickerson, Spencer, and Steenrod [14].

References

1. R. C. Buck, *Advanced Calculus*, 2nd ed., McGraw-Hill, 1965.
2. E. Cartan, Sur certaines expressions différentielles et sur le problème de Pfaff, *Annales École Normale* **16** (1899), 239–332 (*Oeuvres*, Part II, Vol. I, pp. 303–397).
3. E. Cartan, *Leçons sur les invariants intégraux*, Chap. VII, Hermann, 1922.
4. E. Cartan, *Les Systèmes Différentiels Extérieurs et leurs Applications Géométriques*, Hermann, 1945.
5. A. Cauchy, Sur les intégrales qui s'étendent à tous les points d'une courbe fermée, *C. R. Acad. Sci. Paris* **23** (1846), 251–255 (*Oeuvres* (1), X, pp. 70–74).
6. T. DeDonder, Étude sur les invariants intégraux, *Rendiconti del Circolo Matematico di Palermo* **16** (1902), 155–179.
7. H. Flanders, *Differential Forms*, Academic Press, 1963.
8. C. F. Gauss, Theoria attractionis corporum sphaeroidicorum ellipticorum homogeneorum methodo nova tractata, *Commentationes Societatis Regiae Scientiarium Göttingensis Recentiores* (V) II, 1813 (*Werke* 5, pp. 1–22).
9. G. Green, *An Essay on the Application of Mathematical Analysis to the Theories of Electricity and Magnetism*, London, 1828 (*Green's Mathematical Papers*, pp. 3–115).
10. H. Hankel, Zur allgemeinen Theorie der Bewegung der Flüssigkeiten, *Dieterische Univ. Buchdruckerei*, 1861.
11. E. Kähler, *Einführung in die Theorie der Systeme von Differentialgleichungen*, Leipzig, 1934.
12. J. C. Maxwell, Letter to Stokes, in G. Stokes, *Memoir and Scientific Correspondence* II (ed. J. Larmor), Cambridge, 1907, p. 31.
13. J. C. Maxwell, *A Treatise on Electricity and Magnetism*, I, Oxford, 1873.
14. H. Nickerson, D. Spencer, and N. Steenrod, *Advanced Calculus*, Van Nostrand, 1959.
15. M. Ostrogradsky, Démonstration d'un théorème du calcul integral, publ. in Russian in *Istoriko-Matematicheskie Issledovania* **16** (1965), 49–96.
16. M. Ostrogradsky, Note sur la théorie de la chaleur, *Mémoires de L'Acad. Imp. des Sciences de St. Petersburg* (6) **1** (1831), 129–133.
17. M. Ostrogradsky, Sur le calcul des variations des intégrales multiples, *J. Reine Angew. Math.* **15** (1836), 332–354.
18. H. Poincaré, *Les Méthodes Nouvelles de la Mécaniques Céleste*, Vol. III, Gauthier-Villars, 1899, p. 10; Dover, 1957.
19. S. Poisson, Memoire sur l'équilibre et le mouvement des corps élastiques, *Mémoires de l'Académie Royale des Sciences de l'Institut de France* **8** (1829), 357–571.
20. B. Riemann, *Grundlagen für eine allgemeine Theorie der Functionen einer veränderlichen complexen Grösse*, Göttingen, 1851 (*Werke*, pp. 13–48).

21. F. Sarrus, Mémoire sur les oscillations des corps flottans, *Annales de Mathématiques Pures et Appliquées (Nîmes)* **19** (1828), 185–211.

22. G. Stokes, *Mathematical and Physical Papers*, Vol. 5, Cambridge Univ. Press, 1905, p. 320.

23. P. Tait, On Green's and other allied theorems, *Trans. Royal Society of Edinburgh* (1869–70), 69–84.

24. A. Taylor and W. Mann, *Advanced Calculus*, 2nd ed., Xerox, 1972.

25. W. Thomson and P. Tait, *Treatise on Natural Philosophy*, I, Oxford, 1867.

26. H. Vogt, *Eléments de Mathématiques Supérieures*, Paris, 1925.

27. V. Volterra, Delle variabili complesse negli iperspazi, *Rendiconti della R. Accad. der Lincei* **5** (1889), 158–165 (*Opere*, Vol. I, pp. 403–411).

28. A. Yushkevich, *Istoriya Matematiki v Rossii do 1917 goda*, Moscow, 1968, p. 290.

The Mathematical Collaboration of M. L. Cartwright and J. E. Littlewood

SHAWNEE L. McMURRAN AND JAMES J. TATTERSALL

American Mathematical Monthly **103** (1996), 833–845

Balthasar van der Pol's experiments on electrical circuits during the 1920s and 1930s opened an interesting chapter in the history of dynamics. The need for advancements in radio technology made van der Pol's work pertinent and his research stimulated mathematical interest in non-linear oscillators. In particular, van der Pol's work caught the attention of Cambridge mathematicians M. L. Cartwright and J. E. Littlewood. Topology and Poincaré's transformation theory provided a key to analyzing behavior of non-linear oscillators and dissipative systems. Resulting mathematical techniques have played a significant role in the development of the modern theory of dynamical systems and chaos. In addition, non-linear oscillator theory has led to the development of radio, radar, and laser technology.

The collaboration between Cartwright and Littlewood began just before World War II and lasted approximately ten years. They published four joint papers, and individually published several other papers based on joint work. Their collaboration produced some of the earliest rigorous work in the field of large parameter theory. They were among the first mathematicians to recognize that topological and analytical methods could be combined to efficiently obtain results for various problems in differential equations, and their results helped inspire the construction of Smale's horseshoe diffeomorphism ([28], pp. 63–80). Their names can be included with those mathematicians, including Levinson, Lefschetz, Minorsky, Liapounov, Kryloff, Bogolieuboff, Denjoy, Birkhoff, and Poincaré, whose work provided an impetus to the development of modern dynamical theory.

Cartwright knew G. H. Hardy at Oxford before she met Littlewood. Cartwright joined a special group when she began to attend Hardy's Friday class at New College in January 1928. The participants included Gertrude Stanley, John Evelyn, E. H. Linfoot, L. S. Bosanquet, Frederick Brand, and Tirukkannapuram Vijayarhagavan. Nearly all completed their D. Phils by the summer of 1928. The universal feeling among the participants was that they were studying under a very great man who had not yet been recognized fully. Cartwright appreciated Hardy's style and philosophy of mathematics. She recalled that he took immense trouble with his students whether they were good, bad, or indifferent. Once, when she had produced an obviously fallacious result, Hardy ([10], p. 23) remarked, "Let's see, there's always hope when you get a sharp contradiction." Hardy became Cartwright's initial thesis advisor and she finished with E. C. Titchmarsh when Hardy went on leave to Princeton.

Cartwright first met Littlewood in June 1930 when he went to Oxford to examine her for her Doctor of Philosophy degree. The following October she went to Girton College, Cambridge on a three-year research fellowship. She attended a series of courses by Littlewood on the theory of functions and his seminar that took place in his rooms in Trinity at 5:00 P.M. preceded by a nice tea.

In 1931 Hardy assumed the Sadleirian Chair of Mathematics at Cambridge. Cartwright asked Hardy if he would be offering a seminar similar to the Friday evening sessions she had enjoyed at Oxford. He replied that he would probably come to some arrangement with Littlewood. Soon after, the Lecture List announced a Hardy-Littlewood class. Cartwright [16] recalled that while Littlewood was speaking at the first Hardy-Littlewood class, Hardy came in late, helped himself liberally to tea and began to ask questions. It seemed as if he were trying to pin Littlewood on details, whereas Littlewood was trying to illustrate the main point while taking the details for granted. Littlewood told Hardy that he was not prepared to be heckled and Cartwright does not recall them ever being present together at any subsequent class. Thenceforth, Hardy and Littlewood alternated classes. Littlewood usually did the speaking on his turns and Hardy often invited others to speak during the classes. Eventually, Littlewood ceased to participate, though the class continued to be held in his rooms. The class became known as "the Hardy-Littlewood Conversation Class at which Littlewood was *never* present." Littlewood intermittently held his own unlisted lecture class, which Hardy recommended to several of his students.

During her first years at Cambridge, Cartwright occasionally wrote or spoke with Littlewood on topics pertaining to his courses. She recalls his manner as somewhat unconventional; he was always more ready to talk out of doors. She was occasionally able to catch him on the telephone after her late tea and before his early dinner. He never discussed problems with her at a blackboard, but he would draw imaginary figures on a wall as they walked and talked. Once they began to collaborate, nearly all of their collaboration was done by letter with occasional short discussions of particular points. He never came to her rooms and she does not remember going to his. Most of the letters Littlewood sent Cartwright would have something on the back, often galley proofs.

Cartwright suspects that Littlewood may have had some unspoken rules for their collaboration comparable to those he had with Hardy. Harold Bohr ([1], p. xxviii) noted that the following four "axioms of collaboration" formed a basis for the Hardy-Littlewood relationship.

1. When one wrote to the other, it was completely indifferent whether what they wrote was right or wrong.

2. When one received a letter from the other, he was under no obligation whatsoever to read it, let alone answer it.

3. Although it did not really matter if they both simultaneously thought about the same detail, still, it was preferable that they should not do so.

4. It was quite indifferent if one of them had not contributed the least bit to the contents of a paper under their common name.

Bohr was of the opinion that, "seldom, or never, was such an important and harmonious collaboration founded on such negative axioms."

Cartwright gives Hardy credit for the formulation of these rules. She contends that such a precise formulation was typical of him. When she questioned Littlewood about these axioms, he replied that the agreement between Hardy and himself was unwritten. He never personally informed Cartwright about the rules, but she surmises that his policy regarding their collaboration was similar. Cartwright was once told that when preparing some material to send to her, Littlewood "withdrew part of what he had written as being contrary to the rules of collaboration" [16]. She suspects that Littlewood may have broken the first rule by criticizing some of her mistakes.

According to Cartwright [15], one memorable episode between herself and Littlewood concerned an interesting problem that he had presented in his theory of functions class. In particular, Does the function $f(z) = \sum_{n=1}^{\infty} a_n z^n$, analytic for $|z| < 1$, which takes no value more than p times in $|z| < 1$, satisfy

$$\left|f\left(re^\theta\right)\right| < A(p)(1-r)^{-2p}$$

for $0 < r < 1$, where $A(p)$ is a constant depending only on p and $a_1, a_2, \ldots, a_p$? Littlewood gave Cartwright the impression that this was an exciting problem and he would be interested in any significant progress she could make toward its solution.

Meanwhile, in one of Edward Collingwood's classes, Cartwright had learned of Ahlfors' Distortion Theorem. In her bath one night she thought she saw how to apply it to Littlewood's problem. When she later settled down to examine the problem more carefully, she fell into the "usual trap." The proof for the case $p = 1$ was known and she tried to prove the case for $p > 1$ using a modification of that proof rather than her original idea. She sent this attempt to Littlewood. In reply she received a note with the snake in Figure 1. In his note Littlewood explained the common error she had made in her proof.

Figure 1.

J. E. Littlewood

Returning to her original idea, Cartwright sent a new proof to Littlewood using Ahlfors' work. Some time later, while punting with the Vice-Mistress of Girton on the River Cam near Trinity College, Cartwright spotted Littlewood on a nearby bank. As she had received no reply regarding her revised proof, she asked him if he had read her manuscript. He replied, "Have I got to read all that?" She convinced him that he should, and he was impressed with her work.

The polished version of her proof was published [3] and Littlewood referred to the paper in his *Lectures on the Theory of Functions*. Cartwright believes that her work on this problem prompted Littlewood and Hardy to write a joint letter to Girton resulting in a Faculty Assistant Lectureship and the renewal of her research fellowship for a fourth year.

In January 1938, the Radio Research Board of the Department of Scientific and Industrial Research issued a memorandum requesting the "really expert guidance" of pure mathematicians with "certain types of non-linear differential equations involved in the technique of radio engineering" [27]. A copy was sent to the London Mathematical Society. Radio engineers wanted an analysis of the solutions to some very objectionable looking differential equations occurring in connection with radar.

The problems that concerned the radio engineers arose from the use of vacuum tubes (thermionic valves) used to control the flow of electricity in the circuitry of transmitters and receivers. Transistors have replaced most vacuum tubes in transmitters; however, vacuum tubes are still sometimes used when very high voltages are present. When encased in ceramic they are very radiation hard and are used in missiles. The type of problem arising from vacuum tubes had great influence on the mathematics of control theory and airplane construction.

The existing techniques available for the explicit analytical solutions of linear differential equations with constant coefficients were the mainstays of early radio engineers. Linear differential equations were used to approximate physical systems and behavior of systems was inferred from the solutions. According to the memorandum, the need had "arisen for a more complete understanding of the actual behavior of certain assemblages of electrical apparatus, including thermionic valves." Non-linear differential equations were required to obtain a closer approximation of the actual physical system.

Radio engineers hoped mathematicians could provide them with a theory for non-linear differential equations that had the same relative simplicity as that for linear differential equations. At the time, this request could not be considered completely naive. Although interest in celestial mechanics inspired considerable research on non-linear conservative oscillators before 1920, analogous techniques were unavailable for the dissipative systems arising in radio research. Physicists and engineers encountering these systems in their experimental work had done most of the available research. Except for the work of Liapounov and Poincaré, there had been little systematic study of non-linear differential equations by mathematicians.

Although engineers could compute numerical solutions to their systems, these solutions were of little value when one wanted to know how solutions varied with the parameters. During the early twentieth century numerical solutions were less easy to obtain. Moreover, a large number of numerical solutions are necessary to determine how solutions vary with the parameters of the physical system. The engineers hoped that an analytical approach might provide a more efficient way to analyze their systems.

Unfortunately, mathematical analyses of even the simplest equations representing the physical systems were quite complicated and had few practical applications. The memo acknowledged that it was worthwhile to determine if this were the case in order to prevent the "waste of time and energy spent in pursuit of a will-o'-the-wisp." In the end, complications and variations in the valves themselves convinced radio engineers that experimental methods were more effective than mathematical analysis.

Cartwright [13] says, "Although I myself have helped to develop the general theory and settle certain theoretical problems, I do not think that I have ever produced a result useful for any specific practical problem when it was needed." Though the engineers were unable to apply many of the theories developed by mathematicians such as Cartwright and Littlewood, new problems were arising for automatic control mechanisms involving systems of equations sufficiently similar to those of the radio work for some theory still to be applicable. Resulting mathematical techniques allowed for a headway in the analysis of dissipative systems where progress had previously been rather slow. Such analysis provided part of the foundation to the development of modern dynamical theory.

One of the basic building blocks of early twentieth-century circuitry was the triode oscillator illustrated in Figure 2. The objective of the Radio Research Board was to determine which parameter values of the circuit would lead to periodic or almost periodic solutions. It was also essential to determine how the frequency of these oscillations varied with the parameters of the circuit.

The problems appeared interesting to Cartwright. Using the references provided, she began her investigation working back to the research by Appleton and van der Pol in the 1920s. An important paper by van der Pol [29] contained eighty-seven references and was an excellent starting point for her research. The emphasis of van der Pol's article was on phenomena that linear equations could not explain. He referred to several biological and physical research papers such as Volterra's work on interacting species, E. W. Brown's work on the pendulum, and Poincaré's work on celestial mechanics. Cartwright [14] was surprised that van der Pol did not refer to Birkhoff's work, Bendixson's work, or Poincaré's paper "Sur les courbes definies par les équations différentielles." She says that the well-known ideas of Poincaré on curves and Bendixson on limit cycles have since become fundamental concepts when dealing with several types of non-linear equations. In addition, she and Littlewood used techniques inspired by Birkhoff's applications of transformation theory to dynamical systems. Her

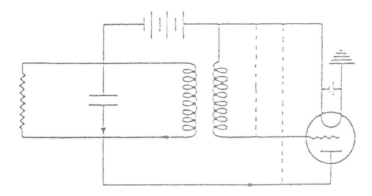

Figure 2.

opinion of van der Pol was that, like many authors, herself included, he did not read all the works to which he referred.

It is interesting that Cartwright would concern herself with these applied problems, since dynamics was a subject that had not appealed to her at Oxford. Nevertheless, Cartwright intuitively recognized the topological undertones of the problems. She says [11], "The mathematical problems which have interested me most are apt to turn into topological problems or problems of topological dynamics."

Although she initially knew very little of the dynamical side, Cartwright found the work of the radio engineers to be much more interesting and suggestive than that of mechanical engineers. Radio engineers wanted their systems to oscillate in an orderly way. Similar equations were studied in astronomy and celestial mechanics, but the cycles there are measured in days and years. The speed of radio oscillations, whose cycles are measured in fractions of a second, enables experiments to suggest the behavior of stable solutions quickly and efficiently.

While exploring the literature, Cartwright found several intriguing problems, which she brought to Littlewood's attention. At the time, Cartwright was in the habit of showing Littlewood anything that she thought would interest him. Littlewood dismissed several simple problems, but he found some of van der Pol's conjectures worth investigating.

Cartwright presented Littlewood with the material she had collected because she thought that he might help her understand the dynamics of the situation. He solved or suggested methods of solutions for most of the problems straightaway, but he found certain problems more interesting. Littlewood often thought in dynamical terms, perhaps because of his earlier work in ballistics during the First World War. She says [13] that he often referred to solutions as "trajectories", as if they were the paths of missiles fired from antiaircraft guns. He certainly seemed to understand the dynamical aspects of the radio work. Van der Pol, recalling a discussion with Littlewood, told Cartwright [16] in "tones of delighted surprise" that all of Littlewood's methods corresponded to physical concepts. Cartwright and Littlewood corresponded with Colebrook, Appleton, and van der Pol in order to gain a better understanding of the situation.

The first problem attacked by Cartwright and Littlewood was that of the amplitude of the stable periodic solution of van der Pol's equation without forcing term:

$$\ddot{x} - k(1-x^2)\dot{x} + x = 0. \qquad (1)$$

Van der Pol had obtained graphical solutions for the cases $k = 0.1, 1$, and 10 (see Figure 3). Van der Pol's graphs appear to show convergence to a periodic solution with amplitude 2 in all three cases. Littlewood showed that the amplitude was not 2 if k was small and positive. For k small and positive, equation (1) has one periodic solution of amplitude $2 + O(k)$. Cartwright and Littlewood [8] succeeded in showing that when k is large all solutions of equation (1), except the trivial solution $x = 0$, converge to a periodic solution whose amplitude tends to 2 as $k \to \infty$. The method they used was elementary, but difficult.

Cartwright and Littlewood next attacked the problem of whether van der Pol's equation with a forced oscillation,

$$\ddot{x} - k(1-x^2)\dot{x} + x = bk\lambda \cos \lambda t, \qquad (2)$$

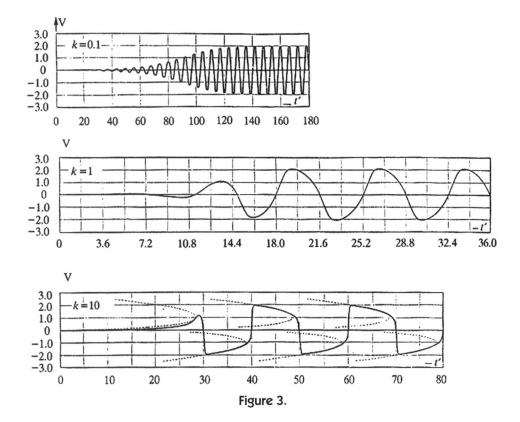

Figure 3.

could have two stable periodic solutions with different periods when k was large. Van der Pol had suggested the phenomenon after he and van der Mark found experimentally that increasing and decreasing the relative frequency λ of the driving force produced two subharmonics of different orders for the same values of the parameters (if a differential equation with a forcing function of period P has a periodic solution with period nP, where n is a positive integer, then that oscillation is called a *subharmonic* of order n).

Cartwright and Littlewood focussed a great deal of their attention on van der Pol's equation. They found the equation interesting and suggestive since it seemed to be the simplest type of equation likely to have two stable periodic oscillations with periods prime to one another. They agreed that tackling a really difficult problem first in its simplest form is best.

Most of the early research of Cartwright and Littlewood concentrated on equations with two or more stable periodic solutions. Both considered this problem to be the most analytically interesting problem investigated during their collaboration [14]. Their topological interpretation of this problem led to their paper [26] on fixed-point theorems. They left what Littlewood considered the "duller stuff" until later. However, when they explored this material they found that it was more difficult than they had anticipated. In Littlewood's words to Cartwright [11], "All details have a nasty way of ramifying into difficulties." Their results were published in a paper [25] dealing with second-order equations with positive damping, which was much corrected by Littlewood.

Although Cartwright and Littlewood produced some interesting and innovative mathematics, their collaboration was not always harmonious [2]. Cartwright became irritated with Littlewood's incessant changes, even though her own technique was to "polish, polish, polish" before publishing. Cartwright's sense of history was very good and her historical perspective adds an excellent dimension to their papers. When they separated, they agreed that Littlewood would finish the hardest problem and Cartwright would complete the other bits and pieces. Her results appear in a series of papers ([4], [5], [6], [7], [8]) published in the late 1940s and early 1950s.

A significant portion of the joint work of Cartwright and Littlewood was published under her name alone. This was the case even if the paper was really based on a Littlewood manuscript to which Cartwright had added a bit or filled in the gaps. Littlewood would not let her put his name to any paper not actually written by him. She had to say it was based on joint work with him. There was one exception, the paper on fixed-point theorems. Most of the problems they attacked came from the literature of engineering. Their paper on fixed-point theorems was an exception as it developed from their treatment of van der Pol's equation.

From 1931 Littlewood suffered from, and was treated for, recurring bouts of mental depression. Cartwright says that his mind was so full of ideas that he found it difficult to rest. She recalls that he seemed exhausted whenever he had completed a collaborative paper and he occasionally complained about their joint papers. Cartwright appreciated the special effort

Figure 4. Dame Mary Cartwright
(Courtesy of the Mistress and Fellows of Girton College)

made by Littlewood to have their first paper published in order to support her candidacy for the Royal Society. This joint paper [24] was published in 1945 and was intended as a preliminary survey of their results regarding van der Pol's equation with forcing term and k large. The full proofs of the results presented in the paper were not published until twelve years later. By that time Littlewood had been cured of his depression. Cartwright recalls that in 1958, although he was reluctant, she persuaded Littlewood to go to London to accept the prestigious Copley Medal that the Royal Society had awarded him. She treasures a note from him beginning, "All right, you win" [16]. Littlewood probably could not have been convinced to make such a social appearance before his cure.

Cartwright insisted that the two papers published in 1957, those that gave the full proofs of the results announced in their paper of 1945, be published under his name alone. She may have felt that her contributions had not been significant. However, as Littlewood was leaving for a trip to America, he left some final details of the proof corrections to her. In each manuscript, he affirms that the work was based on joint efforts.

Cartwright and Littlewood's analysis of the van der Pol equation showed that the existence of two stable sets of subharmonics of different orders led to complex topological structures. At the boundary between the domains of attraction of the two sets there exists an intricate fine structure of highly non-stable trajectories. They concluded that the boundary was most likely an indecomposable continuum that, according to Littlewood, is the "dirtiest" thing one can say about a set of points. Except for its instability, this structure closely resembles that of strange attractors. The homeomorphism of the plane defined by following the trajectories through one period maps the set onto itself. Cartwright and Littlewood found this phenomenon to be of great interest since it indicated that the non-stable periodic solutions and subharmonics of the equation corresponded to fixed points of the set under the homeomorphism or an iterate of it. They surmised that existence of these fixed points could be deduced from a suitable fixed-point theorem. This led them to investigate fixed point theorems for continua invariant under a diffeomorphism of the plane.

The fixed-point theorems needed by Cartwright and Littlewood could not be obtained by the methods of existing algebraic fixed-point theory because of the complexity of the continua involved. Fixed-point theorems for plane continua that are acyclic but otherwise unrestricted were required. However, they were able to use some of Birkhoff's results for analytic transformations of the plane into itself with complicated invariant curves. Even reasonable regularity arguments would not allow them to rule out singular invariant curves. Birkhoff had shown the existence of an analytic transformation of the plane into itself with complicated invariant curve K. Cartwright and Littlewood used such a transformation to show existence of stable periodic subharmonic motions, unstable motions, and recurrent motions each associated with some subset of K. They were among the earliest mathematicians to associate such transformations with a simple differential equation.

At the time of Cartwright and Littlewood's investigations, Lefschetz, Levinson, Minorsky, and others in the United States also began to explore some of the problems raised by the study of electrical circuits, in particular the boundedness of solutions to the non-linear differential equations associated with the circuits. Like the English pair, government requests also prompted the Americans. However, whereas the focus of Cartwright and Littlewood was on certain special problems, the Americans were striving for a general and more easily handled mathematical theory. The American emphasis was on amplitude and period. The work of Cartwright and Littlewood emphasized frequencies and parameters. Cartwright [14] speculates that perhaps this difference occurred because engineers became more interested in non-linearities that were not necessarily either small or large.

Meanwhile, Kryloff, Bogoliuboff, and Mitropolsky of the Kiev school were employing averaging methods to analyze equations such as van der Pol's. Until the end of the nineteenth century, techniques of averaging were used mainly in celestial mechanics. Van der Pol promoted the application of averaging techniques to equations arising in electronic circuit theory. However, rigorous proofs of validity of the method were not available until Fatou provided the first proof of asymptotic validity in 1928. Although they used some foreign research results, Cartwright and Littlewood might have made better use of the work of both the Americans and the Russians if not for the isolation due to the war.

In America, Levinson had obtained results for a piecewise linear equation that were similar to the 1945 results of Cartwright and Littlewood. Cartwright [14] says that a monograph [21] of Levinson's, which combined the work of Kryloff, Bogolieuboff, Denjoy, and Birkhoff, considerably influenced the work of herself and Littlewood. Levinson's ideas provided the foundations of a general topological approach to non-autonomous periodic second-order differential equations and stimulated the interest of Cartwright and Littlewood in the mappings suggested by their research on the forced van der Pol's equation.

In 1947 Cartwright wrote to Lefschetz at Princeton trying to get a copy of Minorsky's report on non-linear vibrations. Minorsky was preparing the report for the Office of Naval Research and Lefschetz was listed as head of the project. She included some results of herself and Littlewood in her correspondence. She received the first part of the report and was surprised that it had been classified as "Restricted" and not to be divulged to unauthorized persons. She also received a letter from Lefschetz expressing great interest in her research with Littlewood.

Lefschetz and Mina Rees invited Cartwright to the United States from January to June 1949 to lecture on non-linear differential equations. She spent three weeks at Stanford with Minorsky, one week at UCLA with John Curtis, and the rest at Princeton. Cartwright [6] gave a series of lectures at Princeton that provides a detailed overview of her joint work with Littlewood. While at Princeton, she was officially a consultant under the Office of Naval Research, not a professor, at the time, Princeton did not have women professors. Cartwright got on well with all those she met, but Bochner was not on speaking terms with Lefschetz so she never saw him. She also learned that American academicians tended to be a bit casual. John Tukey told her to feel free to put her feet on the table in the seminar room. Wearing a skirt, not trousers, she refrained. Cartwright also learned that if Lefschetz stopped asking questions of a visiting lecturer for five minutes, he was asleep.

Cartwright suspects that Littlewood might not have read much of the literature from which their problems were extracted. She recalled that Littlewood had once advised her not to pay too much attention to the existing literature on a problem. He surmised [16], "If previous writers had failed to solve it, it was probably because they had tried a wrong method." Consequently, Cartwright and Littlewood approached their problems using methods that were quite different from those of the radio engineers. Their innovative approach led Cartwright and Littlewood to the conclusion that using both topological and analytical methods was indispensable in the study of differential equations. According to Cartwright [12],

> We should widen our conception of topological methods to include all those now claimed as such by topologists and combine them with analytical methods to obtain results more quickly; we need a clarification of existing topological results in terms of differential equations . . . .

Cartwright [17] also stressed that the topological interpretation of a problem can give one valuable insight into the qualitative behavior of solutions, even when one is unfamiliar with or does not plan to use topological methods in their analysis.

Cartwright [7] found it remarkable that the experimental results of van der Pol and van der Mark guided her and Littlewood to so many unsolved problems in topology. Several papers ([22], [23], [25], [8]) that came from the collaboration of Cartwright and Littlewood are among the earliest fully rigorous works in large parameter theory (i.e., relaxation oscillations). Work such as theirs clearly provided an impetus to the development of the modern theory of dynamical systems. The subject has expanded significantly since their work was completed. Modern techniques have enabled contemporary mathematicians to carry through a more thorough analysis of the asymptotic behavior exhibited by solutions to the forced van der Pol equation (see, for example, [20], [18], 19]). Nonlinear oscillation theory has led to the development of radio, radar, and laser technology, and investigations into the van der Pol equation have played a vital role in both the theory of relaxation oscillations and bifurcation theory.

References

1. H. Bohr, *Coll. Math. Works*, Vol. 1, Dansk Mathematisk Forening, 1952.

2. B. Bollobás (ed.), *Littlewood's Miscellany*, Cambridge, 1986.

3. M. L. Cartwright, Some inequalities in the theory of functions, *Math. Annalen* **111** (1935), 98–118.

4. M. L. Cartwright, Forced oscillations in nearly sinusoidal systems, *J. Inst. Elec. Engrs.* **95** (1948), 88–96.

5. M. L. Cartwright, On nonlinear differential equations of the second order III. The equation $\ddot{x} - k(1 - x^2)\dot{x} + x = pk\lambda \cos(\lambda t + \alpha)$, k small and near 1, *Proc. Cambridge Phil. Soc.* **45** (1949), 495–501.

6. M. L. Cartwright, Forced oscillations in nonlinear systems, *Contributions to the Theory of Nonlinear Oscillations. Annals of Math. Studies* 20, Princeton, 1950, pp. 149–241.

7. M. L. Cartwright, Forced oscillations in nonlinear systems, *J. Research Nat. Bur. Standards* **45** (1950), 514–518.

8. M. L. Cartwright, Van der Pol's equation for relaxation oscillations, *Contributions to the Theory of Nonlinear Oscillations II; Annals of Mathematical Studies* 2, Princeton, 1952, pp. 3–18.

9. M. L. Cartwright, Non-linear vibrations: A chapter in mathematical history, *Math. Gazette* **36** (1952), 81–88.

10. M. L. Cartwright, *Religion and the Scientist* (ed. M. Stockwood), SCM Press, 1959.

11. M. L. Cartwright, From non-linear oscillations to topological dynamics, *J. London Math. Soc.* **39** (1964), 193–201.

12. M. L. Cartwright, Topological problems of nonlinear mechanics, *Konferenz über Nichtlineare Schwingungen, 1964, Berlin, Abh. Deutsch. Akad. Wiss. Berlin. K1. Mathematik Physik, und Technik* 1, 1965.

13. M. L. Cartwright, Mathematics and thinking mathematically, *Amer. Math. Monthly* **77** (1970), 20–28.

14. M. L. Cartwright, Some points in the history of the theory of nonlinear oscillations, *Bull. Inst. Math. Appl.* **10** (1974), 329–333.

15. M. L. Cartwright, Some exciting mathematical episodes involving J.E.L., *Bull. Inst. Math. Appl.* **12** (1976), 201–202.

16. M. L. Cartwright, John Edensor Littlewood, FRS, FRAS, Hon FIMA, *Bull. Inst. Math. Appl.* **14** (1978), 87–90.

17. M. L. Cartwright, E. T. Copson, and J. Grieg, Non-linear vibrations, *Advancement Sci.* **6** (1949), 64–75.

18. J. Guckenheimer, Dynamics of the van der Pol equation, *IEEE Trans. Circuits and Systems* **27** (1980), 983–989.

19. J. Guckenheimer, Symbolic dynamics and relaxation oscillations, *Physica D. Nonlinear Phenomena* **1** (1980), 227–235.

20. M. Levi, Qualitative analysis of the periodically forced relaxation oscillations, *Mem. Amer. Math. Soc.* **32** (1981), 244.

21. N. Levinson, Transformation theory of nonlinear differential equations of the second order, *Annals of Math.* **45** (1942), 723–737.

22. J. E. Littlewood, On nonlinear differential equations ..., *Acta Math.* **97** (1957), 267–308.

23. J. E. Littlewood, On nonlinear differential equations ..., *Acta Math.* **98** (1957), 1–110.

24. J. E. Littlewood and M. L. Cartwright, On nonlinear differential equations of the second order I: the equation $\ddot{y} + k(1-y^2)\dot{y} + y = b\lambda k \cos(\lambda t + a)$, k large, *J. London Math. Soc.* **20** (1945), 180–189.

25. J. E. Littlewood and M. L. Cartwright, On nonlinear differential equations of the second order II. The equation $\ddot{y} + kf(y)\dot{y} + g(y,k) = p(t) = p_1(t) + kp_2(t)$; $k > 0$, $f(y) \geq 1$, *Annals Math.* **48** (1947), 472–494; Addendum **50** (1949), 504–505.

26. J. E. Littlewood and M. L. Cartwright, Some fixed point theorems, *Annals Math.* **54** (1951), 1–37.

27. Radio Research Board, A note on certain types of nonlinear differential equations involved in radio engineering, *Math. Gazette* **22** (1938), 217–218.

28. S. Smale, *Differential and Combinatorial Topology*, Princeton, 1965.

29. B. van der Pol, The nonlinear theory of electric oscillations, *Soc. Inst. Radio Eng.* **22** (1934), 1051–1086.

Dr. David Harold Blackwell, African-American Pioneer

NKECHI AGWU, LUELLA SMITH, AND AISSATOU BARRY

Mathematics Magazine **76** (2003), 3–14

Dr. David Blackwell is an African-American educational pioneer and eminent scholar in the fields of mathematics and statistics, whose contributions to our society extend beyond these fields. This paper highlights his significant contributions and the personal, educational, and professional experiences that groomed and nurtured him for leadership as a civic scientist. We hope this account of Dr. Blackwell's life will enhance the literature on African-American achievers, and motivate students majoring in, or considering careers in mathematics and statistics, particularly those from under-represented groups.

The education of David Blackwell

Early childhood

It is April 24, 1919, an era of heightened segregation and racial discrimination in the United States. Welcome to Centralia, Illinois, a small town community on the Mason-Dixon line, with a population of about 12,000 people, and very few African-American families [13]. Witness the birth of David Harold Blackwell. He was to be the eldest of four children born to Grover Blackwell, a hostler for Illinois Central Railroad, and Mabel Johnson Blackwell, a full-time homemaker. His two younger brothers, J. W. and Joseph, and his younger sister Elizabeth would follow soon after.

During his early childhood, David had a grandfather and an uncle living in Ohio who were influential to his cognitive development. His grandfather, whom he had never met, was a school teacher and later a storekeeper. He endowed David with a large library of books. From this library, David read and enjoyed many books, including his first algebra book. His uncle had been home-schooled by his grandfather, because of worries about the effects of racism on his son at school. He impressed David with his ability to add three columns of numbers expeditiously, in a one-step process.

David attended Centralia public schools for the first ten years of his schooling, from 1925 to 1935. His parents enrolled him in integrated schools in his southern Illinois locality, which also had racially segregated schools: whites-only schools and blacks-only schools. However, being at an integrated school, David was unaware of, or unaffected by, issues of racial discrimination. He attributes this to the fact that his parents shielded their children as much as possible from the effects of racism, and to the fact that he experienced few encounters where race was an issue.

High school education

In high school, David developed a strong interest in games such as checkers and in geometry, but was not particularly interested in algebra and trigonometry. He pondered over questions of whether the player with the first move in these games had a higher probability of winning. He states this about geometry ([3], p. 20):

> Until a year after I finished calculus, it was the only course I had that made me see that mathematics is really beautiful and full of ideas. I still remember the concept of a helping line. You have a proposition that looks quite mysterious. Someone draws a line and suddenly it becomes obvious. That's beautiful stuff. I remember the proposition that the exterior angle of a triangle is the sum of the remote angles. When you draw that helping line it is completely clear.

Dr. David Harold Blackwell (1919–)
"Find something that you like.
It is more important than how
much money you make." [1]

Fortunately for David, he had teachers who nurtured his mathematical interests. His geometry teacher got him to love mathematics by helping him to see the beauty of the subject. A teacher named Mr. Huck formed a mathematics club where he would challenge students with problems from the *School Science and Mathematics* journal. Whenever a student came up with a good possible solution Mr. Huck would send the solution to the journal under the student's name. David's solutions got published once in the journal and he was identified three times there as having correct solutions to problems, which gave him great joy. This was something that further motivated his interest in mathematics. Consequently, long before he was admitted to college, David had decided to major in mathematics. He states ([3], p. 21), "I really fell in love with mathematics . . . It became clear that it was not simply a few things that I liked. The whole subject was just beautiful."

Undergraduate education

David graduated from high school in 1935, at the age of 16. He promptly enrolled at the University of Illinois in Champaign-Urbana, a campus with no black faculty at that time. His intention was to earn a Bachelor's degree and become an elementary school teacher. This decision was motivated by the scarcity of jobs at that time and the fact that a good friend of his father, with a strong influence on the school board in a southern Illinois town, had promised to get him hired upon graduation. However, because his decision to become an elementary school teacher was based primarily on the need for employment after graduation rather than a keen interest, he kept postponing his education courses. After a time, they were no longer necessary, due to a change in his career decision.

David's career goal to become an elementary school teacher changed in his combination junior/senior year when he took a course in elementary analysis. This course really sparked his interest in advanced level mathematics. It motivated him to consider a career that would require graduate level education in mathematics. He now set his sights on teaching at the college or high school level. He began to pursue activities that would facilitate his career goals and groom him for leadership, such as serving as president for the mathematics club at this university. His parents, who were not college educated folks, left it to him to make the hard core decisions about his college education and career. However, they supported him in every way they could and encouraged him to work hard to achieve his goals.

At the end of his freshman year, David learned that his father was borrowing money to finance his college education. A young man with strength of character, he decided to spare his father this financial ordeal by taking responsibility for supporting his college education by working as a dishwasher, a waiter, and a cleaner for equipment in the college entomology lab. In spite of having to work his way through college, David facilitated his college education by taking summer classes and passing proficiency exams, which allowed him to skip courses. Thus, in 1938, he graduated with a Bachelor's degree in mathematics within three years of admission to college.

Graduate education

David continued on for graduate study from 1938 to 1941, at Champaign-Urbana, working to pay for his education as usual. In his last two years of graduate study, while he was a doctoral student, he was awarded fellowships from the university. David has mixed feelings about the motivations of the university officials in offering him these fellowships. He says this about the issue ([3], p. 21):

One of my fellow graduate students told me that
I was going to get a fellowship. I said, "How do

you know?" He said, "You're good enough to be supported, either with a fellowship or a teaching assistantship, and they're certainly not going to put you in the classroom." That was funny to me because fellowships were the highest awards; they gave one the same amount of money and one didn't have to work for it. I have no doubt, looking back on it now, that race did enter into it.

In 1939, David earned a master's degree in mathematics, proceeding on for doctoral studies with some trepidation. He was confident that he could handle the mathematics course work and read research papers. However, he was unsure about whether he would be successful in writing a thesis. Being a determined young man, he took on this challenge, bearing in mind that he had the option of high school teaching in the event that he was not successful in completing the doctoral program.

David's thesis advisor was Joseph Doob. He was a probability and statistics professor at Champaign-Urbana, renowned for his contributions to martingale theory. David states [1], "Joseph Doob had the most important mathematical influence on me. I studied his work carefully and learnt a lot from it. I admired him and tried to emulate him." This statement captures the significance of documenting the contributions and biographies of pioneers, innovators, and leaders in any field of study.

Ironically, David had never met Doob prior to approaching him to become his thesis advisor. His decision to appeal to Doob was based on the recommendation of a peer mentor, Don Kibbey, a teaching assistant in whom he placed a great deal of confidence; Doob was Don Kibbey's dissertation advisor at that time. He was also the dissertation advisor to Paul Halmos, a mathematician who contributed immensely to the development of measure theory and who was a significant peer mentor to David in this area while they were both students of Doob.

In 1941, at the age of 22, within five years of graduation from high school, David earned a doctorate in mathematics. He is the seventh African-American to earn a Ph.D. in this field. His dissertation is titled, "Some Properties of Markoff's Chains". It led to his first set of publications: "Idempotent Markoff chains", "The existence of anormal chains", and "Finite non-homogeneous chains" ([4], [5], [6]). David credits the main idea in his thesis to his advisor Joseph Doob. In doing so, he shows how important a thesis advisor is in helping students to identify appropriate research questions.

Post-doctoral education

Upon completion of his doctoral program, David was awarded a Rosenwald Post-doctoral Fellowship for a year at the Institute of Advanced Study (IAS) at Princeton University. His exposure at the IAS was the beginning of a stellar career as a renowned mathematician, statistician, and educator.

His acceptance at the IAS was not devoid of the hurdles of racism. At that time, it was customary for Princeton to appoint IAS members as visiting fellows. However, when the administrators at Princeton, particularly the president, realized that David was a black man they profusely objected to his acceptance at the IAS. Princeton had never admitted a black student nor hired a black faculty member, and the administrators wanted to maintain the status quo. Upon the insistence and threats of the IAS director, the administrators of Princeton later withdrew their objections to David's acceptance at the IAS. At the time David accepted the Rosenwald Fellowship, he was unaware of the racial controversy that took place between the IAS director and the president of Princeton. He discovered the exact details several years later during the prime of his professional career. Thus, he was shielded from the marring effects of racism on his acceptance of the Rosenwald Fellowship and his stay at the IAS.

At the IAS, there were two mathematicians in particular who influenced David's post-doctoral education, Samuel Wilks and John von Neumann, a renowned Hungarian-American mathematician credited with initiating the development of game theory. David developed a keen interest in statistics by auditing Samuel Wilks's course. Wilks was a mathematician renowned for his work in developing the field of mathematical statistics. He was a founding member of the Institute of Mathematical Statistics, an international professional and scholarly society devoted to the development, dissemination, and application of statistics and probability. (Much later, in 1955, David would serve as president of this organization.)

Another important mathematician who took an interest in David was John von Neumann. He encouraged David to meet with him to discuss his thesis. David avoided this meeting for several months because he did not think that the great John von Neumann was genuinely interested, or had the time to listen to him discuss his thesis. This turned out to be a flawed assumption, for von Neumann was indeed interested in mentoring students. When David and von Neumann finally met to discuss his thesis, von Neumann spent about 10 minutes listening to David's explanation about his thesis and asking him related questions. Afterwards, he took the liberty to explain to David other simpler techniques that he could have used for his thesis problem. The time David spent with von Neumann discussing his thesis, seeing firsthand that he was willing to mentor students, certainly impressed young David. Throughout his professional career, even at the height of success, we see him mentoring students and other young professionals.

Professional career: scholar, teacher, and administrator

We only have to examine the humble beginnings of David's professional career to understand some of the negative consequences of racism, and other forms of discrimination, on society. David was a young African-American pioneer with genius, integrity, and strength of character, whose work was of interest to world-class mathematicians of this period. Yet when he completed his post-doctoral education, the only universities he applied to for a faculty position were Historically Black Colleges and Universities (HBCUs), because he could envision himself nowhere else. He states [10]:

> It never occurred to me to think about teaching in a major university since it wasn't in my horizon at all—I just assumed that I would get a job teaching in one of the black colleges. There were 105 black colleges at that time, and I wrote 105 letters of application ... I eventually got three offers, but accepted the first one I got. From Southern University.

From 1942 to 1943, David was an instructor at Southern University in Baton Rouge, Louisiana. In 1943, he accepted an instructor position for a year, at another HBCU, Clark College, in Atlanta, Georgia. In 1944, at the end of his term at Clark College, David still envisioned himself as a faculty member at an HBCU. He accepted a tenure-track position as an assistant professor at Howard University, Washington D.C., the premier HBCU at that time, where he was one of the Mathematics Department's four faculty members. At Howard he was a generalist, teaching all mathematics courses right up to the master's degree level, which was the highest degree program in the department.

David stayed at Howard for ten years, from 1944 to 1954, rising through the ranks from Assistant Professor to Associate Professor in 1946, and finally to the position of Professor and Chairman of the Mathematics Department in 1947. In spite of the heavy teaching loads of at least 12 hours per week, and heavy administrative duties at these HBCUs, he had over 20 publications by the time he left Howard. He had also earned a strong reputation as an excellent teacher and innovative scholar in probability, statistics, and game theory.

Interestingly, although David enjoyed his work as a mathematics faculty member at Howard, it was not Howard but the larger mathematics community and professional networking that was the springboard for his professional success. He says [10], "I was teaching at Howard and the mathematics environment was not really very stimulating, so I had to look around beyond the university for whatever was going on in Washington that was interesting mathematically." However, Howard should be given some credit. The administrators understood the importance of professional meetings and supported David financially and otherwise to allow him to attend them. This illustrates how important it is for students and young professionals to attend professional meetings and participate in professional organizations.

David credits Abe Girshick for initiating his professional success in statistics. He attended a meeting sponsored by the Washington Chapter of the American Statistical Association. There he listened to an interesting lecture by Girshick on sequential analysis. The lecture involved a discussion of Wald's equation, a concept David found to be unbelievable. Thus, after the meeting, David constructed a counter-example to this equation, which he mailed to Girshick. His counter-example turned out to be wrong. However, it resulted in an invitation from Girshick to David to meet with him in his office to discuss it. This meeting was the beginning of a wonderful relationship for both men and several years of collaboration, which culminated years later in a classic mathematics book, *Theory of Games and Statistical Decisions* [9]. It also resulted in several publications by David, including his favorites: "On an equation of Wald" [7] (a proof with much weaker constraints of the equation he found to be unbelievable) and "Bayes and minimax solutions of sequential decision problems" [8].

According to David, Abe Girshick was his most influential mentor in the field of statistics. He took time off to work in collaboration with Girshick at the RAND Corporation and Stanford University, California, while he was still a faculty member at Howard. The RAND Corporation began as Project RAND, started by the Air Force in 1946 to conduct long-range studies in intercontinental warfare by means other than ground armies. David worked as a mathematician at the RAND Corporation in Santa Monica, California, from 1948 to 1950 during the summer periods, and as a Visiting Professor at Stanford, from 1950 to 1951. These were the most significant times for him. His work during this period resulted in breakthroughs that set the stage for world recognition.

David's work in game theory blossomed at the RAND Corporation while he was collaborating with Abe Girshick and other colleagues. World War II had promoted an interest in the theory of games depicting duels. The theory of duels deals with two-person, zero-sum games.

Imagine two persons initially standing $2n$ paces apart, each with a gun loaded with a single bullet. They are advancing towards each other. At every step forward each person has to decide whether to shoot or hold fire without any prior knowledge of what the other person's decision will be. A strategy certainly involves how many of the possible n steps have been taken already, knowledge of one's own shooting ability, and some guess about one's opponent. Firing too soon means the shooter might miss; firing too late might mean the shooter may have been shot. To simplify the theory, we assume that the game always ends with one person having been shot.

David explored different variations on the basic theory of duels. For instance, if the intention is to kill one's opponent, then the optimal number of steps before firing may be different than it would be if all one wants is to stay alive. It also might make a difference if both guns have silencers, so one might not know that the opponent has fired and missed. His work in the theory of duelling led to significant developments in game theory and earned him a reputation as a pioneer in this area. He developed a game-theoretic proof of the Kuratowski Reduction Theorem, which was ground-breaking in that it connected the fields of topology and game theory, an achievement that gives him great pleasure.

David did not explore beyond two-person, zero-sum games. He attributes his reluctance to do so to the extreme complexity of other types of games and to the fact that the best mathematical response for certain games may have a negative social, psychological, or economical response. This had to do with the *sure thing principle*, which was formulated by Jimmie Savage, one of David's mentors at the RAND Corporation. One way of stating it is this: Suppose you have to choose between two alternatives, A and B, and you think that the outcome depends on some unknown situation X or Y. If knowing that X was the case would lead you to choose A over B and if knowing that Y was the case would still lead you to choose A over B, then, even if you do not know whether X or Y was the case, you should still choose A over B. It was thought that the arms race arising from the Cold War showed the sure thing principle at work.

Suppose that the U.S. and the Soviet Union both operate on the sure thing principle. They have to choose between arming (alternative A) or disarming (alternative B) without knowing whether the other nation is going to arm (situation X) or disarm (situation Y). The sure thing principle indicates that the best mathematical strategy is for both nations to continue arming themselves in order to stay ahead or at par with the opposing nation. This leads to the depletion of valuable resources that each nation could have spent on other important areas of development. This is like the well-known prisoner's dilemma, because both nations are actually losing when they use the sure thing principle. The winning strategy would be for both nations to disarm, a situation that is unlikely to happen due to mistrust between the two nations who both fear that the other will double-cross them if they cooperate. David says [1], "I started working on this particular game where the sure thing principle led to behavior that was not best. So, I stopped working on it." Here we see David as a moral scientist.

David's work with the RAND Corporation led him to an avid study of the works of Thomas Bayes. By a stroke of fate, an economist at the RAND Corporation asked David's mathematical opinion on how to apportion the Air Force research budget over a period of five years between immediate developmental and long-range research. The appropriate proportion is dependent on the probability of a major war within the budget period. If this probability is high, then the budget emphasis will be on immediate developmental research, and if it is low, the emphasis will shift to long-range research.

David gave a mathematically correct but unhelpful answer. He indicated that, in this situation, we are dealing with a unique event and not a sequence of repeated events, so the probability of occurrence of a major war within the five-year period is either 0 or 1, and is unknown until the five-year period has elapsed. The economist remarked, in a manner that intrigued David, that this was a common answer of statisticians. It caused him to ponder the problem, and to discuss it with Jimmie Savage on his visit to the RAND Corporation several weeks later. His discussion with Savage left David with a completely new approach to statistical inference—the *Bayesian approach*.

The Bayesian approach to statistical inference considers probability as the right way to deal with all degrees of uncertainty, and not just the extremes of impossibility and certainty, where the probability is 0 or 1. As a basic example, consider this: Even though we cannot observe the same five-year period repeatedly and deduce the probability of a war, we still may be able to make inferences about this probability and base decisions on our estimates of it. In more sophisticated applications, statisticians develop utility functions based on underlying probability distributions; decision-makers attempt to maximize utility.

Since David developed an appreciation for the Bayesian approach, all his statistical works have incorporated it. Thus, he credits Jimmie Savage as the second most influential person in terms of his statistical thinking.

The years at Berkeley

In 1954, David accepted a visiting position for one year at the University of California, Berkeley. In the following year, he accepted a position as a full Professor at this university, and remained there until his retirement in 1988. It is noteworthy to point out that in 1942, much to David's surprise at the time, he was interviewed for a faculty position at Berkeley. However, he was not surprised or disappointed when he was not offered the position. The reason given by the university for not hiring him was that they had decided to appoint a woman, due to the war and the draft. Nevertheless, destiny prevailed. David finally ended up at Berkeley 12 years later, during the period of civil rights gains. African-Americans were now beginning to enjoy more career opportunities and better employment practices.

Shortly after David's arrival at Berkeley, the Mathematics Department there was divided to make its Statistics Laboratory, headed by Jerzy Neyman, into a separate department of its own. For four years, from 1957 to 1961, David was the chair of the Department of Statistics, succeeding Neyman, the person who had interviewed him in 1942 for a possible faculty position at Berkeley. Neyman turned out to be a good friend. He had a personal influence on David through his warmth, generosity, and integrity.

David enjoyed his stint as chair of the department, but he admits that he did not miss the responsibilities of that position. He sees the primary goal of administrative leadership as creating an environment where the workers are happy. He states ([3], p. 30), "When I was department chairman, I soon discovered that my job was not to do what was right but to make people happy." The success of his leadership at Berkeley shows that it was a winning strategy to build coalitions in which people enjoy working together.

David also provided leadership at Berkeley in other administrative capacities. He was the Assistant Dean of the College of Letters and Science from 1964 to 1968 at a time of serious strife at the university. He was also the Director of the University of California Study Center for the United Kingdom and Ireland, from 1973 to 1975.

While at Berkeley, David continued his scholarly work on the mathematics of competition and cooperation. Interestingly, although he accomplished many innovations while still a faculty member at Howard, he did not gain world recognition until he was a faculty member at Berkeley. Also interesting is the fact that his scholarship was not motivated by doing research for its own sake, but by attempting to understand the problems that intrigued him.

A caring teacher

Surprisingly, even at Berkeley while David was at the peak of his research productivity, he taught probability and statistics courses at all levels, from elementary to graduate courses. He states ([3], p. 26), "There is beauty in mathematics at all levels of sophistication and all levels of abstraction." This statement highlights a very important quality that characterizes talented teachers: they are able to convey the beauty of their subject regardless of the level of mastery of the students.

David is very modest about his ability as a teacher. He sees himself as a good teacher for certain students, but not necessarily for all of them; he recognizes that there are some styles of teaching where he may not excel. He states [1], "People have different learning styles, abstract, concrete, visual, hearing, spatial, and so on. So it is necessary for teachers to reflect these learning styles in their teaching if they would like their students to appreciate the beauty of what they are teaching." Many students evidently found David caring and approachable, since he served as the dissertation advisor to at least 53 students at Berkeley, a very high number.

David is a dynamic scholar and teacher who feels most comfortable when he is around students or those willing to learn and share. He is ever willing to jump to the blackboard to illustrate examples. From his conversations with colleagues and with others to whom he has granted interviews, his excitement with mathematics surfaces when he begins to ponder its beauty, how he fell in love with geometry, or how much pleasure it gave him to be challenged by a difficult proof of a theorem.

World recognition: the leader and civic scientist

David's world recognition as an eminent scholar, educator, and leader in our society is illustrated through his numerous awards, honors, and positions of leadership in professional organizations. His honorary Doctor of Science degrees alone illustrate his widespread recognition. He has received 12 honorary Doctor of Science degrees: from the University of Illinois in 1966; Michigan State University in 1969; Southern Illinois University in 1971; Carnegie-Mellon University in 1980; the National University of Lesotho in 1987; Amherst College and Harvard University in 1988; Howard University, Yale University, and the University of Warwick in 1990; Syracuse University in 1991; and the University of Southern California in 1992.

Equally amazing are his extensive leadership roles and honors in the profession, which speak to his dynamism as a civic scientist, a role in which David emerged full-fledged after he left Howard University. In 1954, he gave the invited address in probability at the International Congress of Mathematicians in Amsterdam. This address is credited with spurring Berkeley to offer him a visiting professorship. From 1959 to 1960, he was a visiting lecturer for the Mathematical Association of America in a program to enhance undergraduate mathematics education. In 1965, he was elected to the National Academy of Sciences. In 1968, he was elected to the American Academy of Arts and Sciences. By this time, David had published at least 60 books and papers.

From 1968 to 1971, David served as the vice president of the American Mathematical Society. From 1972 to 1973, he was chairman of the Faculty Research Lecture Committee. In 1973, he was president of the International Association for Statistics in the Physical Sciences. In 1974, he was the W. W. Rouse Ball Lecturer at the University of Cambridge in the United Kingdom. From 1975 to 1978, he was president of the Bernoulli Society for Mathematical Statistics and Probability. From 1975 to 1977, he was vice president of the International Statistical Institute. In 1976, he was elected Honorary Fellow of the Royal Statistical Society. In 1977, he gave the Wald Lecture for the Institute of Mathematical Statistics. In 1978, he was vice president of the American Statistical Association. Additionally, David has given the Rietz Lecture for the Institute of Mathematical Statistics and he has served on the Board of Directors of the American Association for the Advancement of Science.

In fact, the Wald and Rietz Lectures of the Institute of Mathematical Statistics were instrumental in establishing his reputation as an effective and charismatic lecturer. Noteworthy is the fact that David was among a select few chosen to be filmed by the American Mathematical Society and the Mathematical Association of America, lecturing on mathematical topics accessible to undergraduate students.

The year 1979 was a wonderful one for David. He was awarded the John von Neumann Theory Prize by TIMS/ORSA, which today has become INFORMS, the Institute for Operations Research and Management Sciences. This was a significant honor, given that John von Neumann was one of his earliest professional mentors. The purpose of this prize is to recognize a scholar (or more than one, in cases of joint work) who has made fundamental contributions to theory in operations research and management sciences. Although recent work is not overlooked, the award is usually given for work that has stood the test of time. The criteria for the prize are broad, and include significance, innovation, depth, and scientific excellence. In addition to a cash award and medallion, the citation reads [12]:

The John von Neumann Theory Prize for 1979 is awarded to David Blackwell for his outstanding work in developing the theory of Markovian decision processes, and, more generally, for his many contributions in probability theory, mathematical statistics, and game theory that have strengthened the methodology of operations research and management sciences. In the area of Markovian decision processes Blackwell, in a remarkable series of papers published between

1961 and 1966, put the theory of dynamic programming on a rigorous mathematical footing. He introduced new techniques of analysis and established conditions for the existence of optimal and stationary optimal policies. Particularly noteworthy are his studies of the effect of varying the discount rate and his introduction of the important concepts of positive and negative dynamic programs. Virtually all of the subsequent developments in this field are based on these fundamental papers. In other areas, Blackwell's early work with Arrow and Girshick helped lay the foundations for sequential analysis, and his subsequent book with Girshick systematized the whole field of decision theory, to the great benefit of a generation of mathematical statisticians. The famous Rao-Blackwell theorem on statistical estimation led to a practical method for improving estimates, now known as "Rao-Blackwellization". An elegant and important form of the renewal theorem is due to Blackwell, as is a beautiful characterization of the information content of an experiment. In game theory, he initiated the study of duels (with Girshick) and later made several deep contributions to our understanding of sequential games and the role of information therein.

David, the trailblazer, did not relax after receiving the John von Neumann Theory Prize. He continued scaling the frontiers of twentieth-century developments in mathematics and statistics as a leader. In 1986, he was awarded the R. A. Fisher Award from the Committee of Presidents of Statistical Societies. Upon his retirement in 1988, David received the Berkeley Citation. This is one of the highest honors given to a faculty member at Berkeley, for exemplary service to the university and outstanding achievement in one's field. David received this citation for his work in game theory, Bayesian inference, and information theory, for authoring the classic book, *Theory of Games and Statistical Decisions* [9], and for induction into the American Academy of Arts and Sciences and the National Academy of Sciences. By the time of his retirement, he had well over 90 books and papers published on dynamic programming, game theory, measure theory, probability theory, set theory, and mathematical statistics.

A tribute to David's immense contributions is the long list of lecture series and publications in his honor. The book, *Statistics, Probability and Game Theory, Papers in Honor of David Blackwell* [11], is a compilation of 26 papers edited by T. S. Ferguson, L. S. Shapley, and J. B. MacQueen. These papers treat topics related to his significant contributions in probability, statistics, gambling, game theory, Markov decision processes, set theory, and logic. The editors say this about the man honored by the volume: "It is the mark of an outstanding scientist to be influential in a variety of fields."

Another honor in this category is the Mathematical Sciences Research Institute (MSRI) conference and prize in honor of David Blackwell and Richard A. Tapia, distinguished mathematical scientists who have inspired more than a generation of African-American and Hispanic-American students and professionals in the mathematical sciences. The prize is awarded every second year to a mathematical scientist who has contributed significantly to his or her field of expertise, and who has served as a role model for mathematical scientists and students from under-represented minority groups or contributed in significant ways to the addressing of the problem of the under-representation of minorities in mathematics.

Yet another honor in this category is the David Blackwell Lecture of the National Association of Mathematicians (NAM). This lecture is given annually at the MathFest, the popular summer meeting of the MAA. Its goal is to highlight the contributions of minorities in the mathematical community and to stimulate their professional growth.

Family life and personal tidbits

On December 27, 1944, David married a wonderful woman by the name of Ann Madison. He says [1], "The best thing I ever did in life was to get married to my wife". Thus, it is poignantly clear that Ann played a very supportive role in her husband's successes, and in ensuring the stability and enhancement of their family.

David and Ann have eight children, three sons and five daughters, Ann, Julia, David, Ruth, Grover, Vera, Hugo, and Sara. Notably, none of their children have exhibited any interest in mathematics, nor in a related field, an issue that is viewed positively by David. In response to a question about his children, he says this [1]: "No, they have no particular mathematical interests at all. And I'm rather glad of that. This may sound immodest, but they probably wouldn't be as good at it as I am. People would inevitably make comparisons."

On David's off time, when he's not in a classroom filled with students, or writing fascinating papers on mathematical or statistical topics, or engaging in other professional commitments, you can find him with his wife on their 40-acre property in Northern California, listening to music and enjoying themselves. He might say his dream is to sit beneath a tree and sip a martini, but in reality he is more active, and finds himself planting trees or doing yard work. Many of his good friends are professional colleagues with whom he works and collaborates to advance the fields of mathematics and statistics.

There was a time when David's family home did not have telephone service. One of his children had made quite an expensive long distance telephone call, so David and his wife decided to have the telephone disconnected for a month. During that period, he realized the advantage of not having a telephone—peace and tranquility. One month led to three months, but eventually, the advantages of the telephone won over its cons.

David enjoys playing on the computer. However, these are not trivial games that he plays. He says [10], "I have a little computer at home, and it is a lot of fun just to play with. In fact I'd say that I play with this computer here in my office at least as much as I do serious work with it." He admits to attempting to use his computer to set up a program to take the square root of a positive definite matrix, to minimize functions with five variables, and to look at curves. Perhaps this kind of play is at odds with the image of the mathematician who sits down to prove a theorem, but many of us enjoy this kind of fun.

An example to remember

We learn a great deal from examples. If we know that someone has gone through a situation similar to ours, it helps us to analyze our situation in a more confident manner and to make better decisions. If we want students to make well-informed decisions concerning their educational and professional lives, we must provide them with examples of pioneers, innovators, and leaders, both in all fields of study and in all parts of our nation's history. The information about that successful person motivates students, and gives them the courage to tread similar paths. In the words of a student at Borough of Manhattan Community College (BMCC), City University of New York (CUNY), who was acquainted with Dr. David Harold Blackwell only through reading about him for a research project in an Introduction to Statistics class [2]:

David Blackwell's life has influenced me with the struggles that he has had to endure. He started out just wanting to teach elementary school, but his love of mathematics and his natural talent for mathematics has taken him so much further. This is an inspiration to me for I too love what I do and wish to go further. He has shown me to persevere in the face of adversity. I am happy to have learned so much about this truly incredible man.

In addition, it is important to appreciate the contributions and accomplishments of persons from under-represented groups in any field of study, in order to promote justice, equity, and diversity. This is an avenue for teaching cultural sensitivity and cooperation with people of different cultures, and a way to motivate students from these groups to similar or greater heights of success.

Dr. David Harold Blackwell is one of the world's most accomplished thinkers in the fields of mathematics and statistics. Of great significance is that he is one of the African-American masters in these fields. He is a dynamic educator with a reputation as "one of the finest lecturers in the field" [10]. He is a civic scientist and leader whose life history will certainly motivate others to follow his example. His example can also motivate us to develop the necessary mentoring programs and practices to open the doors of opportunity for all students, especially for students from under-represented groups, in the fields of mathematics and statistics.

By examining the conditions under which this mathematician rose to success, we can learn a lot about leadership, humility, strength of character, and passion for one's field. We can learn that mentoring, professional development, and active participation in professional meetings and organizations are vital opportunities. Providing them helps us to nurture students and encourage them to consider careers in mathematics and scientific fields and to groom young professionals in these disciplines. We can also learn about the social and psychological consequences of any form of discrimination on society.

Dr. David Harold Blackwell was not overly concerned about financial status when he decided to major in a career in mathematics. He had cultivated an appreciation for the subject and had a passion for examining and understanding issues that intrigued him. This passion led him to make ground-breaking innovations in the fields of mathematics and statistics. Our society has benefited from the vast contributions of this most renowned African-American thinker. Unbelievable for a man that thought his story in life was to be an elementary teacher.

References

1. N. Agwu et al., *Interviews with David Blackwell*, Women Research Project, Borough of Manhattan Community College, 1998–1999.

2. N. Agwu et al., *Biographies of Statisticians, MAT 150—Introduction to Statistics Research Project*, Borough of Manhattan Community College, 1998–1999.

3. D. J. Albers and G. L. Alexanderson (eds.), *Mathematical People*, Birkhäuser, 1985.

4. D. H. Blackwell, Idempotent Markoff chains, *Ann. of Math.* **43** (1942), 560–567.

5. D. H. Blackwell, The existence of anormal chains, *Bull. Amer. Math. Soc.* **51** (1945), 465–468.

6. D. H. Blackwell, Finite non-homogeneous chains, *Ann. of Math.* **46** (1945), 594–599.

7. D. H. Blackwell, On an equation of Wald, *Ann. Math. Stat.* **17** (1946), 84–87.

8. D. H. Blackwell, K. J. Arrow, and M. A. Girshick, Bayes and minimax solutions of sequential decision problems, *Econometrica* **17** (1949), 213–244.

9. D. H. Blackwell and M. A. Girshick, *Theory of Games and Statistical Decisions*, Wiley, 1954.

10. M. H. DeGroot, A Conversation with David Blackwell, *Statistical Science* **1** (1986), 40–53.

11. T. S. Ferguson, L. S. Shapley, and J. B. MacQueen (eds.), *Statistics, Probability and Game Theory, Papers in Honor of David Blackwell*, Institute of Mathematical Statistics Lecture Notes—Monograph Series 30, 1996.

12. Institute for Operations Research and Management Sciences (INFORMS), *John von Neumann Theory Prize Winners, 1975–2002*, http://www.informs.org/Prizes/vonNeumannDetails.html#1979.

13. S. W. Williams, *Mathematicians of the African Diaspora*, http://www.math.buffalo.edu/mad/PEEPS/blackwell.david.html, 2002.

Afterword

For more details on the work of Cauchy, the best book is Judith Grabiner's own work, *The Origins of Cauchy's Rigorous Calculus* [8]. A slightly more recent work on nineteenth-century analysis in general is Umberto Bottazzini, *The Higher Calculus: A History of Real and Complex Analysis from Euler to Weierstrass* [1].

It is interesting to compare Cauchy's work with the contemporary work of Bolzano. Bolzano's important paper on the intermediate value theorem is available in [17], while [18] is the complete collection of his mathematical works. For more discussion of the work of Bolzano, see I. Grattan-Guinness [9]. Grattan-Guinness claims that Cauchy took the central ideas of his definitions of continuity and convergence from Bolzano. But H. Freudenthal and H. Sinaceur disagree [7], [19]. In particular, the latter article claims that Bolzano and Cauchy represented two different mathematical traditions.

It is also interesting that Cauchy's definition of convergence had been essentially given by José Anastácio da Cunha (1744–1787), a Portuguese scholar, in a comprehensive textbook written in Portuguese in 1782. Unfortunately, even though the book was translated into French in 1811, it apparently was little noticed. For more on da Cunha, see the articles A. J. Franco de Oliveira [6], João Filipe Queiró [16], and A. P. Youschkevitch [21].

Near the beginning of his article on functions, Kleiner notes that there was a long "prehistory" of functions, but does not discuss this at all. One source that does look at the notion of "function" in Greek times is Olaf Pedersen, *A Survey of the Almagest* [15]. There are more detailed discussions of some of Kleiner's ideas in Ivor Grattan-Guinness, *From the Calculus to Set Theory* [10]. And William Dunham's *The Calculus Gallery* [6] explores some of the ideas of Weierstrass, Baire, and Lebesgue, especially giving examples of some of the "pathological functions."

Besides the two detailed articles by Michael Bernkopf mentioned in Steen's bibliography, other works that give more details on various aspects of his story include Thomas Hawkins, *Lebesgue's Theory of Integration: Its Origins and Development* [11] and Leo Corry, *David Hilbert and the Axiomatization of Physics (1898–1918)* [4]. Another article dealing with Hilbert's work in physics is by A. S. Wightman [20]. Finally, there is a very comprehensive five-volume study of the history of quantum mechanics in the 1920s: J. Mehra and H. Rechenberg, *The Historical Development of Quantum Mechanics* [14].

D. M. Cannell has written a biography of George Green: *George Green. Mathematician and Physicist 1793–1841. The Background to His Life and Work* [2]. This work contains a comprehensive bibliography.

Two works on Kovalevsky are Roger L. Cooke, *The Mathematics of Sofya Kovalevskaya* [3] and Ann Hibner Koblitz, *A Convergence of Lives. Sofia Kovalevskaia: Scientist, Writer, Revolutionary*

[12]. The second book deals in considerable detail with Kovalevskaya's life, while the first emphasizes her mathematical work.

More information on the work of Cartwright and Littlewood is found in another article by McMurran and Tattersall [13].

References

1. Umberto Bottazzini, *The Higher Calculus: A History of Real and Complex Analysis from Euler to Weierstrass*, Springer, 1986.
2. D. M. Cannell, *George Green. Mathematician and Physicist 1793–1841: The Background to His Life and Work*, The Athlone Press, 1993.
3. Roger L. Cooke, *The Mathematics of Sofya Kovalevskaya*, Springer, 1984.
4. Leo Corry, *David Hilbert and the Axiomatization of Physics (1898–1918)*, Kluwer, 2004.
5. William Dunham, *The Calculus Gallery*, Princeton Univ. Press, 2005.
6. A. J. Franco de Oliveira, Anastácio da Cunha and the Concept of Convergent Series, *Archive for History of Exact Sciences* **39** (1988), 1–12.
7. H. Freudenthal, Did Cauchy Plagiarize Bolzano? *Archive for History of Exact Sciences* **7** (1971), 375–392.
8. Judith Grabiner, *The Origins of Cauchy's Rigorous Calculus*, MIT Press, 1981.
9. Ivor Grattan-Guinness, Bolzano, Cauchy and the 'New Analysis' of the Early Nineteenth Century, *Archive for History of Exact Sciences* **6** (1970), 372–400.
10. Ivor Grattan-Guinness, *From the Calculus to Set Theory*, Duckworth, 1980.
11. Thomas Hawkins, *Lebesgue's Theory of Integration: Its Origins and Development*, Chelsea, 1975.
12. Ann Hibner Koblitz, *A Convergence of Lives. Sofia Kovalevskaia: Scientist, Writer, Revolutionary*, Birkhäuser, 1983.
13. Shawnee McMurran and James Tattersall, Cartwright and Littlewood on van der Pol's equation, in *Harmonic Analysis and Nonlinear Differential Equations* (Contemporary Mathematics 208) American Mathematical Society, 1997, 265–276.
14. J. Mehra and H. Rechenberg, *The Historical Development of Quantum Mechanics*, Springer, 1982–1987.
15. Olaf Pedersen, *A Survey of the Almagest*, Odense Univ. Press, 1974.
16. João Filipe Queiró, José Anastácio da Cunha: A Forgotten Forerunner, *The Mathematical Intelligencer*, **10** (1988), 38–43.
17. Steve Russ, A Translation of Bolzano's Paper on the Intermediate Value Theorem, *Historia Mathematica* **7** (1980), 156–185.
18. Steve Russ, *The Mathematical Works of Bernard Bolzano*, Oxford Univ. Press, 2004, pp. 251–278.
19. H. Sinaceur, Cauchy et Bolzano, *Revue d'histoire des sciences* **26** (1973), 97–112.
20. A. S. Wightman, Hilbert's Sixth Problem: Mathematical Treatment of the Axioms of Physics, in *Mathematical Developments Arising from Hilbert's Problems*, American Mathematical Society, 1976, pp. 147–240.
21. A. P. Youschkevitch, J. A. da Cunha et les fondements de l'analyse infinitésimale, *Revue d'Histoire des Sciences* **26** (1973), 3–22.

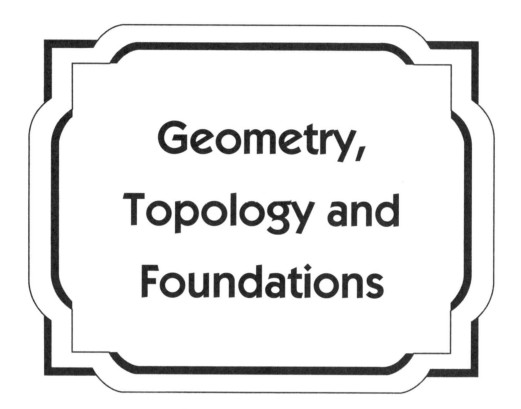

Foreword

One of the most important aspects of geometry in the nineteenth century was the development of non-Euclidean geometry, and this chapter begins with two brief studies of aspects of its development. In the first, George Bruce Halsted reviews volume VII of Gauss's *Werke* and concludes from a study of many of Gauss's letters first published in that volume that Gauss's ideas on the subject had no influence on the independent discoveries of János Bolyai and Nikolai Lobachevsky, or on the earlier publication by Ferdinand Karl Schweikart (1780–1859). In the second article, Florence P. Lewis gives us a whirlwind tour through the history of the parallel postulate, from Proclus to Bolyai and Lobachevsky. She then proposes how this history, and its effect in how mathematicians understood the nature of a *postulate*, could affect the teaching of geometry in schools. In particular, she emphasizes that one reason for the study of geometry is its role in "training the mind."

Another major aspect of nineteenth-century geometry was the development of projective geometry. That development began in the seventeenth-century work of Girard Desargues and Blaise Pascal, but, as Julian Lowell Coolidge notes, it then "dragged along" for about a century. In his article, Coolidge takes up the story in the nineteenth century, summarizing the work of such mathematicians as Jean-Victor Poncelet, Michal Chasles, Jacob Steiner, and Johann Karl Christian von Staudt. But then Coolidge remarks that by the end of the nineteenth century, the field of synthetic projective geometry was "pretty much worked out." Dan Pedoe discusses two particular aspects of projective geometry: the development of homogeneous coordinates by August Ferdinand Möbius and others, and the recognition of the importance of the principle of duality, by Poncelet and Joseph Gergonne. In particular, he deals with the controversy between the two men relating to this development.

Although topology had its origins in certain specific problems that came up in the eighteenth and nineteenth centuries, such as the Königsberg bridges problem and Euler's polyhedron formula, the subject only began to have a systematic development late in the nineteenth century. Three papers in this chapter deal with specific aspects of topology. First, Julian Fleron recalls that the study of point-set topology on the real line began with two questions, the conditions under which a function could be integrated (under some reasonable definition of integration), and the uniqueness of the expression of a function in terms of its Fourier series. These questions led to the study of the point sets at which a discontinuity of the function would still allow it to be integrated or to have a unique Fourier series. As Fleron shows, the Cantor set was essentially discovered by the English mathematician Henry Smith in 1875, a few years before Cantor himself found it, with the first discovery linked to integrability and the second linked to Fourier series.

Connectedness in topology is the subject of R. L. Wilder's article. Wilder traces the changes in the definition and understanding of this concept from the work of Bolzano in the mid-nineteenth century to that of N. J. Lennes and F. Riesz early in the twentieth century. Similarly, Peter Hilton gives a history of the notions of homology and homotopy, beginning with the work of Poincaré and continuing through the remarkable contributions made at Göttingen in the 1920s and in Moscow in 1935.

This chapter concludes with several papers dealing with the foundations of mathematics. First, H. C. Kennedy traces the development of the modern meaning of an *axiom*, concentrating on the work of Pasch and Peano. Then Joseph Dauben discusses Charles Sanders Peirce's ideas on infinite sets, ideas that were related to those of Cantor and Dedekind in Europe. Although the latter work grew out of questions in analysis, Peirce's work was more closely related to philosophy. Interestingly, Peirce attempted to make mathematical sense out of the notion of an *infinitesimal*, an idea that was not to come to fruition until the work of Abraham Robinson many years later.

Ivor Grattan-Guinness discusses the developments in logic in the 1920s and 1930s. He shows that, although *logicism* was the most fully developed philosophy of mathematics in 1918, having its roots in Whitehead and Russell's monumental *Principia Mathematica* of 1910–1913, two other philosophies, *intuitionism* and *formalism*, both grew to prominence over the next twenty years.

Finally, David Fowler looks at Dedekind's theorem that $\sqrt{2} \times \sqrt{3} = \sqrt{6}$. Dedekind himself invented the notion of "cuts", and then claimed to be able to prove this result. But there were other notions of the real numbers available at the time, and Fowler considers how Dedekind's result could be proved under other assumptions on the nature of the set of real numbers.

Gauss and the Non-Euclidean Geometry

GEORGE BRUCE HALSTED

American Mathematical Monthly **7** (1900), 247–252

A Review of Gauss's *Werke*, Vol. VII

We are so accustomed to the German professor who does, we hardly expect the German professor who does not. Such, however, was Schering of Göttingen, who so long held possession of the papers left by Gauss.

Schering had planned and promised to publish a supplementary volume, but never did, and only left behind him at his death certain preparatory attempts thereto, consisting chiefly of excerpts copied from the manuscripts and letters left by Gauss. Meantime these papers for all these years were kept secret and even the learned denied all access to them.

Schering dead, his work has been quickly and ably done, and here we have a stately quarto [1] of matter supplemental to the first three volumes, and to the fourth volume with the exception of the geodetic part.

Of chief interest for us is the geometric portion (pp. 159–452), edited by just the right man, Professor Stäckel of Kiel.

One of the very greatest discoveries in mathematics since ever the world began is, beyond peradventure, the non-Euclidean geometry. By whom was this given to the world in print? By a Hungarian, John Bolyai, who made the discovery in 1823, and by a Russian, Lobachevski, who had made the discovery by 1826. Were either of these men prompted, helped, or incited by Gauss, or by any suggestion emanating from Gauss? No, quite the contrary.

Our warrant for saying this with final and overwhelming authority is this very seventh volume of Gauss's works, just now at last put in evidence, published to the world.

The geometric part opens (p. 159), with Gauss's letter of 1799 to Bolyai Farkas the father of John (Bolyai János), which I gave years ago in my Bolyai as demonstrative evidence that in 1799 Gauss was still trying to prove Euclid's the only non-contradictory system of geometry, and also the system of objective space. The first is false; the second can never be proven.

But both these friends kept right on working away at this impossibility, and the more hot-headed of the two, Farkas, finally thought he had succeeded with it, and in 1804 sent to Gauss his *Göttingen Theory of Parallels*. Gauss's judgment on this is the next thing given (pp. 160–162). He shows the weak spot. "Could you *prove*, that $dkc = ckf = fkg$, etc., then were the thing perfect. However, this theorem is indeed true, only difficult, with already presupposing the theory of parallels, to prove rigorously." Thus in 1804 instead of having or giving any light, Gauss throws his friend into despair by intimating that the link missing in his labored attempt is true enough but difficult to prove without *petitio principii*.

Of course we now know it is *impossible* to prove. Anything is impossible to prove which is the equivalent of the parallel postulate. Yet both the friends continue their strivings after this impossibility.

In this very letter Gauss says: "I have indeed yet ever the hope, that those rocks sometime, and indeed before my end, will allow a through passage."

Farkas in 1808, December 27, writes to Gauss:

Oft thought I, gladly would I, as Jacob for Rachel serve, in order to know the parallels founded, even if by another.

Now just as I thought it out on Christmas night, while the Catholics were celebrating the birth of the Saviour in the neighboring church, yesterday wrote it down, I send it to you enclosed herewith. Tomorrow must I journey out to my land, have no time to revise, neglect I it now, maybe a year is lost, or indeed find I the fault, and send it not, as has already happened with hundreds, which I as I found them took for genuine. Yet it did not come to writing those down, probably because they were too long, too difficult, too artificial, but the present I wrote off at once. As soon as you can, write me your real judgment.

This letter Gauss never answered, and never wrote again until 1832, a quarter of a century later, when the non-Euclidean geometry had been published by both Lobachevski and Bolyai János.

This settles now forever all question of Gauss having been of the slightest or remotest help or aid to young János, who in 1823 announced to his father Farkas in a letter still extant, which I saw at the Reformed College in Maros-Vásárhely, where Farkas was professor of mathematics, his discovery of the non-Euclidean geometry as something undreamed of in the world before.

This immortal letter, a charming and glorious outpouring of pure young genius, speaks as follows:

> Temesvár, 3 Nov. 1823.
> My dear and good father. I have so much to write of my new creations, that it is at the moment impossible for me to enter into great detail, so I write you only on a quarter of a sheet. I await your answer to my letter of two sheets; and perhaps I would not have written you before receiving it, if I had not wished to address to you the letter I am writing to the Baroness, which letter I pray you to send her.
>
> First of all I reply to you in regard to the binomial....
>
> Now to something else, so far as space permits. I intend to write, as soon as I have put it into order, and when possible to publish, a work on parallels. At this moment it is not yet finished, but the way I have hit upon promises me with certainty the attainment of the goal, if it in general is attainable. It is not yet attained, but I have discovered such magnificent things that I am myself astonished at them. It would be damage eternal if they were lost. When you see them, my father, you yourself will acknowledge it. Now I cannot say more of them, only so much: *that from nothing I have created a wholly new world*. All that I have hitherto sent you compares to this only as a house of cards to a castle.
>
> P. S. I dare to judge absolutely and with conviction of these works of my spirit before you, my father; I do not fear from you any false interpretation (that certainly I would not merit),

which signifies that, in certain regards, I consider you as a second self.

In his autobiography János says:

> First in the year 1823 did I completely penetrate through the problem according to its essential nature, though also afterward further completions came thereto. I communicated in the year 1825 to my former teacher, Herrn Johann Walter von Eckwehr (later imperial-royal general), a written paper, which is still in his hands. On the prompting of my father I translated my paper into Latin, in which it appeared as *Appendix* to the *Tentamen* in 1832.

So much for Bolyái.

The equally complete freedom of Lobachevski from the slightest idea that Gauss had ever meditated anything different from the rest of the world on the parallels I showed in [2].

Passing on to the next section (pp. 163–164), in the new volume of Gauss, we find it important as showing that in 1805 Gauss was still a baby on this subject. It is an erroneous pseudo-proof of the impossibility of what in 1733 Saccheri had called "hypothesis anguli obtusi." To be sure, Saccheri himself thought he had proven this hypothesis inadmissable, so that Gauss blundered in good company; but his pupil Riemann in 1854 showed that this hypothesis gives a beautiful non-Euclidean geometry, a new universal space, now justly called the space of Riemann.

Passing on, we find that in 1808, Schumacher writes:

> Gauss has led back the theory of parallels to this, that if the accepted theory were not true, there must be a constant *a priori* line given in length, which is absurd. Yet he himself considers this work still not conclusive.

Again, with the date April 27, 1813, we read:

> In the theory of parallels we are even now not farther than Euclid was. This is the *partie honteuse* (shameful part) of mathematics, which soon or late must receive a wholly different form.

Thus in 1813 there is still no light.

In April 1816, Wachter on a visit to Göttingen had a conversation with Gauss whose subject was what he calls the anti-Euclidean geometry. On December 12, 1816, he writes to Gauss a letter which shows that this anti-Euclidean geometry, as he understands it, far from being the non-Euclidean geometry of Lobachevski and Bolyai János, was a monstrous conglomerate blunder.

The letter as here given by Stäckel (pp. 175–176), is as follows:

... Consequently the anti-Euclidean or your geometry would be true. However the *constant* in it remains undetermined: *why* may perhaps be made comprehensible by the following ... The result of the foregoing may consequently be so expressed:

The Euclidean geometry is false; but nevertheless the true geometry must begin with the same eleventh Euclidean axiom or with the assumption of lines and surfaces which have the property presumed in that axiom. Only instead of the straight line and plane are to be put the great circle of that sphere described with infinite radius together with its surface. From this comes indeed the one inconvenience, that the parts of this surface are merely symmetric, not, as with the plane, congruent; or that the radius out on the one side is infinite, on the other imaginary. Only it is clear how that inconvenience is again overbalanced by many other advantages which the construction on a spherical surface offers: *so that probably also then even*, if the Euclidean geometry were true, the necessity no longer indeed exists, to consider the plane as an infinite spherical surface, though still the fruitfulness of this view might recommend it.

Only, as I thought through all this, as I had already fully settled myself about the result, in part since I believed I had recognized the ground (la métaphysique) of that indeterminateness necessarily inherent in geometry—also even the complete indecision in this matter, then, if that proof against the Euclidean geometry, as I could not expect, were not to be considered as stringent —; in part, while yet not to consider as lost all the many previous researches in plane geometry: but to be used with few modifications, and if still also the theorems of solid geometry and mechanics might have approximate validity, at least to a quite wide limit, which perhaps yet could be more nearly determined; I found this evening—just while busied with an attempt to find an entrance to your transcendental trigonometry, and while I could not find in the plane sufficing, determinate functions thereto, going on to space constructions, to my no small delight the following *demonstration for the Euclidean parallel-theory* ... Just in the idea to conclude I remark still, that the above proof for the Euclidean parallel theory is fallacious ... Consequently has here also the hope vanished, to come to a fully decided result, and I must content myself again with the above cited. Withal I believe I have made upon that way at least a step toward your transcendental trigonometry, since I, with aid of the spherical trigonometry, can give the ratios of all constants, at least *by construction of the right-angled triangle*. I yet lack the actual reckoning of the base of an isosceles triangle from the side, to which I will seek to go from the equilateral triangle.

If Gauss's transcendental trigonometry were as sad a hodgepodge as the anti-Euclidean geometry here explained by Wachter, it is fortunate that nothing was ever given about it but its name. *Requiescat in pace.*

Yet Gauss writes, April 28, 1817:

Wachter has printed a little piece on the foundations of geometry. Though Wachter has penetrated farther into the essence of the matter than his predecessors, yet is his proof not more valid than all others.

We come now to an immortal epoch, that of the discovery of the real non-Euclidean geometry by Schweikart, and his publication of it under the name of Astralgeometry.

On the twenty-fifth of January, 1819, Gerling writes to Gauss:

Apropos of parallel-theory I must tell you something, and execute a commission. I learned last year, that my colleague Schweikart (prof. juris, now Prorector) formerly occupied himself much with mathematics and particularly also had written on parallels. So I asked him to lend me his book. While he promised me this, he said to me, that now indeed he perceived how errors were present in his book (1808) (he had, for example, used quadrilaterals with equal angles as a primary idea), however that he had not ceased to occupy himself with the matter, and was now about convinced, that without some datum the Euclidean postulate could not be proved, also that it was not improbable to him, that our geometry is only a chapter of a more general geometry.

Then I told him how you some years ago had openly said, that since Euclid's time we had not in this really progressed; yes, that you had often told me, how you through manifold occupation with this matter had not attained to the proof of the absurdity of such a supposition. Then when he sent me the book asked for, the enclosed paper accompanied it, and shortly after (end of December) he asked me orally, when convenient to enclose to you this paper of his, and to ask you in his name to let him know when convenient your judgement on these ideas of his. The book itself has, apart from all else, the advantage that it contains a copious bibliography of the subject; which he also, as he tells me, has not ceased still further to add to.

Now comes (pp. 180–181) the precious enclosure, dated Marburg, December, 1818, which, though so brief, may fairly be considered the first *published* [not printed] treatise on non-Euclidean geometry. It is a pleasure to give this here in English for the first time.

The Non-Euclidean Geometry of 1818
by Schweikart

There is a two-fold geometry—a geometry in the narrower sense — the Euclidean; and an astral science of magnitude.

The triangles of the latter have the peculiarity, that the sum of the three angles is not equal to two right angles.

This presumed, it can be most rigorously proven:

1. That the sum of the three angles in the triangle is *less* than two right angles;

2. that this sum becomes ever smaller, the more content the triangle encloses;

3. that the altitude of an isosceles right-angled triangle indeed ever increases, the more one lengthens the side, that it however cannot surpass a certain line, which I call the *constant*.

Squares have consequently the following form:

Is this constant *for us* half the earth's axis (as a consequence of which each line drawn in the universe from one fixed star to another, which are ninety degrees apart from one another, would be a tangent of the earth-sphere), so is it in relation to the spaces occurring in daily life infinitely great.

The Euclidean geometry holds good only under the presupposition, that the constant is infinitely great. Only then is it true, that the three angles of every triangle are equal to two right angles; also this can be easily proven if one takes as given the proposition, that the constant is infinitely great.

Such is the brief declaration of independence of this hero. Nor was Schweikart's courage and independence without farther issue. Under his direct influence his own nephew Taurinus developed the real non-Euclidean trigonometry and published it in 1825 with successful application to a number of problems.

Moreover this teaching of Schweikart's made converts in high places. In the letter of Bessell to Gauss of February 10, 1829 (p. 201) he says:

Through that which Lambert said, and what Schweikart disclosed orally, it has become clear to me, that our geometry is incomplete, and should receive a correction, which is hypothetical and, if the sum of the angles of the plane triangle is equal to a hundred and eighty degrees, vanishes.

That were the *true* geometry, the Euclidean the *practical*, at least for figures on the earth.

The complete originality and independence of Schweikart and of Lobachévski is recognized as a matter of course in the correspondence between Gauss and Gerling, who writes (p. 238):

The Russian steppes seem therefore indeed a proper soil for these speculations, for Schweikart (now in Königsberg) invented his 'Astral-Geometry' while he was in Charkow.

This fixes the date of the first conscious creation and naming of the non-Euclidean geometry as between 1812 and 1816.

Gauss adopts and uses for himself this first name, Astralgeometry (p. 226 (1832); p. 232 (1841)).

At length the true prince comes. On February 14, 1832, Gauss receives the profound treatise of the young Bolyai János, the most marvellous two dozen pages in the history of thought. Under the first impression Gauss writes privately to his pupil and friend

Gerling of the ideas and results as "mit grosser eleganz entwickelt." He even says "I hold this young geometer von Bolyai to be a genius of the first magnitude."

Now was Gauss's chance to connect himself honorably with the non-Euclidean geometry, already independently discovered by Schweikart, by Lobachevski, by Bolyai János.

Of two utterly worthless theories of parallels Gauss had already given extended notices in the *Göttingische gelehrte Anzeigen* (pp. 170–174, 183–185).

To this marvel of János, Gauss vouchsafed never one printed word.

As Stäckel gently remarks, this certainly contributed thereto, that the worth of this mathematical gem was first recognized when John had long since finished his earthly career.

The 15th of December, 1902, will be the centenary of the birth of Bolyai János. Should not the learned world endeavor to arouse the Magyars to honor Hungary by honoring then this truest genius her son?

References

1. Carl Friedrich Gauss, *Werke*, Band VII (ed. P. Stäckel), Göttingen, 1900.
2. G. Halsted, *Science* **9** (232), 813–817.

History of the Parallel Postulate

FLORENCE P. LEWIS

American Mathematical Monthly **27** (1920), 16–23

Like the famous problems of construction, Euclid's postulate concerning parallels is a thought that links the ages. Its history is a long story with a dramatic climax and far-reaching influence on modern mathematical and general scientific thought. I wish to recall briefly the salient features of the story, and to state what seem to me its suggestions in regard to the teaching of elementary geometry.

Euclid's fifth postulate (called also the eleventh or twelfth axiom) states: "If a straight line falling on two straight lines makes the interior angles on the same side less than two right angles, the two straight lines if produced indefinitely meet on that side on which are the angles less than two right angles" [2]. The earliest commentators found fault with this statement as being not self-evident. Concerning the meaning of *axiom*, Aristotle says: "That which it is necessary for anyone to hold who is to learn anything at all is an axiom"; and "It is ignorance alone that could lead anyone to try to prove the axiom." Without going into the difficult question of the precise distinction to the Greek mind between axiom and postulate, we may take it that the character of being indisputable pertained to each. Postulates stating that a straight line joining any two points can be drawn, that a circle can be drawn with given center and radius, or that all right angles are equal, were accepted, while the postulate of parallels was scrutinized and admitted at best with reluctance.

Proclus, writing in the fifth century A.D., gives some of the reasons for this attitude, and we may surmise others. The postulate makes a positive statement about a region beyond the reach of possible observation or geometrical intuition. Proclus insinuates that those who "suppose they have ground for instantaneous belief" are "yielding to mere plausible imaginings"; the conclusion is "plausible but not necessary" (see [1] and [2]). The converse is proved in Proposition 27, Book I of Euclid's *Elements*, and there seems to be no reason why this proposition should be more or less self-evident than its converse. The fact that the two lines continually approach each other (even the meaning of this phrase requires further elucidation) was not a convincing argument to the Greek geometer who was acquainted with the relation of the hyperbola to its asymptote. The form of statement of the postulate is long and awkward compared with that of the others, and its obviousness thereby lessened. There is evidence that Euclid himself endeavored to prove the statement before putting it down as a postulate; for in some manuscripts it appears not with the others but only just before Proposition 29, where it is indispensable to the proof. If the order is significant, it indicates that the author did not at first intend to include this among the postulates, and that he finally did so only when he found that he could neither prove it nor proceed without it.

Most of the early geometers appear to have attacked the problem. Proclus quotes and criticizes several proofs, and gives one of his own. He instances one writer who even attempted to prove the falsity of the statement, the argument being similar to those used in Zeno's paradoxes. The common opinion, however, seems to have been that the postulate stated a truth, but that it ought to be proved. Euclid had proved two sides of a triangle greater than the third, which is far more obvious than this. If the statement was true it should be proved in order to convince the doubters; if false, it should be removed. In no case should it be retained among the fundamental presuppositions. Sir Henry Savile (1621) and the Italian Saccheri (1733) refer to it as a blot or blemish on a work that is otherwise perfect, and this expresses the common attitude of mathematicians until the first quarter of the nineteenth century.

Early attempts at proof usually took the form of a change in the definition of parallels, or the substitution, conscious or unconscious, of a new assumption. Neither of these methods resulted in satisfaction to any but their inventors; for the definitions usually concealed an assumption, and the new postulates were no more obvious than the old. Posidonius, quoted by Proclus, defines parallels as lines everywhere equidistant. This begs the question; surely such parallels do not meet, but may there not be in the same plane other lines, not equidistant, which also do not meet? The definition involves also the assumption, that the locus of points in a plane at a given distance from a straight line is a straight line, and this was not self-evident. (It should be noted that even the meaning of the criterion suggests several questions of logic. If two lines are so placed that perpendiculars to one of them from points on the other are equal, will the same statement hold when the roles of the two lines are reversed? Will a perpendicular to one of two non-intersecting lines necessarily be perpendicular to the other? Of course the answers to these questions are closely bound up with the very postulate under discussion [note by original editor]).

Ptolemy says that two lines on one side of a transversal are no more parallel than their extensions on the other side; hence if the two angles on one side are together less than two right angles, so also are the two angles on the other side, which is impossible since the sum of the four angles is four right angles. This is another way of saying that through a point but one parallel to a given line can be drawn, which is exactly Euclid's postulate. Proclus himself assumes (with some concealment) that if a line cuts one of two parallels it cuts the other, which is again postulate 5.

Even as late as the close of the eighteenth century we find this argument advanced by one Thibault, and attributed also to Playfair. Let a line segment with one end A at a vertex of a triangle be rotated through the exterior angle. Translate it along the side until A comes to the next vertex and repeat the process. We finally arrive at the original position and must therefore have rotated through $360°$. Hence the sum of the interior angles of a triangle is $180°$; and, since Legendre had satisfactorily proved that this proposition entails Euclid's postulate of parallels, the latter is at last demonstrated. The fact that the same process could equally well be carried out with a spherical triangle, in which the angle-sum is not $180°$, might have given him pause. The assumption that translation and rotation are independent operations is in fact equivalent to Euclid's postulate. Heath [2] gives a long and instructive list of these substitutes. In the course of centuries the minds of those interested became clear on one point: they did not wish merely to know whether it was possible to substitute some other assumption for Euclid's, though this question has its interest; they wished to know primarily whether exactly his form of the postulate was logically deducible from his other postulates and established theorems. To change the postulate was merely to re-state the problem.

After certain Arabs and Persians had had their say in their day, the curtain rises on the Italian Renaissance of the sixteenth century, where the problem was attacked with great vigor. French and British assailants were not lacking. The first modern work devoted entirely to the subject was published by Cataldi in 1603. When the eighteenth century took up the unfinished business of proving the parallel postulate, we find most of the giants of those days attacking the enemy of geometers with an even keener sense that without victory there could be no peace. Yet d'Alembert toward the close of the century could still refer to the state of the theory of parallels as "the scandal of elementary geometry." Klügel in 1763 examined thirty demonstrations of the postulate. He was perhaps the first to express doubt of its demonstrability. Lagrange, according to De Morgan, in about 1800, when in the act of presenting to the French Academy a prepared memoir on parallels, interrupted his reading with the exclamation, "Il faut que j'y songe encore," [I must dream about it again.] and withdrew his manuscript.

While the results of these investigations were on the whole negative, certain positive and valuable results were nevertheless obtained. The relation between the parallel postulate and the angle sum of a triangle was clearly brought out. Legendre proved that if in a single triangle the angle sum is two right angles, the postulate holds. Other equivalents are of interest. John Wallis and Laplace wished to assume: There exists a figure of arbitrary size similar to any given figure. Gauss could proceed rigorously provided he could prove the existence of a rectilinear triangle whose area is greater than any previously assigned area. W. Bolyai could have succeeded with the assumption that a circle can be passed through any three points not in a straight line. It must be borne in mind, moreover, that few mathematical questions have served so well as whetstones on which to sharpen the critical powers of mankind.

The work of the Italian priest Saccheri deserves notice because his method is that which finally brought the discussion to a close. Though published in 1733 his results did not become well known until after 1880, and therefore had little influence on other investigations. Legendre's *Réflexions*, published a hundred years later, covered much of the same ground without advancing quite so far. The title of Saccheri's work is *Euclides ab omni Naevo Vindicatus* [Euclid Vindicated of every Flaw]. His plan was to prove the postulate by assuming its contradictory and showing that an inconsistency followed. He succeeded in proving that, according as in one triangle the angle sum is greater than, equal to, or less than two right angles, the same holds in every triangle, and that accordingly Euclid's postulate or one of its contradictories will hold. He makes three hypotheses which were recognized later to correspond to the elliptic, Euclidean and hyperbolic geometries. But at the end of his work, in order to exhibit a contradiction when Euclid's postulate is denied, he is forced to make use of a somewhat vague and unacceptable assumption about "the nature of a straight line."

Gauss's activity in connection with the parallel postulate is of especial interest because of its psychologic aspect. It is difficult for us to picture a mathematician hesitating to publish a discovery for fear of the outcry that its publication might produce—perhaps not many would be displeased to awaken an echo; yet this is believed by some to have been the attitude of Gauss. Though he was keenly interested and thought deeply on the subject of parallels, he published nothing; he feared, as he said, "the clamor of the Bœotians." When forced to write a letter on the subject, he begs his correspondent to keep silence as to the information imparted. In 1831 he writes in a letter: "In the last few weeks I have begun to put down a few of my Meditations [on parallels] which are already to some extent forty years old. These I had never put in writing, so that I have been compelled three or four times to go over the whole matter afresh in my head. Also I wished that it should not perish with me."

It is only when we call to mind the unrivalled place of honor held by Euclidean geometry among branches of human knowledge—a respect no doubt enhanced by the prominence given it in Kant's *Critique of Pure Reason*—that we realize the uncomfortable position of one who even appeared to attack its validity. Gauss's meditations were leading him through tedious and painstaking labors to the conclusion that Euclid's fifth postulate was not deducible from his other postulates. The minds of those not conversant with the intricacies of the problem might easily rush to the conclusion that Euclid's geometry was therefore untrue, and feel the whole structure of human learning crashing about their ears.

Between 1820 and 1830, the conclusion toward which Gauss tended was finally made sure by the invention of the hyperbolic non-Euclidean geometry by Lobachevsky and Johann Bolyai, working simultaneously and independently. The question—is Euclid's fifth postulate logically deducible from his other postulates?—is answered by showing that the denial of this postulate while all the others are retained leads to a geometry as consistent as Euclid's own. The method, we recall, was that used by Saccheri, whose intellectual conservatism alone prevented his reaching the same result. The famous postulate is only one of three mutually exclusive hypotheses which are logically on the same footing. Thus was Euclid "vindicated" in an unexpected manner. Knowingly or not, the wise Greek had stated the case correctly, and only his followers had been at fault in their efforts for improvement. To quote Heath [2]: "We cannot but admire the genius of the man who concluded that such an hypothesis, which he found necessary to the validity of his whole system, was really indemonstrable."

Thus in some sense the problem of the parallel postulate was laid to rest, but its spirit marches on. If the fifth postulate could without logical error be replaced by its contradictory, could the other postulates be similarly treated? What is the nature of a postulate or axiom? What requirements should a satisfactory system of axioms fulfill? Are we sure that accepted proofs will bear as keen scrutiny as that to which proofs of the postulate have been subjected? The facing of these questions has brought us to the modern critical study of the foundations of geometry. It has been realized that if geometry is to continue to enjoy its reputation for logical perfection, it should at least try to deserve it. The edge of criticism, sharpened on the parallel postulate, is turned against the whole structure. Out of this movement has grown the critical examination of the foundations of algebra, of projective geometry, of mechanics, of logic itself; and the end is not in sight.

One obvious result of this critical study is that geometrical axioms are not necessary truths, but merely presuppositions: they are the hypotheses on which the whole body of theorems rests. It is essential that a system of axioms should be consistent with each other, and desirable that they be non-redundant, and complete. No one has found Euclid's system inconsistent, and redundancy would be a crime against elegance rather than against logic. But on the score of completeness Euclid is far from giving satisfaction. He not infrequently states conclusions which could be arrived at only by looking at a figure, i.e., by space intuition; but we are all familiar with cases where space intuition misleads (for example, in the fallacious proof that all triangles are isosceles), and if we accept it as a guide how can we be sure that our intuitions will always agree? In constructing an equilateral triangle Euclid says, "From point C where the two circles meet, draw ····." Perhaps they do meet—but not on the basis of anything previously stated. In dropping a perpendicular from a point to a line he says a certain circle will meet the line twice. Why should it not cut thrice or not at all? In another proof he says that a certain line will lie within a certain angle. *I see* that it does, but I do not see it proved. We are told in the midst of a proof to bisect an angle of a triangle and produce the bisector to meet the opposite side. How do we know it will meet? Because it is not a parallel. And probably it is not parallel because it is inside the triangle. How do we know it is inside; or, being inside, that it must get outside? When have these terms been defined? You may answer: It is not necessary to define them because everyone with common sense knows inside from outside without being told. "Who is so dull as not to perceive ····?" says Simson, one of Euclid's apologists. This may be granted. But it must be pointed out that common sense knows that two straight lines cannot enclose space, yet this is given prominence as a axiom; or that a straight line is the shortest distance between two points, yet this is proved as a proposition. To state in words what distinguishes the inside from the outside of a polygon is not easy. The word *between* is likewise difficult of definition. But the modern geometer imbued with the critical spin feels it necessary to define such terms, and what is more, he finds a way of doing it.

Hilbert's *Grundlagen der Geometrie* [3], published in 1899, is a classic product of this movement. It presupposes no space concepts at all, but only such general logical terms as "corresponds to", "associated with", "determined by". Contrary to tradition, it does not begin by defining terms. The first sentence is: "I think of three systems of things which I call points, lines and planes." Note the unadorned simplicity of the concept *things*. The axioms serve as definitions. They state, in non-spatial terms, relations between these "things"; that is to say, the points, lines and planes are such things as have such and such relations. "That is all ye know on earth, and all ye need to know." Twenty-one axioms are found necessary, as against Euclid's meager five. The whole work could be read and comprehended by a being with no space intuitions whatever. We could substitute the names of colors or sounds for points, lines and planes, and get on equally well. The ideal of making a thing "so plain that a blind man could see it" is literally realized. And the age-old ideal of a body of proved propositions, close-knit together by unassailable logic, is immeasurably nearer realization.

Although to the best of my knowledge no one has yet had the hardihood to invite a child of fourteen to consider "three systems of things", the modern critical movement is not without bearing on problems of teaching. I wish, with proper humility, to put forward a few ideas on this subject.

If it is true that our traditional formal geometry, taken directly or indirectly from Euclid, is not the logically perfect thing we had imagined, and if its modern perfected descendant is so abstract that not even the most rationalistic of us would venture to force it on beginners, why not acknowledge these facts and bravely face anew the question of how we can best make the study of elementary geometry serve its proclaimed purpose of training the mind? I would suggest two lines along which progress might be made.

First, by sacrificing the ideal of non-redundancy in our underlying assumptions we could save time and stimulate interest by arriving more quickly at propositions whose truth is not immediately evident and which could be presented as subjects for investigation. Must we, because Euclid did it, prove that the base angles of an isosceles triangle are equal? A child that has cut the triangle out of paper and folded it over knows as much as any proof can teach him. If to treat the proposition in this way is repugnant to the teacher's logical conscience, let him privately label it "axiom" or "postulate", and proceed, even though this proposition could have been proved. *The place to begin producing arguments is the place where the truth of the proposition is even momentarily in doubt.* One statement which presents itself with a question mark and is found after investigation to be true or to be false is worth ten obviously true statements proved with all the paraphernalia of hypothesis, conclusion, step one and step two, with references. The only apparent reason for proving in the traditional way the theorems on the isosceles triangle and congruent triangles is in order to familiarize the student with the above-mentioned paraphernalia. This brings me to my second suggestion.

Formal geometry has been looked upon as a complete and perfect thing to which the learner can with profit play the sedulous ape. Yet I sometimes think that by emphasizing too early the traditional form of presentation of geometrical argument, and paying too little attention to the psychology of the learner, we may have corrupted some very good minds. "I wish to prove ... ," says the student; meaning, I wish to prove something stated and accepted as true in advance of argument. Should we not prefer to have our students say, "I wish to examine ... , to understand, to find out whether ... , to discover a relation between ... , to invent a means of doing ... "? What better slogan could prejudice desire than "I wish to prove"? The conscienceless way in which college debaters collect and enumerate arguments regardless of the issues involved is another aspect of the same evil. A student said not long ago, "The study of mathematics would be good fun if we did not have to learn proofs." It had never been brought home to her that mathematical reasoning is not a thing to be acquired, like a knowledge of Latin verbs, but a thing to be participated in like any other form of exercise. Another said, "I cannot apply my geometry because all we did in school was to learn the proofs and pass the examinations."

In the midst of a proof the student hesitates and says, "I am sure this is the next step, but I cannot recall the reason for it." The step and its reason would occur simultaneously to a mind that had faced the proposition as a problem and thought it through. I should like to see in every text an occasional page of exercises to prove or disprove. And if formal proofs must be printed in full, by all means let some of the proofs be wrong.

When the student has thus halted with one foot in the air in this progress from step to step down the printed page, on what does the ability to proceed depend? On the ability to quote something: to quote, usually, a single statement—compact, authoritative, triumphantly produced. Surely it is bad training that leads the mind to expect to find support for its surmises in a form so simple; and the temptation to substitute ability to quote in place of the labor of finding out the truth may be a real danger. What wonder if the mind so trained quotes Washington's Farewell Address or the Monroe Doctrine and feels that its work is done?

I do not mean that formal proof should never be given. It has its place as an exercise in literary composition; for it deals with the form in which thought is expressed. We should, however, take every possible precaution to see that the thought is first there to be expressed, lest the form be mistaken for the substance. Just how this is to be brought about I am not prepared to discuss, although I suspect that drawing and measurement in the early stages of study, problems of construction and investigation, and the total absence of complete proofs from the printed page, would help. I wish merely to state my belief that only in so far as we succeed in these aims shall we succeed in making geometry really train the mind.

> It can be done, said the butcher, I *think*;
> It must be done, I am sure.

One point further. Perhaps we are a little too modest about the importance of having our students retain something of the subject matter of the courses we teach. Evidently it is here that memory, based on understanding, should rightly be used. I sometimes think we might in some way collectively take out insurance against a student's arriving at the junior year in college in the belief that two triangles are similar whenever they have a side in common.

References

1. R. Bonola, *Non-Euclidean Geometry*, Open Court, 1912.

2. Euclid, *Elements* (ed. T. Heath), Cambridge Univ. Press, 1908.

3. David Hilbert, *Grundlagen der Geometrie*, Teubner, 1899.

The Rise and Fall of Projective Geometry

J. L. COOLIDGE

American Mathematical Monthly **41** (1934), 217–228

1 The early period

The subject of projective geometry occupied an important position on the mathematical stage during a large portion of the nineteenth century. In recent years it has moved considerably towards the wings. Why did it appear? Why was it prominent? Why is it now moving aside? These are pertinent questions which perhaps it is worth our while to consider.

First of all, what is projective geometry anyway? It is sometimes defined as that branch of geometrical science which deals with those properties of figures which are unaltered by radial projection from plane to plane or space to space, no matter what the number of dimensions involved. This definition is at once too large and too restrictive. The fact that the projective plane has the connectivity 1 or that, looked upon as an assemblage of points, it has the power of the continuum—this fact is invariant under projection, but is not usually looked upon as a projective property. On the other hand, the properties of a figure which are conserved by projection are invariant under the wider group of linear transformations. The proper use of the adjective *projective* would seem to be to describe those properties of figures which are invariant under the general linear transformation but not under all transformations of a wider group. The subject is vast. We can at present deal only with the most significant steps in its development.

At the outset we must make an important distinction. Projective properties were discovered and used long before any one grasped the idea of the projective group. The earliest and most fundamental projective invariant is the cross ratio of four collinear points. We do not know who first discovered this. By a very curious coincidence we have a statement of the corresponding theorem in spherical geometry which is earlier than any we have of the plane theorem. This occurs in Menelaus' *Spherica*. The modern texts of Menelaus are based on various Hebrew manuscripts which differ widely one from another. Thus, Halley's reconstruction does not give this particular theorem at all. The most recent and complete study of Menelaus is by Björnbo, from which it seems safe to assume that Menelaus was cognizant of the theorem in spherical geometry. But who discovered the simpler theorem in the plane? We find it in the second volume of Pappus who wrote long after Menelaus' time, but this portion of Pappus deals with Euclid's lost book of porisms, which suggests that the invariance of the cross ratio was known to Euclid. Chasles has elaborated an ingenious reconstruction of Euclid's work carrying to their logical conclusions the thirty-eight porisms given by Pappus. The reconstruction of this book has been in fact quite a favorite sport with geometers, and there has been not a little speculation as to what a porism is anyway. I am personally somewhat skeptical as to the ultimate importance of such conjectures. One certain thing is that Pappus, and probably Euclid who wrote six hundred years earlier, knew that the cross ratio is invariant.

After this first advance, the subject of projective geometry rested for the substantial period of twelve centuries. The first writer to return thereto was Johannes Werner who wrote in 1522. His rare work has excited a good deal of curiosity among people who have never seen it. He starts out by determining a parabola as a section of a cone as the Greeks did, but keeps his cone by him, instead of discarding it. He treats the hyperbola in the same way. He reaches a few interesting properties of these two curves in this fashion, but really finds nothing which is not much more simply found in the work of Apollonius. There is no sign that he had heard of a cross ratio.

For that we must wait another century and turn to the work of a much better known writer, Girard Desargues. When I say that Desargues was better known, I mean he is better known today. His work was little esteemed by his contemporaries and immediate followers, with the very important exception of Pascal. This is not surprising. His style and nomenclature are weird beyond imagination. Fortunately, his editor Poudra has given the explanation of seventy-one strange, frequently botanical, terms which this writer affects. Who today could guess that a *palm* meant a *straight line*? When there are points thereon (a straight line with no points would be more novel), it becomes a *trunk*. A *tree* is a line with three pairs of points of an involution. The *stump* is the point which is the mate of the infinite point. The stump is *engaged* when the involution is elliptic.

The figure which Desargues finds most significant is three pairs of an involution, which he gives by equations of the form

$$OA \times OA' = OB \times OB' = OC \times OC'.$$

The harmonic set is called an involution of four points. Menelaus' theorem, which is attributed to Ptolemy, comes next and thereby it is proved that an involution is projected into an involution. Next come cones and cylinders, the latter being called cones with an infinite vertex.

The polar theory then appears, derived presumably from the special case of the circle. Desargues treats a diameter as the polar of an infinite point. He carries the whole polar theory much further than did Apollonius, giving the quadrangle construction, the properties of a self-conjugate triangle, and the polar theory with regard to a surface. Moreover, he derives two very famous theorems which bear his name. The first says that the intersections of a line with the pairs of the opposite sides of a complete quadrangle and with any conic circumscribed to that quadrangle, are pairs of an involution. The essentially new element here is including the intersections with the conic. The quadrangle property was known to Pappus. The second theorem is that which tells us that if corresponding pairs of vertices of two triangles be collinear with the fixed point, the intersections of corresponding pairs of sides lie on a line.

The most famous pupil of Desargues was Blaise Pascal who also used the method of projection. There remains very little of what he actually wrote, and there is a tendency at present to reduce the previous somewhat exaggerated estimates of the importance of his work. There is, however, no doubt that he discovered the *hexagramma mysticum*. There seems no doubt also that Pascal and Desargues were both aware of the invariance of the cross ratio.

Another writer of this period who had a positive passion for conic sections was La Hire. His first venture was in 1673 when he wrote a little book on planoconics. He defined each conic separately by sums or differences of distances and wrote a text which would not be a bad introduction to put into the hands of a student today, except for the omission of the focus-directrix property which was known to Pappus. His great work on conics, however, was a splendid folio which saw the light in 1685. It contains over three hundred theorems of a projective sort, as well as an appendix showing that all three hundred and sixty-four theorems of Apollonius can be proved by La Hire's method of projection. The book is an attempt to collate all known material connected with conic sections and very nearly succeeds. There is ample evidence that he knew some of Desargues' work, though, curiously enough, he makes no mention of the great theorems which bear that writer's name. I have the impression that he added little to our knowledge of projective geometry as such, and that his main object was to prove the superiority of his method over the ancient method of Apollonius or the modern ones of Descartes.

The subject of projective geometry dragged along for about a century after La Hire's time. A few theorems were discovered by Newton, MacLaurin, and Braickenridge. The next writer deserving attention is Carnot. He, like La Hire, was possessed of the desire of overcoming the apparent increase in generality given by the algebraic methods of Descartes. His basic idea, which was much esteemed by subsequent writers—I suppose for its obscurity—seems to have been this. We establish geometrical relations in the simplest case where all quantities involved are positive, but no further restrictions imposed. It is then assumed that these relations are identities which are unaltered when the figure is replaced by what is called a *correlative*, that is to say, the figure obtained from the first by continuous deformation. If one is willing to accept such an axiom, of course it is very convenient. Carnot also wrote about the theory of transversals, giving generalizations of theorems of Menelaus and Ceva.

There is just one other writer who deserves to be in this group—Brianchon. He is especially skillful in handling the theory of polar reciprocation, deducing from Pascal's theorem the hexagon theorem which bears his name.

2 The great period

A characteristic feature of the work of Brianchon and his predecessors was that they saw isolated theorems and ingenious methods for solving particular problems. The only general conception was that of beating Descartes. They spoke of the geometry of the ruler and sought the solution of geometrical questions which had but one answer, without going deeply into the question of what plane figures were unaltered by projection. The great advance marked by putting the question in this latter way was due to Jean-Victor Poncelet whose classic *Treaté de Propriétés Projectives des Figures* was commenced in 1813 under the depressing conditions of a Russian military prison at Saratow.

The fundamental task, according to Poncelet, is to study the graphic properties of figures which he defines as those that do not involve distances or angles. These latter he regards as contingent whereas the graphic ones are unaltered by a central projection. In looking for such invariants, he develops the theory of harmonic separation at great length, though, curiously enough, the invariance of the general cross ratio escapes him.

In his second chapter Poncelet makes a bold attack on the problem of imaginary points in pure geometry with a courage and thoroughness far ahead of anything shown by his predecessors. The need for this first appears in connection with conic sections. If we connect the center of a central conic with the pole of a given line, we have the diameter conjugate to the line, which passes midway between the intersections of conic and line in case these intersections exist.

Suppose that we have an ellipse with a set of parallel chords. The conjugate diameter will bisect these chords and meet the ellipse in two points A and B. There is just one hyperbola having double contact with the ellipses at A and B whose asymptotes are conjugate diameters of the ellipse. The diameter in question bisects not only parallel chords of the ellipse but parallel chords of the hyperbola. Each curve is called a *supplementary* of the other, the chords of one being called *ideal chords* of the other. In this way Poncelet avoids using imaginary points which he defines lamely enough as something which, originally real, becomes impossible or inconstructible when we pass from the first figure to a correlative in the sense of Carnot. He gives a long discussion of supplementaries and ideal chords and closes with a discussion of the line at infinity. He makes quite casually the fundamental remark that two coplanar circles should not be looked upon as completely independent figures, but as having two immovable, infinite points in common. This is one of the basic principles of modern geometry here announced for the first time.

Later Poncelet allows imaginary projections throwing real chords into ideal ones. This is essentially Carnot's correlative idea. Poncelet calls it the principle of continuity. His statement is essentially this. If a theorem be true for an infinite number of real cases in a figure about which we make no particular restriction as to reality, it is true in the complex case also. This amounts analytically to saying that if we have an algebraic identity

$$f(a_1, a_2, \ldots, a_n) \equiv 0$$

which holds for all real values of the variables, it holds equally when the variables are complex, but whereas such an algebraic statement is precise, the geometric one is dangerously vague and may lead to error. Another weakness of Poncelet's theory of ideal chords is that it is not easily carried over to curves of higher order, a difficulty we shall return to later. But when all is said and done, the supplementaries and ideal chords give us something really tangible when dealing with imaginary elements, a great advance over the work of all previous and many subsequent writers.

In the second section Poncelet gives the general projective theory of conics and straight lines, Desargues' involution, harmonic separation, and so on. He notes the analogy between projection from plane to plane and the transformation of a plane into itself by means of a central similitude. This he generalizes into a transformation by homology, where corresponding points are collinear with a fixed point and corresponding lines are concurrent on a fixed line. This again he generalizes to three dimensions as he generalizes the idea of harmonic separation.

The last important topic taken up by Poncelet is polar reciprocation. Special applications are made to the study of algebraic curves. He gives in an appendix a long account of a rather sordid controversy between himself and Gergonne on this subject [see the following article by Pedoe in this volume]. Poncelet used his methods as a very convenient tool and discovered useful theorems thereby. Gergonne saw that the conic or quadric of reciprocation was of altogether subordinate importance. What was involved was a very deep geometric principle, but he never bothered to deduce many particular results.

I have paid special attention to Poncelet because he saw far deeper into the essential questions than any of his predecessors. He placed the subject in its right aspect, and one would expect that numerous followers would follow along his path to splendid results. Such was the case, but, curiously enough, only one of the followers of major importance was a countryman of his, Michel Chasles. This enterprising and ingenious writer enriched geometrical science by several fruitful ideas. On the other hand, he was decidedly uncritical as a historian of mathematics both of his own work and of that of others. In his much vaunted study of the history of geometrical methods, he makes the interesting admission that he has neglected German writers because he did not know their language.

The greatest contribution that Chasles made to projective geometry, and it was certainly very great, was in developing the theory of cross ratio of points, lines, or planes. He discovered the theorem which subsequently played a vital role in the work of Steiner, that four points of a conic determine at any fifth point four lines with fixed cross ratio. Another fundamental idea was *homography* or the general linear transformation of the plane. Central projection and the homology of Poncelet are special cases of this.

If Chasles neglected the work of the German geometers, the reverse was not the case, for, contemporary with him, there sprang up across the Rhine a school of German geometers who closely watched the work of their French neighbors. The first of these in the synthetic field was Jacob Steiner whose *Systematische Entwickelung der Abhängigkeit geometrischer Gestalten von einander* appeared in Berlin in 1832. This is based on three fundamental principles.

(a) Points, lines, and planes are the essential data of geometry. Other figures must be constructed from these in definite fashion.

(b) The principle of duality, which appears at the outset and is carried through consistently, much of the work being in double column.

(c) A fundamental concept is that of projective fundamental forms, that is, ranges of points, pencils of lines, and pencils of planes.

Right here is a slip which is hard to account for. I cannot see that Steiner ever really gives a good definition of what he means by *projective*. A pencil of lines and the range they cut on a transversal are called projective, and they are said to remain so when the one or the other is moved about without disturbing distances or angles. This is clumsy and unsatisfactory. Essentially, what he means by projectivity is a one-to-one relation where corresponding cross ratios and senses of description are the same. He shows that such a relation between two fundamental one-dimensional forms is determined when the fate of three elements is known. The proof is unsatisfactory as there is no proper way to handle questions of limits and continuity. There are no less than eighty pages, mostly in double column, devoted to questions of this sort.

The next form of figure to receive his attention is the *cone*, that is to say, the cone with circular sections. Construction of a conic by means of projective pencils comes in naturally here. There is one logical slip; at least I cannot see any proof that every conic so constructed can be cut from a circular cone. Except for this defect, we have here the best possible approach to the study of the conics from the point of view of bringing out their projectively invariant properties. No wonder that all the simplest projective properties appear immediately in excellent form. The same methods are applied to the study of the regulus, that is to say, rulings of a surface of the second order. It is rather curious that he does not at once proceed to a study of the cubic space curve which is reached by the intersections of corresponding planes of three pencils.

The most characteristic feature of Steiner's work is given by the adjective in its title, *systematische*. He has a consistent and uniform method for treating a variety of figures. He handles it beautifully. He is usually considered the greatest of the German school of projective geometers. It is my own feeling that in originality and power he falls far below his distinguished successor, Johann Karl Christian von Staudt. This deep thinker perceived two essential weaknesses in the synthetic geometry of his predecessors.

(a) The basis of projective relations was the cross ratio. This is projectively invariant but, as previously given, was based on distances and angles which are not in themselves unalterable.

(b) What are imaginary points anyway? What can be said about them, except that they are imaginary?

Let us give an outline of his method for developing geometry so as to meet these difficulties. In his first book *Geometrie der Lage*, published in 1847, he starts with points, lines, and planes as fundamental objects. Desargues' two-triangle theorem appears early, leading to the configurations of a complete quadrilateral and quadrangle. It never occurred to him that we really need a proof that the diagonals of a complete quadrilateral are not concurrent. Next we have harmonic separation and the fundamental definition that two one-dimensional forms are projective when their members are in one-to-one correspondence and a harmonic set corresponds to a harmonic set. Here at last we have projectivity defined with no relation to distance. The fact that von Staudt was not able to carry the thing through rigorously does not detract from the originality of the idea. The failure was right here. Sooner or later we have to prove that a projective transformation of a one-dimensional form into itself that leaves three elements in place, leaves all elements in place. Von Staudt could not do this as he had no axiom of continuity. Subsequent writers have filled the void and shown how cross ratios can be defined by successive harmonic constructions, exactly in the way that distances are defined by successively laying down a fixed length or aliquot part thereof.

Von Staudt defines a *collineation* of the plane as a transformation of point to point and line to line and shows that this is projective in that it carries harmonic elements into harmonic elements. He goes further, however, and defines a *correlation*, let us say in the plane, as a correspondence of point to line and points on a line to lines on a point. Poncelet reached such a transformation by polar reciprocation with respect to a conic. Von Staudt exactly reversed the process. Suppose that we have a correlation which is involutory. It is easy to show that such things exist. Then if there be any point lying on the corresponding line, it is easy to show that there is a whole curve of points each of which lies on its line and this is defined as a conic. This definition is clearly self-dual and leads at once to the classic theorems of Pascal, Brianchon, Desargues, and Steiner. The method has the advantage that it is immediately applicable in three or more dimensions. He did not make the mistake of supposing that if we have an involutory correlation in three dimensions, the points lying in the corresponding planes necessarily generate a quadric surface. He was familiar enough with the null-system usually ascribed to Möbius.

Von Staudt was acutely conscious that the treatment of imaginary elements in pure geometry was extremely unsatisfactory. Poncelet's system of ideal chords and supplementaries was the only contribution to the subject that had any real substance. He set to work to remedy this defect in truly heroic fashion. Suppose that on a straight line we have an elliptic involution. A point and its mate in the involution trace the straight line *in the same sense*. "Very well," says von Staudt, "we will define an elliptic point involution and a sense of description as an imaginary point. The same involution with the contrary sense shall be defined as the *conjugate* imaginary point. We can define an imaginary plane as an elliptic involution in a pencil of planes with a sense of description. Reverse the sense and we get the conjugate plane. An imaginary line is defined as the system of points in two planes which are not both real."

These definitions of von Staudt are certainly revolutionary. It was a bold step to define as an imaginary point something that is made up of an infinite number of real points. Von Staudt could not foresee the analogy to Dedekind's definition of an irrational number as a split in the real number system. What he did do was to show by the most careful reasoning that the new elements thus introduced obeyed just the laws of the old ones. Two of his points determine one of his lines which lies completely in any one of his planes through the two points, etc. It is true that his work has the weakness that sense of description is an intuitive concept which he is not able to define and about which he has no definite mathematical axioms. This difficulty can be overcome by methods subsequently invented. His treatment, which is found in his subsequent *Beiträge zur Geometrie der Lage*, is a marvellous piece of careful geometry.

But von Staudt did not rest even here. Chasles had based his projective geometry on the idea of the cross ratio which is defined in terms of distances. Von Staudt perceived that this was a blemish which he undertook to correct. Four collinear points, four lines or planes of a pencil are called a *throw*. The value of this is defined as unchanged by a double interchange in the pairs of elements, and two throws connected by a train of projective transformations are defined as an equivalent. He proceeds to develop an algebra of throws. If we wish to add two throws, we transform them projectively into two others with three common elements and find the sum by two successive harmonic constructions. In the same way the product of two throws is to be found by elimination between two with three common elements. There follows a very careful demonstration that in this algebra the usual commutative, associative, and distributive laws are obeyed. If a harmonic set be defined as a throw having the value of -1, we can show that any real throw of points is equal to the cross ratio in the usual sense.

But von Staudt does not stop with real throws. He considers complex ones as well. He fixes his attention on an imaginary line of the second sort which has no real points. Let us take four complex points of $ABCD$ on such a line and connect them with their conjugates $A'B'C'D'$ by four real skew lines. Three of these will determine a ruled quadric surface including the given complex line and its conjugate. The necessary and sufficient condition that the throw of the original four points should be real is that the four lines should be generators of one quadric. A set of points on such an imaginary line with the property that the cross ratio of any four is real, was defined as a *chain*. Such a set of points can be projectively transformed into the real points of a real line. This theory was further developed by Lüroth and later by our own Veblen and Young.

3 The gradual decline

The work of von Staudt marks the close of the great period in the history of synthetic geometry. A definite geometrical method had been elaborated. The work of subsequent geometers consisted principally in applying to it such problems as lent themselves readily to treatment. For fifty years the devotees of this branch of science were many and enthusiastic. Progress consisted, however, chiefly in extending the methods already discovered to new problems rather than in finding new methods.

An exception to this rule must be made to cover the case of geometrical transformations. Poncelet had a clear grasp of the idea of central projection and the cognate idea of homology. Chasles constructed the general projective transformation. Steiner developed something essentially different in the form of a quadratic transformation by means of a skew projection. Here a point-to-point transformation is established between two planes by means of projecting lines that do not go through a fixed point but do intersect two skew lines. A straight line in one of the planes will correspond to a conic in the other that passes through three fixed points. Another type of quadratic transformation is the projective generalization of circular inversion, where corresponding points are collinear with a fixed point and conjugate with regard to a fixed conic. A better form was developed by Seydewitz, where a point in the plane is carried to the single point conjugate to it with regard to a pencil of conics. Many other forms of one-to-one geometric transformations were discovered. A complete account will be found in the four volumes of Sturm. There are, however, many pitfalls in the path of one who would handle geometrical transformations by means of pure geometry alone, especially when he becomes involved in the complicated singularities of higher curves and surfaces. It is safer to treat the whole question algebraically as a part of the great subject of Cremona transformations.

It was inevitable that students of synthetic geometry should be forced sooner or later to face the question of applying their favorite methods to the study of general algebraic curves and algebraic surfaces, and that forces upon us the question of whether synthetic methods really lend themselves well to the study of such curves or surfaces. And at this point I am forced to confess that it seems to me the answer must be frankly negative, though able geometers have striven to prove the contrary.

One of these was Cremona, whose long memoir *Introduzione alla teoria geometrica della curve piane* was written in 1862. When we examine this work in detail we are forced to acknowledge its defects. The author speaks freely of real and complex points without ever saying just what he means by the latter. It is true that we can, if we wish, lay down a system of axioms for the projective geometry of the complex plane, but Cremona never bothered to do anything of the sort. In his time real points were supposed to be given us by nature and no man had a right to speak of complex points, which nature certainly did not provide, without saying what he meant thereby. Cremona's various proofs involve a good deal of credulity on the part of the reader. For instance, he has assumed without a shadow of proof that the number of intersections of two plane curves is a function of their orders alone and independent of whether they be reducible or irreducible. There is a similar optimism about the number of parameters on which a curve of order n depends. The whole work is shot through with doubtful assumptions. What the writer really does is to use a few essential theorems about the general equation of degree n and then ungratefully discard it.

The two great difficulties have now been mentioned—a sound theory of complex points and something definite to replace the theory of the general equation. Von Staudt, as mentioned above, conquered the first difficulty once and for all in most brilliant fashion, but his method was so cumbersome that in practice nobody could be expected to use it. The only way to avoid the second obstacle was by means of mathematical induction. The most ambitious attempt to use this process was made in 1886 by Kötter in his *Theorie der algebraischen ebenen Curven* (Berlin 1887). His favorite implement is the general involution.

We may lead up to this as follows. Suppose that we have a projective transformation of one of our fundamental forms, let us say the straight line, into itself, whereby the points $ABC\ldots$ correspond to $B'A'C'\ldots$. There will be a pair of self-corresponding points, either a real pair or a conjugate imaginary pair in the von Staudt sense. Now let all of these points remain fixed, except C' which traces the whole line. These pairs of self-corresponding points will trace that involution which is determined by the pairs AA' and BB' and the pairs of this are projectively related to the range traced by C'. Suppose now that we know what we mean by a projective relation between a range of points and the groups of an involution of order $n-1$. Suppose that we have such an involution with the group

$$A_1 A_2, \ldots, A_{n-1},$$
$$B_1 B_2, \ldots, B_{n-1},$$
$$C_1 C_2, \ldots, C_{n-1}$$

projectively related to the points $B'A'C'$, that everything remain fixed except C' which traces the line. The groups of n, real or imaginary, self-corresponding points will trace the involution of order n determined by the two groups

$$A' A_1 A_2, \ldots, A_{n-1}, \quad B' B_1 B_2, \ldots, B_{n-1}, \ldots.$$

In this way we get the general involution. Its establishment involves some delicate considerations of continuity and analysis situs as well as the von Staudt theory. When all is ready, Kötter establishes a projective relation between a pencil of curves of order r and one of order $n-r$, and thus generates a curve of order n.

This is the general idea of Kötter's ambitious attempt. It is a remarkable piece of geometry, also a very difficult one, so that I personally marvel at the thought that I once read it through. I do not wonder that the Berlin Academy awarded it a prize in 1886 or that few have had the temerity to read the thing since. It seems to me that there is really a better approach which consists in generalizing von Staudt's conception of a polar. Suppose that we know a good deal about a thing which we call a curve of order $n-1$, as well as a pencil and a two-parameter net of such curves, and that we know what we mean by the first polar of a point with regard to such a curve. It may be defined, with Guccia, as the locus of points of contact of tangents from a point P to curves of order $n-1$, linearly dependent on a given curve, and $n-1$ arbitrary lines through P. Suppose further that we have a two-parameter set of curves of order $n-1$ which correspond to the points of the plane in such a way that the first polar of P with regard to the curve corresponding to Q is identical with the first polar of Q with regard to the curve corresponding to P. Then, the locus of points which lie on the curves which correspond to them is a curve of order n, and the curve of order $n-1$ corresponding to a given point is its first polar with regard to the new curve. I think that this is a better approach than Kötter's. I once carried it out a short distance in an article in the *Circolo Matematico di Palermo, Rendiconti* (vol. XI, 1915) in ignorance of the fact that the same idea had been previously developed by Thieme (Die Definition der geometrischen Gebilde, *Zeitschrift für Mathematik und Physik*, vol. 24, 1879). And yet I now doubt whether it would be worth while to follow such a lead, for the algebraic approach is so much easier and more satisfactory. It is for this reason that the latest advances in projective geometry, before it finally flickered out, lay in a closer examination of the fundamental postulates.

We have seen that von Staudt's great work had certain intuitive elements such as the sense of description on a line, continuity, separation, etc. It never occurred to either him or any contemporary that the diagonals of a complete quadrilateral might be concurrent. All of these points were cleaned up by a number of writers who dealt with the postulates of projective geometry. At the risk of creating offense by citing one at the expense of others, I would mention the two-volume work of Veblen and Young. Here have we not only a rigorous postulational basis for projective geometry, but the obvious step is taken of extending it to n dimensions. There is a discussion of nets of rationality which are essentially the same thing as generalizations of the von Staudt chain, of finding projective geometries and other strange systems obtained by varying the axioms. It is not too much to say that this is the last great work dealing with this field.

It is hard to escape the sad conclusion that the field of synthetic projective geometry is pretty much worked out. This gloomy foreboding can only be put forth with all possible caution. There have been so many times when the geometrical field, even the most elementary field, has appeared to be completely exhausted and then surprisingly new and attractive results have appeared. Elementary geometry, involving nothing more complicated than a little trigonometry, made far more progress in the nineteenth century than in the sixteen centuries preceding, but the analogy is not perfect. Most of the theorems about the straight line, circle, and sphere which we can demonstrate at all, find their easiest proofs by the most elementary means. Such is not the case with projective geometry. An algebraic handling is almost always the easiest and generally the most adaptable.

Until and unless some totally new principle is discovered, the subject of synthetic projective geometry is no longer a very fruitful field for original research. On the other hand, it would be a disaster to the whole geometric fabric if a time ever came when synthetic methods were completely abandoned. Not only have they a permanent beauty which no one who has ever studied them can forget, but they afford an invaluable insight into the inner significance of geometrical science and an invaluable training for any geometer. If Plato wrote over the gate of the Academy, "Let none ignorant of geometry presume to enter here," surely we may write today in the same spirit, "Let none ignorant of the fundamentals of synthetic projective geometry presume to the title of geometer."

Notes on the History of Geometrical Ideas

DAN PEDOE

Mathematics Magazine **48** (1975), 215–217, 274–277

1 Homogeneous coordinates

It is agreed, even by those who disparage them (see [6], p. 712) that barycentric coordinates, first introduced by August Ferdinand Möbius [4] in 1827, were the first homogeneous coordinates systematically used in geometry. The Möbius idea, in plane geometry for example, is to attach masses p, q and r, respectively, to three non-collinear points A, B and C in the plane under consideration, and then to consider the centroid $P = pA + qB + rC$ of the three masses. The point P necessarily lies in the plane, and varies as the ratios $p : q : r$ vary. As Möbius points out:

> And conversely, given any point P in the plane, the ratios $p : q : r$ are always and uniquely determinable.

It will be noted that Möbius was using position vectors for his points in 1827, and reading of the text shows that he developed all the techniques of homogeneous coordinates known nowadays, changing the simplex of reference, if necessary, and so on. For a more accessible account, see Section 4.2 of [5]. Nobody has suggested that there is a better system of coordinates for projective geometry.

But new ideas are not always easily accepted. All the same, it is strange nowadays to read some of Cauchy's criticisms (XI of [4]). He says:

> only by deeper study can one decide whether the advantages of this method outweigh its difficulties ... One should be quite sure that he is making a considerable advance in science before introducing so many new terms and requiring readers to follow studies which confront them with so much strangeness.

It should be said at this point that the Möbius work is beautifully lucid and unpedantic. As Möbius' editor, Richard Baltzer, points out, Cauchy's next remark shows that Cauchy, like most reviewers, had not read Möbius very carefully. Cauchy says:

> One must suppose that the author of the barycentric calculus was unaware of the general theory of reciprocity which Gergonne has established between the properties of systems of points and systems of lines.

In fact Möbius, in Chapters IV and V of the third section of his book, gives a very clear statement, with applications, of the *principle of reciprocity* and the perhaps more general *principle of duality*, and these discoveries were made independently of Poncelet and Gergonne.

Gauss, writing to Schumacher rather later, in 1843, also confessed that he found the new ideas of Möbius difficult, and commits himself to a rather heavy philosophical statement about new ideas in general. This can also be read in the introduction to the Möbius work (XII, [4]). But the editor Baltzer, although an advocate for Möbius, makes a statement which I have been unable to verify. He says that although the Möbius coordinates are of special significance historically, they were soon overtaken by homogeneous coordinates and the introduction of homogeneous equations (IX, [4]). While the Möbius work was being printed, says Baltzer, Plücker, using ideas of Gergonne, with some points of contact with Lamé and Bobillier, was writing his *Entwicklungen* [8] in which the homogeneous coordinates of a point with respect to three lines and four planes occur more and more.

It is true that Möbius never writes down an equation for his lines or his conics. Everything is treated *parametrically*, as we say nowadays. I have read Plücker carefully, in an attempt to confirm Baltzer's assertion, and I find that although Plücker conceives the idea of the homogeneous coordinates $[u, v, w]$ of a line $uX + vY + w = 0$ at a very early stage in his two-volume work, there is a very definite psychological block with regard to the homogeneous coordinates of a point. These are mentioned for the first time in Section 416 of the second volume, where he says of the equation $Ux + Vy + Wz = 0$:

> This new form of the equation of a straight line, which is homogeneous in the three variables y, x, and z, seems to me to be preferable in regard to elegance and symmetry in all those situations where no elimination is involved.

It is significant that Plücker rules out the use of homogeneous coordinates in elimination, and in fact he devotes nearly 200 pages to the conic considered as a line-locus in his second volume, and only about 140 pages to the conic considered as a point-locus in his first volume, and there is no systematic use of homogeneous point-coordinates in these two volumes.

What Baltzer may refer to is Plücker's very ingenious use of what we call *abridged notation* nowadays. The simple, but seminal, idea behind this is that if $U = 0$ and $V = 0$ are the equations of two curves, then $U + kV = 0$, for all values of k, is a curve which passes through the intersections of the two given curves. When the curves are lines, Plücker uses, in essence, the trilinear coordinates of a point in proving his many theorems about triangles. Trilinear coordinates are the distances of a point from the sides of a given triangle in two dimensions, and the sides of a given tetrahedron in three dimensions, and they are a form of homogeneous coordinates.

But there is no doubt that Plücker had an enormous influence on the development of analytical geometry, and there is little that one can add to his extensive treatment of circles, for example. He even uses inversion. Möbius avoids circles, and as we know, they are not too easy to deal with in homogeneous coordinates. Laguerre had not yet arrived on the scene, and the analytic treatment of the circular points at infinity is very helpful in this connection. But Möbius, besides solving many delightful problems in a masterly way, is the only author I have ever encountered who considers the different types of conic which can be drawn through five given points in a plane. His approach is a mixture of notions of convexity and the use of theorems already derived by his methods, and he says:

> Given five points chosen arbitrarily in a plane, the chance that the unique conic which can be drawn through these points is a hyperbola rather than an ellipse is $\sqrt{\infty} : 1$.

This is, of course, a pre-Cantorian statement, but it embodies a definite theorem. Does it come under the heading of *elementary* or *advanced* problems? Treated algebraically, the theorem gives information about the range of a certain polynomial in 10 variables.

2 The principle of duality

In projective geometry we have a remarkable method which produces another, usually distinct theorem, called a *dual* theorem, from any given theorem, by a simple transliteration of the terms involved in the statement of the given theorem. The "dictionary" for the transliteration depends on the space we are working in. We shall restrict ourselves in this note to the plane, in which case, to obtain a dual theorem, we interchange *point* and *line*, *lying on* and *passing through*, *collinear* and *concurrent*, and *intersection* and *join*. The Desargues theorem on perspective triangles dualizes to the converse theorem, which can be proved from the direct theorem, so that we say the Desargues theorem is a *self-dual* theorem, and the Pappus theorem dualizes to a theorem which looks different from the given theorem, but can also be proved by the use of the Pappus theorem, so that we say the Pappus theorem is also self-dual. These facts are of fundamental importance in the foundations of geometry, and enable us to assert that the so-called *principle of duality* is, in fact, a *theorem of duality*. Karl Menger ([3], pp. 201, 211) writes both fundamental theorems in self-dual forms, so that the transliteration given above does not essentially change the statement of the theorem.

The richness of the applications of duality is enormously increased by the fact that the dual of a conic, considered as a set of points, is a conic considered as a set of tangents (see Section 78.1 in [5]). Thus the Pascal and Brianchon theorems are dual theorems for a conic.

The statements in the first paragraph above make no mention of coordinates. If we use homogeneous coordinates (x, y, z) for points in a projective plane, the equation of a line is $uX + vY + wZ = 0$, and we can introduce the homogeneous coordinates $[u, v, w]$

of a line. The condition that the point (x, y, z) should be incident with the line $[u, v, w]$ is

$$ux + vy + wz = 0,$$

and the symmetry of this relation between point-coordinates (x, y, z) and line- (or *tangential-*) coordinates $[u, v, w]$ is the basis for the theorem of duality, since in any relation involving points, lines and incidences we can replace (x, y, z) by $[x, y, z]$, and $[u, v, w]$ by (u, v, w), and the incidence relations are undisturbed.

We note that the line $[x, y, z]$ is the polar line of the point (x, y, z) with respect to the conic

$$X^2 + Y^2 + Z^2 = 0, \qquad (1)$$

and so the theorem of duality can be interpreted as a *theorem of reciprocal polars*. We also note that any polarity of a plane involves a conic as the locus of points which lie on their corresponding polars. For the polar p of a point P is given by the equation $p = \mathbf{A}P$, where $\mathbf{A}$ is a symmetric matrix, and P lies on p if and only if $P^T \mathbf{A} P = 0$. The conic, as in (1), may not contain real points but this in no way invalidates its use as a *deus ex machina*. It is interesting to find that George Salmon ([10], p. 126) says:

> The principle of duality may be established independently of the method of reciprocal polars by showing ... that all the equations we employ admit of a two-fold interpretation.

Salmon then introduces tangential coordinates for planes (he is working in three dimensions) as we introduced them for lines above. But this is making unconscious use of the god in the machine, the quadric surface

$$X^2 + Y^2 + Z^2 + T^2 = 0. \qquad (2)$$

Veblen and Young ([11], p. 29) say:

> The principle of duality was first stated explicitly by Gergonne (1826), but was led up to by the writings of Poncelet and others during the first quarter of the 19th century.

Controversy as to priority clouded a large part of Poncelet's life. In a supplement to the second edition (1866) of the second volume of his great *Traité des Propriétés Projectives des Figures* [9], first published in 1822, Poncelet devotes 132 large pages in small print to diatribes against the many geometers who seemed not to be on his side in this controversy. He attacks Gergonne, of course, but also Plücker and Möbius, each of whom claimed independent discovery of the principle of reciprocal polars; but Poncelet reserves very special blasts for Chasles, remarking acidly that Cremona had called Chasles *the Archimedes*, and that de Jonquières had named him *the La Fontaine* of modern geometry! Poncelet also has sideswipes at Cayley, remarks with surprise that Sylvester writes good French, and even brings in Salmon, referring to the professor of divinity as *un journaliste philosophe*, with a reference to a footnote on p. 241 of the 1862 edition of Salmon's *Analytical Geometry of Three Dimensions*, which I have not been able to locate.

Amidst all this smoke there is little fire. However, Poncelet does accuse Gergonne (II of [9], p. 375) of stating that the number of tangents or tangent planes common to two curves or three surfaces is equal to the product of the degrees of their equations, which is certainly incorrect. In fact, Poncelet asserts that the principle of duality, as enunciated and used by Gergonne and his supporters, cannot deal with metrical theorems either. It should be stated at this point that the method of reciprocal polars, called *point reciprocation*, when applied to circles, using a circle as fundamental conic, produces theorems about conics and their foci or, conversely, produces theorems about circles from given theorems about conics and their foci ([2]; see also [7]). The principle of duality can hardly transform theorems in this way.

Positive statements about the Poncelet-Gergonne rivalry are rare, but before we discuss one made by Coxeter, an acid comment made by Plücker is worth noting (F. N. VI, vol. II [8]). Plücker is commenting on the alleged superiority of the Gergonne principle of duality:

> Above all, every thought which does not immediately arise from the nature of the case, but is taken over from a metaphysical abstraction, easily exercises a dictatorial rule over us, which always carries its own punishment, imposing bonds on unrestricted modes of thought.

Coxeter, in an exercise ([1], p. 75), asserts that Gergonne must be regarded as the victor in his historic contest with Poncelet. Coxeter first shows that the Desargues configuration is self-polar (p. 74), points in the configuration mapping onto lines in the configuration in a certain polarity, and then considers the Pappus configuration. It turns out that this is self-polar if and only if three lines in the configuration are concurrent, which would involve a specialization of the Pappus configuration. Coxeter then says:

Since the general Pappus configuration is self-dual without being self-polar, the old controversy between Poncelet and Gergonne is settled in the latter's favour.

I feel that this is a judgement against which history must appeal, that the criteria used are not valid, and seem to depend, in the final analysis, on a semantic misunderstanding. If we call the principle of reciprocal polars (P), and the principle of duality (G), then application of (P) to any theorem in the projective plane, using the fundamental conic (1), the ground field being that of the complex numbers, produces a theorem identical with that obtained by applying (G), and no more specialized, as far as the figure goes, than that obtained by applying (G). For any specialization of the dual figure, obtained by applying (P), which involves the lines p_i and the points P_i, implies a specialization of the initial figure involving the points p_i and the lines P_i.

On the other hand, the application of (P) produces theorems which cannot be produced by (G), as we have already remarked. It is possible that (G) may operate outside the regions of projective geometry over the complex numbers, as a general property of the space (*une propriété générale de l'étendue*), to quote the Gergonne phrase which disturbed Plücker, where (P) may have no validity, there being no conics, for instance, and further discussions on this ancient but fascinating controversy are surely to be welcomed.

References

1. H. S. M. Coxeter, *The Real Projective Plane*, Cambridge Univ. Press, 1955.
2. C. V. Durell, *Projective Geometry*, Macmillan, 1945.
3. K. Menger (and L. M. Blumenthal), *Studies in Geometry*, Freeman, 1970.
4. A. F. Möbius, *Gesammelte Werke*, Band I, S. Hirzel, 1885.
5. D. Pedoe, *A Course of Geometry for Colleges and Universities*, Cambridge Univ. Press, 1970.
6. D. Pedoe, Thinking geometrically, *Amer. Math. Monthly* **77** (1970), 711–721.
7. D. Pedoe, The most elementary theorem of Euclidean geometry, *Math. Magazine* **49** (1976), 40–42, 261.
8. J. Plücker, *Analytisch-Geometrische Entwicklungen*, Vol. I, 1828, Vol. II, Baedeker, 1831.
9. J. V. Poncelet, *Traité des Propriétés Projectives des Figures*, 2nd ed., Vol. I, 1865, Vol. II, Paris, 1866.
10. G. Salmon, *A Treatise on the Analytical Geometry of Three Dimensions*, Longmans, 1928.
11. O. Veblen and J. W. Young, *Projective Geometry*, Vol. I, 1938, Vol. II, Ginn, 1946.

A note on the history of the Cantor Set and Cantor function

JULIAN F. FLERON

Mathematics Magazine **67** (1994), 136–140

A search through the primary and secondary literature on Cantor yields little about the history of the Cantor set and Cantor function. In this note, we would like to give some of that history, a sketch of the ideas under consideration at the time of their discovery, and a hypothesis regarding how Cantor came upon them. In particular, Cantor was not the first to discover "Cantor sets". Moreover, although the original discovery of Cantor sets had a decidedly geometric flavor, Cantor's discovery of the Cantor set and Cantor function was neither motivated by geometry nor did it involve geometry, even though this is how these objects are often introduced (see for example [24]). In fact, Cantor may have come upon them through a purely arithmetic program.

The systematic study of point set topology on the real line arose during the period 1870–1885 as mathematicians investigated two problems:

1. conditions under which a function could be integrated,
2. uniqueness of trigonometric series.

It was within the framework of these investigations that the two apparently independent discoveries of the Cantor set were made; each discovery was linked to one of these problems.

Bernhard Riemann (1826–1866) spent considerable time on the first question, and suggested conditions he thought might provide an answer. Although we will not discuss the two forms his conditions took ([16], pp. 17–18), we note that one of these conditions is important as it eventually led to the development of measure-theoretic integration (see [16], p. 28). An important step in this direction was the work of Hermann Hankel (1839–1873) during the early 1870s. Hankel showed, within the framework of Riemann, that the integrability of a function depends on the nature of certain sets of points related to the function. In particular, "a function, he [Hankel] thought, would be Riemann-integrable if and only if it were *point-wise discontinuous* ([16], p. 30)," meaning, in modern terminology, that for every $\sigma > 0$ the set of points x at which the function oscillated by more than σ in every neighborhood of x was nowhere dense. Basic to Hankel's reasoning was his belief that sets of the form $\{1/2^n\}$ were prototypes for all nowhere dense subsets of the real line. Working under this assumption Hankel claimed that all nowhere dense subsets of the real line could be enclosed in intervals of arbitrarily small total length (i.e., had zero outer content) ([16], p. 30). As we shall see, this is not the case (see also [20]).

Although Hankel's investigation into the nature of certain point sets would become extremely important,

> as was the case with Dirichlet and Lipschitz, it was the inadequacy of his understanding of the possibilities of infinite sets—in particular, nowhere dense sets—that led him astray. It was not until it was discovered that nowhere dense sets can have positive outer content that the importance of negligible sets in the measure-theoretic sense was recognized

([16], p. 32). The discovery of such sets, nowhere dense sets with positive outer content, was made by H. J. S. Smith (1826–1883), Savilian Professor of Geometry at Oxford, in a paper [22] of 1875. After an exposition of the integration of discontinuous functions, Smith presented a method for constructing nowhere dense sets that were much more "substantial" than the set $\{1/2^n\}$. Specifically, he observed the following:

Let m any given integral number greater than 2. Divide the interval from 0 to 1 into m equal parts; and exempt the last segment from any subsequent division. Divide each of the remaining $m - 1$ segments into m equal parts; and exempt the last segments from any subsequent subdivision. If this operation be continued *ad infinitum*, we shall obtain an infinite number of points of division P upon the line from 0 to 1. These points lie in loose order ... ([22], p. 147).

In modern terminology Smith's 'loose order' is what we refer to as nowhere dense. Implicit in Smith's further discussion is the assumption that the exempted intervals are open, so the resulting set is closed. Today this set would be known as a general Cantor set, and this seems to be the first published record of such a set.

Later in the same paper, Smith shows that by dividing the intervals remaining before the nth step into m^n equal parts and exempting the last segment from each subdivision we obtain a nowhere dense set of positive outer content. Smith was well aware of the importance of this discovery, as he states, "the result obtained in the last example deserves attention, because it is opposed to a theory of discontinuous functions, which has received the sanction of an eminent geometer, Dr. Hermann Hankel" ([22], p. 149). He continues by explaining the difficulties in the contemporary theories of integration that his examples illuminate.

It is interesting to note that an editor's remark at the conclusion of Smith's paper states "this paper, *though it was not read*, was offered to the society and accepted in the usual manner." [Emphasis added. It is possible that "not read" simply meant that the paper was not presented at a meeting of the London Mathematical Society. However, in weighing the significance of this note, one must consider that in Vols. 3–10 of the *Proceedings of the London Mathematical Society* (1871–1879), and perhaps even further, no other paper was similarly noted.] In fact, this paper went largely unnoticed among mathematicians on the European continent and unfortunately Smith's crucial discoveries lay unknown. It took the rediscovery, almost a decade later, of similar ideas by Cantor to illuminate the difficulties of contemporary theories of integration and to begin the evolution of measure-theoretic integration.

Georg Cantor (1845–1918) came to the study of point-set topology after completing a thesis on number theory in Berlin in 1867. He began working with Eduard Heine (1821–1881) at the University of Halle on the question of the uniqueness of trigonometric series. This question can be posed as follows:

If for all x except those in some set P we have

$$\frac{1}{2}a_0 + \sum_{n=1}^{\infty}(a_n \cos(nx) + b_n \sin(nx)) = 0,$$

must all the coefficients a_n and b_n be zero?

Heine answered the question in the affirmative "when the convergence was *uniform in general* with respect to the set P, which is thus taken to be finite" ([16], p. 23), meaning, by definition, that the convergence was uniform on any subinterval that did not contain any points of the finite set P.

Cantor proceeded much further with this problem. In papers [1] and [2] of 1870–1871, he removed the assumption that the convergence was "uniform in general" and began to consider the case when P was an infinite set. In doing so he began to look at what we now consider the fundamental point-set topology of the real line. In a paper [3] of 1872, Cantor introduced the notion of a *limit point* of a set that he defined as we do today, calling the limit points of a set P the *derived set*, which he denoted by P'. Then P'' was the derived set of P', and so on. Cantor showed that if the set P was such that $P^{(n)} = \emptyset$ for some integer n and the trigonometric series

$$\frac{1}{2}a_0 + \sum_{n=1}^{\infty}(a_n \cos(nx) + b_n \sin(nx)) = 0,$$

except possibly on P, then all of the coefficients had to be zero. Cantor's work on this problem was "decisive" ([18], p. 49), and doubly important as his derived sets would play an important role in much of his upcoming work.

In the years 1879–1884 Cantor wrote a series of papers entitled *Über unendliche, lineare Punktmannichfaltigkeiten* [5–10], that contained the first systematic treatment of the point-set topology of the real line. [In addition, these papers contained many other topics that had far-reaching implications (see [13], [14]), including Cantor's investigation of higher-order derived sets that marked the "beginnings of Cantor's theory of transfinite numbers" ([16], p. 72).] It is the introduction of three terms in this series that concerns us most here. In the first installment of this series Cantor defines what it meant for a set to be *everywhere dense* (literally, "überall dicht"), a term whose usage is still current. He gives a few examples, including the set of numbers of the form $2^{2n+1}/2^m$, where n and m

are integers, and continues by noting the relationship between everywhere dense sets and their derived sets. Namely, $P \subseteq (\alpha, \beta)$ is everywhere dense in (α, β) if [and only if] $P' = (\alpha, \beta)$ ([5], pp. 2–3). In the fifth installment of this series Cantor discusses the partition of a set into two components that he terms *reducible* and *perfect* ([9], p. 575). His definition of a perfect set is also still current: a set P is *perfect* provided that $P = P'$.

After introducing the term *perfect* in the fifth installment, Cantor states that perfect sets need not be everywhere dense ([9], p. 575). In the footnote to this statement Cantor introduces the set that has become known as the *Cantor (ternary) set*: the set of real numbers of the form

$$x = \frac{c_1}{3} + \cdots + \frac{c_\nu}{3^\nu} + \cdots,$$

where c_ν is 0 or 2 for each integer ν. Cantor notes that this is an infinite perfect set with the property that it is not everywhere dense in any interval, regardless of how small the interval is taken to be. We are given no indication of how Cantor came upon this set.

During the time Cantor was working on the 'Punktmannichfaltigkeiten' papers, others were working on extensions of the Fundamental Theorem of Calculus to discontinuous functions. Cantor addressed this issue in a letter [11] dated November 1883, in which he defines the Cantor set, just as it was defined in the paper [9] of 1883 (which had actually been written in October 1882). However, in the letter he goes on to define the *Cantor function*, the first known appearance of this function. It is first defined on the complement of the Cantor set to be the function whose values are

$$\frac{1}{2}\left(\frac{c_1}{2} + \cdots + \frac{c_{\mu-1}}{2^{\mu-1}} + \frac{2}{2^\mu}\right)$$

for any number between

$$a = \frac{c_1}{3} + \cdots + \frac{c_{\mu-1}}{3^{\mu-1}} + \frac{1}{3^\mu} \quad \text{and}$$

$$b = \frac{c_1}{3} + \cdots + \frac{c_{\mu-1}}{3^{\mu-1}} + \frac{2}{3^\mu},$$

where each c_ν is 0 or 2. Cantor then concludes this section of the letter by noting that this function can be extended naturally to a continuous increasing function on $[0, 1]$. That serves as a counter-example to Harnack's extension of the Fundamental Theorem of Calculus to discontinuous functions, which was in vogue at the time (see, for example, [16], p. 60). We are given no indication of how Cantor came upon this function.

There are two other topics that interested Cantor that we would like to mention, because they are indicative of Cantor's facility with arithmetic constructions and it is possibly within this setting that Cantor came upon the Cantor set and Cantor function. First, Cantor spent some time in the mid-1870s considering the possible existence of a bijective correspondence between a line and a plane, a question most of his contemporaries had dismissed as absurd. In 1877, in a letter to Richard Dedekind (1831–1916), Cantor explained that he had found such a correspondence. This 'correspondence' can be expressed as follows.

Let (x_1, x_2) be a point in the unit square, and let $0.x_{1,1}x_{1,2}x_{1,3}\ldots$ and $0.x_{2,1}x_{2,2}x_{2,3}\ldots$ be decimal expressions of x_1 and x_2, respectively. Map the point (x_1, x_2) to the point on the real line whose decimal expansion is $0.x_{1,1}x_{2,1}x_{1,2}x_{2,2}\ldots$. (See for example [13], p. 187.)

Dedekind pointed out that there was a problem with this approach. The decimal expansions of rationals are not unique, so to avoid duplication we must not allow expansions of some type, say expansions that contain infinite strings of zeros. However, by disallowing expansions with infinite strings of zeros, the irrational number 0.11010201010201010102 ... could never be obtained under Cantor's correspondence.

This reasoning does however give us an injection of $[0, 1] \times [0, 1]$ into $[0, 1]$. It is trivial to find an injection of $[0, 1]$ into $[0, 1] \times [0, 1]$. These two facts, together with the Schroeder-Bernstein Theorem (if there are injections of the set A into the set B and B into A, respectively, then there is a bijective correspondence between A and B (see, e.g., [21]), allow us to conclude that there is a bijective correspondence between $[0, 1]$ and $[0, 1] \times [0, 1]$. However, set theory was in its infancy during the period in question and it would be 20 years before E. Schroeder and Felix Bernstein independently proved the theorem that bears their names ([14], pp. 172–173) and occasionally Cantor's name as well (e.g. [17], [23]). So this was not an option for Cantor.

Instead, Cantor needed to explicitly exhibit a bijection. To do this he modified his previous approach to use continued fractions [4]. Denote the continued fraction

$$\cfrac{1}{a_1 + \cfrac{1}{a_2 + \cfrac{1}{a_3 + \cdots}}} \quad by [a_1, a_2, a_3, \ldots],$$

where $a_1, a_2, a_3, \ldots > 0$ are integers.

Since a continued fraction is infinite if and only if it represents an irrational number, in which case the representation is unique (see e.g. [15]), Cantor could set up the correspondence

$$([a_{1,1}, a_{1,2}, \ldots], [a_{2,1}, a_{2,2}, \ldots], \ldots [a_{n,1}, a_{n,2}, \ldots])$$
$$\leftrightarrow [a_{1,1}, a_{2,1}, \ldots a_{n,1}, a_{1,2}, a_{2,2}, \ldots a_{n,2}, \ldots]$$

between n-tuples of irrationals in $(0,1)^n = (0,1) \times (0,1) \times \cdots \times (0,1)$ and irrationals in $(0,1)$. This avoids the difficulties of the previous approach and gives a bijective correspondence between $([0,1]\backslash\mathbb{Q})^n$ and $[0,1]\backslash\mathbb{Q}$. Cantor then took great lengths to prove there was a bijective correspondence between $[0,1]$ and $[0,1]\backslash\mathbb{Q}$. Repeated application of this fact combined with the previous correspondence gives a bijective correspondence between $[0,1]^n$ and $[0,1]$.

Secondly, it is known that Cantor studied binary expansions. In fact:

Cantor recognized that the power of the linear continuum, denoted by $\mathfrak{o}$, could be represented as well by [the power of] the set of all representations:

$$x = \frac{f(1)}{2} + \cdots + \frac{f(\nu)}{2^\nu} + \cdots,$$

where $f(\nu) = 0$ or 1 [for each integer ν] (see [13], p. 209).

There is, so it seems, no substantive evidence about how Cantor came upon the Cantor set and Cantor function. However, given Cantor's route into point-set topology, his arithmetic introduction of the Cantor set and Cantor function, and his facility with arithmetic methods, as we have just illustrated, it is feasible that it is within the arithmetic framework of binary and ternary expansions that Cantor came upon the Cantor set and Cantor function.

References

1. G. Cantor, Beweis, daß eine für jeden reellen Werth von x durch eine trigonometrische Reihe gegebene Function $f(x)$ sich nur auf eine einzige Weise in dieser Form darsteuen läßt, Part 1, *Crelle J. Math.* **72** (1870), 139–142; reprinted in [12], pp. 80–83.

2. G. Cantor, Beweis, daß eine für jeden reellen Werth von x durch eine trigonometrische Reihe gegebene Function $f(x)$ sich nur auf eine einzige Weise in dieser Form darsteuen läßt, Part 2, *Crelle J. Math.* **73** (1871), 294–296; reprinted in [12], pp. 84–86.

3. G. Cantor, Über die Ausdehnung eines Satzes aus der Theorie der trigonometrischen Reihen, *Math. Ann.* **5** (1872), 123–132; reprinted in [12], pp. 92–102.

4. G. Cantor, Ein Beitrag zur Mannigfaltigkeitslehre, *Crelle J. Math.* **84** (1878), 242–258; reprinted in [12], pp. 119–133.

5. G. Cantor, Über unendliche, lineare Punktmannichfaltigkeiten, Part 1, *Math. Ann.* **15** (1879), 1–7; reprinted in [12], pp. 139–145.

6. G. Cantor, Über unendliche, lineare Punktmannichfaltigkeiten, Part 2, *Math. Ann.* **17** (1880), 355–358; reprinted in [12], pp. 145–148.

7. G. Cantor, Über unendliche, lineare Punktmannichfaltigkeiten, Part 3, *Math. Ann.* **20** (1882), 113–121; reprinted in [12], pp. 149–157.

8. G. Cantor, Über unendliche, lineare Punktmannichfaltigkeiten, Part 4, *Math. Ann.* **21** (1883), 51–58; reprinted in [12], pp. 157–164.

9. G. Cantor, Über unendliche, lineare Punktmannichfaltigkeiten, Part 5, *Math. Ann.* **21** (1883), 545–591; reprinted in [12], pp. 165–209.

10. G. Cantor, Über unendliche, lineare Punktmannichfaltigkeiten, Part 6, *Math. Ann.* **23** (1884), 453–488; reprinted in [12], pp. 210–246.

11. G. Cantor, De la puissance des ensembles parfaits de points, *Acta Math.* **4** (1884), 381–392; reprinted in [12], pp. 252–260.

12. G. Cantor, *Gesammelte Abhandlungen Mathematischen und Philosophischen Inhalts* (ed. F. Zermelo), Springer-Verlag, 1980.

13. J. W. Dauben, The development of Cantorian set theory, in I. Grattan-Guinness, *From the Calculus to Set Theory: 1630–1910*, Gerald Duckworth, 1980.

14. J. W. Dauben, *Georg Cantor: His Mathematics and Philosophy of the Infinite*, Princeton Univ. Press, 1990.

15. G. H. Hardy and E. M. Wright, *An Introduction to the Theory of Numbers*, 5th ed., Clarendon Press, 1989, pp. 129–140.

16. T. Hawkins, *Lebesgue's Theory of Integration: Its Origins and Development*, Chelsea, 1975.

17. K. Hrbacek and T. Jech, *Introduction to Set Theory*, Marcel Dekker, 1984, p. 72.

18. W. Purkert, Cantor's Philosophical Views, in D. E. Rowe and J. McCleary, *The History of Modern Mathematics*, Vol. 1: *Ideas and Their Reception*, Academic Press, 1989.

19. W. Purkert and H. J. Ilgauds, *Georg Cantor: 1845–1918*, Birkhäuser Verlag, 1987.

20. H. L. Royden, *Real Analysis*, Macmillan, 1988.

21. G. F. Simmons, *Introduction to Topology and Modern Analysis*, R. E. Krieger, 1983, pp. 28–30.

22. H. J. S. Smith, On the integration of discontinuous functions, *Proc. London Math. Soc.* (1) **6** (1875), 140–153.

23. C. Takeuti and W. M. Zaring, *Introduction to Axiomatic Set Theory*, Springer-Verlag, 1982, p. 86.

24. R. L. Wheeden and A. Zygmund, *Measure and Integral: An Introduction to Real Analysis*, Marcel Dekker, 1977.

Evolution of the Topological Concept of "Connected"

R. L. WILDER

American Mathematical Monthly **85** (1978), 720–726

Introduction

The purpose of this paper is to trace the evolution of one of the most basic concepts in topology, viz., that of *connected* (not to be confused with *simply connected*). Like many other mathematical concepts of a fundamental nature (e.g., continuous function), it had only an intuitive meaning (such as *connected figure* in geometry) until the increasingly subtle demands of analysis and topology forced formulation of a satisfactory definition. The latter was not achieved, as one might expect, until a number of definitions had been proposed—each sufficient within its mathematical context, but quite insufficient as the configurations studied became more general and abstract.

We try to clear up, incidentally, the existing confusion regarding the actual authorship of the definition ultimately adopted. Not surprisingly, we uncover a "multiple." For several years, European topologists considered F. Hausdorff to be the prime originator of the definition, apparently because their knowledge of set theory and fundamental topological notions was usually derived from his classic *Grundzüge der Mengenlehre* [5], published in 1914. However, by the time of publication of his 1944 *Mengenlehre* [6], which was a third edition of the *Grundzüge*, Hausdorff had discovered Lennes's earlier version of the same definition (see below), and called attention thereto in a note at the end of his book.

Thereafter the definition was commonly called the "Lennes–Hausdorff definition" of connected. Many modern textbooks on topology seem to have adopted the term "Hausdorff–Lennes Separation Condition", or "Hausdorff–Lennes condition" for the type of separation involved in the definition. Possibly this received stimulus from the use of the term by S. Lefschetz in his American Mathematical Society Colloquium volume entitled *Algebraic Topology* [10]. On page 15, Lefschetz speaks of the "so-called Hausdorff–Lennes separation condition."

In his classic work on topology [9], Kuratowski states in a footnote (p. 127) that the definition of connected "originates from" C. Jordan's *Cours d'Analyse* of 1893, and also cites Lennes's work. A justification for Kuratowski's statement is offered below.

W. Sierpiński, in the 1952 English edition of his work on general topology [19], attributes the definition to Hausdorff. However, in his *Foundations of Point Set Theory* ([13], p. 378), R. L. Moore attributes the definition to Lennes. In my own book, *Topology of Manifolds* [20], I cited Schoenflies, Lennes, and Hausdorff, the Schoenflies definition being the same, although independently arrived at, as the definition of Jordan which was cited by Kuratowski.

Without further citing of literature, it seems fair to conclude that little attention was paid to United States journals during the early part of the present century, since Lennes' definition was published in both the *Bulletin of the American Mathematical Society* and the *American Journal of Mathematics* in 1906 and 1911, respectively. Perhaps, too, the same should be said about the Hungarian journals, for nowhere in the topological literature cited above is the name of F. Riesz mentioned in connection with the definition of connected, although the same definition as that given by Lennes was published by him in 1906 (in Hungarian) and in 1907 (in German).

The evolution

Unquestionably the roots of the concept of connected lie in the notion of the continuous, but more specifically the linear continuum, which goes back as far as the Greeks, who struggled to clarify the notion in the light of Zeno's paradoxes. The history of this, so far as it is known, is already adequately covered in the literature. Similar remarks hold for the contributions of the medieval mathematicians and philosophers, especially of the scholastic tradition, whose influence on both Bolzano and Cantor have been widely commented upon.

Bolzano's contribution

Although the theory of proportion given by Eudoxus (and reproduced in Euclid's *Elements*) has been credited by some as the equivalent of Dedekind's definition of the real continuum, it seems not to have figured in the analysis of the early part of the nineteenth century. During the latter period, there arose a growing stress for a proper basis for establishing the *location theorem* of algebra, using only arithmetic (as opposed to geometric) means. This theorem asserts that if a real polynomial $f(x)$ is negative for $x = a$ and positive for $x = b$, then it is zero at some value between a and b. This led Bernard Bolzano to offer a proof of the theorem in 1817 [1]. A casual reading of Bolzano's works convinces one that he had a remarkable intuitive knowledge of the structure of the real continuum as it is understood today. Along with this, he evidently conceived of the notion of a general continuum. Consider the following definition (given in his *Paradoxien* [2], p.129):

> ... a continuum is present when, and only when, we have an aggregate of simple entities (instances or points or substances) so arranged that each individual member of the aggregate has, at each individual and sufficiently small distance from itself, at least one other member of the aggregate for a neighbor. When this does not obtain, when so much as a single point of the aggregate is not so thickly surrounded by neighbors as to have at least one at each individual and sufficiently small distance from it, then we call such a point *isolated*, and say for this reason that our aggregate does not form a continuum.

Curiously, the motivation for this definition, according to Bolzano's own testimony, lay in the paradoxes that plagued the philosophical and mathematical conceptions of time, space, and "substance." Bolzano reasoned that, by establishing a suitable characterization of the abstract structural pattern common to all these concepts, the paradoxes could be explained. The analogy with the Greek dilemma and the efforts to resolve it is striking.

Now the *Paradoxien* was written toward the end of Bolzano's life and published posthumously in 1851, while the proof of the location theorem, cited above, was published some 34 years previously. But the motivation for the latter was strictly mathematical in that it was to free analysis of its notorious reliance on the geometric aspects of continuity. There can be little doubt, however, that Bolzano's development of his intuition of the continuous in the latter work was contributory to his philosophical conception of time, space, and substance as continua. And it seems to represent the first attempt at a mathematical formulation of the topological notion of connected. Since, as was to be the case for over a half-century thereafter, the definition of *connected* was tied to that of *continuum*, it would perhaps be more proper to term it the mathematical progenitor of the notion of continuum. However, the time for consideration of point sets having no compactness properties had not arrived in mathematics, and there is little doubt that the intuitive notion which Bolzano (and after him Cantor) was trying to make precise was equivalent, in its context, to that which led later to the "unrestricted" notion of topological connectedness.

Cantor's contribution

Cantor, who was familiar with Bolzano's work, saw clearly that the property used by Bolzano was insufficient to make precise the intuitive notion of continuum. In a paper often called the *Grundlagen* ([3], Section 10), he pointed out, for example, that sets consisting of several separated continua satisfy Bolzano's condition. Moreover, he recognized intuitively that the compactness properties now associated in topology with the notion of continuum had not been required in Bolzano's definition, and pointed out that the complement of an *isolated* point set in n-dimensional coordinate space E^n, $n \geq 2$, is a continuum according to Bolzano. (It is curious that Cantor did not point out that the set of rational points in E^1, the real line, is also a continuum by Bolzano's definition.) He also rejected enlisting the concepts of time or space as aids in exploring the mathematical notion of continuum, deeming the relationship quite the reverse.

According to Cantor, a continuum in E^n must possess two basic properties, namely, the property of being *perfect* and that of being *connected*. In modern terms, a point set in E^n is perfect if it is closed and dense-in-itself—i.e., each of its points is a limit point of it. (Cantor's precise definition utilized the notion of *derivative* of a point set, a point set P being perfect if it coincides with its first derivative P'.) In this connection he pointed out the insufficiency of requiring a point set to be only perfect in order that it be a continuum by giving his classical example of a totally disconnected perfect set—the *Cantor ternary set*, now often called simply the *Cantor set*. (See [3], p. 590.) He then defined *connected* as follows:

> A point set T is connected if for every two of its points t and t', and arbitrary given positive number ε, there always exists a *finite* number of points $t_1, t_2, \ldots, t_n$ of T such that the distances $tt_1, t_1t_2, \ldots, t_nt'$ are all smaller than ε.

Then any perfect and connected subset of E^n is a continuum, according to Cantor, who pointed out in a footnote ([3] p. 590) that no special dimension was implied in the definition; a line, surface, solid, etc., are all continua. Incidentally, for bounded subsets of E^n, this is equivalent to the modern definition of a continuum.

The most important aspect of Cantor's definition of continuum is his separation of the two concepts *perfect* and *connected*, thus identifying for the first time the latter as an independent property. At the time, however, topology was virtually non-existent as a field of study, and it could not be expected that sets having the sole property of connectedness would receive any attention. And as already implied above, Cantor's definition of connected was quite adequate for the study of continua.

C. Jordan's contribution

The next noteworthy step in the evolution of the concept of connected is found in C. Jordan's *Cours d'Analyse* [7]. Apparently Jordan was not familiar with Cantor's definition of ten years earlier, since he makes no mention of it. Following a discussion of closed sets (but using the word *perfect* (*parfait*) instead of *closed*), establishing the notion of distance (*écart*) between them, and defining sets as separated when the distance between them is greater than zero, he gives a definition of what he calls *un seul tenant*— in modern terms *component* — of a bounded closed set, to wit: a bounded and closed set of points has a single component if it cannot be decomposed into several closed separated sets.

> One sees easily that the distinctive character of such a set is the following: *For arbitrary ε, one can intercalate, between any two of its points p, p', a chain of intermediate points of the set such that the distance between consecutive points is less than ε.*

It is this statement that Jordan italicized, not the preceding definition.

This coincides, of course, with Cantor's definition of connected and is proved a necessary and sufficient condition for a bounded and closed set to consist of a single component ([18], p. 26). It is then simple to prove ([7], p. 27) that a subset of the real line which forms a single component and contains two numbers, a and b, must contain every number between a and b. This corollary of Bolzano's theorem seems to have been the chief motive for Jordan's definition of component.

Schoenflies' contribution

In 1904, A. Schoenflies published the first of his fundamental researches into the *topological* aspects of point set theory [18]. He was aware of Cantor's definition of connected, which he cited ([18], pp. 208–209), but went on to comment that even though the concept of *distance* formed a primitive geometric notion for the axiomatic basis of his work, it was preferable to give a purely set-theoretic definition of connected, whereupon he gives the following:

> A perfect set is called *connected* if it is not decomposable into [at least two non-empty] subsets, each of which is perfect.

This is, for bounded sets, the equivalent of Jordan's definition of *un seul tenant* which, according to the accompanying remarks, became known to Schoenflies only after he had announced his own version. Stating that Jordan introduced the definition only to derive Cantor's formulation of the concept (with which he operated thereafter), Schoenflies observes that "connectedness is an important and fundamental property for Analysis Situs as a whole." This statement represents an important step forward in the evolution of the connectedness concept. Whereas Cantor only separated the notion from the other properties of a continuum, Schoenflies now elevated it to the position of a *fundamental property* of topology, and went on to prove its invariance and to study the property especially in the context of plane topology.

Despite Schoenflies' recognition of the fundamental character of connectedness, his view was still limited, in that he expressed the opinion that, while the definition was formulated for perfect sets, it could equally well be stated for (merely) closed sets; but "since for closed sets which are not perfect, connectedness cannot come into question, it is sufficient to limit the definition to perfect sets" ([8], p. 173)! Thus, while making an important step forward, Schoenflies made another step backward.

The work of W. H. and G. C. Young

Although chronologically the work of W. H. and G. C. Young virtually coincides with that of Lennes and Riesz, to be discussed below, it is interpolated here as a kind of capstone to the work already described, as well as being of intrinsic interest for its adumbration of later work in the theory of connectedness.

The classic book [22] of W. H. and G. C. Young, intended as the "first attempt at a systematic exposition" of Cantor's ideas on the theory of sets, introduces a definition of connected in terms of regions. In modern terms, a *region* was a connected open set with or without an arbitrary set of its boundary points; the Youngs defined it as generated by successively overlapping triangles:

> A set of points such that describing a region in any manner round each point and each limiting point of the set as internal point, these regions always generate a single region, is said to be a *connected* set provided it contains more than one point. Hence if a set is connected the set got by closing it is connected, and *vice versa* ([22], p. 34).

From this definition, the Youngs prove: A connected set cannot be divided into closed components (= subsets) without common points. Conversely a set which cannot be divided into closed components without common points is, if closed, a connected set. We recall that this proposition was used by Jordan and others to define connected in the case where the set in question is closed.

The Lennes and Riesz definitions

It is remarkable that throughout the period discussed above—from the time of Bolzano to 1905, over half a century—the notion of connected was confined to closed sets; and this in spite of the fact that Cantor divorced the notion from closure in his definition of continuum. On the other hand, it is not surprising, since attention was devoted exclusively either to the real continuum or to the subsets of Euclidean space (usually the plane), and the only non-closed sets of importance were of a special character, such as the set of rationals or open segments on the real line, and the circular or triangular regions of the plane.

Of course, Jordan, and following him Schoenflies, had proposed definitions of connected which virtually begged for generalization to non-closed sets. And this step was finally taken in 1905–06 by both N. J. Lennes and F. Riesz. Lennes gave his definition at a meeting of the American Mathematical Society in December 1905, and it was published in the abstract of his talk the following year in the *Bulletin* of that Society [11]. Riesz's definition was presented to the Hungarian Academy of Sciences on January 22, 1906, and published later the same year [15]. Here was clearly a "multiple"—a case of independent invention by more than one investigator.

Lennes's definition reads [11]:

> A set of points is connected if in every pair of complementary subsets at least one subset contains a limit point of points in the other set.

This is stated in such a fashion that it is meaningful in any space in which limit point is defined (although undoubtedly the author's thinking, like that of most topologists of the time, was of Euclidean spaces).

Riesz's definition has several remarkable features. In the first place, it is given in the context of an essay [16] devoted to the relations between the "physical continuum" and the "mathematical continuum". In defining the physical continuum, Riesz uses the relation *unterscheidbar* between space points, not the topological notion of limit point. The definition proceeds as follows:

> Das physikalische Kontinuum heisst zusammenhängend, wenn es nicht in zwei Teilmengen zerlegt werden kann, dass jedes Element der einen Teilmenge unterscheidbar sei von jedem Elemente der anderen Teilmenge. The physical continuum is called connected if it cannot be split into two subsets so that each element in one subset is distinguishable from each element in the other subset.

Notice the striking resemblance to the Jordan-Schoenflies definition, although Riesz makes no reference to the latter. However, there is conclusive evidence in previous papers of Riesz's that he was familiar with Jordan's *Cours d'Analyse* and hence, probably, with Jordan's definition.

In the second place, the definition of *connected* for topological spaces (he does not use the latter term) is given initially for what he calls a "mathematical continuum", which, in modern terms, is an abstract topological space defined by four axioms adumbrative of such later systems as were given by Hausdorff and Kuratowski. The definition reads as follows:

> Das mathematische Kontinuum heisse zusammenhängende wenn es nicht in zwei offene Teilmengen zerlegt werden kann, die Komplementarmengen für einander sein. The mathematical continuum is called connected if it cannot be split into two open subsets, which are complements of each other.

For *subsets* of such a space he then distinguishes two degrees of connectedness: a set is called *connected* if it cannot be decomposed into two subsets whose closures are disjoint; it is called *absolutely connected* if for every decomposition of it into two [non-empty] subsets, there exists at least one element which belongs to one subset and is a limit point of the other. It is the second of these, of course—i.e., absolutely connected—that is the modern definition of connected.

Actually, the "multiple" which occurred when these two definitions were given was joined by a third, viz., F. Hausdorff's definition. Apparently when giving his definition in his book [5] of 1914, Hausdorff was unaware of either Lennes' or Riesz' definitions. On the other hand, Hausdorff did proceed, in this book, to study some of the properties of connected sets as topological concepts in their own right.

However, the first paper devoted to the study of connected sets was not published until 1921; we refer here to the classic paper "Sur les ensembles connexes" of B. Knaster and C. Kuratowski [8]. This paper was significant in the evolution of the concept of connectedness because:

1. it established the fact that connected spaces lacking compactness properties have a variety of interesting topological properties;

2. it gave impetus to a host of studies, both in topology and in the logical foundations of set theory;

3. it gave the ultimate emphasis to Schoenflies' statement, quoted above, concerning the fundamental character of connectedness.

Concluding remarks

From an evolutionary point of view, the development of the concept of connectedness proves to be a revealing case study. Its roots, as in the case of many other mathematical concepts, are embedded in the contemplation of physical time, space, and "substance". At the hands of Cantor it finally split off from philosophical and physical considerations to become a part of mathematical theory. But it was not easily divorced from the concept of *continuum* within which it was first formulated—a consequence of its mathematical environment, which consisted chiefly of the study of curves and surfaces, examples of what Cantor called *continua* in the mathematical sense. This was a case of the operation of "environmental stress", in that the mathematical environment worked to confine the notion within a restricted area.

It failed to find its proper place in mathematical theory until Schoenflies pointed out its invariance under topological transformations, as well as its independent status as a topological property. But although Schoenflies, who discovered essentially the same definition that Jordan had given over a decade earlier, made an important step forward, topology had still not grown much beyond the study of configurations whose compactness properties made the Jordan-Schoenflies definition quite adequate. Indeed, so much so, that when the Youngs wrote their classic *The Theory of Sets of Points* during the decade between Jordan and Schoenflies, they seem to have deliberately phrased their definition of *region* (allowing a region to include boundary points freely) so that the Cantor definition would be preserved (see the remark above concerning the Young definition).

Lennes's generalization of these definitions was apparently a result of his consideration of non-closed sets. In the paper [12] giving in detail the results announced in the 1906 abstract [11], he first defined connectedness for open sets (in Euclidean space) by using broken lines, an open set U being connected if every pair of points a, b in U are joined by a broken line lying wholly in U ([12], p. 293). He observed that by the Cantor definition of connected, the union of the interior and exterior of a planar circle forms a connected set; moreover, that "if from the ordinary continuum in space of any dimensions any set whatever which is nowhere dense is removed, the residue would form a connected set" ([12], p. 303). He thereupon gave the form of the definition now generally accepted, remarking that it "applies in cases where the former does not." (In other words, it renders sets connected which our intuition tells us should be connected and rules out those that, like the complement of the circle in the plane, should not be termed connected.)

One of the remarkable features of Riesz's definition, as we have already noticed, is that it was given in the context of an abstract topological space. This aspect of Riesz's work seems also to have been generally unnoticed for some time, despite the fact that Fréchet called attention ([4], Note B) to Riesz's abstract space axioms as they were later presented at the International Congress in Rome, 1908 [17]. In any event, his definition, although agreeing with Lennes's, achieves thereby its most general character, freed from all metric considerations.

Although the lack of diffusion from one country to another, which characterized earlier periods in mathematics, had begun to subside, the period during which the topological concept of connectedness was developed still shows considerable lack of diffusion. Lennes's and Riesz's work, both published in reputable journals during the first decade of the century, was generally unknown until the journal *Fundamenta Mathematicae* commenced publication in 1920. The occurrence of a three-member multiple during the first quarter of the present century is quite noteworthy.

One further comment: One of the noteworthy features of the Knaster and Kuratowski article [8] was its presentation of paradoxical examples of connected sets having no compactness properties. I have pointed out elsewhere [21] the contribution that paradox can make to the development of mathematical concepts. The examples given by Knaster and Kuratowski proved a great stimulation to the study of connectedness. There ensued a sizable literature devoted to the concept, and in recent years this has engendered interesting questions in the foundations of set theory.

References

1. B. Bolzano, *Rein analytischer Beweis des Lehrsatzes, dass zwischen je zwei Werten die ein entgegengesetzes Resultat gewähren, wenigstens eine reele Wurzel der Gleichung liege* (ed. F. Prihonsky), Prag, 1817.

2. B. Bolzano, *Paradoxes of the Infinite* (translation of the *Paradoxien* by F. Prihonsky), Routledge and Kegan Paul, 1950.

3. G. Cantor, Über unendliche lineare Punktmannigfaltigkeiten, 5 Fortsetzung, *Math. Ann.* **21** (1883), 545–591.

4. M. Fréchet, *Les Espaces Abstraits*, Gauthier-Villars, 1928.

5. F. Hausdorff, *Grundzüge der Mengenlehre*, von Weit, 1914.

6. F. Hausdorff, *Mengenlehre*, Dover, 1944.

7. C. Jordan, *Cours d'Analyse*, Vol. I, 2nd ed., Gauthier-Villars, 1893.

8. B. Knaster and C. Kuratowski, Sur les ensembles connexes, *Fund. Math.* **2** (1921), 206–255.

9. C. Kuratowski, *Topology*, Vol. 2, Academic Press, 1968.

10. S. Lefschetz, *Algebraic Topology*, American Mathematical Society, 1942.

11. N. J. Lennes, Curves in non-metrical analysis situs, *Bull. Amer. Math. Soc.* **12** (1905–06), 284, abstract #10.

12. N. J. Lennes, Curves in non-metrical analysis situs with applications in the calculus of variations, *Amer. J. Math.* **33** (1911), 287–326.

13. R. L. Moore, *Foundations of Point Set Theory*, American Mathematical Society, 1962.

14. F. Riesz, *Oeuvres Complètes*, Vol. 1, Gauthier-Villars, 1960.

15. F. Riesz, A térfogálom genesise, *Math. u. Phys. Lapok* **15** (1906), 97–122 and **16** (1907), 145–161 (paper A6 in [14]).

16. F. Riesz, Die Genesis des Raumbegriffes, *Math. u. Naturwiss. Berichte aus Ungarn* **24** (1907), 309–353 (paper A7 in [14]).

17. F. Riesz, Stetigkeitsbegriff und abstrakt Mengenlehre, *Atti del IV Congresso Internazional dei Matematici, (Roma 1908)*, Vol. 2, p. 18.

18. A. Schoenflies, Beiträge zur Theorie der Punktmengen I, *Math. Ann.* **58** (1904), 195–238.

19. W. Sierpiński, *General Topology* (transl. C.C. Krieger), Univ. Toronto Press, 1952.

20. R. L. Wilder, *Topology of Manifolds*, American Mathematical Society, 1949.

21. R. L. Wilder, Hereditary stress as a cultural force in mathematics, *Historia Math.* **1** (1974), 29–46.

22. W. H. Young and G. C. Young, *The Theory of Sets of Points*, Cambridge Univ. Press, 1906.

A Brief, Subjective History of Homology and Homotopy Theory in this Century

PETER HILTON

Mathematics Magazine **61** (1988), 282–291

I have recently been recalling that about twenty-five years ago, when I first came to settle in this country, I was invited to participate in the celebration of the opening of the Mathematics Building, Van Vleck Hall, at the University of Wisconsin. On that occasion I learned a new American word, namely "banquet", which has a totally different meaning in the United States from the meaning that it has in Britain. But more importantly, I must recall the immense respect I felt for some of the after-dinner speakers who were able to make the recounting of an event last much longer than the event itself. So I'm very conscious of the fact that in attempting to recount to you the history of algebraic topology in this century, I must not make the recounting of this history last longer than the history. In fact, I must telescope it very dramatically, one might almost say, abruptly. So I apologize in advance that much of the treatment will be necessarily very superficial. I would like to start off with the first epoch which is up to 1926. And here the inspiration for homology theory comes from the work of Poincaré.

Poincaré, during a period earlier than the one I'm thinking of, had already invented or discovered, according to your philosophy, the *fundamental group*. But he published a series of papers in which he was studying what we would call *algebraic varieties*, the configuration of points in higher-dimensional Euclidean space given by polynomial equalities and inequalities; and he was looking again at what we might call *vector fields* and generalizations of vector fields on such varieties. He was led through this study to look at what we would now call the *homology* of these varieties. In particular, he saw the significance for the solution of such vector field problems of what were then called the Betti numbers, which determine essentially the number of holes the configuration had. As a simple example, let us take the torus (Fig. 1), which is of course an algebraic variety and has two very conspicuous one-dimensional holes. These are cycles which do not bound anything in the torus. Formally, the torus is itself a two-dimensional hole and any given point constitutes a zero-dimensional hole. Problems about the solutions of differential equations on the torus are different from problems relating, for example, to the sphere, because the sphere has a different one-dimensional Betti number from that of the torus.

Poincaré also realized that there was a further subtlety, that is to say, there was a phenomenon which today is described as the phenomenon of the *torsion coefficients*. Essentially, what one means by torsion coefficients can be demonstrated by the real projective plane, which I can represent by a circular disk, with diametrically opposite points identified on the boundary of the disk (Fig. 2). It is not realizable in three-dimensional space, let alone two, so I make no apology for having to represent it this way on the blackboard. Now if we look at the path from P back to P, that is a cycle. That cycle does not in itself bound anything. But if you repeat the cycle, then

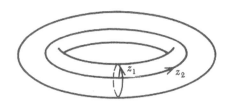

Figure 1. The torus, with its two basic one-dimensional cycles or holes, z_1 and z_2.

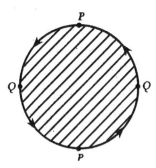

Figure 2. The real projective plane, with its basic one-dimensional cycle $z_1 = PQP$, such that z_1 does not bound the disk, but $2z_1$ does.

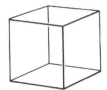 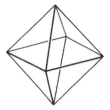

Figure 3. The cube and the octahedron, equivalent to the topologist but not to the geometer

twice the cycle bounds the disk. This exemplifies the phenomenon of the existence of cycles which do not themselves bound, but multiples of them bound.

I want to stress at this point that we are dealing with numbers. These Betti numbers are numbers and in each dimension there are Betti numbers. The torsion coefficients are also numbers.

Other names that one should associate with this period of the great pioneers are J. W. Alexander (one usually associates the name of Alexander with knots but he did many other things that I will refer to), and Veblen (he wrote one of the great books, *Analysis Situs*, which was the early name for topology—the essential difference between the words analysis situs and topology is a choice between the Latin and Greek cultures; they essentially mean the same thing). The word *topology*, Veblen felt, had been pre-empted, because the word topology exists outside mathematics. Also, one should mention here the name of Brouwer, and I am most happy to do so because I understand Karel de Bouvère spoke here on Brouwer. In particular, with Brouwer one associates the idea of the degree of a map, which I shall be referring to later. I should also say that van Kampen and Lefschetz did pioneering work. These will come back into the story, even though I list them in this very early period. To go into any detail on the contribution of any one of them would certainly occupy the entire lecture. So what I really want to do is proceed to what I regard as one of the golden years, 1927.

In regard to Lefschetz, however, I would like to make a remark of a personal nature, because it seems to me that, of all of the mathematicians, it is to Lefschetz one should give the credit for introducing systematically the notion of a polyhedron in a generalized sense. When I began a very enjoyable collaboration with Jean Pedersen, I discovered rather early on that we were separated by the misfortune of a common language, in which we used precisely the same terms but with different meaning. In particular, this word 'polyhedron' was almost responsible for the breaking up of a beautiful friendship. What does 'polyhedron' mean and what does it mean to classify polyhedra? To the topologist today, this is such a standard term, meaning the underlying topological space of a much more general type of combinatorial structure than that which is admitted by the geometer, the combinatorial geometer; and the classification of polyhedra, according to the topologist, is by homeomorphism generally, possibly by combinatorial equivalence, but again in a sense different from that used by the geometer (see Fig. 3). I think we can say that our trouble stems from Lefschetz. It's a very key notion, this idea of homeomorphisms between polyhedra; it raised many questions, more questions than it answered, as any good mathematical notion will.

So now, with this very brief introduction to that early period, I want to discuss those years 1926–27. The reason I pick out these years is because these are the years when Alexandroff and Hopf were in Göttingen. Both of them were there as guests, a splendid example of the efficacy of having mathematical guests! Alexandroff was from the Soviet Union and Hopf, at that time, was from Berlin. They were both of them very much impressed with the work of Lefschetz; and, in particular, they were discussing the Lefschetz fixed-point theorem. They recognized that this was in some way, which they began to put their fingers on, a generalization of the Euler-Poincaré characteristic. In connection with these Betti numbers which Poincaré had established, the so-called *Euler characteristic* was a topological invariant. The Lefschetz fixed-point theorem, expressed in admittedly somewhat clumsy notation, they saw to be closely related. Indeed, the Lefschetz fixed-point theorem, applied to the identity map from a space into itself, seemed to give the Euler-Poincaré characteristic.

But, very significantly, there was also Emmy Noether in Göttingen; she would not have been there but for Hilbert's insistence. He felt that Göttingen was a place for mathematicians and not for sexism—which, in those days, was a very special point of view. Emmy Noether recognized that what Alexandroff and Hopf were talking about, and what Lefschetz had talked about, should not be thought of as numbers but should be thought of as Abelian groups. So really one should credit Emmy Noether, not with the discovery of these topological invariants, but with understanding their mathematical place. Thus Emmy Noether recognized the homology groups, and that the Betti numbers and torsion coefficients were merely numerical invariants of isomorphism classes of finitely-generated Abelian groups. If you take a finitely-generated Abelian group A, then it can be written as the direct sum of a free Abelian group F and a family of cyclic groups A/h_i, where $h_1|h_2|\cdots$. Moreover, the rank of the free Abelian part F and these numbers h_i are invariants of the group A. If A is the homology group in dimension d, say, then its rank is the dth Betti number, and the h_i are the torsion coefficients. Now you see that this is an enormous improvement over just a consideration of the numbers, because this immediately gives you the opportunity of adopting a far more dynamic approach to the whole question of homology; for then it is not simply that with a polyhedron you associate Betti numbers and torsion coefficients, but with the polyhedron you associate homology groups. But, with Emmy Noether's improvement, there is very much more.

There is the natural question: how do you transform one polyhedron to another? How do you map one polyhedron into another? In the first place you have the idea of a simplicial map of the polyhedron and that induces a homomorphism of homology groups. I said, you remember, that the rank of F is an invariant—but F itself is not. A homomorphism from one Abelian group to another does not send the free part into the free part, nor does it preserve the nice little pieces, the Z/h_i. Things can get very much mixed up. The free part can go into the finite part. So you can only begin to understand the transformation of homology from this group-theoretical point of view.

This was an enormous advance both conceptually and dynamically, in terms of the real understanding of what you have. It also posed the very obvious question: simplicial maps induce homomorphisms of homology groups. What can one say in general about merely continuous functions of the underlying polyhedra? Now I have not undertaken to talk about the complete history of the *Hauptvermutung*, which would again occupy a lot of time. So I must leave on one side such questions about the different combinatorial structures on the polyhedron, but what this throws into prominence immediately is the question: can you approximate to any continuous map of the underlying topological spaces by a nice combinatorial transformation of the polyhedra? The answer is yes; that is the *simplicial approximation theorem*, which dates from about this time.

And so the picture begins to emerge of what homology is all about, and we agree first to look at this restricted class of spaces, the polyhedra and the simplicial maps between them, as a combinatorial structure. Here you can say you can do some forgetting, so that the polyhedra can be just thought of as topological spaces—we can forget their combinatorial structure—and the simplicial maps are then simply continuous functions.

From the polyhedra and the simplicial maps you can construct, first, what are called *chain complexes* and *chain maps*. That means you look at the chains on the space. They are the linear combinations of the simplexes, the generalized triangles into which you have subdivided your space to make it a polyhedron. And finally you take the homology groups of the chain complex and get the homology groups of the space. You define first chain complexes and the chain maps, then homology groups and homomorphisms. For the polyhedron, you pick any polyhedral structure on the space, and you prove that the resulting homology groups only depend on the underlying space. That is the essential statement of the topological invariance of the homology groups. You go back from an arbitrary continuous function to a simplicial map by an approximation. You have choices of approximations but, whatever approximation you use, the homomorphism induced on homology groups is the same. This diagram here (Fig. 4), where I have jumped many, many decades, shows essentially what these people were moving towards as they were elucidating the ideas of homology. Central to these ideas is the fact that the homology groups are really homotopy invariants—that is homotopic continuous maps induce the same homology homomorphism. A nice example is furnished by maps f of the n-sphere S^n to itself. Now $H_n S^n$ is cyclic infinite, so the induced homology homomorphism in dimension n is really just an integer, called the *degree* of f. The Brouwer-Hopf Theorem asserts that the degree is the unique invariant of the homotopy class of f.

I should also mention here the first person in this account who, as far as I know, is still alive: Vietoris, an Austrian mathematician, living in Innsbruck. Vietoris

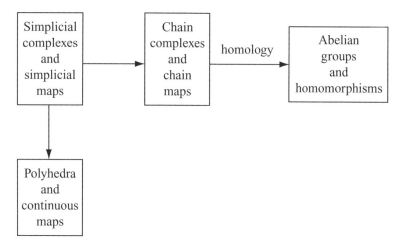

Figure 4. Scheme of the homology theory of polyhedra. Note that there are crucial notions of *homotopy* associated with continuous maps, simplicial maps and chain maps.

is one of the two who should be credited with seeing that you could define homology groups without the space having to be the underlying space of a polyhedron. With just the notion of a topological space, intrinsically defined, Vietoris defined homology groups (it even appears that Vietoris, independently of Emmy Noether, recognized the importance of the group concept). So there are homology groups of arbitrary spaces. He did it by means of coverings of the space by open sets.

The other person we credit here is a Czech mathematician, Čech. They did it independently, as far as I can see, but they did it differently. They both did it by considering coverings of a space by open sets; the open sets in Čech's definition behaved like the vertices of a simplicial complex. According to Čech, then, a finite number of open sets span a simplex if and only if they have a non-empty intersection. Thus he saw these open sets as the vertices of a complex, so, from the present-day point of view, what Čech did was to study the *nerve* of a covering, which is an abstract simplicial complex which has a chain complex and homology groups attached.

Vietoris did something different. He took the open sets and said that a collection of points of the space constitute (in some sense) a simplex, if they all lie in one of the sets of the covering. These two points of view we now see as being, in a very real sense, dual to each other. The first person who made the duality of the Vietoris-Čech definitions precise was Hugh Dowker, a very fine Canadian mathematician, who, I regret to say, died recently.

Also I should mention, in this connection, Mayer, who is known, and always will be known—as Heine is known as part of Heine-Borel—as part of Mayer-Vietoris. There is the *Mayer-Vietoris sequence* in homology. Hopf also credits Mayer, independently of Emmy Noether, with recognizing that groups were involved in the definition of homology, in a paper he published in 1929. But if I go on a bit from that time, I should mention that in 1932 there appeared a book by Alexandroff, which was very influential and in 1935 there was the great book of Alexandroff and Hopf. There was a Volume I but there was never a Volume II. (The reason Volume II never appeared was the advent of cohomology.) This was an extremely influential book and was a sort of bible for the study of algebraic topology. It was a very beautifully written work. The original was, of course, in German, but even if you do not understand German, it was easier to understand than most mathematical books written in English. The purpose of the treatment was to make the subject crystal clear. The pictures were beautiful.

Many things happened in 1935, which was very much a golden year. There was an international meeting in Moscow, in the summer. Hopf sent to Moscow his young student, Stiefel, who had begun a study of the existence of solutions of differential equations from the homological point of view and had come up with a certain idea which we now call *characteristic classes*. In Moscow at the meeting Stiefel read his paper, and in the audience was Hassler Whitney. (There is another living mathematician!) Hassler Whitney came up to Stiefel after and said: "This is

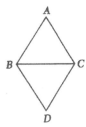

Figure 5. The boundary of BC is $C - B$. The coboundary of BC is $ABC + DBC$.

remarkable, it is almost exactly what I have been doing"—in his study of (what we would now call) *fiber spaces*. Alexandroff, too, said that these fiber spaces are what they (the Russian school) had been doing. There they were called twisted products. So there was an extraordinary confluence of ideas revealed at this Moscow meeting. We now talk of the *Stiefel-Whitney classes*. These are thought of as characteristic cohomology classes on a real differentiable manifold. They are completely understood now, but were then in their infancy.

Also, this was the year when cohomology emerged and it emerged with the understanding that in cohomology you have a ring structure. Now cohomology was very slow to emerge for the simple reason that Emmy Noether's point of view was simply not understood by topologists. Despite the obvious advantages of the algebraic viewpoint, topologists continued to think exclusively geometrically. And I put the emphasis on the word 'exclusively'. That is what was wrong. And it is a remarkable fact about Hopf, the greatest of them all, that he could never feel comfortable with the idea of cohomology.

The reason goes back to the idea of chains to which I have already (inadequately) referred. A *chain* is a linear combination of simplexes. For example, if you have some sort of triangulation, then you think of a chain as involving, in the one-dimensional case, the edges with certain multiplicities. Hopf always thought of it as some sort of path that was traced out on the polyhedron. And the current attitude towards homology and cohomology was this: if I, for example, look at a particular edge, this oriented edge, the two vertices of the edge, one with coefficient $+1$ and one with coefficient -1, that is, with this edge I will associate its boundary. The other thing I can do is associate, with this edge, all the triangles of which it is a side, that is, its coboundary (Fig. 5). So in one case you lower the dimension and in the other case you raise it. The lowering of the dimension is *homology*, the raising of the dimension is *cohomology*.

That was the current point of view and, indeed, under Alexandroff's influence, for many, many years the Russians continued to talk about lower and upper homology. But the point of view is flawed because, in the sense of linear algebra, cohomology is dual to homology—that is to say, if we think of 1-chains as linear combinations of edges, we should think of 1-cochains as functions of those edges. So we should distinguish between an edge and the function that takes on the value 1 on the edge. It is exactly the difference between a basis element of a vector space and the associated basis element of its dual. So cochains have to be thought of as functions on the simplexes. And although Hopf recognized that point of view he could not adopt it, because to him homology was all about geometry. The idea that you were looking at functions which take values in an arbitrary Abelian group and which are defined on the edges or triangles of a polyhedron was a point of view that was totally uncongenial to him.

Cohomology was a very long time emerging because it was incorrectly regarded. There was also the feeling that it should be possible to introduce a multiplicative structure into cohomology which is not present in homology. Many attempts were made to do this. Alexander, whom I have mentioned, was one of the pioneers here and the attempt was finally successful in 1935. Of course, it enriches enormously the structure and adds to the discrimination that you get through homology theory, because you could very well have two topological spaces such that the Abelian group structure in cohomology is the same but in fact they have different multiplicative structures. If you take the torus, and on the other hand the following configuration (Fig. 6), then the homology of the latter has the same additive structure as that of the torus, but it is distinguished from the torus by the multiplicative structure of its cohomology. The two one-dimensional cocycles on the torus multiply together to give the torus, but in the other configuration they multiply together to give zero. So this was the great refinement of cohomology theory.

It's interesting in this connection that Hopf, for whom I have enormous respect, didn't feel comfortable with this. Hopf had earlier seen, in the case of manifolds, that if you have a mapping f of manifolds, an m-dimensional manifold M mapped to an n-dimensional manifold N, then you have intersection

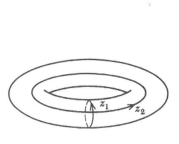 

Figure 6. The torus and the "fake torus" consisting of two circles and a sphere joined at a point. The product $z_1 z_2$ on the torus gives the torus itself, but $z_1 z_2 = 0$ on the fake.

rings of the manifolds. An *intersection ring* means that you take two cycles and look at their intersection in general position on the manifold.

Hopf proceeded from there to define something that he called the *Umkehrshomomorphismus*, the backward homomorphism. The backward homomorphism was something that was induced by f which would go from the homology of N in dimension $n - p$ to the homology of M in dimension $m - p$ for arbitrary p. He defined it very carefully and he gave it this name, the *Umkehrshomomorphismus*. What is it? Well, under the duality present in a manifold, the $(n - p)$th homology of N is essentially the same as the pth cohomology of N, as N is n-dimensional, and the $(m - p)$th homology of M is the same as the pth cohomology of M. So this homomorphism is simply the induced map in cohomology theory. Hopf defined this backward homomorphism and drew attention to the strange way that it changed the dimension in homology and went in the wrong direction. Of course it goes in the wrong direction because cohomology is contravariant, being based on dual vector spaces. You have a linear map from one vector space to another and the dual map maps the dual vectors back in the other direction. Hopf, with his wonderful, wonderful insight, got the right idea but stopped short of clarifying it in that way.

But having made the only criticism one could possibly make of Hopf's mathematics, let us move on to one of his tremendous contributions, also in 1935. In December of 1935 there was a meeting in Geneva and at this meeting Élie Cartan drew attention to the remarkable property of the classical Lie groups. He pointed out that all the classical Lie groups had the property that their Betti numbers—he was still talking about Betti numbers—were just like those of products of odd-dimensional spheres. So if you took any one of the classical Lie groups,—that is, the series of orthogonal groups, the unitary groups, and the symplectic groups—each one of those groups behaved, in a way, like a Cartesian product of odd-dimensional spheres. And this had simply been a matter of, you might say, empirical observation; that is to say, Cartan knew them all and their Betti numbers, and he just looked at them all and it was true for them all. But there was no explanation. He challenged people to produce an explanation. These facts, by the way, go back to Brouwer, Pontrjagin, and Ehresmann. And then he asked the next natural question, you take the five exceptional Lie groups, is it also true for them?

Why do the classical Lie groups always look like products of odd-dimensional spheres? Hopf thought about this and he came up with the answer. In his answer he pointed out that you needed very little of the structure of the classical Lie groups. Essentially all you needed is to say that you have a topological space, together with a continuous multiplication on that space which has a two-sided identity. So we assume, of course, a compact space so that our homology will be finitely generated, and a multiplication with a two-sided identity. That is all!

This has always struck me as being a piece of incredible genius, because at the time that Hopf was working, there were known exactly two examples of this phenomenon which were not Lie groups. One was the seven-dimensional sphere—and it is surely somewhat unexciting to be told that the seven-dimensional sphere behaves like a product of odd-dimensional spheres! And the other is the real projective seven-dimensional space. It is almost equally unexciting because the real projective seven-dimensional space is an orientable manifold of which the seven-sphere is a two-sheeted covering. Obviously from the point of view of Betti numbers it behaves exactly like the seven-dimensional sphere. So there was no interesting example—but Hopf gave this as the explanation, and this was the birth of the whole theory of *Hopf algebras* which is now a tremendous industry. Today we have infinitely many examples of the so-called Hopf manifolds which are not even topological groups. So as you see, much happened in 1935.

It is now time to talk of the homotopy groups. There is a beautiful trade-off between the homology groups and the homotopy groups. The homology groups are terribly difficult to define, but once you have defined them they are very easy to calculate. The homotopy groups are terribly easy to define but essentially impossible to calculate. The homotopy groups generalize the fundamental group. For the fundamental group you look at just the homotopy classes of maps of a circle into your space and, for the higher homotopy groups, you map spheres.

Actually, Hurewicz should not be credited with the actual invention of the homotopy groups. Really the credit for the invention should go to Čech. At a meeting in Vienna in 1931, Čech gave a paper in which he described certain groups from the homotopy point of view. He had no applications of these groups. Moreover, he had only one theorem, that they were commutative. And he was persuaded by people, and we know that Alexandroff played a role here, that they could not be interesting, because it was thought that any information that could be obtained from Abelian groups must come from the homology.

Hurewicz redefined the homotopy groups and immediately gave important applications in a series of four notes which were intended as preliminary publications. In that series of four papers he showed the significance of what we now call *obstruction theory*. Essentially, as Steenrod was later to remark and codify, the basic problem you are facing in topology can so often be represented in the following way. You have a configuration X and you have a configuration Y, you have a subspace of X called L, and a mapping g from L into Y. The question is—can that continuous function g be extended to X? It's amazing how many questions inside and outside topology can be reduced to that. What Hurewicz showed was that this type of question could be answered in terms of certain obstructions which are cohomology classes of X modulo L, with coefficients in the homotopy groups of Y. (In fact you usually cannot answer the question because you cannot calculate the obstructions!) The whole of obstruction theory was made absolutely systematic by Eilenberg. Hurewicz showed the significance of the homotopy groups and there is one great theorem that provides the link between homotopy and homology called the *Hurewicz Isomorphism*. Hurewicz pointed out that there was always a homomorphism going from the nth homotopy group to the nth homology group. If the space is such that the first $n-1$ homotopy groups vanish ($n \geq 2$), then this is an isomorphism. Thus the first place where the Hurewicz homomorphism is interesting it is an isomorphism. This generalizes the classical result already known essentially to Poincaré, the case $n=1$ where the first homology group is the fundamental group Abelianized. For the higher dimensions the homotopy groups are already Abelian, so you don't have to Abelianize them.

And that is the best that can be said of Alexandroff's point of view; that is, the first place where the homotopy groups come into play they are just homology groups. After that, their divergence is very significant, so you can say that homology and homotopy are proceeding together essentially as complementary concepts.

And if I could just make one remark about what I regard as my own small contribution to this evolution, it would be my work with Eckmann. We showed that, though the homology and homotopy groups are essentially different, the method of construction of cohomology groups and of homotopy groups can be regarded as dual manifestations of exactly the same process. That is to say, the actual structure of cohomology theory can be mirrored by the structure of homotopy theory. Of course the results we get provide a vital link between the two theories.

One name I have not mentioned is Henry Whitehead. So let me say that what Henry Whitehead did involved a very beautiful idea. To go back to the beginning of homology, a topological space admitting homology was originally endowed with a combinatorial structure. Vietoris and Čech freed it of this combinatorial structure by defining homology on an arbitrary topological space. Homotopy theory was originally defined for an arbitrary topological space, and what Henry Whitehead did was to impose a combinatorial structure on the space and show how this combinatorial structure on the space could, in fact, lead you to insights into its homotopy groups. These results first appeared in papers he wrote before the war and he then rewrote them afterwards. Whitehead said about his pre-war work that he shared with Karl Marx the property of being frequently quoted and never read. And after the war he tried to deal with this by recasting a lot of his work in algebraic language. I was his first student after the Second World War; so I came under his influence in that period and that probably accounts a great deal for my early taste.

Question-and-answer session

Question: When I studied algebraic topology I never realized that Emmy Noether had anything to do with it. Is this commonly known?

Answer: No. Hopf was very clear about this, about her tremendous contribution, this wonderful insight that she had. He said that he would never have realized that, in doing homology, they were talking about Abelian groups until she pointed this out. He saw the significance of the algebraic viewpoint in homology, but his difficulty was with cohomology. Hopf said that there was this wonderful atmosphere in Göttingen when they all got together and talked. Emmy Noether listened and then she came back and said, "Well, what you're really talking about is Abelian groups." They had these Betti numbers and they told her that the Betti numbers were for manifolds, but she said that they were talking about Abelian groups. And then she said that once you're talking about groups you must be talking about homomorphisms, but in the earlier proofs of invariance, you don't find any use of that fact.

There is a good book by Andrew Wallace on homology theory. He gives, in a tiny little lemma tucked away, the fact that a map between spaces induces a homomorphism of their homology groups and that the composite of two maps induces the composite of the homomorphisms. Well, that's what topological invariance is all about, and the lemma makes it plain that you're transferring one theory into another, you're going from topology to algebra. That explains what you are doing in homology, and that is the point that Emmy Noether really clarified. Let me just tell one little story since this is Pólya country. I was present in Zürich on the occasion of Pólya's eightieth birthday and was invited to his birthday party. He and Hopf were discussing Emmy Noether and one of the two of them insisted that Emmy Noether was very ugly and the other hotly denied it.

Question: Why did some topologists fail to adopt the algebraic methods? Did they just feel that it would not yield worthwhile results?

Answer: It was the feeling that all of topology came from geometry and to some extent from analysis. Poincaré was, in the broad sense, an analyst. I think the idea of algebra as a tool in the hands of a geometer was a peculiar and strange idea. Now it seems much more natural but, in those days, it was very strange. In 1935, such a great mathematician as Élie Cartan was still just thinking in terms of Betti numbers. So this point of view was so foreign to them, this was what really held them up.

Consider, for example, this business with cohomology and the contortions that many mathematicians went into with the 'upper boundary', with things always going wrong! They could see that though you got a map of chains from, say, a simplicial map, it didn't commute with the upper boundary, and Hopf had the idea that you had to think of a map going back the other way in some way. How? They just did not think in algebraic terms of cochains as functions on chains. As a further example, although it was clear to Alexandroff and to Hopf that it was natural, in connection with the Lefschetz fixed-point theorem, to think in terms of homology with rational coefficients, they thought that homology with rational coefficients was just a means of getting rid of the torsion. They didn't think of it as a way of getting a vector space structure which is the way we would think of it now. Take rational coefficients, you get a nice rational vector space. So we can just talk about the dimension of the vector space. They just did not have these ideas.

And I should say, too, that Henry Whitehead, in a paper before the war, introduced the key idea of an exact sequence in homotopy, but the way he wrote it out was extraordinarily obscure to us. Instead of just writing that, in the sequence, the kernel of a homomorphism is the same as the image of the preceding homomorphism, he has complicated statements because he distinguishes between the homotopy of maps into X and the homotopy of maps into a subspace A. In each case he wrote down what exactness meant, and so had two inclusions, one going one way, one going the other.

All of these algebraic ideas were a very long time in the making because the people doing homology and homotopy theory were not algebraists and the algebraists didn't take any interest. The only pure algebraist who took any interest was Emmy Noether.

Question: Now it's very common that you use the techniques of one branch of mathematics to solve problems in another. Was algebraic topology the first place where this happened?

Answer: Yes, in a systematic way, unless you would say that it was already being done in analytic number theory. There you are using methods of classical analysis in order to get results in number theory. There is a sense there though that you are not doing anything but the analysis. In topology, it's a sort of wedding of the two methods simultaneously—it's not the abandonment of one for the other but it's the establishing of links between the two, and in a sense it's gone so much further now because from topology one is led into the construction of new algebraic structures—and you can get algebraic results from the topology as well. By the use of covering spaces you can get theorems from combinatorial group theory that are terribly hard without the topological methods. So we have a two-way application. Cohomology theory has now spread over the whole of mathematics through differential equations, differential operators and so forth. And in algebraic geometry, of course, homology theory has become a basic tool.

The Origins of Modern Axiomatics: Pasch to Peano

H. C. KENNEDY

American Mathematical Monthly **79** (1972), 133–136

The modern attitude toward the undefined terms of an axiomatic mathematical system is that popularized by Hilbert's remark: "One must be able to say at all times—instead of points, straight lines, and planes—tables, chairs, and beer mugs" ([20], p. 57). This view was not widely accepted before the twentieth century, and even in 1959 the well-known James and James *Mathematics Dictionary* gave "A self-evident and generally accepted principle" as first meaning of the term "axiom", although this may only be meant as a reflection of the view universally accepted before the developments in geometry in the nineteenth century. The change in attitude appears to be due to internal pressures within mathematics (what R. L. Wilder ([22], p. 170) has called "hereditary stress"). These include the flowering of projective geometry and, especially, the discovery of the non-Euclidean geometries, i.e., of the possibility of a geometry based on axioms, one of which is the negation of one of Euclid's axioms. The transition from viewing an axiom as "a self-evident and generally accepted principle" to the modern view took place in the second half of the nineteenth century and can be found in the very brief period from 1882 to 1889, from Pasch's *Vorlesungen über neuere Geometrie* [13] to Peano's *I Principii di Geometria, Logicamente Esposti* [15].

Already in 1882, Pasch showed a shift in interest from the theorems to the axioms from which the theorems are derived, when he insisted that everything necessary to deduce the theorems must be found among the axioms ([13], p. 5). Pasch was concerned that his axiom set be complete, i.e., that it furnish a basis for rigorous proofs of the theorems. ("The father of rigor in geometry is Pasch", wrote Hans Freudenthal ([5], p. 619).) There is also a strong hint of the modern attitude, as expressed in Hilbert's remark about "tables, chairs, beer mugs" in his statement:

> In fact, provided the geometry is to be truly deductive, the process of inference must be entirely independent of the meaning of the geometrical terms, just as it must be independent of the figures ([13], p. 98).

Hilbert's remark was made to a few friends in the waiting room of a railway station in 1891 but was not published until 1935 ([10], p. 403). The exposition of the axioms in his famous *Grundlagen der Geometrie* [7] begins:

> Let us consider three distinct systems of things. The things composing the first system, we will call points, and designate them by the letters $A, B, C, \ldots$" ([8], p. 3).

The viewpoint is quite clear—but he was not the first to publish this view. Pasch has already been mentioned, and Hans Freudenthal, in a study of geometrical trends at the turn of the century, says: "Hilbert had in this view too at least one forerunner, namely G. Fano,..." ([4], p. 14). He refers to Fano's statement:

> As basis for our study we assume an arbitrary collection of entities of an arbitrary nature; entities which, for brevity, we shall call points, and this quite independently of their nature ([3], p. 108).

Somewhat surprisingly Freudenthal overlooks Peano's monograph of 1889, even though it is cited in Fano's article, perhaps because Fano says that Peano's work was based on that of Pasch. Peano's work was indeed based on his reading of Pasch, but there are important innovations, and one of them is the explicit statement of the modern attitude toward the undefined terms of an axiomatic mathematical system. The first line of his exposition is: "The sign **1** is read **point**," and in his commentary he says: "We thus have a category of entities, called points. These entities are not defined. Also, given three points, we consider a relation among them, indicated by $c \in ab$, and this relation is likewise undefined. The reader may understand by the sign **1** any category whatever of entities, and by $c \in ab$ any relation whatever among three entities of that category, ... " ([15]; [18], p. 77). We find in this statement explicit acceptance of the axiomatic view. (It should be noted that Peano's view was purely methodical. As we have indicated elsewhere ([12], p. 264), he was not a member of what came to be called the 'formalist' school.)

E. W. Beth has noted ([2], p. 82):

Since the publication of D. Hilbert's *Grundlagen der Geometrie* (1899), it has become customary to require every set of axioms to be (1) *complete*, (2) *independent*, and (3) *consistent*.

Again, it was Hilbert who popularized this custom, but that these properties of an axiom set are desirable was already accepted by Peano and others. The property of consistency is indeed a *sine qua non*, but as the consistency of Euclid's axioms was never doubted, it was only with the advent of non-Euclidean geometry that attention was focused on this property, and it was not until 1868 that a consistency proof was found by E. Beltrami [1]. The property of independence can be reduced to that of consistency; we often say that Beltrami proved the independence of Euclid's parallel postulate, but this reflects a later view, that of Peano who developed this technique into a general method.

Peano's acceptance of the goal of an independent set of axioms is indicated in his *I Principii di Geometria*:

This ordering of the propositions clearly shows the value of the axioms, and we are morally certain of their independence ([15]; [18], p. 57).

In a similar remark about his axioms for the natural numbers, published earlier that year, Peano later wrote:

I had moral proof of the independence of the primitive propositions from which I started, in

their substantial coincidence with the definitions of Dedekind ([17]; [19], p. 243).

It was only in 1891, however, after he had separated the 'famous five' from the postulates dealing with the symbol =, that he showed their absolute independence ([16]; [19], p. 87).

Hermann Weyl wrote of Hilbert ([20], p. 264):

It is one thing to build up geometry on sure foundations, another to inquire into the logical structure of the edifice thus erected. If I am not mistaken, Hilbert is the first who moves freely on this higher 'metageometric' level: systematically he studies the mutual independence of his axioms and settles the question of independence from certain limited groups of axioms for some of the most fundamental geometric theorems. His method is the construction of models: the model is shown to disagree with one and to satisfy all other axioms; hence the one cannot be a consequence of the others.

This method was, as we have seen, already used systematically by Peano, although one would not learn this from reading Hilbert. In the *Grundlagen der Geometrie* there is no mention of Peano. The only Italian mentioned is G. Veronese, and the reference is to a German translation of his work. Nor does Hilbert mention Peano even in his presentation of postulates for the real numbers [9]. Indeed (without naming him) he labels Peano's development of the real numbers the "genetic method", while reserving the label "axiomatic method" for his own presentation!

A word more may be said about the originality of Peano's work. In contrast with Hilbert, Peano always tried to place his work in the historical evolution of mathematics, to see it as a continuation and development of the work of others. Furthermore, he was scrupulously honest (although sometimes mistaken) in assigning priority of discovery. Thus in *I principii di Geometria* he praises Pasch's book and indicates precisely to what extent his treatment coincides with that of Pasch, and where it differs. On the other hand, Peano's discovery of the postulates for the natural numbers was entirely independent of the work of Dedekind, contrary to what is often supposed. Jean van Heijenoort says ([6], p. 83):

Peano acknowledges that his axioms come from Dedekind,

referring the reader to the statement of Peano:

> The preceding primitive propositions are due to Dedekind ([16]; [19], p. 86).

Hao Wang says ([21], p. 145):

> It is rather well known, through Peano's own acknowledgement... that Peano borrowed his axioms from Dedekind...,

and he gives a reference to Jourdain ([11], p. 273), which in turn refers to the same passage of Peano just quoted. Since Peano had already written in *Arithmetices Principia*:

> Also quite useful to me was a recent work: R. Dedekind, *Was sind und was sollen die Zahlen*, Braunschweig, 1888 ([14]; [18], p. 22),

the conclusion of these authors would seem justified. In fact, Peano was only acknowledging Dedekind's priority of publication.

The exact story was given in 1898 when Peano wrote:

> The composition of my work of 1889 was still independent of the publication of Dedekind just mentioned; before it was printed I had moral proof of the independence of the primitive propositions from which I started, in their substantial coincidence with the definitions of Dedekind. Later I succeeded in proving their independence ([17]; [19], p. 243).

We see from this that the reference to Dedekind's work was added to the preface of *Arithmetices Principia* just before the pamphlet went to press, and we have an explanation of how Dedekind's work was "useful".

Ironically, the very modesty of Peano and his desire to see his work as in the mainstream of the evolution of mathematics have contributed to the lack of recognition of his originality. As for clarity, while giving much credit to Peano, Constance Reid ([20], p. 60) says of Hilbert that in the *Grundlagen der Geometrie* he

> attempted to present the modern point of view with even greater clarity than either Pasch or Peano.

What could be clearer than: "The reader may understand by the sign **1** any category whatever of entities"? Let the reader compare for himself the clarity of Dedekind's presentation of the foundations of arithmetic with that of Peano. There can be no doubt that the famous five axioms for the natural numbers are rightly called *Peano's Postulates*.

References

1. E. Beltrami, Saggio di interpretazione della geometria noneuclidea, *Giorn. Mat. Battaglini* **6** (1868), 284–312.

2. E. W. Beth, *The Foundations of Mathematics*, North-Holland, 1959.

3. G. Fano, Sui postulate fondamentali delta geometria in uno spazio lineare a un numero qualunque di dimensioni, *Giorn. Mat. Battaglini* **30** (1892), 106–132.

4. Hans Freudenthal, Die Grundlagen der Geometrie um die Wende des 19. Jahrhunderts, *Math.-Phys. Semester-ber.* **7** (1961), 2–25.

5. Hans Freudenthal, The main trends in the foundations of geometry in the 19th century, *Logic, Methodology and Philosophy of Science*, Stanford Univ. Press, 1962.

6. Jean van Heijenoort (ed.), *From Frege to Gödel: A Source Book in Mathematical Logic, 1879–1931*, Harvard Univ. Press, 1967.

7. David Hilbert, *Grundlagen der Geometrie*, Teubner, 1899.

8. David Hilbert, *The Foundations of Geometry* (tr. E. J. Townsend), Open Court, 1902.

9. David Hilbert, Über den Zahlbegriff, *Jber. Deutsch. Math.-Verein.* **8** (1900), 180–184.

10. David Hilbert, *Gesammelte Abhandlungen*, Vol. 3, Springer-Verlag, 1935.

11. P. E. B. Jourdain, The development of the theories of mathematical logic and the principles of mathematics, *Quart. J. Pure Appl. Math.* **43** (1912), 219–314.

12. H. C. Kennedy, The mathematical philosophy of Giuseppe Peano, *Philos. Sci.* **30** (1963), 262–266.

13. Moritz Pasch, *Vorlesungen über neuere Geometrie*, Teubner, 1882.

14. G. Peano, *Arithmetices Principia, Nova Methodo Exposita*, Bocca, 1889.

15. G. Peano, *I Principii di Geometria, Logicamente Esposti*, Bocca, 1889.

16. G. Peano, Sul concetto di numero, *Rivista di Matematica* **1** (1891), 87–102, 256–267.
17. G. Peano, Sul §2 del Formulario t. II: Aritmetica, *Rivista di Matematica* **6** (1896–99), 75–89.
18. G. Peano, *Opere Scelte*, Vol. 2, Edizioni Cremonese, 1958.
19. G. Peano, *Opere Scelte*, Vol. 3, Edizioni Cremonese, 1959.
20. Constance Reid, *Hilbert*, Springer-Verlag, 1970.
21. Hao Wang, The axiomatization of arithmetic, *J. Symb. Logic* **22** (1957), 145–158.
22. R. L. Wilder, *Evolution of Mathematical Concepts*, Wiley, 1968.

C. S. Peirce's Philosophy of Infinite Sets

JOSEPH W. DAUBEN

Mathematics Magazine **50** (1977), 123–135

> I fear I might seem to talk gibberish to you, so different is your state of mental training and mine ([31], p. 109).
>
> Charles Sanders Peirce

American mathematics, like American science generally in the nineteenth century, remained underdeveloped, depended heavily upon European models, and made few independent and recognized contributions of its own. Though presidents like Jefferson might take a pedagogical interest in mathematics and its teaching, and while Garfield, in fact, discovered an interesting variation on the many proofs of the Pythagorean theorem, American mathematics generally remained without support, either institutional or financial, until late in the century (see [28] and [37]).

Despite the lack of incentives to pursue a mathematical career in America, there were nevertheless some who made important contributions to mathematics in the United States. One of the most interesting of these, Charles Sanders Peirce, made fundamental discoveries, largely independent of his European counterparts, in set theory and mathematical logic. This paper explores the nature and significance of Peirce's contributions.

Though his study of continuity led him to produce results paralleling in some ways the contributions of Georg Cantor and Richard Dedekind in Germany, Peirce's work was dramatically different in its origins, inspiration, and ultimate mathematical character. In order to understand the fate of Peirce and his mathematical studies of continuity and infinity, it is necessary first to say something about the status of American science in general, and of American mathematics in particular, in the nineteenth century. Following a brief sketch of the major developments of set theory, largely in the hands of Cantor and Dedekind, we shall then turn to consider Peirce, his mathematics, and finally the reasons why his genius and multitudinous insights did not exert more influence upon his contemporaries than they did.

Mathematics in America in the nineteenth century

Alexis de Tocqueville, in assessing the status of science in America in the early nineteenth century, remarked that Americans found it easier to borrow their science from Europe than to pursue it earnestly themselves. "I am convinced", he wrote [41], "that if the Americans had been alone in the world... they would not have been slow to discover that progress cannot long be made in the application of the sciences without cultivating the theory of them." Mathematics, as the handmaiden in particular to astronomy and physics, was as essential, but just as neglected, as the other basic sciences in America until well past mid-century. In large measure de Tocqueville found the unusual combination of democracy and economic opportunities responsible for America's indifference to basic science. By this he meant that the egalitarian ideal encouraged the idea that anyone, with hard work, could transform the nation's national resources into personal fortune. Thus if anything was sought in science, it was only the immediate means by which nature might be exploited. Europe had its monarchies and aristocracies to encourage pure science, but de Tocqueville was certain that it could flourish equally as well in the United States, if only constituted authorities here would give it encouragement and support.

Thus while utility was highly praised for reasons that were religious, political and entrepreneurial, little interest was paid to abstract study which seemed to offer no evidence of immediate usefulness [38].

Mathematics was no exception. In fact, most of America's mathematicians until the end of the century were individuals of means, and the case of Josiah Willard Gibbs is illustrative. Gibbs taught at Yale University for many years without pay. In a country where "success" was more often than not associated with financial prosperity, it is no wonder that scientists in America played virtually no role in government or public affairs, unlike their European counterparts. Even as late as 1902, the mathematician C. J. Keyser of Columbia University, could write [21] that:

I know personally of six young men, five of whom have relinquished the pursuit of science and the (sixth) of whom told me only yesterday that he seriously contemplated doing so, all of them, for the reason that, as they allege, the university career furnishes either not at all, or too tardily, a financial competence and consequent relief from practical condemnation to celibacy.

As for the American government, it consistently refused to support any national organization for science until after the Civil War, and it was only through the bequest of an Englishman that the Smithsonian Institution was finally established (for the Smithsonian, see [10] and [20]; for American science and the Civil War, see [39], [16] and [40]). Often, in fact, Americans were better known abroad than at home. Nothing reflects so poorly upon the state of American science even towards the end of the century than a story J. J. Thomson [40] once told:

When a great university was founded in 1887, the newly elected President came over to Europe to find professors. He came to Cambridge and asked me if I could tell him of anyone who would make a good professor of molecular physics. I said, "You need not come to Europe for that; the best man you could get is an American, Willard Gibbs." "Oh", he said, "you mean Wolcott Gibbs", mentioning a prominent chemist. "No, I don't", I said, "I mean Willard Gibbs", and told him something of Gibbs' work. He sat thinking for a minute or two and then said, "I'd like you to give me another name. Willard Gibbs can't be a man of much personal magnetism or I should have heard of him".

Mathematicians in America could blame their lack of status on apathy and indifference. One mathematician assessing mathematical productivity in the United States echoed de Tocqueville's words when he noted that "educational and scientific activity shall come to be generally understood, and especially in proportion as we learn to value the things of mind, not merely for their utility, but for their spiritual worth" ([21], p. 346; see also [27] for American indifference towards mathematics). The assessment was C. J. Keyser's, and in part he blamed the low level of productivity by American mathematicians before the turn of the century upon their isolation, saying that "in general there was no suspicion that, on the other side of the Atlantic, mathematics was a vast and growing science" ([21], p. 347).

Throughout the nineteenth century, mathematics, like the sciences generally, became more complex, more technical, more specialized. More formal training was required, more professionalization, and at first Americans clearly followed the pattern de Tocqueville had described in 1835. If Americans wanted to learn the newest techniques, to study the latest theories, they went to Europe, many to Germany, and to such centers for mathematics as Berlin and Göttingen. In rare instances, Europeans came to America. Perhaps no one was more influential for the future of American mathematics in this capacity than was J. J. Sylvester, whose arrival in America accompanied another significant development ([27], p. 463, and [28], pp. 30–32). As a counterpart to the growing specialization of European science, advanced training at the graduate level was regarded increasingly as imperative. Consequently, following the examples of the great European schools like the *École Polytechnique* and the University of Berlin, America's first graduate school was founded in Baltimore.

In 1876 Johns Hopkins University opened, and one of its first great attractions was the English mathematician J. J. Sylvester. Both he and Johns Hopkins exerted a tremendous influence upon American mathematics, in part through the *American Journal of Mathematics*, edited by Sylvester and begun at Johns Hopkins in 1878. It was perhaps symptomatic that not only was the journal originally edited by an Englishman, but that to a great extent the journal's articles were written by foreigners. Since Johns Hopkins was primarily a graduate school, it served to encourage graduate studies at other U.S. universities. Two similar institutions were founded by the end of the century: Clark University in Worcester, Massachusetts, in 1889; and the University of Chicago in 1892, which was extremely influential in the Midwest.

But the single strongest factor in organizing and promoting mathematical research in America was the *American Mathematical Society* (see [21], pp. 350–352 and [28], pp. 30–32). Its precursor, The New York

Mathematical Club, had been founded in 1888, and at first was little more than a small group meeting at Columbia University. But soon the Mathematical Club became the New York Mathematical Society, publishing a monthly bulletin. In 1894 this organization was again transformed, becoming the American Mathematical Society with membership then at nearly four hundred. Bi-monthly meetings were held, summer meetings were scheduled around the country, and soon sections scattered from coast to coast were established. The first was the Chicago section, chaired by E. H. Moore, on April 24, 1897; five years later, in 1902, the San Francisco Section was chaired by Irving Stringham; and shortly thereafter, in 1906, the Southwestern Section was established in St. Louis by E. R. Hedrick.

In terms of its societies, journals, and publishing mathematicians, mathematics in America had come a long way from its status early in the century, when arithmetic was still taught in the first year at Harvard College, and only became an entrance requirement in 1816. By the end of the nineteenth century, in many respects if not completely, the words of the French mathematician Laisant were in large measure true. As he surveyed the progress Americans had made in mathematics, he concluded [22] that:

> Mathematics in all its forms and in all its parts is taught in numerous universities, treated in a multitude of publications, and cultivated by scholars who are in no respect inferior to their fellow mathematicians of Europe. It is no longer an object of import from the old world, but it has become an essential article of national production, and this production increases each day both in importance and in quantity.

Before turning to assess the character and significance of the contribution made by C. S. Peirce to American mathematics, in particular to various aspects of set theory, it is necessary to survey briefly the state of that art in Europe at the time. Above all, until about 1895 at least, the development of transfinite set theory was almost exclusively the work of the German mathematician Georg Cantor.

Georg Cantor (1845–1918)

Georg Cantor, creator of transfinite set theory, published a theorem in 1872 which brought him to the attention of the mathematical world, and which also marked the beginning of his work in set theory. His theorem established the uniqueness of functional representations by trigonometric series over domains from which certain infinite sets of points could be excepted [2]. The only restriction limited the set of exceptional points to first species sets P, ones for which the nth derived set P^n was empty for some finite value of n.

But in order to provide a satisfactory foundation for his proof, Cantor discovered that he was obliged to introduce a rigorous construction of the real numbers. He did so in terms of equivalence classes of infinite sequences of rational numbers subject to the Cauchy criterion for convergence, and was also led to formulate his Axiom of Continuity, which postulated the equivalence of the arithmetic and geometric continuums. In the same year, 1872, Richard Dedekind (1831–1916) published his construction of the real numbers in terms of his famous "cuts", and he did not fail to acknowledge the similar work Cantor had done in his paper on trigonometric series [12].

Cantor's unexpected results seem to have spurred his interest in the properties of continuous domains in general, and late in 1873 he discovered that the set of all real numbers was non-denumerable ([3]; see also the correspondence [9]). In 1879 he finally managed to publish a startling proof showing that any continuous space of n dimensions could be mapped (though not continuously) in a one-to-one fashion onto the real line. Cantor was so unprepared for this discovery that it prompted one of his most oft-quoted remarks: "I see it, but I don't believe it" ([9], p. 34).

Cantor's first systematic presentation of his transfinite numbers was published in 1883. His *Grundlagen einer allgemeinen Mannichfaltigkeitslehre* [4] was as much philosophy as it was mathematics, a combination that was also to be characteristic of much of C. S. Peirce's research. In the *Grundlagen* Cantor introduced his transfinite ordinal numbers. He began by identifying two principles of generation. The first produced new ordinals by the successive addition of units, hence

$$1, 2, 3, \ldots, n, n+1, \ldots.$$

The second principle was called upon to introduce a new number representing the totality of all ordinals generated by the first principle, when such a sequence continued without coming to an end. For example, though it was not permissible to think of a last of all natural integers, one could posit a least number coming after *all* the natural numbers. This number Cantor defined as ω, and it represented the totality of all the positive integers in their natural order. It was then possible to apply the first principle of generation to produce higher transfinite ordinals: $\omega+1, \omega+2, \ldots, \omega+n, \ldots.$ When this sequence continued without end, Cantor again called upon his second principle of generation to produce the least number following all those of the form $\omega + n$, namely the ordinal number 2ω, and so on. Later Cantor would define the second number class of such transfinite ordinals as the class $\mathcal{Z}(\aleph_0)$, the totality of all order types α of well-ordered sets of cardinality $\aleph_0$. The power of this second number class $\mathcal{Z}(\aleph_0)$ Cantor denoted by the second transfinite cardinal number $\aleph_1$ [7].

Cantor, of course, was not the only mathematician interested in the properties of infinite sets. In 1888, Richard Dedekind published a small pamphlet, *Was sind und was sollen die Zahlen* [13], in which he introduced, among other things, the distinction between finite and infinite collections that has since become a classic: a system S is said to be *infinite* when it is similar to a proper part of itself; in the contrary case S is said to be a *finite* system. This definition is of particular interest for it seems that Peirce had been led to an equivalent distinction, even earlier, and for very different reasons.

In 1891 Cantor published his famous method of diagonalization, whereby it was possible to generate an unending sequence of sets of greater and greater cardinality [6]. In 1874 Cantor had shown only that the set of real numbers was non-denumerable. The result of 1891 was considerably more powerful, and impressively general, for he was able to show that for any exponent $\aleph$, the power $2^{\aleph} > \aleph$. If $\aleph_0$ be taken as the cardinality of the denumerable set of natural numbers, then $2^{\aleph_0}$ was a set of greater cardinality representing the set of all real numbers. Moreover, Cantor could show that there were sets of cardinality even greater than the real numbers — for example the set of all single-valued functions on the interval $(0, 1)$.

Between 1895 and 1897 Cantor's most ambitious and influential work appeared in the *Mathematische Annalen*: his "Beiträge zur Begrundung einer transfiniten Mengenlehre" ([7] and [8]). In Part I (1895) he presented his general theory of the order types of simply-ordered sets like the rationals taken in their natural order (type η), and the reals (type θ); in Part II (1897) he defined his transfinite cardinal numbers in terms of well-ordered sets, and explored in detail basic arithmetic properties of his transfinite numbers. He also took the opportunity to condemn the doctrine of infinitesimals in the recently published work of the Italian mathematician Giuseppe Veronese. Cantor had always been an ardent opponent of infinitesimals, and at one point called them the "infinitary cholera bacillus of mathematics" (see a letter of Cantor's cited in [25]). Early in his career Cantor rejected the idea of infinitesimals, and when Mittag-Leffier asked if there might not be other kinds of numbers between the rational and real numbers, Cantor responded with an emphatic "no" (see [26], p. 234, and [5]). In 1887 Cantor published a proof of their logical impossibility, based not surprisingly upon the Archimedean character of what he called linear numbers, and some years later Peano published a similar proof against infinitesimals in his own journal, the *Rivista di Matematica* [30]. To have admitted infinitesimals, it might be added, would have complicated greatly Cantor's continuum hypothesis, which asserted that the cardinality of the set of all real numbers, $2^{\aleph_0}$, was equal to that of his second number class, in other words, $2^{\aleph_0} = \aleph_1$. To allow infinitesimals in addition to the rationals and irrationals would have made this conjecture concerning the power of the continuum considerably more complicated.

By the end of the century, the status of Cantor's work was brought dramatically into question by discovery of the paradoxes of set theory. While Burali-Forti was the first to publish his paradox of the largest ordinal number, Cantor had discovered the paradoxes of both a largest ordinal and cardinal number even earlier, probably as early as 1895. Cantor sketched a proof for Dedekind in 1899, in which he concluded that it was a direct consequence of the paradoxical nature of the unending sequence of *all* cardinals that the continuum must be a set whose cardinality was one of Cantor's transfinite alephs ([43], pp. 443–450; also see [19]).

But in 1897 Burali-Forti drew a very different conclusion from his study of the collection of all ordinal numbers. Such a collection, he argued, must have an ordinal number δ greater than any ordinal in the collection. But if the set contained all ordinal numbers, then it must contain δ, and Burali-Forti was forced to the contradictory conclusion that $\delta > \delta$. From this he did not draw comfort, as did Cantor, in alleging that this was the key to solving much deeper problems of set theory. Instead, Burali-Forti concluded that mathematicians could only agree to abandon any hope of strict comparability between transfinite numbers [1]. In 1902 Bertrand Russell constructed a strictly logical paradox and shocked Frege by showing that there were certain antinomies inherently part of logic, and consequently of mathematics as a form of structured reason (see Frege's reaction in [18], Vol. II, 443–450).

In considering the paradoxes of set theory, in particular those of the greatest ordinal and cardinal numbers, Peirce agreed with Bertrand Russell that these were, properly speaking, questions of logic. The basic business of mathematics, for Peirce, was the formation of hypotheses ([31], pp. 41, 73, 769, 785, 875, 916). In terms of set theory, this meant the determination of what grades of multitude between infinite collections were mathematically possible. And, as we shall see, Peirce drew from the paradox of the largest cardinal number a principle which he felt might help to resolve the question of the *true* nature of continuity.

Charles Sanders Peirce (1839–1914)

C. S. Peirce, the son of Benjamin Peirce, was born in Cambridge, Massachusetts, on September 10, 1839 (for studies of Peirce's life and works, see [17], [29], [42]). Benjamin Peirce, a professor of mathematics and natural philosophy at Harvard University, was careful to direct his son's schooling, and saw that young Charles had as rigorous a scientific education as he and the private schools of Boston could provide. When Peirce graduated from Harvard with an Sc. B. in chemistry, in 1863, he did so *summa cum laude*. But Peirce was not to go on immediately to devote himself to the study of pure science. His interests ran more to the study of method and logic, and in hopes of gaining more experience in the nature and method of scientific investigation, he joined the U.S. Coast Survey. For more than thirty years Peirce remained with the Survey, and in addition to working on the nautical almanac, he conducted numerous pendulum experiments, was a special assistant in gravity research, and devoted a good deal of time to the observation of solar eclipses.

As for teaching, Johns Hopkins University made it possible for Peirce to lecture in logic from 1879 to 1884, and some of his earliest work of relevance to set-theoretic problems dates from this period. In fact, in 1881, Peirce published a paper in the *American Journal of Mathematics*, "On the Logic of Number" [33], in which (he was later always proud to emphasize) he had characterized the difference between finite and infinite sets well before Dedekind had done so in 1888. Peirce asserted that Dedekind's *Was sind und was sollen die Zahlen* was doubtless influenced by his own paper, because Peirce had sent Dedekind a copy (see [31], pp. 883 and 1117, and [32], Vol. 3, 564). But the most interesting feature of Peirce's entire approach to mathematics was not the way in which it was like the research then being conducted in Europe, but in the ways it was unlike the approaches taken by Georg Cantor and Richard Dedekind to the problems of continuity and infinity.

Cantor, as we have seen, was motivated to study the continuum of real numbers as a result of his early study of the representation theorem for trigonometric series. Similarly, Dedekind's characterization of the continuum and his introduction of the now famous "Dedekind cut" to define the real numbers was also inspired by analysis. In trying to teach the basic elements of the differential calculus, particularly theorems involving limits, Dedekind realized that geometric intuition, though a guide, was not rigorously satisfactory. And so he turned to produce a purely arithmetic study of continuity and the irrational numbers (see Dedekind's discussion in [12], pp. 1–3).

Peirce, on the contrary, took an entirely different approach. His inspiration was not analysis, and his interests were not in probing the foundations of mathematics in order to provide a certain, unshakable beginning from which function theory could proceed without difficulty. Instead, Peirce was led to study the mathematics of infinity, infinitesimals, and continuity as a result of his interests in logic and philosophy. In this difference lies the key to understanding why Peirce differed so markedly from Cantor and Dedekind in his approach to the problems of continuity and the infinite.

Peirce's first publication describing the difference between finite and infinite classes appeared in 1881, while he was lecturing in logic at Johns Hopkins. His paper ([32], Vol. 3, 256) began with a definition (though insufficient) of continuity:

> A continuous system is one in which every quantity greater than another is also greater than some intermediate quantity greater than that other.

But since the rationals would be continuous under this definition, Peirce's description is inadequate, although it does represent a necessary feature of any continuum. Peirce, however, was only beginning his study of continuity at the time; the most interesting feature of his paper appeared towards the end, where he offered a distinction between finite and infinite collections. He announced that a set was finite if no one-to-one correspondence could be found between the set and any proper subset. Peirce's favorite example (([31], p. 772; [32], Vol. 3, 288) was a syllogism which appeared in numerous equivalent forms throughout his mathematical and logical writings:

> Every Hottentot kills a Hottentot,
> No Hottentot is killed by more than one Hottentot,
> Hence, every Hottentot is killed by a Hottentot.

The syllogism is true only if the set of Hottentots is finite. The form of the syllogism, Peirce noted, was due to De Morgan, who called it the syllogism of transposed quantity [15]. Thus Peirce's interest in the infinite was inspired by studies in logic and the consequences one might draw from the *syllogism of transposed quantity*. In keeping with his interests, and in pushing his study of quantity, both finite and infinite, as far as he could, Peirce decided that a perfectly logical definition of continuity was needed, and in 1897, when he published "The Logic of Relatives" [33] in *The Monist*, he wrote that:

> A perfectly satisfactory logical account of the conception of continuity is required. This involves the definition of a certain kind of infinity, and in order to make that quite clear, it is requisite to begin by developing the logical doctrine of infinite multitude. This doctrine still remains, after the works of Cantor, Dedekind, and others, in an inchoate condition. For example, such a question remains unanswered as the following: is it, or is it not, logically possible for two collections to be so multitudinous that neither can be put into a one-to-one correspondence with a part or the whole of the other? To resolve this problem demands, not a mere application of logic, but a further development of the conception of logical possibility.

But what did Peirce mean by the need to define a certain kind of infinity before the concept of continuity could be accounted for logically? What was "inchoate" about the work of Cantor and Dedekind? Why was the comparability of cardinals, in Peirce's view, impossible to establish without developing further the concept of logical possibility? What, in fact, did Peirce mean by logical possibility? The answers to all these questions hinge on Peirce's view of the infinite, and upon a very important discovery, one he apparently made independently of Georg Cantor, and one for which Peirce was always, and justifiably so, very proud.

Peirce proved (though exactly when he did so for the first time is unclear), that the power of the set of all subsets of a given set is always greater than the power of the original set itself. In other words, for any exponent $\aleph$, $2^\aleph > \aleph$, a result, as Peirce put it, of "prime importance" (see [31], pp. 51, 777, 785; also [32], Vol. 4, 200–212). Beginning from the smallest infinite set, the set of all integers, he concluded that it was always possible to produce increasingly larger sets of greater and greater power. Peirce designated the collection of all integers "denumerable". The set of all real numbers comprised what he called the "first abnumeral" or the "primipostnumeral" multitude. The set of all subsets of the real numbers produced the "second abnumeral", or the "secundo-postnumeral" multitude, and so on. Since one could always form from such sets the set of all subsets, Peirce noted that there could be no maximal multitude. But Peirce also commented that as for the second abnumeral, mathematics could offer no example of such a multitude, and added that in fact mathematics had no occasion to consider multitudes as great as this (which suggests that Peirce was unaware of, or did not fully appreciate Cantor's paper [6]), a comment that is somewhat puzzling, as we shall see, particularly in light of his construction of infinitesimals and his assertion that continua were greater in power than *any* postnumeral multitude.

In a letter to his friend E. H. Moore, Peirce commented ([31], p. 925) upon the significance of his discovery that there was no maximum multitude:

> Here we have a hint about continuity... The continuum is a General. It is a General of relation. Every General is a continuum vaguely defined.

By a "General" Peirce seems to have meant that which was neither discrete nor definite, "General" as opposed to particular or to something completely specified. But what did Peirce see in all this to help solve the mystery of continuity?

Peirce argued that if a continuum did not contain all the points that it possibly could, then there would be gaps or discontinuities present ([31], pp. 880–881). Thus it was a problem of the utmost importance to determine what the maximum possible multitude *was* in order to determine the power of continua. But Peirce had already shown that there could be *no* such greatest possible multitude, that the process of forming power sets was unending, and that consequently it was a process that remained potential, indefinite. Similarly, if the continuum were to contain the maximum possible multitude of points, it had to be correspondingly potential, indefinite.

Since Peirce regarded the essence of multitudes to be their definiteness, thus making it possible to determine their powers or cardinalities, it was reasonable for him to assert that since the collection of all abnumerals was unending, potentially infinite, thus entirely indefinite, it could not properly be called a set. Or, in Peirce's terminology, it could not be called a *multitude*. Likewise, the elements constituting a continuum could not be regarded as comprising a set or multitude of objects. Thus the complete determination of continua was impossible, for Peirce regarded the concept as intrinsically potential, essentially general ([31], pp. 62 and 981).

Consequently, Peirce was led to reject Cantor's view that the geometric continuum was somehow made up of a multitude of points. Peirce realized that there were two features of such continua that had to be considered: one involved quantity, the other involved order. Cantor had published his analysis of what he called the simply-ordered type θ of the continuum of real numbers in 1895 [7]. Peirce, however, disagreed on two counts. The collection of points comprising any continuum must be infinitely larger than any Cantor contemplated because, Peirce claimed, the real numbers $\mathbb{R}$ as defined by Cantor were grossly insufficient to account for the geometric continuum ([31], p. 122). This was so, he argued, because of his discovery that the set of abnumerals was unending. Since the line must contain all points possible, and since the set $\mathbb{R}$ corresponded to the multitude represented by $2^{\aleph_0}$, it could not possibly, as a completed multitude, account for the nature of continuity. As for the question of the order of elements constituting a continuum, Peirce suggested that something like the proper idea was approximated if between every pair of rational numbers one inserted a sequence of irrational numbers. Between any two irrationals of this sequence one could pack yet another such sequence, and so on, without end. Thus Peirce thought it was possible that between any two points of the continuum, however close, one could always pack sets of points of higher and higher multitude. The continuum would eventually be "cemented together", and not by virtue of discrete points ([31], p. 98).

Peirce illustrated his case by imagining a series of photographs ([31], p. 59). No matter how close the intervals, no motion will appear in any of them. But our perception of motion in time shows that time must be more than a succession of instants. More than single, isolated point "instants" must be present to our consciousness. What is this something more? Peirce claimed that

1. in a sensible time there exists room for *any* multitude of distinct instants.
2. The instants are so close together they merge and cannot be distinct.

And this view of time and its continuity greatly influenced Peirce's view of the continuum and the logical status of continuity. Just as parts of time merged to lose their identity, so too the points of a line. If continuity consisted of nothing but a special type of serial order, he argued that two continuous lines thought to intersect might actually "slip through" each other, this presumably being possible wherever spaces between one object and its "next" in the serial ordering on each line might occur. This certainly suggests that Peirce had a very different way of thinking about continuity and serial order than did Cantor.

Peirce, as he put it himself, took the very word *continuity* to mean that the instants of time or the points of a line were everywhere "welded together" ([31], pp. 61–62). As evidence of this, he seems to have been content to argue that its justification was nothing more than common sense. Peirce concluded that the instants of time did not constitute a multitude (or set, whose elements had to be distinct and definite). In fact, the collection of instants in any continuum of time had to be *more* than any multitude.

To illustrate his idea of continuity, Peirce outlined a procedure which he called interpolation on the unit interval, and which involved decimal representations given with only the digits 0 and 1. At step (I) there was only one interpolation, at step (II) there were 2, at step (III) there were four, at step (IV) there were 8, and at step (N) there would be 2^N. Schematically:

```
       0.000...                                                    1.000...
(I)                              .100...
(II)              .010...                         .110...
(III)   .001...          .011...         .101...         .111...
```

Here Peirce found, as he put it, a "premonition of continuity". In carrying out this procedure, a non-denumerably infinite number of interpolations would be made. Somehow this process was to help Peirce explain how the continuum was able, in his words, to "stick together" ([31], p. 88). No finite collection could ever "stick together" in any order, nor could any denumerably infinite collection if considered in any well-ordered form. The difficulty in describing the continuity of the real line, Peirce believed, reduced to the fact that numbers *per se* could never account for continuity ([31], 3:93–94 and [32], 3:568). Numbers expressed nothing but the order, he believed, of discrete objects. Nothing discrete could possibly be multitudinous enough to account for the continuum.

For example, Peirce noted that in supposing the countable collection of the set of rational points to be completely present on the line, one was in effect supposing also the collection of irrational points, in the sense that the irrationals could be considered as interpolated between the rationals ([31], 3:94, 3:.880–881). Hence the denumerable set of rationals carried with it the existence of a non-denumerable collection of points—namely, the first abnumeral multitude of irrational numbers. In exactly the same way, said Peirce, the system of irrational points on a line led to a secundo-postnumeral collection of points interpolated between the irrationals. This secundo-postnumeral collection for Peirce involved his infinitesimals, and it is possible to understand the role they played more clearly by returning for a moment to his diagram of decimal interpolations.

In a letter to M. F. C. S. Schiller of 1906, Peirce explained that by a Leibnizian infinitesimal of the first order he meant an assumed quantity smaller than any finite positive quantity ([31], 3.989). It was the *first* quantity after the sequence $0.1, 0.01, 0.001, \ldots$. Peirce believed it was impossible to prove that there was no such quantity. In fact, he believed his infinitesimals were given an imprimatur of sorts by nature, since he took their existence to be necessary for physics ([31], pp. 898 and 957). To support this claim of the physical reality of infinitesimals, Peirce referred to the experience of memory. The perception of the flow of time must extend, he said, beyond a single instant. Yet Peirce could not see how such phenomena could be satisfactorily explained unless time were believed to be strictly infinitesimal. Moreover, there were reasons from physics which also established the necessary existence of infinitesimal magnitudes.

In a letter written in 1908 to P. E. B. Jourdain ([31], pp. 123–124, [32], Vol. 3, p. 570), Peirce struck up a theme which Cantor had sketched, but never developed, in a short article of 1885. Cantor had conjectured that in order to explain satisfactorily the phenomena of nature, one had to suppose two sorts of monads, material and aetherial. He conjectured that the power of the set of all material monads was denumerable, of cardinality $\aleph_0$, while the set of aetherial monads was equipotent with the second cardinal, $\aleph_1$.

Peirce applied this idea in order to explain how matter could act on the brain to produce thought. Peirce posited two fluids, one vortex of matter-monads, and an infinitely subtler vortex of soul-monads, where the diameters of the soul-monads were to be taken as infinitesimals of infinite order, which Peirce felt was quite appropriate for the character of soul-monads.

This theory was given the name of the "introvortical theory" ([32], Vol. 3, 896). It all helped to convince Peirce that with physical counterparts for his logical infinitesimals, there was pragmatic justification for arguing their validity.

Peirce's most important reason for insisting that his infinitesimals were acceptable involved their self-consistency. Logically, being non-contradictory, there was no reason not to admit them into mathematics. Though Peirce did not undertake a careful arithmetic investigation of the properties of his infinitesimals, nor did he undertake any investigation of non-archimedean systems in general, he was a proto-proponent of non-standard analysis in believing that there was perhaps more to understanding the nature of continuity than the rationals and irrationals together could manage to explain.

To make Peirce's point as clear as possible, it may be helpful, in closing, to sketch his reasons for arguing that the rationals and irrationals together were insufficient to account for the nature of continuity. While Cantor and Dedekind regarded the irrational numbers as completing the rationals, and thus conferring completeness on the real numbers, Peirce saw the relation between rationals and irrationals somewhat differently. He concluded that there was a kind of "nextness" in the reals which actually constituted a breach of continuity ([31], pp. 125, 121–122). Suppose, he argued, that we do have a clear idea of a sequence of real numbers. If we have a clear idea of their order, it can be assumed that any set of objects sufficiently multitudinous can be similarly ordered. Assume each of these objects to be replaced with a sequence of points, similar in order to the real numbers on the open interval $(0, 1)$. But there is no reason to stop here, and Peirce went on to replace each point of such series by yet other series, and so on, without end. In his own words ([31], p. 126):

The result is, that we have altogether eliminated points. We have a series of series of series, *ad infinitum*. Every part, however closely designated, is still a series and divisible into further series. There are no points in such a line. There is no exact boundary between any parts.

It would be easier to interpret the significance Peirce attached to his infinitesimals for mathematics had he ever commented explicitly upon Cantor's and Dedekind's axioms of continuity, which hypothesized the equivalence of the standard Archimedean arithmetic continuum and the geometric continuum. But all we have is Peirce's allegation that the real numbers were insufficient to account for the continuity of space and time. As for analysis, however, it never had need to consider multitudes greater in magnitude than that of the first abnumeral, which meant that the set of real numbers was enough for the interests of analysis and presumably for all of mathematics ([31], p. 85).

Peirce wanted to explore the logical boundaries of the possible in terms of the infinitely large and the infinitely small, and found no logical contradictions or constraints in either conception. Since his interests were primarily philosophical, it is perhaps easier to understand why he never submitted his ideas to a more searching mathematical analysis than he did.

Ultimately Peirce's infinitesimals remained vague rather than rigorously defined mathematical entities. He never suggested how they might be useful in analysis. But then Peirce's interests had never been inclined towards analysis. From the very beginning he had been inspired by the purely logical implications of the syllogism of transposed quantity, and the logic of relations. Thus, unlike Cantor, he was not concerned to develop the arithmetic properties of his ideas, since their existence as numbers was not for him of great consequence. He was interested in illuminating a deep philosophical problem of long standing, namely that of the continuum, and he felt that conceptually he had found an approach to the subject that was the most satisfying of all.

Finally, how can we now draw all of these ideas together in order to interpret Peirce's claim that "the continuum is a General", meaning that it could not be defined as a set in Cantor's sense of a collection of distinct elements, and Peirce's statement that "Infinity is nothing but a peculiar twist given to generality"? (See Peirce's letter to William James, [32], Vol. 8, 268.) Peirce took generality to involve the potential infinite, in that anything not general, anything specific, was completed in some sense or other. He also took infinity to be a potentiality, something never completed, and thus it too reposed upon generality. Reasoning about such matters, he noted with perhaps too little emphasis, was always exceedingly puzzling.

Almost all metaphysicians, and even mathematicians, he added, had fallen into pitfalls concerning the infinite where the "ground was spongy" ([31], p. 79). Trouble inevitably stemmed from quantifiers like "all and every", but if one were always certain to determine *how* objects in question were to be selected, he believed that erroneous reasoning could easily be avoided. The trouble with the collection of all abnumerable multitudes was simply that it could never be considered as completed. It was self-generating, without end. Such collections were so great that they were no longer discrete, and not being complete, no definite determination of their magnitude could be given. They were potential and as such, general. In the same way the line was general, indeterminate, since between any of its points, Peirce imagined that more could be packed representing supermultitudinous collections. To quote Peirce directly ([31], p. 95):

> Such supermultitudinous collections stick together by logical necessity. Its constituent individuals are no longer distinct, or independent. They are not subjects but phases expressive of the properties of the continuum.

Peirce offered a graphic illustration. Suppose a collection of blades were to cut the line. So long as they did not comprise a supermultitudinous collection, the line would be cut up into bits, each of which was still a line. Peirce therefore urged mathematicians to discard all analytical theories about lines, and recommended that they begin from his view, a synthetic point of view ([31], p. 96). By a simple mental experiment, he believed he had shown that the line refused to be cut up into points. But even if mathematicians refused to accept his arguments, Peirce insisted that only one thing mattered: his idea of continuity and of infinitesimals involved no contradiction. In closing, he warned ([31], pp. 87–89):

> I am careful not to call supermultitudinous collections multitudes. Multitudes imply an independence of the individuals of one another which is not found in supermultitudinous sets. Here the elements are cemented together, they become indistinct.

Here Peirce had reached the potentiality, the generality that he had earlier said was essential if one were to understand properly the structure of continuity. For Peirce, the essence of continuity depended upon the supermultitudinousness of the elements of the line, and their "intrinsic arrangement which is inseparable from the particular grade of multitude in which those phenomena of cohesion are found" ([31], pp. 98–99).

It is now possible to see what Peirce meant when he wrote to Paul Carus, editor of *The Monist*, to say that at last he had seen where Cantor had gone wrong ([31], p. 792). Continuity could not come from any collection of points because points were discrete, determined, and if anything, points represented *discontinuities* when removed from the line. In summarizing his view for Judge Francis Russell in 1909, Peirce interpreted the essence of continuity in terms of the potential and completely general nature of the ideas involved. "As to a straight line not having any definition proper, it is demonstrable that it cannot be, properly speaking, defined" ([31], p. 981).

If we may take "properly speaking" to mean "mathematically speaking", then there was not much Peirce could hope to offer in such a view to his mathematical colleagues. But it is also true that he did not feel mathematics needed to go so far, apparently, in the analysis of the continuum as he had gone. Peirce was interested in pushing the logical consequences of his ideas to their ultimate conclusions for the sake of philosophy, but analysis, he seemed to say, could stop at something less.

For Peirce, such ideas justified themselves as a matter of instinct, of common sense ([31], p. 56). He had always held that his intuitive understanding of the continuum, or the continuity of space and time, were the ultimate guides in his analysis of continuity. Nothing could have been further from the aims of Weierstrassian analysis, which sought to reject all such intuitions and aimed to base mathematics upon more certain foundations of arithmetic. Weierstrass had constructed examples of everywhere continuous, nowhere differentiable functions to show the inadequacy of intuition. But Peirce was not convinced, and once even commented that Weierstrassian mathematics, in showing a distrust of intuition, betrayed an ignorance of fundamental principles of logic ([31], p. 968). Peirce followed his intuitions as far as they would carry him, and it may have been this feature of his thinking, as much as his interest in philosophical and metaphysical arguments, that prevented his being more readily accepted by those of his contemporaries even aware of his work.

Conclusion

Peirce once said that it was the business of science and mathematics to guess. For mathematics this reduced to the fabrication of hypotheses to be tested for logical self-consistency ([31], p. 893). If no contradictions could be deduced, then the hypothesis, or the mathematical theory in question, stood as acceptable. This was the basis upon which Peirce argued most persuasively for the respectability of his ideas concerning infinitesimals and continuity. Kepler, to Peirce's mind, was the greatest guesser the history of science had ever seen. But in terms of American mathematics at the turn of the century, Peirce may easily have been an equally impressive guesser, producing the mathematical hypothesis of infinitesimals.

Kepler lacked sufficient mathematical techniques and a theory of gravitation to establish a convincing explanation of his laws, deficiencies Isaac Newton would later remedy. In much the same way, Peirce lacked sufficient techniques to produce a rigorous theory of non-Archimedean systems. But his hypotheses were eventually vindicated, in the twentieth century, by mathematicians like Schmieden, Laugwitz, Robinson, and Luxemburg (see [36], [23], [35], [24]). Perhaps until the middle of this century, only a mathematician as interested in philosophy as was Peirce, and as isolated from the prevailing assumptions of established mathematics of the nineteenth century, could have pursued the problem of continuity in the way he did. Out of touch with much of European mathematics, Peirce considered infinitesimals with an unprejudiced eye to affirm not only their existence, but to argue as well that the arithmetic continuum of real numbers was only a very incomplete picture of the actual richness of any continuum. Working in obscurity, penury, and isolation, he nevertheless saw, if only a glimmer, what later generations might be more willing to accept. In light of current research in non-standard analysis, it is now possible to consider, more rigorously than did Peirce, alternatives to the traditional nineteenth-century view of the standard Archimedean continuum.

References

1. C. Burali-Forti, Una questione sui numeri transfiniti, *Rend. Circ. Mat. Palermo* **11** (1897), 154–164.

2. G. Cantor, Über die Ausdehnung eines Satzes aus der Theorie der trigonometrischen Reihen, *Math. Ann.* **5** (1872), 123–132.

3. G. Cantor, Über eine Eigenschaft des Inbegriffes aller reellen algebraischen Zahlen, *J. Reine Angew. Math.* **77** (1874), 258–262.

4. G. Cantor, *Grundlagen einer allgemeinen Mannichfaltigkeitslehre. Ein mathematisch–philosophischer Versuch in der Lehre des Unendlichen*, Teubner, 1883.

5. G. Cantor, Mitteilungen zur Lehre vom Transfiniten, *Zeit. für Phil. und philosophische Kritik* **91** (1887), 81–125; **92** (1888), 240–265.

6. G. Cantor, Über eine elementare Frage der Mannigfaltigkeitslehre, *Jahresbericht der Deutschen Matematiker–Vereinigung* **1** (1891), 75–78.

7. G. Cantor, Beiträge zur Begründung der transfiniten Mengenlehre (Part I), *Math. Ann.* **46** (1895), 481–512.

8. G. Cantor, Beiträge zur Begründung der transfiniten Mengenlehre (Part II), *Math. Ann.* **49** (1897), 207–246.

9. G. Cantor and R. Dedekind, *Briefwechsel Cantor–Dedekind* (ed. E. Noether and J. Cavaillès), Hermann, 1937.

10. L. Carmichael, *James Smithson and the Smithsonian Story*, Putnam, 1965.

11. I. B. Cohen, Science and the Civil War, *Tech. Review* **48** (1946).

12. R. Dedekind, *Stetigkeit und irrationale Zahlen*, Vieweg, 1892; in [14].

13. R. Dedekind, *Was sind und was sollen die Zahlen?*, in [14].

14. R. Dedekind, *Essays on the Theory of Numbers* (transl. W. Beman), Dover, 1965.

15. A. De Morgan, On the syllogism and the logic of relatives, *Cambridge Philo. Trans.* **10** (1860), 331–358.

16. A. H. Durpee, *Science in the Federal Government: A History of Policies and Activities to 1940*, Harvard Univ. Press, 1957.

17. C. Eisele, Charles Sanders Peirce, Vol. X, pp. 482–488, in *Dictionary of Scientific Biography* (ed. C. C. Gillespie), Scribner's, 1974.

18. G. Frege, *Grundgesetze der Arithmetik, begriffsschriftlich abgeleitet*, Verlag Hermann Pohle, 1903.

19. I. Grattan-Guinness, The rediscovery of the Cantor-Dedekind correspondence, *Jahresbericht der Deutschen Mathematiker-Vereinigung* **76** (1974), 104–139.

20. W. Karp, *The Smithsonian Institution*, Smithsonian Inst., 1965.

21. C. J. Keyser, Mathematical productivity in the United States, *Educ. Review* 24 (1902), 356.

22. C. A. Laisant, *La mathématique philosophie-enseignement*, Carré and Naud, 1898.

23. W. A. J. Luxemburg, *Non-Standard Analysis*, Lecture Notes, Cal. Tech., 1962 and 1964 (rev.).

24. W. A. J. Luxemburg and A. Robinson, *Contributions to Non-Standard Analysis*, North-Holland, 1972.

25. H. Meschkowski, Aus den Briefbüchern Georg Cantors, *Arch. Hist. Exact Sci.* **2** (1965), 505.

26. H. Meschkowski, *Probleme des Unendlichen: Werk und Leben Georg Cantors*, Vieweg, 1967.

27. G. A. Miller, American mathematics, *Popular Science Monthly* **79** (1911).

28. G. A. Miller, *Historical Introduction to Mathematical Literature*, Macmillan, 1921.

29. M. G. Murphey, *The Development of Peirce's Philosophy*, Cambridge Univ. Press, 1961.

30. G. Peano, Dimostrazione dell'impossibilità infinitesimi constanti, *Riv. Mat.* **2** (1892), 58–62.

31. C. S. Peirce, *The New Elements of Mathematics by Charles S. Peirce* (ed. C. Eisele), Mouton, The Hague, 1976. All page references are to Vol. 3.

32. C. S. Peirce, *Collected Papers of Charles Sanders Peirce* (eds. C. Hartshorne and P. Weiss), Harvard Univ. Press, 1960. Citations to this work are by volume and paragraph number.

33. C. S. Peirce, On the logic of number, *Amer. J. Math.* **4** (1881), 85–95; in [31], pp. 252–288.

34. C. S. Peirce, The logic of relatives, *The Monist* **7** (1897), 161–217; in [31], p. 526.

35. A. Robinson, *Non-Standard Analysis*, North-Holland, 1966.

36. C. Schmieden and D. Laugwitz, Eine Erweiterung der Infinitesimalrechnung, *Math. Z.* **69** (1958), 1–39.

37. D. E. Smith and J. Ginsburg, *A History of Mathematics in America before 1900*, Mathematical Association of America, 1934.

38. R. H. Shryock, American indifference to basic science during the nineteenth century, *Archives Internationales d'Histoire des Sciences* **29** (1948–1949), 3–18.

39. D. J. Struik, *Yankee Science in the Making*, Little Brown, 1948.

40. J. J. Thomson, *Recollections and Reflections*, Macmillan, 1937.

41. Alexis de Tocqueville, *Democracy in America* (transl. H. Reeve), D. Appleton, 1904.

42. P. P. Wiener and F. H. Young (eds.), *Studies in the Philosophy of Charles Sanders Peirce*, Cambridge, 1952.

43. E. Zermelo, *Georg Cantor: Gesammelte Abhandlungen mathematischen und philosophischen Inhalts*, Springer, 1932.

On the Development of Logics between the two World Wars

I. GRATTAN-GUINNESS

American Mathematical Monthly **88** (1981), 495–509

1 Introduction

Logic is a disparate topic, occurring in almost any field of human activity without appearing to have much character of its own. Traditionally it was associated largely with methods of reasoning and regarded as encapsulated in the principles of syllogistic logic. It was the concern mostly of philosophers and developed in the context of rather general questions. During the second half of the nineteenth century there was especial concern with connections with psychology. For example, some authors maintained that psychology is a descriptive theory concerned with how we think, while logic is a normative discipline about how we ought to think (see [74] for the case of Mill).

Mathematics began to play a significant role in logic with Boole's work around 1850 on Boolean algebra. From the technical point of view his work increased the scope of reasoning; and his contemporary De Morgan and successors Peirce and Schröder moved still further beyond the confines of syllogistic logic when they developed a theory of relations in their algebraic logic. Otherwise, however, the generality of concern remained; for example, Boole saw his own logic as concerned with the workings of the mind.

A different tradition was instituted around 1880 with the mathematical logic of Frege and Peano, and its development by Russell and Whitehead. Not only is the logic itself rather different in form; the motivations lie in specific questions in the foundations of mathematics, in contrast to the general concerns indicated above for other logicians. There were two principal motivations: the foundations of arithmetic; and the formal language required to express mathematical analysis in the style of Weierstrass, with especial concern for the set theory of Cantor. In Russell and Whitehead *both* problems are treated, because they followed the principle (though not the methods) of Cantor in seeking a foundation for arithmetic in set theory.

All these developments belong to the pre-history of my topic, and I shall not dwell on them here (see [11], [12], [31], [34], [20], and [7]), while better known but less reliable sources include [49], [3] and [83]; [90] contains English translations, with commentaries, of many major papers up through 1918. I must indicate now, however, that there was already in that period something of an overthrow of the old tradition by the new one. This paper, for example, is almost entirely concerned with the consequences and further advance of mathematical logic. I shall conclude this introductory section by indicating the topics to be discussed and also those which I have avoided or set aside.

Section 2 concerns the development of logicism, which had been expounded in Whitehead and Russell's *Principia Mathematica* of 1910–1913 [96]. Discussion is confined to Wittgenstein, Ramsey, and Quine, who made substantial revisions in, or use of, logicism. Section 3 charts the progress of Hilbert's formalism in the 1920s, including the blow dealt to it by Gödel's incompletability theorem of 1931. Section 4 surveys the interest taken in recursion and computability by Church, Kleene, and Turing, especially as an extension of the proof method used by Gödel in the proof of his 1931 theorem. Section 5 deals with the intuitionism of Brouwer, including some remarks on his polemics with Hilbert. Section 6 covers certain extensions to the traditional conception of mathematical logic that were proposed and/or rejected by

Hilbert, Skolem, and Zermelo. Infinitary and second-order logics were the principal candidates for attention. Section 7 selects some of the topics studied by Polish logicians, concentrating on the comments on logicism by Chwistek, Leśniewski's logical systems, Tarski's contributions to semantics and related topics, and Łukasiewicz's advocacy of many-valued logics. Finally, in Section 8 I compare the situation around 1939 with that in 1918.

Even this range of topics far from exhausts the developments in logic between the two world wars. For reasons of both space and prudence I have omitted topics such as inductive logic, phenomenological logic, and the role played by logic in the development of quantum mechanics. I have also touched only *very* lightly on related topics in mathematics which bore also on logic in various ways: abstract algebra, model theory, and especially axiomatic set theory and transfinite arithmetic, although in some sections I comment on the views held on the relationship between logic, mathematics, and set theory. Throughout I have had to elide many of the fine distinctions that in fact played some role in the developments, and also omit many of the historical details. For this I crave the indulgence of specialists.

Note that the general history of inter-war logic has not been written. Dumitriu [22] and Guillaume [35] provide brief notices of various aspects, while Hermes [39] surveys the period 1890–1965 at similar speed. Mangione [61] is much more substantial. Some aspects of developments in the 1930s are summarized *passim* in Mostowski [66]. Van Heijenoort [90] also contains English translations, with commentaries, of major papers up to 1931. Church [16] contains historical notes, and [15] provides a detailed bibliography of symbolic logic up to that time; a more extensive bibliography up to recent times is in preparation by J. M. B. Moss.

2 Logicism and its critics

The philosophical position expounded in Whitehead and Russell's three-volume *Principia Mathematica* of 1910–13 was that pure mathematics is derivable solely from logical principles and by logical processes of reasoning. By '*logic*' they understood a form of mathematical logic comprising calculi of propositions and predicates (or propositional functions), including multi-order quantification over individuals and predicates, and certain rules of inference. This logic was intended to embrace set theory. By 'pure' they intended to emphasize the logical form

$p \to q$ for mathematical propositions. By 'mathematics' they seem to have intended to cover the whole subject, though the details were confined to finite and transfinite arithmetic, the foundations of real-variable analysis, and various other topics preliminary to a fourth volume on geometry that Whitehead never completed. To avoid the paradoxes of set theory the logical system included a theory of types. Unfortunately, in order to express the required mathematics they had to introduce the *axiom of reducibility*, which asserted that to any predicate of an object a there exists a logically equivalent predicate which presupposes only objects of the type of a. The "logical" status of this axiom was, to say the least, not obvious.

I shall confine myself in this section to later considerations of the broad principles of logicism, rather than the various technical improvements that were rendered. Russell's conception of logic was all-embracing in that, to use later terminology, it included indiscriminately both logical and metalogical topics. Thus it was, strictly speaking, *impossible* for Russell to talk about his logic. Among other drawbacks, this situation threatens logicism with circularity; for one might assert that pure mathematics is derivable solely from logic, but also specify logic so as to ensure that pure mathematics is encompassed.

This problem was tackled by some of Russell's followers, who tried to define, or characterize, logic in a manner independent of logicism (see [32]). For example, Wittgenstein announced in the preface to his *Tractatus* that his purpose was "to set a limit... to the expression of thoughts" ([97], p. 3). In his work he outlined the idea of treating complex propositions as truth-functions of elementary propositions, the truth-functionality to be expressed by truth tables ([97], Prop. 4.31). Logical propositions are characterized as *tautologies* (6.1), that is, propositions whose truth-value is *true* under all truth-values of their component propositions (4.46). Mathematics "is a logical method" (6.2), and so presumably also has tautologous status.

In his introduction to *Tractatus*, Russell seized on Wittgenstein's remark about setting a limit to the expression of thoughts, and admitted the possibility "that there may be another language dealing with the structure of the first language, and having itself a new structure, and that to this hierarchy of languages there may be no limit" ([97], p. xxii). It is curious that when beginning to prepare the second edition of *Principia Mathematica* a year later, Russell did not start from this important remark, which was made at the time when the recognition of metalogic as independent of logic was far from widely known. (Frege's distinction of object- and meta-language is the most explicit case of the period.) In his new material he used other suggestions of Wittgenstein, especially the idea of truth-functionality. It made his system much more extensional (that is, based on regarding collections as composed of their members rather than as defined by (intensional) properties), although the extent to which extensionality is taken does not seem to be clear.

This move towards extensionality was followed with enthusiasm by Ramsey, who had read the proofs of Russell's new material. In his own work, especially [71], Ramsey defined universal and existential quantification as infinite conjunctions and disjunctions of propositions, respectively; for example,

$$\text{Definition:} \quad (\exists x)\phi x = \phi a \vee \phi b \vee \cdots .$$

He took all predicates as extensional, and in these terms re-structured the type theory of *Principia Mathematica* without having to use the axiom of reducibility.

Ramsey also distinguished the paradoxes into mathematical ones, concerned with sets and numbers, and semantic ones, which deal with notions such as definability; and he divided type theory correspondingly into *simple* and *ramified* parts. Since his time it has become customary to regard semantic paradoxes as irrelevant to logicism.

Logicism also attracted the attention of the *Vienna Circle* of philosophers, especially Carnap, who wrote one of the first books [9] in German on logicism. In his *Der Logische Aufbau der Welt* [8], which outlined much of the philosophical programme of the Circle, he made considerable use of Russell's logical techniques, including the Russell/Wittgenstein view of extensionality. Interestingly, in his preface to the English edition of the book, he admitted that his use of extensionality "is unclear in some points" ([10], p. ix).

Another Vienna Circle member took up *Principia Mathematica*; for Gödel's incompletability theorem of 1931 showed that the Russellian logicist programme cannot be executed (see the next section). Thereafter the question of the relationship between mathematics, logic, and set theory was obviously wide open. Quine's work has been particularly influential in this area; he devised various logical systems of the scope of *Principia Mathematica* without espousing logicism. Two of his early systems were the so-called *New Foundations* [67], in which type theory is replaced by a stratification of formulae; and *Mathematical Logic* [69], where the paradoxes were avoided by following von Neumann's idea of denying that certain objects can be members of sets. In these and other systems some status was assigned to set theory which is autonomous of logic. No definitive view of the distinction between logic and set theory has been laid down (see [70]); but Russell's logicist standpoint, that set theory is *part* of logic, was abandoned.

Russell himself wrote little on logicism after the second edition of *Principia Mathematica*. Whitehead sketched a revised logico-arithmetical system in [95], based on the notion of instantiation of objects; despite many readings, I have not been able to see how far it can be developed. During the inter-war period logicism fell quite substantially in reputation; although it influenced logicians both as an early example of a comprehensive mathematico-logical system and as a source of techniques, the philosophical position itself won few followers.

I conclude this section with an event of our period which was unknown at the time but which is of especial historical interest. The first logicist was Frege, although he did not include "all" mathematics in his viewpoint. His work was not widely known during the inter-war years, although Scholz hoped to publish his manuscripts. In one of these texts from 1925 [23], Frege abandoned logicism entirely, on the grounds that logic alone could not provide objects for which properties such as equality or set-membership can be appraised. The rejection of a life-long position is a rare achievement in mankind. It reveals a special kind of greatness.

3 Formalism and its fate

Hilbert strongly advocated the use of axiomatics in mathematics in his early years. These were the 1890s, when such a view was far from widespread among mathematicians (see [11]). In the 1900s he applied the approach to the foundations of logic and arithmetic; but he was very unclear on the distinction between a formal system and its interpretation, so that his early papers make peculiar reading. Perhaps for these reasons he seems to have set the work aside; but he resumed his interests during the First World War (see [42]), and he developed his ideas much further in the 1920s to espouse *formalism*, in which a mathematical theory is axiomatized and treated as a string of symbols for the purpose of studying, in metamathematics, properties such as consistency, completeness, and the existence (or not) of decision procedures (that is, procedures to show that any well-formed formula of a formal system is provable or not).

Here is a simple example of his approach, taken from [43]. He gave an axiomatization of arithmetic, roughly like the Peano axioms without the induction axiom, together with the *modus ponens* rule of inference. He proved as a lemma that a provable formula need not contain the conditional connective $\to$ more than twice. He then showed that the system was consistent, by showing that an equation $\alpha = \beta$ and its negation $\alpha \neq \beta$ are not provable.

During the 1920s Hilbert and others, especially his assistants Ackermann and Bernays, obtained results on consistency and completeness, and he and Ackermann published a textbook [46] on mathematical logic outlining their principal interests. Among specific results, von Neumann [92] showed in 1927 that a certain part of first-order arithmetic was consistent; Gödel [27] proved in 1930 the completeness of the first-order predicate calculus. In the same year Herbrand studied various properties equivalent to the provability of a formula in such a system (see [[37]]), and stated (with a defective proof) a "fundamental theorem" with the aid of which he could appraise (in [37] and also [38]) the consistency of, and decision procedures for, various logical and arithmetical systems. He also proved the deduction theorem for the first-order predicate calculus ([37], p. 108), which states that if a formula B is provable from A in the calculus, then $(A \to B)$ is also provable. This theorem has become one of the most widely used results in metamathematics.

Metamathematics can itself be studied in meta-metamathematics, and so a hierarchy of theories is erected. Although Hilbert confined his detailed studies to (first-level) metamathematics, his hope may have been that each tier in the hierarchy would become successively simpler in content and assumption, so that at some suitably high tier uncontroversial assumptions ($0 \neq 1$, say) would guarantee its consistency, and thereby transmit the guarantee down through the lower levels. By this means he could establish the consistency of mathematical theories, and thus show them to be cleansed of paradoxes.

Unfortunately, Gödel's incompletability theorem [28] showed that this conception of the relationship between tiers was incorrect. Gödel defined an axiomatic system P for first-order arithmetic, using a logical system like that of *Principia Mathematica* and the Peano axioms for arithmetic, and proved that P was incomplete in that it contains a proposition A for which neither A nor not-A is provable. As a corollary he then showed that a proof of the consistency of P could not be expressed within P, but would require a formally richer system.

Although Gödel stated that the corollary "do[es] not contradict Hilbert's formalist standpoint", on the grounds that there may be finitary proofs of consistency not expressible within P ([28], p. 615), such a possibility is very unlikely, and it was soon recognized that Hilbert's hopes for a consistency proof for mathematical theories along the lines described above, and belief that truth is equivalent to deducibility, must be abandoned. The incompletability theorem itself also seems to have affected logicism, for P could be converted to a system like *Principia Mathematica*, which was thus also shown to be incomplete.

While Gödel's theorem rebuffed Hilbert's hopes, it did not detract from interest in metamathematics; indeed, it led to interest in non-finitary consistency proofs. Of particular note was Gentzen's proof of the consistency of first-order arithmetic in [26], in which he made use of transfinite induction. Gentzen made other important contributions to metamathematics in his dissertation [25]. He proved a "principal theorem", which has some similarity to Herbrand's "fundamental theorem" mentioned above and was also of value in producing consistency proofs and seeking for decision procedures. He also recast the predicate calculus in a manner which is now called *natural deduction* and stands closer to heuristic reasoning than to traditional axiomatizations of the calculus. The contributions of Gentzen and Herbrand exercised a marked influence on the development of post-Gödelian metamathematics. Indeed, metamathematics became one of the chief interests of logicians in the inter-war period.

4 Recursion and computability: American logic

In order to prove his incompletability theorem Gödel had to design both his formal system P and his metamathematical concepts so that his meta-statement of the incompletability of P could be expressed in its own formal language. Since P was a formulation of first-order arithmetic, the metamathematics had to be expressible in arithmetical terms. This led Gödel to his process of the "arithmetization of metamathematics", in which he developed a theory of what are now called *primitive recursive functions*. (The process of arithmetization requires the distinction between a formal system and its interpretation to be made very carefully. Professor J. Barkley Rosser, who contributed to the early studies of Gödel's theorem, has told me in reminiscence that not until this theorem was published did logicians realize *how* carefully they had to make the distinction.)

Recursion had already been noted in the 1920s, especially in the metamathematical studies discussed in the previous section; Skolem [81] is also a notable contribution. But Gödel's treatment was the most systematic hitherto, and his theorem suggested that there may be similar limits to recursion. These researches were actively pursued in the 1930s. Many of the principal papers are reprinted in [21], and a detailed account of the results is provided by Kleene ([51], part 3).

We take as initial functions the successor function ($\phi(x) = x+1$), the constant function ($\phi(x) = K$), and the identification function ($\phi(\{x_i\}) = x_r$). A primitive recursive function is defined as obtainable from a finite number of uses of the initial functions and of schemata given by

$$\phi(x) = \theta\left(\{\psi_i(x)\}\right)$$
and $\quad \phi(0) = k, \qquad \phi(y+1) = \psi\left(y, \phi(y)\right).$

I have stated the definition for a function of one variable; generalizations to functions of several variables are obvious.

Examples of recursive functions that cannot be obtained by the process of primitive recursion had been known since the late 1920s, and extended definitions were proposed. One, due to Kleene [50], defined a function as *general recursive* if it were obtained from the processes of primitive recursion and also an evaluation procedure of the form (for functions of one variable)

$$\phi(x) = \text{the smallest value of } y \text{ such that } \psi(x, y) = 0.$$

ψ is a given function, and at least one value of y is assumed to exist. The function is not reducible to the others because no upper bound is set on the value allowed for y.

Church proposed as a thesis that general recursion be taken as the definition of the effective calculability of a number-theoretic function, and proved theorems of a generalized Gödelian type on the lack of a decision procedure for first-order arithmetic (see [14]). In the same year Turing proposed as a thesis that computability be defined as the operations that can be executed by a *Turing machine*, which is the conception of a computer reduced to essentials (see [89]).

It was shown that general recursion was equivalent to computability for number-theoretic functions. Thus an interesting connection between logic and computing was established, and has continued in various forms ever since. Both of these notions were shown to be equivalent to a concept called λ-*definability*, introduced by Church as part of his proposal (initiated in [13]) for a logic without variables. This proposal led to the early development of combinatory logic (see [19]).

All the work described in this and the last section was basically inspired by Hilbert's programme for metamathematics, and in their treatise [47] he and Bernays gave a detailed account of the developments. Naturally many of the results were obtained by Hilbert and his colleagues; but Americans also played a prominent role, especially Church, Kleene, Post, Rosser, and Gödel (resident in America from 1933).

Since that time America has been a leading country for logic, and saw the founding in 1936 of the Association for Symbolic Logic. This organization is still the only international organization for the subject, and its *Journal of Symbolic Logic*, also founded in 1936, was the first journal devoted exclusively to logic and related topics. By 1939 the Association had achieved a membership of around 200, which was also around the number of pages published in each volume of the *Journal*. It was founded soon after the establishment in America of the Philosophy of Science Association in 1934, together with the journal *Philosophy of Science*. The Association was then strongly influenced by Vienna Circle émigrés resident in America, and so showed some interest in logic and its applications, after the spirit of Carnap's employment of techniques from *Principia Mathematica* that I noted in Section 2.

A similar recognition of institutional change may be noted in the *Jahrbuch über die Fortschritte der Mathematik*, the leading journal between the two world wars for mathematical reviews. It had established a section for set theory before 1918, containing both the abstract and the point-set branches, and also a rather vague section called *Philosophy* where logic was usually covered. But in Volume 61, for 1935 and published in 1936, point-set theory was placed within *Real functions* or *Topology*, *Philosophy* was abandoned, and a new section entitled *Foundations of mathematics: Abstract set theory* was established. With Volume 65, for 1939 and published in 1943, this section had become *Foundations of mathematical logic* with the sub-sections *Philosophical*, *Logic*, and *Foundational arithmetic and algebra*; and abstract and point set theory had been reunited again as a section under *Analysis*. Other minor changes occurred until the final Volume 68 for 1942, published in 1944; but the sub-section for logic was retained. The *Zentralblatt für Mathematik* allowed for *Foundational questions, logic* from its inception in 1931, and *Mathematical Reviews* included *Foundations* (initially *Foundations of analysis*) when it began to appear in 1940.

5 The emergence of intuitionism

The intuitionistic philosophy of mathematics is associated primarily with Brouwer's rejection of the law of excluded middle as a valid logical principle. He introduced his ideas in the 1900s, and first largely applied them to Cantorian set theory; but around 1918 he began to develop them in more detail as a general viewpoint for both logic and mathematics.

Brouwer advocated that mathematical proofs be restricted to constructive processes. Thus there was a direct clash with Hilbert's formalism; for while Hilbert assumed the law of excluded middle as part of his means of proving the consistency of mathematical theories, Brouwer felt that

> the (contentual) justification of formalistic mathematics by means of the proof of its consistency contains a vicious circle, since this justification rests upon . . . the (contentual) correctness of the principle of excluded middle ([6], p. 491).

He came to reject Cantor's theory of the actual infinite; but Hilbert supported Cantor, and used metamathematical techniques to try to prove Cantor's *continuum hypothesis*. Brouwer also tried to replace the classical formulations of mathematical analysis with intuitionistic forms (see, for example, [5]).

However, Brouwer did not help his cause by the notoriously obscure formulation of his basic notions (for helpful explanations, see [90], pp. 446–457) and dogmatic assertions, such as mathematics being independent of both language and logic. The discussion between him and Hilbert was largely unprofitable, partly for these reasons and partly for a clash of personalities.

These clashes came to a head at the International Congress of Mathematicians, held at Bologna in 1928. Brouwer, a Germanophile although born Dutch, shared the sentiments of some Germans that their country should not participate. However, Hilbert led the German delegation; and in the following year he removed Brouwer from the editorial board of *Mathematische Annalen* (see [72], pp. 184–188). The whole issue undoubtedly involved ideological considerations; but some aspects of the early work of Hilbert and Brouwer suggest that, concerning Brouwer *himself*, there may have been purely personal elements. Hilbert's early papers of 1904 on metamathematics are scarcely intelligible, for want of a careful distinction between symbols and their referents. Brouwer's thesis of 1907, by contrast, makes the distinction very clearly (see [4], pp. 61, 70, 101), although he does not develop a *theory* of "mathematics of the second order". Hilbert will probably not have read this Dutch thesis; but in later papers Brouwer reported communicating these ideas in conversation with Hilbert in 1909 (see, for example, [4], p. 410). When Hilbert started again on metamathematics in 1917, the distinction is clearly made. Brouwer, an expert at nurturing grudges, could well have felt (with or without justice) that his ideas were being used without acknowledgement, and sought some kind of personal revenge at an opportune moment. Their personal relations had become further complicated by other personal factors (for example, Weyl's switch from formalism to a type of intuitionism).

Formalism attracted much the greater support in the inter-war period, although several logicians wrote on various aspects of intuitionistic logic and mathematics, and Weyl was an early convert to a form of intuitionism (see especially [93]). Heyting was the most prominent follower of intuitionism, publishing [40] the first textbook presenting a form of intuitionistic logic.

Although intuitionism owes its origins chiefly to Brouwer, it also suffered between the wars from his perplexing presentations and polemic style. But it has been one of the most interesting and far-reaching criticisms of the foundations of mathematics, and has continued to claim the attention of logicians to this day.

6 Beyond first-order and finitary logics

The accounts in Sections 2–5 have been of specific topics and developments. In this section I wish to draw attention to an undercurrent in inter-war logic, a matter which was not quite a topic in the usual sense but an *issue* which arose and fell at various times, especially in the 1920s: the extension of logic beyond its first-order scope and finitary expression. To discuss the issue fully, I would have to go rather deeply into the development of model theory and axiomatic set theory, which would both imbalance and unduly extend this paper; I refer to [29] and [65] for fuller accounts.

I mentioned in Section 2 that Ramsey urged the extensional interpretation of quantification in terms of infinite conjunctions and disjunctions. Whether wittingly or not, he was following the traditional interpretations of quantifications in algebraic logic. Schröder, in his theory of relatives, wrote quantifiers as $\prod_i a_{ij}$ and $\sum_j a_{ij}$, for example, where a_{ij} is (in effect) a relation between variables i and j (see [30]) and $\prod$ and $\sum$ are intended to convey the idea of infinite conjunctions and disjunctions of propositions.

During the First World War Löwenheim wrote a paper [57] on the first-order predicate calculus following Schröderian principles. He used infinitely long expressions in logic following the Schröderian interpretation of quantification, and noted the conversion of formulae into *normal forms* (where the quantifiers are all collected together, in certain orders, at the front of the formula). He also proved a form of the *Löwenheim-Skolem theorem*, which states essentially that every satisfiable formula of the calculus is satisfiable in a finite or denumerable domain.

Löwenheim's paper stimulated various studies in the 1920s. Skolem dealt in more detail with normal forms in [79]. The reduction of formulae to normal forms became a very useful technique in both mathematical logic and metamathematics, as developed by Hilbert and others in the ways described in Section 3. One feature was the fact that to a given normal form there is a dual form. Here again influence from algebraic logic is evident; for duality was often used by Schröder, whereas it is largely absent from the predicate calculi of Frege and Russell (see [30], p. 121). Skolem produced his own proof of the Löwenheim-Skolem theorem, using only denumerably long expressions rather than the non-denumerably infinite lengths employed by Löwenheim. Later he gave a new, but rather weaker, proof of the theorem, using only finitely long expressions, and also pointed out as a consequence what is now known as *the Skolem paradox*: that Zermelo's axiomatization of set theory can be satisfied in a denumerable model even though the theory allows for non-denumerably infinite sets (see [80]).

Skolem's new proof, unlike his old one, avoided using the axiom of choice. This axiom had emerged in the 1900s in the context of Cantorian set theory and mathematical analysis, and its non-constructive character and occasional avoidability led to a lively controversy over its forms and necessity (see [77], [31], and [64]). By the end of the First World War the controversy had largely died down, although Skolem noted the continuing reservations over the axioms—reservations which he did not share—at the end of [80]. Another mathematician who had no qualms was Hilbert, who now put it to a new use; the interpretation of quantification.

In [44] Hilbert introduced what he called the *transfinite axiom*, which asserts that if an operator τ on the predicate $A(x)$ selects the value c of x for which $A(c)$ is provable, then $A(y)$ is provable for all values of y. He advocated the use of this axiom to define universal quantification in a finitary form, in contrast to the infinitary interpretation as infinite conjunctions of propositions; existential quantification could also be defined by applying τ to not-$A(x)$. Later he restated his transfinite axiom in terms of the axiom of choice in a way which in effect interprets the existential quantification $(\exists x)A(x)$ in terms of using a transfinite choice function to select a (or the) value of x which satisfies A (see [45]).

At this time Hilbert also urged the extension of logic from first-order to second-order predicate calculus (he used the terminology *restricted* and *extended functional calculus*), where quantification is allowed over predicates A as well as individuals x, for the purpose of expressing arithmetical notions in logical form (see, for example, [46], Ch. 4). In this respect (though not in philosophy, of course) he was moving towards the logicists' practice, for they had defined arithmetic notions in their systems and freely allowed second-order quantification. Indeed, they had admitted quantification to arbitrarily high orders; in *Principia Mathematica* type theory in effect separated out the various levels of quantification.

Thus Hilbert's position was to be in favor of second-order logic but opposed to infinitary logic. Skolem, reluctant to rely on the concepts and techniques of set theory, was opposed to both these forms of logic; but in the 1930s Zermelo advocated them both. He supported second-order logic as indispensable for the formalization of axiomatic set theory; and he used infinitary logic in a sketched theory of infinitely long proofs (the fullest account is in [99]), from which he hoped to be able to show that all true formulae are "provable" in his extended sense of proof.

Zermelo's view seems to clash with Gödel's incompletability theorem, which was published after he began to develop his ideas. The two men spoke in September 1931 at the annual meeting of the *Deutsche Mathematiker-Vereinigung* on their very different viewpoints, and they corresponded soon afterwards (in letters published in [33]). In their letters they basically re-stated their positions; in particular, Gödel continued to wish to use only finitary proofs.

Zermelo's ideas were too vague for practical development, and infinitary logics gained little interest until well after the Second World War. Second-order logic maintained some supporters, especially Hilbert, although the relationship with first-order logic was not widely discussed. The considerable differences between the two logics are often not realized even today.

7 The rise of Polish logic

At the end of Section 4 I noted the emergence of America in the mid-1930s as an important center for logic. The other—indeed, the first—major national development between the wars was the rise to prominence of Polish logicians in the 1920s. The principal figures were Adjukiewicz, Chwistek, Kotarbiński, Leśniewski, Łukasiewicz, Stupecki, Sobociński, Tarski, and Wajsberg, although there are many other significant names that one could adjoin.

The rise of Poland to prominence—a position which it has retained to this day—is one of the most remarkable features of the history of twentieth-century logic; and its early development is still more interesting because of the simultaneous emergence of a school of Polish mathematicians (see [53]). Banach, Kuratowski, Lindenbaum, Mazurkiewicz, and Sierpiński were perhaps the most prominent mathematicians; and since many of them were strongly interested in algebra and set theory, their work overlapped in part with that of the logicians. Indeed, until 1928 Łukasiewicz and Leśniewski were members of the editorial board of the Polish journal *Fundamenta Mathematicae*, launched in 1920 and devoted to set theory and related topics, and some logical papers appeared there. Otherwise the logicians published in various Polish academy and society journals, and also sometimes abroad.

In this section I can indicate only some principal interests; a more detailed account is provided in [48]. A selection of English translations of Tarski's works will be found in [84], and of other authors in [62]. At the time many of their works were published in French or German (or occasionally English), though some appeared only in Polish.

Naturally, in the early 1920s *Principia Mathematica* was a major interest. Chwistek's contributions were rather similar to those of Ramsey, which I noted in Section 2; indeed, he anticipated Ramsey to some extent (see [32]). In a two-part paper [17]—rejected by *Fundamenta Mathematicae* (see [32]) but published by the Polish Mathematical Society—he divided Russell's type theory into *simple* and *branched* (that is, ramified) parts, although he was less clear than Ramsey on the distinction between kinds of paradox. He also tried to re-construct type theory while avoiding the axiom of reducibility, although his new system is harder to comprehend than is Ramsey's; for example, the extent of his commitment to extensionality is less clear. His work is of some note as an early attempt to bridge the gap between the prevailing philosophies of mathematics; for while he constructed a system of scope comparable with that of *Principia Mathematica*, he also tried to specify his formal system in formalist style.

Russell also influenced Leśniewski, although here it was Russell's paradox (of the set of all sets which do not belong to themselves) which was the prime source of inspiration (see [82]). Leśniewski came eventually to construct three logical systems (see [58] and [75]). *Ontology* is a modernized version of traditional logic; in its structural aspects it includes a theory of classes and relations. *Mereology* is a study of the part-whole relationship between objects. Both these systems took as their logical basis *protothetic*, a propositional calculus assuming only equivalence as a primitive connective but also using quantification over propositions.

This *equivalential calculus* was one of Tarski's early contributions to logic; in fact, it was the subject of his doctoral dissertation 1923. But his main work lay in the growing interest in metalogic, especially concerning semantical questions. Here he showed some influence from Leśniewski, who in the early 1920s laid emphasis on semantic categories (see [55]) and on the role of definitions in formal theories and their formulation as equivalences. Tarski's most celebrated work in this area is his treatise [87] on the semantic definition of truth as the property of a sentence in a language. He also broadened Hilbert's conception of metamathematics into a study of "the methodology of the deductive sciences" (see [86]), where the notion of logical consequence and Herbrand's deduction theorem (for which Tarski claimed priority) played major roles. In his text-book [88] he laid great emphasis on the theory of deduction.

Tarski also worked with his teacher Łukasiewicz on many-valued logics. This topic received little attention between the wars (see the historical outline and bibliography in [73]); in fact, the Poles were the logicians most active in it, with Łukasiewicz as the chief protagonist. After early work on the propositional calculus, especially as presented in *Principia Mathematica*, he extended the method of truth-tables to three-valued logic, and in a joint paper [60] with Tarski he explored many-valued logical systems. Łukasiewicz also introduced the Polish system of notation for logic, in which certain conventions are employed to avoid the use of brackets or dots in formulae.

As in America, this burgeoning interest led to organizational consequences in the mid-1930s. The periodical *Studia Philosophica* was launched in 1936 to cover both logic and philosophy; it contained Tarski's German translation of his treatise [87] on truth in its first volume. In 1939 *Collectanea Logica* was planned, with a first volume on proof, as a forum for publishing Polish logic; but the print-shop was destroyed in the bombing of Warsaw, and only off-prints of some papers that had already been posted to foreign logicians and philosophers have survived.

8 Conclusions and comparisons

In this final section I compare the situation around 1918 with that in 1939. I do so from seven points of view, the first three illustrating contrasts, the next three similarities.

(a) Philosophies of mathematics

In 1918 logicism was the most fully developed philosophy of mathematics (and *Principia Mathematica* the best-known logical system), and it held the dominant position. But during the next decade intuitionism and especially formalism grew to prominence. The disputes between the three schools gave the impression that they were both mutually inconsistent and also exhaustive of possible views. But during the 1930s this impression dissolved, as each school faced great difficulties or was revealed as inadequate in some ways.

A good impression of the three major philosophies of mathematics just prior to Gödel's 1931 theorem may be obtained from the papers by Carnap on logicism, Heyting on intuitionism, and von Neumann on formalism which were delivered to a Vienna Circle symposium on the foundations of mathematics in 1930. They were published in the Vienna Circle's journal *Erkenntnis* (1931), 91–121; English translations are contained in [1], pp. 31–54.

From that time until today there was no *dominant* philosophy of mathematics, or logical system. In particular, logicism, having been center stage in 1918, was very much in the wings from the 1930s on. Hilbert was arguably the most influential figure during the inter-war years.

(b) The status of metalogic

In 1918 metalogic was little recognized as independent of (object-level) logic. But the rise of metamathematics with Hilbert, the work on model theory by Skolem and others, and the study of problems in semantics (especially by the Poles) brought metalogic to the forefront of concern by the middle 1920s, where it has remained ever since.

(c) A logical revolution?

In 1918 logic was still an occasional interest, pursued by a few figures mostly working in isolation. Much of the writing in logic of that time was still discursive stuff written by philosophers for philosophers, with little symbolic content (although, in the work of the phenomenologist Husserl, much penetrating philosophical thought). But during the 1920s the subject was largely taken over by mathematicians, or at least by people with substantial mathematical training, and most major papers appeared in mathematical journals.

Thus, can we speak of a revolution in logic? The word *revolution* is used so often by historians of science that I am disinclined to add myself to their voices; but we can certainly point to a substantial change of emphasis during this change of tradition. Among theories of scientific revolution, I am generally attracted to Kuhn's theory of normal science; but in this case the theory does not seem to function well. There was not really a normal logic in the period before Frege, since various traditions were active and only the Boolean algebra made substantial use of mathematics. Nor was there a new paradigm afterwards, since the controversies between logicism, formalism, and intuitionism, already nascent in the birth of the latter two doctrines in the 1900s, prevented any of the three becoming a new 'normal' logic. Perhaps we can point to a change of paradigm in the much weaker, more general, sense of a conversion from general considerations, often involving psychology, to relatively specific concerns using mathematical techniques.

(d) Professional standing

There were signs of professionalisation, especially in America and Poland: the Association for Symbolic Logic and its journal, *Studia Philosophica*, and so on. In addition, quite a number of text-books were published, espousing one or another of the new developments in logic, and I believe that logic was being taught at university level rather more widely in some countries. However, I would not wish to exaggerate the extent of professionalisation; even today, logic nestles under the wings of larger subjects. For example, there seems to have been little science policy applied to logic between the wars—though this was to change during the Second World War, when logic was applied to crypto-analysis and some logicians became code-breakers.

(e) Non-classical logics

These logics did not progress very far between the wars; intuitionistic logic aroused the most attention. The interest in many-valued logics was rather slight (see [76]); even their bearing on quantum mechanics was apparently not given prominence until Birkhoff and von Neumann [2]. Modal logics seem also to have been rather unpopular (see [98]), although they have their origins in [56].

(f) Interest in the history of logic

Through the inter-war period interest in the history of logic, both ancient and relatively modern, maintained a steady low level. Of the major figures discussed above, Łukasiewicz showed the deepest historical concern and inspired among some of his compatriots an interest that has continued to this day. Church's bibliography [15] of symbolic logic, which graced the early pages of the *Journal of Symbolic Logic*, is an important historical aid. Editions of the works of Hilbert and Peirce (and also of Cantor and Dedekind) were prepared during the early 1930s.

(g) Mathematicians and logic

I mentioned in (e) that mathematicians largely took over logic in the 1920s. The main areas of contact between mathematics and logic were, of course, in the relationship between arithmetic, set theory, and predicate calculi. In addition, abstract algebra furnished appropriate structures for aspects of model theory; and then there was lattice theory, which sprang to prominence in the 1930s after lying dormant in Schröder's algebraic logic among other places (see [63]), and which furnished Birkhoff and von Neumann with the quantum logic that I mentioned in (e).

One could certainly add other topics to this list; but nevertheless the main impression is that contacts between logic and mathematics—especially "working" mathematics—were rather scattered between the wars (and, indeed, both before and after that period). While mathematicians emphasize the need for rigorous proofs and exact definitions, and take a *general*

interest in problems of axiomatization and proof, they are often reluctant to study these "logical" questions *explicitly*; even the most exact mathematics rarely meets logicians' standards (see [18]). A good example is provided by the French, who always produce a remarkable number of great mathematicians but who show little interest in logic. Herbrand, by far the greatest French logician between the wars, had difficulty in obtaining a panel to judge his doctoral thesis [37], and he published it in Poland.

The mathematician's attitude to logic is best understood by considering the position of set theory. It became clear between the wars that the attempt of logicism to embrace set theory within logic was forlorn, although the line of division between the two fields was not clearly or definitively drawn. Now mathematicians were interested in set theory much more than in logic; and even then they tended to draw on the set-*topological* aspects of the subject, as a source of definitions, techniques, and theorems, rather than the *axiomatic* set theory of Zermelo and others, where most of the contacts with logic were made. This situation is well exemplified by mathematical analysis, where set topology continued to play a significant role but axiomatic set theory and logic were largely absent—indeed, logic was much less prominent than it had been in the 1890s and 1900s, when Peano brought the new mathematical logic into analysis (at the time Peano's group made Italy the leading country for mathematical logic, but Italian interest between the wars was at a low level) and Whitehead and Russell were working out the details of their logicist programme. Ziehen made a perceptive judgement and prediction at the beginning of our period [100], with which most inter-war mathematicians and logicians could agree:

> Set theory is no part of logic but its favored-daughter discipline, from whose inspirations it has many more results to await.

References

1. P. Benacerraf and H. Putnam, *Philosophy of Mathematics: Selected Readings*, Oxford, 1964.
2. G. Birkhoff and J. von Neumann, The logic of quantum mechanics, *Ann. Math.* **37** (1936), 823–843; or [91], Vol. 4, pp. 105–125.
3. G. D. Bowne, *The Philosophy of Logic 1880–1908*, The Hague, 1966.
4. L. E. J. Brouwer, *Collected Works*, Vol. 1 (ed. A. Heyting), Amsterdam, 1975.

5. L. E. J. Brouwer, Über Definitionsbereiche von Functionen, *Math. Ann.* **97** (1927), 60–75; or [4], pp. 390–405.

6. L. E. J. Brouwer, Intuitionistische Betrachtungen über den Formalismus, *Proc. Kon. Akad. Wetens. Amsterdam* **31** (1927) 374–379; or [4], pp. 409–414.

7. R. Bunn, Developments in the foundations of mathematics, 1870–1910, *From the calculus to set theory: 1630–1910, An Introductory History* (ed. I. Grattan-Guinness), London, 1980, pp. 220–255.

8. R. Carnap, *Die Logische Aufbau der Welt*, Weltkreis, 1928; English transl. 1967.

9. R. Carnap, *Abriss der Logistik...*, J. Springer, 1929.

10. R. Carnap, *The Logical Structure of the World*, Univ. California Press, 1967.

11. J. Cavaillès, *Méthode Axiomatique et Formalisme*, Hermann, 1937.

12. J. Cavaillès, *Remarques sur la Formation de la Théorie Abstraite des Ensembles*, 2 vols., Hermann, 1938.

13. A. Church, A set of postulates for the foundation of logic, *Ann. Math.* **33** (1932), 346–366; **34** (1933), 839–864.

14. A. Church, An unsolvable problem of elementary number theory, *Amer. J. Math.* **58** (1936), 345–363; or [21], pp. 88–107.

15. A. Church, A bibliography of symbolic logic, *J. Symb. Logic* **1** (1936), 121–218; **3** (1938), 178–212. (Addenda in later volumes.)

16. A. Church, *Introduction to Mathematical Logic*, Princeton, 1952.

17. L. Chwistek, The theory of constructive types (Principles of logic and mathematics), *Ann. Soc. Polon. Math.* **2** (1924), 9–48; **3** (1925), 92–141.

18. J. Corcoran, Gaps between logical theory and mathematical practice, *The Methodological Unity of Science* (ed. M. Bunge), Springer, 1973, pp. 23–50.

19. H. B. Curry and R. Feys, *Combinatory Logic*, Vol. 1, North-Holland, 1968.

20. J. W. Dauben, *Georg Cantor...*, Princeton Univ. Press, 1979.

21. M. Davis (ed.), *The Undecidable...*, Raven Press, 1965.

22. A. Dumitriu, *History of Logic*, Vol. 4, Tunbridge Wells, 1978.

23. G. Frege, Erkenntnisquellen der Mathematik..., *Nachgelassene Schriften* (ed. H. Hermes and others), Hamburg, 1969, pp. 286–294. English transl. *Posthumous writings*, Oxford 1979, pp. 267–274.

24. G. Gentzen, *Collected Papers* (ed. M. E. Szabo), North-Holland, 1969.

25. G. Gentzen, Untersuchungen über das logische Schliessen, *Math. Zeitschrift* **39** (1935), 176–210; 405–431; English transl. [24], pp. 68–131.

26. G. Gentzen, Die Widerspruchsfreiheit der reinen Zahlentheorie, *Math. Ann.* **112** (1936), 493–565; English transl. [24], pp. 132–213.

27. K. Gödel, Die Vollstandigkeit der Axiome des logischen Funktionkalkuls, *Monats. Math. Phys.* **31** (1930), 349–360; English transl. [90], pp. 582–591.

28. K. Gödel, Über formal unentscheidbare Sätze der Principia Mathematica und verwandte Systeme. I, *Monats. Math. Phys.* **38** (1931), 173–198; English transl. [21], pp. 4–38 and [90], pp. 596–616.

29. W. Goldfarb, Logic in the twenties: the nature of the quantifier, *J. Symb. Logic* **44** (1979), 351–368.

30. I. Grattan-Guinness, Wiener on the logics of Russell and Schröder..., *Ann. Sci.* **32** (1975), 103–132.

31. I. Grattan-Guinness, *Dear Russell – Dear Jourdain*, Columbia Univ. Press, 1977.

32. I. Grattan-Guinness, On Russell's logicism and its influence, 1910–1930, *Proceedings of the 3rd International Wittgenstein Symposium, 1978*, Kirchberg am Wechsel, 1979, pp. 275–280.

33. I. Grattan-Guinness, In memoriam Kurt Gödel:..., *Hist. Math.* **6** (1979), 294–304.

34. I. Grattan-Guinness, Georg Cantor's influence on Bertrand Russell, *Hist. Phil. Logic* **1** (1980), 61–93.

35. M. Guillaume, Axiomatique et logique, *Abrégé d'Histoire des Mathématiques 1700–1900* (ed. J. Dieudonné), Vol. 2, Paris, 1978, pp. 315–430.

36. J. Herbrand, *Ecrits Logiques* (ed. J. van Heijenoort), Univ. Press de France, 1968; English transl. D. Reidel, 1971.

37. J. Herbrand, Recherches sur la théorie de la démonstration, *Prace Towarz. Nauk. Warzaw* **3** (1930); [36], or pp. 35–153.

38. J. Herbrand, Sur le problème fondamental de la logique mathématique, *Spraw. Posted. Towarz. Nauk. Warsaw* **3** (1931), 12–56; or [36], pp. 167–207.

39. H. Hermes, Zur Geschichte der mathematischen Logik und Grundlagenforschung in den letzten fünfundsiebzig Jahren, *Jber. Deutsch. Math.-Ver.* **68** (1966), 75–96.

40. A. Heyting, *Mathematische Grundlagenforschung. Intuitionismus. Beweistheorie*, J. Springer, 1934.

41. D. Hilbert, *Gesammelte Abhandlungen*, Vol. 3, J. Springer, 1935.

42. D. Hilbert, Axiomatisches Denken, *Math. Ann.* **78** (1918), 405–415; [41], pp. 146–156; English transl. in *Phil. Math.* **7** (1971), 1–12.

43. D. Hilbert, Neubegründung der Mathematik, *Abh. Math. Sem. Hamburg Univ.* **1** (1922), 157–177; or [41], pp. 157–177.

44. D. Hilbert, Die logischen Grundlagen der Mathematik, *Math. Ann.* **88** (1923), 151–165; or [41], pp. 178–191.

45. D. Hilbert, Über das Unendliche, *Math. Ann.* **95** (1926), 161–190; English transl., [90], pp. 367–392.

46. D. Hilbert and W. F. Ackermann, *Grundzüge der Theoretischen Logik*, J. Springer, 1928; English transl. New York, 1950.

47. D. Hilbert and P. Bernays, *Grundlagen der Mathematik*, 2 vols., J. Springer, 1934–1939.

48. Z. A. Jordan, *The Development of Mathematical Logic and of Logical Positivism in Poland between the Two Wars*, McCall, 1967, pp. 346–406.

49. J. Jörgensen, *A Treatise of Formal Logic*, 3 vols., Oxford Univ. Press, 1931.

50. S. C. Kleene, General recursive functions of natural numbers, *Math. Ann.* **112** (1936), 727–742; or [21], pp. 236–253.

51. S. C. Kleene, *Introduction to Metamathematics*, Von Nostrand, 1952.

52. W. Kneale and M. Kneale, *The Development of Logic*, Oxford, 1962.

53. K. Kuratowski, *A Half-century of Polish Mathematics...*, Pergamon, 1980.

54. S. Leśniewski, *Collected Works*, Kluwer, 1992.

55. S. Leśniewski, Grundzüge eines neuen Systems der Grundlagen der Mathematik, *Fund. Math.* **14** (1929), 1–81.

56. C. I. Lewis, *A Survey of Symbolic Logic*, Univ. California Press, 1918.

57. L. Löwenheim, Über Möglichkeiten in Relativkalkül, *Math. Ann.* **76** (1915), 447–470; English transl. [90], pp. 228–251.

58. E. C. Luschei, *The Logical Systems of Leśniewski*, North-Holland, 1962.

59. Jan Łukasiewicz, *Selected Works* (ed. L. Borkowski), North-Holland, 1970.

60. J. Łukasiewicz and A. Tarski, Untersuchungen über den Aussagenkalkül, *C. R. Soc. Sci. Lett. Varsovie*, **23** (1930), 30–50; English transl. [59], pp. 131–152 or [84], pp. 38–59.

61. C. Mangione, La logica del ventesimo secolo, *Storia del pensiero scientifico* (ed. L. Geymonat), Vol. 6, 1972, pp. 469–682; Vol. 7, 1976, pp. 299–433.

62. S. McCall (ed.), *Polish Logic 1920–1939*, Oxford, 1967. [See review by W. A. Pogorzelski in *J. Symb. Logic* **35** (1970), 442–446.]

63. H. Mehrtens, *Die Entstehung der Verbandstheorie*, Gerstenberg, 1979.

64. G. H. Moore, Beyond first-order logic: the historical interplay between mathematical logic and axiomatic set theory, *Hist. Phil. Logic* **1** (1980), 95–137.

65. G. H. Moore, *Zermelo's Axiom of Choice: Its Origins, Development, and Influence*, Springer-Verlag, 1981.

66. A. Mostowski, *Thirty Years of Foundational Studies...*, Acta Philosophica Fennica, 1965.

67. W. van O. Quine, New foundations for mathematical logic, *Amer. Math. Monthly* **44** (1937), 70–80.

68. W. van O. Quine, *From a Logical Point of View*, Harvard Univ. Press, 1961, Ch. 4.

69. W. van O. Quine, *Mathematical Logic*, Harvard Univ. Press, 1940.

70. W. van O. Quine, *Set Theory and its Logic*, Harvard Univ. Press, 1969.

71. F. P. Ramsey, The foundations of mathematics, *Proc. London Math. Soc.* **25** (1926), 338–384.

72. C. Reid, *Hilbert*, Springer-Verlag, 1970.

73. N. Rescher, *Many-Valued Logic*, McGraw Hill, 1969.

74. J. Richards, Boole and Mill: differing perspectives on logical psychologism, *Hist. Phil. Logic* **1** (1980), 19–36.

75. V. F. Rickey, A survey of Leśniewski's logic, *Studia Logica* **36** (1977), 407–426.

76. P. Rutz, *Zweiwertige und mehrwertige Logik. Ein Beitrag zur Geschichte und Einheit der Logik*, Ehrenwirth, 1973.

77. W. Sierpiński, L'axiome de M. Zermelo et son rôle dans la théorie des ensembles et l'analyse, *Bull. Acad. Sci. Cracovie, Cl. Sci. Math. Nat. A* (1918), 97–152; *Oeuvres Choisies*, Vol. 2 (ed. S. Hartman and A. Schinzel), Warsaw, 1975, pp. 208–255.

78. T. Skolem, *Selected Works in Logic* (ed. J. E. Fenstad), Scandinavian Univ. Books, 1970.

79. T. Skolem, Logisch-kombinatorische Untersuchungen..., *Videns. Skr. I Mat. Naturv. Kl.*, **4** (1920); English transl. (part), [78], pp. 252–263.

80. T. Skolem, Einige Bemerkungen zur axiomatischen Begründung der Mengenlehre, *Matematikerkongressen i Helsingfors den 4–7 Juli 1922...*, Helsinki, 1923, pp. 217–232; English transl. [78], pp. 290–301.

81. T. Skolem, Begründung der elementaren Arithmetik durch die recurrierende Denkweise, *Videns. Skr. I. Mat. Naturv. Kl.* **6** (1923); English transl. [78], pp. 302–333.

82. B. Sobociński, L'analyse de l'antinomie Russellienne par Leśniewski, *Methodus* **1** (1949), 94–107; 220–228, 308–316; **2** (1950), 237–257.

83. N. I. Styazhkin, *From Leibniz to Peano: a Concise History of Mathematical Logic*, Cambridge, 1970.

84. A. T. Tarski, *Logic, Semantics, Metamathematics* (tr. J. H. Woodger), Oxford, 1956.

85. A. T. Tarski, O wyrazie pierwotnym logistyki, *Prz. Filoz.* **26** (1923), 68–89. English transl. [84], pp. 1–23.

86. A. T. Tarski, Fundamentale Begriffe der deduktiven Wissenschaften. I, *Monats. Math. Phys.* **37** (1930), 361–404; English transl. [84], pp. 60–109.

87. A. T. Tarski, *Pojęcie Prawdy Wjęzykach Nauk Dedukcyjnej*, Warsaw, 1933; English transl. [84], pp. 152–278.

88. A. T. Tarski, *O Logice Matematycznej i Metodzie Dedukcyjnej*, Warsaw, 1935; English transl., Oxford U. Press, 1941.

89. A. M. Turing, On computable numbers, with an application to the Entscheidungsproblem, *Proc. London Math. Soc.* **42** (1936), 230–265 [with corrections in **43** (1937), 544–546]; or [21], pp. 116–154.

90. J. van Heijenoort (ed.), *From Frege to Gödel: A Source Book in Mathematical Logic 1879–1931*, Cambridge, 1967.

91. John von Neumann, *Collected Works* (ed. A. H. Taub), 6 vols., Pergamon, 1961–1963.

92. John von Neumann, Zur Hilbertschen Beweistheorie, *Math. Zeitschrift* **26** (1927), 1–46; or [91], pp. 256–300.

93. Hermann Weyl, *Philosophie der Mathematik und Naturwissenschaften*, Hermann, 1927; English transl., Princeton Univ. Press, 1952.

94. A. N. Whitehead, Indication, classes, numbers, validation, *Mind* **43** (1934), 281–297; 543; or [95], pp. 227–240.

95. A. N. Whitehead, *Essays in Science and Philosophy*, Rider, 1948.

96. A. N. Whitehead and B. Russell, *Principia Mathematica*, 3 vols., Cambridge Univ. Pres, 1910–1913, 1925–1927.

97. L. Wittgenstein, *Tractatus Logico-Philosophicus*, London, 1922. Revised English transl., Routledge, 1961.

98. J. J. Zeman, *Modal Logic*, Oxford, 1972.

99. Ernst Zermelo, Grundlagen einer allgemeinen Theorie der mathematischen Satzsysteme. (Erste Mitteilung), *Fund. Math.* **25** (1935), 136–146.

100. T. Ziehen, *Das Verhältnis der Logik zur Mengenlehre*, Berlin, 1917.

Dedekind's Theorem: $\sqrt{2} \times \sqrt{3} = \sqrt{6}$

DAVID FOWLER

American Mathematical Monthly **99** (1992), 725–733

1 Dedekind's theorem

When the young Richard Dedekind, newly arrived at the Zurich Polytechnik (now the ETH), had to give for the first time the introductory calculus course, it had repercussions that were eventually to spread far beyond his class of students. He tells us in the introduction to his *Stetigkeit und irrationale Zahlen* [3] how his search for a satisfactory foundation for the calculus led him, on November 24, 1858, to his construction of the real numbers. (It was a Wednesday.) His immediate objective was to make precise and therefore, he argued, arithmetical the previously vague geometrical appeals to what we now call completeness, and every modern treatment develops lovingly and in detail this crucial property of the real numbers. But in the body of his essay, which is still the most lucid available account of his construction, he points to an equally fundamental achievement. After having described how to define addition, he goes on to say ([3], p. 22):

> Just as addition is defined, so can the other operations of the so-called elementary arithmetic be defined, viz., the formation of differences, products, quotients, powers, roots, logarithms, and in this way we arrive at real proofs of theorems (as, e.g. $\sqrt{2} \times \sqrt{3} = \sqrt{6}$), which to the best of my knowledge have never been established before.

He then elaborates this opinion in letters to Lipschitz of 1876 [3], and repeats and emphasizes it at the end of the introduction to his later *Was sind und was sollen die Zahlen?* (*What Are Numbers and What Should They Be?* is a better translation of this than the pusillanimous *The Nature and Meaning of Numbers* of [31]; and Dedekind throughout used the word *Stetigkeit*, continuity, to denote our completeness.)

This contribution is often slighted or even completely overlooked in descriptions of the reals, so my objective here is to celebrate his achievement by illustrating some of the problems that lie in the way of some alternative interpretations of what I shall call Dedekind's theorem, $\sqrt{2} \times \sqrt{3} = \sqrt{6}$, and then discuss briefly the wider historical issue of the evolving idea of the real numbers since antiquity.

The illustrations will be of two sharply contrasted types. The first group will be arithmetical, in the spirit of Dedekind's approach. For simplicity, consider the non-negative numbers. Take a half-infinite line, with left end-point labelled 0 and another distinguished point labelled 1, and somehow describe a labelling system for the points of the line. Abstracting from this, the set of real numbers will then *be* the set of all possible labels, so the labels will determine what we then *conceive* as the points of the line, and the properties of these labels will *determine* the geometrical properties of the line. Throughout, we suppose that we have available the integers and their arithmetic and order, but nothing more; and we want to try to extend this arithmetic and order structure to the set of all labels. Dedekind's insight was to use cuts in the rational numbers as labels, and to see that the features of arithmetic, order, and completeness are easy to define on these cuts.

Dedekind's cogent objections to referring other than like this to the "points" of the line, or, in his terms, to "extensive magnitudes", are that it is "not scientific" ([3], p. 1), since these points or magnitudes are "nowhere carefully defined" (p. 9; also see pp. 36–8); he calls his labels "numbers" and describes his procedure as "arithmetical". My underlying historical point in this first group of examples is to emphasize that carefully defined arithmetic—addition, multiplication, etc.—for models of the number line was far from obvious before Dedekind, and it would have been difficult to satisfy his requirement that "I demand that arithmetic should be developed out of itself" (p. 10). Behind these examples is also my dissent from an opinion that is almost universally held but is rarely articulated, and it is another indication of Dedekind's insight and lucidity that he brings it into the open:

> According to my view, the notion of the ratio between two magnitudes of the same kind can be clearly defined only after the introduction of irrational numbers (p. 10 footnote; the translation has 'two numbers', but 'two magnitudes' is clearly meant by the German original).

The first examples (in Sections 2, 3, and 5) show that there is no difficulty in defining the idea of the ratio of, for example, the diagonal and side of a square, and describing the order structure on these kinds of ratios; the problems arise precisely in trying to do, and prove results about, their arithmetic. And the final example (in Section 6) will abandon Dedekind's programme and will formulate his theorem in a geometrical model, Euclidean style, bypassing the need to talk about ratios.

The article [2] is highly recommended for details of Dedekind's intellectual and personal life.

2 The continued fraction representation

We use the so-called Euclidean algorithm (or *anthyphairesis*) to generate a labelling system. Let X_0 be any point on the half-infinite line, and write x_0 for any line congruent to $0X_0$, x_1 for 01. Now express

$$x_0 = n_0 x_1 + x_2 \quad \text{with } x_2 < x_1,$$
$$x_1 = n_1 x_2 + x_3 \quad \text{with } x_3 < x_2, \text{ etc.},$$

where if at any stage there is no remainder, the process terminates. (In some circumstances, it is more illuminating to conceive of the terminating case as finishing with an additional infinite term.) Thus far, this represents a process of decomposing $0X_0$ and 01 into subintervals, and we can use the n_i to label the point X_0; let us write $X_0 = [n_0, n_1, n_2, \ldots]$. Purely geometrical arguments show us that if $\sqrt{n}$ denotes the side of the square equal to n times the square on 01, then

$$\sqrt{2} = [1, \bar{2}], \; \sqrt{3} = [1, \overline{1, 2}], \; \text{and} \; \sqrt{6} = [2, \overline{2, 4}],$$

where the bar denotes an indefinitely repeating period. (Details of three different kinds of such proofs are given in [6], Chapter 3.)

We now reverse this way of looking at things, and define the set of all real numbers to be the set of all such terminating or non-terminating sequences of integers $[n_0, n_1, n_2, \ldots]$, with $n_i \in \mathbb{Z}, n_i \geq 1$ if $i \geq 1$ and, if the sequence terminates with an n_K for which $K \geq 1$, then $n_K \geq 2$. Relaxing the description of the process to

$$x_0 = n_0 x_1 + x_2 \quad \text{with} \quad x_2 \leq x_1, \text{ etc.}$$

eliminates this last condition on terminating expansions but introduces an innocent ambiguity,

$$[n_0, n_1, n_2, \ldots n_K] = [n_0, n_1, n_2, \ldots n_{K-1}, 1]$$
where $K \geq 0$, and if $K \geq 1$ then $n_K \geq 2$.

The order structure is easily described: lexicographic in the even-indexed terms, and reverse lexicographic in the odd-indexed terms, when terminating expansions have been put in standard notation ($n_K \geq 2$) with an infinite term adjoined. The final ingredient for the statement of Dedekind's theorem is a description of multiplication, but here a classic account of continued fractions describes how the situation is generally perceived ([8], p. 20):

> For continued fractions there are no practically applicable rules for arithmetical operations; even the problem of finding the continued fraction for a sum from the continued fractions representing the addends is exceedingly complicated, and unworkable in computational practice.

In fact, there is a simple algorithm for arithmetic, an elaboration of the procedure for evaluating the convergents, discovered in the 1970s by R. W. Gosper but never published conventionally by him! It is described in [6], pp. 114–116 and 354–360, where it is illustrated for the evaluation of

$$[1, 2, 2, 2, 2, 2, 2, \ldots] \times [1, 2, 1, 2, 1, 2, 1, \ldots]$$
$$= [2, 2, 4, 2, 4, \ldots].$$

But I cannot see how to go on to construct a direct proof of Dedekind's theorem using it.

I have argued (see [6]) that anthyphairesis may have been one of the early Greek ways of defining ratio, but anybody who knows anything about continued fractions would suspect that this approach to the real numbers and Dedekind's theorem might be unfruitful. Let us now try a much more promising and more familiar approach.

3 The decimal representation

Again, write x_0 for any line congruent to $0X_0$, x_1 for 01, and

$$x_0 = n_0 x_1 + x_2 \quad \text{with } x_2 < x_1$$
$$10x_2 = n_1 x_1 + x_3 \quad \text{with } x_3 < x_1$$
$$10x_3 = n_2 x_1 + x_4 \quad \text{with } x_4 < x_1$$

Clearly $0 \leq n_i \leq 9$ if $i \geq 1$; and we have constructed a decimal expansion of x, traditionally written $x = n_0.n_1 n_2 n_3 \cdots$.

There is again no difficulty in describing the order structure, and so no difficulty in using decimal expansions to describe the ordered set of real numbers. But once again we have problems with arithmetic. Under pressure from calculations like $\frac{4}{9} + \frac{5}{9}$ and $3 \times \frac{1}{3}$, expressed decimally, we are led to consider decimal expansions ending in strings of nines and allow identities like $0.999\ldots = 1.000\ldots$. (These do not arise from the algorithm as described above, but can occur if we modify it to

$$x_0 = n_0 x_1 + x_2 \quad \text{with} \quad x_2 \leq x_1, \text{ etc.,}$$

and we now must decide whether to allow $n_i = 10$ or not.) This opens Pandora's box, letting out the confusion of non-unique representations and indefinitely long carry, but it is difficult to conceive of any kind of decimal manipulation without these complicating features in some form or another.

Many mathematicians have a touching and naive belief that arithmetical operations on decimals pose no problems; or they pretend to believe this, as in some circumstances the most scrupulously honest among us may sometimes pretend to believe in Father Christmas (see, e.g., [1], pp. 26, 47); or perhaps they have never considered the question to be problematic.

Of course arithmetic with terminating decimal expansions is straightforward, since it is only arithmetic in $\mathbb{Z}$, represented decimally and slightly modified to accommodate the notation of the decimal point. But an example first shown to me by Christopher Zeeman shows how we again encounter problems long before we reach Dedekind's theorem. Let those who believe that an algorithm for decimal multiplication exists use it to evaluate the first non-zero digit of the expansion of the product of non-terminating periodic numbers

$$1.222\ldots \times 0.818181\ldots.$$

Is the answer 9, or 1, or 9 or 1? (For more discussion and another surprising example of Zeeman's, see [5].) The analogous problem with Gosper's algorithm for continued fraction arithmetic also concerns terminating expansions: in evaluating expressions like $\sqrt{2} \times \sqrt{2}$, the output from the algorithm will be of the form $[2, n_1]$ or $[1, 1, n_2]$, where n_1 and n_2 increase indefinitely as the algorithm struggles to evaluate them and move on to the next term. In fact, n_1 and n_2 here are infinite.

There is some evidence that mathematicians of the early nineteenth century and before, and especially those who were moving towards the developing ε-δ arithmetized analysis, were aware of the fundamental and messy problems with decimal arithmetic. For example, Cauchy's celebrated *Cours d'Analyse* has a long appended Note 1 ("Sur la théorie des quantités positives et négatives") in which he defines arithmetical operations on 'numbers' (also called 'quantities') in rather vague terms of manipulations of rational approximations; whilst in its Note 3 ("Sur la résolution numérique des équations"), he describes a proof of the intermediate value theorem in terms of what is, in effect, a decimal algorithm, though expressed there to any base. But, vague though his account often is, Cauchy does not fudge the issue by describing arithmetic in terms of terminating decimal expansions, and then pretend that he has described arithmetic in general.

Many Babylonian clay tablets containing arithmetical tables and problems of some sophistication expressed throughout to base 60, and which date from around 2000 B.C. onwards, have been found and edited. Division appears to have been handled by reciprocation followed by multiplication; but most of the reciprocal tables only contain entries for those numbers whose reciprocals terminate (i.e., numbers whose only factors are 2, 3 and 5). So Babylonian mathematicians also seem to have had a proper caution about the problem of arithmetic with non-terminating radix fractions.

4 The unit fraction representation

Egyptian mathematics tends to be viewed with amazement by mathematicians of today because of its practice of expressing rational numbers as sums of

different unit fractions:
$$\frac{p}{q} = n_0 + \frac{1}{n_1} + \frac{1}{n_2} + \cdots \frac{1}{n_K},$$
with $\quad 1 < n_1 < n_2 < \cdots < n_K$.

I do not think that I need to belabor the opinion that this makes a very unpromising base on which to attempt to state, let alone prove, Dedekind's theorem. What does need belaboring is that this same practice is found throughout Greek texts; see [6], Chapter 7 for details, and for a discussion of the evidence that leads me to argue that we have no good grounds for arguing that early Greek mathematics and commercial practice had anything corresponding to our common fractions p/q and their arithmetic! Unit fraction expressions are then found in Greek, Arabic, and Italian texts up to the sixteenth century; for example, astronomical texts continue to use the Egypto-Greek unit fractions side-by-side with the Babylonian sexagesimal numbers.

We can, incidentally, generate a class of unit fraction expansions using another variant of the subtraction algorithm. Write

$$x_0 = n_0 x_1 + x_2 \quad \text{with } x_2 < x_1,$$
$$x_1 = n_1 x_2 + x_3 \quad \text{with } x_3 < x_2,$$
$$x_2 = n_2 x_3 + x_4 \quad \text{with } x_4 < x_3, \text{ etc.}$$

We then get
$$\frac{x_0}{x_1} = n_0 + \frac{1}{n_1} - \frac{1}{n_1 n_2} + \frac{1}{n_1 n_2 n_3} - \cdots.$$

Or, by overshooting from the second step onwards,
$$x_1 = n_1 x_2 - x_3 \quad \text{with } x_3 < x_2, \text{ etc.,}$$

we can eliminate all the negative signs. While there is absolutely no evidence that anyone in antiquity used any such algorithm, similar kinds of expressions do appear in Arabic mathematics from the twelfth century onwards. They were described by Fibonacci in his *Liber Abaci* of 1202, and then persisted up to the sixteenth century, known as the *practica Italiano*. They correspond to ascending continued fractions, and sometimes have more general numerators:

$$n_0 + \frac{m_1}{n_1} + \frac{m_2}{n_1 n_2} + \frac{m_3}{n_1 n_2 n_3} + \cdots$$
$$= n_0 + \frac{m_1 + \dfrac{m_2 + \dfrac{m_3 + \cdots}{n_3}}{n_2}}{n_1},$$

though occasionally there is an additional complication of numerators that build up in a similar way to the denominators, and so do not correspond as closely to continued fractions. In their notation,

$$\frac{m_3 m_2 m_1}{n_3 n_2 n_1} \quad \text{(a variant of } practica\ Italiano\text{)}$$
$$= \frac{m_1}{n_1} + \frac{m_1 m_2}{n_1 n_2} + \frac{m_1 m_2 m_3}{n_1 n_2 n_3},$$

and all such expressions are usually written backwards like this, presumably a vestige of their Arabic origins.

Lagrange, in [10], gave a unified treatment of the three subtraction algorithms I have described so far, and others, expressed in terms of approximations. He noted the connection of the first with continued fractions and ended with an oblique reference to ascending continued fractions, which he attributed to Lambert. Ascending continued fractions appear sporadically thereafter, but they are much more simply handled in terms of series.

5 The Eudoxian representation

I start with the celebrated *Elements V*, Definition 5, attributed to Eudoxus:

> [Four] magnitudes are said to be in the same ratio, the first to the second and the third to the fourth, when, if any equi-multiples whatever be taken of the first and third, and any whatever of the second and fourth, the former equi-multiples alike exceed, are alike equal to, or alike fall short of, the latter equi-multiples respectively taken in corresponding order.

and a passage from Heath's note on this that has spawned or reinforced endless confusion:

> Max Simon remarks (*Euclid und die sechs planimetrischen Bücher*, p. 110), after Zeuthen, that Euclid's definition of equal ratios is word for word the same as Weierstrass' definition of equal numbers. So far from agreeing in the usual view that the Greeks saw in the irrational no *number*, Simon thinks it is clear from Book V that they possessed a notion of number in all its generality as clearly defined as, nay almost identical with, Weierstrass' conception of it. Certain it is that there is an exact correspondence, almost coincidence, between Euclid's definition of equal ratios and the modern theory of irrationals due to Dedekind ([4], Vol. ii, p. 124).

The first two sentences of this second quotation represent what is, for me, an ugly disease of much scholarship: the repetition of incorrect, misleading, or meaningless stereotyped verbal formulae backed up by a liturgical parade of names. As far as I am aware, Weierstrass' description of number looks nothing like *Elements V*, Def. 5; and, far from finding "numbers in all generality" in the *Elements*, the consensus of serious investigations of the *Elements* uncovers little or no trace there of proto-real numbers; indeed, almost the only numbers found there are the positive integers, of which the unit has an ambiguous status.

As remarked in the previous section, I go even further and argue, in [6], Chapter 7, that we find nowhere in early Greek mathematics (i.e., up to and including Archimedes) any convincing evidence for an understanding of the rational numbers, such as we derive from manipulations of common fractions p/q. So, as concerns the final sentence, while we can now easily translate V, Def. 5 into Dedekind's definition of a cut, the mathematical contexts of the two definitions are even more strikingly different than the correspondence between their formulations.

Elements V is about proportionality, the equivalence relation of equality between ratios. If, as here, we are only given this equivalence relation, we can now play the formal trick of taking the equivalence classes it defines, and refer to them as ratios; but this is a late nineteenth-century device, at the earliest. However, behind Book V of the *Elements*, especially when set in the context of Eudoxus' interest in cyclical calendars (for more details of this and what follows below, see the discussion in [6], pp. 121–130), we may be able to detect another procedure that we could try to use to describe the set of reals. This will involve leaving the subtraction algorithms of the previous examples and passing to addition.

Each point x of the positive line will generate a characteristic pattern in the way the points $x, 2x, 3x, \ldots$ interlace with the integer points $1, 2, 3, \ldots$. We can describe this, for example by specifying how many points of $\{x, 2x, 3x, \ldots\}$ lie in $[0, 1)$, how many in $[1, 2)$, etc. The labels this generates for the rational points of the line were described, in the 1870s, by Christoffel and H. J. S. Smith, and their descriptions can be extended to generate the labels for $\sqrt{n}$.

Abstracting from this, we may describe the set of reals to be all possible patterns that arise in this way. First, we need to characterize these patterns; then, define the order structure on them; finally, define arithmetic with them. The first problem is solved by an ingenious algorithm of Zeeman, closely related to the idea of rotation numbers; see [11] and [12].

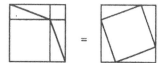

Figure 1. Pythagoras' theorem

The order structure is described in *Elements V*, Definition 7. But I have no idea if any direct definition of their arithmetic is known or accessible. So, once again, and at yet another place, the attempt to formulate and prove Dedekind's theorem is unsuccessful.

6 A geometrical description

I finish this cycle of formulations of Dedekind's theorem with a purely geometrical version, based on a naive model of Euclidean geometry in which figures are manipulated by congruence transformations and equality is interpreted in terms of scissors-and-paste operations. This goes against Dedekind's approach, but it corresponds to what is found in much, though not all, of Euclid's *Elements*. For example, the so-called Pythagoras' theorem is a statement about decomposing and reassembling squares, and Fig. 1 gives a proof which, though not found in the *Elements*, may have been excised from between Propositions 8 and 9 of Book II.

We again make a geometrical definition of $\sqrt{n}$ as the side of a square equal to n concatenated copies of the unit square; this can be constructed, for example, by making repeated use of Pythagoras' theorem, as in Fig. 2. We now define multiplication geometrically: if n and m denote natural numbers, a and b lines, and A and B regions in the plane, then na or nA will denote n concatenated copies of a or A; $a.b$ will denote the rectangle with adjacent sides a and b; $a.B$ will

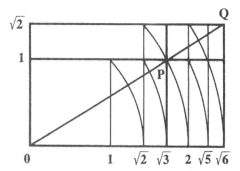

Figure 2. Geometrical formulation of Dedekind's theorem.

Figure 3. Elements I43.

denote the rectangular prism with base B and height a; and $A.B$ is not defined. Again this corresponds to what we find in the *Elements*. With these definitions, our original version of Dedekind's theorem is not well-posed—a rectangle cannot be compared with a line—but we can immediately adjust its formulation to $\sqrt{2}.\sqrt{3} = \sqrt{6}.1$.

A proof will now consist of an argument about a figure involving these two rectangles. Many different figures are possible; Fig. 2 illustrates one straightforward construction, using nothing more than has been described above; and the equality of the rectangles $\sqrt{2}.\sqrt{3}$ and $\sqrt{6}.1$ therein will be equivalent to the collinearity of $0, P$ and Q. (This is the easy converse of *Elements I*, 43, which shows that, in Fig. 3, the shaded parallelograms are equal.) Curiously, a proof now requires arguments based on similarity—that the construction of P, starting from the line 01, is the same as the construction of Q, starting from $0P\sqrt{2}$, and hence the triangles $0P\sqrt{3}$ and $0Q\sqrt{6}$ are similar—and so seems to depend on the Euclidean nature of the geometry. I remark, in passing, that Euclid's version of the parallel postulate is not actually expressed in terms of parallels, but is closer to expressing the possibility of constructing similar triangles of any given size.

I know of no explicit reference to any formulation of Dedekind's theorem in early Greek mathematics; whether or not it is there implicitly is a delicate historical matter.

7 Mathematics and the real numbers

Mathematical thinking today is built on our intuitions of the real numbers, but I have tried here to illustrate how it may sometimes distort the past when we interpret their mathematics in these terms. Here are some related opinions about other related past mathematical developments:

(a) Early Greek mathematics, up to the time of Archimedes, does not seem to be arithmetized. However it is interpreted arithmetically thereafter, for example in the metrical geometry of Heron and the astronomy of Ptolemy, and many modern descriptions are now set against some assumed background of a developing idea of rational and real numbers, For example, the lurid stories of the discovery and effect of incommensurability, which is so damaging to a naive arithmetical mathematics based on the rationals, are found in later commentators but are surprisingly absent from our earlier evidence; they may, I think, be part of a later overlay. (There is a discussion of the historical evidence concerning the discovery of incommensurability in [6], 294–308.)

(b) Babylonian mathematics and astronomy is highly arithmetized, but it does not have the deductive structure we now associate with Greek and our mathematics. One possibly fruitful line of interpretation, which I have not seen explored anywhere, might be to develop the distinction proposed, by Knuth [9], between mathematical and algorithmic thinking, and see if it applies to this Babylonian material, especially the later Babylonian astronomy of the Seleucid period. In the grandest sweep of history, is it possible that the paradigm of deductive mathematics, which has dominated our view since the fourth century BC, may have run its course, and may now be giving way to an older algorithmic style, which is now flourishing in a changed environment of automatic computation, the appeal of experimental mathematics, economic and political pressures, shifts in the school curriculum, and the increasing specialization and inaccessibility of much of mathematics today?

(c) Western mathematics since the seventeenth century has owed a lot of its power to the way it has been successfully and comprehensively arithmetical, though it managed to ignore the basic problems with a precise description of its underlying arithmetic until the nineteenth century. (This simple picture must be filled in with details of the excursions into infinitesimal and infinite numbers and nonstandard analysis, which fit well with the approach to arithmetical models of the line I described in Section 1, and the mathematico-algorithmic hybrid of constructive mathematics.) I believe that the dramatic and explosive growth of symbolism in the seventeenth century—for there is no symbolism before about 1600, apart from numerals and a few things that are better described as abbreviations—may be connected with a new fluency in arithmetized thinking, which itself may owe a lot to the popularization of decimal fractions at the end of the fifteenth century. Stevin, for example, was a thorough-going arithmetizer: he published, in 1585, the first popularization of decimal fractions in the West (both in Dutch, *De Thiende*, and French, *La Disme*); in 1594, he described an algorithm for finding the decimal expansion of the root of any polynomial, the same algorithm we find later in Cauchy's proof of the intermediate value theorem to which I referred above; and he argued vigorously for an arithmetical understanding of the *Elements*, including its notorious Book *X*. But this is another story, part of which is described in [7].

References

1. L. Bers, *Calculus*, Holt, Reinhart and Winston, 1969.

2. K.-R. Bierman, Dedekind, *Dictionary of Scientific Biography* iv, Scribner's, 1970–1978.

3. R. Dedekind, *Essays on the Theory of Numbers: Continuity and Irrational Numbers & The Nature and Meaning of Numbers* (transl. W. W. Beman, of *Stetigkeit und irrationale Zahlen* (1872) and *Was sind und was sollen die Zahlen?* (1888), Open Court, 1901, repr. Dover, 1963; all page references refer to this translation. The German originals are reprinted in his *Gesammelte Mathematische Werke* (ed. R. Fricke, E. Noether, and O. Ore), 3 vols., 1930–1932, repr. Chelsea, 1969, in Vol. iii, 315–391, which also contains his letters of June 10 and July 27, 1876, to Lipschitz on this topic, 468–479.

4. Euclid, *The Thirteen Books of Euclid's Elements* (transl. and ed. T. L. Heath), 2nd ed., 3 vols., Cambridge Univ. Press, 1926; repr. Dover, 1956.

5. D. H. Fowler, 400 years of decimal fractions, *Mathematics Teaching* **110** (1985) 20–21; and 400.25 years of decimal fractions, *Mathematics Teaching* **111** (1985), 30–31.

6. D. H. Fowler, *The Mathematics of Plato's Academy*, Oxford Univ. Press, 1987; corrected paperback repr., 1991.

7. D. H. Fowler, An invitation to read Book X of Euclid's Elements, *Historia Mathematica* **19** (1992), 233–264.

8. A. Ya. Khinchin, *Continued Fractions* (transl. Scripta Technica Inc. of *Cepnye Drobi* (1936)), Chicago Univ. Press, 1964.

9. D. E. Knuth, Algorithmic thinking and mathematical thinking, *Amer. Math. Monthly* **92** (1985), 170–81.

10. J. L. Lagrange, Essai d'analyse numérique sur la transformation de fractions, *Journal de l'Ecole Polytechnique* 2 (prairial an VI [= 28 May–18th June, 1799]), reprinted in *Oeuvres* vii, 291–313.

11. C. Series, The geometry of Markoff numbers, *Mathematical Intelligencer* **7** (1983), 20–29.

12. E. C. Zeeman, Gears from the Greeks, *Proceedings of the Royal Institution* **58** (1986), 137–156.

Afterword

Halsted's review is of Volume VII of Gauss's *Werke*, but in Volume VIII, the editors published Gauss's manuscript material on non-Euclidean geometry. These show that Gauss anticipated much of the work of Bolyai and Lobachevsky, as well as of Schweikart. The existence of these materials does not, however, challenge Halsted's conclusion that Gauss had no influence on the work of any of these men. Roberto Bonola's *Non-Euclidean Geometry* [2] is still a good source for details on Gauss's work; the English edition also contains Halsted's translations of the fundamental articles on non-Euclidean geometry by Bolyai and Lobachevsky.

More recent works on the history of non-Euclidean geometry include Jeremy Gray's *Ideas of Space: Euclidean, non-Euclidean and Relativistic* [10] and *János Bolyai, Non-euclidean Geometry and the Nature of Space* [11] and B. A. Rosenfeld's *A History of Non-Euclidean Geometry* [17]. The latter work, in particular, deals in great detail with the work of "certain Arabs and Persians" that Lewis skips over. In fact, it is now clear that some of the Islamic work on the parallel postulate was read in Europe in the seventeenth century and affected the European developments on the subject. Jeremy Gray has also written a historically-based textbook in geometry, *Worlds out of Nothing: A Course in the History of Geometry in the 19th Century* [12], that covers in detail not only non-Euclidean geometry but also projective geometry and many other topics. Julian Lowell Coolidge himself expanded on his article on projective geometry in *A History of Geometrical Methods* [4], first published in 1940.

A general history of point-set topology is Jerome Manheim's *The Genesis of Point-Set Topology* [15], which also contains a history of set theory in its relationship with topology. As to algebraic topology, the most detailed history is Jean Dieudonné's *A History of Algebraic and Differential Topology, 1900–1960* [8], in which the author attempts to use modern terminology to represent older ideas and concentrates heavily on the algebraic side of the story. For the prehistory of homotopy theory, see R. Van den Eynde's article [18]; this article is one of forty in I. M. James's *History of Topology* [13], a work that covers most of the major topics in the field.

An article closely related to Hilton's is by Solomon Lefschetz [14], originally published in 1970 and reprinted in James's volume. Another history of the homology concept is by Maja Bollinger [1]. Auguste Dick's biography, *Emmy Noether, 1882–1935* [7] contains details of Noether's influence on the development of algebraic topology in Göttingen. It also reproduces Pavel Sergeiivich Aleksandrov's address in her memory that describes some of the atmosphere there in the 1930s. Hassler Whitney has written on the international topology conference in Moscow in 1935 [19], and Hilton mentions that, in 1988, Leopold Vietoris was still alive. In fact, Vietoris lived until 2002, dying at the age of 110!

Afterword

W. S. Contro [5] covers some of the same ground as H. C. Kennedy, as he considers both Pasch's and Peano's approaches to axiomatization. Joseph Brent has written an excellent biography of Charles Sanders Peirce: *Charles Sanders Peirce: A Life* [3], which includes information on his logical work, and Peirce also appears in the masterful volume of Karen Parshall and David Rowe: *The Emergence of the American Mathematical Research Community, 1876–1900: J. J. Sylvester, Felix Klein, and E. H. Moore* [16]. Parshall and Rowe in particular deal with Peirce's tenure at Johns Hopkins University, where he lectured on logic for several years.

For more information on logic, a good source, with many bibliographic references, is Part 5 of Ivor Grattan-Guinness's *Companion Encyclopedia of the History and Philosophy of the Mathematical Sciences* [9]. A good treatment of the major philosophies of mathematics can be found in Chapter 7 of *The Mathematical Experience*, by Philip Davis and Reuben Hersh [6].

References

1. Maja Bollinger, Geschichtliche Entwicklung des Homologie-begriffs, *Arch. Hist. Exact Sci.* **9** (1972), 94–166.
2. Roberto Bonola, *Non-Euclidean Geometry: A Critical and Historical Study of Its Development* (transl. H. S. Carslaw), Dover, 1955.
3. Joseph Brent, *Charles Sanders Peirce: A Life*, Indiana Univ. Press, 1998.
4. Julian Lowell Coolidge, *A History of Geometrical Methods*, Dover, 1963.
5. W. S. Contro, Von Pasch zu Hilbert, *Arch. Hist. Exact Sci.* **15** (1975–1976), 283–295.
6. Philip Davis and Reuben Hersh, *The Mathematical Experience*, Birkhäuser, 1981.
7. Auguste Dick, *Emmy Noether, 1882–1935*, Birkhäuser, 1981.
8. Jean Dieudonné, *A History of Algebraic and Differential Topology. 1900–1960*, Birkhäuser, 1989.
9. Ivor Grattan-Guinness (ed.), *Companion Encyclopedia of the History and Philosophy of the Mathematical Sciences*, (2 vols.), Routledge, 1994.
10. Jeremy Gray, *Ideas of Space: Euclidean, Non-Euclidean and Relativistic*, Clarendon Press, 1989.
11. Jeremy Gray, *Worlds out of Nothing: A Course in the History of Geometry in the 19th Century*, Springer, 2007.
12. I. M. James (ed.), *History of Topology*, Elsevier, 1999.
13. S. Lefschetz, The early development of algebraic topology, in [13], pp. 531–560.
14. Jerome Manheim, *The Genesis of Point-Set Topology*, Macmillan, 1964.
15. Karen H. Parshall and David E. Rowe, *The Emergence of the American Mathematical Research Community, 1876–1900: J. J. Sylvester, Felix Klein, and E. H. Moore*, American Mathematical Society, 1994.
16. B. A. Rosenfeld, *A History of Non-Euclidean Geometry*, Springer, 1988.
17. R. Van den Eynde, Development of the concept of homotopy, in [13], pp. 65–102.
18. Hassler Whitney, Moscow 1935: Topology Moving Toward America, in *A Century of Mathematics in America* (Peter Duren ed.), Part I, American Mathematical Society, 1988.

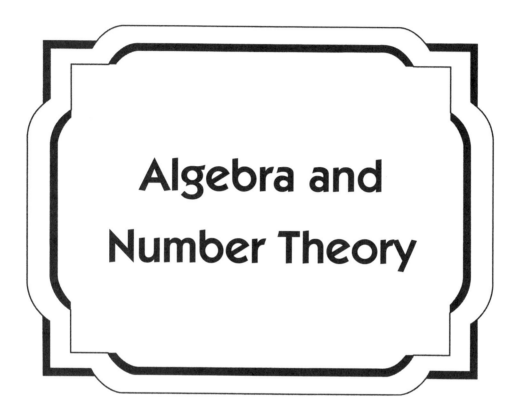

Foreword

Algebra in 1800 meant the solving of equations. By 1900, the term was beginning to encompass the study of various mathematical structures—sets of elements with well-defined operations, satisfying certain specified axioms—and by the 1930s, under the influence of Emmy Noether, this change was complete. This chapter explores some aspects of this change in the notion of algebra over that time period.

The nineteenth century also saw a change in number theory, not so much in the subject matter but in the methods of proof. From the inception of number theory, one proved results about positive integers by using positive integers. But in the nineteenth century, mathematicians generalized their results by beginning to look at other "numbers", beginning with the Gaussian integers. Then, in a more striking change, they began to consider proofs about numbers using techniques from real and complex analysis. Several articles in this chapter explore these changes.

The first two articles in this chapter discuss aspects of the history of *quaternions*, the first algebraic system to be studied that did not satisfy the commutative law. B. L. van der Waerden tells the story of Hamilton's discovery of the quaternions, beginning with his search for a law of multiplication for triples (or vectors in three-dimensional space) that would generalize the law of multiplication for complex numbers, thought of as pairs or as vectors in two-dimensional space. Simon Altmann then continues the story, showing how over many years Hamilton attempted to interpret his quaternions in terms of rotations in space. Ultimately, Hamilton failed in this attempt, and the reasons were only elucidated by Olinde Rodrigues, a French banker whose contributions to mathematics are little known.

The first algebraic concept to be fully abstracted was that of a *group*. Two articles are devoted to the beginnings of that abstract theory. In the first, Josephine Burns considers the work of Augustin Cauchy in developing some of the basic concepts of the theory of groups, although always in the context of permutation groups. Cauchy not only wrote out all the elements (in the form of permutations) of a number of important groups, but also proved one of the first theorems on the way to the Sylow theorems, that if m is the order of a group and p any prime that divides m, then the group contains an element of order p.

Israel Kleiner surveys the history of group theory, tracing its roots to four areas: the solving of equations, especially the fundamental work of Lagrange; number theory, beginning with Carl Friederich Gauss's *Disquisitiones Arithmeticae*; geometry, with the ideas of Felix Klein; and finally, analysis, including especially the work of Sophus Lie. Kleiner then continues with the work of Cauchy, Galois and others in studying permutation groups in detail, the work of Kronecker in developing what is now called the fundamental theorem of Abelian groups, and the work of Klein and Lie in investigating the theory of transformations. Kleiner finally looks at the eventual emergence of abstraction in group theory, beginning in the 1880s, although with some of this abstraction accomplished "prematurely" by Arthur Cayley.

One particular aspect of finite group theory that has been important of late is the theory of *finite simple groups*. In an article published in 1976, Gallian takes us through the work on their classification up to that time. Interestingly, at the time he wrote, new sporadic groups had recently been discovered, and it was not clear whether such groups would continue to appear. As we know now, the sporadic groups discovered by the mid-1970s were the last ones, and the classification of finite simple groups was complete by the early 1980s.

Another field of algebra that was studied in detail in the nineteenth century was *linear algebra*, although it was not axiomatized until the twentieth century. But Desmond Fearnley-Sander argues that the true creator of this subject was Hermann Grassmann, in his two versions of the *Ausdehnungslehre* (1844 and 1862). We see how Grassmann not only developed the theory of finite-dimensional vector spaces and proved most of the important theorems, but also developed what we now call the *exterior algebra* of a vector space, a concept that was applied at the end of the nineteenth century to develop the theory of differential forms. Fearnley-Sander also shows how Grassmann was successful in one of his basic aims: providing an easy-to-use symbolic language for geometry.

At the beginning of the nineteenth century, Gauss's *Disquisitiones Arithmeticae* discussed the basics of number theory, not only giving a proof of the law of quadratic reciprocity, but also introducing various new concepts that provided early examples of groups and matrices. Gauss's study of higher reciprocity laws soon led to his study of the *Gaussian integers*, complex numbers of the form $a + bi$ where a and b are ordinary integers. Attempts to generalize the properties of these integers to integers in other number fields led Ernst Kummer to the realization that some of the most important of these properties, such as unique factorization, fail to hold. The search to recover this property is discussed by Israel Kleiner, as he summarizes the work of Gauss and Kummer and also considers in detail Richard Dedekind's invention of the theory of ideals. Reinhard Laubenbacher and David Pengelley consider Gotthold Eisenstein's geometric proof of the law of quadratic reciprocity, a result proved by Gauss in the *Disquisitiones* and several times later.

The next three articles deal with aspects of the theory of numbers, especially the change in the nature of proof. First, Charles Small surveys the history of *Waring's problem*, which asks whether for each exponent $k > 1$, there is an integer $g(k)$ such that every positive integer can be expressed as the sum of at most $g(k)$ kth powers. Waring himself claimed that $g(2) = 4$, $g(3) = 9$, and $g(4) = 19$. Next, Larry Goldstein surveys the history of the *prime number theorem*, that the number $\pi(x)$ of prime numbers not exceeding x can be approximated by the function $x/\ln(x)$. It was Legendre who first conjectured a result of this sort, and it was Gauss who collected extensive numerical data to come up with his own similar conjecture. Later on, Riemann showed that the prime number theorem was closely related to what became known as the *Riemann hypothesis*, that all the non-trivial zeros of the Riemann zeta function have real part equal to $\frac{1}{2}$. As it turned out, however, the truth of the Riemann hypothesis was not necessary for a proof of the prime number theorem, which was accomplished by Hadamard and de la Vallée Poussin in 1896. Their proofs, and other work related to the prime number theorem, are discussed in detail in the article by Paul Bateman and Harold Diamond.

One of the major successes of twentieth-century number theory was the proof of Fermat's last theorem, the famous conjecture that $x^n + y^n = z^n$ has no non-trivial integer solutions for $n > 2$. Fernando Gouvêa surveys Andrew Wiles's initial announced proof of this result, a proof announced by surprise at the end of a series of lectures Wiles gave at a number theory conference in Cambridge in June 1993. Gouvêa outlines the history of attempts to prove the theorem and the advances in the theory of elliptic functions and modular forms on which Wiles based his proof.

Interspersed among the historical articles in this chapter are four primarily biographical articles. First, we have the paper by Karen Parshall and Eugene Seneta on J. J. Sylvester, concentrating mostly on his efforts to build an international reputation for himself. Then, we have Tony Rothman's article on Evariste Galois, in which the author debunks many of the myths surrounding the last few years of Galois's life and gives his own hypothesis about the fatal duel. Next is an article by G. H. Hardy on Ramanujan, discussing the close relationship between the two and his opinions on the Indian mathematician's brilliance. The chapter also includes Clark Kimberling's biographical sketch of Emmy Noether, containing long quotations from several people—notably Pavel Alexandroff—who wrote memorials to her after her death.

Hamilton's Discovery of Quaternions

B. L. VAN DER WAERDEN

Mathematics Magazine **49** (1976), 227–234

Introduction

The ordinary complex numbers $a + ib$ (or, as they were formerly written, $a + b\sqrt{-1}$) are added and multiplied according to definite rules. The rule for multiplication reads as follows:

First multiply according to the rules of high school algebra:
$$(a + ib)(c + id) = ac + adi + bci + bdi^2$$
and then replace i^2 by -1:
$$(a + ib)(c + id) = (ac - bd) + (ad + bc)i.$$

Complex numbers can also be defined as couples (a, b). The product of two couples (a, b) and (c, d) is defined as the couple $(ac - bd, ad + bc)$. The couple $(1, 0)$ is called 1, the couple $(0, 1)$ is called i. Then we also have the result
$$i^2 = (0, 1)(0, 1) = (-1, 0) = -1.$$

By means of this definition the *imaginary unit* $i = \sqrt{-1}$ loses all of its mystery: i is simply the couple $(0, 1)$.

The *quaternions* $a + bi + cj + dk$ which William Rowan Hamilton discovered on the 16th of October 1843, are multiplied according to fixed rules, in analogy to the complex numbers; that is to say:
$$i^2 = j^2 = k^2 = -1,$$
$$ij = k, \quad jk = i, \quad ki = j,$$
$$ji = -k, \quad kj = -i, \quad ik = -j.$$

They can also be defined as quadruples (a, b, c, d). Quaternions form a *division algebra*; that is, they cannot only be added, subtracted, and multiplied, but also divided (excluding division by zero). All rules of calculation of high school school algebra hold; only the commutative law $AB = BA$ does not hold since ij is not the same as ji.

How did Hamilton arrive at these multiplication rules? What was his problem and how did he find the solution? We are accurately informed about these matters in documents and papers which appear in the third volume of Hamilton's collected *Mathematical Papers* [3]:

1. Through an entry in Hamilton's *Note Book* dated 16 October 1843 ([3], pp. 103–105);

2. Through a letter to John Graves of the 17th of October 1843 ([3], pp. 106–110);

3. Through a paper in the *Proceedings of the Royal Irish Academy* 2 (1844), 424–434, presented on the 13th of November 1843 ([3], pp. 111–116);

4. Through the detailed Preface to Hamilton's *Lectures on Quaternions*, dated June 1853 ([3], pp. 117–155, in particular, pp. 142–144);

5. Through a letter to his son Archibald which Hamilton wrote shortly before his death, that is shortly before the 2nd of September 1865 ([3], pp. xv–xvi).

We can follow exactly each of Hamilton's steps of thought through all of these documents. This is a rare occurrence in which we can observe what flashed across the mind of a mathematician as he posed the problem, as he approached the solution step by step, and then through a lightning stroke so modified the problem that it became solvable.

A brief history of complex numbers

Expressions of the form $A + \sqrt{-B}$ had already been encountered in the Middle Ages in the solution of quadratic equations. They were called "impossible solutions" or *numeri surdi*: absurd numbers. The negative numbers too were called "impossible." Cardan used numbers $A + \sqrt{-B}$ in the solution of equations of the third degree in the *casus irreducibilis* in which all three roots are real. Bombelli showed that it was possible to calculate with expressions such as $A + \sqrt{-B}$ without contradiction, but he did not like them: he called them "sophistical" and apparently without value. The expression "imaginary number" stems from Descartes.

Euler had no scruples about operating altogether freely with complex numbers. He proposed formulas such as

$$\cos \alpha = \frac{1}{2}(e^{i\alpha} + e^{-i\alpha}).$$

The geometric representation of complex numbers as vectors or as points in a plane stems from Argand (1813), Warren (1828) and Gauss (1832).

The first named, Argand, defined the complex numbers as directed segments in the plane. He took the basis vectors 1 and i as mutually perpendicular unit vectors. Addition is the usual vector addition, with which Newton made us familiar (the parallelogram law of velocities or of forces). The length of a vector was denoted at that time by the term *modulus*, the angle of the vector with the positive x-axis as the *argument* of the complex number. Multiplication of complex numbers, according to Argand, then takes place so that the moduli are multiplied and the arguments are added. Independently of Argand, Warren and Gauss also represented complex numbers geometrically and interpreted their addition and multiplication geometrically.

"Papa, can you multiply triples?"

Hamilton knew and used the geometric representation of complex numbers. In his published papers, however, he emphasized the definition of complex numbers as the couple (a, b) which followed definite rules for addition and multiplication. Related to that, Hamilton posed this problem to himself:

> To find how number-triplets (a, b, c) are to be multiplied in analogy to couples (a, b).

For a long time Hamilton had hoped to discover the multiplication rule for triplets, as he himself stated. But in October 1843 this hope became much stronger and more serious. He put it this way in a letter to his son Archibald ([3], p. xv):

> the desire to discover the law of multiplication of triplets regained with me a certain strength and earnestness....

In analogy to the complex numbers $a + ib$, Hamilton wrote his triplets as $a + bi + cj$. He represented his unit vectors $1, i, j$ as mutually perpendicular "directed segments" of unit length in space. Later Hamilton himself used the word *vector*, which I also shall use in the following. Hamilton then sought to represent products such as $(a + bi + cj)(x + yi + zj)$ again as vectors in the same space. He required, first, that it be possible to multiply out term by term; and second, that the length of the product of the vectors be equal to the product of the lengths. This latter rule was called the *law of the moduli* by Hamilton.

Today we know that the two requirements of Hamilton can be fulfilled only in spaces of dimensions 1, 2, 4 and 8; this was proved by Hurwitz [5]. Therefore Hamilton's attempt in three dimensions had to fail. His profound idea was to continue to 4 dimensions, since all of his attempts in 3 dimensions failed to reach the goal.

In the previously mentioned letter to his son, Hamilton wrote about his first attempt:

> Every morning in the early part of the above-cited month [October 1843], on my coming down to breakfast, your brother William Edwin and yourself used to ask me, "Well, Papa, can you multiply triplets?" Whereto I was always obliged to reply, with a sad shake of the head, "No, I can only add and subtract them."

From the other documents we learn more precisely about Hamilton's first attempts. To fulfill the law of the moduli, at least for the complex numbers $a + ib$, Hamilton set $ii = -1$, as for ordinary complex numbers, and similarly so that the law would also hold for the numbers $a + cj$, $jj = -1$. But what was ij and what was ji? At first Hamilton assumed $ij = ji$ and calculated as follows:

$$(a + ib + jc)(x + iy + jz)$$
$$= (ax - by - cz) + i(ay + bx)$$
$$+ j(az + cx) + ij(bz + cy).$$

Now, he asked, what is one to do with ij? Will it have the form $\alpha + \beta i + \gamma j$?

First attempt: The square of ij had to be 1, since $i^2 = -1$ and $j^2 = -1$. Therefore, wrote Hamilton, in this attempt one would have to choose $ij = 1$ or $ij = -1$. But in neither of these two cases will the law of the moduli be fulfilled, as calculation shows.

Second attempt: Hamilton considered the simplest case

$$(a + ib + jc)^2$$
$$= a^2 - b^2 - c^2 + 2iab + 2jac + 2ijbc.$$

Then he calculated the sum of the squares of the coefficients of $1, i$, and j on the right-hand side and found

$$\left(a^2 - b^2 - c^2\right)^2 + (2ab)^2 + (2ac)^2$$
$$= \left(a^2 + b^2 + c^2\right)^2.$$

Therefore, he said, the product rule is fulfilled if we set $ij = 0$. And further: if we pass a plane through the points $0, 1$ and $a + ib + jc$, then the construction of the product according to Argand and Warren will hold in this plane: the vector $(a + bi + cf)^2$ lies in the same plane and the angle which this vector makes with the vector 1 is twice as large as the angle between the vectors $a + bi + cj$ and 1. Hamilton verified this by computing the tangents of the two angles.

Third attempt: Hamilton reports that the assumption $ij = 0$, which he made in the second attempt, subsequently did not appear to be quite right to him. He writes in the letter to Graves ([3], p. 107):

> Behold me therefore tempted for a moment to fancy that $ij = 0$. But this seemed odd and uncomfortable, and I perceived that the same suppression of the term which was *de trop* might be attained by assuming what seemed to me less harsh, namely that $ji = -ij$. I made therefore $ij = k$, $ji = -k$, reserving to myself to inquire whether k was 0 or not.

Hamilton was entirely right in giving up the assumption $ij = 0$ and taking instead $ij = -ji$. For example, if $ij = 0$, then the modulus of the product ij would be zero, which would contradict the law of the moduli.

Fourth attempt: Somewhat more generally, Hamilton multiplied $a + ib + jc$ and $x + ib + jc$. In this case the two segments which are to be multiplied also lie in one plane, that is, in the plane spanned by the points $0, 1$, and $ib + jc$. The result of the multiplication was

$$ax - b^2 - c^2 + i(a+x)b + j(a+x)c + k(bc - bc).$$

Hamilton concluded from this calculation ([3], p. 107) that:

> the coefficient of k still vanishes; and $ax - b^2 - c^2$, $(a + x)b$, $(a + x)c$ are easily found to be the correct coordinates of the *product-point*, in the sense that the rotation from the unit line to the radius vector of a, b, c being added in its own plane to the rotation from the same unit-line to the radius vector of the other factor-point x, b, c conducts to the radius vector of the lately mentioned product-point; and that this latter radius vector is in length the product of the two former. Confirmation of $ij = -ji$; but no information yet of the value of k.

The leap into the fourth dimension

After this encouraging result Hamilton ventured to attack the general case: "Try boldly then the general product of two triplets, ..." ([3], p. 107.) He calculated

$$(a + ib + jc)(x + iy + jz)$$
$$= (ax - by - cz) + i(ay + bx) + j(az + cx)$$
$$+ k(bz - cy).$$

In an exploratory attempt he set $k = 0$ and asked: Is the law of the moduli satisfied? In other words, does the identity

$$(a^2 + b^2 + c^2)(x^2 + y^2 + z^2)$$
$$= (ax - by - cz)^2 + (ay + bx)^2 + (az + cx)^2$$

hold?

No, the first member exceeds the second by $(bz - cy)^2$. But this is just the square of the coefficient of k, in the development of the product $(a + ib + ic)(x + iy + jz)$, if we grant that $ij = k$, $ji = -k$, as before.

And now comes the insight which gave the entire problem a new direction. In the letter to Graves ([3], p. 108), Hamilton emphasized the insight:

> And here there dawned on me the notion that we must admit, in some sense, a *fourth dimension* of space for the purpose of calculating with triplets.

This fourth dimension appeared as a "paradox" to Hamilton himself and he hastened to transfer the paradox to algebra ([3], p. 108):

> ... or transferring the paradox to algebra, [we] must admit a *third* distinct imaginary symbol k, not to be confounded with either i or j, but equal to the product of the first as multiplier, and the second as multiplicand; and therefore [I] was led

to introduce *quaternions* such as

$$a + ib + jc + kd, \text{ or } (a, b, c, d).$$

Hamilton was not the first to think about a multidimensional geometry. In a footnote to the letter to Graves he wrote:

> The writer has this moment been informed (in a letter from a friend) that in the *Cambridge Mathematical Journal* for May last [1843] a paper on Analytical Geometry of n dimensions has been published by Mr. Cayley, but regrets he does not yet know how far Mr. Cayley's views and his own may resemble or differ from each other.

"This moment" can in this connection only mean the same day in which he wrote the letter to Graves. In the *Note Book* of the 16th of October 1843 there is no mention of the paper by Cayley. Hamilton therefore appears to have arrived at the concept of a 4-dimensional space independently of Cayley.

After Hamilton had introduced $ij = -ji = k$ as a fourth independent basis vector, he continued the calculation ([3], p. 108):

> I saw that we had probably $ik = -j$, because $ik = iij$, and $i^2 = -1$; and that in like manner we might expect to find $kj = ijj = -i$.

From the use of the word "probably" it can be seen how cautiously Hamilton continued. He scarcely trusted himself to apply the associative law $i(ij) = (ii)j$ because he was not yet certain if the associative law held for quaternions. Likewise, Hamilton could have used the associative law to determine ki:

$$ki = -(ji)i = -j(ii) = (-j)(-1) = j.$$

Instead he applied a conclusion by analogy. He wrote ([3], p. 108):

> ...from which I thought it likely that $ki = j, jk = i$, because it seemed likely that if $ji = -ij$, we should have also $kj = -jk, ik = -ki$.

Finally k^2 had to be determined. Hamilton again proceeded cautiously:

> And since the order of multiplication of these imaginaries is not indifferent, we cannot infer that k^2, or $ijij$, is $= +1$, because $i^2 \times j^2 = (-1)(-1) = +1$. It is more likely that $k^2 = ijij = -iijj = -1$.

This last assumption $k^2 = -1$, asserts Hamilton, is also necessary if we wish to fulfill the *law of the moduli*. He carried this out and concluded ([3], p. 108):

> My assumptions were now completed, namely,
>
> $$i^2 = j^2 = k^2 = -1$$
> $$ij = -ji = k$$
> $$jk = -kj = i$$
> $$ki = -ik = j.$$

And now Hamilton tested if the law of the moduli was actually satisfied.

> But I considered it essential to try whether these equations were consistent with the law of moduli ... without which consistence being verified, I should have regarded the whole speculation as a failure.

He therefore multiplied two arbitrary quaternions according to the rules just formulated

$$(a, b, c, d)(a', b', c', d') = (a'', b'', c'', d''),$$

calculated (a'', b'', c'', d'') and formed the sum of the squares

$$(a'')^2 + (b'')^2 + (c'')^2 + (d'')^2$$

and found to his great joy that this sum of squares actually was equal to the product

$$(a^2 + b^2 + c^2 + d^2)(a'^2 + b'^2 + c'^2 + d'^2).$$

In Hamilton's letter to his son we learn even more about the external circumstances which befell him at this flash of insight. Immediately after the previously cited words, "No, I can only add and subtract them." Hamilton continued ([3], pp. xx–xxvi):

> But on the 16th day of the same month [October 1843]—which happened to be a Monday and a Council day of the Royal Irish Academy—I was walking in to attend and preside, and your mother was walking with me, along the Royal Canal, to which she had perhaps been driven; and although she talked with me now and then, yet an under-current of thought was going on in my mind, which gave at last a result, whereof it is not too much to say that I felt at once the importance. An electric circuit seemed to close; and a spark flashed forth, the herald (as I foresaw immediately) of many long years to come of definitely directed thought and work, by myself if spared, and at all events on the part of others, if I should ever be allowed to live long enough distinctly to communicate the discovery. I pulled

out on the spot a pocket-book, which still exists, and made an entry there and then. Nor could I resist the impulse—unphilosophical as it may have been—to cut with a knife on a stone of Brougham Bridge, as we passed it, the fundamental formula with the symbols i, j, k;

$$i^2 = j^2 = k^2 = ijk = -1,$$

which contains the solution of the Problem, but of course as an inscription, has long since mouldered away.

The entry in the pocket book is reproduced on the title page of [3]: it contains the formulas

$$i^2 = j^2 = k^2 = -1$$
$$ij = k, \quad jk = i, \quad ki = j$$
$$ji = -k, \quad kj = -i, \quad ik = -j.$$

I assume as likely that before his walk Hamilton had already written on a piece of paper the result of the somewhat tiresome calculation which showed that the sum of squares

$$(ax - by - cz)^2 + (ay + bx)^2 + (az + cx)^2$$

still lacked $(bz - cy)^2$ compared with the product $(a^2 + b^2 + c^2)(x^2 + y^2 + z^2)$. What then happened immediately before and during that remarkable walk along the Royal Canal, he described again on the same day in his *Note Book*, as follows:

I believe that I now remember the order of my thought. The equation $ij = 0$ was recommended by the circumstances that

$$(ax - y^2 - z^2)^2 + (a + x)^2(y^2 + z^2)$$
$$= (a^2 + y^2 + z^2)(x^2 + y^2 + z^2).$$

I therefore tried whether it might not turn out to be true that

$$(a^2 + b^2 + c^2)(x^2 + y^2 + z^2)$$
$$= (ax - by - cz)^2 + (ay + bx)^2 + (az + cx)^2,$$

but found that this equation required, in order to make it true, the addition of $(bz - cy)^2$ to the second member. This *forced* on me the non-neglect of ij, and *suggested* that it might be equal to k, a new imaginary.

By underscoring the italicized words *forced* and *suggested*, Hamilton emphasized that he was concerned with two entirely different facts. The first was a compelling logical conclusion, which came immediately out of the calculation: it was not possible to set ij equal to zero, since then the law of the moduli would not hold. The second fact was an insight which came over him in a flash at the canal ("an electric circuit seemed to close, and a spark flashed forth"); that is, that ij could be taken to be a new imaginary unit.

After the insight was once there, everything else was very simple. The calculations $ik = iij = -j$ and $kj = ijj = -i$ could be made easily enough by Hamilton in his head. The assumptions $ki = -ik = j$ and $jk = -kj = i$ were immediate. And k^2 could be easily calculated too: $k^2 = ijij = -iijj = -1$.

And so during his walk Hamilton also discovered the rules of calculation which he entered into the pocket book. The pocket book also contains the formulas for the coefficients of the product

$$(a + bi + cj + dk)(\alpha + \beta i + \gamma j + \delta k),$$

that is,

$$a\alpha - b\beta - c\gamma - d\delta$$
$$a\beta + b\alpha + c\delta - d\gamma$$
$$a\gamma - b\delta + c\alpha + d\beta$$
$$a\delta + b\gamma - c\beta + d\alpha$$

as well as the sketch for the verification of the fact that in the sum of the squares of these coefficients all mixed terms (such as $ad\alpha\delta$) cancel and only $(a^2 + b^2 + c^2 + d^2)(\alpha^2 + \beta^2 + \gamma^2 + \delta^2)$ remains. In the *Note Book* of the same day, everything was again completely restated.

Octonions

The letter to Graves in which Hamilton announced the discovery of quaternions was written on the 17th of October 1843, one day after the discovery. The seeds, which Hamilton sowed, fell upon fertile soil, since in December 1843 the recipient John T. Graves already found a linear algebra with 8 unit elements $1, i, j, k, l, m, n, o$, the algebra of *octaves* or *octonions*. Graves defined their multiplication as follows ([3], p. 648):

$$i^2 = j^2 = k^2 = l^2 = m^2 = n^2 = o^2 = -1$$
$$i = jk = lm = on = -kj = -ml = -no$$
$$j = ki = ln = mo = -ik = -nl = -om$$
$$k = ij = lo = nm = -ji = -ol = -mn$$
$$l = mi = nj = ok = -im = -jn = -ko$$
$$m = il = oj = kn = -li = -jo = -nk$$
$$n = jl = io = mk = -lj = -oi = -km$$
$$o = ni = jm = kl = -in = -mj = -lk.$$

In this system the *law of the moduli* also holds:

$$\left(a_1^2 + \cdots + a_8^2\right)\left(b_1^2 + \cdots + b_8^2\right) = \left(c_1^2 + \cdots + c_8^2\right). \quad (1)$$

Hamilton answered on the 8th of July 1844 ([3], p. 650). He noted to Graves that the associative law $A \cdot BC = AB \cdot C$ clearly held for quaternions but not for octaves.

Octaves were re-discovered by Cayley in 1845; because of this they are also known as *Cayley numbers*. Graves also made an attempt with 16 unit elements but it was unsuccessful. It could not succeed since we know today that identities of the form (1) are only possible for sums of 1, 2, 4 and 8 squares. I should like to close with a brief comment about the history of these identities.

Product formulas for the sums of squares

It is likely that the *law of the moduli* for complex numbers was already known to Euler:

$$\left(a^2 + b^2\right)\left(c^2 + d^2\right) = (ac - bd)^2 + (ad + bc)^2.$$

A similar formula for the sum of 4 squares

$$(a_1^2 + \cdots + a_4^2)(b_1^2 + \cdots + b_4^2) = (c_1^2 + \cdots + c_4^2)$$

was discovered by Euler; the formula is stated in a letter from Euler to Goldbach on May 4th 1748 [4]. The formula (1) for 8 squares, which Graves and Cayley proved by means of octonions, was previously found by Degen (1818) [6]. Degen erroneously thought that he could generalize the theorem to 2^n squares.

The problem, which started with Hamilton, reads: can two triplets (a, b, c) and (x, y, z) be so multiplied that the law of the moduli holds? In other words: is it possible so to define (u, v, w) as bilinear functions of (a, b, c) and (x, y, z) that the identity

$$(a^2 + b^2 + c^2)(x^2 + y^2 + z^2) = (u^2 + v^2 + w^2) \quad (2)$$

results?

The first to show the impossibility for this identity was Legendre. In his great work *Théorie des Nombres* he remarked on page 198 that the numbers 3 and 21 can easily be represented rationally as sums of three squares: $3 = 1 + 1 + 1, 21 = 16 + 4 + 1$, but the product $3 \times 21 = 63$ cannot be so represented, since 63 is an integer of the form $8n + 7$. It follows from this that an identity of the form (2) is impossible, to the extent that it is assumed that (u, v, w) are bilinear forms in (a, b, c) and (x, y, z) with rational coefficients. If Hamilton had known of this remark by Legendre he would probably have quickly given up the search to multiply triplets. Fortunately he did not read Legendre: he was self-taught.

The question for which values of n a formula of the kind

$$\left(a_1^2 + \cdots + a_n^2\right)\left(b_1^2 + \cdots + b_n^2\right) = \left(c_1^2 + \cdots + c_n^2\right)$$

is possible, was finally decided by Hurwitz in 1898. With the help of matrix multiplication he proved (in [5]) that $n = 1, 2, 4$ and 8 are the only possibilities. For further historical accounts the reader may refer to [1] or [2].

References

1. C. W. Curtis, The four and eight square problem and division algebras, *Studies in Modern Algebra* (ed. A. A. Albert), Prentice-Hall, 1963, pp. 100–125.

2. L. E. Dickson, On quaternions and their generalizations and the history of the eight square theorem, *Ann. Math.* **20** (1919), 155–171.

3. W. R. Hamilton, *The Mathematical Papers of Sir William Rowan Hamilton*, Vol. III: *Algebra* (ed. H. Halberstam and R. E. Ingram), Cambridge Univ. Press, 1967.

4. W. R. Hamilton, *Correspondence Mathématique et Physique*, Vol. I (ed. P. H. Fuss), St. Petersburg, 1843.

5. A. Hurwitz, Ueber die Composition der quadratischen Formen von beliebig vielen Variabeln, *Nachr. der königlichen Gesellschaft der Wiss. Göttingen* (1898), 309–316; in *Mathematische Werke*, Bd. II, Basel, 1932, pp. 565–571.

6. C. P. Degen, Adumbratio Demonstrationis Theorematis Arithmeticae maxime generalis, *Mémoires de l'Académie de St. Petersbourg* **8** (1822), 207.

Hamilton, Rodrigues, and the Quaternion Scandal

SIMON L. ALTMANN

Mathematics Magazine **62** (1989), 291–308

Some of the best minds of the nineteenth century—and this was the century that saw the birth of modern mathematical physics—hailed the discovery of quaternions as just about the best thing since the invention of sliced bread. Thus James Clerk Maxwell, the discoverer of electromagnetic theory, wrote ([31], p. 226):

> The invention of the calculus of quaternions is a step towards the knowledge of quantities related to space which can only be compared, for its importance, with the invention of triple coordinates by Descartes. The ideas of this calculus, as distinguished from its operations and symbols, are fitted to be of the greatest use in all parts of science.

Not everybody, alas, was of the same mind, and some of the things said were pretty nasty (Lord Kelvin, letter to Hayward, 1892; see [38], Vol. II, p. 1070):

> Quaternions came from Hamilton after his really good work had been done; and, though beautifully ingenious, have been an unmixed evil to those who have touched them in any way, including Clerk Maxwell.

Such robust language as Lord Kelvin's may now be largely forgotten, but the fact remains that the man in the street is strangely averse to using quaternions. Side by side with matrices and vectors, now the *lingua franca* of all physical scientists, quaternions appear to exude an air of nineteenth-century decay, as a rather unsuccessful species in the struggle-for-life of mathematical ideas. Mathematicians, admittedly, still keep a warm place in their hearts for the remarkable algebraic properties of quaternions, but such enthusiasm means little to the harder-headed physical scientist.

This article will attempt to highlight certain problems of interpretation as regards quaternions which may seriously have affected their progress, and which might explain their present parlous status. For claims were made for quaternions which quaternions could not possibly fulfil, and this made it difficult to grasp what quaternions are excellent at, which is handling rotations and double groups. It is essentially the relation between quaternions and rotations that will be explored in this paper, and the reader interested in double groups will find this question fully discussed in my recent book [1].

The men involved: Hamilton and Rodrigues

It is not possible to understand the quaternions' strange passage from glory to decay unless we look a little into the history of the subject, and the history of quaternions, more perhaps than that of any other nineteenth-century mathematical subject, is dominated by the extraordinary contrast of two personalities, the inventor of quaternions, Sir William Rowan Hamilton, Astronomer Royal of Ireland, and Olinde Rodrigues, one-time director of the Caisse Hypothécaire (a bank dedicated to lending money on mortgages) at the Rue Neuve-Saint-Augustin in Paris ([4], p. 107).

Hamilton was a very great man indeed; his life is documented in minute detail in the three volumes of Graves [15]; and a whole issue in his honour was published in 1943, the centenary of quaternions, in the *Proceedings of the Royal Irish Academy*, Vol. A50 and, in 1944 in *Scripta Mathematica*, Vol. 10. There are also two excellent new biographies ([23], [33]) and numerous individual articles (see, e.g. [26]). Of Hamilton, we know the very minute of his birth, precisely midnight, between 3 and 4 August, 1805, in Dublin. Of Olinde Rodrigues, despite the excellent one-and-only published article on him by Jeremy Gray [16], we know next to nothing. He is given a mere one-page entry in the Michaud *Biographie Universelle* [32] as an "economist and French reformer". So little is he known, indeed, that Cartan ([6], p. 57) invented a non-existent collaborator of Rodrigues by the surname of Olinde, a mistake repeated by Temple ([37], p. 68). Booth [4] calls him *Rodrigue* throughout his book, and Wilson ([41], p. 100) spells his name as *Rodriques*.

Nothing that Rodrigues did on the rotation group—and he did more than any man before him, or than anyone would do for several decades afterwards—brought him undivided credit; and for much of his work he received no credit at all. This Invisible Man of the rotation group was probably born in Bordeaux on 16 October 1794, the son of a Jewish banker, and he was named Benjamin Olinde, although he never used his first name in later life. The family is often said to have been of Spanish origin, but the spelling of the family name rather suggests Portuguese descent (as indeed asserted by the *Enciclopedia Universal Illustrada Espasa-Calpe*). He studied mathematics at the École Normale, the École Polytechnique not being accessible to him owing to his Jewish extraction. He took his doctorate at the new University of Paris in 1816 with a thesis which contains the famous *Rodrigues formula* for Legendre polynomials, for which he is mainly known [14].

The next 24 years or so until, out of the blue, he wrote the paper on rotations which we shall discuss later, are largely a blank as far as Rodrigues's mathematics is concerned. But he did lots of other things. The little that we know about Rodrigues relates to him mainly as a paranymph of Saint-Simon, the charismatic Utopian Socialist, whom he met in May 1823, two months after Saint-Simon's attempted suicide. So, we read ([40], p. 30) that the banker Rodrigues helped the poor victim in his illness and destitution, and supported him financially until his death in 1825. That Rodrigues must have been very well off we can surmise from Weill's reference to him as belonging to high banking circles, on a par with the wealthy Laffittes ([40], p. 238). After Saint-Simon died, with Rodrigues by his bedside, the latter shared the headship of the movement with Prosper Enfantin, an old friend and disciple of Saint-Simon. Thus he became *Pére Olinde* for the acolytes. But the union did not last very long: in 1832 Rodrigues repudiated Enfantin's extreme views of sexual freedom and he proclaimed himself the apostle of Saint-Simonism. In August of that year he was charged with taking part in illegal meetings and outraging public morality, and was fined fifty francs [4]. Neither of the two early historians of Saint-Simonism, Booth and Weill, even mention that Rodrigues was a mathematician: the single reference to this is that in 1813 he was Enfantin's tutor in mathematics at the École Polytechnique. Indeed, all that we know about him in the year 1840 when he published his fundamental paper on the rotation group is that he was "speculating at the Bourse" ([4], p. 216).

Besides his extensive writings on social and political matters, Rodrigues published several pamphlets on the theory of banking and was influential in the development of the French railways. He died in Paris almost forgotten, however [32]. Even the date of his death is uncertain: 26 December 1850, according to the *Biographie Universelle* [32], or 17 December 1851, according to Larousse [27]. Sébastien Charléty ([9], pp. 26, 294), although hardly touching upon Rodrigues in his authoritative history of Saint-Simonism, gives 1851 as the year of Rodrigues's death, a date which most modern references seem to favour.

Hamilton survived Rodrigues by fourteen years and had the pleasure, three months before his death in 1865, to see his name ranked as that of the greatest living scientist in the roll of the newly created Foreign Associates of the American National Academy of Sciences. And quite rightly so: his achievements had been immense by any standards. In comparison with Rodrigues, alas, he had been born with no more than a silver-plated spoon in his mouth: and the plating was tarnishing. When he was three the family had to park various children with relatives and William was sent to his uncle, the Rev. James Hamilton, who ran the diocesan school at Trim. That was an intellectually explosive association of child prodigy and eccentric pedant: at three William was scribbling in Hebrew and at seven he was said by an expert at Trinity College, Dublin, to have surpassed the standard in this language of many Fellowship candidates. At ten, he had mastered ten oriental languages, Chaldee, Syriac, and Sanscrit amongst them plus, of course, Latin and Greek and various European languages. This is, at least, the received wisdom on Hamilton and it may contain an element of legend: it is pretty clear, e.g., that his knowledge of German was not all that strong in later life and the veracity of the reports on these linguistic feats is disputed by O'Donnell [33]. Mathematics—if one does not count mental arithmetic, at which he was prodigious—came late but with a bang when, at seventeen, reading on his own Laplace's *Mécanique Céleste*, he found a mistake in it which he communicated to the President of the Irish Academy. His mathematical career was already set in 1823 when, still seventeen, he read a seminal paper on caustics before the Royal Irish Academy.

From then on Hamilton's career was meteoric: Astronomer Royal of Ireland at 22, when he still had to take two quarterly examinations as an undergraduate, knight at 30. Like Oersted, the Copenhagen pharmacist who had stirred the world in 1820 with his discovery of the electromagnetic interaction, Hamilton was a Kantian and a follower of the *Naturphilosophie* movement then popular in Central Europe. For Hamilton "The design of physical science is ... to learn the language and to interpret the oracles of the universe" (Lecture on Astronomy, 1831, see [15], Vol. I, p. 501). He discusses in 1835, prophetically (because of the later application of quaternions in relativity theory), "Algebra as the Science of Pure Time". He writes copiously both in prose and in stilted verse, engages in a life-long friendship with Wordsworth, and goes to Highgate in the spring of 1832 to meet Coleridge, whom he visits and with whom he corresponds regularly in the next few years, the poet praising him for his understanding "that Science ... needs a Baptism, a regeneration in Philosophy" or Theosophy ([15], Vol. I, p. 546).

The discovery of quaternions

Hamilton had been interested in complex numbers since the early 1830s and he was the first to show, in 1833, that they form an algebra of couples (See [22], Vol. III.) I shall review briefly his ideas so as to lead the way to quaternions, but, here and hereafter, I shall use my own notation in order to avoid ambiguities. First, we define imaginary units, 1 and $\mathbf{i}$ with the well-known multiplication rules in Table 1. Then the elements of the algebra are the complex numbers $\mathbb{A} = a1 + A\mathbf{i}$, with a and A real.

Of course, to say that they form an algebra merely means that the formal rules of arithmetical operations are valid for the objects so defined. Thus, given $\mathbb{A}$ and a similarly defined $\mathbb{B}$, their product is

$$\mathbb{AB} = ab - AB + \mathbf{i}(aB + bA). \tag{1}$$

	1	i
1	1	i
i	i	−1

Table 1. Multiplication Table of the Quaternion Units

We can now write the complex numbers $\mathbb{A}$ and $\mathbb{B}$ as couples (or ordered pairs)

$$\mathbb{A} = [\![a, A]\!], \quad \mathbb{B} = [\![b, B]\!], \tag{2}$$

and their product is also a couple:

$$\mathbb{AB} = [\![ab - AB, aB + bA]\!]. \tag{3}$$

Hamilton also recognized that the real number a can be written as the complex couple

$$a = [\![a, 0]\!]. \tag{4}$$

For the next ten years Hamilton's mind was occupied, if not obsessed, with two problems. On the one hand, Hamilton tried to extend the concept of the complex number as a couple in order to define a triple, with one real and two imaginary units. This however, not even he could do. On the other hand, the concept of a vector was beginning to form in his mind. It must be remembered that in the 1830s not even the word *vector* existed, although people were playing about, in describing forces and such other quantities, with concepts that we would recognize today as at least vector-like. It is pretty clear that during this gestation period, as a result of which Hamilton would eventually invent the notion of vector, he had built up in his mind a picture of the addition and of some form of multiplication of vectors, but there was an operation which baffled him in the extreme: coming down the stairs for breakfast, Hamilton often could hear his elder son asking: "Father, have you now learned how to divide vectors?". Out of this preoccupation Hamilton was to invent the most beautiful algebra of the century, but he was also to feed the fever that eventually led him to corrupt his own invention.

We must now come to Monday, 16 October 1843, one of the best documented days in the history of mathematics and which, by one of those ironies of fate, happened to be the 49th birthday of Olinde Rodrigues, whose work, however ignored, was to give a new meaning to Hamilton's creation. Hamilton's letter to his youngest son, of 5 August 1865 ([15], Vol. II, p. 434), is almost too well known, but bears brief repetition. The morning of that day Hamilton, accompanied by Lady Hamilton, was walking along the Royal Canal in Dublin towards the Royal Irish Academy, where Hamilton was to preside at a meeting. As he was walking past Broome Bridge (referred to as Brougham Bridge by Hamilton and called by this name ever since), Hamilton, in a real flash of inspiration, realized that three, rather than two, imaginary units were needed, with the following properties:

$$\mathbf{i}^2 = \mathbf{j}^2 = \mathbf{k}^2 = -1, \quad \mathbf{ij} = \mathbf{k}, \quad \mathbf{ji} = -\mathbf{k}, \quad (5)$$

and cyclic permutation. As everyone knows, and de Valera was to do almost one century later on his prison wall, Hamilton carved these formulae on the stone of the bridge: poor Lady Hamilton had to wait. Armed now with four units, Hamilton called the number

$$\mathbb{A} = a\mathbf{1} + A_x\mathbf{i} + A_y\mathbf{j} + A_z\mathbf{k}, \quad (6)$$

where the coefficients here are all real, a *quaternion*. Thus were quaternions born and baptized: it was entered on the Council Books of the Academy for that day that Mr. W. R. Hamilton was given leave to read a paper on quaternions at the First General Meeting of the Session, 13 November 1843.

One of the various falsehoods which have to be dispelled about quaternions is the origin of their name, since entirely unsupported sources are often quoted, in particular Milton, *Paradise Lost*, Vol. 181 ([28], p. 70) and the *Vulgate*, Acts 12:4 ([37], p. 46). Of course, we know that Milton was a favourite poet of Hamilton at 24 ([15], Vol. I, p. 321), and to suggest that he was not aware of Acts and the apprehension of Peter by a quaternion of soldiers would be absurd. (These references appear in fact in Dr. Johnson's *Dictionary*, which was familiar to every schoolboy of the time.) But no one with the slightest acquaintance with Hamilton's thought would accept the obvious when the recondite will do. In his *Elements of Quaternions* ([21], p. 114) we find our first clue: "As to the *mere word*, quaternion, it signifies primarily (as is well known), like its Latin original, *Quaternio* or the Greek noun $\tau\epsilon\tau\rho\alpha\chi\tau\upsilon\varsigma$, a Set of Four". The key word here is *tetractys*, and there is evidence for this coming from Hamilton's closest, perhaps his only real pupil, P. G. Tait, who, writing in the *Encyclopaedia Britannica* (see article on Quaternions in the XIth edition) says: "Sir W. R. Hamilton was probably influenced by the recollection of its Greek equivalent the Pythagorean Tetractys..., the mystic source of all things...". That Tait very much believed in this is supported by the unattributed epigraph in Greek in the title page of his own treatise on quaternions [36]: these are verses 47 and 48 of *Carmen Aureum* (*Golden Song*), a Hellenistic Pythagorean poem much in vogue in the Augustan era, the full text of which appears in Diehl ([11], p. 45). Of course, the concept of the tetractys of embodying, as we shall see, multiple layers of meaning in a single word, must have attracted Hamilton: for Pythagoras, having discovered that the intervals of Greek music are given by the ratios $2:1, 3:2$ and $4:3$ made it appear that *kosmos*, that is, order and beauty, flow from the first four digits, $1, 2, 3, 4$, the sum of which gives the perfect number 10, and is symbolized by the sacred symbol, the tetractys:

$$\begin{array}{c} 0 \\ 0\ 0 \\ 0\ 0\ 0 \\ 0\ 0\ 0\ 0 \end{array}$$

(A famous depiction of the tetractys can be seen in the *School of Athens*, the fresco by Raphael at the Vatican where, anachronistically, the sacred symbol is given in Latin numerals in the figure held in front of Pythagoras.) The Pythagoreans used to take an oath by the tetractys, as recorded by Sextus Empiricus (see ([24], p. 233):

> The Pythagoreans are accustomed sometimes to say "All things are like number" and sometimes to swear this most potent oath: "Nay, by him that gave to us the tetractys, which contains the fount and root of ever-flowing nature."

That the tetractys exercised the imagination of Hamilton, there is no doubt: besides the cryptic footnote in the *Elements*, already quoted, we find Augustus De Morgan (with whom Hamilton entertained a very copious correspondence) acknowledging on 27 December 1851 a sonnet from Hamilton (apparently lost) on the tetractys. It is tempting to speculate that Hamilton might have been introduced to the tetractys by Coleridge, who called it "the adorable tetractys, or tetrad" (see [2], p. 252), and who referred to it many times.

In praise of Hamilton: the algebra of quaternions

In comparison with the binary form (2) of a complex number, the quaternion (6) can also be written as a couple of a real number a and a vector $\mathbf{A}$ of components A_x, A_y, A_z (as already said, we use modern rather than historical notation):

$$\mathbb{A} = [\![a, \mathbf{A}]\!], \quad \mathbf{A} = (A_x, A_y, A_z). \quad (7)$$

Just as for the complex numbers, in order to multiply two such objects we need the multiplication table of the quaternion units, which follows at once from equation (5); it is given in Table 2.

	1	i	j	k
1	1	i	j	k
i	i	−1	k	−j
j	j	−k	−1	i
k	k	j	−i	−1

Table 2. Multiplication table of the quaternion units

Consider now a second quaternion $\mathbb{B}$; write both $\mathbb{A}$ and $\mathbb{B}$ as in equation (6) and, on using the table, their product follows at once in the same manner as that in (3):

$$\mathbb{A}\mathbb{B} = [\![ab - \mathbf{A} \cdot \mathbf{B}, a\mathbf{B} + b\mathbf{A} + \mathbf{A} \times \mathbf{B}]\!]. \quad (8)$$

Although Hamilton did not give names or symbols for these operations, it is here that the scalar and vector products of two vectors appear for the first time in history. We can now go back to Table 2 and reflect a little about why Hamilton made the product **ij** non-commutative. Not only was this the first time that a non-commutative product appeared in mathematics, but this was a true stroke of genius. Remember that Hamilton wanted to divide vectors: he never really achieved this (neither was it worth trying) but the point is that, because of this, he was after a *division algebra*, i.e., one in which the quotient of an element of the algebra by any other non-null element always exists. Now, a necessary condition for a division algebra is this: the product of two elements of the algebra must vanish if and only if one of the factors is the null element. Table 2 is designed so that this happens, as can easily be verified from the resulting multiplication rule, given by (8). A counterexample will be instructive. Suppose we take **ij** and **ji** as equal in Table 2, and, similarly, for the other products. Then, under the new multiplication rules, it is very easy to verify that the product of the two non-null elements

$$\mathbb{A} = \mathbf{i} + \mathbf{j}, \quad \mathbb{B} = -\mathbf{i} + \mathbf{j}, \quad (9)$$

vanishes.

Hamilton's everlasting monument (see [30]) is his construction of objects which, except for commutativity, obey the same algebra as that of the real and complex numbers and which therefore, like them, form a division algebra: and Hamilton was aware of this—although he could not foresee that his brainchild was going to receive at the hands of Frobenius in 1878 the supreme accolade of being proved to be the only possible algebra, in addition to the real and complex numbers, with this property.

The trouble starts

Now back to 16 October 1843. That same evening Hamilton wrote a long and detailed draft of a letter to his friend John Graves, first published by A. J. Mc-Connell ([29] and included in [22], Vol. III). Next day a final letter was written and sent, later published in the *Philosophical Magazine* [17]. The November report to the Irish Academy was published almost at the same time [18]. We can thus follow almost hour by hour Hamilton's first thoughts on quaternions. Although in the morning of the glorious day he had been led to the discovery through the algebra of the quaternions, by the evening (and in this he acknowledges the influence of Warren [39]), he had been able to recognize a relation between quaternions and what we now call rotations. And in this, sadly, we cannot but see the germ of the canker that eventually consumed the quaternion body. Three separate themes, ever present in Hamilton's mind, contributed to this infirmity.

As regards the first theme: as in (4), Hamilton identified a real number with a real quaternion:

$$a = [\![a, \mathbf{0}]\!]. \quad (10)$$

Nothing wrong here, but it invited Hamilton to go on and identify a *pure quaternion* (a quaternion with a null scalar part) with a *vector*, a word which he invented for this purpose in 1846 ([19], p. 54):

$$\mathbf{A} = [\![0, \mathbf{A}]\!]. \quad (11)$$

As the inventor of the vector he was entitled to call this object anything he wanted, but the problem is that by this time people were already thinking about forces and such like objects very much as we think of vectors today and that the identification of Hamilton's 'vectors' with what they had in mind created a great deal of confusion. The apparently innocent convention (11) entails in fact two serious problems. That something here was a serious worry must have been evident for decades, as Klein ([25], p. 186), one of the leading nineteenth-century geometers, implied himself. Yet, the first explicit statement to the effect that something is wrong here which I have been able to find is as recent as 1958 by Marcel Riesz ([34], p. 21):

> Hamilton and his school professed that the quaternions make the study of vectors in three-space unnecessary since every vector can be considered as the vectorial part ... of a quaternion ... this interpretation is grossly incorrect since the vectorial part of a quaternion behaves

with respect to coordinate transformations like a bivector or *axial* vector and not like an ordinary or *polar* vector.

However damning this statement is, it is only half the story, since the pure quaternion (11) is not anything like a vector at all: we shall see that it is a *binary rotation*, that is, a rotation by π. The left-hand side of (11) should be written as $\mathbb{A}$ and carefully distinguished from the vector **A**. The fact that neither Hamilton, nor his successors to the present day, introduced any notational distinction between these two objects is the source of extraordinary confusion, as we shall soon witness ourselves.

Hamilton's second theme was closely connected to his first and has already been mentioned: he wanted to find a definition of the quotient of two vectors and however grateful we must be for this obsession, which has given us the last possible division algebra, we shall soon see that it led Hamilton to an interpretation of quaternions and of their operations which is not right.

Hamilton's capacious mind could not be at rest until he understood not just the formalities of his work but also what went on behind the scenes, and he had to understand the physical or geometrical meaning of equating the square of the imaginary or quaternion unit, $\mathbf{i}^2$, with -1. This was his third everlasting theme, for which he took a cue from Argand, who had observed in 1806 that the imaginary unit **i** rotates what we would now call a vector in the Argand plane by $\pi/2$, which made it possible to visualize the relation in question. From that point of view, in fact, $\mathbf{i}^2$ should be a rotation by π which, duly enough, multiplies each vector of the plane by the factor -1. For this reason, Hamilton always identified the quaternion units with *quadrantal rotations*, as he called the rotations by $\pi/2$ (see [20], p. 64, art. 71). Clifford ([10], p. 351) associates himself with this interpretation which he presents with beautiful clarity. The sad truth is that, however appealing this argument is, to identify the quaternion units with rotations by $\pi/2$ is not only not right, but it is entirely unacceptable in the study of the rotation group: we shall see, in fact, that they are nothing else except binary rotations.

Quaternions and rotations: the first steps

Already during the first day of his creation Hamilton knew what he had to do in order to define rotations. Since rotations must leave the lengths of vectors invariant, and since for Hamilton a vector was a particular case of a quaternion, the first thing he had to do is to define the *norm* or length $|\mathbb{A}|$ of a quaternion. He defined for this purpose the *conjugate quaternion*

$$[\![a, \mathbf{A}]\!]^* = [\![a, -\mathbf{A}]\!]. \quad (12)$$

The norm is now defined as follows:

$$\mathbb{A}\mathbb{A}^* = [\![a, \mathbf{A}]\!][\![a, -\mathbf{A}]\!] = [\![a^2 + A^2, 0]\!]$$
$$= a^2 + A^2 = |\mathbb{A}|^2. \quad (13)$$

A quaternion of unit norm is called *a normalized quaternion* and, although Hamilton also considered more general quaternions, these will be the only quaternions which we shall need for the purposes of this article. It was easy for Hamilton to prove that *the product of two normalized quaternions is a normalized quaternion.* (See, e.g. [1], p. 208.) We are now ready to accompany Hamilton in performing an extraordinary piece of legerdemain.

An optical illusion: the rectangular rotation

A rotation is an operation which transforms a unit, that is, a normalized position vector **r** (a vector with its tail at the origin), into another unit position vector $\mathbf{r}'$. If we go along with Hamilton and identify the vector **r** with the pure quaternion $\mathbb{R}$ equal to $[\![0, \mathbf{r}]\!]$, the latter is clearly normalized. In order to achieve a rotation and keep the normalization requirement of $\mathbb{R}$, all that we need is to act on $[\![0, \mathbf{r}]\!]$ with a normalized quaternion. (See the italicized statement in the previous paragraph.) Let us, therefore, choose for this purpose the quaternion

$$\mathbb{A} = [\![\cos\alpha, \sin\alpha\,\mathbf{n}]\!], \quad |\mathbf{n}| = 1, \quad (14)$$

which is clearly normalized. (Here **n** is a unit position vector.) This is not the end of the story, however, because we must require that the product $\mathbb{A}\mathbb{R}$ be not only normalized, but also a pure quaternion $\mathbb{R}'$ of the form $[\![0, \mathbf{r}']\!]$ which Hamilton would then identify with the rotated position vector $\mathbf{r}'$. This is what Hamilton envisioned on the same evening of Creation Day, and he also realized that for this idea to work it was the necessary that the vector **n**, which he called the *axis of the quaternion*, be normal to the vector **r**. (This is why this is called the *rectangular transformation*.) To verify that this works is child's play on using the quaternion multiplication rule (8).

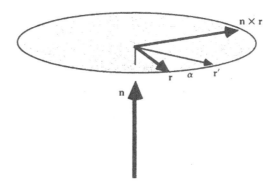

Figure 1. The rectangular transformation. The vector $\mathbf{r}'$ is defined in equation (15).

Given that $\mathbf{r} \cdot \mathbf{n} = 0$, then

$$\begin{aligned}
\mathbb{A}\mathbb{R} &= [\![\cos\alpha, \sin\alpha\mathbf{n}]\!][\![0, \mathbf{r}]\!] \\
&= [\![0, \cos\alpha\mathbf{r} + \sin\alpha(\mathbf{n}\times\mathbf{r})]\!] \\
&= [\![0, \mathbf{r}']\!] = \mathbb{R}'.
\end{aligned} \quad (15)$$

If we briefly avert our gaze while Hamilton rewrites this equation as

$$\mathbb{A}\mathbf{r} = \mathbf{r}', \quad (16)$$

then the job is done—that is, the quaternion $\mathbb{A}$ transforms the unit position vector $\mathbf{r}$ into another unit position vector $\mathbf{r}'$ and, therefore, has rotated $\mathbf{r}$ into $\mathbf{r}'$. What is more: it is clear from Figure 1 that the angle of rotation is α. Thus Hamilton identified quaternion (14) with a rotation around the axis $\mathbf{n}$ of the quaternion by the angle α of rotation of the quaternion.

All this is so wonderfully convincing that it is difficult to believe that there is anything wrong here. Moreover, Hamilton immediately obtained confirmation that one of his themes was coming out all right, for it was clear to him that the quaternion units in this picture are quadrantal rotations, as it is immediate from (14): on comparing this equation with, say, the quaternion unit $\mathbf{k}$, given by $[\![0, \mathbf{k}]\!]$, one can see at once that the rotation angle to be associated with $\mathbf{k}$ must be $\pi/2$. His other theme was also coming out well here, since from (16) the quaternion $\mathbb{A}$ can be considered as the quotient of the vector $\mathbf{r}'$ by the vector $\mathbf{r}$. This picture of quaternions was thus so near Hamilton's heart that in his *Lectures on Quaternions* ([20], p. 122), and ever after, the primary definition of a quaternion which he used was

The quotient of two vectors, or the operator which changes one vector into another,

as later adopted by the *Oxford English Dictionary*, and this definition became the core of the quaternion dogma, thus causing endless damage. We shall see, in fact, that a quaternion can never operate on a vector, as (16) implies, and that this equation must always be understood as the quaternion product in (15).

The comical transformation
(this heading contains a misprint)

The conical transformation was the means by which nature began to make its protest against Hamilton. Even accepting that a pure quaternion is a vector, we must ensure, in order to have a rotation, that the transform of a normalized pure quaternion is another normalized pure quaternion. Thus, a general rotation cannot be written as $\mathbb{A}\mathbb{R}$, because this product is not a pure quaternion unless, as we have seen, the axis of $\mathbb{A}$ is normal to $\mathbf{r}$. Hamilton and his colleagues, therefore, searched for a quaternion transformation of a pure quaternion $\mathbb{R}$ under a normalized quaternion $\mathbb{A}$ which would always produce a normalized pure quaternion $\mathbb{R}'$. The result of this search was the following transformation:

$$\mathbb{A}\mathbb{R}\mathbb{A}^* = \mathbb{R}'. \quad (17)$$

It is, in fact, quite easy by means of (8) to verify that the left-hand side of this equation is normalized and pure. With a little bit of geometry (see [1], p. 214), and assuming that $\mathbb{A}$ is given by (14), it can be proved that $\mathbf{r}$, $\mathbf{n}$, and $\mathbf{r}'$ are related as shown in Figure 2, i.e., that the vector $\mathbf{r}$ is rotated around $\mathbf{n}$ by the angle 2α.

There are two problems here: first, the form of (17) has nothing at all to do with that of (15), so that it is no longer possible to say that the quaternion operates on a vector transforming it into another vector,

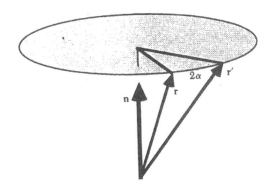

Figure 2. The conical transformation

or even less that it is the quotient of two vectors. The second problem is this: which is the angle of rotation to be associated with a quaternion (14), α, as in the rectangular transformation, or 2α, as it turns out in the case of a general rotation? It is, perhaps, significant that Hamilton obtained (17) (writing the conjugate quaternion as the reciprocal, as it is valid for a normalized quaternion) but did not publish it for some time. Cayley [7] was the first to go into print, although in his collected papers ([8], Vol. I, p. 586, note 20) he concedes priority to Hamilton. Cayley notices that the components of $\mathbf{r}'$ are

> precisely those given for such transformations by M. Olinde Rodrigues... It would be an interesting question to account, *a priori* for the appearance of these coefficients here.

Let us see what Hamilton has to say about this ([20], p. 271, my italics):

> The SYMBOL OF OPERATION $q(\ )q^{-1}$, where q may be called (as before) the operator quaternion, while the symbol (suppose r) of the operand quaternion is conceived to occupy the place marked by the parentheses... 'can be regarded as' a conical transformation of the operand round the axis of the operator, through *double the angle thereof*.

It is clear that Hamilton, rather than accepting the result of the more general transformation (17) to recognize that the angle of rotation of the quaternion (14) is 2α, gives greater weight to the transformation (16) and keeps talking of the angle α in (14) as the angle of rotation. Naturally, whereas (16) had the shape that he expected, the form of (17), as Cayley stated, could not be explained. It is, perhaps, because of this that, although Figure 2 is nothing else than the most general rotation of a vector, Hamilton refers to it with the *ad hoc* name of *conical rotation*, as if it were a particular case of the transformation of a vector. It had, instead, been Rodrigues who had recognized, three years before Hamilton's invention of quaternions, that the angle α in (14) is not the rotation angle but only half of it. But his paper, which had puzzled Cayley, was almost certainly never read by Hamilton and it was never again quoted by any of the major quaternionists. As for Cayley's question, it was probably never answered until 1986. (See [1], p. 214.)

The Rodrigues programme

Hamilton constructed quaternions as an algebra, whence the elements of the algebra were given a dual role as operators (rotations) and operands (vectors). This was very lucidly explained by Clifford [10], but it must be clearly appreciated that, as we have already asserted, the status of vectors in this scheme is highly dubious, of which more later. Be that as it may, in Hamilton's approach rotations become subservient to the algebra, which opens the door to a variety of misinterpretations.

Historically, however, a treatment of rotations and quaternions had been going on for some years before 1843, quite independently of Hamilton and taking a diametrically opposed view to his. This treatment was entirely geometric, and because it tried to do a simple job in a simple way it was clear and precise and it was entirely successful; but it was largely ignored by everyone.

Let us consider rotations of a unit sphere with fixed center about various axes. The first problem which

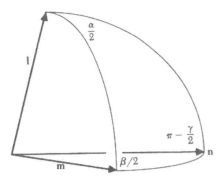

Figure 3. The product of the rotations $R(\alpha\mathbf{l})$ and $R(\beta\mathbf{m})$ is the rotation $R(\gamma\mathbf{n})$.

arises is whether, if we apply one rotation after another, the net result is a rotation of the sphere around some unique axis by some unique angle. Euler [12] proved algebraically that this is so, but he did not provide either a geometric or a constructive solution (i.e., a solution in which the axis and angle of the resultant rotation are determined geometrically or algebraically).

It was the paper by Rodrigues in 1840 [35] which solved all these aspects of the problem. In its §8 he describes most clearly, without a figure, a geometric construction which, given the angles and axes of two successive rotations, determines the orientation of the resultant axis of rotation and the geometrical value of the angle of rotation. This construction is usually called in the literature the *Euler construction*, although Euler had nothing to do with it. Not only was this construction ignored by the quaternionists, but it is not even mentioned in modern books on the rotation group. Although Hamilton himself rediscovered, geometrically ([20], p. 328), the results of the Rodrigues construction, this is not a theorem to which either he or his commentators paid much attention (see [1], pp. 19–20).

Let us represent a rotation around the axis $\mathbf{p}$ by the angle ϕ with the symbol $R(\phi\mathbf{p})$. Then, if we use the Rodrigues construction for the following product of rotations,

$$R(\alpha\mathbf{l})R(\beta\mathbf{m}) = R(\gamma\mathbf{n}), \quad (18)$$

it turns out that the axes $\mathbf{l}$, $\mathbf{m}$, and $\mathbf{n}$ form a spherical triangle with the angles shown in Figure 3. (Remember that in the left-hand side of (18) the rotation around $\mathbf{m}$ is applied first and it is followed by that around $\mathbf{l}$: this is the usual convention for reading operators.)

What is very remarkable about this very simple triangle is that the angles of the rotations appear in it as *half-angles*, and this is the first time that half-angles occur in the study of rotations. Their importance is absolutely crucial, as we shall see, and yet they were ignored by Euler and were never considered by Hamilton or his followers: it took more than forty years before their significance was appreciated. One would shirk nowadays at the solution of the spherical triangle in Figure 3 (see [1], p. 157), but mathematical training in France on surveying and such was very good and Rodrigues was able to obtain quite easily expressions for the angle and axis of the resultant rotation—that is, the one on the right-hand side of (18)—in terms of those of the factors which appear on its left. The following are Rodrigues's formulae exactly as he gave them, except that I have introduced vector notation, which was non-existent in his time:

$$\cos\frac{\gamma}{2} = \cos\frac{\alpha}{2}\cos\frac{\beta}{2} - \sin\frac{\alpha}{2}\sin\frac{\beta}{2}\mathbf{l}\cdot\mathbf{m}, \quad (19)$$

$$\sin\frac{\gamma}{2}\mathbf{n} = \sin\frac{\alpha}{2}\cos\frac{\beta}{2}\mathbf{l} + \cos\frac{\alpha}{2}\sin\frac{\beta}{2}\mathbf{m}$$
$$+ \sin\frac{\alpha}{2}\sin\frac{\beta}{2}\mathbf{l}\times\mathbf{m}. \quad (20)$$

These formulae immediately suggest that a rotation $R(\alpha\mathbf{l})$ can be represented by a couple of a scalar and a vector (although this notation was not used by Rodrigues):

$$R(\alpha\mathbf{l}) = \left[\!\!\left[\cos\frac{\alpha}{2}, \sin\frac{\alpha}{2}\mathbf{l}\right]\!\!\right], \quad (21)$$

so that the product (18) is written as follows

$$\left[\!\!\left[\cos\frac{\alpha}{2}, \sin\frac{\alpha}{2}\mathbf{l}\right]\!\!\right]\left[\!\!\left[\cos\frac{\beta}{2}, \sin\frac{\beta}{2}\mathbf{m}\right]\!\!\right]$$
$$= \left[\!\!\left[\cos\frac{\gamma}{2}, \sin\frac{\gamma}{2}\mathbf{n}\right]\!\!\right], \quad (22)$$

with the parameters on the right-hand side of this equation being given by (19) and (20). It can immediately be seen that the multiplication rule for the couples so defined is identical with the multiplication (8) of Hamilton's quaternions! Rodrigues's couples are, therefore, quaternions, but the difference in parametrization between (21) and (14) is profound. We see at once that the conical transformation, which gives the angle of rotation as twice the angle which appears in the quaternion was right, and that Hamilton committed a serious error of judgement in basing his parametrization on the special case of the rectangular transformation.

Simple as the distinction is, the consequences are dramatic, and never more so than when we consider pure quaternions. From (21), it is clear that for a quaternion to be pure the angle of rotation must be π—that is, a pure quaternion is nothing other than a binary rotation:

$$[\![0, \mathbf{r}]\!] = R(\pi \mathbf{r}). \qquad (23)$$

Thus, it is entirely wrong ever to identify a pure quaternion with a vector, as Hamilton had done in (11). This simple fact will exorcise all the demons so far lurking into our story. Before we do this, we must mention that the (four) rotation parameters in (21) are called the *Euler-Rodrigues parameters* in the literature. The reasons for this are entirely disreputable (see [1], p. 20), since Euler never came near them: in particular, he never used half-angles which, as demonstrated by Rodrigues, are an essential feature of the parametrization of rotations.

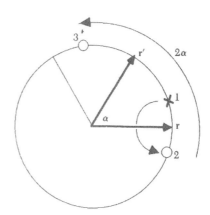

Figure 4. Product of a rotation by 2α around the axis $\mathbf{n}$ normal to the plane of the drawing, with a binary rotation about the axis $\mathbf{r}$.

The resolution of the paradoxes

Although quaternions are always rotations and never vectors, they allow us to mark points in space very much as a position vector does. Consider the unit sphere centered and fixed at the origin. A rotation of it determines a single point of the sphere which is called the *pole of the rotation*. This is the point of the sphere which is left invariant by the rotation and such that from outside it the rotation is seen as positive (counter-clockwise). If we want to mark a point in space by means of rotation poles, it is sensible to use always binary rotations for this purpose, since, as it follows from (23), these are the nearest things to vectors that we can get within the quaternion algebra. (It should be stressed that this is purely a matter of convenience: the pole of any arbitrary rotation is just as good to denote a point of the unit sphere and thus to masquerade as a vector.) Let us now look again at (15) with the quaternion $\mathbb{A}$ in it given by

$$\mathbb{A}\mathbb{R} = \mathbb{R}', \quad \mathbb{A} = [\![\cos\alpha, \sin\alpha \mathbf{n}]\!]. \qquad (24)$$

If we compare the left-hand side of this line with (18) and (21), it says this: the product of a rotation by 2α around the axis $\mathbf{n}$, times a binary rotation around the axis $\mathbf{r}$, is a binary rotation around the axis $\mathbf{r}'$:

$$R(2\alpha \mathbf{n}) R(\pi \mathbf{r}) = R(\pi \mathbf{r}'). \qquad (25)$$

The angle of rotation, which we already know, and the orientation of the axis $\mathbf{r}'$ can, of course, be obtained from the quaternion multiplication rules (as given, e.g., in (19) and (20)) but it will be instructive to obtain an independent geometrical verification, since this will show up the paradox involved in Hamilton's interpretation of the rectangular transformation. We do this in Figure 4.

In order to multiply rotations we transform the unit sphere, whose intersection with the plane of the drawing is shown in Figure 4. The rotation axis $\mathbf{n}$ is perpendicular to and above the plane of the drawing. The rotation axis $\mathbf{r}$ is in the plane of the drawing and thus, as it must be in the rectangular transformation, is normal to $\mathbf{n}$. A point above the plane of the drawing is represented with a cross and those below with a circle. In order to multiply the two rotations on the left of (25) we start with the point 1 above the plane of the drawing. The first rotation to act on it (remember to read the left-hand side of (25) from right to left) is a rotation by π around $\mathbf{r}$ which takes it into the point 2 below the plane of the drawing. The rotation around $\mathbf{n}$ by 2α takes 2 to 3. Thus, the two combined operations take the point 1 above the drawing to the point 3 below the drawing, which is the effect of a binary rotation around the axis $\mathbf{r}'$. Notice that the angle between the axes $\mathbf{r}$ and $\mathbf{r}'$ is α and that this angle *is not the angle of rotation*.

We can now see how Hamilton's optical illusion was performed. If in (24) we identify the quaternions $\mathbb{R}$ and $\mathbb{R}'$ with their corresponding vectors $\mathbf{r}$ and $\mathbf{r}'$, Figure 4 now reads as the rotation of $\mathbf{r}$ into $\mathbf{r}'$ by α. Incidentally, the correct reading of Figure 4 as stating that a rotation axis by 2α and a perpendicular binary axis determine another perpendicular binary axis at an angle α to the first one is so fundamental in crystallography that the whole of this science would collapse like a pack of cards if it were not true.

What about the conical transformation? I cannot go into all the details of the theory, but a sketch will suffice. We must accept the following result (see [1], p. 215). If we take a pole of a binary rotation **r** and we rotate this pole about an axis **n** by an angle α, the new pole thus obtained, **r**′, is the pole of another binary rotation given in the following form:

$$R(\alpha\mathbf{n})R(\pi\mathbf{r})R(-\alpha\mathbf{n}) = R(\pi\mathbf{r}'). \quad (26)$$

Because the corresponding quaternions must multiply in the same manner, we get immediately (17), since the inverse and conjugate quaternions are identical in our case. We must remember, though, to use the Rodrigues parametrization of the quaternion, as in (21), and not Hamilton's (14):

$$\mathbb{A}\mathbb{R}\mathbb{A}^* = \mathbb{R}', \quad \mathbb{A} = \left[\!\left[\cos\frac{\alpha}{2}, \sin\frac{\alpha}{2}\mathbf{n}\right]\!\right]. \quad (27)$$

In this case it is heuristically possible to substitute **r** and **r**′ for the quaternions $\mathbb{R}$ and $\mathbb{R}'$, with the poles of the binary rotations masquerading successfully as position vectors. This substitution, however, must never be done anywhere else. In particular, one must never attempt to operate with a quaternion on a vector, as is shown by the disastrous results of the crude interpretation of the rectangular transformation.

We must now discuss again the significance of the quaternion units. Because they are pure quaternions they must now be identified with binary rotations (rotations by π). This, for Hamilton, must have been absurd: the relation $\mathbf{i}^2$ equal to -1 must still be satisfied. But the product of two rotations by π about the same axis is a rotation by 2π. This is clearly the identity operation, i.e., one which does not change any vector, whereas we are now saying that it is equal to -1, i.e., that it changes the sign of all vectors in space.

I believe that this is the reason Hamilton was forced to accept his parametrization, since this agreed with his picture of quaternion units as quadrantal rotations. Rodrigues, practical man as bankers must be, knew better than to worry about this strange result of his geometry—he did not carry, like Hamilton, all the world's problems on his shoulders. Nature and history, alas, were playing games with Hamilton. How was he to know that Cartan was going to discover in 1913 [5] objects (spinors) which are indeed multiplied by -1 under a rotation by 2π, exactly as Rodrigues's parametrization requires? Moreover, when the topology of the rotation group became understood in the 1920s through the work of Hermann Weyl, it became natural to accept that the square of a binary rotation multiplies the identity by -1, and thus behaves like the quaternion units. Though this should have shown the enormous importance of quaternions in the rotation group, they were by that time somewhat discredited, so that other much less effective parametrizations of the rotation group were in universal use.

It must be stressed that the Rodrigues approach to rotations, by emphasizing their multiplication rules and by regarding them entirely as operators, fully reveals the group properties of the set of all orthogonal rotations, the full orthogonal group $SO(3)$, as it is now called. The set of all normalized quaternions (in the Rodrigues parametrization) is a group homomorphic to $SO(3)$, and it is its *covering group*. Although I cannot go into the mathematical significance of this statement, its practical importance in quantum mechanics, e.g., can be easily understood: it permits the study of the transformation properties of the wave functions of the electron spin. It is for this purpose that quaternions are superb, because their use in dealing with rotations makes the work not only simpler but also more precise than with any other method.

The decline

Hamilton was still under forty when he discovered quaternions, but he had more than twenty years of very productive research past him and was already showing the signs of having passed his prime. Financial and even sentimental worries are often mentioned, as well as overwork and an increasing consumption of alcohol [33]. I am inclined to believe that a major factor was that Hamilton himself was in some way aware of the internal contradictions of his doctrine and that he could not rest until he could peel off all layers of reality one by one to reach to the core. This was always beyond his power, since he was not prepared to renounce the, for him, essential picture of the quaternion units as quadrantal rotations. Be it as it may, his writing became more and more obscure: even his supporters found his books unreadable. And he himself became more isolated and eccentric.

E. T. Bell ([3], Ch. 19) labelled the last twenty years of Hamilton *The Irish Tragedy*. Lanczos [26] compares them with Einstein's fruitless search for a unified field theory in his own last two decades. The truth is probably somewhere between these two views. For Hamilton suffered the weight of his own greatness: it was not enough for him to have an algebra, it was not enough to have a geometry, he had to "interpret the oracles of the Universe" and the oracles trailed in front of him false clues that no one was to unravel for another three-score years after his death.

If only he had known about spinors! The result, however, was that Hamilton, and to a much greater degree his followers, became dogmatic and intolerant (see [25], p. 182) and that a great deal of sterile discussion ensued.

The last years of Hamilton, despite his immense fame, were not without worries: Continentals were spreading rumours that the great Gauss had actually discovered quaternions but had never bothered to publish. (They were right, as shown by Gauss's notes from 1819, published in 1900; see [13], Vol. VIII, pp. 357–362.) In letters to De Morgan of January 1852 ([15], Vol. II, p. 490; Vol. III, p. 330), Hamilton attacks these allegations. Curiously enough, of Rodrigues, who in 1840 not only had invented quaternions bar their name, but also published his formulae, there was never a word. Who would pay attention to a Socialist banker in matters mathematical?

After Hamilton's death his work began to give fruits but not in the direction which he had expected. His ideas of vectors and of their scalar and vector products were much too important so that people began to try and graft a new skin onto them in order to make these concepts usable. Grassmann in Germany and Heaviside in Britain moved some way in this direction, but one must admit that they were not much more transparent than Hamilton himself. It was left to Willard Gibbs of Yale to produce not only the first coherent picture of vectors and of their operations but also a good and successful working notation. This hardened the response of Hamilton's followers, who adopted a truly Byzantine posture, intent on stopping the flood of rebellion from across the Atlantic. Thus P. G. Tait ([36], p. vi):

> Even Prof. Willard Gibbs must be ranked as one of the retarders of quaternion progress, in virtue of his pamphlet on *Vector Analysis*, a sort of hermaphrodite monster, compounded of the notations of Hamilton and of Grassmann.

The kiss of life for quaternions, alas, much too late, came with the foundation in 1895 of an International Association for Promoting the Study of Quaternions and Allied Systems of Mathematics: an acknowledgement that quaternions were a corpse in need of resuscitation. Alexander Macfarlane, who taught at Texas, became the leading force of the Association, which actually published a Bulletin from 1900 to 1923. The influence of this group extended as far as Japan, where Kimura in 1907 became one of the major influences of the Association. Nothing that they did, however, succeeded in preventing the rise of vectors and the consequent decline of quaternions.

A number of applications of quaternions went on appearing from time to time (see [1], p. 18]). Ironically, however, by the time in the late twenties when quantum mechanics made the study of the rotation group crucial, thus giving the quaternions their real *raison d'être*, they had been submerged for much too long in the murky waters of their battle against vectors to be able to come to the surface again. They are much too useful in this context, though, for their time not to return.

There is a moral to this story: Rodrigues's applied mathematics yields a more accurate picture of the quaternions than that afforded by the pure mathematics of their inventor: it is probably a myth that pure mathematics is either born or can stand entirely on its own, although the aesthetic appeal of pure mathematics makes us often think otherwise.

Epilogue

After this article was communicated, a book was announced which contains new information about Rodrigues. This is the *Dictionnaire du Judaïsme Bordelais aux XVIIIe et XIXe Siècles*, by Jean Cavignac (Archives Départementales de La Gironde, Bordeaux, 1987). This book contains a family tree of Rodrigues, which shows that his great-grandfather, Isaac Rodrigues-Henriques, was born in Spain, around 1689–1691 and died in Bordeaux in 1767. He was indeed a banker but, contrary to previous belief, Olinde's father was an accountant. Surprisingly, the so-far universally accepted date of birth of Rodrigues is not right (so that the extraordinary coincidence with the day of the discovery of quaternions becomes a second-order effect). The correct date is 6 October 1795, and this is now unimpeachable, since his birth certificate is fully transcribed in a paper on Rodrigues by Paul Courteault (*Un Bordelais Saint-Simonien*) which I, like most people so far, had missed, since it was published in an obscure journal (*Revue Philomatique de Bordeaux*, Octobre–Décembre 1925, pp. 151–166). In accordance to this certificate, the date of birth was 14 *vendémiare* in the year IV of the Republic, at 1 p.m. Courteault (and Cavignac), instead, both agree with Michaud's date of death, 26.12.1850. Courteault also gives evidence that, although Olinde tried to enter the École Normale, he did not succeed in so doing, being prevented by his religion, so that how he learnt his advanced mathematics remains an unsolved mystery. Even worse, it appears that Rodrigues did not even attend the local secondary school (Lycée) at Bordeaux, so that we do not yet know anything at all about his formative years. It is of some value, however, that some of the traditional wisdom about this period, as repeated, I am afraid, in my paper, is now known to be worthless.

References

1. S. L. Altmann, *Rotations, Quaternions, and Double Groups*, Clarendon Press, 1986.
2. O. Barfield, *What Coleridge Thought*, Oxford Univ. Press, 1972.
3. E. T. Bell, *Men of Mathematics*, Gollancz, 1937.
4. A. J. Booth, *Saint-Simon and Saint-Simonism, a Chapter in the History of Socialism in France*, Longmans, Green, Reader and Dyer, 1871.
5. E. Cartan, Les groupes projectifs qui ne laissent invariante aucune multiplicité plane, *Bull. Soc. Math. de France* **41** (1913), 53–96.
6. E. Cartan, *Leçons sur la Théorie des Spineurs*, Vol. I, Hermann, 1938.
7. A. Cayley, On certain results relating to quaternions, *Phil. Mag.* **26** (1845), 141–145.
8. A. Cayley, *The Collected Mathematical Papers*, 13 vols. plus Index, Cambridge Univ. Press, 1888–1898.
9. S. Charléty, *Histoire du Saint-Simonisme (1825–1864)*, 2nd ed., Hermann, 1936 (1st ed., 1896).
10. W. K. Clifford, Applications of Grassmann's extensive algebra, *Amer. J. of Math. Pure and Applied* **1** (1878), 350–358.
11. E. Diehl, *Anthologia Lyrica Græca* Fasc. 2, 3rd ed., Teubner, 1940.
12. L. Euler, Formulae generales pro translatione quacunque corporum rigidorum, *Novi. Comm. Acad. Sci. Imp. Petrop.* **20** (1775), 189–207.
13. C. F. Gauss, *Werke*, 12 vols., Königliche Gesellschaft der Wissenschaften, 1863–1929.
14. I. Grattan-Guinness, personal communication, 1983.
15. R. P. Graves, *Life of Sir William Rowan Hamilton*, 3 vols., Hodges, Figgis and Co., 1882–1889.
16. J. Gray, Olinde Rodrigues' paper of 1840 on transformation groups, *Arch. Hist. Exact Sci.* **21** (1980), 375–385.
17. W. R. Hamilton, On quaternions; or a new system of imaginaries in algebra, *Phil. Mag.* **25** (1844), 489–495.
18. W. R. Hamilton, On a new species of imaginary quantities connected with the theory of quaternions, *Proc. Royal Irish Acad.* **2** (1844), 424–434.
19. W. R. Hamilton, On symbolical geometry, *Cambridge and Dublin Mathematical Journal* **1** (1846), 45–57.
20. W. R. Hamilton, *Lectures on Quaternions*, Hodges and Smith, 1853.
21. W. R. Hamilton, *Elements of Quaternions*, 2nd ed. (ed. C. J. Jolly), 2 vols., Longmans, Green and Co., 1899–1901.
22. W. R. Hamilton, *The Mathematical Papers of Sir William Rowan Hamilton*, Vol. I, *Geometrical Optics* (eds. A. W. Conway and J. L. Synge); Vol. II, *Dynamics* (eds. A. W. Conway and A. J. McConnell); Vol. III, *Algebra* (eds. H. Halberstam and R. E. Ingram), Cambridge Univ. Press, 1931, 1940, 1967.
23. T. L. Hankins, *Sir William Rowan Hamilton*, Johns Hopkins Univ. Press, 1980.
24. G. S. Kirk, J. E. Raven, and M. Schofield, *The Presocratic Philosophers*, 2nd ed., Cambridge Univ. Press, 1963.
25. F. Klein, *Vorlesungen über die Entwicklung der Mathematik im 19. Jahrhundert*, 2 vols., Springer, 1926, 1927.
26. C. Lanczos, William Rowan Hamilton—an appreciation, *American Scientist* **55** (1967), 129–143.
27. P. Larousse, *Grand Dictionnaire Universel du XIXe Siècle*, Larousse, Vol. I, 1866; Vol. XIII, no date.
28. A. L. Mackay, *The Harvest of a Quiet Eye. A Selection of Scientific Quotations* (ed. M. Ebison), The Institute of Physics, Bristol, 1977.
29. A. J. McConnell, The Dublin mathematical school in the first half of the nineteenth century, *Proc. Royal Irish Acad.* **A50** (1943), 75–92.
30. C. C. MacDuffee, Algebra's debt to Hamilton, *Scripta Math.* **10** (1944), 25–36.
31. J. C. Maxwell, Remarks on the mathematical classification of physical quantities, *Proc. London Math. Soc.* **3** (1869), 224–232.
32. L.-G. Michaud, *Biographie Universelle Ancienne et Moderne*, Nouvelle edition, Desplaces, Vol. I, 1843; Vol. XXXVI, no date, pp. 288–289.
33. S. O'Donnell, *William Rowan Hamilton. Portrait of a Prodigy*, Boole Press, 1983.
34. M. Riesz, *Clifford Numbers and Spinors*, Lecture Series No. 38, The Institute for Fluid Dynamics and Applied Mathematics, Univ. of Maryland, 1958.
35. O. Rodrigues, Des lois géométriques qui régissent les déplacements d'un système solide dans l'espace, et la variation des coordonnées provenant de ses déplacements considérés indépendamment des causes

qui peuvent les produire, *J. de Mathématiques Pure et Appliquées* **5** (1840), 380–440.

36. P. G. Tait, *An Elementary Treatise on Quaternions*, 3rd ed., Cambridge Univ. Press, 1890.

37. G. Temple, *Cartesian Tensors. An Introduction*, Methuen, 1960.

38. S. P. Thompson, *The Life of William Thomson, Baron Kelvin of Largs*, Macmillan, 1910.

39. J. Warren, On the geometrical representation of the powers of quantities, whose indices involve the square roots of negative quantities, *Phil. Trans. Roy. Soc.* **119** (1829), 339–359.

40. G. Weill, *Saint Simon et son Oeuvre. Un Précurseur du Socialisme*, Perrin et Cie, 1894.

41. E. Wilson, *To the Finland Station. A Study in the Writing and Acting of History*, Secker and Warburg, 1941.

Building an International Reputation: The Case of J. J. Sylvester (1814–1897)

KAREN HUNGER PARSHALL AND EUGENE SENETA

American Mathematical Monthly **104** (1997), 210–222

James Joseph Sylvester—prolific, gifted, flamboyant, egocentric, cantankerous. At the time of his death in London on 15 March 1897, Sylvester's reputation internationally as one of the nineteenth century's principal mathematical figures had long been secure. He had worked hard to assure this. Obviously, he had done much seminal work in building the theory of invariants, and this had contributed to his renown. Yet, Sylvester had felt compelled to establish ties directly with mathematicians at home—but more importantly abroad—in order to make his name known. Was this just a matter of egocentrism, or did other factors contribute to his international focus? What did it take to become an internationally recognized British mathematician in the latter half of the nineteenth century, when first France and then Germany very much set the mathematical standard? Why was this even important? As the centenary of Sylvester's death brings historians and mathematicians together in England to celebrate his life and research, we examine some of the reasons why Sylvester valued the international mathematical arena so highly and how he used it to his advantage during the course of his career.

It is well known that Sylvester, as a Jew, was, like all non-Anglicans, debarred by the Test Acts from taking the Cambridge degree he had earned as Second Wrangler in 1837 and from holding a Cambridge fellowship or professorship. His first position, the professorship of natural philosophy at non-sectarian University College London, was too far from his real interests and expertise to satisfy him, so he gave it up in 1841 after only three years for the uncertain fortunes of a professorship of mathematics in exile far from home (and, he quickly came to think, far from civilization!) at the University of Virginia. He lasted there for four-and-a-half months before resigning over a matter of principle and fleeing northward to New York City and his brother's home. From there, he tried in vain for some eighteen months to secure a new position in the United States—at Columbia College, at Harvard with Benjamin Peirce, in the Washington, D.C. area, and even at the University of South Carolina—before returning to England to resume what he termed the "fruitless and hopeless struggle with an adverce [sic] tide of affairs" [39]. By the close of 1844, though, he proudly reported having "recovered [his] footing in the world's slippery path" [39] thanks to his assumption of the post of actuary and secretary at the Equity and Law Life Assurance Company in London.

During the decade from 1845 to 1855, he prepared for and passed the Bar; he met his mathematical alter ego, Arthur Cayley; and he produced his groundbreaking series of papers in what would come to be known as invariant theory. The next fifteen years found him in his first sustained academic post, the professorship of mathematics at the Royal Military Academy in Woolwich, where he taught drudgerous mathematics to mostly uncaring students and fought with the military authorities over teaching loads destined, he was convinced, to bring the "extinction of my scientific existence" [40]. Sylvester's career trajectory clearly diverged from those guided within the ivy-covered walls of the colleges of Cambridge and Oxford.

As an establishment-outsider, who, unlike Augustus De Morgan, did not even hold a position at the leading anti-establishment institution, how was Sylvester to secure the reputation that his healthy-sized ego demanded and that his manifest mathematical talents warranted? Surely, the British Association for the Advancement of Science or the Royal Society, to which he had been elected in 1839, represented

James Joseph Sylvester, aged 26, by George Patton.
In the private collection of Mr. and Mrs. Alain Enthoven.

avenues toward the establishment of national recognition, and Sylvester presented his work-like accounts of his research on Sturm's theorem for locating the roots of an algebraic equation ([1], I: 59–60 and 429–586)—before both of these organizations throughout his career. In Sylvester's Platonistic view, however, mathematics was a universal endeavor that transcended national boundaries (see [24] for an analysis). Moreover, as Sylvester well realized, the Continent—and particularly France and Germany—dominated mathematical research at mid-century, while England tended to assume a more isolationist posture. It was thus important to make one's work known abroad. This would help assure that credit was given where due, that results published in British journals were not ignored or overlooked outside England, and that British mathematicians effectively contributed to building the eternal edifice of mathematics. As Marin Mersenne had shown in the seventeenth century, establishing an international network of correspondents could be remarkably helpful in achieving these goals.

Sylvester actively sought to forge his own international mathematical connections beginning around

1850. Initially, at least, he seemed to have the greatest number of ties with France, a country which had dominated mathematics and the sciences during the first half of the nineteenth century [14], a country with whose language Sylvester felt at ease, a country with an influential scientific society and mathematicians of the highest repute. In France, Sylvester enjoyed perhaps his most lasting and most intimate mathematical association with Charles Hermite.

Eight years Sylvester's junior, Hermite had come upon the mathematical scene in the 1840s with work on elliptic and hyperelliptic functions that had earned Jacobi's admiration. In 1848, the year after taking his *baccalauréat* and *licence*, he was named *répétiteur* and admissions examiner at the *École Polytechnique*, where he himself had pursued his studies. Rather quickly, he established a reputation as one of the rising stars in French mathematics with his work on quadratic forms and, beginning in 1854, on the theory of invariants. His election to the Paris *Académie des Sciences* in 1856 gives a clear indication of his stature in the French scientific community. Since Hermite's research interests fundamentally overlapped those of Sylvester, it is little wonder that the Englishman sought out the kindred French mathematical spirit. In particular, Sylvester wanted to make sure that Hermite knew of and gave the proper credit to his research.

Charles Hermite, aged 25, by Charles Wittman, from *Oeuvres de Charles Hermite*

After finishing his law studies in 1850 and following the establishment of his close personal friendship and mathematical exchange with Arthur Cayley, Sylvester finally began to come into his own as a researcher. Between 1850 and 1854, he published much of his work on determinants as well as some of his seminal papers on the emergent theory of invariants. The year 1852, in particular, witnessed the publication of his first major contribution to invariant theory, his paper, "On the Principle of the Calculus of Forms" ([1], I: 284–327 and 328–363). Sylvester earnestly desired that this work *not* escape the immediate notice of the French mathematical community, and, to this end, he wrote several letters in 1852 to his correspondent, Irenée-Jules Bienaymé (compare [20]), asking him to distribute offprints to certain key individuals in addition to the *Société Philomatique de Paris*, to which Sylvester had been elected as a corresponding member early that year, and to the *Institut de France*.

According to Bienaymé's handwritten tally [2] in response to Sylvester's first request on 7 February 1852 [31], Michel Chasles, Charles Hermite, Olry Terquem, Eugène Catalan, Joseph Serret, and Joseph Bertrand were to receive copies of the first installment of Sylvester's latest pronouncement on the calculus of forms ([1], I: 284–327) as well as an 1851 paper, "On the General Theory of Associated Forms" ([1], I: 198–202), while Cauchy and the *Institut de France* were to get these together with a paper on the theory of determinants ([1], I: 241–250). Bienaymé dutifully delivered these and received additional requests from Sylvester on 4 June [32] and again on 27 August [33] to deliver more papers to these and selected others. Sylvester's letter of 4 June revealed at least one additional motivation for this general distribution of his work. There, he asked Bienaymé to pass copies of the second installment of his paper on the calculus of forms to Hermite and to the *Société Philomatique* as before, but this time he also wanted a copy to go to the editor of the *Journal des Mathématiques Pures et Appliquées*, Joseph Liouville. As he explained to Bienaymé,

> I wish M. Liouville to have a copy because I am told that M. Eisenstein of Berlin has sent to Liouville's Journal the same kind of matter as is in my Section VI [32].

This was the demonstration of the sufficiency of two particular differential operators that Cayley and Sylvester employed for detecting invariants ([1], I: 351–360), and Sylvester clearly wanted Liouville to know that he and Cayley—and *not* Eisenstein—had discovered the properties of these operators first. Eisenstein did not end up publishing the contested proof [11].

Irenée-Jules Bienaymé (1796–1878) by Jules Franceschi.
Photo courtesy of Arnaud Bienaymé

As Sylvester's delivery list to Bienaymé also indicates, the Englishman's association with Michel Chasles dates from at least the early 1850s. Chasles, who had established his reputation as both a geometer and an historian of mathematics as early as 1837 with the publication of his *Aperçu Historique sur l'Origine et le Développement des Méthodes en Géométrie*, had been named to the chair of higher geometry at the Sorbonne in 1846 and had become a full member of the Paris *Académie des Sciences* in 1851. He was thus another influential member of the French mathematical community and a worthy contact for Sylvester to establish, even though his brand of geometric research was never Sylvester's forte. In 1852, as Sylvester was busy with the early invariant-theoretic work that he was anxious for Bienaymé to deliver to Chasles, Chasles had just published a new book, his *Traité de Géométrie Supérieure* [81], and asked Sylvester to serve as the messenger for two copies to British colleagues [7]. In his letter of 26 August, Chasles also provided Sylvester with an indication that he was starting to make a name for himself in France. Chasles reported that ([7], our translation)

I saw M. Terquem yesterday; he spoke to me with pleasure and enthusiasm of the beautiful theorem that you sent him, and knowing that I was going to write you, he asked me to give you his compliments and to tell you that he was going to publish your communication without delay in his journal.

Olry Terquem was the editor of the *Nouvelles Annales de Mathématiques*, and the submission in question was Sylvester's first publication in a foreign journal, his determinant-theoretic proof of the fact that if $\mathbf{A}$ is an $n \times n$ symmetric matrix, then the roots of the characteristic equation of $\mathbf{A}^p$ for p an integer are the roots of the characteristic equation of $\mathbf{A}$, each raised to the pth power ([1], I: 364–366). Sylvester quickly followed this with two more notes to Terquem's journal ([1], I: 413 and 424–428). Having successfully taken the step of bringing his work directly before the Continental audience by publishing in a foreign journal, Sylvester continued throughout his career to submit his work to periodicals both at home and abroad.

Besides Hermite, Bienaymé, and Chasles, Sylvester also established ties in the 1850s in France with Jean-Victor Poncelet, Jean-Marie-Constant Duhamel, Joseph Serret, and Joseph Bertrand. Relative to Germany, however, his contacts were initially fewer and his relations never as close. One exception to this was Carl Borchardt, who had spent the year from 1846 to 1847 studying in Paris and forging his own links with Chasles, Hermite, and Liouville. Borchardt had earned his doctorate under Jacobi at the University of Königsberg in 1843 and had made a splash in 1846 with his first mathematical publication, a work right up Sylvester's mathematical alley. Borchardt had given a determinant-theoretic argument showing that the Sturm functions that arise from the equation determining the secular disturbances of the planets can be represented as the sum of squares [4]. In subsequent work, he considered related algebraic questions involving symmetric functions and elimination theory, that also actively engaged Sylvester. Given their common interests, it is little wonder that Sylvester and Borchardt began to correspond. It is also little wonder that, in 1852, Sylvester wanted to apprise Borchardt of his latest researches.

On 20 February, just two weeks after he had entrusted Bienaymé with the bundle of offprints for distribution in Paris, Sylvester wrote Borchardt enclosing a copy of one of the same papers, the first installment of his paper "On the Principle of the Calculus of Forms." Burdened by his ongoing preparation of Jacobi's collected works and by his teaching duties as *Privatdozent* at the University of Berlin, Borchardt only responded to Sylvester's letter on 6 April [3]. There, he apologized for the delay in his response and for not having had a chance to read and study Sylvester's latest achievement. More interestingly, he also returned to a topic that had apparently been under discussion in earlier letters between the two men: the integrity of Otto Hesse.

Because the early 1850s were abundantly productive years for Sylvester mathematically, it was then that he actively sought to build and solidify his increasingly deserved reputation. Always touchy about matters of priority, but perhaps most touchy during these early years, Sylvester came to feel that Hesse [18] had stolen a result he had published in the *Philosophical Magazine* in 1841 ([1], I: 75–85). Sylvester, furious, felt that Hesse had consciously failed to credit his work and lambasted him in print in ([1], I: 184–197 on p. 189). In his letter, however, Borchardt offered a very different read on the situation. In his measured and more objective view, Borchardt offered that

[i]f Mr. Hesse had known of your memoir... he would not have committed plagiarism, and I know him too well to believe him capable of it. This changes nothing relative to your priority but much relative to the moral judgment on Mr. Hesse. In mathematics it often happens that, owing to an insufficient knowledge of the literature, results are published as new which have already been obtained earlier by others. Such oversights must certainly be corrected, but if in every such case of this kind one claimed plagiarism, one would not be justified ([3], our translation).

This dispassionate assessment of the situation as well as Sylvester's heated reaction to what he deemed to be Hesse's initial slight reflect a key aspect of reputation-building, the paramount importance of priority. Without it, someone else's reputation may grow incrementally at the expense of one's own hard work and effort. In Sylvester's case, this became even more of an issue when publications serving primarily different national constituencies were involved. If Hesse read the *Philosophical Magazine*, a periodical that published, by and large, mathematical work of lesser quality than its German counterparts, then he had *overlooked* Sylvester's paper. If he—and by extension his countrymen—did *not* read the *Philosophical Magazine*, then all the worse for those British mathematicians trying conscientiously to make their work known to the broader community of mathematicians through that means. At least in this case, the priority dispute was ultimately little more than a tempest in a teapot; just two years later, at a crucial juncture in Sylvester's career, Hesse served as one of the Englishman's hand-picked references.

By 1854, Sylvester had been in his job as actuary and secretary at Equity Law and Life for some nine years. He had also been doing some of his best mathematical research and, understandably, wanted a position more consonant with his interests and training. When the professorship of mathematics at the Royal Military Academy in Woolwich came open, then, he quite naturally took the opportunity to apply. When he lost out on the appointment to a mathematically inferior candidate and felt that the military authorities had misrepresented some of the facts surrounding the election, he wrote to another longtime acquaintance, the former Lord Chancellor, Lord Brougham, expressing his frustration. Not without a certain amount of pride, Sylvester let Brougham know that

> Letters were written or sent in in support of my application and couched in the strongest language in which a recommendation could be clothed from
>
> Sir William Hamilton, Dublin
> Professor Graves, D[itt]o
> Professor Kelland, Edinburgh
> Professor Challis, Cambridge
> The Bishop of Natal
> General Poncelet, Paris
> M. Chasles, Paris
> Rev[eren]d Geo Salmon, Dublin
> Duhamel, Paris
> Serret, Hermite, Bertrand of the Examiners at the Polytechnique
> and many others.
>
> Letters were also written but too late to be sent in by
> the great Lejeune Dirichlet, Berlin
> Professors Peters & Hesse, Königsberg
> Professor Joachimsthal, Halle
> Professor Thomson, Glasgow
>
> and also from distinguished pupils testifying to my teaching powers [34].

As is evident from this roster, Sylvester made important use of the international network he had established, in his attempt to break into the English academic world. This first try proved unsuccessful, but, when the victorious candidate died unexpectedly after only a few months on the job, Sylvester reapplied and won the appointment (see [28] on the changing London mathematical scene). His distinguished list of referees may not have helped him get the post, but it apparently did not hurt. Sylvester had finally broken back into academe after a dozen years away.

The late 1850s and early 1860s found Sylvester hard at work on his ongoing researches in invariant theory and in new, but not unrelated, work in combinatorics. In a series of seven lectures delivered at King's College, London between 6 June and 11 July 1859 (but published only in 1897), he brought together a large number of known and new combinatorial results in an early effort to systematize a theory of partitions ([1], II: 119–175). Prior to delivering the last of these lectures, Sylvester wrote to his friend, Pafnuty Chebyshev, thanking him for the elucidating remarks on some of Euler's work on partitions that enabled him to include a discussion of it in his presentation [38]. Sylvester and Chebyshev had known each other at least since the fall of 1852, when Chebyshev was on a foreign tour to study the scientific and technological advances in evidence in Paris, London, and Berlin. After seeing Bienaymé and others in Paris, he went on to visit Sylvester in London, and the two remained in touch sporadically for many years thereafter. In fact, like Sylvester, Chebyshev also maintained his contact with Hermite. As the Russian mathematician would later recollect,

> [w]e were once sitting in Paris, the three of us: Hermite, Sylvester, and I. Hermite—the leading mathematician of France, Sylvester—the leading mathematician of England, and I ([15], our translation).

Sylvester's growing international network also extended to Italy by the early 1860s to include Enrico Betti, fellow determinant-theorist Francesco Brioschi, and physicist and later Minister of Education, Carlo Matteucci, among others. In the winter of 1862, in fact, Sylvester followed one of his many trips to Paris with an extended scientific tour of the newly unified Italy and described at least part of his itinerary in a letter to Cayley [36]. Here, Sylvester entered into a different role relative to his international network; his interest—as well as that of other foreign mathematicians—in the post-unification Italian scene helped to legitimize the research efforts of the Italians in the latter half of the nineteenth century ([5] and [22]). Sylvester's presence helped the Italians establish *their* reputations; by the early 1860s, *his* reputation, at least in their eyes, was already secure. This was apparently true elsewhere as well.

Late in 1863, Sylvester received what must have been the most significant symbol of his international stature as a mathematician, his election as foreign correspondent to the Paris *Académie des Sciences*. It was his contact, Joseph Bertrand, whom Sylvester chose to communicate one of his first "official" works to the assembled *savants*, and that communication related to his (the first) proof of Newton's rule for isolating pairs of complex roots of polynomial equations ([1], II: 361–362). As Sylvester explained to Bertrand,

> I have proved, not without some difficulty, Newton's rule to the fifth degree inclusive. Messers De Morgan and Cayley have expressed to me their firm conviction that Newton himself never had a proof of this rule, which remains to this day the marvel and the disgrace of Algebra ([30], our translation).

Interestingly, Sylvester announced the result in print first in France, but only after having read the paper before the Royal Society; he later published the full exposition of it in the *Philosophical Transactions* ([1], II: 376–479). In so doing, he was, in a sense, maximizing the publicity for the new result; he gave it to the most important scientific bodies in both his native England and in France. Moreover, after the founding of the London Mathematical Society in 1865, he took yet another opportunity to highlight his work on the rule by lecturing at one of the Society's early meetings on the proof he had subsequently discovered of the general case ([1], II: 498–513).

This mathematical triumph was followed in the late 1860s through the mid-1870s by a troubled period which found Sylvester first casting about for mathematical direction, next in premature retirement from his post at the Military Academy, and then adrift and unemployed in London. All of this changed in 1876, when he accepted the first professorship of mathematics at The Johns Hopkins University in Baltimore, Maryland. He had made a bold transatlantic move as a young man of twenty-eight; in 1876, at the age of sixty-one, he did it again. The second time around the outcome was much more positive.

Sylvester's arrival on American shores marked the beginning of a quarter-century-long process of establishing mathematics at the research level in the United States [27]. By 1877, Sylvester had regained his research footing and had reengaged in his earlier invariant-theoretic researches. In particular, he sought to vindicate the British approach to invariant theory that he and Cayley had developed by providing a British-style proof of Gordan's finiteness theorem of 1868 (on the history of this problem, see [9], [10] and [26]). Despite Cayley's supposed proof in 1856 that the number of irreducible covariants of a binary quintic form is infinite [6], Paul Gordan showed that this number is, in fact, finite for any given binary form [13]. Beginning in the late 1870s, Sylvester worked on and off to supply a British proof of Gordan's theorem. Writing to William Spottiswoode on 19 November 1876, he made his nationalistic and personal motivations crystal clear in announcing what he thought was a proof of Gordan's theorem. "The piratical Germans Clebsch and Gordan who have so unscrupulously done their best to rob us English of all the credit belonging to the discoveries made in the New Algebra will now suffer it is to be hoped the due Nemesis of their misdeeds," Sylvester declared.

> Nothing in Clebsch and Gordan is *really new* but their Cumbrous method of *limiting* (not *determining*) the Invariants to any given form . . . I see a splendid vista of investigations open to me on this subject destined I believe to reduce to annihilation all that the school of Clebsch and Gordan, by aid of methods borrowed by the Germans without acknowledgement from Cayley and myself, have attempted on the subject ([42]; Sylvester's emphasis).

In Sylvester's view, this was a priority issue of the greatest magnitude. Virtually his entire scientific reputation rested on his work in invariant theory, and the Germans had not only failed to give it its due, but also isolated and patched a major hole in the entire British approach. May the better theory win; this battle would be fought in the international mathematical arena.

In 1877, Sylvester was convinced that he had not only won the battle but had in some sense won the invariant-theoretic war against the Germans. In May of that year, Hermite communicated the following announcement to the Paris Academy ([17], p. 975, our translation):

> Baltimore—Since my last communication, please inform the Academy that I have resolved the problem of finding the complete set of *Groundforms* for arbitrary forms in n variables.

This stunning announcement was rather quickly followed by a retraction, but Sylvester persisted in the struggle for the result. In the fall of 1878, he met his adversary head on when he claimed again to have the theorem in its full generality. Once again, though, his claim was false.

Sylvester continued on and off to try to find a British-style proof of Gordan's theorem, and, while he never succeeded, he also never seemed to question the personal and nationalistic motivations behind such a quest, accepting fully the reality of the increasingly international arena [29] in which mathematicians competed. When David Hilbert gave the first inklings in 1888 and 1889 ([19], II: 176–198) of the new invariant-theoretic methods he would bring forth in their fuller glory in 1890 ([19], II: 199–257), Sylvester knew he had lost a major skirmish and conceded defeat gracefully in a letter to Felix Klein. Hilbert "has rendered a very good service to Algebra, in obtaining so simple a proof of Gordan's theorem," Sylvester told Klein, "and I should like to be able to congratulate him on his brilliant invention. What a relief from the previous methods of proof!" [41]. International competition was important, but it was also important that credit be given where due.

It is not hard to imagine, however, that Sylvester got some satisfaction from the fact that, in a real sense, Gordan had been bested as well!

As a professor at The Johns Hopkins University, Sylvester did not seek to shield his students from the realities of international competition like his own with the German invariant-theoretic school. Rather, he worked to instill in them his strong sense of the importance of international, as opposed to merely national, exposure. From his own personal experience, he knew how important the foreign imprimatur was to the establishment of real reputation. In 1881, for example, when Sylvester and his Hopkins students were hard at work on what would become their groundbreaking paper on partition theory ([1], IV: 1–81), one of the students, Fabian Franklin, devised a wonderfully simple, graphical proof of Euler's pentagonal number theorem [12]. Sylvester was so pleased by and excited about this result that he had Franklin write it up for communication through Hermite to the Paris Academy. On 29 April 1881, Hermite wrote to Sylvester offering his own praise for Franklin's proof and giving an indication of the attention that it was getting in France. "It certainly will not be unpleasant for you to hear that I was not the only person to be very interested in Mr. Franklin's very original and ingenious proof," he wrote.

> Mr. Halphen, one of our most eminent young mathematicians, . . . found Franklin's method so remarkable that he lectured on it in one of the recent sessions of the *Société Philomatique*. Please tell Mr. Franklin that his talent is appreciated, as it deserves to be, by the mathematicians of the old world ([16], our translation).

Writing to Cayley perhaps on the very day he received Hermite's letter, Sylvester could not contain his pleasure over this reaction to his student's work. "Hermite," he gushed, "is overflowing with admiration at the beauty of the method" [37].

Following the success of his student, Franklin, and in the wake of the concerted combinatorial research in which he had engaged all of his students [23], Sylvester began to feel the strain of leading America's first research-level program in mathematics. In 1883, after seven-and-a-half years in Baltimore, he resigned from his Hopkins professorship to assume the Savilian Chair of Geometry at Oxford; the repeal of the Universities Test Act in 1871 allowed Sylvester, as a non-Anglican, to hold the Oxford chair. At the age of sixty-nine, he returned home, his efforts at reputation-building having finally yielded what for him was the ultimate prize—the recognition of the Oxbridge academic establishment.

Sylvester's career can hardly be considered typical of mathematicians in Victorian Britain. For one thing, the fact that he was Jewish initially closed many of the usual avenues of a mathematical career to him. Twice this led him to leave Britain in the hopes of greener pastures in the United States, to broaden his focus to an international arena, albeit the wrong one at least in the 1840s. Moreover, it forced him to look beyond the English academic scene for ways to establish his reputation in his chosen field. If he could not have the validation of a prestigious position at an English university, he could at least measure his self-worth in terms of an international renown—hard won and carefully cultivated—that the majority of the mathematical practitioners within the English academic system would never enjoy.

Another aspect of Sylvester that made him atypical of Victorian mathematicians was his ego. It was undeniably large. He *wanted* to be known for his research; he *wanted* his work appreciated; he was often the first person to pronounce his latest theorem "remarkable" or "beautiful." Yet, Sylvester was much more complicated than this simplistic analysis would suggest. He loved mathematics and believed it to be eternal and transcendent. He wanted to make enduring contributions to it. As he explained to Lord Brougham in the aftermath of one of his altercations with the authorities at Woolwich, "I trust . . . to leave a lasting mark on 'The Algebra of the Future' " [35]. This, too, motivated him to make connections with mathematicians internationally as well as nationally. They were all striving toward the same goal, the construction of what Camille Jordan called the "temple of Algebra" [21] or, more generally, the temple of mathematics. And, even if they competed like Sylvester and Gordan, they labored in common cause.

Both of these atypical aspects of Sylvester as Victorian mathematician led, in his case, to the formation, maintenance, and utilization of an international network. Beginning at mid-century, national mathematical research communities—defined in terms of specialized professional associations, specialized journals, venues for the training of future researchers, and the overall emphasis on the production of original research—were under formation in Europe and somewhat later in the United States [25]. Sylvester, also beginning at mid-century, seemed to have a strong sense of the value of a further step in the professional development of mathematics, the internationalization of the field.

References

1. H. F. Baker (ed.), *The Collected Mathematical Papers of James Joseph Sylvester*, 4 vols., Cambridge Univ. Press, 1904–1912; reprint ed., Chelsea, 1973 (all page references to Sylvester's papers refer to the pagination in this edition).

2. Irenée-Jules Bienaymé to J. J. Sylvester, draft of a letter dated 4 April 1852, in the private collection of Arnaud Bienaymé.

3. Carl W. Borchardt to J. J. Sylvester, 6 April 1852, *Sylvester Papers*, Box 2, St. John's College, Cambridge.

4. Carl W. Borchardt, Développements sur l'équation à l'aide de laquelle on détermine les inégalités séculaires du mouvement des planètes, *Journal de Mathématiques Pures et Appliquées* **12** (1847), 50ff, or Carl W. Borchardt, *Gesammelte Werke* (ed. G. Hettner), Georg Reimer, 1888, pp. 15–30.

5. Umberto Bottazzini, Il diciannovesimo secolo in Italia, in Dirk Struik, *Mathematica: Un Profilo Storico*, Il Mulino, 1981, pp. 249–312.

6. Arthur Cayley, A second memoir upon quantics, *Philosophical Transactions of the Royal Society of London* **146** (1856), 101–126, or *The Collected Mathematical Papers of Arthur Cayley* (ed. Arthur Cayley and A. R. Forsyth), 14 vols., Cambridge Univ. Press, 1889–1898, II: pp. 250–275.

7. Michel Chasles to J. J. Sylvester, 26 August 1852, *Sylvester Papers*, Box 3, St. John's College, Cambridge.

8. Michel Chasles, *Traité de Géométrie Supérieure*, Bachelier, 1852.

9. Tony Crilly, The rise of Cayley's invariant theory (1841–1862), *Historia Mathematica* **13** (1986), 241–254.

10. Tony Crilly, The decline of Cayley's invariant theory (1863–1895), *Historia Mathematica* **15** (1988), 332–347.

11. Gotthold Eisenstein, Extrait d'une lettre adressée à M. Charles Hermite, *Journal de Mathématiques Pures et Appliquées* **17** (1852), 473–477, or Gotthold Eisenstein, *Mathematische Werke*, 2 vols., Chelsea, 1975, II: pp. 771–775.

12. Fabian Franklin, Sur le développement du produit infini $(1-x)(1-x^2)(1-x^3)\cdots$, *C. R. Acad. Sci. Paris* **82** (1881), 448–450.

13. Paul Gordan, Beweis dass jede Covariante und Invariante einer binären Form eine ganze Function mit numerische Coefficienten einer endlichen Anzahl soichen Formen ist, *Journ. Reine Angew. Math.* **69** (1868), 323–354.

14. Ivor Grattan-Guinness, *Convolutions in French Mathematics, 1800–1840: From the Calculus and Mechanics to Mathematical Analysis and Mathematical Physics*, 3 vols., Birkhäuser Verlag, 1990.

15. D. A. Grave, My life and scientific activity, *Istoriko-Matematicheskie Issledovaniia* **34** (1993), 219–246 (in Russian).

16. Charles Hermite to J. J. Sylvester, 29 April 1881, *Gilman Papers*, Coll #1 Corresp., Box 16, The Milton S. Eisenhower Library, Special Collections Department, The Johns Hopkins University.

17. Charles Hermite, Études de M. Sylvester sur la théorie algébrique des formes, *C. R. Acad. Sci. Paris* **84** (1877), 974–975.

18. Otto Hesse, Über die Elimination der Variabeln aus drei algebraischen Gleichungen von zweiten Grade mit zwei Variabeln, *Journal für die Reine und Angewandte Mathematik* **28** (1844), 68–96.

19. David Hilbert, *Gesammelte Abhandlungen*, 3 vols., Springer-Verlag, 1932–1935.

20. François Jongmans and Eugene Seneta, The Bienaymé family history from archival materials and background to the turning-point test, *Bulletin de la Société Royale des Sciences de Liège* **62** (1993), 121–145.

21. Camille Jordan to J. J. Sylvester, 13 May 1877, *Sylvester Papers*, Box 2, St. John's College, Cambridge.

22. Erwin Neuenschwander, Der Aufschwung der italienischen Mathematik zur Zeit der politischen Einigung Italiens und seine Auswirkungen auf Deutschland, *Symposia Mathematica* **27** (1986), 213–237.

23. Karen Hunger Parshall, America's first school of mathematical research: James Joseph Sylvester at The Johns Hopkins University 1876–1883, *Arch. Hist. Exact Sci.* **38** (1988), 153–196.

24. Karen Hunger Parshall, Chemistry through invariant theory? James Joseph Sylvester's mathematization of the atomic theory, *Experiencing Nature: Proceedings of a Conference in Honor of Allen G. Debus* (eds. Paul Theerman and Karen Hunger Parshall), Kluwer, 1997, pp. 81–111.

25. Karen Hunger Parshall, Mathematics in national contexts (1875–1900): an international overview, in *Proceedings of the International Congress of Mathematicians: Zürich* (ed. S. D. Chatterji), 2 vols., Birkhäuser Verlag, 1995, II, pp. 1581–1591.

26. Karen Hunger Parshall, Toward a history of nineteenth-century invariant theory, *The History of Modern Mathematics* (eds. David E. Rowe and John McCleary), 2 vols., Academic Press, 1989, I, pp. 157–206.

27. Karen Hunger Parshall and David E. Rowe, *The Emergence of the American Mathematical Research Community, 1876–1900: J. J. Sylvester, Felix Klein, and E. H. Moore*, American Mathematical Society and London Mathematical Society, 1994.

28. Adrian Rice, Mathematics in the metropolis: a survey of Victorian London, *Historia Mathematica* **23** (1996), 376–417.
29. Brigitte Schroeder-Gudehus, Nationalism and internationalism, *Companion to the History of Modern Science* (eds. R. C. Olby *et al.*), Routledge, 1990, pp. 909–919.
30. J. J. Sylvester to Joseph Bertrand, 12 April 1864, Académie des Sciences, Pochette de Séance, 18 April 1864.
31. J. J. Sylvester to Irenée-Jules Bienaymé, 7 February 1852, in the private collection of Arnaud Bienaymé.
32. J. J. Sylvester to Irenée-Jules Bienaymé, 4 June 1852, in the private collection of Arnaud Bienaymé.
33. J. J. Sylvester to Irenée-Jules Bienaymé, 27 August 1852, in the private collection of Arnaud Bienaymé.
34. J. J. Sylvester to Lord Brougham, 9 August 1854, *Brougham Papers*, Sylvester, J. J., 20232, University College London Archives.
35. J. J. Sylvester to Lord Brougham, 8 May, 1863, *Brougham Papers*, Sylvester, J. J., 20253, University College London Archives.
36. J. J. Sylvester to Arthur Cayley, undated, *Sylvester Papers*, Box 10, St. John's College, Cambridge (Tony Crilly has estimated the date of this letter as January or February 1862).
37. J. J. Sylvester to Arthur Cayley, 12 May 1881, *Sylvester Papers*, Box 11, St. John's College, Cambridge.
38. J. J. Sylvester to Pafnuty Chebyshev, 5 July 1859, P. L. Chebysbev, *Polnoe Sobranie Sochinenii*, Vol. 5 *Prochie Sochinenia Biograficheskie Materialy*, AN SSSR, 1951, p. 448.
39. J. J. Sylvester to Joseph Henry, 12 April 1846, *Henry Papers*, M099 #8573, Smithsonian Institution Archives, or Nathan Reingold and Marc Rothenberg (eds.), *The Papers of Joseph Henry*, 6 vols., Smithsonian Institution Press, 1972–1992, VI, pp. 407–410.
40. J. J. Sylvester to Thomas Archer Hirst, 21 March 1863, *London Mathematical Society Papers, Sylvester, J. J.*, University College London Archives.
41. J. J. Sylvester to Felix Klein, 24 November 1889, Klein Nachlass XI, Niedersächsische Staats und Universitätsbibliothek, Göttingen, or Gert Sabidussi, Correspondence between Sylvester, Petersen, Hilbert, and Klein on invariants and the factorisation of graphs 1889–1891, *Discrete Mathematics* **100** (1992), 99–155, on p. 126.
42. J. J. Sylvester to William Spottiswoode, 19 November 1876, *Sylvester Papers*, Box 1, St. John's College, Cambridge.

The Foundation Period in the History of Group Theory

JOSEPHINE E. BURNS

American Mathematical Monthly **20** (1913), 141–148.

Introduction

The earliest group notions

Henri Poincaré has pointed out that the fundamental conception of a group is evident in Euclid's work; in fact, that the foundation of Euclid's demonstrations is the group idea. Poincaré establishes this assertion by showing that such operations as successive superposition and rotation about a fixed axis presuppose the displacements of a group. However much the fundamental group notions were unconsciously used in the work of early mathematicians, it was not until the latter part of the eighteenth century that these notions began to take life and develop.

The foundation period

The period of foundation of group theory as a distinct science extends from Lagrange (1770) to Cauchy (1844–1846), a period of seventy-five years. We find Lagrange considering the number of values a rational function can assume when the variables are permuted in every possible way. With this beginning the development may be traced down through the contributions of Vandermonde, Ruffini, Abbati, Abel, Galois, Bertrand and Hermite, to Cauchy's period of active production (1844–1846). At the beginning of this period group theory was a discovery useful in the theory of equations; at the end it existed as a distinct science, not yet, to be sure, entirely free but so nearly so that this may be called the close of the foundation period.

Lagrange, Ruffini, Galois

Lagrange

The contributions of Lagrange are included in his memoir [7], published at Berlin in 1770–1771. In this paper Lagrange first applies what he calls the "calcul des combinaisons" to the solution of algebraic equations. This is practically the theory of substitutions, and he uses it to show wherein the efforts of his predecessors Cardan, Ferrari, Descartes, Tschirnhaus, Euler and Bézout fail in the case of equations of degree higher than the fourth. He studies the number of values a rational function can assume when its variables are permuted in every possible way. The theorem that the order of the subgroup divides the order of the group is implied but not explicitly proved. The theorem that the order of a group of degree n divides $n!$ is however explicitly stated. Lagrange does not use at all the notation or the terminology of group theory but confines himself entirely to direct applications to the theory of equations. He mentions the symmetric group of degree 4, and the four-group and the cyclic group of order 4 as subgroups of this group. This embodies practically all of the work of Lagrange in the theory of substitution groups.

Ruffini

The first man to follow Lagrange and to make any signal progress was Paolo Ruffini, an Italian, who published in 1799 [9] a number of important theorems in the theory of substitutions. Burkhardt [1] says that several fundamental concepts are implied, if not explicitly stated. Ruffini's "permutation" corresponds to the later accepted term of *group* and to Cauchy's "system of conjugate substitutions." These permutations he classifies as "simple" and "complex." The first are of two sorts, of one cycle or of more than one cycle. The second he divides into three classes which correspond to the modern notions of (1) intransitive, (2) transitive imprimitive, and (3) transitive primitive. He enunciates the following theorems:

1. If a substitution of n letters leaves the value of a rational function invariant, the result of applying the substitution any number of times is that the function is left invariant.

2. The order of a substitution is the least common multiple of the orders of the cycles.

3. There is not necessarily a subgroup corresponding to every arbitrary divisor of the order of the group. (This theorem he established by showing that there is no eight-valued, four-valued or three-valued function on five letters.)

4. A group of degree five that contains no cycle of degree five cannot have its order divisible by five.

Abbati, Abel

From Ruffini to Galois, only two or three contributions of merit were made. Abbati gave the first complete proof that the order of the subgroup divides the order of the group. This he did by putting the substitutions of the group in rectangular array. Abbati also proved that there is no three or four-valued function on n letters when n is greater than five. Cauchy published in 1815 a relatively unimportant memoir [3]. Abel, by using the theory of substitutions to prove that it is impossible to solve algebraically equations of degree higher than the fourth, called attention to this useful instrument.

Galois

Up to the time of Cauchy, Galois had done the most for group theory. His work was written in 1831 and 1832, but not made public until 1846. To Galois is due the credit for the conception of the invariant subgroup, for the notion of the simple group and the extension of the idea of primitivity. Galois first used the term *group* in its present technical sense. Several important theorems are due to him. Among these are the following:

1. The lowest possible composite order of a simple group is 60. [5]

2. The substitutions common to two groups form a group. [6]

3. If the order of a group is divisible by p, a prime, there is at least one substitution of order p.

4. The substitutions of a group that omit a given letter form a group.

Cauchy, the founder of group theory

Terminology

Cauchy's work on groups is found in his *Exercises* [4], published in 1844, and in the series of articles published in the Paris *Comptes Rendus* in 1845–1846 [2]. Cauchy defines *permutation* and *substitution* in the same way, but uses the latter term almost entirely. He uses several devices to denote a substitution, the most common being (x, y, z, u, v, w) where each letter is replaced by the one which follows it and the last by the first. He defines this as a *cyclic substitution*.

In the article written in 1815 [7] he defines the *degree* of a substitution as the first power of the substitution that reduces to identity, but later he also defines *order* in this way and uses the word "order" rather than "degree" in his subsequent work. He uses both the term *unity* and *identical substitution*. A *transposition* is defined and the terms *similar, regular, inverse* and *permutable substitutions* are found with their present day significance. A group he calls a "system of conjugate substitutions," which may be transitive or intransitive. A transitive imprimitive group is "transitive complex." The *order* of a system of conjugate substitutions is the number of substitutions that it contains. The term "divisor indicatif" is also found for the order of a group. Cauchy frequently transfers his definition of order to the theory of equations and speaks of the number of equal values a rational function can assume when the variables are permuted in every possible way. The number of distinct values of such a function is the *index*. In the early article the order of operation is from left to right, but in all his later work he reverses the order and operates from right to left.

The Memoir of 1815

In the paper published in 1815 the one theorem which is of special interest is that the number of distinct values of a non-symmetric function of degree n cannot be less than the largest prime that divides n, without becoming equal to 2. This theorem is proved. He states the special cases, that if the degree of a function is a prime number greater than 2, the number of distinct values cannot be less than the degree; and that if the degree is 6 the number of distinct values cannot be less than 6. He makes special reference to the functions belonging to

1. the intransitive group of degree 6 and order 36;
2. the transitive imprimitive group of degree 6 and order 72;
3. the intransitive group of degree 6 and order 48;
4. the symmetric group of degree 5 considered as an intransitive group of degree 6.

The functions are as follows:

1. $a_1a_2a_3 + 2a_4a_5a_6$,
2. $a_1a_2a_3 + a_4a_5a_6$,
3. $a_1a_2a_3a_4 + a_5a_6$,
4. $a_1a_2a_3a_4a_5 + a_6$.

The Memoir of 1844

There is much more of importance in his 1844 *Exercises* [4]. The first theorem proved is that every substitution similar to a given substitution P is the product of three factors, the extremes of which are the inverse of each other and the middle term of which is P. Conversely, every product of three factors, the first and last of which are the inverse of each other, is similar to the middle term P.

An important formula developed is for the number of substitutions similar to a given substitution. If ω is the number required, n the total number of letters and P the given substitution, composed of f cycles of order a, g cycles of order b, h cycles of order c, etc., and r the number of letters fixed in P, then

$$\omega = \frac{n!}{(f!)(g!)(h!)\cdots(r!)a^f b^g c^h \cdots}.$$

Among other theorems that he proves in this article are the following:

1. If P is a substitution of order i, h any number and θ the highest common factor of h and i, then P^h is of order i/θ.

2. If P is a substitution of order i, the substitutions among the powers of P that are of order i are the powers of P whose indices are prime to i. These substitutions are likewise similar to P, and hence the number of substitutions similar to P among the powers of P are in number equal to the number of numbers less than i and prime to it.

3. Let P be any substitution, regular or irregular; let i be its order and p be any prime factor of i. Then a value for i can always be found such that P^i is a substitution of order p.

4. A substitution and its inverse are always similar.

5. The powers of a cycle constitute the totality of substitutions that transform the given cycle into itself.

6. The order of a system of conjugate substitutions is divisible by the order of each substitution.

7. Two permutable substitutions, with no common power but the identity, together generate a system of conjugate substitutions whose order is the product of the orders of the two substitutions.

8. If $1, P_1, P_2, \cdots, P_{a-1}$ and $1, Q_1, Q_2, \cdots, Q_{b-1}$ are two groups, one of order a and the other of order b, which are permutable and have no common terms but the identity, then the group generated by these groups is of order ab.

9. The converse of (8).

10. If P and Q are two substitutions, one of order a and the other of order b, and if the two series $Q, PQ, P^2Q, \cdots, P^{a-1}Q$ and $Q, QP, QP^2, \cdots, QP^{a-1}$ are made up of the same terms in the same or different order, then the cyclic group generated by P is permutable with the cyclic group generated by Q.

A most important element in establishing Cauchy's claim as the founder of group theory is the proof which he gives of the fundamental theorem that if m is the order of a group and p is any prime which divides m, there is at least one substitution of order p. Galois had stated the theorem but had not proved it. Its importance is due to the fact that it is the first step toward Sylow's theorem which appeared nearly thirty years later.

In this memoir there are a number of specific groups mentioned, although no enumeration of groups of special degrees is attempted. The substitutions of several groups are written out. Among them are

1. the octic,
2. the four-group,
3. the holomorph of the cyclic group of order five,
4. the intransitive group of degree 6 and order 9,
5. the intransitive group of degree 6 and order 16.

In addition to these groups, the substitutions of which are given, several other groups are mentioned, among them

1. the holomorph of order 42,
2. the group of degree 7 and order 21,
3. the holomorph of the cyclic group of order 9.

There is also mentioned a group of degree 9 and order 27, generated by the two substitutions

$$P \equiv x_0 x_3 x_6 \cdot x_1 x_4 x_7 \cdot x_2 x_5 x_8 \quad \text{and}$$
$$Q \equiv x_1 x_2 x_4 x_8 x_5 \cdot x_3 x_6.$$

This is obviously impossible but a little study of his method reveals that he should have used

$$Q \equiv x_1 x_3 x_4 x_8 x_7 x_5 \cdot x_3 x_6,$$

and that in order to get the order of the group he has incorrectly applied the theorems he was announcing and illustrating. The order of the group should be 18. The two theorems, which he was illustrating may be stated as follows:

1. If P is a substitution on n letters, $P = x_0 x_1 x_2 \cdots x_{n-1}$ and if r is any number prime to n and if Q is a substitution derived from P by replacing each letter by another whose subscript is r times its own, then for any values of h and k,

$$Q^k P^h = P^{hr^k} Q^k.$$

2. If in the above theorem we take v to be any divisor of n, distinct from unity, and let R be any substitution that replaces any letter x_l by the letter with the subscript $l + v$, then $Q^k R^h$ generates a group of order vi, where i is the smallest value of k that satisfies the congruence $r^k \equiv i \pmod{n}$.

Since in the present instance r is 2, i must be 6. The order of the group is then $vi = 3 \cdot 6 = 18$, whereas Cauchy seems to take $i = 9$ and derives the order of the group $vi = 3 \cdot 9 = 27$.

The Memoirs of 1845–1846

Some of the material in the memoirs [2] published in the *Comptes Rendus* had already appeared in [4], but is treated more extensively here. There is much, however, in the later articles that is not touched in [4]. It is in the later memoirs that he first defines what he means by a system of conjugate substitutions. The definition is as follows:

> I shall call derived substitutions all that can arise from the given substitutions by multiplying them one or more times by each other or by themselves; and the given substitutions together with all the derived substitutions form what I shall call a system of conjugate substitutions.

Then follow the general theorems:

1. If i is the order of a substitution P and $a, b, c, \ldots$ are the prime factors of i, then the substitution P and its powers form a group generated by the substitutions

$$P^{i/a}, P^{i/b}, P^{i/c}, \ldots.$$

2. If $P, Q, R, S, \cdots$ are the substitutions that leave a given function Ω invariant, they form a group whose order is the number of equal values of Ω when the variables are permuted in every possible way.

3. The order of a transitive group of degree n is n times the order of the subgroup that leaves one letter fixed.

4. If Ω is a transitive function on n letters, and if m is the index of the corresponding transitive group under the symmetric group of degree n, then m will likewise be the index of the subgroup that leaves one letter fixed under the symmetric group of degree $n - 1$.

Intransitive groups

The general subject which Cauchy treats next is that of intransitive groups. He implicitly divides intransitive groups into two classes. The first class includes those in which the systems are independent, that is, in our terminology, those in which the systems of transitivity are united by direct product. The second class includes those in which the systems are dependent, that is, those in which the systems are united by some sort of isomorphism. The concept of isomorphism is however only implied. He states first that an intransitive group is formed by the combination in some way of transitive constituents. He then develops some interesting formulas for the order of intransitive groups. If the systems are independent, then the order M is the direct product of $A, B, C, \cdots$ the orders of the transitive constituents and the index will be $\frac{n!}{ABC\cdots}$, where n is the total number of letters. This same formula holds also when the systems are not independent if $A, B, C, \cdots$ are defined in a special way. These definitions are as follows:

- let A be the order of the first transitive constituent;
- let B be the number of substitutions involving letters in the second system without involving any of the first;
- let C be the number of substitutions involving letters in the third system, without involving any of the first or second, and so on.

With these definitions of $A, B, C, D, \ldots$ the order of the intransitive group is $M = ABC \cdots$.

Imprimitive groups

Cauchy next turns to imprimitive groups. Here he confines himself almost entirely to the consideration of those imprimitive groups which are simply transitive and which have for the subgroup that leaves one letter fixed the direct product of the transitive constituents. In regard to this particular type of imprimitive groups, he gives several theorems which are of interest, even though he does not touch upon the broader and more general principles.

If in a simply transitive group the subgroup that leaves one letter fixed is an intransitive group formed by the direct product of its transitive constituents, the group is imprimitive. He considers the special cases $a > n/2, a < n/2, a = n/2$ in which a is the number of letters in the largest transitive constituent and n is the degree of the group. The only theorems which he enunciates that apply to imprimitive groups in general are:

1. The number of letters in the systems of imprimitivity must be a divisor of the degree.
2. If A is the number of substitutions that permute the variables within the systems and K is the number of ways the k systems can be permuted, then the order of the group is KA^k.

Symmetric groups

Cauchy then considers briefly symmetric groups. One theorem which he states with its corollaries is as follows:

- If a transitive group of degree n has a symmetric subgroup of degree a, where $a > n/2$, then the group is symmetric on n letters.
- For $n > 2$, a transitive group of degree n is symmetric if it contains a symmetric subgroup of degree $n - 1$.
- If $n > 3$, a transitive group of degree n is symmetric if it contains a symmetric subgroup of degree $n - 2$. The special case $n = 4$ is excepted; for the octic group, belonging to the function $\Omega = xy + zu$, is symmetric on two letters, yet not symmetric on all four.
- If $n > 4$, a transitive group of degree n is symmetric if it contains a symmetric subgroup of degree $n - 3$. The case $n = 6$ is excluded. A group requiring this exception is the imprimitive group of degree 6 and order 72 which contains the symmetric group of order 6.
- If $n > 5$, a transitive group of degree n is symmetric if it contains a symmetric subgroup of degree $n - 4$. If $n = 6$ or $n = 8$, this does not hold. Cauchy gives as examples of the case $n = 6$ the imprimitive groups of degree 6 and orders 72 and 48.

Some implied theorems

One of the most important and interesting parts of Cauchy's work on group theory is that in which he develops some complicated formulas from which we may easily deduce the theorem that the average number of letters in the substitutions of a transitive group is $n - 1$. Cauchy himself does not enunciate the theorem, although he brings the proof to the point where only the final statement is necessary to complete it. The formula for the case where the group is simply transitive is as follows:

$$M = 2H_{n-2} + 3H_{n-3} + \cdots + (n-2)H_2 + n,$$

where M is the order of the group, n its degree and H_{n-r} the number of substitutions involving $n - r$ letters.

Further development of these formulas leads to the two conclusions:

1. If G is l-fold transitive of degree n, the number of substitutions involving $n-l+1$ letters is equal to or greater than
$$\frac{n(n-1)\cdots(n-l+2)}{(n-l)l!};$$

2. If G is simply transitive, of degree n, the number of substitutions involving n letters is equal to or greater than $n-1$.

We find here the assertion that if in an l-fold transitive group $l+1$ letters are left fixed, all are left fixed. That this is false is demonstrated by considering the imprimitive group of degree 6 and order 72. It is simply transitive and hence $l=1$; but $l+1=2$ letters may be left fixed without all being so.

The theorem that the order of the holomorph of a cyclic group is the product of the orders of the cyclic group and its group of isomorphisms is found implicitly in Cauchy. He gives two theorems relative to this point.

1. Let $P = x_0 x_1 x_2 \cdots x_{n-1}$ be a substitution of order n. Let r be a primitive root of the modulus n, and I be the smallest of the indices of unity belonging to the base r. Then let Q be the substitution that replaces x_i by $x_{ri}j$. The order of Q will be I and the order of the group will be nI. This I is nothing else than the order of the group of isomorphisms of the cyclic group. If n is a power of a prime p, then $I = n(1-1/p)$. When n is a prime then $I = n-1$.

2. With the same hypothesis as in (1), P^a and Q^b, where a and b are divisors of n and I respectively, generate a group of order nI/ab.

Enumeration of orders

Cauchy was the first to attempt an enumeration of the possible orders of groups. This he did with a fair degree of accuracy up to and including the sixth degree. The enumeration including degree 5 is correct and complete, but several errors occur in the enumeration of those of degree 6. For instance, 150 is given as the index of a group of degree 6 under the symmetric group, although 6! is not a multiple of 150.

Cauchy goes back to his original distinction between imprimitive groups with heads which are direct products and those with heads formed by isomorphisms. He gives as the possible orders of groups of degree 6 with heads the direct products of the transitive constituents 72, 48, 24, 18, 16, and 8. The last two numbers are clearly impossible for there is no transitive group of degree 6 and order 8 or 16. The orders of the groups possible when the head is formed with isomorphisms are given as 6, 12, 4, while there is no imprimitive group of degree 6 and order 4. According to his classification there should also be included in this last enumeration 24 and 18. A complete omission occurs in the list of imprimitive groups since the groups of order 36 with two systems of imprimitivity are not mentioned. Otherwise the enumeration through degree 6 is correct and complete.

Errors of Cauchy

The errors in Cauchy's work on group theory may he divided into two classes. The first and smaller class includes a number of serious errors in logic. The second class includes a large number of minor errors, some typographical, others arising through careless statement. All of the serious errors found have been noted in this paper with the exception of one to which attention has already been called (see [8]). He states an erroneous theorem on imprimitive groups in the following form.

> If G_1 is a subgroup composed of all the substitutions that omit a given letter in a simply transitive group, then G is imprimitive unless all the transitive constituents of G_1 are of the same degree.

Conclusion

In conclusion we may say that the foundation period in the history of group theory includes the time from Lagrange to Cauchy inclusive; that at the beginning of this period group theory was a means to an end and not an end in itself. Lagrange and Ruffini thought of substitution groups only in so far as they led to practical results in the theory of equations. Galois, while broadening and deepening the application to the theory of equations, may be considered as taking the initial step toward abstract group theory. In Cauchy, while a group is still spoken of as the substitutions that leave a given function invariant, and the order of a group is still thought of as the number of equal values which the function can assume when the variables are permuted in every possible way, nevertheless quite as often a group is a system of conjugate substitutions and its relation to any function is entirely ignored.

In Cauchy's work it is to be noted that use is made of all the concepts originated by the earlier writers in substitution theory except those of Galois, a number of which would have proved powerful instruments in his hands, notably the idea of the invariant subgroup of which he makes no explicit use.

Because of the important theorems Cauchy proved, because of the break which he made in separating the theory of substitutions from the theory of equations, and because of the importance that he attached to the theory itself, he deserves the credit as the founder of group theory.

References

1. H. Burkhardt, Die Anfinge der Gruppen Theorie und Paolo Ruffini, *Zeitschrift für Math. u. Phys.* **37** (1892), 122–160.
2. Auguste Cauchy, papers published in *Comptes Rendus* in 1845–1846, *Oeuvres de A. Cauchy*, First series, Vols. 9 and 10.
3. Auguste Cauchy, *Journal de l'École Polytechnique* **10** (1815), 1–112.
4. Auguste Cauchy, *Exercises d'analyse et de physique mathematique*, 1844; reprinted in *Oeuvres Complètes*, Gauthier-Villars, 1882–1974, pp. 171–282.
5. Évariste Galois, *Oeuvres Mathématiques d'Éveriste Galois*, Gauthier-Villars, 1906–1907, pp. 26 ff.
6. Évariste Galois, *Manuscrits de Évariste Galois*, Gauthier-Villars, 1908, pp. 39 ff.
7. Joseph Lagrange, *Reflexions sur la Résolution Algébrique des Equations, Mémoires of the Academy of Science at Berlin*, 1770–1771; in *Oeuvres de Lagrange*, Vol. 3, pp. 204–420.
8. G. A. Miller, *Bibliotheca Mathematica* **10** (1910), 321.
9. Paolo Ruffini, *Teoria Generale Della Equazioni*, Stamperia di S. Tommaso d'Aquino, 1799.

The Evolution of Group Theory: A Brief Survey

ISRAEL KLEINER

Mathematics Magazine **59** (1986), 195–215

This article gives a brief sketch of the evolution of group theory. It derives from a firm conviction that the history of mathematics can be a useful and important integrating component in the teaching of mathematics. This is not the place to elaborate on the role of history in teaching, other than perhaps to give one relevant quotation (C. H. Edwards [11]):

> Although the study of the history of mathematics has an intrinsic appeal of its own, its chief raison d'être is surely the illumination of mathematics itself. For example the gradual unfolding of the integral concept from the volume computations of Archimedes to the intuitive integrals of Newton and Leibniz and finally to the definitions of Cauchy, Riemann and Lebesgue—cannot fail to promote a more mature appreciation of modern theories of integration.

The presentation in one article of the evolution of so vast a subject as group theory necessitated severe selectivity and brevity. It also required omission of the broader contexts in which group theory evolved, such as wider currents in abstract algebra, and in mathematics as a whole. (We will note *some* of these interconnections shortly.) We trust that enough of the essence and main lines of development in the evolution of group theory have been retained to provide a useful beginning from which the reader can branch out in various directions. For this the list of references will prove useful.

The reader will find in this article an outline of the origins of the main concepts, results, and theories discussed in a beginning course on group theory. These include, for example, the concepts of (abstract) group, normal subgroup, quotient group, simple group, free group, isomorphism, homomorphism, automorphism, composition series, direct product; the theorems of J. L. Lagrange, A.-L. Cauchy, A. Cayley, C. Jordan–O. Hölder; the theories of permutation groups and of abelian groups. At the same time we have tried to balance the technical aspects with background information and interpretation.

Our survey of the evolution of group theory will be given in several stages, as follows:

1. Sources of group theory.

2. Development of "specialized" theories of groups.

3. Emergence of abstraction in group theory.

4. Consolidation of the abstract group *concept*; dawn of abstract group *theory*.

5. Divergence of developments in group theory.

Before dealing with each stage in turn, we wish to mention the context within mathematics as a whole, and within algebra in particular, in which group theory developed. Although our story concerning the evolution of group theory begins in 1770 and extends to the twentieth century, the major developments occurred in the nineteenth century. Some of the general mathematical features of that century which had a bearing on the evolution of group theory are:

(a) an increased concern for rigor;

(b) the emergence of abstraction;

(c) the rebirth of the axiomatic method;

(d) the view of mathematics as a human activity, possible without reference to, or motivation from, physical situations.

Each of these items deserves extensive elaboration, but this would go beyond the objectives (and size) of this paper.

Up to about the end of the eighteenth century, algebra consisted (in large part) of the study of solutions of polynomial equations. In the twentieth century, algebra became a study of abstract, axiomatic systems. The transition from the so-called classical algebra of polynomial equations to the so-called modern algebra of axiomatic systems occurred in the nineteenth century. In addition to group theory, there emerged the structures of commutative rings, fields, non-commutative rings, and vector spaces. These developed alongside, and sometimes in conjunction with, group theory. Thus Galois theory involved both groups and fields; algebraic number theory contained elements of group theory in addition to commutative ring theory and field theory; group representation theory was a mix of group theory, non-commutative algebra, and linear algebra.

1 Sources of group theory

There are four major sources in the evolution of group theory. They are (with the names of the originators and dates of origin):

(a) Classical algebra (J. L. Lagrange, 1770)

(b) Number theory (C. F. Gauss, 1801)

(c) Geometry (F. Klein, 1874)

(d) Analysis (S. Lie, 1874; H. Poincaré and F. Klein, 1876)

We deal with each in turn.

(a) **Classical algebra** (J. L. Lagrange, 1770) The major problems in algebra at the time (1770) that Lagrange wrote his fundamental memoir *Réflexions sur la Résolution Algébrique des Équations* concerned polynomial equations. There were *theoretical* questions dealing with the existence and nature of the roots (e.g., Does every equation have a root? How many roots are there? Are they real, complex, positive, negative?), and *practical* questions dealing with methods for finding the roots. In the latter instance there were exact methods and approximate methods. In what follows we mention exact methods.

The Babylonians knew how to solve quadratic equations (essentially by the method of completing the square) around 1600 B.C. Algebraic methods for solving the cubic and the quartic were given around 1540. One of the major problems for the next two centuries was the algebraic solution of the quintic. This is the task Lagrange set for himself in his paper of 1770.

In his paper Lagrange first analyzes the various known methods (devised by F. Viète, R. Descartes, L. Euler, and E. Bézout) for solving cubic and quartic equations. He shows that the common feature of these methods is the reduction of such equations to auxiliary equations—the so-called *resolvent equations*. The latter are one degree lower than the original equations. Next Lagrange attempts a similar analysis of polynomial equations of arbitrary degree n. With each such equation he associates a resolvent equation as follows: let $f(x)$ be the original equation, with roots $x_1, x_2, \ldots, x_n$. Pick a rational function $R(x_1, x_2, \ldots, x_n)$ of the roots and coefficients of $f(x)$. (Lagrange describes methods for doing so.) Consider the different values which $R(x_1, x_2, \ldots, x_n)$ assumes under all the $n!$ permutations of the roots $x_1, x_2, \ldots, x_n$ of $f(x)$. If these are denoted by $y_1, y_2, \ldots, y_k$ then the resolvent equation is given by

$$g(x) = (x - y_1) \cdot (x - y_2) \cdots (x - y_k).$$

(Lagrange shows that k divides $n!$—the source of what we call *Lagrange's theorem* in group theory.) For example, if $f(x)$ is a quartic with roots x_1, x_2, x_3, x_4, then $R(x_1, x_2, x_3, x_4)$ may be taken to be $x_1 x_2 + x_3 x_4$, and this function assumes three distinct values under the 24 permutations of x_1, x_2, x_3, x_4. Thus the resolvent equation of a quartic is a cubic. However, in carrying over this analysis to the quintic, he finds that the resolvent equation is of degree six!

Although Lagrange did not succeed in resolving the problem of the algebraic solvability of the quintic, his work was a milestone. It was the first time that an association was made between the solutions of a polynomial equation and the permutations of its roots. In fact, the study of the permutations of the roots of an equation was a cornerstone of Lagrange's general theory of algebraic equations. This, he speculated, formed "the true principles for the solution of equations." (He was, of course, vindicated in this by E. Galois.) Although Lagrange speaks of permutations without considering a *calculus* of permutations (e.g., there is no consideration of their composition or closure), it can be said that the germ of the group concept (as a group of permutations) is present in his work. For details see [12], [16], [19], [25], [33].

(b) **Number Theory.** (C. F. Gauss, 1801) In the *Disquisitiones Arithmeticae* of 1801 Gauss summarized and unified much of the number theory that preceded him. The work also suggested new directions which kept mathematicians occupied for the entire century. As for its impact on group theory, the *Disquisitiones* may be said to have initiated the theory of finite abelian groups. In fact, Gauss established many of the significant properties of these groups without using any of the terminology of group theory. The groups appear in four different guises: the additive group of integers modulo m, the multiplicative group of integers relatively prime to m (modulo m), the group of equivalence classes of binary quadratic forms, and the group of nth roots of unity. And though these examples appear in number-theoretic contexts, it is as abelian groups that Gauss treats them, using what are clear prototypes of modern algebraic proofs.

For example, considering the non-zero integers modulo p (p a prime), Gauss shows that they are all powers of a single element; i.e., that the group $\mathbb{Z}_p^*$ of such integers is cyclic. Moreover, he determines the number of generators of this group (he shows that it is equal to $\phi(p-1)$, where ϕ is Euler's ϕ-function). Given any element of $\mathbb{Z}_p^*$ he defines the order of the element (without using the terminology) and shows that the order of an element is a divisor of $p-1$. He then uses this result to prove P. Fermat's *little theorem*—namely, that $a^{p-1} \equiv 1 \pmod{p}$ if p does not divide a, thus employing group-theoretic ideas to prove number-theoretic results. Next he shows that if t is a positive integer which divides $p-1$, then there exists an element in $\mathbb{Z}_p^*$ whose order is t—essentially the converse of Lagrange's theorem for cyclic groups.

Concerning the nth roots of 1 (which he considers in connection with the cyclotomic equation), he shows that they too form a cyclic group. In connection with this group he raises and answers many of the same questions he raised and answered in the case of $\mathbb{Z}_p^*$.

The problem of representing integers by binary quadratic forms goes back to Fermat in the early seventeenth century. (Recall his theorem that every prime of the form $4n+1$ can be represented as a sum of two squares $x^2 + y^2$.) Gauss devotes a large part of the *Disquisitiones* to an exhaustive study of binary quadratic forms and the representation of integers by such forms. (A *binary quadratic form* is an expression of the form $ax^2 + bxy + cy^2$, with a, b, c integers.) He defines a composition on such forms, and remarks that if K and K^1 are two such forms one may denote their composition by $K + K^1$. He then shows that this composition is associative and commutative, that there exists an identity, and that each form has an inverse, thus verifying all the properties of an abelian group.

Despite these remarkable insights one should not infer that Gauss had the concept of an abstract group, or even of a finite abelian group. Although the arguments in the *Disquisitiones* are quite general, each of the various types of groups he considers is dealt with separately—there is no unifying group-theoretic method which he applies to all cases. For details see [5], [9], [25], [30], [33].

(c) **Geometry** (F. Klein, 1872). We are referring here to Klein's famous and influential (but see [18]) lecture entitled *A Comparative Review of Recent Researches in Geometry*, which he delivered in 1872 on the occasion of his admission to the faculty of the University of Erlangen. The aim of this so-called *Erlanger Program* was the classification of geometry as the study of invariants under various groups of transformations. Here there appear groups such as the projective group, the group of rigid motions, the group of similarities, the hyperbolic group, the elliptic groups, as well as the geometries associated with them. (The affine group was not mentioned by Klein.) Now for some background leading to Klein's Erlanger Program.

The nineteenth century witnessed an explosive growth in geometry, both in scope and in depth. New geometries emerged: projective geometry, non-Euclidean geometries, differential geometry, algebraic geometry, n-dimensional geometry, and Grassmann's geometry of extension. Various geometric methods competed for supremacy: the synthetic versus the analytic, the metric versus the projective. At mid-century, a major problem had arisen—namely, the classification of the relations and inner connections among the different geometries and geometric methods.

This gave rise to the study of *geometric relations*, focusing on the study of properties of figures invariant under transformations. Soon the focus shifted to a study of the transformations themselves. Thus the study of the geometric relations of figures became the study of the associated transformations. Various types of transformations (e.g., collineations, circular transformations, inversive transformations, affinities) became the objects of specialized studies. Subsequently, the logical connections among transformations were investigated, and this led to the problem of classifying transformations and eventually to Klein's group-theoretic synthesis of geometry.

Klein's use of groups in geometry was the final stage in bringing order to geometry. An intermediate stage was the founding of the first major theory of classification in geometry, beginning in the 1850s, the Cayley-Sylvester Invariant Theory. Here the objective was to study invariants of forms under transformations of their variables. This theory of classification, the precursor of Klein's Erlanger Program, can be said to be *implicitly* group-theoretic. Klein's use of groups in geometry was, of course, explicit. (For a thorough analysis of implicit group-theoretic thinking in geometry leading to Klein's Erlanger Program, see [33].) In the next section (2(c)) we will note the significance of Klein's Erlanger Program (and his other works) for the evolution of group theory. Since the Program originated a hundred years after Lagrange's work and eighty years after Gauss' work, its importance for group theory can best be appreciated after a discussion of the evolution of group theory beginning with the works of Lagrange and Gauss and ending with the period around 1870.

what we call *Lie groups* today. Such a group is represented by the transformations

$$x'_i = f_i(x_1, x_2, \ldots, x_n, a_1, a_2, \ldots, a_n),$$
$$i = 1, 2, \ldots, n,$$

where the f_i are analytic functions in the x_i and a_i (the a_i are parameters, with both x_i and a_i real or complex). For example, the transformations given by

$$x' = \frac{ax + b}{cx + d}, \text{ where } a, b, c, d \text{ are real numbers}$$
$$\text{and } ad - bc \neq 0,$$

define a continuous transformation group.

Sophus Lie (1842–1899

Lie thought of himself as the successor of N. H. Abel and Galois, doing for differential equations what they had done for algebraic equations. His work was inspired by the observation that almost all the differential equations which had been integrated by the older methods remain invariant under continuous groups that can be easily constructed. He was then led to consider, in general, differential equations that remain invariant under a given continuous group and to investigate the possible simplifications in these equations which result from the known properties of the given group (cf. Galois theory). Although Lie did not succeed in the actual formulation of a "Galois theory of differential equations", his work was fundamental in the subsequent formulation of such a theory by E. Picard (1883/1887) and E. Vessiot (1892).

Felix Klein (1849–1925)

(d) Analysis (S. Lie, 1874; H. Poincaré and F. Klein, 1876) In 1874 Lie introduced his general theory of (continuous) transformation groups—essentially

Poincaré and Klein began their work on *automorphic functions* and the groups associated with them around 1876. Automorphic functions (which are generalizations of the circular, hyperbolic, elliptic, and other functions of elementary analysis) are functions of a complex variable z, analytic in some domain D, which are invariant under the group of transformations

$$z' = \frac{az+b}{cz+d}$$

(a, b, c, d real or complex and $ad - bc \neq 0$)

or under some subgroup of this group. Moreover, the group in question must be *discontinuous* (i.e., any compact domain contains only finitely many transforms of any point). Examples of such groups are the modular group (in which a, b, c, d are integers and $ad - bc = 1$), which is associated with the elliptic modular functions, and Fuchsian groups (in which a, b, c, d are real and $ad - bc = 1$) associated with the Fuchsian automorphic functions. As in the case of Klein's Erlanger Program, we will explore the consequences of these works for group theory in section 2(c).

Henri Poincaré (1854–1912)

2 Development of "specialized" theories of groups

In Section 1 we outlined four major sources in the evolution of group theory. The first source—classical algebra—led to the theory of permutation groups; the second source—number theory—led to the theory of abelian groups; the third and fourth sources — geometry and analysis—led to the theory of transformation groups. We will now outline some developments within these specialized theories.

(a) **Permutation groups** As noted earlier, Lagrange's work of 1770 initiated the study of permutations in connection with the study of the solution of equations. It was probably the first clear instance of implicit group-theoretic thinking in mathematics. It led directly to the works of P. Ruffini, Abel, and Galois during the first third of the nineteenth century, and to the concept of a permutation group.

Ruffini and Abel proved the unsolvability of the quintic by building upon the ideas of Lagrange concerning resolvents. Lagrange showed that a necessary condition for the solvability of the general polynomial equation of degree n is the existence of a resolvent of degree less than n. Ruffini and Abel showed that such resolvents do not exist for $n > 4$. In the process they developed a considerable amount of permutation theory. (See [1], [9], [19], [23], [24], [25], [30], [33] for details.) It was Galois, however, who made the fundamental conceptual advances, and who is considered by many as the founder of (permutation) group theory.

Galois's aim went well beyond finding a method for the solvability of equations. He was concerned with gaining insight into general principles, dissatisfied as he was with the methods of his predecessors: "From the beginning of this century," he wrote, "computational procedures have become so complicated that any progress by those means has become impossible" ([19], p. 92).

Galois recognized the separation between *Galois theory* (i.e., the correspondence between fields and groups) and its application to the solution of equations, for he wrote that he was presenting "the general principles and just one application" of the theory ([19], p. 42). "Many of the early commentators on Galois theory failed to recognize this distinction, and this led to an emphasis on applications at the expense of the theory" (Kiernan, [19]).

Galois was the first to use the term *group* in a technical sense—to him it signified a collection of permutations closed under multiplication: "if one has in the

Evariste Galois (1811–1832)

same group the substitutions S and T one is certain to have the substitution ST" ([33], p. 111). He recognized that the most important properties of an algebraic equation were reflected in certain properties of a group uniquely associated with the equation—"the group of the equation." To describe these properties he invented the fundamental notion of *normal subgroup* and used it to great effect. While the issue of resolvent equations preoccupied Lagrange, Ruffini, and Abel, Galois's basic idea was to bypass them, for the construction of a resolvent required great skill and was not based on a clear methodology. Galois noted instead that the existence of a resolvent was equivalent to the existence of a normal subgroup of prime index in the group of the equation. This insight shifted consideration from the resolvent equation to the group of the equation and its subgroups. Galois defines the group of an equation as follows ([19], p. 80):

> Let an equation be given, whose m roots are $a, b, c, \ldots$. There will always be a group of permutations of the letters $a, b, c, \ldots$ which has the following property:

1. that every function of the roots, invariant under the substitutions of that group, is rationally known [i.e., is a rational function of the coefficients and any adjoined quantities].

2. conversely, that every function of the roots, which can be expressed rationally, is invariant under these substitutions.

The definition says essentially that the group of the equation consists of those permutations of the roots of the equation which leave invariant all relations among the roots over the field of coefficients of the equation—basically the definition we would give today. Of course the definition does not guarantee the existence of such a group, and so Galois proceeds to demonstrate it. Galois next investigates how the group changes when new elements are adjoined to the *ground field F*. His treatment is amazingly close to the standard treatment of this matter in a modern algebra text.

Galois's work was slow in being understood and assimilated. In fact, while it was done around 1830, it was published posthumously in 1846, by J. Liouville. Beyond his technical accomplishments, Galois (Wussing, [33])

> challenged the development of mathematics in two ways. He discovered, but left unproved, theorems which called for proofs based on new, sophisticated concepts and calculations. Also, the task of filling the gaps in his work necessitated a fundamental clarification of his methods and their group theoretical essence.

For details see [12], [19], [23], [25], [29], [31], [33].

The other major contributor to permutation theory in the first half of the nineteenth century was Cauchy. In several major papers in 1815 and 1844 Cauchy inaugurated the theory of permutation groups as an autonomous subject. (Before Cauchy, permutations were not an object of independent study but rather a useful device for the investigation of solutions of polynomial equations.) Although Cauchy was well aware of the work of Lagrange and Ruffini (Galois's work was not yet published at the time), Wussing suggests that Cauchy "was definitely not inspired directly by the contemporary group-theoretic formulation of the solution of algebraic equations" [33].

In these works Cauchy gives the first systematic development of the subject of permutation groups. In the 1815 papers Cauchy uses no special name for sets of permutations closed under multiplication. However, he recognizes their importance and gives a name to the number of elements in such a closed set, calling it "diviseur indicatif". In the 1844 paper he defines the concept of a group of permutations generated by certain elements ([22], p. 65):

> Given one or more substitutions involving some or all of the elements $x, y, z, \ldots$ I call the products of these substitutions, by themselves or by any other, in any order, *derived* substitutions. The given substitutions, together with the derived ones, form what I call a *system of conjugate substitutions*.

In these works, which were very influential, Cauchy makes several lasting additions to the terminology, notation, and results of permutation theory. For example, he introduces the permutation notation

$$\begin{pmatrix} x & y & z \\ x & z & y \end{pmatrix}$$

in use today, as well as the cyclic notation for permutations; defines the product of permutations, the degree of a permutation, cyclic permutation, transposition; recognizes the identity permutation as a permutation; discusses what we would call today the *direct product* of two groups; and deals with the alternating groups extensively. Here is a sample of some of the results he proves.

(i) Every even permutation is a product of 3-cycles.

(ii) If p (prime) is a divisor of the order of a group, then there exists a subgroup of order p. (This is known today as *Cauchy's theorem*, though it was stated without proof by Galois.)

(iii) He determined all subgroups of S_3, S_4, S_5, S_6 (making an error in S_6).

(iv) All permutations which commute with a given one form a group (the *centralizer* of an element).

It should be noted that all these results were given and proved in the context of permutation groups. For details see [6], [8], [23], [24], [25], [33].

The crowning achievement of these two lines of development—a symphony on the grand themes of Galois and Cauchy—was Jordan's important and influential *Traité des Substitutions et des Équations Algébriques* of 1870. Although the author states in the preface that

> the aim of the work is to develop Galois's method and to make it a proper field of study, by showing with what facility it can solve all principal problems of the theory of equations,

it is in fact group theory *per se*—not as an offshoot of the theory of solvability of equations—which forms the central object of study.

Camille Jordan (1838–1922)

The striving for a mathematical synthesis based on key ideas is a striking characteristic of Jordan's work, as well as that of a number of other mathematicians of the period (e.g., F. Klein). The concept of a (permutation) group seemed to Jordan to provide such a key idea. His approach enabled him to give a unified presentation of results due to Galois, Cauchy, and others. His application of the group concept to the theory of equations, algebraic geometry, transcendental functions, and theoretical mechanics was also part of the unifying and synthesizing theme. "In his book Jordan wandered through all of algebraic geometry, number theory, and function theory in search of interesting permutation groups" (Klein, [20]). In fact, the aim was a survey of all of mathematics by areas in which the theory of permutation groups had been applied or seemed likely to be applicable. "The work represents ... a review of the whole of contemporary mathematics from the standpoint of the occurrence of group-theoretic thinking in permutation-theoretic form" (Wussing, [33]).

The *Traité* embodied the substance of most of Jordan's publications on groups up to that time (he wrote over 30 articles on groups during the period 1860–1880) and directed attention to a large number of difficult problems, introducing many fundamental concepts. For example, Jordan makes explicit the notions of isomorphism and homomorphism for (substitution) groups, introduces the term *solvable group* for the first time in a technical sense, introduces the concept of a composition series, and proves part of the Jordan–Hölder theorem, namely, that the indices in two composition series are the same (the concept of a quotient group was not explicitly recognized at this time); and he undertakes a very thorough study of transitivity and primitivity for permutation groups, obtaining results most of which have not since been superseded. Jordan also gives a proof that A_n is simple for $n > 4$.

An important part of the treatise is devoted to a study of the *linear group* and some of its subgroups. In modern terms these constitute the so-called *classical groups*—namely, the general linear group, the unimodular group, the orthogonal group, and the symplectic group. Jordan considers these groups only over finite fields, and proves their simplicity in certain cases. It should be noted, however, that he considers these groups as permutation groups rather than groups of matrices or linear transformations (see [29], [33]).

Jordan's *Traité* is a landmark in the evolution of group theory. His permutation-theoretic point of view, however, was soon to be overtaken by the conception of a group as a group of transformations (see (c) below):

> The *Traité* marks a pause in the evolution and application of the permutation-theoretic group concept. It was an expression of Jordan's deep desire to effect a conceptual synthesis of the mathematics of his time. That he tried to achieve such a synthesis by relying on the concept of a permutation group, which the very next phase of mathematical development would show to have been unduly restricted, makes for both the glory and the limitations of the *Traité* ... (Wussing, [33]).

For details see [9], [13], [19], [20], [22], [24], [29], [33].

(b) **Abelian groups** As noted earlier, the main source for abelian group theory was number theory, beginning with Gauss's *Disquisitiones Arithmeticae*. In contrast to permutation theory, group-theoretic modes of thought in number theory remained implicit until about the last third of the nineteenth century. Until that time no explicit use of the term *group* was made, and there was no link to the contemporary, flourishing theory of permutation groups. We now give a sample of some implicit group-theoretic work in number theory, especially in algebraic number theory.

Algebraic number theory arose in connection with Fermat's conjecture concerning the equation $x^n + y^n = z^n$, Gauss's theory of binary quadratic forms, and higher reciprocity laws. Algebraic number fields and their arithmetical properties were the main objects of study. In 1846 G. L. Dirichlet studied the units in an algebraic number field and established that (*in our terminology*) the group of these units is a direct product of a finite cyclic group and a free abelian group of finite rank. At about the same time E. Kummer introduced his *ideal numbers*, defined an equivalence relation on them, and derived, for cyclotomic fields, certain special properties of the number of equivalence classes (the so-called *class number* of a cyclotomic field; in our terminology, the order of the ideal class group of the cyclotomic field). Dirichlet had earlier made similar studies of *quadratic* fields.

In 1869 E. Schering, a former student of Gauss, investigated the structure of Gauss's (group of) equivalence classes of binary quadratic forms. He found certain fundamental classes from which all classes of forms could be obtained by composition. In group-theoretic terms, Schering found a basis for the abelian group of equivalence classes of binary quadratic forms. L. Kronecker generalized Kummer's work on cyclotomic fields to arbitrary algebraic number fields. In a paper in 1870 on algebraic number theory, entitled "Auseinandersetzung einiger Eigenschaften der Klassenzahl idealer complexer Zahlen", he began by taking a very abstract point of view: he considered a finite set of arbitrary "elements", and defined an abstract operation on them which satisfied certain laws—laws which we may take nowadays as axioms for a finite abelian group:

> Let $\theta', \theta'', \theta''', \ldots$ be finitely many elements such that with any two of them we can associate a third by means of a definite procedure. Thus, if f denotes the procedure and θ', θ'' are two (possibly equal) elements, then there exists

a θ''' equal to $f(\theta', \theta'')$. Furthermore,

$$f(\theta', \theta'') = f(\theta'', \theta'),$$
$$f(\theta', f(\theta'', \theta''')) = f(f(\theta', \theta''), \theta''')$$

and if θ'' is different from θ''' then $f(\theta', \theta'')$ is different from $f(\theta', \theta''')$. Once this is assumed we can replace the operation $f(\theta', \theta'')$ by multiplication $\theta' \cdot \theta''$ provided that instead of equality we employ equivalence. Thus using the usual equivalence symbol $\sim$ we define the equivalence $\theta' \cdot \theta'' \sim \theta'''$ by means of the equation $f(\theta', \theta'') = \theta'''$.

Kronecker aimed at working out the laws of combination of "magnitudes", in the process giving an implicit definition of a finite abelian group. From the above abstract considerations Kronecker deduces the following consequences:

(i) If θ is any "element" of the set under discussion, then $\theta^k = 1$ for some positive integer k. If k is the smallest such, then θ is said to "belong to k". If θ belongs to k and $\theta^m = 1$, then k divides m.

(ii) If an element θ belongs to k, then every divisor of k has an element belonging to it.

(iii) If θ and θ' belong to k and k' respectively, and k and k' are relatively prime, then $\theta\theta'$ belongs to kk'.

(iv) There exists a "fundamental system" of elements $\theta_1, \theta_2, \theta_3, \ldots$ such that the expression

$$\theta_1^{h_1} \theta_2^{h_2} \theta_3^{h_3} \cdots \quad (h_i = 1, 2, 3, \ldots, n_i)$$

represents each element of the given set of elements just once. The numbers $n_1, n_2, n_3, \ldots$ to which, respectively, $\theta_1, \theta_2, \theta_3, \ldots$ belong, are such that each is divisible by its successor; the product $n_1 n_2 n_3 \cdots$ is equal to the totality of elements of the set.

The above can, of course, be interpreted as well-known results on finite abelian groups; in particular (iv) can be taken as the basis theorem for such groups. Once Kronecker establishes this general framework, he applies it to the special cases of equivalence classes of binary quadratic forms and to ideal classes. He notes that when applying (iv) to the former, one obtains Schering's result.

Although Kronecker did not relate his implicit definition of a finite abelian group to the (by that time) well-established concept of a permutation group, of which he was well aware, he clearly recognized the advantages of the abstract point of view which he adopted:

> The very simple principles... are applied not only in the context indicated but also frequently elsewhere—even in the elementary parts of number theory. This shows, and it is otherwise easy to see, that these principles belong to a more general and more abstract realm of ideas. It is therefore proper to free their development from all inessential restrictions, thus making it unnecessary to repeat the same argument when applying it in different cases... Also, when stated with all admissible generality, the presentation gains in simplicity and, since only the truly essential features are thrown into relief, in transparency.

The above lines of development were capped in 1879 by an important paper of G. Frobenius and L. Stickelberger entitled "On groups of commuting elements." Although Frobenius and Stickelberger built on Kronecker's work, they used the concept of an abelian group explicitly and, moreover, made the important advance of recognizing that the abstract group concept embraces congruences and Gauss's composition of forms as well as the substitution groups of Galois. (They also mention, in footnotes, groups of infinite order, namely groups of units of number fields and the group of all roots of unity.) One of their main results is a proof of the basis theorem for finite abelian groups, including a proof of the uniqueness of decomposition. It is interesting to compare their explicit, *modern*, formulation of the theorem to that of Kronecker ((iv) above) ([33], p. 235):

> A group that is not irreducible [indecomposable] can be decomposed into purely irreducible factors. As a rule, such a decomposition can be accomplished in many ways. However, regardless of the way in which it is carried out, the number of irreducible factors is always the same and the factors in the two decompositions can be so paired off that the corresponding factors have the same order.

They go on to identify the "irreducible factors" as cyclic groups of prime power orders. They then apply

Georg Frobenius (1849–1917)

their results to groups of integers modulo m, binary quadratic forms, and ideal classes in algebraic number fields. The paper by Frobenius and Stickelberger is (Fuchs, [30])

> a remarkable piece of work, building up an independent theory of finite abelian groups on its own foundation in a way close to modern views.

For details on this section (b), see [5], [9], [24[, [30], [33].

(c) Transformation groups As in number theory, so in geometry and analysis, group-theoretic ideas remained implicit until the last third of the nineteenth century. Moreover, Klein's (and Lie's) explicit use of groups in geometry influenced conceptually rather than technically the evolution of group theory, for it signified a genuine shift in the development of that theory from a preoccupation with permutation groups to the study of groups of transformations. (That is not to imply, of course, that permutation groups were no longer studied.) This transition was also notable in that it pointed to a turn from finite groups to infinite groups.

Klein noted the connection of his work with permutation groups but also realized the departure he was making. He stated that what Galois theory and his own program have in common is the investigation of "groups of changes", but added ([33], p. 191) that to be sure, the objects the changes apply to are different: there [Galois theory] one deals with a finite number of discrete elements, whereas here one deals with an infinite number of elements of a continuous manifold.

To continue the analogy, Klein notes that just as there is a theory of permutation groups, "we insist on a *theory of transformations*, a study of groups generated by transformations of a given type" ([33], p. 191).

Klein shunned the abstract point of view in group theory, and even his technical definition of a (transformation) group is deficient ([33], p. 185):

> Now let there be given a sequence of transformations $A, B, C, \ldots$ If this sequence has the property that the composite of any two of its transformations yields a transformation that again belongs to the sequence, then the latter will be called a group of transformations.

His work, however, broadened considerably the conception of a group and its applicability in other fields of mathematics. Klein did much to promote the view that group-theoretic ideas are fundamental in mathematics ([33], p. 228):

> Group theory appears as a distinct discipline throughout the whole of modern mathematics. It permeates the most varied areas as an ordering and classifying principle.

There was another context in which groups were associated with geometry, namely, *motion-geometry*; i.e., the use of motions or transformations of geometric objects as group elements. Already in 1856 W. R. Hamilton considered (implicitly) *groups* of the regular solids. Jordan, in 1868, dealt with the classification of all subgroups of the group of motions of Euclidean 3-space. And Klein in his *Lectures on the Icosahedron* of 1884 "solved" the quintic equation by means of the symmetry group of the icosahedron. He thus discovered a deep connection between the groups of rotations of the regular solids, polynomial equations, and complex function theory. (In these *Lectures* there also appears the "Klein 4-group".)

Already in the late 1860s Klein and Lie had undertaken, jointly,

> to investigate geometric or analytic objects that are transformed into themselves by *groups of changes*. (This is Klein's retrospective description, in 1894, of their program.)

While Klein concentrated on discrete groups, Lie studied continuous transformation groups. Lie realized that the theory of continuous transformation groups was a very powerful tool in geometry and differential equations and he set himself the task of "determining all groups of... [continuous] transformations" ([33], p. 214). He achieved his objective by the early 1880s with the classification of these groups (see [33] for details). A classification of discontinuous transformation groups was obtained by Poincaré and Klein a few years earlier.

Beyond the technical accomplishments in the areas of discontinuous and continuous transformation groups (extensive theories developed in both areas and both are still nowadays active fields of research), what is important for us in the founding of these theories is that

(i) they provided a major extension of the scope of the concept of a group—from permutation groups and abelian groups to transformation groups;

(ii) they provided important examples of infinite groups—previously the only objects of study were finite groups;

(iii) they greatly extended the range of applications of the group concept to include number theory, the theory of algebraic equations, geometry, the theory of differential equations (both ordinary and partial), and function theory (automorphic functions, complex functions).

All this occurred prior to the emergence of the abstract group concept. In fact, these developments were instrumental in the emergence of the concept of an abstract group, which we describe next. For further details on this section (c), see [5], [7], [9], [17], [18], [20], [24], [29], [33].

3 Emergence of abstraction in group theory

The abstract point of view in group theory emerged slowly. It took over one hundred years from the time of Lagrange's implicit group-theoretic work of 1770 for the abstract group concept to evolve. E. T. Bell discerns several stages in this process of evolution towards abstraction and axiomatization [2]:

> The entire development required about a century. Its progress is typical of the evolution of any major mathematical discipline of the recent period; first, the discovery of isolated phenomena, then the recognition of certain features common to all, next the search for further instances, their detailed calculation and classification; then the emergence of general principles making further calculations, unless needed for some definite application, superfluous; and last, the formulation of postulates crystallizing in abstract form the structure of the system investigated.

Although somewhat oversimplified (as all such generalizations tend to be), this is nevertheless a useful framework. Indeed, in the case of group theory, first came the "isolated phenomena"—e.g., permutations, binary quadratic forms, roots of unity; then the recognition of "common features"—the concept of a finite group, encompassing both permutation groups and finite abelian groups (cf. the paper of Frobenius and Stickelberger cited in section 2(b)); next the search for "other instances"—in our case transformation groups (see section 2(c)); and finally the formulation of "postulates"—in this case the postulates of a group, encompassing both the finite and infinite cases. We now consider when and how the intermediate and final stages of abstraction occurred.

In 1854 Cayley, in a paper entitled "On the theory of groups, as depending on the symbolic equation $\theta^n = 1$," gave the first abstract definition of a finite group. (In 1858 R. Dedekind, in lectures on Galois theory at Göttingen, gave another.) Here is Cayley's definition:

> A set of symbols $1, \alpha, \beta, \ldots$ all of them different, and such that the product of any two of them (no matter in what order), or the product of any one of them into itself, belongs to the set, is said to be a *group*.

Cayley goes on to say that

> These symbols are not in general convertible [commutative], but are associative,

and

> it follows that if the entire group is multiplied by any one of the symbols, either as further or nearer factor [i.e., on the left or on the right], the effect is simply to reproduce the group.

Cayley then presents several examples of groups, such as the quaternions (under addition), invertible matrices (under multiplication), permutations,

Arthur Cayley (1821–1895)

Gauss's quadratic forms, and groups arising in elliptic function theory. Next he shows that every abstract group is (in our terminology) isomorphic to a permutation group, a result now known as *Cayley's theorem*. He seems to have been well aware of the concept of isomorphic groups, although he does not define it explicitly. He introduces, however, the multiplication table of a (finite) group and asserts that an abstract group is determined by its multiplication table. He then goes on to determine all the groups of orders four and six, showing there are two of each by displaying multiplication tables. Moreover, he notes that the cyclic group of order n "is in every respect analogous to the system of the roots of the ordinary equation $x^n - 1 = 0$," and that there exists only one group of a given prime order.

Cayley's orientation towards an abstract view of groups—a remarkable accomplishment at this time of the evolution of group theory—was due, at least in part, to his contact with the abstract work of G. Boole. The concern with the abstract foundations of mathematics was characteristic of the circles around Boole, Cayley, and Sylvester already in the 1840s. Cayley's achievement was, however, only a personal triumph. His abstract definition of a group attracted no attention at the time, even though Cayley was already well known. The mathematical community was apparently not ready for such abstraction: permutation groups were the only groups under serious investigation, and more generally, the formal approach to mathematics was still in its infancy. As M. Kline put it in his inimitable way [21]:

> Premature abstraction falls on deaf ears, whether they belong to mathematicians or to students.

For details see [22], [23], [24], [25], [29], [33].

It was only a quarter of a century later that the abstract group concept began to take hold. And it was Cayley again who in four short papers on group theory written in 1878 returned to the abstract point of view he adopted in 1854. Here he stated the general problem of finding all groups of a given order and showed that any (finite) group is isomorphic to a group of permutations. But, as he remarked ([22], p. 141), this

> does not in any wise show that the best or easiest mode of treating the general problem is thus to regard it as a problem of substitutions; and it seems clear that the better course is to consider the general problem in itself, and to deduce from it the theory of groups of substitutions.

These papers of Cayley, unlike those of 1854, inspired a number of fundamental group-theoretic works.

Another mathematician who advanced the abstract point of view in group theory (and more generally in algebra) was H. Weber. It is of interest to see his "modern" definition of an abstract (finite) group given in a paper of 1882 on quadratic forms ([23], p. 113):

> A system G of h arbitrary elements $\theta_1, \theta_2, \ldots, \theta_h$ is called a group of degree h if it satisfies the following conditions:
>
> I. By some rule which is designated as composition or multiplication, from any two elements of the same system one derives a new element of the same system. In symbols $\theta_r \theta_s = \theta_t$.
>
> II. It is always true that $(\theta_r \theta_s)\theta_t = \theta_r(\theta_s \theta_t) = \theta_r \theta_s \theta_t$.
>
> III. From $\theta \theta_r = \theta \theta_s$ or from $\theta_r \theta = \theta_s \theta$ it follows that $\theta_r = \theta_s$.

Weber's and other definitions of abstract groups given at the time applied to *finite* groups only. They thus encompassed the two theories of permutation groups and (finite) abelian groups, which derived from the two sources of classical algebra (polynomial

Heinrich Weber (1842–1913)

equations) and number theory, respectively. Infinite groups, which derived from the theories of (discontinuous and continuous) transformation groups, were not subsumed under those definitions. It was W. von Dyck who, in an important and influential paper in 1882 entitled "Group-theoretic studies," consciously included and combined, for the first time, all of the major historical roots of abstract group theory—the algebraic, number-theoretic, geometric, and analytic. In von Dyck's own words:

> The aim of the following investigations is to continue the study of the properties of a group in its abstract formulation. In particular, this will pose the question of the extent to which these properties have an invariant character present in all the different realizations of the group, and the question of what leads to the exact determination of their essential group-theoretic content.

Von Dyck's definition of an abstract group, which included both the finite and infinite cases, was given in terms of generators (he calls them "operations") and defining relations (the definition is somewhat long—see [7], pp. 5, 6). He stresses that "in this way all ... *isomorphic* groups are included in a *single* group," and that "the *essence* of a group is no longer expressed by a particular presentation form of its operations but rather by their mutual relations." He then goes on to construct the free group on n generators, and shows (essentially, without using the terminology) that every finitely generated group is a quotient group of a free group of finite rank. What is important from the point of view of postulates for group theory is that von Dyck was the first to require explicitly the existence of an inverse in his definition of a group:

> We require for our considerations that a group which contains the operation T_k must also contain its inverse T_k^{-1}.

In a second paper (in 1883) von Dyck applied his abstract development of group theory to permutation groups, finite rotation groups (symmetries of polyhedra), number-theoretic groups, and transformation groups.

Although various postulates for groups appeared in the mathematical literature for the next twenty years, the abstract point of view in group theory was not universally applauded. In particular, Klein, one of the major contributors to the development of group theory, thought that the "abstract formulation is excellent for the working out of proofs but it does not help one find new ideas and methods," adding ([33], p. 228) that

> in general, the disadvantage of the [abstract] method is that it fails to encourage thought.

Despite Klein's reservations, the mathematical community was at this time (early 1880s) receptive to the abstract formulations (cf. the response to Cayley's definition of 1854). The major reasons for this receptivity were:

(i) There were now several major "concrete" theories of groups—permutation groups, abelian groups, discontinuous transformation groups (the finite and infinite cases), and continuous transformation groups, and this warranted abstracting their essential features.

(ii) Groups came to play a central role in diverse fields of mathematics, such as different parts of algebra, geometry, number theory and several areas of analysis, and the abstract view of groups was thought to clarify what was essential for such applications and to offer opportunities for further applications.

(iii) The formal approach, aided by the penetration into mathematics of set theory and mathematical logic, became prevalent in other fields of mathematics—for example, various areas of geometry and analysis.

In the next section we will follow, very briefly, the evolution of that abstract point of view in group theory.

4 Consolidation of the abstract group *concept*; dawn of abstract group *theory*

The abstract group concept spread rapidly during the 1880s and 1890s, although there still appeared a great many papers in the areas of permutation and transformation groups. The abstract viewpoint was manifested in two ways:

(a) concepts and results introduced and proved in the setting of "concrete" groups were now reformulated and reproved in an abstract setting;

(b) studies originating in, and based on, an abstract setting began to appear.

An interesting example of the former case is a reproving by Frobenius, in an abstract setting, of Sylow's theorem, which was proved by Sylow in 1872 for permutation groups. This was done in 1887, in a paper entitled "Neuer Beweis Sylowschen Satzes". Although Frobenius admits that the fact that every finite group can be represented by a group of permutations proves that Sylow's theorem must hold for all finite groups, he nevertheless wishes to establish the theorem abstractly:

> Since the symmetric group, which is introduced in all these proofs, is totally alien to the context of Sylow's theorem I have tried to find a new derivation of it ...

(For a case study of the evolution of abstraction in group theory in connection with Sylow's theorem see [28] and [32].)

Hölder was an important contributor to abstract group theory, and was responsible for introducing a number of group-theoretic concepts abstractly. For example, in 1889 he introduced the abstract notion of a quotient group (the "quotient group" was first seen as the Galois group of the "auxiliary equation", later as a homomorphic image and only in Hölder's time as a group of cosets), and "completed" the proof of the Jordan–Hölder theorem, namely, that the quotient groups in a composition series are invariant up to isomorphism (see section 2(a) for Jordan's contribution).

In 1893, in a paper on groups of order p^3, pq^2, pqr, and p^4, he introduced abstractly the concept of an automorphism of a group. Hölder was also the first to study simple groups abstractly. (Previously they were considered in concrete cases—as permutation groups, transformation groups, and so on.) As he says ([29], p. 338):

> It would be of the greatest interest if a survey of all simple groups with a finite number of operations could be known.

(By "operations" Hölder meant elements.) He then goes on to determine the simple groups of order up to 200.

Other typical examples of studies in an abstract setting are the papers by Dedekind and G. A. Miller in 1897/1898 on Hamiltonian groups—i.e., non-abelian groups in which all subgroups are normal. They (independently) characterize such groups abstractly, and introduce in the process the notions of the commutator of two elements and the commutator subgroup (Jordan had previously introduced the notion of commutator of two permutations).

The theory of group characters and the representation theory for finite groups (created at the end of the nineteenth century by Frobenius and Burnside/Frobenius/Molien, respectively) also belong to the area of abstract group theory, as they were used to prove important results about abstract groups. See [17] for details.

Although the abstract group *concept* was well established by the end of the nineteenth century, (Wussing, [33])

> this was not accompanied by a general acceptance of the associated method of presentation in papers, textbooks, monographs, and lectures. Group-theoretic monographs based on the abstract group concept did not appear until the beginning of the twentieth century. Their appearance marked the birth of abstract group *theory*.

The earliest monograph devoted entirely to abstract group theory was the book by J. A. de Séguier of 1904 entitled *Elements of the Theory of Abstract Groups* [27]. At the very beginning of the book there is a set-theoretic introduction based on the work of Cantor. B. Chandler and W. Magnus [7] remark that

> De Séguier may have been the first algebraist to take note of Cantor's discovery of uncountable cardinalities.

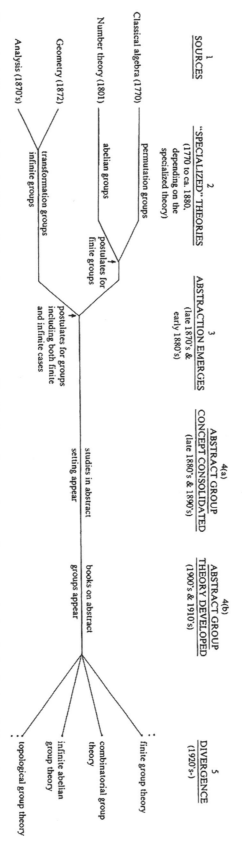

Figure 1. A diagrammatic sketch of the evolution of group theory as outlined in the various sections, and as summarized at the beginning of this section.

Next is the introduction of the concept of a *semi-group* with two-sided cancellation law and a proof that a finite semigroup is a group. There is also a proof, by means of counterexamples, of the independence of the group postulates. De Séguier's book also includes a discussion of isomorphisms, homomorphisms, automorphisms, decomposition of groups into direct products, the Jordan–Hölder theorem, the first isomorphism theorem, abelian groups including the basis theorem, Hamiltonian groups, and finally, the theory of p-groups. All this is done in the abstract, with "concrete" groups relegated to an appendix.

> The style of de Séguier is in sharp contrast to that of Dyck. There are no intuitive considerations ... and there is a tendency to be as abstract and as general as possible ... "
> (Chandler and Magnus, [7]).

De Séguier's book was devoted largely to finite groups. The first abstract monograph on group theory which dealt with groups in general, relegating finite groups to special chapters, was O. Schmidt's *Abstract Theory of Groups* of 1916 [26]. Schmidt, founder of the Russian school of group theory, devotes the first four chapters of his book to group properties common to finite and infinite groups. Discussion of finite groups is postponed to chapter 5, there being ten chapters in all. See [7], [10], [33].

5 Divergence of developments in group theory

Group theory evolved from several different sources, giving rise to various concrete theories. These theories developed independently, some for over one hundred years (beginning in 1770) before they converged (early 1880s) within the abstract group concept. Abstract group theory emerged and was consolidated in the next thirty to forty years. At the end of that period (around 1920) one can discern the divergence of group theory into several distinct theories. Here is the barest indication of some of these advances and new directions in group theory, beginning in the 1920s (with contributors and approximate dates):

(a) *Finite group theory*. The major problem here, already formulated by Cayley (1870s) and studied by Jordan and Hölder, was to find all finite groups of a given order. The problem proved too difficult and mathematicians turned to special cases (suggested especially by Galois theory): to find all simple or all solvable groups (cf. the Feit-Thompson theorem of 1963, and the classification of all finite simple groups in 1981). See [14], [15], [30].

(b) *Extensions of certain results from finite group theory to infinite groups with finiteness conditions*; e.g., O. J. Schmidt's proof, in 1928, of the Remak-Krull-Schmidt theorem. See [5].

(c) *Group presentations (Combinatorial Group Theory)*, begun by von Dyck in 1882, and continued in the twentieth century by M. Dehn, H. Tietze, J. Nielsen, E. Artin, O. Schreier, *et al.* For a full account, see [7].

(d) *Infinite abelian group theory* (H. Prüfer, R. Baer, H. Ulm *et al.*—1920s to 1930s). See [30].

(e) *Schreier's theory of group extensions* (1926), leading later to the cohomology of groups.

(f) *Algebraic groups* (A. Borel, C. Chevalley *et al.*—1940s).

(g) *Topological groups*, including the extension of group representation theory to continuous groups (Schreier, E. Cartan, L. Pontrjagin, I. Gelfand, J. von Neumann *et al.*—1920s and 1930s). See [4].

References

We give references here to *secondary* sources. Extensive references to *primary* sources, including works referred to in this article, may be found in [25] and [33].

1. R. G. Ayoub, Paolo Ruffini's contributions to the quintic, *Arch. Hist. Exact Sci.* **23** (1980), 253–277.

2. E. T. Bell, *The Development of Mathematics*, McGraw Hill, 1945.

3. G. Birkhoff, Current trends in algebra, *Amer. Math. Monthly* **80** (1973), 760–782, and **81** (1974), 746.

4. G. Birkhoff, The rise of modern algebra to 1936, *Men and Institutions in American Mathematics* (eds. D. Tarwater, J. T. White and J. D. Miller), Texas Tech. Press, 1976, pp. 41–63.

5. N. Bourbaki, *Elements d'Histoire des Mathématiques*, Hermann, 1969.

6. J. E. Burns, The foundation period in the history of group theory, *Amer. Math. Monthly* **20** (1913), 141–148; reproduced in this volume.

7. B. Chandler and W. Magnus, *The History of Combinatorial Group Theory: A Case Study in the History of Ideas*, Springer-Verlag, 1982.

8. A. Dahan, Les travaux de Cauchy sur les substitutions. Etude de son approche du concept de groupe, *Arch. Hist. Exact Sci.* **23** (1980), 279–319.

9. J. Dieudonné (ed.), *Abrégé d'Histoire des Mathématiques, 1700–1900*, 2 vols., Hermann, 1978.

10. P. Dubreil, L'algèbre, en France, de 1900 à 1935, *Cahiers du Seminaire d'Histoire des Mathématiques* **3** (1981), 69–81.

11. C. H. Edwards, *The Historical Development of the Calculus*, Springer-Verlag, 1979.

12. H. M. Edwards, *Galois Theory*, Springer-Verlag, 1984.

13. J. A. Gallian, The search for finite simple groups, *Math. Mag.* **49** (1976), 163–179; reproduced in this volume.

14. D. Gorenstein, *Finite Simple Groups: An Introduction to Their Classification*, Plenum Press, 1982.

15. D. Gorenstein, *The Classification of Finite Simple Groups*, Plenum Press, 1983.

16. R. R. Hamburg, The theory of equations in the eighteenth century: The work of Joseph Lagrange, *Arch. Hist. Exact Sci.* **16** (1976/77), 17–36.

17. T. Hawkins, Hypercomplex numbers, Lie groups, and the creation of group representation theory, *Arch. Hist. Exact Sci.* **8** (1971/72), 243–287.

18. T. Hawkins, The Erlanger Programme of Felix Klein: Reflections on its place in the history of mathematics, *Hist. Math.* **11** (1984), 442–470.

19. B. M. Kiernan, The development of Galois theory from Lagrange to Artin, *Arch. Hist. Exact Sci.* **8** (1971/72), 40–154.

20. F. Klein, Development of mathematics in the nineteenth century (transl. from the 1928 German ed. by M. Ackerman), *Lie Groups: History, Frontiers and Applications*, Vol. IX (ed. R. Hermann), Math. Sci. Press, 1979, pp. 1–361.

21. M. Kline, *Mathematical Thought from Ancient to Modern Times*, Oxford Univ. Press, 1972.

22. D. R. Lichtenberg, *The Emergence of Structure in Algebra*, Doctoral Dissertation, Univ. of Wisconsin, 1966.

23. U. Merzbach, *Development of Modern Algebraic Concepts from Leibniz to Dedekind*, Doctoral Dissertation, Harvard Univ., 1964.

24. G. A. Miller, History of the theory of groups, *Collected Works*, Vol. 1, pp. 427–467, Vol. 2, pp. 1–18, and Vol. 3, pp. 1–15, Univ. of Illinois Press, 1935, 1938, and 1946.

25. L. Novy, *Origins of Modern Algebra*, Noordhoff, 1973.

26. O. J. Schmidt, *Abstract Theory of Groups*, W. H. Freeman and Co., 1966. (Translation by F. Holling and J. B. Roberts of the 1916 Russian edition.)

27. J.-A. de Séguier, *Théorie des Groupes Finis, Elements de la Théorie des Groupes Abstraits*, Gauthier Villars, 1904.

28. L. A. Shemetkov, Two directions in the development of the theory of non-simple finite groups, *Russ. Math. Surv.* **30** (1975), 185–206.

29. R. Silvestri, Simple groups of finite order in the nineteenth century, *Arch. Hist. Exact Sci.* **20** (1979), 31–356.

30. J. Tarwater, J. T. White, C. Hall, and M. B. Moore (eds.), *American Mathematical Heritage: Algebra and Applied Mathematics*, Texas Tech. Press, 1981. Has articles by Feit, Fuchs, and MacLane on the history of finite groups, abelian groups, and abstract algebra, respectively.

31. B. L. Van der Waerden, Die Algebra seit Galois, *Jahresbericht der Deutsch. Math. Ver.* **68** (1966), 155–165.

32. W. C. Waterhouse, The early proofs of Sylow's theorem, *Arch. Hist. Exact Sci.* **21** (1979/80), 279–290.

33. H. Wussing, *The Genesis of the Abstract Group Concept*, MIT Press, 1984. (Translation by A. Shenitzer of the 1969 German edition.)

The Search for Finite Simple Groups

JOSEPH A. GALLIAN

Mathematics Magazine **49** (1976), 163–179

At present, simple group theory is the most active and glamorous area of research in the theory of groups and it seems certain that this will remain the case for many years to come. Roughly speaking, the central problem is to find some reasonable description of all finite simple groups. A number of expository papers [36], [42], [45], [47], [49], [79] and books [21], [46], [67] detailing progress on this problem have been written for professional group theorists, but very little has appeared which is accessible to undergraduates. (Only Goldschmidt's proof of the Brauer-Suzuki-Wall theorem [44] comes to mind.) This paper is intended as a historical account of the search for simple groups for readers who are not experts in the subject. It is the hope of the author that the paper may profitably be read by one who is conversant with the contents of Herstein's algebra book [55]. A complete discussion of all important contributions to simple group theory is beyond the scope of this paper.

What are simple groups and why are they important? Évariste Galois (1811–1832) called a group *simple* if its only normal subgroups were the identity subgroup and the group itself. The abelian simple groups are the group of order 1 and the cyclic groups of prime order, while the non-abelian simple groups generally have very complicated structures. These groups are important because they play a role in group theory somewhat analogous to that which the primes play in number theory or the elements do in chemistry; that is, they serve as the "building blocks" for all groups.

These building blocks are called the *composition factors* of the group and may be determined in the following way. Given a finite group G, choose a maximal normal subgroup G_1 of $G = G_0$. Then the factor group G_0/G_1 is simple, and we next choose a maximal normal subgroup G_2 of G_1. Then G_1/G_2 is also simple, and we continue in this fashion until we arrive at $G_n = \{e\}$. The simple groups

$$G_0/G_1, \; G_1/G_2, \ldots, G_{n-1}/G_n$$

are the *composition factors* of G and by the Jordan-Hölder theorem these groups are independent of the choices of the normal subgroups made in the process described.

In a certain sense, a group can be reconstructed from its composition factors and many of the properties of a group are determined by the nature of its composition factors. This, and the fact that many questions about finite groups can be reduced (by induction) to questions about simple groups, make clear the importance of determining all non-abelian finite simple groups.

The narrative which follows is divided into 16 sections which appear in more or less chronological order according to theme. Within a particular section, however, we usually include a number of results which are related to the theme without regard to time. Thus, for example, the section on the odd order problem appears early in the paper but includes results ranging from 1895 to 1963. In this way we hope to emphasize two points:

1. the problems of one generation very often have deep roots in the work of previous generations;

2. there is frequently a large temporal gap between certain results and their subsequent improvements.

Throughout the remainder of this paper we use the term *simple group* to mean a finite non-abelian simple group.

1 The alternating groups and the classical linear groups

Although Galois had formulated the definition of a simple group and had observed that the alternating group (of even permutations) on 5 symbols was simple, the first major results in the theory were due to Camille Jordan (1838–1922). In 1870, Jordan published *Traité des Substitutions*, the first book ever written on group theory [58].

In this book he established the existence of five infinite families of finite simple groups. One of these families, which we denote by A_n, consists of the alternating permutation groups on $n > 4$ symbols. Jordan formed the other four families by using matrices with entries from finite fields. One of these may be described as follows. For $m > 1$, the *special linear group* $SL(m, p^n)$ is the multiplicative group of $m \times m$ matrices of determinant 1 with entries from the field with p^n elements and the *projective special linear group* $PSL(m, p^n)$ is the factor group $SL(m, p^n)/Z(SL(m, p^n))$, where $Z(SL(m, p^n))$, the *center* of $SL(m, p^n)$, is the subgroup of $SL(m, p^n)$ consisting of all scalar matrices with determinant 1. Jordan proved that $PSL(m,p)$ is simple when (m, p) is not $(2, 2)$ or $(2, 3)$. The other three families have been given the names *orthogonal*, *unitary* and *symplectic* groups and, following Hermann Weyl, mathematicians refer to these four families collectively as the *classical simple groups*.

The last three types mentioned above are most easily defined as certain groups of invertible linear transformations of a finite-dimensional vector space V over a finite field modulo the center of the group and in each case the group is obtained by considering those transformations T which leave a non-degenerate form f of V invariant (i.e., f is a certain function from $V \times V$ into the field and $f(Tx, Ty) = f(x, y)$ for all x, y in V). A *symmetric bilinear form* (i.e., a dot product) gives an orthogonal group; *hermitian*, a unitary group; and *skew symmetric bilinear*, a symplectic group. The precise definitions of these groups are not needed here but the reader can find them in [2] and [6]. Jordan introduced these three families as groups of matrices instead of groups of linear transformations and proved they are simple when the field has prime order (except for a few trivial cases).

2 Range problem 1–660

A different approach was taken by Otto Hölder (1859–1937) when in 1892 he initiated what we will call the *range problem*; namely, the complete determination of all simple groups whose orders are in a given range. Here both the existence and the uniqueness questions must be considered; that is, it must be determined which integers in the range are the orders of simple groups and, for each such integer, all possible simple groups of that order must also be determined (up to isomorphism). Hölder [56] proved that the only two simple groups whose orders lie between 1 and 200 are A_5 of order 60 and $PSL(2,7)$ of order 168.

F. N. Cole (1861–1927), the first American-born mathematician to publish in group theory, followed Hölder's lead in 1892 [23] when he examined the integers between 201 and 500 for simple groups. He was not totally successful for he was unable to prove that A_6 was the unique simple group of order 360; nor was he able to show that 432 was not the order of a simple group. He overcame these difficulties [24] a year later, however, when he completed the determination of all the simple groups with orders in the range 1 to 660. In addition to the ones in this range already found by Jordan, Cole discovered one more, $PSL(2, 8)$, having order 504. This provided the first example of a simple group not known to Jordan and the first proof of the simplicity of one of the groups $PSL(m, q)$ with q not prime.

The methods of Hölder and Cole are of interest. The three Sylow theorems rule out 596 of the first 660 integers as possible orders of simple groups. (In fact, they rule out 9431 of the first 10,000 [74].) If G is assumed to be simple and H is a proper subgroup of G, then G is isomorphic to a subgroup of the symmetric group on the cosets of H in G (compare with the proof of Theorem 2.92 in [55]). Thus G with order $|G|$ is represented as a group of permutations on $|G|/|H|$ symbols and it follows that $(|G|/|H|)!$ is a multiple of $|G|$. This last fact is called the *index theorem* and it further reduces the list of integers to be examined to 47. Finally, a combination of the Sylow theorems, the index theorem, and other elementary techniques such as counting elements reduces to 33 the list of those integers from 1 to 660 which require *ad hoc* arguments. Since the theory of permutation groups was much further developed than the theory of abstract groups at that time, these remaining 33 integers were handled with permutation group techniques. An example of a permutation-type argument will be given later.

It is noteworthy that while the proofs of the non-simplicity of a group of order 144 or 180 occupied more than 10 pages of Hölder's paper, the author has had undergraduates [59] who have done this in less

CHRONOLOGY OF THE RANGE PROBLEM

Date	Integers	Individual
1892	1–200	Hölder [56]
1892–93	201–660	Cole [23], [24]
1895	661–1092	Burnside [15]
1900	1093–2000	Ling and Miller [60]
1912	2001–3640 (except 2520)	Siceloff [71]
1922	2520	Miller [65]
1924	3641–6232 (except 5616 and 6048)	Cole [26]
1942	5616 and 6048	Brauer [7]
1963	6233–20,000	Michaels [62]
1972	1–1,000,000 (21 exceptions)	Hall [49], [50]
1975	Hall's exceptions	Beisiegel and Stingl [3]

Table 1. The search for all simple groups of specified orders reveals sporadic progress as new methods made possible sudden bursts of successful analysis on groups of increasingly large order. At present the range problem is completed through groups of order 1,000,000: all simple groups of order less than 1,000,000 are known, but only some of those beyond that order have been discovered.

than 2 pages using only the results found in Herstein [55]. Similarly, using a bit more machinery, three undergraduates from the University of Wisconsin [27] covered all the integers up to 1000 with the exceptions of 720 and the uniqueness question. Their proofs for the cases 144 and 180 require only 12 lines.

3 $PSL(m, p^n)$

Cole's discovery of the simplicity of $PSL(2, 8)$ had far-reaching consequences because that same year E. H. Moore (1862–1932), the first mathematics department chairman of the University of Chicago, used it for the starting point of his investigations which resulted [66] in a proof that the family of groups $PSL(2, p^n)$ are all simple except when $p^n = 2$ or 3. William Burnside (1852–1927) also obtained this result [13] shortly after Moore. Moore's paper, in turn, led his first Ph.D. student, Leonard E. Dickson (1874–1954), to the complete generalization of Jordan's original result when in 1897 he proved [29] that the family of groups $PSL(m, p^n)$ ($m > 1$) consisted of simple groups except when $p^n = 2$ or 3. Dickson called this family a *triply infinite system* since each of $p, n,$ and m may take on infinitely many values. Moore's paper also contains many interesting results on finite fields, the most important of which is that for each prime power p^n there exists a unique field of order p^n (Galois had proved such fields exist in 1830 [41]). In the opinion of E. T. Bell ([4], p. 10) these results on finite fields clearly mark the beginning of abstract algebra in America.

4 Range problem to 1092

In 1895 Burnside ([14], [15]) obtained several powerful arithmetic tests for simple groups. By far the most important of these is the fact that a simple group of even order must be divisible by one of 12, 16 or 56. In proving this result, he showed that an even order simple group cannot have a cyclic Sylow 2-subgroup. This theorem appears to be the first non-simplicity criterion which is based on the structure of the Sylow 2-subgroups. In the past two decades much of the research in simple group theory has dealt with the problem of classifying all simple groups whose Sylow 2-subgroups have a specified structure. (All the Sylow 2-subgroups of a group are isomorphic.) For example, John Walter [77], in a long (110 pages!) and difficult proof, obtained a broad generalization of Burnside's result when he determined all simple groups with abelian Sylow 2-subgroups. Similarly, all simple groups whose Sylow 2-subgroups are dihedral (that is, are groups of symmetries of some regular n-sided polygon) have been determined by Daniel Gorenstein and Walter [48]. The proof of this important result appears in three papers and runs to 160 pages! Commenting on this proof in *Mathematical Reviews* (**32** #7634), John Thompson wrote:

> The techniques of these papers cover the spectrum of finite group theory more thoroughly than any single paper known to the reviewer.

By 1893 the range problem had been completed as far as 660. Since 1092 was the next integer known to be the order of a simple group, Burnside [15] decided to examine the integers between these two. The arithmetical tests of his previous paper disposed of all but 17 of the 432 integers in this range and the Sylow theorems ruled out six more. The remaining 11 integers were considered individually although his proof for the hardest integer in the range, 720, was erroneous and he inadvertently omitted 1008. The efficiency of Burnside's non-simplicity tests is further evidenced by the fact that they dispose of all odd integers up to 2025 and all but 14 of the odd integers less than 9000. As a rule, even integers are much harder to eliminate than odd integers, but Burnside's "12, 16, 56 theorem" alone rules out 3691 of the first 5000 even integers [74].

5 Permutation representations and character theory

In obtaining their results Burnside, Cole, and Hölder utilized permutation representations of groups. Certain permutation groups—transitive, doubly transitive, and primitive—play an especially important role in simple group theory. (A permutation group on a set S is called *transitive* on S if, for each pair a, b of letters of S, there is an element in G which sends a to b; G is called *doubly transitive* on S if, for each pair of distinct letters of S, (a, a') and (b, b'), there is an element in G which sends a to b and a' to b'; see [78], p. 15, for the definition of a *primitive* group.) The reasons for the importance of these groups are that the representation of a group as permutations of the cosets of a subgroup is transitive and that many of the known simple groups can be represented as a doubly transitive (and therefore primitive) permutation group. Thus, a common technique when dealing with a simple group G is to represent it as a transitive, doubly transitive or primitive group and then utilize the theory of these groups to obtain important information about G.

Much effort was devoted in the late 1800s and early 1900s to classifying the transitive and primitive permutation groups of low degree. These results often prove useful in simple group theory. To illustrate, let us consider Siceloff's proof [71] that there is no simple group of order $1188 = 2^2 \cdot 3^3 \cdot 11$. If G were a simple group of order 1188, Sylow's theorem implies that G has 12 subgroups of order 11 which are conjugate in G. If for each element g in G we define T to be the mapping which sends the Sylow 11-subgroup S to the Sylow 11-subgroup $g^{-1}Sg$, we see that G may be viewed as a transitive permutation group on the set of Sylow 11-subgroups of G (see, for example, the proof of 2.92 in [55]). Then letting H denote a subgroup of G which consists of all permutations which have some Sylow 11-subgroup fixed, it follows that $|H| = 3^2 \cdot 11$ (see [78], p. 51) and H is a permutation group on the other 11 Sylow 11-subgroups. By Sylow's theorem H has an element of order 11 and so this element is an 11-cycle. Thus H is a transitive permutation group of degree 11 and order 99. But Cole [25] has shown no such group exists.

A homomorphism from a group into a group of matrices with entries from some field is called a *representation* of the group. If T is a representation of G, the *character* of this representation is the function X from G to the field defined by $X(g) = \text{trace}(T(g))$ for all g in G. There exist numerous arithmetical relations on the characters of a group G which are intimately related to the structure of G. Thus a knowledge of the characters of a group reveals much information about the group itself. The theory of group characters has profoundly influenced the search for simple groups. This theory was developed by Georg Frobenius (1849–1917) in a series of papers beginning in 1896. (The historical background to Frobenius's creation of group characters is detailed in [51] and [52].) Around the turn of the century Issai Schur (1875–1941) and especially Burnside simplified the theory and found many important applications of it. In recent times, character theory has been further developed and refined by Brauer, Suzuki, and Feit.

6 Odd order problem

During the period 1895–1901 much attention was focused, particularly by Burnside, on the possibility of the existence of a simple group of odd order. In his 1895 paper [15], Burnside had shown that there is no simple group of odd order less than 2025. He later extended this to 9000 [16] and then to 40,000 [18]. Numerous arithmetical theorems obtained by Burnside in this period reduced the list of possible odd orders less than 40,000 to 7; these were then eliminated by elementary considerations.

In 1901 Burnside [17] used character theory to prove that a non-solvable transitive permutation group of prime degree is doubly transitive. Since a simple group which has a subgroup of prime index can be represented as a transitive permutation group on the cosets of this subgroup it must be doubly transitive. But the order of a doubly transitive group of degree n is divisible by $n(n-1)$ ([78], p. 20) so Burnside's result shows there is no odd order simple group which has a subgroup of prime index. Burnside

[18] also proved in 1901 that if a simple group has odd order n and p is the smallest prime divisor of n then n is divisible either by p^4 or by both p^3 and a prime factor of $p^2 + p + 1$.

Burnside's efforts convinced him that there were no simple groups of odd order and that the eventual proof of this would involve the use of character theory. In fact, he wrote ([20], p. 503):

> The contrast that these results shew between groups of odd and of even order suggests inevitably that simple groups of odd order do not exist.

He further wrote [17]:

> The results obtained in this paper, partial as they necessarily are, appear to me to indicate that an answer to the interesting question as to the existence or non-existence of simple groups of odd order may be arrived at by further study of the theory of group characters.

The next important step in this direction, however, did not come for more than 50 years. In 1957, Michio Suzuki [72] used character theory to prove that a simple group in which the centralizer of any non-identity element is abelian must have even order. (The *centralizer* of an element x in a group G is the subgroup $C(x) = \{g \in G | gx = xg\}$.) Three years later, in a major work [37], Walter Feit, Marshall Hall, Jr., and John Thompson obtained a broad generalization of Suzuki's result by showing that "abelian" could be replaced by the much weaker condition "nilpotent". (A group is *nilpotent* if all of its Sylow subgroups are normal.) Their proof was similar to Suzuki's and character theory played an important role in it.

Burnside's prophecy was at last fulfilled in 1963 when Feit and Thompson expanded on the ideas of the two papers mentioned above and proved [38] that groups of odd order are solvable. (A finite group is *solvable* if all of its composition factors have prime order; thus, solvable groups are not simple.) The difficulty of this proof and the significance of both the theorem and the methods employed cannot be exaggerated. Concerning one portion of the proof, Suzuki wrote in *Mathematical Reviews* (29 #3538) "... [This 50 page portion] represents one of the highest points ever achieved in the theory of finite groups."

The proof of the "Odd Order Theorem" occupies an entire 255-page issue of the *Pacific Journal of Mathematics*. It proceeds by assuming that there is a group G of minimal odd order which is not solvable. Then every proper subgroup of G is solvable and therefore Philip Hall's extensive work on solvable groups could be brought to bear on the subgroups. Ultimately, they were able to derive a contradiction. For their achievement, Feit and Thompson were awarded the Frank Nelson Cole Prize in Algebra by the American Mathematical Society in 1965. (The Cole Prize is named after the same Cole who had determined the simple groups with orders between 201 and 660, and was established in his honor in recognition of his many years of service to the Society.)

7 Dickson's simple groups

In the period from 1897 to 1905 Dickson made many fundamental contributions to the theory of simple groups. In a series of papers appearing from 1897 to 1899 he extended Jordan's results on the simplicity of the orthogonal, unitary and symplectic groups over fields of prime order to arbitrary finite fields. Much of this work emanated from his Ph.D. dissertation, the first one ever done in mathematics at the University of Chicago.

Whether there exist two non-isomorphic simple groups of the same order had been a long-standing problem by 1899. But Dickson's proof in 1897 that $PSL(3,4)$ is simple provided a possible answer to this question since it and the simple group A_8 both have order 20,160. It was quickly suspected that these two were not isomorphic since A_8 contains elements of orders 6 and 15 while no such elements were known to be in $PSL(3,4)$. At Moore's suggestion, Ida Schottenfels investigated these two groups and proved [70] they were not isomorphic. Shortly thereafter, Dickson showed [30] that there are infinitely many such examples. Since these examples were given by Dickson no others have been found and there is no known triple of non-isomorphic simple groups of equal order. After 20,160 the next known integer for which there is a pair of non-isomorphic simple groups of equal order is 4,585,351,680, and it wasn't until the mid-1960s that 20,160 was shown to be the smallest possible integer for which this can happen.

In his classic book *Linear Groups* Dickson listed all the isomorphisms between the simple groups he knew. For example, A_5, $PSL(2,4)$, and $PSL(2,5)$ are defined differently but are isomorphic. The question of whether Dickson's list of isomorphisms contained all which were possible among the simple groups known to him was not answered until 50 years later when Jean Dieudonné proved [35] that Dickson's list was complete. Dickson also listed all the coincidences in the orders of the simple groups known to him but whether this list was complete was not determined until 1955 when Emil Artin (1898–1962) proved with an elegant number-theoretical study [1] that it was.

Without going into detail we mention that the classical linear groups over the field of complex numbers are Lie groups (because, roughly, they possess a smooth geometric structure) and Wilhelm Killing (1847–1923) and Elie Cartan (1869–1951) proved (1888–1894) that besides the simple Lie groups corresponding to the classical groups there are only five additional simple ones called *exceptional Lie groups*. In two papers in 1901 and 1905 Dickson discovered a new infinite family of finite simple groups by defining analogs over finite fields of one of these exceptional Lie groups ([31], [32]). It is remarkable that no additional new finite simple groups were found until Claude Chevalley and others, 50 years later, were able to show (among other results) that the remaining four exceptional Lie groups also had finite analogs.

8 The Mathieu groups

In 1861 E. Mathieu discovered a family of five transitive permutation groups. This remarkable family has become very important in both the theory of simple groups and coding theory as well as in permutation group theory. In 1895, while determining all transitive permutation groups on 10 or 11 symbols, Cole observed [25] that the smallest Mathieu group (order 7920) is simple and by 1900 G. A. Miller (1863–1951) had shown ([63], [64]) that the other four are also simple. Three of these groups have order less than 1,000,000 and this brought to 53 the number of such simple groups known in 1900. This number would not be enlarged until 1960.

Among all the simple groups known by 1905 the Mathieu groups had the peculiar distinction of being the only ones which were not part of an infinite family of simple groups (such as A_n or $PSL(m, p^n)$). To this date they (and 21 or so other simple groups) still have not been shown to be members of any infinite family of simple groups in a natural way.

9 Range problem to 6232

The determination of all simple groups of order up to 2000 was completed in 1900 when G. Ling and Miller proved [60] that there is no simple group whose order is between 1093 and 2000. It is interesting to note that although there are 908 integers in this range only 28 required any special treatment. For example, Burnside's result on odd order simple groups and his "even, but not divisible by 12, 16 or 56" theorem eliminated all but 118 possibilities. Then the Sylow theorems, the index theorem, and several results of Frobenius on groups with orders of the form $p^a q^b$ reduced the possibilities to 28.

In 1902 Frobenius determined [40] all transitive permutation groups of degree $p + 1$ and order $p\left(p^2 - 1\right)/2$ where p is prime. Since any simple group of order $p\left(p^2 - 1\right)/2$ can be represented as a transitive permutation group on the $p + 1$ conjugate Sylow p-subgroups, it follows from Frobenius's theorem that $PSL(2, p)$ is the only simple group of order $p\left(p^2 - 1\right)/2$. This appears to be the first arithmetical characterization of an infinite family of simple groups. Fifty-six years later this uniqueness theorem was generalized in answer to a question of Artin. Artin had observed that the simple groups $PSL(2, p)$ ($p > 3$) and $PSL(2, 2^n)$, where $2^n + 1$ is prime, have orders which are divisible by a prime whose cube exceeds the order of the group. He conjectured that these were the only such simple groups and Brauer and Reynolds [11] used modular character theory (the character values lie in a field of prime characteristic) to prove this conjecture.

Twelve years after Ling and Miller completed the range problem up to 2000, L. P. Siceloff, at the suggestion of Cole, proved [71] that the only integers between 2001 and 3640 which are orders of simple groups are 2448, 2520, and 3420. All of these were on Dickson's list in 1901. Since 2448 and 3420 have the form $p\left(p^2 - 1\right)/2$ the uniqueness question concerning these integers had been answered affirmatively by Frobenius ten years earlier and it was Miller who ten years afterward showed [65] that A_7 is the unique simple group of order 2520. Thus by 1922 all simple groups of order up to 3640 had been determined.

Again there was a twelve-year hiatus before the exhaustive enumeration of integers in a certain range was continued. In 1924, Cole returned to the problem again and showed [26] that the only integers between 3641 and 6232 which are orders of simple groups were the four on Dickson's 1901 list. Unfortunately Cole's paper was so lacking in detail that its value was diminished. The uniqueness question for $PSL(2, 23)$ (order 6072) had previously been settled affirmatively by Frobenius, and Cole did the same for $PSL(2, 16)$ (order 4080). Eighteen years after Cole's paper Brauer [7] showed, with the use of character theory, that the remaining two integers also corresponded to unique simple groups. So by 1942 all simple groups of order as far as 6232 had been determined.

10 Burnside's $p^a q^b$ theorem

In 1904, Burnside used character theory to prove [19] that every group of order $p^a q^b$, where p and q are primes, is solvable. This theorem represented the final generalization of a large number of special cases which had been established by Sylow, Frobenius, Burnside, Jordan, and Cole; it has become the classic example of the power of character theory. With character theory a simple proof was possible 70 years ago, but a character-free proof of Burnside's theorem, although long sought, has appeared only in the past few years. Thompson had indicated in his fundamental paper [75] on minimal simple groups (see section 14) that a character-free proof of the Burnside theorem could be extracted from that paper and the "odd order paper" [38]. David Goldschmidt [43], in 1970, gave a short character-free proof of the theorem when p and q are odd and Helmut Bender [5] two years later proved the general result without character theory. By combining the arguments of Bender in the odd order case and of Hiroshi Matsuyama [61] in the even order case it is now possible to obtain a short and attractive character-free proof of the $p^a q^b$ theorem.

Despite the early outstanding achievements with character theory by Burnside and Frobenius, others seemed to ignore it as a tool in simple group theory until Brauer brought it to the forefront in the 1940s and 1950s. This is partly explained by the fact that interest in simple groups began to wane around 1905.

11 The Chevalley groups

In 1948 Dieudonné [34] classified all the known simple groups according to their method of construction. This classification scheme modernized the one Dickson had devised in 1901. Whereas Dickson obtained his results by means of complicated matrix calculations which often obscured the underlying ideas, Dieudonné utilized the geometric properties of linear transformations and vector space theory to simplify and clarify Dickson's work. Although Dieudonné's approach to the classical linear groups is much more elegant than Dickson's, his methods still required that each family of simple groups be treated individually and he found no new simple groups. Thus, simple group theory was revitalized in 1955 when Chevalley, in his celebrated paper [22], introduced a new approach which provided a uniform method for investigating three kinds of the classical linear groups and Dickson's simple groups of Lie type as well. In addition to encompassing most of the then-known simple groups, his method also yielded several new infinite families of finite simple groups.

This was accomplished in the following way. With every pair (L, K) where L is a *Lie algebra* over the complex numbers (i.e., an algebra whose associative law is replaced by the Jacobi identity and also satisfies the condition $x^2 = 0$ for all x) and K is a field, a new Lie algebra L_K is constructed. Chevalley was able to associate with every such pair a certain subgroup of the automorphism group of L_K which is simple. With the appropriate choice of L and K these simple groups are those investigated by Jordan, Dickson, and Dieudonné. With other choices of L and K Chevalley obtained his new simple groups (the smallest of these has order $2^{24} 3^6 5^2 7^2 13 \cdot 17$). These groups were the first new simple groups found in more than 50 years. That they were indeed new was established by comparing their orders with the orders of the simple groups which had been known. The formulas for the orders of the Chevalley groups over finite fields were derived by using topological properties of the Lie group of the same type. Artin [1] developed a new classification scheme for the known simple groups which included the Chevalley groups. He used fewer classes than did Dickson and Dieudonné and his method considerably improved theirs (see also [2]).

12 Groups of Lie type

During the period 1958–1959 Chevalley's methods were extended and modified by Robert Steinberg, Jacques Tits, and D. Hertzig (see [21]) to obtain additional new infinite families of simple groups and the classical groups not handled by Chevalley. Shortly thereafter, Suzuki [73], while in the process of classifying a certain type of doubly transitive permutation groups, also discovered another new infinite family. Analyzing the Suzuki groups, Rimhak Ree noticed that when interpreted from a Lie-theoretical point of view, they were closely related to a certain family of Chevalley groups. He then showed that the method of Steinberg could be used to construct the Suzuki groups. This in turn led him to investigate two other similar situations and eventually discover his two families of simple groups ([68], [69]). The Suzuki and Ree groups, together with those of Chevalley and Steinberg, are collectively referred to as the *simple groups of Lie type*. These, together with the alternating groups A_n ($n \geq 5$), account for all but 26 or so of the finite simple groups known to date.

The Suzuki groups are noteworthy for another reason. They provided the first examples of simple groups whose orders are not divisible by 3, and Thompson, in a major classification theorem, has recently shown that these are the only possible such groups. The elements of the Suzuki groups are certain 4×4 matrices with entries from the Galois fields of order 2^{2n+1}, again illustrating the extremely important role that matrix groups over finite fields play in simple group theory.

13 Sporadic simple groups

A simple group which no one has yet been able to fit into an infinite class of simple groups in a natural way is called a *sporadic* simple group. For example, A_n and $PSL(n, q)$ are infinite families of simple groups while the five Mathieu groups are sporadic. In the classification schemes of Dickson, Dieudonné, and Artin only the five Mathieu groups are sporadic and no new ones were discovered until Zvonimir Janko found one of order 175,560 in 1966 [57]. Remarkably, the discoveries of Mathieu and Janko are separated by more than one hundred years. Using other results of Janko, M. Hall proved [49] the existence of a sporadic simple group of order 604,800 in 1967. This brought to 56 the number of known simple groups of order less than 1,000,000, and to 7 the number of sporadic simple groups. Paraphrasing Gorenstein ([47], p. 14) we recount the following anecdote in connection with Hall's simple group:

> Shortly after Hall constructed his group he gave a lecture on it at Oxford. Donald Higman and Charles Sims were in the audience and they were both struck by the fact that one of the Mathieu groups had certain permutation properties analogous to those of a group which Hall had used in his construction. That very same night this observation led them to the construction of a new sporadic simple group!

Hall's method has also led to the discovery of two others by analogous methods. By now there are 26 or so sporadic simple groups and it has been proved that there are exactly 56 simple groups of order less than 1,000,000. (Of late, new sporadic simple groups are being discovered so frequently that it is difficult for one to be sure of their precise number.)

Some of the sporadic simple groups have been discovered in the course of solving certain problems in permutation group theory, while others have turned up as the automorphism group of a distance-transitive graph (see Chapter 4 and the Appendix in [6]). Quite often, two or more sporadic groups are related in some way. Indeed, the Conway .1 simple group contains 12 sporadic simple groups as subgroups! The existence of many of these recently-discovered groups has been verified by means of a permutation representation of the group and extensive use of computers.

An important technique which has led to the discovery of a number of sporadic simple groups involves the notion of the centralizer of an *involution* (i.e., of an element of order 2). This method is employed in the following way. Choose H to be the centralizer of an involution from some known simple group G. Next, assume G^* is any simple group which contains an involution x such that $C(x)$ is isomorphic to H. (By a theorem of Brauer and Fowler [10] only a finite number of such groups can exist, so H "almost" determines G.) Then a great deal of information about G^* can be obtained. With this information it is often possible to show that G^* is isomorphic to G or to some other known simple group. For example Dieter Held [53] has proved such a theorem when $G = A_8$ or A_9. If it cannot be shown that G^* must be isomorphic to some known simple group, the information may be adequate to suggest a method of constructing a new simple group. Held [54] has also been instrumental in accomplishing this. He began with the observation that $PSL(5, 2)$ and the largest Mathieu group possess involutions with isomorphic centralizers, and no other known simple group has this property. Choosing this for H, he was led to three possible configurations for G^*. Ultimately, enough properties of this third group were derived so that Graham Higman and John McKay were able to construct it with the use of a computer.

Similarly, one may proceed by assuming G is a simple group which contains an involution whose centralizer closely resembles the form of a centralizer of an involution from a known simple group. In this case, if the information about G is not self-contradictory it suggests the possible existence of a new simple group and may be sufficient to lead to an actual construction of the group. This is how Janko discovered his simple group of order 175,560. Each member of a family of simple groups discovered by Ree has a centralizer isomorphic to $\mathbb{Z}_2 \times PSL(2, 3^n)$ and has its Sylow 2-subgroups isomorphic to $\mathbb{Z}_2 \times \mathbb{Z}_2 \times \mathbb{Z}_2$.

Janko set out to determine all simple groups which have a centralizer isomorphic to $\mathbb{Z}_2 \times PSL(2, p^n)$), p odd, and with Sylow 2-subgroups isomorphic to $\mathbb{Z}_2 \times \mathbb{Z}_2 \times \mathbb{Z}_2$. Eventually he was able to show that either $p = 3$ and the group is of Ree type or $p^n = 5$. The information he obtained about this latter case led him to write down a pair of 7×7 matrices with entries in the field of order 11 which generated a new simple group.

This method was also recently used independently by Bernd Fischer and Thompson and by Robert Griess to discover a possible new simple group. This object, called the *Monster M*, is not defined in terms of generators and relations. In fact, it has not been defined at all! What Fischer, Thompson and Griess did was to assume that there exists a finite simple group M that satisfies certain hypotheses. They showed, under this assumption, that M would be in fact a new simple group (i.e., not isomorphic to any existing simple group). They also obtained a great deal of other information about M, such as its order, properties of certain subgroups, and a portion of its character table. Thompson has computed the order of M to be

$$808{,}017{,}424{,}794{,}512{,}875{,}886{,}459{,}904{,}961{,}710{,}$$
$$757{,}005{,}754{,}368{,}000{,}000{,}000$$
$$= 2^{46} \cdot 3^{20} \cdot 5^9 \cdot 7^6 \cdot 11^2 \cdot 13^3 \cdot 17 \cdot 19 \cdot 23 \cdot 29 \cdot$$
$$31 \cdot 41 \cdot 59 \cdot 71$$

(hence the name). If there is a simple group satisfying the stipulated hypothesis we should be able to deduce what it must look like, and once we know what it looks like, we should be able to define it. For the Monster M this last step has not yet been accomplished.

A certain *section* of M (i.e., a group of the form H/K, where H and K are subgroups of M and K is normal in H), the *Baby Monster B*, is also a possible new simple group. Fischer has computed the order of B to be

$$4{,}154{,}781{,}481{,}226{,}426{,}191{,}177{,}580{,}544{,}000{,}000$$
$$= 2^{41} \cdot 3^{13} \cdot 5^6 \cdot 7^2 \cdot 11 \cdot 13 \cdot 17 \cdot 19 \cdot 23 \cdot 41 \cdot 47.$$

Of course, neither of the groups M or B has yet been shown to exist.

In addition to the role the centralizers of involutions play in the discovery of new simple groups, these subgroups are important for another reason. There is presently under way a systematic attempt to use the centralizers of involutions as a means for classifying all the known finite simple groups. This program began in 1954 with Brauer's characterization of $PSL(3, q)$, q odd, and is now almost complete. According to Gorenstein ([47], p. 21),

> Probably more individuals have been involved in this effort than in any other single area of simple group theory.

We refer the reader to section 4.4 of [36] for a survey of results of this type.

14 Thompson's *N*-paper

Most simple groups contain other simple groups as subgroups. For example,

$$A_5 \subset A_6 \subset A_7 \subset \cdots .$$

On the other hand, a minimal simple group is one, all of whose proper subgroups are solvable. It follows then that every simple group has a minimal simple group as a section. Minimal simple groups are therefore basic and the complete determination of all such groups would clearly be of great value. In the early 1960s Thompson set out to do just this. Such an endeavor was a natural successor to the Odd Order Theorem since the minimal counterexample G in that proof was a minimal simple group. Actually, Thompson decided to tackle a more general classification problem. The normalizer of a non-identity solvable subgroup of a group G is called a *local* subgroup of G and an N-group is one in which all local subgroups are solvable. Evidently, every minimal simple group is also an N-group.

As early as 1963, Thompson had concluded that, with only finitely many exceptions, the simple N-groups were $PSL(2, q)$ ($q > 3$) and the Suzuki groups. The complete classification of all non-solvable N-groups however, did not come until several years later. The 407-page proof (!!) [75] of this remarkable theorem is spread out over six journal issues during the seven-year period 1968–1974. Describing his approach, Thompson writes ([75], p. 383):

> In a broad way, this paper may be thought of as a successful transformation of the theory of solvable groups to the theory of simple groups. By this is meant that a substantial structure is constructed which makes it possible to exploit properties of solvable groups to obtain delicate

information about the structure and embedding of many solvable subgroups of the simple group under consideration. In this way, routine results about solvable groups acquire great power.

(An essay which outlines the organization of the proof and discusses some of the arguments used is given in [46], pp. 473–480.)

This result has a number of important corollaries, the most important of which is a classification of all minimal simple groups. A few consequences of this corollary are mentioned in the next two sections. For his many profound contributions to simple group theory, Thompson was awarded the Fields Medal—the mathematical equivalent of the Nobel Prize—by the International Congress of Mathematicians in 1970.

15 The $p^a q^b r^c$ problem

One of the corollaries of Thompson's result classifying the minimal simple groups seems to have put the solution of a very difficult, well-known problem within reach. This problem concerns the natural generalization of Burnside's $p^a q^b$ theorem to three primes. Since there are eight known simple groups which have orders divisible by exactly three distinct primes, the logical extension of Burnside's result would be the complete determination of all simple groups with orders of the form $p^a q^b r^c$. The first steps in this direction were taken by Burnside, Frobenius and E. Maillet. For example, Burnside [14] showed that there is no simple group whose order has the form $pq^b r$, where $p < q < r$. Fifty years later Brauer and Hsio Tuan [12] used character theory to show that except for the groups $PSL(2,5)$ and $PSL(2,7)$, the restriction that $p < q < r$ was unnecessary. In 1962 and again in 1968 Brauer ([8], [9]) returned to this problem and determined all simple groups whose orders have the form $p^a q^b r$, where $a = 1$ or 2, or the form $2^a 3^b 5$.

In spite of the fact that Brauer and his predecessors had solved the "three-prime problem" in numerous special cases, the complete solution was far from sight until Thompson proved his result. Specifically, he proved that a simple group whose order is divisible by exactly three distinct primes must have one of

$$PSL(2,4),\ PSL(2,7),\ PSL(2,8),$$
$$PSL(2,17) \text{ or } PSL(3,3)$$

as a section. From this it follows that such a group must have order of the form $2^a 3^b p^c$ where p is 5, 7, 13 or 17. Then, since character theory is a natural tool for analyzing groups whose orders have a prime to the first power only, David Wales [76] used it in conjunction with the N-paper to determine all simple groups (8 of them) whose orders have the form $2^a 3^b p^c$. Finally, Kenneth Klinger and Geoffrey Mason are presently in the midst of showing (they hope) that there are no simple groups with orders of the form $2^a 3^b p^c$ with $c > 1$. Of course, the completion of this work will finish the $p^a q^b r^c$ problem.

16 The range problem to 1,000,000

At the suggestion of Brauer, Sister Michaels, in her 1963 Ph.D. dissertation [62], showed that there was no known simple group in the range 6232 to 20,000. Her work was superseded during the late 1960s and early 1970s when the range problem was taken up by M. Hall ([49], [50]). Using a wide assortment of methods from elementary to advanced, as well as a computer, he succeeded in eliminating all but 21 of the first 1,000,000 integers as possible orders for new simple groups. For the integers not eliminated by elementary considerations, the theory of modular characters and Thompson's result on minimal simple groups played an important role. The character theory yields integer equations which certain parameters of the group must satisfy and a computer was used to make the verifications. Every simple group must have a section which is a minimal simple group, and so the order of any simple group is divisible by the order of a minimal simple group. Hall was able to show that from among Thompson's list of minimal simple groups only

$$PSL(2,5),\ PSL(2,7),\ PSL(2,8),$$
$$PSL(2,13), PSL(2,17), PSL(3,3),$$
$$PSL(2,23) \text{ and } PSL(2,27)$$

could occur as a section of an unknown simple group of order less than 1,000,000. From a result of Gorenstein it then follows that such a group has order divisible by 840 or 2184. Eventually the list was pared down to 1146 integers which required individual consideration. The first paper [49] eliminates all but about 100 of these and the second paper [50] reduces the list to 21.

Then Paul Fong [39] classified all simple groups whose Sylow 2-subgroups have order at most 26 and this reduced Hall's list to 13 integers as possible orders for new simple groups of order less than 1,000,000. Finally, the range problem to 1,000,000 was recently finished when two students of Held, Bert Beisiegel and Volker Stingl [3], eliminated the remaining integers on Hall's list by extending Fong's work as far as the case 2^{10}.

In conclusion we mention that even though the range problem is not the central one in simple group theory, this achievement is a dramatic illustration of how far the theory has progressed in the years since Hölder determined the simple groups of order up to 200.

A Chronological Collection of the Highlights in the Theory of Simple Groups

Date	Mathematician	Achievement
1870	Jordan	Established simplicity of alternating groups and linear groups over fields of prime order.
1892	Hölder	Began range problem.
1895–1900	Cole, Miller	Proved simplicity of Mathieu groups.
1896–1901	Frobenius–Burnside	Developed character theory.
1897–1905	Dickson	Established simplicity of linear groups over arbitrary finite fields. Discovered a family of simple groups of Lie type.
1904	Burnside	Proved $p^a q^b$ theorem.
1954	Brauer	Began the program of characterizing simple groups in terms of centralizers of involutions.
1955	Chevalley	Discovered new approach to simple groups. Discovered new families of simple groups of Lie type.
1958–1961	Steinberg, Tits, Hertzig, Ree	Extended Chevalley's methods and discovered new infinite families of simple groups of Lie type.
1960	Suzuki	Discovered new infinite family of simple groups (only simple groups with orders not divisible by 3).
1963	Feit–Thompson	Proved simple groups have even order.
1965	Gorenstein–Walter	Classified all simple groups with dihedral Sylow 2-subgroups.
1966–1975	Janko, Hall, Higman, Sims, McKay, McLaughlin, Suzuki, Held, Conway, Thompson, Fischer, Lyons, Rudvalis, Wales, O'Nan, Smith	Discovered new sporadic simple groups.
1968	Thompson	Proved N-Theorem. Classified all minimal simple groups.
1969	Walter	Classified all simple groups with Abelian Sylow 2-subgroups.
1971	Thompson	Proved Suzuki groups are the only simple groups with orders not divisible by 3.
1972	Wales	Classified all simple groups with orders of the form $p^a q^b r$.
1975	Hall, Beisiegel–Stingl	Completed range problem to 1,000,000.

The Known Finite Simple Groups: Their Types, Notations and Orders, Listed in Order of Discovery

Date and Discoverer	Type: Name	Group Notation	Order p a prime, $q = p^n$
1870 Jordan	alternating group of degree m	A_m $(m > 4)$	$m!/2$
1870 Jordan	classical linear projective special linear	$L_m(p)$ or $PSL(m,p)$ $(m > 1)$	$d^{-1} p^{m(m-1)/2} \prod_{i=2}^{m}(p^i - 1)$; $d = (m, p-1)$
1870 Jordan	classical linear symplectic	$S_{2m}(p)$ $(m > 1)$	$d^{-1} p^{m^2} \prod_{i=1}^{m}(p^{2i} - 1)$; $d = (2, p-1)$
1870 Jordan	classical linear orthogonal	$O_{2m}(\varepsilon, p)$, $\varepsilon = \pm 1$, $(m > 3)$	$d^{-1} p^{m(m-1)} (p^m - \varepsilon) \times \prod_{i=1}^{m-1}(p^{2i} - 1)$; $d = (4, p - \varepsilon)$
1870 Jordan	classical linear unitary	$U_m(p)$ $(m > 2)$	$d^{-1} p^{m(m-1)/2} \prod_{i=2}^{m}\left(p^i - (-1)^i\right)$ $d = (m, p+1)$
1893 Cole	classical linear $L_2(8)$	$L_2(8)$ or $PSL(2, 8)$	$504 = 2^3 \cdot 3^2 \cdot 7$
1893 Moore	classical linear projective special linear	$L_2(q)$ or $PSL(2, q)$ $(q > 3)$	$d^{-1} q(q^2 - 1)$; $d = (2, q-1)$
1895 Mathieu–Cole	sporadic Mathieu 11	M_{11}	$7920 = 2^4 \cdot 3^2 \cdot 5 \cdot 11$
1897 Dickson	classical linear projective special linear	$L_m(q)$ or $PSL(m, q)$ $(m > 1)$	$d^{-1} q^{m(m-1)/2} \prod_{i=2}^{m}(q^i - 1)$ $d = (m, q-1)$
1897 Dickson	classical linear symplectic	$S_{2m}(q)$ $(m > 1)$	$d^{-1} q^{m^2} \prod_{i=1}^{m}(q^{2i} - 1)$; $d = (2, q-1)$
1898 Dickson	classical linear orthogonal	$O_{2m}(\varepsilon, q)$, $\varepsilon = \pm 1$ $(m > 3)$	$d^{-1} q^{m(m-1)} (q^m - \varepsilon) \times \prod_{i=1}^{m-1}(q^{2i} - 1)$; $d = (4, q^m - \varepsilon)$
1898 Dickson	classical linear unitary	$U_m(q)$ $(m > 2)$	$d^{-1} q^{m(m-1)/2} \prod_{i=2}^{m}\left(q^i - (-1)^i\right)$ $d = (m, q+1)$
1899 Mathieu–Miller	sporadic Mathieu 12	M_{12}	$95{,}040 = 2^6 \cdot 3^3 \cdot 5 \cdot 11$
1900 Mathieu–Miller	sporadic Mathieu 22	M_{22}	$443{,}520 = 2^7 \cdot 3^2 \cdot 5 \cdot 7 \cdot 11$
1900 Mathieu–Miller	sporadic Mathieu 23	M_{23}	$10{,}200{,}960 = 2^7 \cdot 3^2 \cdot 5 \cdot 7 \cdot 11 \cdot 23$
1900 Mathieu–Miller	sporadic Mathieu 24	M_{24}	$244{,}823{,}040 = 2^{10} \cdot 3^3 \cdot 5 \cdot 7 \cdot 11 \cdot 23$
1901, 1905 Dickson	Lie groups of type G_2	$G_2(q)$	$q^6(q^6 - 1)(q^2 - 1)$
1955 Chevalley	Lie Chevalley groups of type E_4	$E_4(q)$	$q^{24}(q^{12} - 1)(q^8 - 1)$ $\times (q^6 - 1)(q^2 - 1)$

Date and Discoverer	Type: Name	Group Notation	Order p a prime, $q = p^n$
1955 Chevalley	Lie *Chevalley groups of type E_6*	$E_6(q)$	$d^{-1}q^{36}(q^{12}-1)(q^9-1)$ $\times(q^8-1)(q^6-1)(q^5-1)$ $\times(q^2-1);\ d=(3,q-1)$
1955 Chevalley	Lie *Chevalley groups of type E_7*	$E_7(q)$	$d^{-1}q^{63}(q^{18}-1)(q^{14}-1)$ $\times(q^{12}-1)(q^{10}-1)(q^8-1)$ $\times(q^6-1)(q^2-1);\ d=(2,q-1)$
1955 Chevalley	Lie *Chevalley groups of type E_8*	$E_8(q)$	$q^{120}(q^{30}-1)(q^{24}-1)$ $\times(q^{20}-1)(q^{18}-1)(q^{14}-1)$ $\times(q^{12}-1)(q^8-1)(q^2-1)$
1959 Steinberg–Tits–Hertzig	Lie *twisted groups of type E_6*	$^2E_6(q^2)$	$d^{-1}q^{36}(q^2-1)(q^5+1)$ $\times(q^6-1)(q^8-1)(q^9+1)$ $\times(q^{12}-1);\ d=(3,q+1)$
1959 Steinberg–Tits–Hertzig	Lie *twisted groups of type D_4*	$^3D_4(q^3)$	$q^{12}(q^2-1)(q^6-1)(q^8+q^4+1)$
1960 Suzuki	Lie *Suzuki groups*	$Sz(q)$ or $^2B_2(q)$, $q=2^{2m+1}$	$q^2(q^2+1)(q-1)$
1961 Ree	Lie *Ree groups of type G_2*	$^2G_2(q)$, $q=3^{2m+1}$	$q^3(q^3+1)(q-1)$
1961 Ree	Lie *Ree groups of type F_4*	$^2F_4(q)$, $q=2^{2m+1}$	$q^{12}(q^6+1)(q^4-1)$ $\times(q^3+1)(q-1)$
1966 Janko	sporadic *Janko*	Ja	$175{,}560 = 2^3 \cdot 3 \cdot 5 \cdot 7 \cdot 11 \cdot 19$
1967 Hall–Janko	sporadic *Hall–Janko*	HaJ	$604{,}800 = 2^7 \cdot 3^3 \cdot 5^2 \cdot 7$
1968 Higman–Sims	sporadic *Higman–Sims*	HiS	$44{,}352{,}000 = 2^9 \cdot 3^2 \cdot 5^3 \cdot 7 \cdot 11$
1969 Hall–Janko–McKay	sporadic *Hall–Janko–McKay*	HJM	$50{,}232{,}960 = 2^7 \cdot 3^5 \cdot 5 \cdot 17 \cdot 19$
1969 McLaughlin	sporadic *McLaughlin*	McL	$898{,}128{,}000 = 2^7 \cdot 3^6 \cdot 5^3 \cdot 7 \cdot 11$
1969 Suzuki	sporadic *Suzuki*	Suz	$448{,}345{,}497{,}600$ $= 2^{13} \cdot 3^7 \cdot 5^2 \cdot 7 \cdot 11 \cdot 13$
1969 Held–Higman–McKay	sporadic *Held–Higman–McKay*	HNM	$4{,}030{,}387{,}200$ $= 2^{10} \cdot 3^3 \cdot 5^2 \cdot 7^3 \cdot 17$
1969 Conway–Thompson	sporadic *Conway's .1 group*	Co_1	$4{,}157{,}776{,}806{,}543{,}360{,}000$ $= 2^{21} \cdot 3^9 \cdot 5^4 \cdot 7^2 \cdot 11 \cdot 13 \cdot 23$
1969 Conway–Thompson	sporadic *Conway's .2 group*	Co_2	$42{,}305{,}421{,}312{,}000$ $= 2^{18} \cdot 3^6 \cdot 5^3 \cdot 7 \cdot 11 \cdot 23$

Date and Discoverer	Type: *Name*	Group Notation	Order p a prime, $q = p^n$
1969 Conway–Thompson	sporadic *Conway's .3 group*	Co_3	495,766,656,000 $= 2^{10} \cdot 3^7 \cdot 5^3 \cdot 7 \cdot 11 \cdot 23$
1969 Fischer	sporadic *Fischer 22*	Fi_{22}	64,561,751,654,400 $= 2^{17} \cdot 3^9 \cdot 5^2 \cdot 7 \cdot 11 \cdot 13$
1969 Fischer	sporadic *Fischer 23*	Fi_{23}	4,089,460,473,293,004,800 $= 2^{18} \cdot 3^{13} \cdot 5^2 \cdot 7 \cdot 11 \cdot 13 \cdot 17 \cdot 23$
1969 Fischer	sporadic *Fischer 24*	Fi_{24}	1,255,205,709,190,661,721,292,800 $= 2^{21} \cdot 3^{16} \cdot 5^2 \cdot 7^3 \cdot 11 \cdot 13 \cdot 17 \cdot 23 \cdot 29$
1971 Lyons–Sims	sporadic *Lyons–Sims*	LyS	51,765,179,004,000,000 $= 2^8 \cdot 3^7 \cdot 5^6 \cdot 7 \cdot 11 \cdot 31 \cdot 37 \cdot 67$
1972 Rudvalis–Conway–Wales	sporadic *Rudvalis*	Rud	145,926,144,000 $= 2^{14} \cdot 3^3 \cdot 5^3 \cdot 7 \cdot 13 \cdot 29$
1973 O'Nan–Sims	sporadic *O'Nan*	$O'N$	460,815,505,920 $= 2^9 \cdot 3^4 \cdot 5 \cdot 7^3 \cdot 11 \cdot 19 \cdot 31$
1974 Fischer	sporadic *Monster*	M or F_1 (possible new simple group)	808,017,424,794,512,875,886,459,904, 961,710,757,005,754,368,000,000,000 $= 2^{46} \cdot 3^{20} \cdot 5^9 \cdot 7^6 \cdot 11^2 \cdot 13^2 \cdot 17 \cdot 19 \cdot 23 \cdot 29 \cdot 31 \cdot 41 \cdot 47 \cdot 59 \cdot 71$
1974 Fischer	sporadic *Baby Monster*	B or F_2 (possible new simple group)	4,154,781,481,226,426,191, 177,580,544,000,000 $= 2^{41} \cdot 3^{13} \cdot 5^6 \cdot 7^2 \cdot 11 \cdot 13 \cdot 17 \cdot 19 \cdot 23 \cdot 31 \cdot 41 \cdot 47$
1974 Fischer–Smith–Thompson	sporadic *Fischer 3 or Thompson group*	E or F_3	90,745,943,887,872,000 $= 2^{15} \cdot 3^{10} \cdot 5^3 \cdot 7^2 \cdot 13 \cdot 19 \cdot 31$
1974 Fischer–Smith	sporadic *Fischer 5 or Harada group*	F or F_5	273,030,912,000,000 $= 2^{14} \cdot 3^6 \cdot 5^6 \cdot 7 \cdot 11 \cdot 19$
1975 Janko	sporadic *Janko 4*	J_4 (possible new simple group)	86,775,571,046,077,562,880 $= 2^{21} \cdot 3^3 \cdot 5 \cdot 7 \cdot 11^3 \cdot 23 \cdot 29 \cdot 31 \cdot 37 \cdot 43$

References

1. E. Artin, The orders of the classical simple groups, *Comm. Pure Appl. Math.* **8** (1955), 455–472.

2. E. Artin, *Geometric Algebra*, Interscience, 1957.

3. B. Beisiegel, Über einfache endliche Gruppen mit Sylow-2-Gruppen der Ordnung höchstens 2^{10}, *Comm. Alg.* **5** (1977), 113–170.

4. E. T. Bell, Fifty years of algebra in America, 1888–1938, *Amer. Math. Soc. Semicentennial Publications* Vol. II, 1938, pp. 1–34.

5. H. Bender, A group theoretic proof of Burnside's $p^a q^b$ theorem, *Math. Zeitschrift* **126** (1972), 327–338.

6. N. Biggs, *Finite Groups of Automorphisms*, Cambridge Univ. Press, 1971.

7. R. Brauer, On groups whose order contains a prime number to the first power I, *Amer. J. Math.* **64** (1940), 401–420.

8. R. Brauer, On some conjectures concerning finite simple groups, *Studies in Mathematical Analysis Related Topics*, Stanford Univ. Press, 1962, pp. 56–61.

9. R. Brauer, On simple groups of order $5 \cdot 3^a 2^b$, *Bull. Amer. Math. Soc.* **74** (1968), 900–903.

10. R. Brauer and K. A. Fowler, Groups of even order, *Ann. Math.* **62** (1955), 565–583.

11. R. Brauer and W. F. Reynolds, On a problem of E. Artin, *Ann. Math.* **68** (1958), 713–720.

12. R. Brauer and H. Tuan, On simple groups of finite order I, *Bull. Amer. Math. Soc.* **51** (1945), 756–766.

13. W. Burnside, On a class of groups defined by congruences, *Proc. London Math. Soc.* **25** (1894), 113–139.

14. W. Burnside, Notes on the theory of groups of finite order, *Proc. London Math. Soc.* **26** (1895), 191–214.

15. W. Burnside, Notes on the theory of groups of finite order (continued), *Proc. London Math. Soc.* **26** (1895), 325–338.

16. W. Burnside, On transitive groups of degree n and class $n-1$, *Proc. London Math. Soc.* **32** (1900), 240–246.

17. W. Burnside, On some properties of groups of odd order, *Proc. London Math. Soc.* **33** (1901), 162–185.

18. W. Burnside, On some properties of groups of odd order (second paper), *Proc. London Math. Soc.* **33** (1901), 257–268.

19. W. Burnside, On groups of order $p^a q^b$, *Proc. London Math. Soc.* **2** (1904), 388–392.

20. W. Burnside, *Theory of Groups of Finite Order*, 2nd ed., Dover, 1955.

21. R. Carter, *Simple Groups of Lie Type*, Wiley, 1972.

22. C. Chevalley, Sur certains groupes simples, *Tohoku Math. J.* **7** (1955), 14–66.

23. F. Cole, Simple groups from order 201 to order 500, *Amer. J. Math.* **14** (1892), 378–388.

24. F. Cole, Simple groups as far as order 660, *Amer. J. Math.* **15** (1893), 303–315.

25. F. Cole, List of the transitive substitution groups of ten and of eleven letters, *Quart. J. Pure Appl. Math.* (1895), 39–50.

26. F. Cole, On simple groups of low order, *Bull. Amer. Math. Soc.* **30** (1924), 489–492.

27. G. Cornell, N. Pelc, and M. Wage, Simple groups of order less than 1000, *J. Undergraduate Math.* (1973), 77–86.

28. C. W. Curtis, The classical groups as a source of algebraic problems, *Amer. Math. Monthly* **74** (1967), 80–91.

29. L. E. Dickson, The analytic representation of substitutions on a power of a prime number of letters with a discussion of the linear group, *Ann. Math.* **11** (1897), 161–183.

30. L. E. Dickson, Proof of the non-isomorphism of the simple Abelian group on $2m$ indices and the orthogonal group on $2m+1$ indices for $m>2$, *Quart. J. Pure Appl. Math.* **32** (1900), 42–63.

31. L. E. Dickson, Theory of linear groups in an arbitrary field, *Trans. Amer. Math. Soc.* **2** (1901), 363–394.

32. L. E. Dickson, A new system of simple groups, *Math. Ann.* **60** (1905), 137–150.

33. L. E. Dickson, *Linear Groups with an Exposition of the Galois Field Theory*, Dover, 1958.

34. J. Dieudonné, *Sur les Groupes Classiques*, Actualités Sci. Indust., **1040**, Hermann, 1948.

35. J. Dieudonné, On the automorphisms of the classical groups, *Mem. Amer. Math. Soc.* **2** (1951).

36. W. Feit, The current situations in the theory of finite simple groups, *Proc. Internat. Congr. Mathematicians Nice, 1970*, Gauthier-Villars, 1971.

37. W. Feit, M. Hall, Jr., and J. G. Thompson, Finite groups in which the centralizer of any non-identity element is nilpotent, *Math. Zeitschrift* **74** (1960), 1–17.

38. W. Feit and J. G. Thompson, Solvability of groups of odd order, *Pacific J. Math.* **13** (1963), 775–1029.

39. P. Fong, private communication.

40. G. Frobenius, Über Gruppen des Grades p oder $p+1$, *Berliner Sitzgsb.* (1902), 351–369.

41. E. Galois, Sur la théorie des nombres, *J. Math. Pures Appl.* **11** (1846), 398–407.

42. G. Glauberman, Subgroups of finite groups, *Bull. Amer. Math. Soc.* **73** (1967), 1–12.

43. D. Goldschmidt, A group theoretic proof of the $p^a q^b$ theorem for odd primes, *Math. Zeitschrift* **113** (1970), 373–375.

44. D. Goldschmidt, Elements of order two in finite groups, *Delta* **4** (1974), 45–58.

45. D. Gorenstein, Some topics in the theory of finite groups, *Rend. Mat. e Appl.* (5) **23** (1964), 298–315.

46. D. Gorenstein, *Finite Groups*, Harper and Row, 1968.

47. D. Gorenstein, Finite simple groups and their classification, *Israel J. Math.* **19** (1974), 5–66.

48. D. Gorenstein and J. Walter, The characterization of finite groups with dihedral Sylow 2-subgroups, *J. Algebra* **2** (1965), 85–151, 218–270, 354–393.

49. M. Hall, A search for simple groups of order less than one million, *Computational Problems in Abstract Algebra* (ed. J. Leech), Pergamon Press, 1969, pp. 137–168.

50. M. Hall, Simple groups of order less than one million, *J. Algebra* **20** (1972), 98–102.

51. T. Hawkins, The origins of the theory of group characters, *Arch. Hist. Exact Sci.* **7** (1971), 142–170.

52. T. Hawkins, New light on Frobenius' creation of the theory of group characters, *Arch. Hist. Exact Sci.* **12** (1974), 215–243.

53. D. Held, A characterization of the alternating groups of degrees eight and nine, *J. Algebra* **7** (1967), 218–237.

54. D. Held, The simple group related to M_{24}, *J. Algebra* **13** (1969), 253–296.

55. I. N. Herstein, *Topics in Algebra*, 2nd ed., Xerox College, 1975.

56. O. Hölder, Die einfachen Gruppen im ersten und zweiten Hundert der Ordnungszahlen, *Math. Ann.* **40** (1892), 55–88.

57. Z. Janko, A new finite simple group with Abelian Sylow 2-subgroups and its characterization, *J. Algebra* **3** (1966), 147–186.

58. C. Jordan, *Traité des Substitutions*, Gauthier-Villars, 1870.

59. R. Lindberg and J. Robinson, *A project in simple group theory*, Senior paper, Univ. of Minn., Duluth, 1973.

60. G. Ling and G. A. Miller, Proof that there is no simple group whose order lies between 1092 and 2001, *Amer. J. Math.* **22** (1900), 13–26.

61. H. Matsuyama, Solvability of groups of order $2^a p^b$, *Osaka J. Math.* **10** (1973), 375–378.

62. E. Michaels, *A study of simple groups of even order*, Ph.D. dissertation, Univ. of Notre Dame, 1963.

63. G. A. Miller, On the simple groups which can be represented as substitution groups that contain cyclical substitutions of a prime degree, *Amer. Math. Monthly* **6** (1899), 102–103.

64. G. A. Miller, Sur plusieurs groupes simples, *Bull. Soc. Math. France* **28** (1900), 266–267.

65. G. A. Miller, The simple group of order 2520, *Bull. Amer. Math. Soc.* **22** (1922), 98–102.

66. E. H. Moore, A doubly-infinite system of simple groups, *Bull. New York Math. Soc.* **3** (1893), 73–78.

67. M. B. Powell and G. Higman (eds.), *Finite Simple Groups*, Academic Press, 1971.

68. R. Ree, A family of simple groups associated with the simple Lie algebra of type (F_4), *Amer. J. Math.* **83** (1961), 401–420.

69. R. Ree, A family of simple groups associated with the simple Lie algebra of type (G_2), *Amer. J. Math.* **83** (1961), 432–462.

70. I. M. Schottenfels, Two non-isomorphic simple groups of the same order 20,160, *Ann. Math.* **1** (1900), 147–152.

71. L. Siceloff, Simple groups from order 2001 to order 3640, *Amer. J. Math.* **34** (1912), 361–372.

72. M. Suzuki, The nonexistence of a certain type of simple groups of odd order, *Proc. Amer. Math. Soc.* **8** (1957), 686–695.

73. M. Suzuki, A new type of simple groups of finite order, *Proc. Nat. Acad. Sci. U.S.A.* **46** (1960), 868–870.

74. P. Telega, *A computer project in simple group theory*, Senior paper, Univ. of Minn., Duluth, 1975.

75. J. G. Thompson, Nonsolvable finite groups all of whose local subgroups are solvable, *Bull. Amer. Math. Soc.* **74** (1968), 383–437; *Pacific J. Math.* **33** (1970), 451–537; *Pacific J. Math.* **39** (1971), 483–534; *Pacific J. Math.* **48** (1973), 511–592; *Pacific J. Math.* **50** (1974), 215–297; *Pacific J. Math.* **51** (1974), 573–630.

76. D. Wales, Simple groups of order $13 \cdot 3^a 2^b$, *J. Algebra* **20** (1972), 124–143.

77. J. H. Walter, The characterization of finite groups with abelian Sylow 2-subgroups, *Ann. Math.* **89** (1969), 405–514.

78. H. Wielandt, *Finite Permutation Groups*, Academic Press, 1964.

79. W. J. Wong, Recent work on finite simple groups, *Math. Chronicle* **1** (1969), 5–12.

A Simple Song

For readers who have difficulty in keeping track of the history of simple group theory we offer the following summary in ballad form. This classic first appeared in print in 1973 in the *American Mathematical Monthly* (p. 1028) where it is claimed to have been found scrawled on a table in Eckhart Library at the University of Chicago. The author, possibly for cause, is claimed to be unknown. The tune is that of "Sweet Betsy from Pike".

Oh, what are the orders of all simple groups?
I speak of the honest ones, not of the loops.
It seems that old Burnside their orders has guessed
Except for the cyclic ones, even the rest.

CHORUS: *Finding all groups that are simple is no simple task.*

Groups made up with permutes will produce
   Some more:
For A_n is simple, if n exceeds 4.
Then, there was Sir Matthew who came into view
Exhibiting groups of an order quite new.

Still others have come on to study this thing.
Of Artin and Chevalley now we shall sing.
With matrices finite they made quite a list
The question is: Could there be others they've missed?

Suzuki and Ree then maintained it's the case
That these methods had not reached the end
   of the chase.

They wrote down some matrices, just four by four,
That made up a simple group. Why not make more?

And then came the opus of Thompson and Feit
Which shed on the problem remarkable light.
A group, when the order won't factor by two
Is cyclic or solvable. That's what is true.

Suzuki and Ree had caused eyebrows to raise,
But the theoreticians they just couldn't faze.
Their groups were not new: if you added a twist,
You could get them from old ones with a flick of the wrist.

Still, some hardy souls felt a thorn in their side.
For the five groups of Mathieu all reason defied;
Not A_n, not twisted, and not Chevalley,
They called them sporadic and filed them away.

Are Mathieu groups creatures of heaven or hell?
Zvonimir Janko determined to tell.
He found out that nobody wanted to know:
The masters had missed 1 7 5 5 6 0.

The floodgates were opened! New groups were the rage!
(And twelve or more sprouted, to greet the new age.)
By Janko and Conway and Fischer and Held
McLaughlin, Suzuki, and Higman, and Sims.

No doubt you noted the last lines don't rhyme.
Well, that is, quite simply, a sign of the time.
There's chaos, not order, among simple groups;
And maybe we'd better go back to the loops.

Genius and Biographers: The Fictionalization of Evariste Galois

TONY ROTHMAN

American Mathematical Monthly **89** (1982), 84–106

1 Introduction

In Paris, on the obscure morning of May 30, 1832, near a pond not far from the pension Sieur Faultrier, Evariste Galois confronted Pescheux d'Herbinville in a duel to be fought with pistols, and was shot through the stomach. Hours later, lying wounded and alone, Galois was found by a passing peasant. He was taken to the Hospital Cochin where he died the following day in the arms of his brother Alfred, after having refused the services of a priest. Had Galois lived another five months, until October 25, he would have attained the age of twenty-one.

The legend of Evariste Galois, creator of group theory, has fired the imagination of generations of mathematics students. Many of us have experienced the excitement of Freeman Dyson who writes ([8], p. 14):

In those days, my head was full of the romantic prose of E. T. Bell's *Men of Mathematics*, a collection of biographies of the great mathematicians. This is a splendid book for a young boy to read (unfortunately, there is not much in it to inspire a girl, with Sonya Kovalevsky allotted only half a chapter), and it has awakened many people of my generation to the beauties of mathematics. The most memorable chapter is called "Genius and Stupidity" and describes the life and death of the French mathematician Galois, who was killed in a duel at the age of twenty.

Dyson goes on to quote Bell's famous description ([1], p. 375) of Galois's last night before the duel:

All night long he had spent the fleeting hours feverishly dashing off his scientific last will and testament, writing against time to glean a few of the great things in his teeming mind before the death which he saw could overtake him. Time after time he broke off to scribble in the margin "I have not time; I have not time," and passed on to the next frantically scrawled outline. What he wrote in those last desperate hours before the dawn will keep generations of mathematicians busy for hundreds of years. He had found, once and for all, the true solution of a riddle which had tormented mathematicians for centuries: under what conditions can an equation be solved?

This extract is likely the very paragraph which has given the greatest impetus to the Galois legend. As with all legends the truth has become one of many threads in the embroidery. E. T. Bell has embroidered more than most, but he is not alone. James R. Newman, writing in *The World of Mathematics* ([16], Vol. 3, p. 1534) notes: "The term *group* was first used in a technical sense by the French mathematician Evariste Galois in 1830. He wrote his brilliant paper on the subject at the age of twenty, the night before he was killed in a stupid duel." From a description in the famed Bullitt archives of mathematics [5] issued by the University of Louisville library, we learn: "Goaded by a 'mignonne' and two of her slattern confederates into a 'duel of honor', Galois was shot and killed at the age of 20." Leopold Infeld, in his biography of Galois [13], invokes a conspiracy theory to explain Galois's death: Galois was considered one of the most dangerous republicans in Paris; the government wanted to get rid of him; a female agent provocateur set him up for the duel with d'Herbinville; *et cetera*. Fred Hoyle, in his *Ten Faces of the Universe* ([11], Chapter 1), attempts a partial inversion of the argument: Galois's ability to carry on complex calculations entirely in his head made him appear distant to others; personal animosities arose with his republican friends; they began to think he was not fully for the cause; Galois in their eyes was the agent provocateur; *et cetera*. All three authors, Bell, Hoyle, and Infeld, invoke a political cause for the duel, with a mysterious coquette just off center.

This article is an attempt to sift some of the facts of Galois's life from the embroidery. It will not be an entirely complete account and will assume the reader is familiar with the story, presumably through Bell's version. Because these authors have emphasized the end of Galois's life, I will do so here. As will become apparent, many of the statements just cited are at worst nonsensical, and at best have no basis in the known facts.

Although a number of the documents presented here are, I believe, translated into English for the first time, it should be emphasized that they are not new, just ignored. There is more known about Galois than recent authors admit. It is my hope that some ambitious historian will find the requisite letter in an attic trunk or a newspaper clipping in the Paris archives to unravel the remaining mysteries.

2 Sources

It is not difficult to trace the story of Galois's brief life through its increasingly embellished incarnations. The primary source of information, containing eye-witness accounts and many relevant documents, is the original study of Paul Dupuy [7], which appeared in 1896. Dupuy was a historian and the Surveillant Général of the Ecole Normale. Bell, Hoyle, and Infeld all cite it as an important reference but never once explicitly quote it. Indeed, Bell acknowledges ([1], p. vii) that his account is based on Dupuy and the documents in Tannery [19] (see below), but it remains unclear how much Bell has read of Dupuy; for while numerous passages are lifted bodily from Dupuy, other important information in the latter is strangely absent. Dupuy's study itself is lacking a number of important letters and documents. Whether Dupuy was unaware of their existence or chose not to publish them I do not know. He also makes a number of minor errors in chronology. In any case, the first lesson is already learned: those who use Dupuy as their sole source of information must make mistakes. Nevertheless, this original biography is much more complete and accurate than the subsequent dilutions and contains more information than a reading of Bell,

Hoyle, or Infeld would even suggest. A translation of Dupuy into English should be undertaken.

Some of the documents not found in Dupuy are contained in Tannery's 1908 edition [19] of Galois's papers. All are contained in the definitive 1962 edition [4] of Bourgne and Azra. This volume contains every scrap of paper known to have been written by Galois, an accurate chronology, facsimiles of some of his original manuscripts, and a number of relevant letters by others. When quoting Galois, I have worked exclusively from this edition.

The memoirs of Alexandre Dumas ([6], Vol. 4, Chapter 204) contain a pertinent chapter, and the *Lettres sur les Prisons du Paris* by François Vincent Raspail ([17], Vol. 2) are the primary source on Galois's months in prison. Some of these letters are quoted by Dupuy and Infeld. Other references will be cited as they appear.

3 Early life and Louis-le-Grand

I will not dwell at length on the first sixteen years of Galois's life, for they are reported with fair accuracy by Bell. This is not surprising; his account approaches that of a somewhat abridged translation of Dupuy. The divergences will set in later. Thus this section and the next may be taken as a rather condensed review and criticism of Bell. Infeld and Hoyle, who concentrate most of their energies on the duel, will be dealt with at the appropriate time.

Evariste Galois was born on October 25, 1811, not far from Paris in the town of Bourg-la-Reine, France. His father was Nicholas-Gabriel Galois, who was then thirty-six years of age, and his mother was Adelaide-Marie Demante. Both parents were highly intelligent and well educated in the subjects considered important at the time: philosophy, classical literature, and religion. Bell points out that there is no record of any mathematical talent on either side of the family. A more neutral statement should perhaps be made: no record exists in favor of or against any such talent. M. Galois did possess the talent for composing rhymed couplets with which he would amuse neighbors. This harmless activity, as Bell notes, would later cause his undoing. Evariste seems to have inherited some of this ability, participating in the fun at house parties. For the first twelve years of his life, Evariste's mother served as his sole teacher, giving him a solid background in Greek and Latin, as well as passing on her own skepticism toward religion.

In 1815, during The One Hundred Days, M. Galois was elected mayor of Bourg-la-Reine. He had been a supporter of Napoleon and, in fact, had been elected chief of the town's liberal party during Napoleon's first exile. After Waterloo, he had planned to relinquish his post to his predecessor, but the latter had left the country. Galois demanded to be either confirmed or replaced, and in the confusion managed to keep his office. He served the new King faithfully, but from this point on he met increasing resistance from the conservative elements of his town. It is probably safe to say that the younger Galois inherited his liberal ideas from his parents.

On October 6, 1823, Evariste was enrolled in the Lycée of Louis-le-Grand, a famous preparatory school (which still exists) in Paris. Both Robespierre and Victor Hugo had studied there. Louis-le-Grand is where Evariste's troubles began, where Infeld's account of his life essentially opens, and where Bell introduces his theme of "Genius and Stupidity", taking on the tone of a blanket condemnation of almost everyone and everything that surrounded Galois. "Galois was no 'ineffectual angel'," Bell writes in his introduction ([1], p. 362), "but even his magnificent powers were shattered before the massed stupidity aligned against him, and he beat out his life fighting one unconquerable fool after another." I believe we will see that the problems ran much deeper than that.

Bell's first liberties with Dupuy are minor. Bell describes Louis-le-Grand as a "dismal horror" and goes on to say "the place looked like a prison and was" ([1], p. 363). Admittedly, Dupuy writes that Louis-le-Grand happened to look like a jail because of its grills, but he then goes on ([7], p. 203) to describe the underlying "passions of work, academic triumph, passions of liberal ideas, passions of memories of the Revolution and the Empire, contempt and hate for the legitimist reaction." Bell, by cutting Dupuy's sentence in half, has begun the slant toward the negative.

At this particular time, there were problems. During Galois's first term, the students, who suspected the new provisor of planning to return the conservative Jesuits to the school, protested by staging a minor rebellion. When required to sing at a chapel service, they refused. When required to recite in class, they refused. When required to toast King Louis XVIII at an official school banquet, they refused. The provisor summarily expelled the forty students whom he suspected of leading the insurrection. Galois was not among those expelled, nor is it known if he was even among the rebels, but we may guess that the arbitrariness of the provisor and the general severity of the school's regime made a deep impression on him.

Nevertheless, Galois's first two years at Louis-le-Grand were marked by a number of successes. He received a prize in the General Concourse and three mentions. At this point we witness the first of Bell's distortions of chronology to give the impression that Galois was misunderstood and persecuted. Galois was asked to repeat his third year because of his poor work in rhetoric. Bell writes, "His mathematical genius was already stirring," and "He was forced to lick up the stale leavings which his genius had rejected" ([1], p. 364; compare with [7], p. 205). I cannot say for certain whether Galois's mathematical genius was already stirring, but it is known that Galois did not enroll in his first mathematics course until *after* he had been demoted.

During this first mathematics course, which he began in February 1827, Galois discovered Legendre's text on geometry, soon followed by Lagrange's original memoirs: *Resolution of Numerical Equations* [= Algebraic Equations], *Theory of Analytic Functions*, and *Lessons on the Calculus of Functions*. Doubtlessly, Galois received his initial ideas on the theory of equations from Lagrange. I do not understand why Bell claims Galois's classwork was mediocre; his instructor, M. Vernier, constantly writes such accolades as "zeal and success", "zeal and progress very marked" ([7], pp. 255–256).

With his discovery of mathematics, Galois became absorbed and neglected his other courses. Before enrolling in M. Vernier's class, typical comments about him had been ([7], pp. 254–255):

Religious Duties—Good   Work—Sustained
Conduct—Good   Progress—Marked
Disposition—Happy   Character—Good, but singular

After a trimester in M. Vernier's class, the comments were:

Religious Duties—Good   Work—Inconstant
Conduct—Passable   Progress—Not very satisfactory
Disposition—Happy   Character—Closed and original

The words "singular", "bizarre", "original", and "closed" would appear more and more frequently during the course of Galois's career at Louis-le-Grand. His own family began to think him strange. His rhetoric teachers would term him "dissipated". Bell discusses these remarks at some length. His use of the indefinite pronoun "they" gives the impression that the entire faculty was aligned against Galois.

A perusal of Dupuy's appendix, however, shows the negative remarks were penned, by and large, by Galois's two rhetoric teachers. Until this point in his life, I believe it fair to say that Galois was somewhat misunderstood by his teachers in the humanities, but not that he was persecuted.

Slightly more serious problems were soon to arise. His mathematics teacher, M. Vernier, constantly implored Galois to work more systematically. His remark on one of Galois's trimester reports makes this clear ([7], p. 256):

> Intelligence, marked progress, but not enough method.

Galois did not take the advice; he took the entrance examination to l'Ecole Polytechnique a year early, without the usual special course in mathematics, and failed. Apparently he did not know some basics. To Galois, his failure was a complete denial of justice. This and subsequent rejections embittered him for life. When we examine some of his later writings, I think it will be evident that he developed not a little paranoia.

Galois did not yet give up. The same year, 1828, saw him enroll in the course of Louis-Paul-Emile Richard, a distinguished instructor of mathematics. Richard encouraged Galois immensely, even proclaimed that he should be admitted to the Polytechnique without examination. The results of such encouragement were spectacular. In April of 1829, Galois published his first small paper, "Proof of a Theorem on Periodic Continued Fractions". It appeared in the *Annales de Gergonne*.

This paper was a minor aside. Galois had also been working on the theory of equations (*Galois theory*). On May 25 and June 1, 1829, while still only 17, he submitted to the Academy his first researches on the solubility of equations of prime degree. Cauchy was appointed referee.

We now encounter a major myth which evidently has its origins in the very first writings on Galois and which has been perpetuated by virtually all writers since. This myth is the assertion that Cauchy either forgot or lost the papers (Dupuy [7], p. 209, Bell [1], p. 368) or intentionally threw them out (Infeld [13], p. 306). Recently, however, René Taton [20] has discovered a letter of Cauchy in the Academy archives which conclusively proves that he did not lose Galois's memoirs but had planned to present them to the Academy in January 1830. There is even some evidence that Cauchy encouraged Galois. The letter and related events will be discussed in more detail later; for now we note only that to hold Cauchy responsible for "one of the major disasters in the history of mathematics", to paraphrase Bell ([1], p. 368), is simply incorrect, and to add neglect by the Academy to the list of Galois's difficulties during this period appears entirely unwarranted.

A truly tragic blow came within a month of the submissions: on the second of July, 1829, Galois's father committed suicide. The reactionary priest of Bourg-la-Reine had signed Mayor Galois's name to a number of maliciously forged epigrams directed at Galois's own relatives. A scandal erupted. M. Galois's good nature could not withstand such an attack and he asphyxiated himself in his Paris apartment "not two steps from Louis-le-Grand." During the funeral, when the same clergyman attempted to participate, a small riot erupted. The loss of his father may explain much of Galois's future behavior. We must wait a few years, until Evariste's second prison term, to see this. In any case, he loved his father dearly, and if an iron link had not already been forged between the Bourbon government and the Jesuits, it had now. (Probably the clearest picture of the relationship between the Jesuits and the Bourbons, one which contains episodes paralleling that of M. Galois's misfortunes, is Stendhal's famous novel *The Red and The Black*.)

But Galois's troubles were not yet over. A few days later, he failed his examination to l'Ecole Polytechnique for the second and final time. Legend has it that Galois, who worked almost entirely in his head and who was poor at presenting his ideas verbally, became so enraged at the stupidity of the examiner that he hurled an eraser at him. Bell records this ([1], p. 369) as a fact, although Dupuy specifically states ([7], p. 211) that it is only an unverified tradition. The examination failures, as well as the misunderstanding of his humanities teachers, left him irrevocably embittered. Bell quotes him as writing ([1], p. 371), "Genius is condemned by a malicious social organization to an eternal denial of justice in favor of fawning mediocrity." I believe Bell constructed this quotation from a passage of Dupuy ([7], p. 217), but Galois did express similar sentiments in his fragmentary essay "Sciences Hiérarchie: Ecoles" and in "Sur l'Enseignment des Sciences" ("Hierarchy is a means for the inferior." ([4], p. 27). In Bell's diatribe against this famous examination, as well as in other accounts of it, the suggestion that the death of Galois's father several days before may have had something to do with the outcome never arises. It is a simple matter for Bell to lay the fault squarely with the examiner's stupidity because Bell has placed the examination before

M. Galois's unfortunate suicide. In this case, Bell is not fully to blame; Dupuy does not date the examination (I assume here, as elsewhere, the chronology in [4], pp. xxvii–xxxi). I do not wish to suggest Galois should not have been failed. I only wish to point out that the examination must have been held under the worst possible conditions.

Thus, Galois's secondary school career ended in a string of minor setbacks and two major disasters. Evariste had not planned to take the Baccalaureate examinations, because the Ecole Polytechnique did not require them. Now, having failed the Polytechnique's entrance examination and having decided to enter the less prestigious Ecole Normale, he was forced to reconsider. "Still persecuted and maliciously misunderstood by his preceptors," in Bell's words ([1], p. 370) "Galois prepared himself for the final examinations." Despite such malice, Galois did very well in mathematics and physics, although less well in literature. He received both a Bachelor of Letters and a Bachelor of Science on the twenty-ninth of December, 1829.

It is interesting to note that, although he has continued to play the role of muckraker of malice, Bell has failed to mention M. Richards distinct cooling toward Galois, on whom he had previously bestowed encomia. After the first trimester of the 1828–1829 academic year, Richard wrote: "This student is markedly superior to all his classmates." After the second: "This student works only in the highest realms of Mathematics." After the third: "Conduct good, work satisfactory."

Because I do not have an accurate date for this report, I cannot propose a specific event as the cause of this obvious change in attitude. Presumably, it occurred in the spring of 1829, shortly before or after Galois's time of troubles began. One could, of course, argue that M. Richard had simply become bored with Galois. Otherwise, it does serve to show that Bell's black-and-white presentation of Galois's preceptors is an oversimplification.

4 L'Ecole Normale

The early months of 1830, which saw Galois officially enrolled as a student at l'Ecole Normale, also witnessed an interesting series of transactions with the Academy. As will be recalled, Galois submitted his first researches to the Academy on May 25 and June 1 of 1829. On January 18, 1830, Cauchy wrote the previously mentioned letter discovered by Taton ([20], p. 134):

I was supposed to present today to the Academy first a report on the work of the young Galois, and second a memoir on the analytic determination of primitive roots in which I show how one can reduce this determination to the solution of numerical equations of which all roots are positive integers. Am indisposed at home. I regret not to be able to attend today's session, and I would like you to schedule me for the following session for the two indicated subjects.

Please accept my homage ...

A.-L. Cauchy

This letter makes it clear that, six months after their receipt, Cauchy was still in possession of Galois's manuscripts, had read them, and very likely was aware of their importance. At the following session on 25 January, however, Cauchy, while presenting his own memoir mentioned above, did not present Galois's work. Taton hypothesizes that between January 18 and January 25, Cauchy persuaded Galois to combine his researches into a single memoir to be submitted for the Grand Prize in Mathematics, for which the deadline was March 1. Whether or not Cauchy actually made the suggestion cannot yet be proved, but in February Galois did submit such an entry to Fourier in his capacity as perpetual secretary of mathematics and physics for the Academy. In any case, there is an additional piece of evidence which attests to Cauchy's appreciation of Galois's work. This is an article ([20], p. 139) which appeared the following year on 15 June, 1831, in the Saint-Simonian journal *Le Globe*. The occasion was an appeal for Galois's acquittal after his arrest following the celebrated banquet at the *Vendanges des Bourgogne*:

Last year before March 1, M. Galois gave to the secretary of the Institute a memoir on the solution of numerical equations. This memoir should have been entered in the competition for the Grand Prize in Mathematics. It deserved the prize, for it could resolve some difficulties that Lagrange had failed to do. *M. Cauchy had conferred the highest praise on the author about this subject.* And what happened? The memoir is lost and the prize is given without the participation of the young *savant*. [Taton's italics]

(My own interpretation of this article is slightly different from that of Taton. Taton writes that the journalist evidently had first-hand information. But note the date: 15 June 1831. In the aftermath of the July revolution, Cauchy fled France during September, almost nine months prior to the article's publication. It is difficult to see when the journalist would have spoken to Cauchy. However, the article appeared in a Saint-Simonian journal. Galois's best friend, Auguste Chevalier, was one of the most active Saint-Simonians. My own suspicion, which I cannot prove, is that the journalist was Chevalier and the information was coming directly from Galois. If this hypothesis is correct, Galois himself is admitting Cauchy's encouragement.)

The misfortune referred to above was the death of Fourier in April. Galois's entry could not be found among Fourier's papers. In Galois's eyes this could not be an accident. "The loss of my memoir is a very simple matter," he wrote ([4], p. xxix). "It was with M. Fourier, who was supposed to have read it and, at the death of this savant, the memoir was lost." It was an unfortunate coincidence; however it was not Fourier's sole responsibility to read the manuscript, for the committee appointed to judge the Grand Prize consisted also of Lacroix, Poisson, Legendre, and Poinsot ([20], p. 138). I mention this because a number of sources give the impression that somehow Fourier either intentionally lost the paper or could not understand it (see, e.g., [14]).

In spite of the setback caused by the loss of his manuscript, April saw the publication of Galois's paper "An Analysis of a Memoir on the Algebraic Resolution of Equations" in the *Bulletin de Ferussac*. In June he published "Notes on the Resolution of Numerical Equations" and the important article "On the Theory of Numbers" ([4], p. xxviii).

In addition to propagating the legend that Cauchy lost the manuscripts, Bell, curiously, does not mention Fourier by name in the preceding misadventure, although Dupuy is explicit on the identity of the Academy's Perpetual Secretary. I do not understand the reason for this omission unless Bell felt it a little too much to "expose" Cauchy, Fourier, and later Poisson, as incompetents. Bell also does not make it clear that the papers listed above (plus a later memoir) constitute what is now called "Galois theory". If this point had been clarified, the claim that Galois had written the theory down on the eve of the duel would be difficult to substantiate or even to suggest.

From this point onward, the scenario of Galois as a passive victim of negligence, misunderstanding, and bad luck begins to break down—if it has not already. More and more he participated in the creation of his own disasters. But this picture does not fit Bell's plan.

Therefore chronology is rearranged, events are omitted, and others invented in increasing quantity, until the end of his account is largely fantasy. The wholesale reordering of events will be especially evident in what follows.

Most important, Bell gives an extremely late start to Galois's political activities. He remarks that had Evariste's teachers at Louis-le-Grand allowed him to study only mathematics he might have lived to be eighty ([1], p. 366). Unlikely. According to Dupuy, one of the reasons Galois had hoped to attend the Polytechnique was to participate in political activities. At l'Ecole Normale he became a "polytechnician in exile". The July revolution of 1830 reared its head. The Director of l'Ecole Normale, M. Guigniault, locked the students in so that they would not be able to fight on the streets. Galois was so incensed at the decision that he tried to escape by scaling the walls. He failed, and in doing so missed the revolution. Afterwards, the Director put the students in the service of the provisional government. Charles X had fled France. He would be followed in September by Cauchy. Louis-Phillipe was the new King.

The events of July, severely abridged here, Bell chronicles accurately. He does fail to mention that Galois probably joined the Society of the Friends of the People, one of the most extreme republican secret societies, within the next month, certainly before December ([7], p. 221). The importance of this omission will be explained after we have filled in the remaining gaps of the narrative.

In December of that year, M. Guigniault was engaging in polemics against students in the pages of several newspapers. Galois saw his chance for attack and jumped into the squabble with a blistering letter to the *Gazette des Ecoles*. It read in part ([4], p. 462; also [7], p. 225; translated in part in [13], Chapter 5):

Gentlemen:

The letter which M. Guigniault inserted yesterday in the *Lycée* on the occasion of one of the articles in your journal has seemed to me very inappropriate. I had thought that you would welcome with eagerness every means to expose this man.

Here are the facts which can be verified by forty-six students.

On the morning of July 28, when many of the students wished to leave the school and fight, M. Guigniault told them on two occasions that he would call the police to reestablish order within the school. The police on the 28th of July!

On the same day, M. Guigniault told us with his usual pedantry: "There are many brave men fighting on both sides. If I were a soldier I would not know what to decide—to sacrifice liberty or LEGITIMACY?"

Here is the man who the next day covered his hat with an immense tricolor cockade. Here are our liberal doctrines!

Galois continues. According to Dupuy, every statement in the letter is accurate. Nonetheless, the result was what might have been anticipated: Galois was expelled. The action was to become official on January 4, but Galois quit school immediately and joined the Artillery of the National Guard, a branch of the militia which was almost entirely composed of republicans. It is interesting that the forty-six students referred to in the letter actually published a reply *against* Galois, but this seems to have been at the "prompting" of M. Guigniault ([7], pp. 227–228).

December was a turbulent month for other reasons. After the Bourbons had fled France, four of their ex-ministers were tried for treason. Popular sentiment called for their execution. The decision to execute or imprison for life was to be announced on December 21. That day, the Artillery of the National Guard was stationed in the quadrangle at the Louvre. Galois was certainly there. The atmosphere was very tense. If the ministers were given a life sentence, the artillerymen had planned to revolt. But the Louvre was soon surrounded by the full National Guard and troops of the line, more trustworthy arms of the military. A distant cannon shot was heard. It signaled the end of the trial and that the ministers had indeed been given imprisonment over execution. The artillerymen and the National Guard readied themselves for bloodshed, but with the arrival at the Louvre of thousands of Parisians, the fighting did not erupt. Over the next few days, the situation in Paris grew calmer with the appearance of Lafayette, who called for peace, and daily proclamations calling for order. On December 31, 1830, the Artillery of the National Guard was abolished by royal decree in fear of its threat to the throne (account based on [3]; also see [13], Chapter 5).

In January 1831 Galois, no longer a student, attempted to organize a private class in mathematics. At the first meeting, about forty students appeared ([7], p. 234), but the endeavor did not last long, evidently because of Galois's political activities. On the 17th of that month, upon the invitation of Poisson, Galois submitted a third version of his memoir to the Academy. Later, in July, Poisson would reject the manuscript (Dupuy does not date this event and its placement in the narrative may have misled Bell). This rejection will be discussed at the proper time, but we should note that by that time Galois would have already been arrested.

If we return to Bell's account, we will now find a totally distorted chain of events. The months after July are missing; Galois still has not joined the Society of Friends of the People. He leaves school in December but has not joined the artillery. The events at the Louvre, which will turn out to have critical importance for the remainder of the story, never take place. Galois attempts to organize his private course in mathematics. Bell writes: "Here he was at nineteen, a creative mathematician of the first rank, peddling to no takers... Finding no students, Galois temporarily abandoned mathematics and joined the Artillery of the National Guard..." ([1], p. 372). According to Bell, Galois submits his paper to Poisson, it is rejected; and this being the "last straw", Galois decides to devote "all his energy to revolutionary politics."

The chronology presented by Bell is thus completely backwards. The impression given by this rearrangement of events is once again that of a misunderstood and persecuted Galois who, surrounded on all sides by idiots, finally gives up and goes into radical politics. By writing that Galois found no students, Bell of course strengthens this impression. A more balanced account clearly requires what is lacking in Bell: a Galois of volition. We may get a better indication of his character and behavior during the spring of 1831 from a letter written on April 18 by the mathematician Sophie Germain to her colleague Libri [9]:

> ...Decidedly there is a misfortune concerning all that touches upon mathematics. Your preoccupation, that of Cauchy, the death of M. Fourier, have been the final blow for this student Galois who, in spite of his impertinence, showed signs of a clever disposition. All this has done so much that he has been expelled from l'Ecole Normale. He is without money and his mother has very little also. Having returned home, he continued his habit of insult, a sample of which he gave you after your best lecture at the Academy. The poor woman fled her house, leaving just enough for her son to live on, and has been forced to place herself as a companion in order to make ends meet. They say he will go completely mad, I fear this is true.

Unfortunately, as Bell observes, Galois was no ineffectual angel.

Before continuing, another historical detail should be mentioned. As an aftermath of the December events at the Louvre and the dissolution of the Artillery of the National Guard, nineteen officers were arrested, having been suspected of planning to deliver their cannons to the people. The charge was conspiracy to overthrow the government. In April, all nineteen were acquitted.

Until now, my criticism has been devoted almost entirely to Bell. Partly, this has been because his account is by far the most famous. There are other reasons as well. Hoyle's short essay, as already mentioned, is purely concerned with Galois's death and thus has little to say concerning the foregoing events. Infeld's account, on the other hand, is of book length. In a single article it would be difficult to debate all salient points. Nonetheless, Infeld has also stated ([13], Afterword) that he is primarily concerned with the events surrounding the duel. It is then reasonable to devote attention here to that aspect of the book. Infeld's work is actually something of a curiosity. The bulk of it is a fictionalized biography, interspersed with real documents and eyewitness accounts. All dates, names, and places are respected. The second part of the biography consists of a lengthy Afterword in which Infeld details exactly what he has invented, what he has not, and what he believes to be true. He also includes a fairly comprehensive bibliography. In my criticisms of Infeld to follow, I only take issue with those points he claims not to have invented. The reader may get the flavor of the author's intent by noting that at Galois's private class in algebra, spoken of earlier, Infeld has stationed two police spies ([13], p. 169).

It might also be noted that, according to James R. Newman's brief remark quoted in the Introduction, Galois at this point in the narrative would be dead.

5 Arrest and prison

And thus we arrive on May 9, 1831. The occasion was the republican banquet at the restaurant *Vendanges des Bourgogne*, where approximately two hundred republicans were gathered to celebrate the acquittal of the nineteen republicans on conspiracy charges. As Dumas says in his memoirs, "It would be difficult to find in all Paris two hundred persons more hostile to the government than those to be found reunited at five o'clock in the afternoon in the long hall on the ground floor above the garden" ([6], p. 331). It is worth quoting Bell's description ([1], p. 372) of this event:

> The ninth of May, 1831, marked the beginning of the end. About two hundred young republicans held a banquet to protest against the royal order disbanding the artillery which Galois had joined. Toasts were drunk to the Revolutions of 1789 and 1793, to Robespierre, and to the Revolution of 1830. The whole atmosphere of the gathering was revolutionary and defiant. Galois rose to propose a toast, his glass in one hand, his open pocket knife in the other. "To Louis-Phillipe"—the King. His companions misunderstood the purpose of the toast and whistled him down. Then they saw the open knife. Interpreting this as a threat against the life of the King, they howled their approval. A friend of Galois, seeing the great Alexander Dumas and other notables passing by the open windows, implored Galois to sit down, but the uproar continued. Galois was the hero of the moment, and the artillerists adjourned to the street to celebrate their exuberance by dancing all night. The following day Galois was arrested at his mother's house and thrown into the prison at Sainte-Pélagie.

Dumas himself describes this event at length in his memoirs. To save space, we quote only a portion ([6], pp. 332–333):

> Suddenly, in the midst of a private conversation which I was carrying on with the person on my left, the name Louis-Phillipe, followed by five or six whistles, caught my ear. I turned around, One of the most animated scenes was taking place fifteen or twenty seats from me.
>
> A young man who had raised his glass and held an open dagger in the same hand was trying to make himself heard. He was Evariste Galois, since killed by Pescheux d'Herbinville, a charming young man who made silk-paper cartridges which he would tie up with silk ribbons.

[This is a literal translation of Dumas. We have not been able to discover exactly to what this occupation refers, but it is a plausible guess that d'Herbinville made what the British call *crackers*, party favors that pop when the ribbons are pulled and contain inspirational messages. They seem to have been invented at about this time.]

Evariste Galois was scarcely 23 or 24 at the time. He was one of the most ardent republicans. The noise was such that the very reason for this noise had become incomprehensible.

All I could perceive was that there was a threat and that the name of Louis-Phillipe had been mentioned; the intention was made clear by the open knife.

This went way beyond the limits of my republican opinions. I yielded to the pressure from my neighbor on the left who, as one of the King's comedians, didn't care to be compromised, and we jumped from the window sill into the garden.

I went home somewhat worried. It was clear this episode would have its consequences. Indeed, two or three days later, Evariste Galois was arrested.

The amusing discrepancies between the two accounts are not entirely difficult to explain. Bell has taken his description from Dupuy, almost word for word, who in turn has based his account on Dumas and the report in the *Gazette des Ecoles* ([7], pp. 234–235). The toasts Bell mentions, as well as the description of the general atmosphere, are found in Dupuy. But Bell has mistranslated: Dupuy writes ([7], p. 235) that "... Dumas et quelques autres passaient par le fenêtre dans le jardin pour ne pas se compromettre..." which, in this context, means "... Dumas and several others jumped through the window into the garden in order not to be compromised." It does not here mean, "Dumas and several others passed by the window in order not to be compromised." One of course wonders how Bell interpreted the clause "in order not to be compromised" in light of his own translation. Why should Dumas be passing by open windows in order not to be compromised? It is difficult to call this carelessness. Bell has also distorted the reason for the banquet. Dupuy clearly states ([7], p. 234) that it was a celebration for the acquittal of the nineteen conspirators. But Bell has not mentioned the trial. For consistency's sake, he must therefore emphasize the obviously revolutionary character of the gathering.

The issue of accuracy becomes more important when we question the most glaring omission in Bell's account: the absence of any mention of Pescheux d'Herbinville. The single sentence in Dumas is the only extant evidence that d'Herbinville was the man who eventually shot Galois. Although Dumas is repeatedly cited by Dupuy, Bell has obviously not read Dumas. If he had, Bell might have seen fit to close the discrepancies in the banquet accounts, in order not to be compromised. On the other hand, Bell claims to have read Dupuy; Dupuy, once again citing Dumas, explicitly names d'Herbinville as Galois's adversary ([7], p. 247). Hoyle is guilty of the same charge; listing Dupuy as a main reference, he relegates d'Herbinville to the ranks of anonymous assassins. Infeld, who does identify d'Herbinville, attempts to prove he was a police agent.

For mathematics, of course, it is not important to know exactly who killed Galois; for historical accuracy, it is. In light of the plethora of theories which have arisen to explain the cause of the celebrated duel, most of which involve police spies, agents provocateurs, and political overtones, the identity of d'Herbinville might be a key piece of information. We will, in fact, find that the only evidence strongly indicates that d'Herbinville was *not* a police agent. D'herbinville and the conspiracy theories will be discussed in greater detail later, but for now let us return to Galois.

Evariste was arrested at his mother's house the day following the banquet, which does indicate that police or informers were at the dinner, although the celebration was open to any subscriber. Galois was held in detention at Sainte-Pélagie prison until June 15, when he was tried for threatening the King's life. Bell's description of this event is highly oversimplified. Indeed, the defense lawyer did claim Galois had actually said, "To Louis-Phillipe, *if he betrays*", but that the noise had been such to drown out the qualifying clause. Nonetheless, the matter took on a less facetious aspect when the prosecutor asked Galois if he really intended to kill the King. Galois replied, "Yes, if he betrays." The prosecutor goes on to ask how Galois "can believe this abandonment of legality on the part of the King", and Galois answers, "Everything makes us believe he will soon turn traitor if he has not done so already." Galois is asked to clarify his remarks and basically repeats what he has already said: "I will say that the trend in government can make one suppose that Louis-Phillipe will betray one day if he hasn't already."

As Dumas aptly remarks, "One understands that with such lucidity in the questions and answers, the discussion did not last long." Apparently moved by Galois's youth, the jury acquitted him within moments. Dumas writes, "I repeat that this is a rude generation, perhaps a bit foolish, but you will recall Beranger's song *Les Fous* [*The Fools* or *The Madmen*] ([6], UG Edition, Volume 2, Chapter 37).

Shortly after this event, the Academy rejected Galois's memoir on the resolution of equations, this time with Poisson as referee. The rejection was written on July 4, although according to Infeld ([13], p. 230) Galois did not receive the letter until October, when he was in prison again (I have found no other source which either corroborates or contradicts Infeld's claim that the rejected manuscript was not received until October, three months after the actual rejection.) By this time, about eight months had passed since he had submitted the paper at Poisson's request. As we will see, Galois did not take the rejection lightly.

The cause for Galois's second arrest was preventative: On Bastille Day, July 14, 1831, he and his republican friend Duchatelet were apprehended dressed in Artillery Guard uniforms and heavily armed. Because the Artillery Guard had been disbanded on the last day of 1830 in fear of its becoming an instrument of the republicans, to wear the uniform was an outright gesture of defiance. It was also illegal. This was the charge brought against Galois, but not until the late date of October 23; he was sentenced to six months in prison. The sentence was confirmed by the court of appeals on December 3. In the meantime, Galois had been sitting in Sainte-Pélagie prison since his arrest in July.

In Bell's fierce diatribe against this arrest he forgets to mention several relevant points. First, he does not seem to comprehend that this was not the Paris of our day but Paris one year after a revolution, when street riots were rampant, assassination attempts not uncommon, and republican activity dangerous (see, e.g., [10]). The "celebration" Bell mentions was a republican demonstration on Bastille Day. Today such a demonstration would be considered patriotic; then it was seditious. This is exactly what the police chief decided when he went on record opposing the demonstration ([7], p. 238). Bell concedes, "True, Galois was armed to the teeth when arrested, but he had not resisted arrest" ([1], p. 378). More precisely, Galois was carrying a loaded rifle, several pistols, and his dagger, a punishable offense even in our more moderate times. To say that he had not resisted arrest may also be inaccurate. The police came to Galois's house to detain him, but Evariste had already decamped.

Galois's predicament was not helped by his friend Duchatelet, who drew a picture on the wall of his cell of the King's head lying next to a guillotine with the inscription "Phillipe will carry his head to your altar, O Liberty!" ([7], p. 238). Part of the delay in bringing Galois to trial was the fact that Duchatelet was tried first.

The point here is not to argue for or against the justice of Galois's arrest. The point is that he was behaving dangerously in a dangerous time. Two forces are clearly at work here: the government's intention to deal harshly with him after his threat of regicide and his own inability to keep out of trouble.

During his stay in prison, a number of events occurred which throw further light on Galois's personality. These incidents were recorded by the republican François Vincent Raspail. Raspail was an early botanist, one of the first to advocate the use of the microscope to examine cell structure in plants. He also had his troubles with the Academy and was sitting next to Dumas at the May 9 banquet. An ardent republican, he refused to receive the Cross of the Legion of Honor from Louis-Phillipe and during the years 1830–1836 spent a total of twenty-seven months in prison [21]. Later in life, Raspail became a famous statesman. He is now remembered by a boulevard and a metro stop in Paris. He lived to be about eighty. One of his many arrests occurred at about the same time that Galois was taken. Raspail recorded the following incidents in several of his letters. Infeld quotes him several times at great length, but never explains who he was.

On July 25, 1831, Raspail wrote ([17], p. 84) that his fellow prisoners had taunted Galois into drinking some liquor, a pastime at which he was apparently a novice:

> To refuse the challenge would be an act of cowardice. And our poor Bacchus has so much courage in his frail body that he would give his life for the hundredth part of the smallest good deed. He grasps the little glass like Socrates courageously taking the hemlock; he swallows it at one gulp, not without blinking and making a wry face. A second glass is not harder to empty than the first, and then the third. The beginner loses his equilibrium. Triumph! Homage to the Bacchus of the jail! You have intoxicated an ingenuous soul, who holds wine in horror.

The scene repeats itself. This time Galois empties a bottle of brandy in a single draught. Galois, drunk, pours out his soul to Raspail in haunting prophecy ([17], p. 89):

> How I like you, at this moment more than ever. You do not get drunk, you are serious and a friend of the poor. But what is happening to my body? I have two men inside me, and unfortunately I can guess which is going to overcome the other. I

am too impatient to get to the goal. The passions of my age are all imbued with impatience. Even virtue has that vice with us. See here! I do not like liquor. At a word I drink it, holding my nose, and get drunk. I do not like women and it seems to me that I could only love a Tarpeia or a Graccha. And I tell you, I will die in a duel on the occasion of some *coquette de bas étage*. Why? Because she will invite me to avenge her honor which another has compromised.

Do you know what I lack, my friend? I confide it only to you: it is someone whom I can love and love only in spirit. I have lost my father and no one has ever replaced him, do you hear me ...

The aftermath of this episode is neither heartwarming nor pleasant; Galois in a delirium attempts suicide ([17], p. 90):

We laid him out on one of our beds. But the fever of intoxication tormented our unhappy friend.... He would fall back senseless only to raise himself with new exaltation, and he foretold sublime things which a certain reserve often rendered ridiculous.

"You despise me, you who are my friend! You are right, but I who committed such a crime must kill myself!"

And he would have done it if we had not flung ourselves on him, for he had a weapon in his hands ...

Several important points need to be made about these passages. Bell, in his account, says only, "Goaded beyond endurance, Galois seized a bottle of brandy, not knowing or caring what it was, and drank it down. A decent fellow prisoner took care of him until he recovered" ([1], p. 374). Thus, the really important parts of the episode, which tell us something about Galois's character and which bear on future events, are omitted altogether.

Later, in attempting to understand the cause of Galois's death, Dupuy remarks ([7], p. 245), "If I credit an allusion of Raspail, Galois lost his virgin heart to *quelque coquette de bas étage*." Bell writes ([1], p. 374): "Some worthless girl [*quelque coquette de bas étage*] initiated him." Here, Bell is taking a conjecture of Dupuy based on a letter of Raspail reporting an utterance of Galois spoken in a delirium a year before the duel as a characterization of real events. This can only be termed fabrication. And it is very likely that this piece of fabrication is responsible for the widespread belief that a prostitute was the cause of Galois's death.

Infeld, in his version of the prison scene, quotes the letters far more fully than Dupuy, but jumps from "Tarpeia and Graccha" to "Do you know what I lack, my friend?" In other words, he omits Gaiois's prophecy that he will die in a duel. He also makes no comment whatsoever on Galois's suicide attempt. This selective presentation and slanting of evidence is characteristic of Infeld's book. He publishes any document or any portion of a document which does not interfere with his stated hypothesis that Galois was killed by the secret police. More obvious examples will be presented later when we discuss the actual circumstances surrounding the duel.

On August 2, Raspail chronicles an interesting series of events which took place after his previous letters. On July 27, the prisoners were invited to attend a mass in memory of those killed during the July revolution a year earlier. Because many of the prisoners were political, the atmosphere was tense and an open riot was expected to erupt at any moment. A few prudent prison leaders defused the situation, and two days passed without violence. At lock-up time on the 29th, a shot was heard throughout the prison, followed by cries of "Help, murder!" The following day, the mystery was clarified. Raspail, quoting the conversation of another prisoner with the prison superintendent, writes ([17], pp. 117–118):

"Here are the facts. I am one of those in the attic room of the bathing pavilion. We were quietly going to bed. The man whose bed is between two casements had his face toward the window while undressing and he was humming a tune."

"At that moment a shot was fired from the garret opposite. We thought our comrade was dead, but he was only unconscious. Not knowing where the shot had come from, nor how serious the wound was, we called for help. For in such a room, open in all directions through six windows, a better-aimed shot would have struck down its man."

The shot, it turned out, came from a garret, across the street, where one of the prison guards lived. Galois was not the man who was at the window and wounded. However, he was in the same room and was later thrown into the dungeon, evidently because he had insulted the superintendent, probably accusing him of having intentionally arranged the shooting. Raspail continues ([17], p. 118; also [7], p. 243) to record the conversation. The prisoner already quoted is talking:

"What? You have no order to seize the guilty man [the guard who fired the shot]? But you have one to throw into the dungeon both the victim of this shameful trap and the witnesses of it? It may sound insolent to say that the administration pays turnkeys to murder prisoners. But what if this insolent statement is true? And I bear witness that no other insolence has come from those who were thrown into the dungeon. This young Galois doesn't raise his voice, as you well know; he remains as cold as his mathematics when he talks to you."

"Galois in the dungeon!" repeats the crowd. "Oh, the bastards! They have a grudge against our little scholar."

"Of course they have a grudge against him. They trick him like vipers. They entice him into every imaginable trap. And then, too, they want an uprising."

An uprising they got. This oblique conversation ends with the superintendent taking to his heels as the prisoners take control of the prison. The situation remains stalemated until late that night when the infantry is called in. The prisoners surrender without violence and remarkably no one is hurt.

I have tried to present this episode in as neutral a tone as possible. Infeld interprets the shot as an assassination attempt on Galois's life, and later cites it in his Afterword as his first piece of evidence that Galois was murdered by the government ([13], p. 308). It is agreed that the moderate government of Louis-Phillipe would have liked to have been rid of all political extremists. But a conspiracy theory presumes that there exists a reason to single out a particular victim. Why Galois over Raspail? A shot was fired in a prison full of political prisoners on the verge of a riot, at night ("lock-up time"), into a room containing an unknown number of men, evidently "aimed" at someone else. Yes, it could have been an attempt to kill Galois. I do not find the evidence compelling.

More compelling is the evidence for the absolute hatred Galois had developed for the Academy, which I feel can only be termed paranoid. And, as is not uncommon with paranoiacs, there was a kernel of justification for the behavior. At some point in October, according to Infeld, Galois was notified of Poisson's rejection of his latest manuscript on the theory of equations. Infeld quotes the following letter (from [2]; see also [13], p. 230), which was originally published by Bertrand. It is not quoted by Dupuy.

Dear M. Galois:

Your paper was sent to M. Poisson to referee. He has returned it with his report, which we quote:

"We have made every effort to understand M. Galois's proofs. His argument is neither sufficiently clear nor sufficiently developed to allow us to judge its rigor; it is not even possible for us to give an idea of this paper.

The author claims that the propositions contained in his manuscript are a part of a general theory which has rich application. Often different parts of a theory clarify each other and can be more easily understood when taken together than when taken in isolation. One should rather wait to form a more definite opinion, therefore, until the author publishes a more complete account of his work."

For this reason, we are returning your manuscript in the hope that you will find M. Poisson's remarks useful in your future work.

François Arago, Secretary to the Academy

Bell, elaborating from Dupuy, states ([1], p. 371) that Poisson found the manuscript "incomprehensible" but "did not state how long it had taken him to reach this remarkable conclusion." I believe that this is an unfair characterization of Poisson's comments. This is the rejection that Bell has occurring before Galois's arrest.

In light of previous events and in light of his character, it is not terribly surprising that Galois reacted violently to what might nowadays be considered an encouraging rejection letter. He gave up all plans to publish his papers through the Academy and decided to publish them privately with the help of his friend Auguste Chevalier. Galois collected his manuscripts and in December, while still in Sainte-Pélagie, penned what surely must be one of the most remarkable documents in the history of mathematics, his *Préface*. The entire *Préface* runs to about five pages. Infeld, to his credit, prints some of it, although he alters and omits certain parts at will. To save space, I here quote only the first page. The full text can be found in Bourgne and Azra ([4], pp. 3–11):

Firstly, you will notice the second page of this work is not encumbered by surnames, Christian names or titles. Absent are eulogies to some prince whose purse would have opened at the smoke of incense, threatening to close once the incense holder was empty. Neither will you see, in characters three times as high as those in the text, homage respectfully paid to some high-ranking official in science, or to some savant-protector, a thing thought to be indispensable (I should say inevitable) for someone wishing to write at twenty. I tell no one that I owe anything of value in my work to his advice or encouragement. I do not say so because it would be a lie. If I addressed anything to the important men of science or of the world (and I grant the distinction between the two at times is imperceptible) I swear it would not be thanks. I owe to important men the fact that the first of these papers is appearing so late. I owe to other important men that the whole thing was written in prison, a place, you will agree, hardly suited for meditation, and where I have been dumbfounded at my own listlessness in keeping my mouth shut at my stupid, spiteful critics: and I think that I can say "spiteful critics" in all modesty because my adversaries are so low in my esteem. The whys and wherefores of my stay in prison have nothing to do with the subject at hand; but I must tell you how manuscripts go astray in the portfolios of the members of the Institute, although I cannot in truth conceive of such carelessness on the part of those who already have the death of Abel on their consciences. I do not want to compare myself with that illustrious mathematician but, suffice to say, I sent my memoir on the theory of equations to the Academy in February of 1830 (in a less complete form in 1829) and it has been impossible to find them or get them back. There are other anecdotes in this genre but I would be ungracious to recount them because, other than the loss of my manuscripts, those incidents do not concern me. Happy voyager, only my poor countenance saved me from the jaws of wolves. Perhaps I have already said too much for the reader to understand why, as much as I would have liked otherwise, it is absolutely impossible for me to embellish or disfigure this work with a dedication.

The remainder of the *Préface* continues in much the same tone ("And thus it is knowingly that I expose myself to the laughter of fools"). Others of his writings are not dissimilar ([4], pp. 21–27). Among his papers is the picture of a bizarre, torsoless figure, captioned by Bourgne and Azra [4] "Riquet à la Houppe." The picture must have been drawn shortly before his death. It may be significant that Riquet à la Houppe was in French folklore a character, short, ugly, disdained by all, but nonetheless very clever.

6 The duel and theories surrounding it

We are almost at the end of this short story. Galois remained in Sainte-Pélagie without further recorded incident until March 16, 1832, when he was transferred to the pension Sieur Faultrier. Ironically enough, this was to prevent the prisoners from being exposed to the cholera epidemic then sweeping Paris. Galois was due to be given his freedom on April 29. From this point on, the historical record is very scanty. On May 25, Galois writes ([4], pp. 468–469) to his friend Chevalier and clearly alludes to a broken love affair:

My dear friend, there is a pleasure in being sad if one can hope for consolation; one is happy to suffer if one has friends. Your letter, full of apostolitic unction, has given me a little calm. But how can I remove the trace of such violent emotions that I have felt?

How can I console myself when in one month I have exhausted the greatest source of happiness a man can have, when I have exhausted it without happiness, without hope, when I am certain it is drained for life?

The letter continues in similar tones. Galois goes on ([4], p. 470) to say that he is disgusted with the world: "I am disenchanted with everything, even the love of glory. How can a world I detest soil me?"

The next few days are a complete blank. On the morning of May 30, the famous duel took place. The previous evening, Galois wrote ([4], p. 471) several well-known letters to his republican friends:

I beg patriots, my friends, not to reproach me for dying otherwise than for my country.

I die the victim of an infamous coquette and her two dupes. It is in a miserable piece of slander that I end my life.

Oh! Why die for something so little, so contemptible?

I call on heaven to witness that only under compulsion and force have I yielded to a provocation which I have tried to avert by every means. I repent in having told the hateful truth to those who could not listen to it with dispassion. But to the end I told the truth. I go to the grave with a conscience free from patriots' blood.

> I would like to have given my life for the public good.
>
> Forgive those who kill me for they are of good faith.

Galois also writes ([4], p. 471) another, similar letter to two republican friends, Napoleon Lebon and V. Delauney:

> My good friends,
>
> I have been provoked by two patriots... It is impossible for me to refuse.
>
> I beg your forgiveness for not having told you.
>
> But my adversaries have put me on my word of honor not to inform any patriot.
>
> Your task is simple: prove that I am fighting against my will, having exhausted all possible means of reconciliation; say whether I am capable of lying even in the most trivial matters.
>
> Please remember me since fate did not give me enough of a life to be remembered by my country.
>
> I die your friend.

We will return to Galois's activities during this last night in due time. For now we discuss a few of the many theories which purport to explain the cause of this celebrated duel. There is perhaps enough in the two letters to raise suspicions of foul play. The attempts to make Galois the victim of royalists, a female agent provocateur, a prostitute, or a government conspiracy doubtlessly stem from these letters, for there is no other direct evidence in existence. Thus, we have the origin of Bell's assertion ([1], p. 375):

> What happened on May 29th is not definitely known. Extracts from two letters suggest what is usually accepted as the truth: Galois had run afoul of political enemies immediately after his release.

The first statement is accurate, the second is not. Dupuy certainly believes the exact opposite, as will be seen shortly. Dupuy does mention that Alfred Galois, unjustifiably in his view, did maintain that his older brother was murdered. Because Bell "followed" Dupuy exclusively, one can only conclude that he took Alfred's position and termed it widely accepted or that he invented the whole thing.

Although Bell may have invented the theory, or merely propagated it to previously unattained heights, he is not its chief advocate. Infeld goes further. He assumes the "infamous coquette" was a female agent provocateur who set up Galois for the duel with a police agent. Infeld's evidence is by admission circumstantial. In addition to the bullet episode at Sainte-Pélagie it consists of the following ([13], pp. 308–311): the police were known to have used spies; the police broke up a meeting of the Society of Friends of the People the night before Galois's funeral; Police Chief M. Gisquet wrote in 1840 that Galois "had been killed by a friend"; police spies were unmasked in 1848, at which time a claim appeared in a journal that Galois "had been murdered in a so-called duel of honor"; Galois's brother Alfred always maintained that Evariste had been murdered; Galois was abandoned by his adversaries and his seconds and found by a peasant.

It should be noted that this evidence is consistent and does not contradict known facts. However, necessity does not follow from consistency. The bullet episode has already been discussed. It is true that the police used spies and that they were unmasked in 1848. We will return to this point below. Infeld does not mention that the newspapers announced Galois's funeral before the fact and explicitly named him as a member of both the Artillery of the National Guard and the Friends of the People. In any event, his membership in these organizations must have been widely known. One must weigh for oneself whether it is remarkable that police knew of republican meetings. Infeld finds it suspicious that the police chief, eight years after the fact, knew Galois had been "killed by a friend". He does not find it suspicious that Dumas knew more—precisely who that friend was. Dupuy feels that Alfred's position was the result of justifiable anger over his brother's death and points out some unlikely details Alfred attributed to the duel, such as stating that Evariste would have fired into the air. The assertion that Galois was abandoned to die, another of Alfred's claims, is also open to dispute. Dupuy mentions ([7], pp. 247–248) that one of the witnesses went to Galois's mother the following day to explain what had happened. He feels, then, it was more likely that the witnesses were searching for a doctor when the peasant happened along. This explanation may be weak; nonetheless Infeld fails to mention that Mme. Galois was informed.

The remarks above are admittedly as circumstantial as the evidence. There is, however, more concrete evidence which weighs very heavily against the political conspiracy theorists: the identity of Pescheux d'Herbinville. More is known about him than his anonymity. He was, in fact, one of the nineteen republicans who were acquitted on charges of conspiring to overthrow the government in the trial spoken of earlier. Is there any reason to suspect that d'Herbinville was a police agent? The historian Louis Blanc, in his exhaustive *History of Ten Years*, writes ([3], p. 431):

> The trial gave rise to highly interesting scenes. In the sittings of the 7th of April, the president having reproached M. Pescheux d'Herbinville, one of the accused, with having had arms by him and with having distributed them, "Yes," replied the prisoner, "I have had arms, a great many arms, and I will tell you how I came by them." Then, relating the part he had taken in the three days, he told how, followed by his comrades, he had disarmed posts, and sustained glorious conflicts; and how, though not wealthy, he had equipped national guards at his own cost. There still burned in the hearts of the people some of the fire kindled by the revolution of July; such recitals as this fanned the embers. The young man himself, as he concluded his brief defense, wore a face radiant with enthusiasm and his eyes filled with tears.

In addition, Blanc mentions ([3], p. 431) the appearance of General Lafayette during the trial:

> The old general came to give his testimony in favor of the accused, almost all of whom he knew, and all saluted him from their places with looks and gestures of regard.

D'Herbinville, it seems, was one of the heroes of the hour. After the acquittal, the crowd pulled his coach through the streets of Paris "amid shouts of rapturous applause".

Bell, by not mentioning d'Herbinville at all, relieves himself of the difficulty of explaining why Galois should be killed in a political duel with a fellow republican or why d'Herbinville should be considered a political enemy. Infeld is in a more difficult position. Having acknowledged d'Herbinville's existence, he must explain why neither Dumas nor Blanc, both republicans, nor evidently the extremely liberal Lafayette (assuming he knew d'Herbinville personally), nor, one would gather from Blanc's account, any republican in Paris, ever held any suspicions that d'Herbinville was an agent. Infeld talks at length about the 1848 unmasking of the police spies, but he does not mention the following extract from Dupuy ([7], p. 247):

> Pescheux was certainly not a "false-brother": all the men who acted as police agents during the reign of Louis-Phillipe were revealed in 1848 when Caussidière became chief of police, as witness Lucien de la Hodde [Hodde was a "republican" who was unmasked as a spy in 1848]. If Pescheux were suspect, he would certainly not have been nominated as curator of the palace of Fontainebleu. It is absolutely necessary to discard the idea of police intervention and of a framed assassination.

Thus there are some serious difficulties with the political enemies scenario. Infeld gets around this problem in characteristic fashion: in his bibliography he cites both Blanc and Dupuy as primary sources but quotes neither. In his Afterword, Infeld goes so far as to admit, "There is no reason to believe Pescheux d'Herbinville was a police agent." But then he goes on to say ([13], p. 310):

> I believe there is enough circumstantial evidence to prove that the intervention of the secret police sealed Galois's fate. I do not believe it is possible to fit all the known facts without assuming Galois was murdered.

It is left as an exercise for the reader to form a rebuttal to this statement. But in order to see just how far "known facts" can be stretched, we turn to Hoyle's version of the event. He writes ([11], p. 14):

> Such are the bare bones of the story of the life and death of Evariste Galois. The classical biography of Galois [he then references Dupuy], in an attempt to add flesh to these bones, suggests that he was done to death by royalist enemies, as does E. T. Bell in his book *Men of Mathematics*. There are dark hints that the release from prison was but a device for encompassing his death, a necessary preliminary to his being matched against a highly skilled assailant in royalist pay. But why should Galois feel it critical to his honor that he should accept the challenge of a right-wing agent, especially if the agent were a known marksman? Gallic logic suggests on account of a girl . . .

We first note the complete misrepresentation of Dupuy's position. If Hoyle is challenging Bell, and admittedly this is unclear, it seems to be on the extremely naive assumption that Galois would have known his opponent was a right-wing agent. Hoyle then goes on to dispose of the "infamous coquette" and propose his own theory ([11], p. 15):

It is possible that the "infamous coquette" was the source of a purely personal quarrel, but it is the normal biological rule among mammals that sexual quarrels between two males cease as soon as one side seeks "accommodation". It is the normal rule that either party to such fights can simply walk away, which is just what Galois seems to have attempted to do.

The more likely possibility is that Galois's habit of working mathematical problems in his head, his ability to think in parallel, caused serious animosities, and perhaps suspicions, to develop during the six months of imprisonment. There may have been suspicions that Galois was not wholly for the "cause", or even that he was an *agent provocateur*...

Lincoln's remark comes to mind: "You can fool some of the people some of the time..." To suggest as Hoyle does that any republican in Paris suspected Galois after his expulsion from l'Ecole Normale, his Artillery activities, his threat to the King, his arrests, trials, sentencings, resentencings, and prison activities borders on the fantastical. This is in addition to the fact that two or three thousand republicans later attended the funeral of this supposed agent provocateur. One might equally well claim that Lenin had been suspected of being a Menshevik.

As to Hoyle's bio-sociological theories, he is contradicted by the historical record. The greatest Russian poet, Alexander Pushkin, was killed in 1837 at the age of 37 in a duel over his wife. England's Lord Camelford was killed in a duel over a prostitute. As late as 1838 members of the American legislature were engaging in similar duels. Toward the end of the eighteenth century, during election season, approximately 23 duels *per day* were fought in Ireland *alone*, unlikely just for political reasons. In the last decades of the nineteenth century, Paris newspapers carried notices of the daily duels and their terms. These practices continued until World War I. The cause of such "affairs of honor" ranged from geese, to insults, to politics, to women (see, e.g., [15], "Duels and Ideals", and [18], Chapter 1). Dupuy himself mentions that nothing was more common at the time in question than duels between republicans, and I think that one may safely infer from his remarks that no one paid the slightest attention to them ([7], p. 247).

However, argument by analogy is generally a weak policy when dealing with a specific case, and Hoyle's expansive pronouncements on the sexual behavior of mammals bring to mind further evidence with which anyone wishing to invoke a political cause for Galois's death must contend. This evidence consists primarily of two fragmentary letters written to Galois by one Mademoiselle Stéphanie D, who is none other than the "infamous coquette" over whom the duel was fought. Most authors have assumed her identity to be an absolute mystery and that she, like d'Herbinville, is an anonymous casualty of history. Dupuy apparently was unaware of the letters or chose not to publish them. Bell and Hoyle never mention her name. Infeld calls her Eve Sorel (perhaps inspired by Stendhal). This is a strange state of affairs, for the letters were published in Tannery's 1908 edition of Galois's papers. Tannery does not affix a name to the author of these fragments; it is left for the 1962 edition of Bourgne and Azra to attempt an identification. One can understand why Bell and Infeld did not mention her name since Tannery did not provide it. Hoyle does not have such an excuse, his book being published in 1977. One cannot understand why these letters are never mentioned by anyone, especially by Bell and Infeld who cite Tannery as a major source for Galois's manuscripts.

The letters, as they exist, are copies made by Galois himself on the back of one of his papers ([4], pp. 489–491). The copies contain gaps, which may indicate he had previously torn up the originals and could not completely reconstruct them. More likely, Galois purposely omitted any incriminating or personally distasteful segments. I say this because some words in the French versions are broken in half; one generally does not remember only half a word. Galois has certainly obliterated Stéphanie's last name in a fit of anger. Due to the fragmentary nature of these letters their translation has proved difficult and may be uncertain in places. Where impossible to translate we have allowed the original French to stand.

Letter I:

Please let us break up this affair. I do not have the wit to follow a correspondence of this nature but I will try to have enough to converse with you as I did before anything happened. Here is Mr. the *en a qui doit vous qu'a* me and do not think about those things which did not exist and which never would have existed.

Mademoiselle Stéphanie D
14 May 183–

Letter II:

I have followed your advice and I have thought over what has happened on whichever denomination it may have happened between us. In any case, Sir, be assured there never would have been more. You're assuming wrongly and your regrets have no foundation. True friendship exists nearly only between people of the same sex particularly of friends. full in the vacuum that the absence of all feeling of this kind ... my trust ... but it has been very wounded ... you have seen me sad you have asked the reason; I answered you that I had sorrows that one had inflicted upon me. I had thought that you would take this as anyone in front of whom one drops a word for these one is not The calm of my thoughts leaves me to judge the persons that I usually see without much reflection: this is the reason that I rarely regret having been wrong in my judgment of a person. I am not of your opinion *les sen plus que les a exiger ni se* thank you sincerely for all those who you would bring down in my favor.

These are highly tantalizing morsels, but is there anything else known about the author? Indeed there is. C. A. Infantozzi has examined the original of the first letter [12]. With the help of a magnifying glass he was able to discern Stéphanie's full signature under Galois's erasures: Stéphanie Dumotel. Further archival investigation by Infantozzi shows she was Stéphanie-Félicie Poterin du Motel, daughter of Jean-Louis Auguste Poterin du Motel, a resident physician at the Sieur Faultrier, where Galois stayed the last months of his life. In 1840 Stéphanie married Oscar-Theodore Barrieu, a language professor. Any presumption that she was a prostitute must at this point be discarded as a complete figment of Bell's imagination.

The establishment of Stéphanie's identity unfortunately does not conclusively establish what in actuality did occur. From Stéphanie's second letter it is not difficult to infer that Galois took some song of sorrows on her part too seriously and himself provoked the duel. On face value she certainly seems an unwilling participant in whatever transpired. On the other hand, we have a curious passage from Dupuy, once again not quoted by the other authors. During the course of his researches, Dupuy had asked Galois's cousin if he knew the cause of the duel ([7], p. 246).

His cousin, M. Gabriel Demante, writes me that at a last meeting [with Stéphanie?] Galois found himself in the presence of a supposed uncle and a supposed fiancé, each of whom provoked the duel.

It is difficult to say in which direction this passage points, but in weighing its importance, one should keep in mind Dupuy's own skepticism of anything the Galois family said concerning the duel.

With this passage, all the evidence pertaining to the duel which I have found to date has been presented. One can read the circumstantial evidence as Infeld does to arrive at a conspiracy. No known facts conclusively refute this interpretation. But it must be re-emphasized that there is absolutely no direct evidence that such is the case. Furthermore, there is the testimony of several men, two of whom were republicans, that d'Herbinville was not a police agent. In addition, there is the identity of Stéphanie, who was simply the daughter of a physician who happened to live and work at the pension where Galois was staying. To suggest she was an agent provocateur somehow planted there to entrap Galois becomes a baroque, if not byzantine, invention.

If one chooses to reject the conspiracy theory, a fairly consistent picture of a personal quarrel emerges. We have Galois's unhappy letter to Chevalier of May 25. His famous cry, "I am the victim of an infamous coquette and her two dupes", may mean exactly what it says, with suitable allowances for Galois's usual withering tone. The excerpt from Dupuy quoted above is certainly consistent with "two dupes". Galois himself writes, "Forgive those who kill me for they are of good faith", i.e., they are not political enemies. And we must remember, in his own eyes, Galois was exceedingly honest. If he felt any treachery were involved, we can be sure he would have said as much. The two letters from Stéphanie are perhaps the strongest argument for a personal quarrel. Those dissenting could of course take the extreme position that she was a very good actress.

In addition, there is the more difficult question of psychology. Galois's writings are at times unquestionably violent, and equally violent erasures are preserved on his manuscripts. He was arrested twice for dangerous actions which might have easily been avoided by a more prudent individual, or perhaps in a more prudent age. Raspail writes that Galois attempted suicide in prison, and of course there is Galois's own prophecy, not inconceivably self-fulfilling, that he would be killed in an affair of honor. It is not terribly difficult to believe that such a troubled young man in such a turbulent time could have ended his life in a duel.

In this scenario, the role of Pescheux d'Herbinville admittedly remains unclear. Was he a "supposed" fiancé or a real fiancé whom Galois's cousin took for supposed? Did Galois's cousin invent this epithet, or was d'Herbinville simply involved in a stupid quarrel? I have no answers to these questions. The point I wish to make now for the interested historian is that, although in 1982 Galois will have been dead for 150 years, the investigation of his death has been closed prematurely. D'Herbinville should be traced to see if any letters exist which might shed some light on the matter; a perusal of the standard biographical encyclopedias has failed to reveal any further information on d'Herbinville. If he was in prison with Galois, a background for the quarrel might be established. Letters of Stéphanie or her husband might be extant and could conceivably mention the duel. Dupuy remarks cryptically that Raspail as well as all the republicans knew the cause of the duel. Raspail became a famous politician. Perhaps there is a clue in his correspondence.

These avenues are still open for those who are interested. They have been neglected only because of the intentional or unintentional omission of information by those who have previously written on Galois. We will return to the question of scholarship after disposing of the remaining myths concerning the night before the duel.

7 The last night

We saw in the introduction how Bell all but states outright that Galois committed his theory of equations to paper the night before he was shot. James R. Newman repeats this as an assertion, and the vision of the doomed boy, sitting by candlelight, feverishly bringing group theory into the world seems to be the major myth which most scientists harbor concerning Galois. This is again due to Bell's embellishment of Dupuy, who in this instance is sufficiently romantic of his own accord. But as has already been detailed at great length, Galois had been submitting papers on the subject since the age of 17. The term *group*, used in the sense of *group of permutations*, is used in all of them. During the night before the duel, in addition to the letters already quoted, Galois wrote ([4], p. 173) a long letter to his friend Chevalier. He begins:

My Dear Friend,

I have made some new discoveries in analysis.

The first concern the theory of equations, the others integral functions.

In the theory of equations I have researched the conditions for the solvability of equations by radicals; this has given me the occasion to deepen this theory and describe all the transformations possible on an equation even though it is not solvable by radicals.

All this will be found here in three memoirs.

Galois then goes on to describe and elucidate the contents of the memoir which was rejected by Poisson, as well as subsequent work. Galois had indeed created a field which would keep mathematicians busy for hundreds of years, but not "in those last desperate hours before the dawn". During the course of the night he annotated and made corrections on some of his papers. He comes across a note that Poisson ([4], p. 48) had left in the margin of his rejected memoir:

The proof of this lemma is not sufficient. But it is true according to Lagrange's paper, No. 100, Berlin 1775.

Galois writes directly beneath it:

This proof is a textual transcription of that which we gave for this lemma in a memoir presented in 1830. We leave as an historic document the above note which M. Poisson felt obliged to insert. (Author's note.)

A few pages later ([4], p. 54), Galois scrawls next to a theorem:

There are a few things left to be completed in this proof. I have not the time. (Author's note.)

Galois penned this famous inscription only once during the course of the night. It is unfortunate he tarnished some of the romance by including his parenthetical "Author's note". Galois ends ([4], p. 185) his letter to Chevalier with the following request:

In my life I have often dared to advance propositions about which I was not sure. But all I have written down here has been clear in my head for over a year, and it would not be in my interest to leave myself open to the suspicion that I announce theorems of which I do not have complete proof.

Make a public request of Jacobi or Gauss to give their opinions not as to the truth but as to the importance of these theorems.

After that, I hope some men will find it profitable to sort out this mess.

I embrace you with effusion.   E. Galois

And that was the end. The funeral was to be held on June 2. During the previous evening, the police broke up a meeting of the Society of Friends of the People on the pretext that the republicans were planning a demonstration for Galois's funeral. Thirty of those present were arrested. The next day, two or three thousand republicans were present at the services. Galois's body was interred in a common burial ground of which no trace remains today.

Later, Evariste's brother Alfred and his devoted friend Chevalier would laboriously re-copy the mathematical papers and submit them to Gauss, Jacobi, and others. By 1843 the manuscripts had found their way to Liouville, who, after spending several months in the attempt to understand them, became convinced of their importance. He published the papers in 1846.

There exist many fragments which indicate that Galois carried on his mathematical researches, not only while in prison, but right up until the time of his death. The fact that he could work through such a turbulent life is testimony to the extraordinary fertility of his imagination. There is no question that Galois was a great mathematician who developed one of the most original ideas in the history of mathematics. The invention of legends does not make him any greater.

8 Harsher words

The account of Galois's life given here has not been entirely complete. There are more documents, letters, and events. No doubt I will shortly be exposed for having selectively presented evidence. The purpose of this paper, however, has not been one of completeness, nor entirely one of biography. No, the purpose has been to show that something is wrong. Two highly respected physicists and an equally well-known mathematician have invented history.

Bell's account, by far the most famous, is also the most fictitious. It is a myth devoid of such complications as a protagonist who is faulted as well as gifted. It is myth based on the stereotype of the misunderstood genius whom the conservative hierarchy is out to conquer—as if the befuddled hierarchy is generally organized well enough for persecution. It is a myth based on a misunderstanding of the method by which a scientist works—as if a great theory could be written down coherently in a single night.

It is unclear how far one can go in forgiving Bell. Surely all his mistakes could not result from a poor knowledge of French. No, I believe consciously or unconsciously Bell saw his opportunity to create a legend. The details which are absent in his account, such as Dumas at the banquet, such as d'Herbinville, such as the suicide attempt and Raspail, are those details which lend a concreteness and a humanness to Galois's life which a legend must not have. Unfortunately, if this was Bell's intent, he succeeded. After hearing of my investigation, physicists and mathematicians all open conversations with me with the same question: "Did Galois really invent group theory the night before he was killed?" No, he didn't.

Infeld presents far more details. He is not interested in making Galois a legend. He does intend to make Galois a hero of the people. Politics is the guiding principle for Infeld. His book might be termed the proletarian interpretation of Galois; certainly, parts of it read like the local Workers' Party publication. Infeld is very good at covering his tracks. To delete a phrase here, a paragraph there, a counter-argument in between, is all that is necessary to create conspiracy from chaos.

As to Hoyle's motives, we can only take him at his word: He describes at length how as a child he was taught arithmetic by his mother, how he became proficient at mathematics, and how school for him became an excruciating bore. Hoyle was forced to learn to "think in parallel" in order to fool the teacher into believing he paid attention in class. He then writes, "I mention these personal details because I believe they cast some light on the mysterious death of the French mathematician Evariste Galois." Further comment seems unnecessary.

Dupuy seems to have less of a vested interest. I assume he included all the documents known to him at the time. If not, then he too should be scrutinized more carefully. He does seem *a priori* unwilling to accept conspiracy theory.

At the very least, the three twentieth-century authors are guilty of distorting Dupuy's account and even falsifying it. In each case the story of Galois has been used to put a stamp of approval on the author's personal theories. Indeed, all history is interpretative. But if we do not approve, we understand the liberty: Galois, like Einstein, has passed into the public domain. No act or anecdote attributed to him is too outrageous to be given consideration. There is a closer analogy from farther afield. The Russian composer Reinhold Glière once wrote a symphony, his third, which ran well over an hour. Stokowski—the story goes—worked with Glière to edit the score down to manageable length. Since then, every conductor presents his own edition. I do not know if I have ever heard the original.

The investigations of Galois discussed here have told us less about the man than about his biographers. The misfortune is that the biographers have been scientists. Because they appreciate his genius a century after its undisputed establishment, anyone who did not recognize it at the time is condemned. "In all the history of science," writes Bell, "there is no completer example of the triumph of crass stupidity over untamable genius." "Is it possible to avoid the obvious conclusion," asks Infeld, "that the regime of Louis-Phillipe was responsible for the early death of one of the greatest scientists who ever lived?" The underlying assumption is apparent: Galois was persecuted because he was a genius, and all scientists, to a greater or lesser degree, understand that genius is not tolerated by mediocrity. From this point of view, a genius must be recognized as such, even when standing drunk on a banquet table with a dagger in his hand. Anyone who does not recognize him becomes a fool, an assassin, or a prostitute. This is a presumption of the highest arrogance. Scientists should not be so enamored of themselves.

References

1. E. T. Bell, *Men of Mathematics*, Simon and Schuster, 1937.
2. J. Bertrand, La vie d'Evariste Galois par P. Dupuy, *Eloges Académiques* (1902), 329–345.
3. L. Blanc, *History of Ten Years*, Chapman and Hall, 1844.
4. R. Bourgne and J. P. Azra (eds.), *Ecrits et Mémoires Mathématiques d'Evariste Galois: Edition Critique Intégrale de ses Manuscrits et Publications*, Gauthier-Villars, 1962.
5. *Checklist of the Bullitt Collection of Mathematics*, University of Louisville, 1979.
6. A. Dumas, *Mes Mémoires*, Gallimard, 1967; also the Union Générale d'Editions.
7. P. Dupuy, La vie d'Evariste Galois, *Annales de l'Ecole Normale* **13** (1896), 197–266.
8. F. Dyson, *Disturbing the Universe*, Harper and Row, 1979.
9. C. Henry, Manuscrits de Sophie Germain, *Révue Philosophique* **8** (1879), 631.
10. T. E. B. Howarth, *Citizen King: The Life of Louis-Phillipe*, Eyre and Spottiswoode, 1961.
11. F. Hoyle, *Ten Faces of the Universe*, W. H. Freeman, 1977.
12. C. A. Infantozzi, Sur la mort d'Evariste Galois, *Révue d'Histoire des Sciences* **21** (1968), 157.
13. L. Infeld, *Whom the Gods Love: The Story of Evariste Galois*, Whittlesey House, 1948.
14. L. Lieber, *Galois and the Theory of Groups*, The Science Press, 1932.
15. C. Mackay, *Extraordinary Popular Delusions and the Madness of Crowds*, Noonday Press, 1932.
16. J. R. Newman, *The World of Mathematics*, Simon and Schuster, 1956.
17. F. V. Raspail, *Lettres sur les Prisons de Paris*, Paris, 1839.
18. R. Shattuck, *The Banquet Years*, Vintage Books, 1968.
19. J. Tannery (ed.), *Manuscrits d'Evariste Galois*, Gauthier-Villars, 1908.
20. R. Taton, Sur les relations scientifiques d'Augustin Cauchy et d'Evariste Galois, *Révue d'Histoire des Sciences* **24** (1971), 123.
21. D. B. Weiner, *Raspail: Scientist and Reformer*, Columbia University Press, 1968.

Hermann Grassmann and the Creation of Linear Algebra

DESMOND FEARNLEY-SANDER

American Mathematical Monthly **86** (1979), 809–817

> There is a way to advance algebra as far beyond what Vieta and Descartes have left us as Vieta and Descartes carried it beyond the ancients... We need an analysis which is distinctly geometrical or linear, and which will express *situation* directly as algebra expresses *magnitude* directly.
>
> Leibniz ([17], p. 382)

1 Introduction

From Pythagoras to the mid-nineteenth century, the fundamental problem of geometry was to relate numbers to geometry. It played a key role in the creation of field theory (via the classic construction problems), and, quite differently, in the creation of linear algebra. To resolve the problem, it was necessary to have the modern concept of real number; this was essentially achieved by Simon Stevin, around 1600, and was thoroughly assimilated into mathematics in the following two centuries. The integration of real numbers into geometry began with Descartes and Fermat in the 1630s, and achieved an interim success at the end of the eighteenth century with the introduction into the mathematics curriculum of the traditional course in analytic geometry. From the point of view of analysis, with its focus on functions, this was entirely satisfactory; but from the point of view of geometry, it was not: the method of attaching numbers to geometric entities is too clumsy, the choice of origin and axes irrelevant and (in view of Euclid) unnecessary.

Leibniz, in 1679, had mused upon the possibility of a universal algebra, an algebra with which one would deal directly and simply with geometric entities. The possibility is already suggested by a perusal of Euclid. For example, if D is a point in the side BC of a triangle ABC, then

$$\frac{BD}{BC} = \frac{ABD}{ABC};$$

this ancient theorem begs to be proved by simply multiplying numerator and denominator on the left by A. The geometric algebra of which Leibniz dreamed, and in which the concept of real number is thoroughly assimilated, was created by Hermann Grassmann in the mid-nineteenth century.

Grassmann looked on geometry as it might well be considered today but is not, as being applied mathematics. In his view, there is a part of mathematics, linear algebra, that is applicable to a part of the physical world, chalk figures on a blackboard or objects in space; and geometry, as the business of relating the two, does not belong to mathematics pure and simple. One may say without great exaggeration that Grassmann invented linear algebra and, with none at all, that he showed how properly to apply it in geometry. Linear algebra has become a part of the mainstream of mathematics, though Grassmann gets scant credit for it; but its application to geometry, affine and Euclidean, is remembered only in a half-baked version in which the notion of vector is all important and the notion of point is unnecessary.

The reason is that there are other applications of linear algebra which are of greater practical importance (though they are no more interesting) than geometry—namely, within mathematics, to function spaces, and, within physics, to forces and other vectorial entities. [And yet, oddly enough, Grassmann's geometry is better suited to physics, since, for example, it distinguishes the notions of (polar) vector and axial vector; modern physics texts (such as Feynman [9]), in attempting to explain the physical distinction between the two types of entity, are handicapped by the fact that in the accepted model both have the same mathematical representation.]

2 Grassmann's life

A biography of Grassmann, by Friedrich Engel, may be found in the *Collected Works* ([14], III.2) and also, more briefly, in Michael Crowe's scholarly work [6], the main modern source in English of which I am aware.

Hermann Gunther Grassmann was born in Stettin in 1809, lived there most of his life, and died in 1877. He was one of twelve children and, after marrying at the age of 40, fell short of his father in himself siring only eleven. He spent three years in Berlin studying theology and philology. He had no university mathematical training, nor did he ever hold a university post, though he repeatedly sought one. His life was spent as a schoolteacher.

His major mathematical works are *A Theory of Tides*, which is a kind of thesis written in 1840 in the hope of improving his status as a teacher and which was unpublished until the appearance of the *Collected Works* between 1894 and 1911; a book known briefly as the *Ausdehnungslehre* (literally "Theory of Extension"), which was published in 1844 and almost totally ignored, though it was drawn to the attention of Möbius, Gauss, Kummer, Cauchy, and others; and the *Ausdehnungslehre* of 1862, which was a new work on the same subject, rather than a new edition, and which met an equally cold reception. (Both are included in [14].) His many papers include important contributions to physics as well as to mathematics. He also wrote textbooks in mathematics and languages, edited a political journal for a time, and produced a translation of the *Rig Veda* and a huge commentary on it which, according to *Encyclopaedia Britannica*, is still used today. For his work in philology he received, in the last year of his life, an honorary doctorate. His achievements in mathematics were virtually unrecognized, and it has taken a century for their importance to become clearly visible.

While I intend to devote my attention to his linear algebra and geometry, there are two other contributions of Grassmann to mathematics which may be mentioned. In an arithmetic text [13] published in 1861, he defined the arithmetic operations for integers inductively and he proved their properties—commutativity, associativity, distributivity. He thus anticipated in its most important aspects Peano's treatment [20] of the natural numbers, published 28 years later. Peano generously acknowledges this, but in the naming game by which History distributes fame to the creators of mathematics Peano is a winner, Grassmann a loser. Dedekind, who published a similar development [7] of the natural numbers in 1888, makes no mention of Grassmann.

A feature of Grassmann's work, far in advance of the times, is the tendency toward the use of implicit definition—in which a mathematical entity is characterized by means of its formal properties rather than being obtained by an explicit construction. For example, in the *Ausdehnungslehre* of 1844 he comes very close indeed to the abstract notion of a (not necessarily associative) ring; what is lacking is the language of set theory. This is the second contribution I wanted to mention. Incidentally, the first formal definition of a ring was given by Fraenkel [11] in 1915.

3 The invention of linear algebra

From the beginning, Grassmann distinguished linear algebra, as a formal theory independent of any interpretation, from its application in geometry. However, in the first *Ausdehnungslehre*, the algebra is intermixed with its geometric interpretation—indicating, very interestingly, how he came upon the ideas. Those who did read his work in the late nineteenth century found it easier to follow the 1862 *Ausdehnungslehre*, in which, in modern style, the full development of the mathematical theory precedes its application, and in outlining his linear algebra I shall mainly follow the latter work; it should be borne in mind, though, that some of the ideas have their origin as much as two decades previously.

The definition of a linear space (or vector space) came into mathematics, in the sense of becoming widely known, around 1920, when Hermann Weyl [23] and others published formal definitions. In fact, such a definition had been given thirty years previously by Peano [19], who was thoroughly acquainted with Grassmann's mathematical work. Grassmann did not put down a formal definition—again, the language was not available—but there is no doubt that he had the concept. Beginning with a collection of "units" $e_1, e_2, e_3, \ldots$, he effectively defines the free linear space which they generate; that is to say, he considers formal linear combinations $\sum \alpha_i e_i$, where the α_i are real numbers, defines addition and multiplication by real numbers by setting

$$\sum \alpha_i e_i + \sum \beta_i e_i = \sum (\alpha_i + \beta_i) e_i$$

and
$$\alpha\left(\sum \alpha_i e_i\right) = \sum (\alpha \alpha_i) e_i,$$
and formally proves the linear space properties for these operations. (At the outset, it is not clear whether the set of units is allowed to be infinite, but finiteness is implicitly assumed in some of his proofs.) He then develops the theory of linear independence in a way which is astonishingly similar to the presentation one finds in modern linear algebra texts.

He defines the notions of subspace, independence, span, dimension, join and meet of subspaces, and projections of elements onto subspaces. He is aware of the need to prove invariance of dimension under change of basis, and does so. He proves the Steinitz Exchange Theorem, named for the man who published it [21] in 1913 (and who, incidentally, defined a linear space in terms of "units" in the same way Grassmann did). Among other such results, he shows that any finite set has an independent subset with the same span and that any independent set extends to a basis, and he proves the important identity

$$\dim(U+W) = \dim U + \dim W - \dim(U \cap W).$$

He obtains the formula for change of coordinates under change of basis, defines elementary transformations of bases, and shows that every change of basis (equivalently, in modern terms, every invertible linear transformation) is a product of elementaries.

4 Products

In a paper [12] published in 1855, Grassmann defines a product of elements of a linear space by setting

$$\left(\sum \alpha_i e_i\right)\left(\sum \beta_j e_j\right) = \sum \alpha_i \beta_j e_i e_j,$$

and he proves distributivity. (In this paper the scalars are explicitly allowed to be complex.) If the $e_i e_j$ are themselves linear combinations of the e_i, we have here the concept of an algebra. Instead of following this path (though he did later observe that the algebra of quaternions is a special case), Grassmann singles out particular products by "equations of condition"

$$\sum \xi_{ij} e_i e_j = 0,$$

and, observing as a disadvantage of this notion that it lacks invariance under change of basis, he proceeds to characterize those products whose conditioning equations are invariant under various substitutions.

Grassmann's declared motive for publishing this paper was to claim priority for some results that had been published by Cauchy. The interesting story is related by Engel. In 1847 Grassmann had wanted to send a copy of the *Ausdehnungslehre* to Saint-Venant (to show that he had anticipated some of Saint-Venant's ideas on vector addition and multiplication), but, not knowing the address, Grassmann sent the book to Cauchy, with a request to forward it. Cauchy never did so. And six years later Cauchy's paper [4] appeared in *Comptes Rendus*. Grassmann's comment was that, on reading this, "I recalled at a glance that the principles which are there established and the results which are proved were exactly the same as those which I published in 1844, and of which I gave at the same time numerous applications to algebraic analysis, geometry, mechanics and other branches of physics." An investigating committee of three members of the French Academy, including Cauchy himself, never came to a decision on the question of priority.

In the two *Ausdehnungslehren*, Grassmann singles out for special attention those products, which he calls *linear products*, for which the conditioning equations are invariant under change of basis. He shows that, apart from two trivial cases, there are only two possible types of product: it must be the case that either

$$e_i e_j = e_j e_i \qquad (1)$$

or

$$e_i e_j = -e_j e_i \text{ (and, in particular, } e_i^2 = 0) \qquad (2)$$

for all i and j. Whereas in the 1855 paper he does not consider higher-order products, in the *Ausdehnungslehren* he examines in detail products of the second type, extended, by imposing associativity, to allow multiplication of products of the original simple units (such a product being called a *compound unit*) and, by imposing distributivity, to allow multiplication of linear combinations of compound units (such a combination being called a *form*). Condition (2) entails certain relations among compound units of higher order (for example, that $e_1 e_3 e_2 = -e_1 e_2 e_3$), but it is assumed that no other relations hold. Dimension plays a role here, since, for example, in the three-dimensional case we have only a single independent third-order unit, say $e_1 e_2 e_3$, and this forces independence of the three units $e_1 e_2$, $e_2 e_3$, and $e_3 e_1$ of order 2, because

$$\xi_1 e_1 e_2 + \xi_2 e_2 e_3 + \xi_3 e_3 e_1 = 0$$

implies, when we multiply by e_3, that $\xi_1 e_1 e_2 e_3 = 0$ and hence $\xi_1 = 0$ (and, similarly, $\xi_2 = 0 = \xi_3$).

This multiplication is nowadays called *exterior multiplication*. Half a century later, in treating Grassmann's ideas at length in his *Universal Algebra* [24], A. N. Whitehead explicitly excludes forms of mixed degree like $e_1 + e_1 e_2$ on the rather metaphysical ground that they are "meaningless"; he thus admits arbitrary products of linear combinations of simple units, but not arbitrary linear combination of products. Whitehead's objection itself would have been meaningless to Grassmann and, although he never explicitly brings in such forms, neither, so far as I can see, does he explicitly exclude them. If I am right about this, then Grassmann has the full exterior algebra, while Whitehead's presentation (like many modern treatments of exterior products) restricts consideration to a graded linear space (a far less tidy structure than an algebra).

The full development of exterior algebra as Grassmann did it (in particular the essential invariance under change of basis) is complicated and must be omitted. Perhaps the most important fact is that elements $a_1, a_2, \ldots, a_k$ of the original linear space are linearly independent if and only if $a_1 a_2 \cdots a_k \neq 0$; Grassmann proves this in the modern way and gives the following application (with notation precisely as I have written it) to a system of n linear equations in n unknowns,

$$\alpha_1^{(1)} x_1 + \alpha_2^{(1)} x_2 + \cdots + \alpha_n^{(1)} x_n = \beta^{(1)}$$
$$\alpha_1^{(2)} x_1 + \alpha_2^{(2)} x_2 + \cdots + \alpha_n^{(2)} x_n = \beta^{(2)}$$
$$\cdots$$
$$\alpha_1^{(n)} x_1 + \alpha_2^{(n)} x_2 + \cdots + \alpha_n^{(n)} x_n = \beta^{(n)}.$$

Introducing (independent) units $e^{(1)}, e^{(2)}, \ldots, e^{(n)}$ and quantities

$$a_1 = \alpha_1^{(1)} e^{(1)} + \alpha_1^{(2)} e^{(2)} + \cdots + \alpha_1^{(n)} e^{(n)}$$
$$a_2 = \alpha_2^{(1)} e^{(1)} + \alpha_2^{(2)} e^{(2)} + \cdots + \alpha_2^{(n)} e^{(n)}$$
$$\cdots$$
$$a_n = \alpha_n^{(1)} e^{(1)} + \alpha_n^{(2)} e^{(2)} + \cdots + \alpha_n^{(n)} e^{(n)}$$

and

$$b = \beta^{(1)} e^{(1)} + \beta^{(2)} e^{(2)} + \cdots + \beta^{(n)} e^{(n)},$$

we have

$$b = x_1 a_1 + x_2 a_2 + \cdots + x_n a_n,$$

and so $b a_2 a_3 \cdots a_n = x_1 a_1 a_2 \cdots a_n, \ldots$. Thus if $a_1 a_2 \cdots a_n \neq 0$ we have the unique solution

$$x_1 = \frac{b a_2 a_3 \cdots a_n}{a_1 a_2 a_3 \cdots a_n}, \quad x_2 = \ldots, \quad \ldots$$

(Equivalently, x_1 is the number obtained by dividing the (non-zero) determinant of the matrix obtained from $\alpha_j^{(i)}$ by replacing each $\alpha_1^{(i)}$ by $\beta^{(i)}$.) This is Cramer's rule [5]. The same elegant derivation (but without the double subscript-superscript notation) is given in the first *Ausdehnungslehre*. It is one of the techniques that occurs in the above-mentioned paper of Cauchy.

5 Inner products

Grassmann derives the concept of inner product from that of exterior product in a very interesting way. Working in the algebra generated by the simple units $e_1, e_2, \ldots, e_n$ (subject, as always, to (2)), he defines the *supplement* $|E$ of a compound unit E as $+1$ or -1 times the product of those of the simple units that are not factors of E, the sign $+$ or $-$ being chosen in such a way that

$$E|E = +e_1 e_2 \cdots e_n.$$

For example, in the three-dimensional case,

$$|e_1 = e_2 e_3 \quad \text{and} \quad |e_1 e_3 = -e_2.$$

The supplement is extended to linear combinations of compound units of the same order by linearity:

$$\left| \sum \alpha_j E_j = \sum \alpha_j | E_j. \right.$$

(The map $A \to |A$ is nowadays called the *Hodge star-operator*.) If E and F are forms of the same order, then necessarily $E|F$ is a multiple of $e_1 e_2 \cdots e_n$; indeed, since the forms of order n make up a one-dimensional linear space, one may identify them with the scalars, and this Grassmann does, setting $e_1 e_2 \cdots e_n = 1$. Although $E|F$ makes sense for forms E and F of different order, it is only in the case where both orders are the same (in particular, where they are both 1) that $E|F$ is a number. Noting that $E|F$ is linear in both E and F, Grassmann calls $E|F$ the *inner product* of E and F. Its restriction to the original linear space (the space of forms of order 1) is indeed an inner product in the modern sense, since, as he shows,

$$e_i | e_j = \begin{cases} 1 \text{ if } i = j \\ 0 \text{ if } i \neq j \end{cases},$$

and hence
$$\sum \alpha_i e_i | \sum \beta_j e_j = \sum \alpha_i \beta_i.$$

He calls $\sqrt{a|a}$ the *numerical value* of a. The notion of a complete orthonormal set is introduced, and it is shown that such a set must be independent and that in theorems involving the inner product the original system of units may be replaced by any such set. The notions of orthogonal complement and orthogonal projection are investigated; the *Gram-Schmidt Process* is at least implicitly involved in this.

6 Linear transformations

For the linear transformation that carries the basis elements $e_1, e_2, \ldots, e_n$ to $b_1, b_2, \ldots, b_n$, respectively, Grassmann writes
$$Q = \frac{b_1, b_2, \ldots, b_n}{e_1, e_2, \ldots, e_n},$$
and considers Q to be a generalized quotient. While there are obvious problems with this notation, it does have a certain elegance; for example, if $b_1, b_2, \ldots, b_n$ are independent then the inverse of Q is
$$\frac{e_1, e_2, \ldots, e_n}{b_1, b_2, \ldots, b_n}.$$

He effectively obtains the matrix representation of Q by showing that it may be written
$$Q = \sum \alpha_{r,s} E_{r,s}$$
where
$$E_{r,s} = \frac{0, \ldots, 0, e_s, 0, \ldots, 0}{e_1, \ldots, e_r, \ldots, e_n}.$$

The *determinant* of Q is defined to be the number
$$\frac{b_1 b_2 \cdots b_n}{e_1 e_2 \cdots e_n},$$
eigenvalues and *eigenvectors* are introduced (though different terms are used), and the fact that the eigenvalues are roots of the characteristic polynomial is demonstrated as follows. Suppose that
$$Qx = \rho x$$
where $x = \sum \xi_i e_i \neq 0$. Writing $c_i = (\rho - Q)e_i$, we see that $\sum \xi_i c_i = 0$; thus $c_1, c_2, \ldots, c_n$ are dependent and hence
$$[(\rho - Q)e_1][(\rho - Q)e_2] \cdots [(\rho - Q)e_n] = 0.$$
Since $e_1 e_2 \cdots e_n \neq 0$, this is equivalent to the vanishing of the determinant of the linear map $\rho - Q$. It is shown that the eigenvectors corresponding to distinct eigenvalues are independent and the *spectral theorem* for a symmetric Q is proved.

Turning to a general linear transformation Q, Grassmann shows, in effect (by constructing an appropriate basis), that the whole space may be decomposed as the direct sum of invariant subspaces W_ρ, where ρ ranges through the characteristic roots of Q, and each W_ρ is the kernel of $(Q - \rho)^k$, k being the algebraic multiplicity of ρ. While the *spectral theorem* had been proved by Weierstrass [22] in 1858 (and, for the case of n distinct eigenvalues, by Cauchy [3] in 1829), it appears that Grassmann was the first to prove the latter result which is sometimes called the *primary decomposition theorem*; it goes part of the way to obtaining Jordan's canonical form, published in 1870 [16] (the remaining step being to reduce the nilpotent maps $Q - \rho : W_\rho \to W_\rho$).

7 Geometry

In the *Ausdehnungslehre* of 1844, Grassmann describes the geometric considerations which led him to the theory that we now call linear algebra. Musing on the formula
$$AB + BC = AC, \tag{3}$$
which one might find in old geometry texts, to describe a relationship between lengths that holds for collinear points A, B and C, with B between A and C, he realized that the formula remains valid regardless of the order of the three collinear points, *provided that one sets*
$$BA = -AB; \tag{4}$$
for example, if C lies between A and B then (3) follows from the fact that
$$AB = AC + CB = AC - BC.$$

Over many years Grassmann carefully investigated the consequences of (4), which is the special property that defines an exterior algebra. His development of geometry is complicated, and we shall give here an oversimplified presentation that brings one rapidly to the heart of the matter; those who regard it as criminal to attempt a modern paraphrase of old mathematics should read no further. To avoid (merely notational) complications we shall, like Grassmann, restrict ourselves to three dimensions.

Beginning with the basic material of geometry, numbers and points, we permit them to be combined by formal operations of sum and product, assuming the elementary algebraic rules for such operations, but subject to the condition that (4) holds for all points A and B and also that

$$\alpha A = A\alpha \qquad (5)$$

for all real numbers α and points A. From (4) we deduce that every point A has $A^2 = 0$; thus the square of a point is a number, and we explicitly assume that it is not a point:

$$0 \text{ is not a point.} \qquad (6)$$

We need a rule by which to interpret geometrically the entities which occur in this formal algebra. For a pair of points A and B, and positive real numbers α and β with $\alpha + \beta = 1$, write

$$P = \alpha A + \beta B.$$

Then immediately we have

$$AP = \beta AB, \ PB = \alpha AB \text{ and } AP + PB = AB;$$

these formulas suggest that P should be interpreted as the unique point which divides the line segment from A to B in the ratio β to α. We do so interpret P; it is here, via the bijection $\alpha \to \alpha A + \beta B$ between $[0, 1]$ and the line segment, that numbers enter geometry, and *all* other geometric interpretations follow from this one. To give an immediate example, the interpretation of $P = \alpha A + \beta B$ with $\alpha \leq 0$ and $\beta > 0$ and $\alpha + \beta = 1$ is forced by the fact that, equivalently,

$$B = \left(-\frac{\alpha}{\beta}\right) A + \left(\frac{1}{\beta}\right) P,$$

where $-\alpha/\beta \geq 0$, $1/\beta > 0$ and $(-\alpha/\beta) + 1/\beta = 1$. The *line* through A and B is the set of all $\alpha A + \beta B$ with $\alpha + \beta = 1$, and accordingly we assume that

if A and B are points and $\alpha + \beta = 1$

then $\alpha A + \beta B$ is a point. (7)

To sum up, we consider a ring Ω, in which the number 1 is a unit element and which is generated by $R \cup P$, where R is the set of real numbers and P is a set whose elements are called *points*, subject to the conditions (4), (5), (6), and (7); equivalently, Ω may be regarded as an algebra generated by P. Dimensionality comes in with the assumption that

Ω is generated, as an algebra, by four elements

of P but not by three, (8)

and, finally, ensuring non-triviality of multiplication, that

there exist points A, B, C, D with $ABCD \neq 0$. (9)

Existence and uniqueness of such a structure are proved in [8]. We then have a model for the geometry of three-dimensional space; the multiplication is an exterior product and Grassmann's abstract theory may now be brought to bear in a situation where geometric interpretation is possible.

The difference of two points is a *vector*; here the interpretation is forced by the identity

$$B - A = C - D \iff \frac{1}{2}(A + C) = \frac{1}{2}(B + D).$$

The sum of a vector X and a point A is a point, since

$$A + X = B \iff X = B - A.$$

And a product of a number and a vector is a vector, since for $X = B - A$ we have $\alpha X = P - A$, where $P = (1 - \alpha)A + \alpha B$; this also entails that αX is to be interpreted as having the same direction as X and α times its length.

Here is a classic theorem which with just these ideas becomes trivial: if, in a triangle with vertices A, B, and C, the points D and E, respectively, divide the side from A to B and the side from A to C in equal ratios, then the line segment from D to E is parallel to the one from B to C and the ratio of their lengths is the appropriate number. The proof is two lines:

$$D = \alpha A + \beta B, \ E = \alpha A + \beta C$$
$$\Rightarrow D - E = \beta(B - C).$$

The hypotheses (8) and (9) imply that the space of all fourth-order forms is one-dimensional: if A, B, C, and D are independent points, then any product $A'B'C'D'$ is a multiple of $ABCD$ and one may show that the ratio must be interpreted as the ratio of the oriented volumes of the associated tetrahedra. In particular, $ABCD = A'B'C'D'$ means that the two tetrahedra have the same orientation and volume. This in turn forces one to interpret $ABC = A'B'C'$ as meaning that the associated triangles are coplanar and have the same orientation and area; and $AB = A'B'$ as meaning that the associated line segments lie on the same line and are equal in length and direction.

We may now prove the converse of the above result about triangles in exactly the way that it was done by Euclid. If the line segments from D to E and from B

to C are parallel, then
$$\frac{DB}{AB} = \frac{DBC}{ABC} = \frac{EBC}{ABC} = \frac{EC}{AC},$$
where we have used the fact that for a suitable real α
$$DBC = [E + \alpha(B - C)]BC = EBC.$$

Again, here is the proof that the diagonal of a parallelogram bisects it:
$$D = C + A - B \quad \Rightarrow \quad ABD = -ABC.$$

These examples give an indication, if no more, of the power of Grassmann's geometric interpretation of linear algebra. They are results of affine geometry, but by introducing an inner product one gets equally transparent algebraic proofs of the theorems of Euclidean geometry and trigonometry. (Incidentally, Grassmann, with his inclination toward abstraction and generality, displays little interest in proving known results of geometry, and these proofs are mine; many other examples are given by Forder [10].)

8 Contemporary and later developments

Though in 1844 he had been unaware of their work, Grassmann later acknowledged that in some respects his theory was anticipated, in particular, by the concept of vector addition of Bellavitis [1] and others, and by the barycentric calculus of Möbius [18]. But no one had approached the elegant simplicity of the formula
$$(C - B) + (B - A) = C - A$$
which, for Grassmann, forces the interpretation of the sum of two vectors; and Möbius had not exploited the full possibilities of his notation
$$P = \alpha A + \beta B + \gamma C,$$
which, even today, in his brilliant [2], is dismissed by Boyer as inferior to the homogeneous coordinate representation $P = (\alpha, \beta, \gamma)$. The key to the difference between Möbius and Grassmann is that, whereas for Möbius the equation $\alpha A + \beta B = \alpha' A + \beta' B$ entails merely that $\alpha : \beta = \alpha' : \beta'$, for Grassmann it implies that $\alpha = \alpha'$ and $\beta = \beta'$; the one concept is appropriate to projective geometry, the other to Euclidean.

The story of Hamilton's invention of the quaternions [15] in 1843 and of the subsequent influence of Hamilton and of Grassmann in the emergence of vector analysis is well told by Crowe. But vector analysis in Crowe's sense of the term is a subject that has ceased to exist, or should have; it has been absorbed by linear algebra. It would not be easy to estimate the relative influences of the two men in the development of linear algebra as we know it, but there is no doubt that Grassmann in his own work came far closer to it than Hamilton or any of his contemporaries. Even in those cases where forerunners may be discerned, his results, and especially his methods, were highly original. All mathematicians stand, as Newton said he did, on the shoulders of giants, but few have come closer than Hermann Grassmann to creating, single-handedly, a new subject.

9 Conclusion

The genius of Descartes revealed itself in his decision to drop the ancient convention that a product of line segments is a rectangle. In Grassmann's geometry a product of line segments is again a higher-dimensional object. It is a return to Euclid, but to Euclid with a difference, the difference that had been dreamed of by Leibniz. But Grassmann's geometry (as distinct from his linear algebra) has been largely forgotten. Perhaps this is because, at the moment when it might have been remembered, Hilbert, with his immense prestige, closed the book of Euclid by showing that Euclid's program, rigorously carried through, was too tedious for anyone to bother with. Grassman's way is not tedious; properly done, it is simple, direct, and powerful, and perhaps the book should be opened again.

I conclude with a quotation (as translated in ([6], p. 89) from the preface to the 1862 *Ausdehnungslehre*:

I remain completely confident that the labor I have expended on the science presented here and which has demanded a significant part of my life as well as the most strenuous application of my powers, will not be lost. It is true that I am aware that the form which I have given the science is imperfect and must be imperfect. But I know and feel obliged to state (though I run the risk of seeming arrogant) that even if this work should again remain unused for another seventeen years or even longer, without entering into the actual development of science, still that time will come when it will be brought forth from the dust of oblivion and when ideas now dormant will bring forth fruit. I know that if I also fail to gather around me (as I have until now desired in vain) a circle of scholars, whom I could fructify with these ideas, and whom I could stimulate to develop and enrich them further, yet there will come a time when these ideas, perhaps in a new form, will arise anew and will enter into a living communication with contemporary developments. For truth is eternal and divine.

References

1. Giusto Bellavitis, Saggio di applicazioni di un nuovo methodo di geometria analytica (Calcolo delle equipollenze), *Ann. Sci. Regno Lombardo–Veneto* (1835), 246–247.
2. Carl B. Boyer, *History of Analytic Geometry*, Scripta Mathematica, 1956.
3. Augustin Cauchy, Sur l'équation à l'aide de laquelle on détermine les inégalités séculaires des mouvements des planètes, *Exercices de Mathématiques*, Paris, 1829.
4. Augustin Cauchy, Sur les clefs algébriques, *C. R. Acad. Sci. Paris* **36** (1853), 70–75 and 129–136.
5. Gabriel Cramer, *Introduction à l'Analyse des Lignes Courbes Algébriques*, Geneva, 1750.
6. Michael Crowe, *A History of Vector Analysis*, Notre Dame, 1967.
7. Richard Dedekind, *Was Sind und Was Sollen die Zahlen?*, Braunschweig, 1888.
8. Desmond Fearnley-Sander, Affine geometry and exterior algebra, *Houston J. Math.* **6** (1980), 53–58.
9. Richard P. Feynman, Robert B. Leighton, and Matthew Sands, *Lectures on Physics*, Addison-Wesley, 1963.
10. Henry George Forder, *The Calculus of Extension*, Cambridge, 1941.
11. A. A. Fraenkel, Über die Teiler der Null und die Zerlegung von Ringen, *J. Reine Angew. Math.* **145** (1915), 139–176.
12. Hermann Grassmann, Sur les différents genres de multiplication, *Crelle's Journal* **49** (1855), 123–141.
13. Hermann Grassmann, *Lehrbuch der Arithmetik*, Stetten, 1861.
14. Hermann Grassmann, *Gesammelte Mathematische und Physikalische Werke*, 3 vols. in 6 parts, Teubner, 1894–1911.
15. William Rowan Hamilton, On quaternions or on a new system of imaginaries in algebra, *Phil. Mag.* 3rd Ser. (1844), 10–13.
16. Camille Jordan, *Traité des Substitutions et des Équations Algébriques*, Gauthier-Villars, 1870.
17. Gottfried Wilhelm Leibniz, *Philosophical Papers and Letters* I (transl. and ed. Leroy E. Loemker), Chicago Univ. Press, 1956.
18. August Ferdinand Möbius, *Der Barycentrische Calcul*, Barti, 1827.
19. Giuseppe Peano, *Calcolo Geometrico Secondo l'Ausdehnungslehre di H. Grassmann Preceduto dalle Operazioni della Logica Deduttiva*, Bocca, 1888.
20. Giuseppe Peano, *Arithmetices Principia, Nova Methoda Exposita*, Bocca, 1889.
21. E. Steinitz, Bedingt konvergente Reihen und konvexe Systeme, *J. Reine Angew. Math.* **143** (1913), 128–175.
22. Karl Weierstrass, Über ein die homogenen Funcktionen zweiten Grades betreffendes Theorem, nebst Anwendung desselben auf die Theorie der kleinen Schwingungen, *Monatsh Akad. Berlin* (1858), 207–220.
23. Hermann Weyl, *Raum, Zeit und Materie*, J. Springer, 1923.
24. Alfred North Whitehead, *A Treatise on Universal Algebra with Applications*, Cambridge, 1898.

The Roots of Commutative Algebra in Algebraic Number Theory

ISRAEL KLEINER

Mathematics Magazine **68** (1995), 3–15

Introduction

The concepts of field, commutative ring, ideal, and unique factorization are among the fundamental notions of commutative algebra. How did they arise? In large measure, from three central problems in number theory: *Fermat's Last Theorem*, *reciprocity laws*, and *binary quadratic forms*. In this paper we will describe how this happened.

To put the issues in a broader context, these three number-theoretic problems were instrumental in the emergence of algebraic number theory—one of the two main sources of the modern discipline of commutative algebra. The other source was algebraic geometry. It was in the setting of these two subjects that many of the main concepts and results of commutative algebra evolved in the nineteenth century. Thus commutative algebra can be said to have been well developed before it was created.

Commutative algebra is nowadays understood to be the study of commutative noetherian rings. The first book on the subject, dating from the 1950s, was Zariski and Samuel's *Commutative Algebra*. In the first decades of the twentieth century, what we understand as commutative algebra was known as *Ideal Theory*, a name which reflects the sources of the subject.

To set the scene for our story, a word about mathematics, and especially algebra, in the nineteenth century. The period witnessed fundamental transformations in mathematics—in its concepts, its methods, and in mathematicians' attitude toward their subject. These included a growing insistence on rigor and abstraction; a predisposition for founding general theories rather than focusing on specific problems; an acceptance of non-constructive existence proofs; the emergence of mathematical specialties and specialists; the rise of the view of mathematics as a free human activity, neither deriving from, nor dependent on, nor necessarily applicable to concrete setting; the rise of set-theoretic thinking; and the re-emergence of (a new variant of) the axiomatic method. As for algebra, its subject-matter and methods changed beyond recognition. Algebra as the study of solvability of polynomial equations gave way to algebra as the study of abstract structures defined axiomatically. While in the past algebra was on the periphery of mathematics, it now (the nineteenth and early twentieth centuries) became one of its central concerns. Moreover, algebra began to penetrate other mathematical fields (e.g. geometry, analysis, logic, topology, number theory) to such an extent that in the early decades of the twentieth century one began to speak of the algebraization of mathematics ([3], p. 135 and [23]).

Algebraic number theory and unique factorization

Algebraic number theory arose (as noted) from the investigation of three fundamental problems: Fermat's Last Theorem, reciprocity laws, and binary quadratic forms. Although these problems had their roots in the seventeenth and eighteenth centuries, they began to be intensively studied only in the nineteenth century. The strategy that began to emerge was to embed the domain $\mathbb{Z}$ of integers, in terms of which these problems were formulated, in domains of (what came to be known as) algebraic integers. Let me explain.

Fermat's Last Theorem states that $x^n + y^n = z^n$ has no non-trivial integer solutions for $n > 2$. It suffices to consider $n = 4$ and $n = p$, an odd prime.

Fermat proved the conjecture for $n = 4$ in the early seventeenth century, and Euler proved it for $n = 3$ in the eighteenth (Euler's proof had a gap). In the early nineteenth century, Dirichlet and Legendre proved Fermat's conjecture for $n = 5$, but the case $n = 7$ defied the efforts of the best mathematicians, including Dirichlet and Gauss. A new approach was called for. It was provided by Lamé.

On March 1, 1847, Lamé excitedly announced before the Paris Academy of Sciences that he had proved Fermat's Last Theorem. His basic idea was to factor the left-hand side of $x^p + y^p = z^p$ as follows:

$$x^p + y^p = (x+y)(x+\omega y)(x+\omega^2 y)\cdots(x+\omega^{p-1}y), \quad (1)$$

where ω is a primitive pth root of 1. Since $x^p + y^p = z^p$, the product of the factors on the right-hand side of (1) is a pth power. Lamé claimed that if these factors are relatively prime, then each is a pth power. From this a contradiction can be derived using Fermat's method of infinite descent. If the factors $x+y, x+\omega y, \ldots, x+\omega^{p-1}y$ are not relatively prime, then, upon division by a suitable a, we obtain the relatively prime factors

$$\frac{x+y}{a}, \frac{x+\omega y}{a}, \ldots, \frac{x+\omega^{p-1}y}{a},$$

and the proof, Lamé asserted, proceeds analogously.

The essential idea of Lamé's approach is contained in the following method for finding all primitive Pythagorean triples (x, y, z). Factoring the left-hand side of $x^2 + y^2 = z^2$ we get $(x+yi)(x-yi) = z^2$. It can be shown that $x + yi$ and $x - yi$ are relatively prime (since $(x, y, z) = 1$), and since their product is a square, each is a square. In particular, $x + yi = (a + bi)^2 = (a^2 - b^2) + 2abi$, hence $x = a^2 - b^2$, $y = 2ab$, and it follows that $z = a^2 + b^2$. This is the standard formula yielding all Pythagorean triples.

Following Lamé's presentation, Liouville, who was in the audience, took the floor and pointed to what seemed to him to be a gap in Lamé's proof. The gap, he said, was Lamé's contention that if a product of relatively prime factors is a pth power, then each of the factors must be a pth power. This result is of course true for the integers, Liouville noted, but it required a proof for the complex entities Lamé was dealing with.

Liouville's observation went to the head of what was wrong with Lamé's proof. Specifically, to prove Lamé's claim one had to show that unique factorization holds in the domain

$$D_p = \{a_0 + a_1\omega + \cdots + a_{p-1}\omega^{p-1} : a_i \in \mathbb{Z}\}$$

of so-called cyclotomic integers. About three months after Lamé's presentation Kummer wrote to Liouville that D_p is in general not a unique factorization domain (UFD). (The first p for which it is not is $p = 23$.)

It is important to note that although Lamé's proof was false (it is correct for those p for which D_p is a UFD), his overall strategy proved fundamental for subsequent approaches to Fermat's Last Theorem. In particular, he proved the theorem for $n = 7(!)$, a feat that (we recall) eluded the likes of Dirichlet and Gauss.

We turn now to *reciprocity laws*. It is no easy matter to define what a reciprocity law is; see [10], [22], [30]; for our purposes, the descriptions that follow suffice. Gauss was the first to define formally the notion of *congruence*. Thus it was natural for him to pose the question of solvability of the congruence

$$a_0 + a_1 x + a_2 x^2 + \cdots + a_n x^n \equiv 0 \pmod{n}.$$

The problem in this generality proved intractable (it is so to this day), and so Gauss considered a simpler case, namely $a_0 + a_1 x + a_2 x^2 \equiv 0 \pmod{n}$. Several reductions show that it suffices to consider the congruence $x^2 \equiv q \pmod{p}$, where both p and q are odd primes (the case of even primes has to be considered separately). Gauss proved that there is a *reciprocity relation* between the solvability of $x^2 \equiv q \pmod{p}$ and $x^2 \equiv p \pmod{q}$. Specifically, $x^2 \equiv q \pmod{p}$ is solvable if, and only if, $x^2 \equiv p \pmod{q}$ is solvable, unless $p \equiv q \equiv 3 \pmod{4}$. In the latter case, $x^2 \equiv q \pmod{p}$ is solvable if, and only if, $x^2 \equiv p \pmod{q}$ is not. This is the celebrated quadratic reciprocity law, one of the "jewels of number theory", proved in Gauss' *Disquisitiones Arithmeticae* of 1801.

What about higher reciprocity laws? That is, is there a *reciprocity relation* between the solvability of $x^m \equiv q \pmod{p}$ and $x^m \equiv p \pmod{q}$, for $m > 2$? Gauss opined that such laws cannot even be properly conjectured within the context of natural numbers ([29], p. 105), adding that ([17], p. 108)

The previously accepted laws of arithmetic are not sufficient for the foundations of a general theory... Such a theory demands that the domain of higher arithmetic be endlessly enlarged.

A prophetic statement, indeed. Gauss was calling (in modern terms) for the founding of an arithmetic theory of algebraic numbers. In fact, Gauss himself began to enlarge the domain of arithmetic by introducing what came to be known as the *Gaussian integers*,

$$\mathbb{Z}[i] = \{a + bi : a, b \in \mathbb{Z}\},$$

and showing that they form a UFD. This he did in two papers in 1829 and 1831, in which he used $\mathbb{Z}[i]$ to formulate the law of biquadratic reciprocity.

In the 1831 paper Gauss also introduced the geometric representation of complex numbers as points in the Euclidean plane. At about the same time, Jacobi and Eisenstein (as well as Gauss in unpublished papers) formulated the cubic reciprocity law. Here one needed to consider the domain

$$\mathbb{Z}[\rho] = \{a + b\rho : a, b \in \mathbb{Z}\},$$

where ρ is a primitive cube root of 1. $\mathbb{Z}[\rho]$ was also shown to be a UFD. The search was on for higher reciprocity laws. But as in the case of Fermat's Last Theorem, here too one needed new methods to deal with cases beyond the first few, since unlike $\mathbb{Z}[i]$ and $\mathbb{Z}[\rho]$, other domains of higher arithmetic needed to formulate such laws were not UFDs.

We now turn to the third number-theoretic problem related to the rise of algebraic number theory—namely, representation of integers by *binary quadratic forms*. (This problem is, in fact, intimately connected to reciprocity laws; see [10].) An (integral) binary quadratic form is an expression of the form

$$f(x,y) = ax^2 + bxy + cy^2, \quad a,b,c \in \mathbb{Z}.$$

The major problem of the theory of quadratic forms was: Given a form f, find all integers m that can be represented by f; that is, for which $f(x,y) = m$. This problem was studied for specific f by Fermat, and intensively investigated by Euler. For example, Fermat considered the representation of integers as sums of two squares. It was, however, Gauss in the *Disquisitiones* who made the fundamental breakthroughs and developed a comprehensive and beautiful theory of binary quadratic forms. Most important was his definition of the *composition* of two forms and his proof that the (equivalence classes of) forms with a given *discriminant* $D = b^2 - 4ac$ form a commutative group under this composition.

The *idea* behind composition of forms is simple: if forms f and g represent integers m and n, respectively, then their composition $f*g$ should represent the product mn. The *implementation* of this idea is subtle and "extremely difficult to describe" ([13], p. 334). Attempts to gain conceptual insight into Gauss's theory of composition of forms inspired the efforts of some of the best mathematicians of the time, among them Dirichlet, Kummer, and Dedekind. The key idea here, too, was to extend the domain of higher arithmetic and view the problem in a broader context.

Here is perhaps the simplest illustration. If m_1 and m_2 are sums of two squares, so is $m_1 m_2$. Indeed, if $m_1 = x_1^2 + y_1^2$ and $m_2 = x_2^2 + y_2^2$, then

$$m_1 m_2 = (x_1 x_2 - y_1 y_2)^2 + (x_1 y_2 + x_2 y_1)^2.$$

In terms of the composition of quadratic forms this can be expressed as

$$f(x_1, y_1) * f(x_2, y_2) = f(x_1 x_2 - y_1 y_2, x_1 y_2 + x_2 y_1),$$

or $f * f = f$, where $f(x,y) = x^2 + y^2$. But even this "simple" law of composition seems mysterious and *ad hoc*, until one introduces the Gaussian integers, which make it transparent ($\bar{\alpha}$ denotes the conjugate of α):

$$(x_1^2 + y_1^2)(x_2^2 + y_1^2)$$
$$= (x_1 + y_1 i)(x_1 - y_1 i)(x_2 + y_2 i)(x_2 - y_2 i)$$
$$= (x_1 + y_1 i)(x_2 + y_2 i)\overline{(x_1 + y_1 i)(x_2 + y_2 i)}$$
$$= [(x_1 x_2 - y_1 y_2) + (x_1 y_2 + x_2 y_1)i]$$
$$\quad \times \overline{[(x_1 x_2 - y_1 y_2) + (x_1 y_2 + x_2 y_1)i]}$$
$$= (x_1 x_2 - y_1 y_2)^2 + (x_1 y_2 + x_2 y_1)^2.$$

In general, $ax^2 + bxy + cy^2 = m$ can be written as

$$\frac{1}{a}\left(ax + \frac{b + \sqrt{D}}{2}y\right)\left(ax + \frac{b - \sqrt{D}}{2}y\right) = m,$$

where $D = b^2 - 4ac$ is the discriminant of the quadratic form. We have thus expressed the problem of representation of integers by binary quadratic forms in terms of the domain

$$R = \left\{\frac{u + v\sqrt{D}}{2} : u, v \in \mathbb{Z}, u \equiv v \pmod{2}\right\}.$$

Specifically, if we let

$$a = \alpha, \quad \frac{b + \sqrt{D}}{2} = \beta,$$

then

$$ax^2 + bxy + cy^2 = \frac{1}{a}(\alpha x + \beta y)(\bar{\alpha}x + \bar{\beta}y)$$
$$= \frac{1}{a} N(\alpha x + \beta y),$$

where N denotes the *norm*. Thus to solve $ax^2 + bxy + cy^2 = m$ is to find $x, y \in \mathbb{Z}$ such that $N(\alpha x + \beta y) = m$. In fact, Kummer noted in the 1840s that the entire theory of binary quadratic forms can be regarded as the theory of "complex numbers" (Kummer's terminology) of the form $x + y\sqrt{D}$ ([6], p. 585). And since such domains of "complex numbers" did not, in general, possess unique factorization, the development of their arithmetic theory became an important goal. (Gauss' theory of binary quadratic forms may, in fact, be seen as the first major (implicit) attempt to deal with non-unique factorization.)

It is time to take stock. We have seen that in dealing with central problems in number theory—namely Fermat's Last Theorem, reciprocity laws, and binary quadratic forms—it was found important to formulate them as problems in domains of (what came to be called) *algebraic integers*. This often transformed additive problems into multiplicative ones (e.g., $x^3 + y^3 = z^3$) was changed to
$$(x+y)(x+\rho y)(x+\rho^2 y) = z^3,$$
ρ a primitive cube root of unity. Now, multiplicative problems in number theory are, in general, much easier to handle than additive problems—provided that one has the (multiplicative) machinery to do it, namely unique factorization in the domain under consideration. Thus the study of unique factorization in domains of algebraic integers became the major problem of a newly emerging subject—algebraic number theory. Number theory had lost its innocence (which, at the appropriate time, is of course no bad thing). After more than 2000 years in which number theory meant the study of properties of the (positive) integers, its scope became enormously enlarged. One could no longer use the term "integer" with impunity. The term had to be qualified—a rational (ordinary) integer, a Gaussian integer, a cyclotomic integer, a quadratic integer, or any one of an infinite species of other algebraic integers. Of course the real difficulty was not the multiplicity of types of integers, but the fact that, in general, domains of such integers no longer possessed unique factorization.

Could one regain the paradise lost by establishing unique factorization in *some* sense for these domains? An affirmative answer was given by Dedekind and Kronecker in the last third of the nineteenth century. We will describe the work of Dedekind, who introduced the notions of ideal and prime ideal, and showed that every nonzero ideal in an arbitrary domain of algebraic integers is a unique product of prime ideals. (An ideal P is *prime* if $xy \in P$ implies $x \in P$ or $y \in P$.) This result generalizes the theorems on unique factorization of elements into primes in (for example) $\mathbb{Z}$, $\mathbb{Z}[i]$, and $\mathbb{Z}[\rho]$. But it was preceded by very significant contributions by Kummer to the problem of unique factorization, which inspired both Dedekind and Kronecker (in different ways). We begin with a sketch of Kummer's contribution.

Kummer and ideal numbers

We recall that the domains of cyclotomic integers
$$D_p = \{a_0 + a_1\omega + \cdots + a_{p-1}\omega^{p-1} : \\ a_i \in \mathbb{Z}, \omega \text{ a primitive } p\text{th root of } 1\}$$
were central in the study of Fermat's conjecture. They also proved important in the investigation of higher reciprocity laws. The two problems are, in fact, related, and progress on one was often linked with progress on the other (see [18], p. 203). Gauss contended that if he were able to derive higher reciprocity laws, then Fermat's Last Theorem would be one of the less interesting corollaries. That is one prediction where Gauss proved to be wrong. A general reciprocity law (due to Artin) was obtained in the 1920s, but Fermat's problem was not resolved until Andrew Wiles's proof in the mid-1990s. In any case, Wiles's proof does by no means follow from Artin's reciprocity law.

Both Fermat's problem and higher reciprocity laws were of great interest to Kummer (the latter apparently more than the former), and to make serious progress on them it was essential to establish unique factorization (of some type) in the domains D_p. This Kummer accomplished in the 1840s. As he put it in a letter to Liouville, unique factorization in D_p "can be saved by the introduction of a new kind of complex numbers that I have called *ideal complex numbers*" ([3], p. 101). Kummer then showed that every element in the domain of cyclotomic integers is a unique product of *ideal primes*.

Kummer's theory was vague and computational. In fact, the central notion of ideal number was only *implicitly* defined in terms of its divisibility properties (see [13] or [14]). Kummer noted that in adopting the implicit definition he was guided by the idea of *free radical* in chemistry, a substance whose existence can only be discerned by its effects ([9], p. 35).

The following example (even if not of a cyclotomic domain) may give a sense of Kummer's theory of ideal numbers. Let $D = \{a + b\sqrt{5}i : a, b \in \mathbb{Z}\}$. This is a standard example (due to Dedekind) of a domain in

which factorization is not unique. For example,
$$6 = 2 \times 3 = (1+\sqrt{5}i)(1-\sqrt{5}i),$$
where 2, 3, and $1 \pm \sqrt{5}i$ are primes (indecomposables) in D ([1], p. 250). To restore unique factorization of $6 \in D$, adjoin the *ideal numbers* $\sqrt{2}$, $(1+\sqrt{5}i)/\sqrt{2}$, and $(1-\sqrt{5}i)/\sqrt{2}$. These are, in fact, *ideal primes*.

We then have
$$6 = 2 \times 3 = \sqrt{2} \times \sqrt{2} \times \frac{1+\sqrt{5}i}{\sqrt{2}} \times \frac{1-\sqrt{5}i}{\sqrt{2}},$$
and
$$6 = (1+\sqrt{5}i)(1-\sqrt{5}i)$$
$$= \sqrt{2} \times \frac{1+\sqrt{5}i}{\sqrt{2}} \times \sqrt{2} \times \frac{1-\sqrt{5}i}{\sqrt{2}};$$
that is, the decomposition of 6 into primes is now unique. Furthermore, although the choice of the ideal primes $\sqrt{2}$, $(1 \pm \sqrt{5}i)/\sqrt{2}$ seems to have been *ad hoc*, it will come to seem natural after ideals are introduced.

A cruder but perhaps more revealing example is the "domain"
$$H = \{3k+1 : k = 0, 1, 2, \ldots\}$$
$$= \{1, 4, 7, 10, 13, \ldots\},$$
introduced by Hilbert. (This is not an integral domain, but its structure as a multiplicative semigroup illustrates the ideas involved.) Here
$$100 = 10 \times 10 = 4 \times 25,$$
where 4, 10, and 25 are primes of H. We can eliminate the non-uniqueness of this factorization by adjoining the "ideal primes" 2 and 5. Then
$$100 = 10 \times 10 = 2 \times 5 \times 2 \times 5$$
and
$$100 = 4 \times 25 = 2 \times 2 \times 5 \times 5.$$

The notion of adjoining elements to a given mathematical structure in order to obtain a desired property is commonplace and important in various areas of mathematics. Thus we adjoin $\sqrt{-1}$ to the reals to obtain algebraic closure, and adjoin *ideal points* (points at infinity) to the euclidean plane to obtain symmetry (duality) between points and lines. The adjunction of ideal numbers to the cyclotomic integers should be seen in the same light.

Kummer's ideas were brilliant but difficult and not clearly formulated. The fundamental concepts of ideal number and ideal prime were not intrinsically defined. "Today," notes Edwards, "few mathematicians would find [Kummer's] definition [of ideal primes] acceptable" ([16], p. 6). Most importantly, Kummer's decomposition theory was devised only for cyclotomic integers (he mentions, without elaboration, that an analogous theory could be developed for algebraic integers of the form $x + y\sqrt{D}$). What was needed was a decomposition theory that would apply to arbitrary domains of algebraic integers. This called for a fundamentally new approach to the subject, provided (independently, and in different ways) by Dedekind and Kronecker. We will focus (as mentioned) on Dedekind's formulation, which is the one that has generally prevailed.

Dedekind and ideals

Dedekind's work [11] in 1871 was revolutionary in several ways: in its formulation, its grand conception, its fundamental new ideas, and its modern spirit. Its main result was that every non-zero ideal in the domain of integers of an algebraic number field is a unique product of prime ideals. Before one could state this theorem one had, of course, to define the concepts in its statement, especially "the domain of integers of an algebraic number field" and "ideal." It took Dedekind about 20 years to formulate them.

The number-theoretic domains studied at the time, such as the Gaussian integers, the integers arising from cubic reciprocity, and the cyclotomic integers, are all of the form
$$\mathbb{Z}[\theta] = \{a_0 + a_1\theta + \cdots + a_n\theta^n : a_i \in \mathbb{Z}\},$$
where θ satisfies an appropriate polynomial with integer coefficients. It was therefore tempting to define the domains to which Dedekind's theorem would apply as objects of this type. These were, however, the wrong objects, Dedekind showed. For example, he proved that Kummer's theory of unique factorization could *not* be extended to the domain $\mathbb{Z}[\sqrt{3}i] = \{a + b\sqrt{3}i : a, b \in \mathbb{Z}\}$ ([14], p. 337), and, of course, Dedekind's objective was to try to extend Kummer's theory to *all* domains of algebraic integers. One had to begin the search for the appropriate domains, Dedekind contended, within an *algebraic number field*—a finite field extension
$$\mathbb{Q}(\alpha) = \{q_0 + q_1\alpha + \cdots + q_s\alpha^s : q_i \in \mathbb{Q}\}$$
of the rationals, α an algebraic number (the notion of *algebraic number* was well known at the time, but not that of *algebraic integer*). He showed that

$\mathbb{Q}(\alpha)$ is closed under the four algebraic operations of addition, subtraction, multiplication, and division, and then defined axiomatically (for the first time) the notion of a *field* ([3], p. 118):

> A system A of real or complex numbers is called a *field* if the sum, difference, product, and quotient of any two numbers of A belongs to A.

This procedure was typical of Dedekind's *modus operandi*: He would distill from a concrete object (in this case $\mathbb{Q}(\alpha)$) the properties of interest to him (in this case closure under the four algebraic operations) and proceed to define an abstract object (in this case a field) in terms of those properties. This is, of course, standard practice nowadays, but it was revolutionary in Dedekind's time. Dedekind used it again and again, as we shall see.

To come back to the domain of algebraic integers to which Dedekind's result was to apply: He defined it to be the set R of all elements of $\mathbb{Q}(\alpha)$ that are roots of *monic* polynomials with integer coefficients. (*All* the elements of $\mathbb{Q}(\alpha)$ are roots of polynomials, not necessarily monic, with integer coefficients.) He showed that these elements behave like integers—they are closed under addition, subtraction, and multiplication. They are the *integers* of the algebraic number field $\mathbb{Q}(\alpha)$. Dedekind did not, however, motivate this basic definition of the domain of algebraic integers, a fact lamented by Edwards ([14], p. 332):

> Insofar as this is the crucial idea of the theory, the genesis of the theory appears, therefore, to be lost.

This is not to say, of course, that the notion of the domain of algebraic integers cannot be motivated. The algebraic integers of $\mathbb{Q}(\alpha)$ are to $\mathbb{Q}(\alpha)$ what the rational integers (the elements of $\mathbb{Z}$) are to $\mathbb{Q}$. Now, the elements of $\mathbb{Q}$ can be thought of as roots of *linear* polynomials with integer coefficients (viz. $p(x) = ax + b$). Among these, the integers are the roots of *monic* linear polynomials (viz. $q(x) = x + b$). If we extend this analysis to polynomials of arbitrary degree over $\mathbb{Z}$, their roots in $\mathbb{Q}(\alpha)$ yield all of $\mathbb{Q}(\alpha)$, while the roots of the monic polynomials among them yield the domain of integers of $\mathbb{Q}(\alpha)$. Moreover, just as $\mathbb{Q}$ is the field of quotients of $\mathbb{Z}$, so $\mathbb{Q}(\alpha)$ is the field of quotients of *its* domain of integers. (See also [28], pp. 264–265, for a discussion, from a different perspective, of the issue of motivation of algebraic integers.)

Having defined the domain R of algebraic integers of $\mathbb{Q}(\alpha)$ in which he would formulate and prove his result on unique decomposition of ideals, Dedekind considered, more generally, sets of integers of $\mathbb{Q}(\alpha)$ closed under addition, subtraction, and multiplication. He called them *orders*. (The domain R of integers of $\mathbb{Q}(\alpha)$ is the largest order.) Here, then, was another algebraic first for Dedekind—an essentially axiomatic definition of a commutative ring, albeit in a concrete setting.

The second fundamental concept of Dedekind's theory, that of ideal, derived its motivation and name from Kummer's ideal numbers. Dedekind wanted to characterize them *internally*, within the domain D_p of cyclotomic integers. Thus, for each ideal number σ, he considered the set of cyclotomic integers divisible by σ. These, he noted, are closed under addition and subtraction, as well as under multiplication by all elements of D_p. Conversely, he proved (and this is a difficult theorem) that every set of cyclotomic integers closed under these operations is precisely the set of cyclotomic integers divisible by some ideal number τ. Thus there is a one-one correspondence between ideal numbers and subsets of the cyclotomic integers closed under the above operations. Such subsets of D_p Dedekind called *ideals*. These subsets, then, characterized ideal numbers internally, and served as motivation for the introduction of ideals in arbitrary domains of algebraic integers.

Dedekind defined them abstractly as follows ([14], p. 343:

> A subset I of the integers R of an algebraic number field K is an *ideal* of R if it has the following two properties:
>
> i. if $\beta, \gamma \in I$, then $\beta \pm \gamma \in I$.
> ii. if $\beta \in I, \mu \in R$, then $\beta\mu \in I$.

Dedekind then defined a prime ideal—perhaps the most important notion of commutative algebra—as follows: an ideal P of R is *prime* if its only divisors are R and P. Given ideals A and B, A was said to *divide* B if $A \supseteq B$. In later versions of his work Dedekind showed that A divides B if, and only if, $B = AC$ for some ideal C of R (see [7], p. 122). Having defined the notion of prime ideal, Dedekind proved his fundamental theorem that every nonzero ideal in the ring of integers of an algebraic number field is a unique product of prime ideals.

Here is another way of introducing ideals, in the spirit of *Kronecker's* work on establishing unique factorization in domains of algebraic integers. It is perhaps best illustrated with an example. Let $D = \{a + b\sqrt{5}i : a, b \in \mathbb{Z}\}$. As we noted, D is not a UFD. Since uniqueness of factorization in a domain is equivalent to the existence of the greatest common divisor (g.c.d.) of any two non-zero elements in the domain ([16], p. 199), it follows that one way to re-establish unique factorization in D is to introduce g.c.d.s in D. (This is Kronecker's motivation for the introduction of divisors (ideals); see [5], p. 125, or [14], p. 353.) How do we introduce the g.c.d. of, say, 2 and $1 + \sqrt{5}i$, which appear in the factorizations of 6 into primes, $6 = 2 \times 3 = (1 + \sqrt{5}i) \times (1 - \sqrt{5}i)$? We look at $\mathbb{Z}$ (or any UFD) for guidance.

If $a, b \in \mathbb{Z}$, the g.c.d. of a and b can be found among the elements of the set

$$I = \{ax + by : x, y \in \mathbb{Z}\}.$$

In fact, I is a principal ideal of $\mathbb{Z}$,

$$I = \langle d \rangle = \{dm : m \in \mathbb{Z}\},$$

and g.c.d.$(a, b) = d$. (An ideal J of a ring R is *principal* if it is generated by a single element $a \in R$. We write $J = \langle a \rangle$.) In like manner, if D were a UFD, then g.c.d.$(2, 1 + \sqrt{5}i)$ would be found among the elements of

$$P = \{2\alpha + (1 + \sqrt{5}i)\beta : \alpha, \beta \in D\}.$$

Moreover, we would have $P = \langle s \rangle$ (the principal ideal of D generated by some $s \in D$) and

$$\text{g.c.d.}(2, 1 + \sqrt{5}i) = s.$$

Since D is not a UFD, there is no $s \in D$ such that $P = \langle s \rangle$. But why not have all of P represent (capture) g.c.d.$(2, 1 + \sqrt{5}i)$? Indeed, this is Kronecker's idea. (Dedekind would have argued for an "ideal element" $\hat{s}$ ($\hat{s} \notin D$) to describe P, so that P would be the set of elements in D divisible by $\hat{s}$; we exhibit such an $\hat{s}$ below.) We note that P is an ideal of D, though not a principal ideal. Similarly we let

$$Q = \langle 3, 1 + \sqrt{5}i \rangle = \{3\alpha + (1 + \sqrt{5}i)\beta : \alpha, \beta \in D\}$$
$$\text{and} \quad R = \langle 3, 1 - \sqrt{5}i \rangle$$

(note that $\langle 2, 1 - \sqrt{5}i \rangle = \langle 2, 1 + \sqrt{5}i \rangle$).

We then easily verify that

$$P^2 = \langle 2 \rangle, \quad PQ = \langle 1 + \sqrt{5}i \rangle, \quad QR = \langle 3 \rangle,$$
$$\text{and } PR = \langle 1 - \sqrt{5}i \rangle,$$

where we must recall that if A and B are ideals of a ring R, their product is the ideal

$$AB = \left\{ \sum a_i b_i : a_i \in A, b_i \in B, \text{ a finite sum} \right\}.$$

Returning now to the factorizations

$$6 = 2 \times 3 = (1 + \sqrt{5}i) \times (1 - \sqrt{5}i),$$

they yield the following factorizations of ideals:

$$\langle 6 \rangle = \langle 2 \rangle \langle 3 \rangle = P^2(QR) \text{ and}$$
$$\langle 6 \rangle = \langle 1 + \sqrt{5}i \rangle \langle 1 - \sqrt{5}i \rangle = (PQ)(PR) = P^2 QR.$$

One can readily verify that the ideals P, Q, R are prime. Thus the *ideal* $\langle 6 \rangle$ (if not the *element* 6) has been factored uniquely into prime ideals. Paradise regained. See [5], [14].

Let us compare this factorization of $\langle 6 \rangle$ into *prime ideals* with the factorization of 6 into *ideal primes* (à la Kummer) that we gave earlier:

$$6 = 2 \times 3 = \sqrt{2} \times \sqrt{2} \times \frac{1 + \sqrt{5}i}{\sqrt{2}} \times \frac{1 - \sqrt{5}i}{\sqrt{2}}$$

and

$$6 = (1 + \sqrt{5}i)(1 - \sqrt{5}i)$$
$$= \sqrt{2} \times \frac{1 + \sqrt{5}i}{\sqrt{2}} \times \sqrt{2} \times \frac{1 - \sqrt{5}i}{\sqrt{2}}.$$

Performing some eighteenth-century symbolic callisthenics, we obtain the following. Since

$$P^2 = \langle 2 \rangle, \quad P \sim \sqrt{2}$$

(where "$\sim$" stands for "corresponds to", "captures", "represents"). In fact, P is the ideal consisting of all elements of D divisible by the ideal number $\sqrt{2}$ — that is, such that the quotient is an algebraic integer ([7], p. 235). This would have been Dedekind's way of introducing P (that is, $\hat{s} = \sqrt{2}$). We also have $PQ = \langle 1 + \sqrt{5}i \rangle$, hence $PQ/P \sim (1 + \sqrt{5}i)/\sqrt{2}$, so that the ideal Q corresponds to the ideal number $(1 + \sqrt{5}i)/\sqrt{2}$. And since $PR = \langle 1 - \sqrt{5}i \rangle$,

$$PR/R \sim (1 - \sqrt{5}i)/\sqrt{2},$$

hence $R \sim (1 - \sqrt{5}i)/\sqrt{2}$. This removes the mystery associated with our earlier introduction of the ideal numbers $\sqrt{2}$ and $(1 \pm \sqrt{5}i)/\sqrt{2}$. (How might Dedekind have restored unique factorization to the "domain" $H = \{1, 4, 7, 10, 13, \ldots\}$ we considered earlier?)

Dedekind's legacy

Dedekind's *Supplement X* [11] to Dirichlet's *Zahlentheorie* was the culmination of 70 years of investigations of problems related to unique factorization. It created, in one swoop, a new subject—algebraic number theory. It introduced, albeit in a concrete setting, some of the most fundamental concepts of commutative algebra, such as field, ring, ideal, prime ideal, and module (the last we have not discussed). These became basic in algebra and beyond. *Supplement X* also established one of the central results of algebraic number theory, namely the representation of ideals in domains of integers of algebraic number fields as unique products of prime ideals. The theorem was soon to play a fundamental role in the study of algebraic curves. It would also serve as a model for decomposition results in algebra.

As important as his concepts and results were Dedekind's methods. In fact, "his insistence on philosophical principles was responsible for many of his important innovations" ([14], p. 349). One of his philosophical principles was a focus on intrinsic, conceptual properties over formulas, calculations, or concrete representations. Another was the acceptance of nonconstructive procedures (definitions, proofs) as legitimate mathematical methods. Dedekind's great concern for teaching also influenced his mathematical thinking. His two very significant methodological innovations were the use of the axiomatic method and the institution of set-theoretic modes of thinking. See [15], [16].

To illustrate, compare Dedekind's and Kronecker's definitions of a field. (Recall that these two mathematicians developed essentially the same theory, at about the same time, but used entirely different methods and approaches.) Dedekind defined a field axiomatically (even if in a concrete setting), as the set of all real or complex numbers satisfying certain properties. Kronecker did not, in fact, define a field in the way *we* think of a definition. He *described* it, calling it a *domain of rationality*. It consisted of all rational functions in the quantities $R', R'', R''', \ldots$ with integer coefficients, $Q(R', R'', R''', \ldots)$, in current notation). He put no restrictions on these quantities—they could be indeterminates or roots of algebraic equations. In this sense, Kronecker's field concept is more general than Dedekind's. On the other hand, the elements $R', R'', R''', \ldots$ had to be given explicitly, so that (for example) the field of all algebraic numbers would not be considered a domain of rationality, whereas it qualified as a field under Dedekind's definition. Here was a foreshadowing of the formalist-intuitionist controversy, to emerge in full force in the early decades of the twentieth century.

The axiomatic method was just beginning to surface after 2000 years of near dormancy. Dedekind was instrumental in pointing to its mathematical power and pedagogical value. In this he inspired (among others) David Hilbert and Emmy Noether. His use of set-theoretic formulations (recall, for example, his definition of an ideal as the set of elements of a domain satisfying certain properties), including the use of the completed infinite—taboo at the time—preceded by about 10 years Cantor's seminal work on the subject. Undeniably, it justifies the description of his 1871 memoir as "the 'birthplace' of the modern set-theoretic approach to the foundations of mathematics" ([15], p. 9). Edwards's tribute (especially as it comes from one who is a great admirer of Kronecker's approach to the subject) is fitting ([14], p. 20):

> Dedekind's legacy ... consisted not only of important theorems, examples, and concepts, but of a whole *style* of mathematics that has been an inspiration to each succeeding generation.

Postscript

What of the three fundamental number-theoretic problems that gave rise to the development of algebraic number theory? In the 1871 memoir Dedekind gave a very satisfactory explanation, in terms of ideals, of Gauss's theory of binary quadratic forms, in particular of the composition of such forms. Specifically, with each quadratic form $f(x, y) = ax^2 + bxy + cy^2$ with discriminant $D = b^2 - 4ac$, Dedekind associated the ideal I of integers of the quadratic field $\mathbb{Q}(\sqrt{D})$ generated by a and $(-b + \sqrt{D})/2$, that is, $I = \langle a, (-b + \sqrt{D})/2 \rangle$. He showed that this establishes a one-one correspondence between quadratic forms with discriminant D and ideals of the domain of integers of $\mathbb{Q}(\sqrt{D})$, and that under this correspondence composition of forms corresponds to multiplication of ideals (see [1]). Although this result made Gauss's composition of forms transparent, it did not end interest in the arithmetic study of binary quadratic forms. In particular, the representation of integers by specific forms and the determination of the class number of forms are problems of current interest. See [10].

Various reciprocity laws beyond the biquadratic were formulated and proved in the nineteenth century, especially by Eisenstein, Kummer, and Hilbert. Kummer's work relied on his results on unique factorization of cyclotomic integers, Hilbert's initiated class-field theory (see [10], [18]). But these efforts were not entirely satisfactory. In fact, one of Hilbert's 23 problems presented at the 1900 International Congress of Mathematicians was to find a general reciprocity law.

Artin solved the problem in the 1920s with his celebrated reciprocity law—a centerpiece of class-field theory. It establishes (in one of its incarnations, and roughly speaking) a connection between the Galois group of an abelian extension of an algebraic number field and the arithmetical properties of the ground field (see [10]). A far cry from the quadratic, cubic, and biquadratic reciprocity laws! See [22], [30].

Having instituted unique factorization in the domains D_p of cyclotomic integers, Kummer proved Fermat's Last Theorem for *regular* primes. (A prime p is *regular* if it does not divide the class number h of D_p; equivalently, if for any ideal I of D_p, if I^p is a principal ideal then so is I.) He then showed that all primes $p < 100$ (with the exception of 37, 59, and 67) are regular. Thus Fermat's Last Theorem was established for all primes $p < 100$ (about 10 years later Kummer disposed of the three remaining cases). Given that Gauss and Dirichlet failed to prove Fermat's theorem even for $n = 7$, this was quite an accomplishment. It showed the power of Kummer's theory of ideal numbers in cyclotomic domains. Much progress was made on Fermat's Last Theorem following Kummer's work (see [13] and [26]). In 1976, through the use of powerful computers along with powerful abstract mathematics, Fermat's Last Theorem was shown to hold for all primes $p < 125,000$ (see [26]). In another direction, as a corollary of his pioneering work on abelian varieties, Faltings showed in 1983 that for each fixed $n > 2$ the equation $x^n + y^n = z^n$ has at most finitely many solutions (See [4], pp. 41–42). Wiles's very recent proof of Fermat's Last Theorem has its roots in Taniyama's 1955 conjecture about elliptic curves. It took 30 years to tie the conjecture to Fermat's Last Theorem. The high point of these developments was Ribet's 1987 proof that a special case of the Taniyama conjecture implies Fermat's Last Theorem. Wiles devoted the past six years to proving this special case of Taniyama's conjecture. See [12], [27].

Emmy Noether used to say modestly of her work that it can already be found in Dedekind's. It would not be amiss to say that much of the subject matter of this paper originated in, or was inspired by, Gauss's *Disquisitiones Arithmeticae* of 1801. A new subject, algebraic number theory, came into being. It embodied fundamental concepts whose consequences reached far beyond number theory. Yet some of the major number-theoretic problems of that period are still with us today, two centuries later. Of course, the tools that have been brought to bear on their study are recent, and they are powerful. Gauss, Kummer, and Dedekind would likely have found them initially inscrutable, but in time wondrous. Mathematics is alive and well (if any evidence were needed).

References

1. W. W. Adams and L. J. Goldstein, *Introduction to Number Theory*, Prentice-Hall, 1976.

2. M. F. Atiyah and I. G. Macdonald, *Introduction to Commutative Algebra*, Addison-Wesley, 1969.

3. I. G. Bashniakova and A. N. Hudakov, The theory of algebraic numbers and the beginnings of commutative algebra, *Mathematics in the Nineteenth Century* (ed. A. N. Kolmogorov and A. P. Yushkevich), Birkhäuser, 1992, pp. 86–135.

4. S. Bloch, The proof of the Mordell Conjecture, *Math. Intell.* **6:2** (1984), 41–47.

5. E. D. Bolker, *Elementary Number Theory: An Algebraic Approach*, W. A. Benjamin, 1970.

6. N. Bourbaki, Historical note, *Commutative Algebra*, Addison-Wesley, 1972, pp. 579–606.

7. H. Cohn, *Advanced Number Theory*, Dover, 1980.

8. M. J. Collison, The origins of the cubic and biquadratic reciprocity laws, *Arch. Hist. Exact Sci.* **17** (1977), 63–69.

9. G. Cornell, Review of two books on algebraic number theory, *Math. Intell.* **5:1** (1983), 53–56.

10. D. A. Cox, *Primes of the Form $x^2 + ny^2$: Fermat, Class Field Theory, and Complex Multiplication*, Wiley, 1989.

11. R. Dedekind, Supplement X, *Vorlesungen über Zahlentheorie* (L. Dirichlet), Braunschweig, 1871, 1879, 1894.

12. K. Devlin, F. Gouvêa and A. Granville, Fermat's Last Theorem, a theorem at last, *Focus* **13:4** (1993), 3–4.

13. H. M. Edwards, *Fermat's Last Theorem: A Genetic Introduction to Algebraic Number Theory*, Springer-Verlag, 1977.

14. H. M. Edwards, The genesis of ideal theory, *Arch. Hist. Exact Sci.* **23** (1980), 321–378.

15. H. M. Edwards, Dedekind's invention of ideals, *Bull. Lond. Math. Soc.* **15** (1983), 8–17.

16. H. M. Edwards, Mathematical ideas, ideals, and ideology, *Math. Intell.* **14:2** (1992), 6–19.

17. E. Grosswald, *Topics from the Theory of Numbers*, 2nd ed., Birkhäuser, 1984.

18. K. Ireland and M. Rosen, *A Classical Introduction to Modern Number Theory*, Springer-Verlag, 1982.

19. I. Kaplansky, Commutative rings, *Proc. Conf. on Commutative Algebra* (ed. J. W. Brewer and E. A. Rutter), Springer-Verlag, 1973, pp. 153–166.

20. I. Kleiner, A sketch of the evolution of (noncommutative) ring theory, *L'Enseign. Math.* **33** (1987), 227–267.

21. M. Kline, *Mathematical Thought from Ancient to Modern Times*, Oxford, 1972.

22. E. Lehmer, Rational reciprocity laws, *Amer. Math. Monthly* **85** (1978), 467–472.

23. A. F. Monna, *L'Algébrisation de la Mathématique: Réflexions Historiques*, Comm. Math. Inst. Rijksuniversiteit, 1977.

24. H. Pollard and H. G. Diamond, *The Theory of Algebraic Numbers*, 2nd ed., Mathematical Association of America, 1975.

25. W. Purkert, Zur Genesis des abstrakten Körperbegriffs, I, II, *Naturwiss., Techn. u. Med.* **8** (1971), 23–37, **10** (1973), 8–20.

26. P. Ribenboim, *13 Lectures on Fermat's Last Theorem*, Springer-Verlag, 1979.

27. K. Ribet, Wiles proves Taniyama's conjecture; Fermat's Last Theorem follows, *Notices Amer. Math. Soc.* **40** (1993), 575–576.

28. H. Stark, *An Introduction to Number Theory*, MIT Press, 1978.

29. A. Weil, Two lectures on number theory, past and present, *L'Enseign. Math.* **20** (1974), 87–110.

30. B. F. Wyman, What is a reciprocity law?, *Amer. Math. Monthly* **79** (1972), 571–586.

Eisenstein's Misunderstood Geometric Proof of the Quadratic Reciprocity Theorem

REINHARD C. LAUBENBACHER AND DAVID J. PENGELLEY

College Mathematics Journal **25** (1994), 29–34

The *quadratic reciprocity theorem* has played a central role in the development of number theory, and formed the first deep law governing prime numbers. Its numerous proofs from many distinct points of view testify to its position at the heart of the subject. The theorem was discovered by Euler, and restated by Legendre in terms of the symbol now bearing his name, but was first proven by Gauss. The eight different proofs Gauss published in the early 1800s, for what he called the *fundamental theorem*, were followed by dozens more before the century was over, including four given by Gotthold Eisenstein in the years 1844–1845. Our aim is to take a new look at Eisenstein's geometric proof, in which he presents a particularly beautiful and economical adaptation of Gauss's third proof, and to draw attention to all the advantages of his proof over Gauss's, most of which have apparently heretofore been overlooked.

It is hard to imagine today the sensation caused by Eisenstein when he burst upon the mathematical world. In the autumn of 1843, at age twenty, this self-taught mathematician had barely received his high school certificate and entered the Friedrich-Wilhelms University of Berlin, when he produced a flood of publications, instantly making him one of the leading mathematicians of the early nineteenth century. On July 14, 1844, Gauss wrote to C. Gerling:

> I have recently made the acquaintance of a young mathematician, Eisenstein from Berlin, who came here with a letter of recommendation from Humboldt. This man, who is still very young, exhibits *very* excellent talent, and will certainly achieve great things [4].

In 1844 Eisenstein contributed no less than 16 of the 27 mathematical articles in Volume 27 of *Crelle's Journal*, and by his third semester as a student he had received an honorary doctorate from Breslau [9]. Both Gauss and the great scientist and explorer Alexander von Humboldt made great efforts, for the most part in vain, to obtain recognition and financial security for the impoverished Eisenstein. Gauss wrote to Humboldt that Eisenstein's talent was "that which nature bestows upon only a few in each century" [3]. He did obtain a position as a Privatdozent (unsalaried lecturer) at the University in Berlin, and was eventually admitted to the Berlin Academy of Sciences in early 1852. But by then his lifelong poor health had seriously deteriorated, and later that same year he died, aged 29, of tuberculosis. Gotthold Eisenstein stands with Abel and Galois as another nineteenth-century mathematical genius with a tragic and short life [3], [9].

Eisenstein's geometric proof appeared in *Crelle's Journal* under the title "Geometrischer Beweis des Fundamentaltheorems für die quadratischen Reste" [5]. It is intimately connected to Gauss's third proof (published in [6] and translated in [10]). Many expositions of Eisenstein's proof, beginning with [1], [2], have observed only one of its three geometric aspects, and have overlooked the other important differences between the two proofs. The result has been a failure to recognize and fully appreciate all the ways in which Eisenstein greatly streamlines and illuminates Gauss's proof, and thereby reveals its essence. For instance, Gauss's third proof is based on a result known as *Gauss's lemma*, and Eisenstein was particularly pleased with a shortcut he found to avoid the technicalities involved in applying it:

> I did not rest until I freed my geometric proof ... from the Lemma on which it still depended, and

it is now so simple that it can be communicated in a couple of lines ([7], pp. 173–174).

We believe that the elegance of Eisenstein's proof deserves wide appreciation, and we present it here along with a comparison to Gauss's third proof.

Eisenstein's proof

To set the stage, we recall a few consequences of the fact that the residue classes modulo a prime p form a field $\mathbb{Z}_p$. Fermat's little theorem, that $b^{p-1} \equiv 1 \pmod{p}$ for any integer b not divisible by p, holds because the non-zero residue classes form a (cyclic) group of order $p-1$ under multiplication. When p is odd, the squaring map $x \mapsto x^2$ has kernel $\{-1, 1\}$, so its image, the squares or *quadratic residues* modulo p, form a subgroup of order $(p-1)/2$ and the non-residues form its coset. The quadratic character of a residue class $b \in \mathbb{Z}_p$ is specified by the *Legendre symbol*:

$$\left(\frac{b}{p}\right) = 1 \quad \text{if } b \text{ is a quadratic residue mod } p$$
$$= -1 \quad \text{if not.}$$

From $\left(b^{(p-1)/2}\right)^2 = 1$, it follows that $b^{(p-1)/2} = \pm 1$ for any $b \in \mathbb{Z}_p^*$. But if $b = c^2$, then

$$b^{(p-1)/2} = c^{p-1} = 1,$$

so the quadratic residues are all roots of the polynomial $x^{(p-1)/2} - 1$. Since this polynomial can have no more than $(p-1)/2$ roots in the field $\mathbb{Z}_p$, we conclude that its roots are exactly the quadratic residues. That is, we have *Euler's criterion*:

$$\left(\frac{b}{p}\right) \equiv b^{(p-1)/2}$$

for any integer b not divisible by p.

The quadratic reciprocity theorem compares the quadratic character of two primes with respect to each other.

Quadratic reciprocity theorem. *If p and q are distinct odd primes, then*

$$\left(\frac{p}{q}\right)\left(\frac{q}{p}\right) = (-1)^{(p-1)(q-1)/4}.$$

Here is Eisenstein's proof, closely following both his own language and notation (which he conveniently and successfully abuses).

Proof: Consider the set $a = 2, 4, 6, \ldots, p-1$. Let r denote the remainder (mod p) of an arbitrary multiple qa. Then it is apparent that the list of numbers $(-1)^r r$ agrees with the list of numbers a, up to multiples of p. For clearly each of the numbers $(-1)^r r$ has even least positive residue, and if there were duplication among these residues, e.g.

$$(-1)^{qa} qa \equiv (-1)^{qa'} qa',$$

then $a \equiv \pm a'$. Since the as are distinct, it follows that $a + a' \equiv 0$, which cannot occur since $0 < a + a' < 2p$ and $a + a'$ is even.

Thus,

$$q^{(p-1)/2} \prod a \equiv \prod r \pmod{p}$$
$$\text{and } \prod a \equiv (-1)^{\sum r} \prod r \pmod{p},$$

from which it follows that

$$q^{(p-1)/2} \equiv (-1)^{\sum r} \pmod{p}.$$

Recalling Euler's criterion that

$$\left(\frac{q}{p}\right) \equiv q^{(p-1)/2} \pmod{p},$$

this produces

$$\left(\frac{q}{p}\right) = (-1)^{\sum r}, \qquad (1)$$

so that one may focus solely on the parity of the exponent. Clearly

$$\sum qa = p \sum \left[\frac{qa}{p}\right] + \sum r, \qquad (2)$$

where $[\ ]$ is the greatest integer function. Since the elements a are all even, and p is odd, it follows that $\sum r \equiv [qa/p] \pmod{2}$, and thus

$$\left(\frac{q}{p}\right) = (-1)^{\sum [qa/p]}.$$

(Here Eisenstein remarks that since up to this point q need not have been an odd prime, but merely a number relatively prime to p, one can easily obtain the *Ergänzungssatz*

$$\left(\frac{2}{p}\right) = (-1)^{(p^2-1)/8}$$

from the above formula. This we leave as an exercise for the reader.)

Eisenstein now uses a geometric representation of the exponent in this last equation to transform it twice while retaining its parity. This exponent is precisely the number of integer lattice points with even abscissas lying in the interior of the triangle ABD in the Figure (note that no lattice points lie on the line AB).

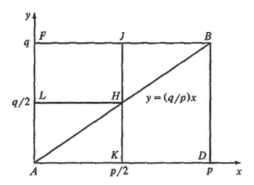

Consider an even abscissa $a > p/2$. Since the number of lattice points on each abscissa in the interior of the rectangle $ADBF$ is even, the number $[qa/p]$ of lattice points on the abscissa below AB has the same parity as the number of lattice points above AB. This in turn is the same as the number of points lying below AB on the odd abscissa $p - a$. This one-to-one correspondence between even abscissas in triangle BHJ and odd abscissas in AHK now implies that $\sum [qa/p] \equiv \mu \pmod 2$, where μ is the number of points inside triangle AHK, and thus

$$\left(\frac{q}{p}\right) = (-1)^\mu.$$

Reversing the roles of p and q yields

$$\left(\frac{p}{q}\right) = (-1)^\nu,$$

where ν is the number of points inside the triangle AHL. Since the total number of points inside both triangles is simply $(p-1)/2 \cdot (q-1)/2$, one may now conclude that

$$\left(\frac{p}{q}\right)\left(\frac{q}{p}\right) = (-1)^{\nu+\mu} = (-1)^{(p-1)(q-1)/4}.$$

Even the normally modest Eisenstein ([7], p. 174) could not restrain his pleasure with this proof:

> How lucky good Euler would have considered himself, had he possessed these lines about seventy years ago.

Eisenstein versus Gauss

Gauss himself considered his third proof to be the most direct and natural of his demonstrations. In introducing it he said:

> For a whole year this theorem tormented me and absorbed my greatest efforts until at last I obtained a proof... Later I came across three other proofs which were built on entirely different principles... I do not hesitate to say that until now a natural proof has not been produced. I leave it to the authorities to judge whether the following proof which I have recently been fortunate enough to discover deserves this description ([10], p. 113).

While Eisenstein essentially follows the same outline as Gauss, each feature of his approach displays great clarity and insight, and offers an elegant view while shortening the path taken by Gauss.

Gauss's third proof begins with his lemma, which says that

$$\left(\frac{q}{p}\right) = (-1)^\alpha, \qquad (3)$$

with α obtained as follows: Let

$$\mathcal{A} = \left\{1, 2, \ldots, \frac{p-1}{2}\right\} \quad \text{and}$$

$$\mathcal{B} = \left\{\frac{p+1}{2}, \frac{p+3}{2}, \ldots, p-1\right\}.$$

Then α is defined to be the number of least positive residues of the set $q\mathcal{A}$ which lie in $\mathcal{B}$.

Instead of using Gauss's lemma, Eisenstein derives equation (1), with the algebraic expression $\sum r$ in the exponent, which is then more easily converted into the key equation

$$\left(\frac{q}{p}\right) = (-1)^{\sum [(qa)/p]}, \qquad (4)$$

common to both proofs, than is equation (3). While Eisenstein's algebraic exponent is easily transformed into the exponent in (4) via (2), Gauss must establish a number of technical properties of the greatest integer function and apply them to relate α to the exponent in (4). Eisenstein's use of the set $a = 2, 4, 6, \ldots, p-1$, as opposed to Gauss's $\mathcal{A}$, allows him to count the same elements as Gauss's lemma, but via the expression $\sum r$, leading quickly to (4):

> The main difference between my argument and that of Gauss is that I do not divide the numbers less than p into those less than $p/2$ and those greater than $p/2$, but rather into even and odd ones [7].

Eisenstein now applies his two clever geometric transformations to convert the exponent $\sum [qa/p]$ into the number of lattice points in the triangle AHK (mod 2). After doing the same for (p/q), yielding the number of lattice points in AHL, the proof is completed simply by counting the lattice points in the rectangle $AKHL$. (Most modern expositions of Eisenstein's proof present only this final counting argument, replacing his two geometric transformations by algebra.) Gauss, on the other hand, in essence performs the same two transformations, and counting, without availing himself of the geometric presentation. He actually counts the lattice points using algebraic properties of the greatest integer function. This makes the remainder of his proof lengthy and non-intuitive, and forces him to consider separate cases, depending on the congruence classes of p and q $(\mod 4)$. (For a more detailed comparison, see [8].)

References

1. P. G. Bachmann, *Niedere Zahlentheorie*, Teubner, 1902–1910; republished by Chelsea, 1968.

2. O. Baumgart, Über das Quadratische Reciprozitätsgesetz, *Zeitschrift für Mathematik und Physik* **30** (1885), Historisch-literarische Abtheilung, 169–277.

3. K.-R. Biermann, Gotthold Eisenstein: Die Wichtigsten Daten seines Lebens und Wirkens, *Mathematische Werke; Gotthold Eisenstein,* Chelsea, 1975, pp. 919–929.

4. K.-R. Biermann, *Carl Friedrich Gauß*, Verlag C. H. Beck, 1990, p. 177.

5. G. Eisenstein, Geometrischer Beweis des Fundamentaltheorems für die quadratischen Reste, *Crelle's Journal* **28** (1844), 246–249.

6. C. F. Gauss, *Commentationes Societatis Regiae Scientiarum Göttingensis* 16 (1808), Göttingen; also *Werke*, Band 2, Göttingen, 1876, pp. 1–8.

7. A. Hurwitz and F. Rudio (eds.), *Briefe von G. Eisenstein an M. Stern*, supplement to *Zeitschrift für Mathematik und Physik* **40** (1895), 169–203.

8. R. Laubenbacher and D. Pengelley, Gauss, Eisenstein, and the "third" proof of the quadratic reciprocity theorem: Ein kleines Schauspiel, *Mathematical Intelligencer* **16** (1994), 67–72.

9. F. Rudio (ed.), Eine Autobiographie von Gotthold Eisenstein. Mit Ergänzenden Biographischen Notizen, *Zeitschrift für Mathematik und Physik* **40** (1895), 143–168.

10. D. E. Smith, *A Source Book in Mathematics*, Dover, 1959, pp. 112–118.

Waring's Problem

CHARLES SMALL

Mathematics Magazine **50** (1977), 12–16

What follows is a non-scholarly survey of the history of Waring's problem. Although a few easy things are proved along the way, the paper is mostly concerned with telling stories—in other words, quoting many beautiful theorems without proof. The proofs, for the most part, involve hard-core analysis, and are difficult. Anyone wishing to pursue the subject should examine chapters 20 and 21 of Hardy and Wright [4] and then [1] and [2]. Ellison's paper [2] provides a much more scholarly and detailed version of the story, with many proofs and an extensive bibliography; the present informal version should serve a useful role as an introduction to [1] and [2].

Waring's problem began with Edward Waring, who published a book entitled *Meditationes Algebraicae* in 1770, in which, among other things, the following remarkable assertion occurs:

> Every number is the sum of 4 squares; every number is the sum of 9 cubes; every number is the sum of 19 biquadrates (4th powers); and so on.

(Here, and throughout, number means natural number, possibly 0.)

The assertion for squares is much older: it is hinted at in Diophantus (roughly third century AD) and stated explicitly by Bachet in 1621. Fermat claimed to have a proof in 1641, but in 1770 when Waring's book appeared, the 4-squares theorem was a well-known "fact" for which no proof was known. It was proved later in the same year by Lagrange, to the chagrin of Euler, who had tried unsuccessfully to find a proof.

There have been many other proofs for the 4-squares theorem since Lagrange; here's one that is particularly short and sweet. Suppose that we are given a number n which we want to write as a sum of four squares. Clearly, we can throw out any squares that divide n: if $n = x^2 n'$ and $n' = a^2 + b^2 + c^2 + d^2$ then $n = (xa)^2 + (xb)^2 + (xc)^2 + (xd)^2$. Thus we may assume that $n = p_1 \ldots p_n$ is a product of distinct primes p_i. Then, by the Chinese Remainder Theorem, we know that $\mathbb{Z}_n$ breaks up as the corresponding direct product $\mathbb{Z}_{p_1} \times \cdots \times \mathbb{Z}_{p_n}$ of finite fields. Now in each $\mathbb{Z}_{p_i}$, -1 is a sum of two squares. (In fact it's easy to prove, more generally, that in any finite field every element is a sum of two squares.) Thus -1 is a sum of two squares in $\mathbb{Z}_n$, so we can find integers c, d, m such that $-1 = c^2 + d^2 - mn$.

Now consider the matrix

$$A = \begin{pmatrix} n & c + di \\ c - di & m \end{pmatrix}$$

over $\mathbb{Z}[i]$, where $i = \sqrt{-1}$. Then $\det A = 1$ and $n > 0$. Under these conditions it can be shown (we indicate one method in the next paragraph) that $A = BB^*$ for some 2×2 matrix B over $\mathbb{Z}[i]$, where $*$ denotes conjugate transpose. The upper left entry in BB^* yields the desired expression for n:

$$\begin{pmatrix} n & - \\ - & - \end{pmatrix} = \begin{pmatrix} x + yi & z + wi \\ - & - \end{pmatrix} \cdot \begin{pmatrix} x - yi & - \\ z - wi & - \end{pmatrix}.$$

In other words, $n = x^2 + y^2 + w^2 + z^2$, a sum of 4 squares.

We can be sure that $A = BB^*$ for some 2×2 complex matrix B for well-known reasons ([3], §72): A represents a positive transformation. But we really need a matrix B over $\mathbb{Z}[i]$. We can get it by a proof using induction on $c^2 + d^2$. This crucial observation is due to George Bergman, and it is what makes this proof work so nicely.

The assertions for cubes and biquadrates, and the implied assertions for higher powers, originate with Waring. The first question, no doubt, is what those implied assertions are: what does Waring's phrase

Power	Integer	Minimal expression as sums of kth powers	Number of kth powers required
2	7	$7 = 2^2 + 3 \cdot 1^2$	$1 + 3 = 4$
3	23	$23 = 2 \cdot 2^3 + 7 \cdot 1^3$	$2 + 7 = 9$
4	79	$79 = 4 \cdot 2^4 + 15 \cdot 1^4$	$4 + 15 = 19$
5	223	$223 = 6 \cdot 2^5 + 31 \cdot 1^5$	$6 + 31 = 37$

Table 1. Waring's Problem concerns efficient ways of expressing integers as sums of kth powers. When $k = 2$ it can always be done with four squares (including 0 if necessary); when $k = 3$ it requires 9 cubes, when $k = 4$ it requires at least 19 biquadrates (or fourth powers), and when $k = 5$ it requires 37 fifth powers. The sequences $7, 23, 79, 223, \ldots$ and $4, 9, 19, 37, \ldots$ form a pattern discovered by Euler: the number of kth powers needed to express the $(k-1)$st number in the first sequence is given by the corresponding number in the second sequence.

"and so on" (& *sic deinceps*) mean? One might observe that $2 \cdot 4 + 1 = 9$ and $2 \cdot 9 + 1 = 19$ and wonder if Waring meant to assert that, for each k, every number is a sum of $s(k)$ kth powers, where $s(k)$ is the appropriate term in the sequence defined by $s(2) = 4$, $s(n + 1) = 2s(n) + 1$. It turns out, as we shall see below, that there is a better candidate for such a sequence in which the next term after 19 is 37 rather than 39. But the most likely interpretation of what Waring had in mind is something weaker than this.

Let us define $g(k)$ to be the smallest r such that every number is the sum of r kth powers, and put $g(k) = \infty$ if no such r exists. Thus, Lagrange's theorem is that $g(2) \leq 4$ (and therefore $g(2) = 4$, since 7 is not a sum of three squares). In this language, the interpretation of Waring's assertion has usually been:

$$g(2) = 4, \ g(3) = 9, \ g(4) = 19,$$
$$\text{and} \quad g(k) < \infty \text{ for all } k.$$

Waring's problem, then, is to prove that $g(k) < \infty$ for all k, and then to compute $g(k)$.

Of course, the general problem is much harder than the case $k = 2$ considered above. Indeed, it is one of those nasty gems, like Fermat's Last Theorem, which begins with a simply-stated assertion about natural numbers, and leads quickly into deep water. One difference between the two is that Fermat's Last Theorem is still a problem whereas Waring's problem is mostly a theorem: Fermat's conjecture that $x^n + y^n = z^n$ has no solution in positive integers when $n > 2$ remains unproved, while Waring's problem, at least in its original form, is 99% solved. [*Editors' note*: This paper was written well before Andrew Wiles's proof in the mid-1990s of Fermat's Last Theorem for all exponents, while the proof of the finiteness of $g(k)$ for all exponents had been accomplished much earlier; the determination of the value for all $g(k)$ remains tantalizingly incomplete, however.]

To get a feeling for the solution, let's look for numbers that cannot be written as sums of just a few kth powers. In other words, we're looking for lower bounds for $g(k)$. (Since we don't know, *a priori*, whether $g(k)$ is finite, it seems prudent to begin by seeking *lower* bounds!) The first obvious choice is $n = 2^k - 1$; clearly n requires n kth powers; hence $g(k) \geq 2^k - 1$. A trick found by Euler will let us do better:

Divide 3^k by 2^k, writing the result as Euclid taught us:

$$3^k = q \cdot 2^k + r, \ 0 \leq r < 2^k.$$

Now consider $n = q \cdot 2^k - 1$. Since $n < q \cdot 2^k \leq 3^k$, we can use only 1^k and 2^k to write n as a sum of kth powers; and since $n < q \cdot 2^k$ we can use only $(q - 1)$ 2^ks; the difference has to be made up with 1s. Thus

$$n = (q - 1) \cdot 2^k + (2^k - 1) \cdot 1^k$$

is a minimal decomposition of n as a sum of kth powers; the number of kth powers involved is $q - 1 + 2^k - 1 = q + 2^k - 2$, which we will call $\bar{g}(k)$. Thus $q \cdot 2^k - 1$ is not the sum of fewer than $\bar{g}(k)$ kth powers, so that $g(k) \geq \bar{g}(k)$ for all k.

This result takes on some life if we tabulate (in Table 3) some values of $\bar{g}$. To do this it is convenient

r			Largest known integer requiring r cubes
9	239	=	$2 \cdot 4^3 + 4 \cdot 3^3 + 3 \cdot 1^3$
8	454	=	$7^3 + 4 \cdot 3^3 + 3 \cdot 1^3$
7	8042	=	$16^3 + 12^3 + 2 \cdot 10^3 + 6^3 + 2 \cdot 1^3$

Table 2. Nine cubes suffice to solve Waring's problem for $k = 3$: every integer can be expressed as a sum of at most 9 cubes. Dickson proved that 239 is the largest number requiring as many as nine cubes. Numerical evidence going back to Jacobi suggests that 454 is the largest number requiring eight cubes, and that 8042 is the largest requiring seven cubes.

k	2	3	4	5	6	7	8	9	$\cdots$
q	2	3	5	7	11	17	25	38	$\cdots$
$\bar{g}(k)$	4	9	19	37	73	143	279	548	$\cdots$
$g(k)$	4	9	19?	37	73	143	279	548	$\cdots$
$G(k)$	4	≥ 4, ≤ 7	16	≥ 6, ≤ 23	≥ 9, ≤ 36	≥ 8, ≤ 137	≥ 32, ≤ 163	≥ 13, ≤ 190	$\cdots$

Table 3. Waring's problem deals with the representation of integers as sums of kth powers. In this table, $g(k)$ is the smallest number with the property that every number is the sum of $g(k)$ kth powers. Euler found a simple lower bound for $g(k)$, called $\bar{g}(k)$, given by the formula $\bar{g}(k) = q + 2^k - 2$, where $q = [(3/2)^k]$. This table indicates that for small $k \neq 4$ Euler's bound is in fact the correct value for $g(k)$; that it might be correct for all values of k is a major unsolved problem known as the *Ideal Waring Theorem*. The final line in this table reflects the relative lack of knowledge concerning the related number $G(k)$, which is defined as the smallest number such that every *sufficiently large number* is the sum of $G(k)$ kth powers.

to express q as $[(3/2)^k]$, and then use $\bar{g}(k) = q + 2^k - 2$. Euler's procedure shows that $7 (= 2 \cdot 2^2 - 1)$ requires 4 squares, $23 (= 3 \cdot 2^3 - 1)$ requires 9 cubes, $79 (= 5 \cdot 2^4 - 1)$ requires 19 biquadrates, ... And Euler knew the procedure, in general, by 1772.

The general theorem—that $g(k)$ is finite for all k—was proved in 1909 by Hilbert. By that time it had been shown that $g(3) = 9$ (Wieferich, 1909), that $g(4) \leq 53$ (Liouville, 1859), and that $g(k)$ was finite for $k = 5, 6, 7, 8$, and 10.

Hilbert's proof is extremely complicated. The essential ingredient is a fantastic identity that had been conjectured by Hurwitz:

$$\left(X_1^2 + X_2^2 + \cdots + X_n^2\right)^k = \sum_{j=1}^{M} \lambda_j \left(a_{1j}X_1 + \cdots + a_{nj}X_n\right)^{2k}.$$

The assertion here is that for each n and k there exist positive rational numbers λ_j and integers a_{ij}, where

$$1 \leq i \leq n, \ 1 \leq j \leq M, \quad \text{and}$$
$$M = (2k+n)!/(2k)!(n-1)!,$$

making the identity in the indeterminates $X_1, \ldots X_n$ true. Hilbert proved the existence of such identities by estimating huge multiple integrals associated with certain convex bodies; in the original paper, a 25-fold integral is evaluated. The proof has since been simplified, but it remains an existence proof for the numbers $g(k)$, and sheds no light on their actual value.

Once Hilbert's theorem is known, the problem "reduces" to computation of $g(k)$. The general question was taken up by Hardy and Littlewood, in a long series of papers published in the 1920s under the title "On Some Problems of Partitio Numerorum" I, II, ..., VIII. *Partitio Numerorum* has since become additive number theory, in approximately the same sense that *analysis situs* has become topology. Additive number theory is now a vast and flourishing subject, which grew from Waring's problem and generalizations of it. Hardy and Littlewood developed a powerful analytic method for handling such questions, and they used it to give an entirely different proof of Hilbert's theorem. Their proof was no easier than Hilbert's, but it did offer some hope of leading to actual computation of the values of $g(k)$.

The Hardy-Littlewood method was improved in many important respects by Vinogradov, and the method which evolved from their combined work is still a dominant force in additive number theory. The Russians call it the Hardy–Littlewood–Vinogradov method, but elsewhere it is usually known as the *circle method*, because it involves integrating a complicated function around a circle in the complex plane. The circle is divided into *major arcs* and *minor arcs* whose precise identification depends on the particular aspect of the problem being considered. Because the integrand behaves in a complicated way near the contour of integration, delicate analysis is required to estimate the integrals on those arcs.

The Hardy–Littlewood–Vinogradov method was sufficiently refined by about 1935 to allow Dickson to determine $g(k)$ for nearly all k. Recall the definition of q, r and $\bar{g}(k)$:

$$3^k = q \cdot 2^k + r, \ 0 \leq r < 2k,$$
$$q = \left[\left(\frac{3}{2}\right)^k\right], \quad \bar{g}(k) = q + 2^k - 2.$$

Let us call k *good* if $q + r \leq 2^k$, and *bad* otherwise. What Dickson showed was this:

If k is good and not equal to 4, then $g(k) = \bar{g}(k)$. If k is bad, then $g(k) = \bar{g}(k) + q' - t$, where $q' = \left[\left(\frac{4}{3}\right)^k\right]$ and $t = 1$ if $2^k < qq' + q + q'$, and 0 otherwise.

(Since $2^k \leq qq' + q + q'$, t can equal 0 only in case $2^k = qq' + q + q'$.) This determines $g(k)$ completely, except for $k = 4$.

It is, of course, a matter of simple arithmetic to determine if a given k is good or not. In particular, the reader can check that all $k \leq 9$ are good, so that if we leave aside $k = 4$ we can add $g(k)$ to our previous table (see the fourth line of Table 3). A computer study in 1964 showed that all $k \leq 200,000$ are good, and a difficult theorem due to Mahler (1957) shows that at most finitely many k are bad; it is not known whether *any* bad k actually exist. This is certainly a tantalizing state of affairs: the conjecture that all k are good is a simple-looking assertion about the natural numbers that is (except for $k = 4$) equivalent to the so-called *ideal Waring theorem*—namely, the assertion that the rather trivial lower bound $\bar{g}(k)$ found by Euler in 1772 is actually the correct value of $g(k)$ for all k.

To be perfectly accurate, we should note that Dickson was not the only person involved in the proof of this key result. Dickson proved it in 1935 for $k \geq 7$, aside from a few cases that were filled in during the 1940s by various people (including Dickson). For $k = 6$ it is due to Pillai (1940), and for $k = 5$ to Chen (1964). For $k = 3$ and $k = 2$ it is, as we have seen, classical and pre-classical, respectively. For $k = 4$ Euler's construction yields, as we have seen, $19 = \bar{g}(4) \leq g(4)$. The upper bounds have decreased over the years, the best known currently being $g(4) \leq 22$ (H. E. Thomas, 1974). There is little doubt that further computation will show $g(4) = 19$. It has been, since 1940, literally a matter of computation; for Auluck proved in that year that all numbers greater than

$$c = 10^{10^{88.39}}$$

are sums of 19 biquadrates!

The theorem of Dickson *et al.* solves Waring's problem in its original form (except for the ambiguity about $g(4)$ and the question of whether any bad numbers exist). However, there is another aspect which opens a whole new range of problems; in fact it is an essential part of the method of proof in the Hardy-Littlewood-Vinogradov-Dickson solution of Waring's problem.

To describe it, let us go back to the case of cubes. Recall that Wieferich, in 1909, had proved that $g(3) = 9$. In the same year, Landau observed that by tracing through Wieferich's arguments very carefully it was possible to prove more: only finitely many numbers actually required 9 cubes because all sufficiently large numbers are sums of 8 cubes. Back in 1851, Jacobi had tabulated for us all numbers up to 12000, writing each as a sum of as few cubes as possible. He found, in that range, exactly 2 numbers that require 9 cubes (23 and 239), 15 numbers that require 8 cubes (the largest is 454), and 121 that require 7 cubes (the three largest are 5306, 5818 and 8042). On the basis of the increasingly large gaps between these numbers (5306, 5818, 8042, and 12000, where he stopped) Jacobi concluded that it was very likely that all numbers greater than 8042 are sums of 6 cubes or fewer. In fact, numbers requiring 6 cubes become sparse toward the end of Jacobi's table, and he almost permits himself to conjecture that all sufficiently large numbers are sums of 5 cubes.

These observations suggest a general definition: let $G(k)$ denote the smallest r such that every sufficiently large number (i.e., every number with perhaps finitely many exceptions) is a sum of r kth powers. Then $G(k) \leq g(k)$ for all k; and $G(2) \geq 4$ since $7 + 8n$ is never a sum of 3 squares for any n (because 7 is not a sum of 3 squares modulo 8). Therefore $G(2) = g(2) = 4$.

The Hardy-Littlewood-Vinogradov proof leads to upper bounds for $G(k)$. If one finds such an upper bound, $\bar{G}(k)$, and if it is less than or equal to the lower bound $\bar{g}(k)$ for $g(k)$, one can use it to compute $g(k)$, by extracting from the proof a specific n_0 such that all $n > n_0$ are sums of $\bar{G}(k)$ kth powers. This is in fact how $g(k)$ is determined. For $k = 4$, the appropriate n_0 was Auluck's number c; it is now known that every number less than 10^{140} or greater than $10^{1408.3}$ is a sum of 19 biquadrates, but the gap is still too large, as yet, to permit complete calculations.

Jacobi's conjecture for cubes—that $G(3) \leq 6$, or perhaps even $G(3) \leq 5$—remains unproved; the best that is known is a theorem of Linnik (1942), according to which $G(3) \leq 7$. As mentioned above, Landau had proved $G(3) \leq 8$ in 1909; and Dickson showed in 1939 that in fact 23 and 239 are the only numbers that require 9 cubes. In the other direction, we know that $G(3) \geq 4$ because of a striking result due to Maillet and Hurwitz: $G(k) \geq k + 1$ for all $k > 1$. The proof, which is elegant and astonishingly easy, can be found in Hardy and Wright ([4], Chapter 21.6). (The special case $G(3) \geq 4$ can also be proved by working modulo 9.)

Only a little more is known about $G(k)$ in the direction of explicit lower or upper bounds for specific k, although there are many asymptotic theorems comparing $G(k)$ with known functions for large k. We have mentioned $G(2) = 4$ and $4 \leq G(3) \leq 7$. For biquadrates, we can show that $G(4) \geq 16$. To begin, we note that 15 requires 15 biquadrates modulo 16, so that $G(4) \geq 15$. Using this it is easy to show that if n requires 16 biquadrates, so does $16n$. Since 31 requires 16 biquadrates, this implies $G(4) \geq 16$. In fact, Davenport showed in 1939 that $G(4) = 16$, and $G(2)$ and $G(4)$ are the *only* values of $G(k)$ known exactly.

Finally, let us mention briefly some generalizations. The most obvious, perhaps, is to ask Waring's question in other rings or fields: is every element which is a sum of kth powers a sum of g kth powers, for some g depending only on k? There is a substantial literature on such questions, for example in algebraic number fields and their rings of integers; see e.g., [5].

Another possibility is to remain in the natural numbers and generalize the question: given an interesting subset S of numbers, is every number (respectively, every sufficiently large number) a sum of r elements of S, for some bound r depending only on S? For example, if S is the set of kth powers, the appropriate r is $g(k)$ (respectively, $G(k)$). The question in this generality is the province of additive number theory, and the literature on it is vast (see, for example, the bibliography of [6]: 28 dense pages, single-spaced!).

One interesting choice for S is the set of primes; this leads to Goldbach's problem. Another is the set of values of some fixed integer polynomial, more general than X^k. We close with a rather special case, where there is a particularly striking theorem (quoted incorrectly in [2]).

Suppose we have a sequence $2 \leq n_0 \leq n_1 \leq n_2 \ldots$ and we ask: is there, for each j, a bound $r = r(j)$ with the property that every sufficiently large number can be written in the following form?

$$x_j^{n_0} + x_{j+1}^{n_1} + \cdots + x_{j+r}^{n_r}.$$

(If all $n_j = k$, the appropriate bound is $G(k) - 1$.) The answer was announced without proof by Freĭman in 1949, and proved by Scourfield in 1960: such bounds r_j exist if and only if $\sum 1/n_j$ diverges! The "only if" is relatively easy; the "if" uses the full Hardy–Littlewood–Vinogradov machine.

References

1. H. Davenport, *Analytic Methods for Diophantine Equations and Diophantine Inequalities*, Ann Arbor Publishers, 1962.

2. W. J. Ellison, Waring's problem, *Amer. Math. Monthly* **78** (1971), 10–36.

3. P. R. Halmos, *Finite-Dimensional Vector Spaces* (2nd ed.), Van Nostrand, 1958.

4. G. H. Hardy and E. M. Wright, *An Introduction to the Theory of Numbers*, Oxford Univ. Press, 4th ed., (1960); reprinted with corrections, 1968.

5. J.-R. Joly, Sommes de puissances d-ièmes dans un anneau commutatif, *Acta Arith.* **17** (1970), 37–114.

6. H.-H. Ostmann, *Additive Zahlentheorie*, two vols., Springer Ergebnisse, 1956.

A History of the Prime Number Theorem

L. J. GOLDSTEIN

American Mathematical Monthly **80** (1973), 599–615

The sequence of prime numbers, which begins

$$2, 3, 5, 7, 11, 13, 17, 19, 23, 29, 31, 37, \cdots,$$

has held untold fascination for mathematicians, both professionals and amateurs alike. The basic theorem which we shall discuss in this lecture is known as the *prime number theorem* and allows one to predict, at least in gross terms, the way in which the primes are distributed. Let x be a positive real number, and let $\pi(x) = $ the number of primes $\leq x$. Then the prime number theorem asserts that

$$\lim_{x \to \infty} \frac{\pi(x)}{x/\log(x)} = 1, \qquad (1)$$

where $\log x$ denotes the natural log of x. In other words, the prime number theorem asserts that

$$\pi(x) = \frac{x}{\log x} + o\left(\frac{x}{\log x}\right), \quad (x \to \infty), \qquad (2)$$

where $o(x/\log x)$ stands for a function $f(x)$ with the property

$$\lim_{x \to \infty} \frac{f(x)}{x/\log x} = 0.$$

Actually, for reasons which will become clear later, it is much better to replace (1) and (2) by the following equivalent assertion:

$$\pi(x) = \int_2^x \frac{dy}{\log y} + o\left(\frac{x}{\log x}\right). \qquad (3)$$

To prove that (2) and (3) are equivalent, it suffices to integrate

$$\int_2^x \frac{dy}{\log y}$$

once by parts to get

$$\int_2^y \frac{dy}{\log y} = \frac{x}{\log x} - \frac{2}{\log 2} + \int_2^x \frac{dy}{\log^2 y}. \qquad (4)$$

However, for $x \geq 4$,

$$\int_2^x \frac{dy}{\log^2 y} = \int_2^{\sqrt{x}} \frac{dy}{\log^2 y} + \int_{\sqrt{x}}^x \frac{dy}{\log^2 y}$$
$$\leq \sqrt{x} \cdot \frac{1}{\log^2 2} + x \cdot \frac{1}{\log^2(\sqrt{x})} \qquad (5)$$
$$= o\left(\frac{x}{\log x}\right),$$

where we have used the fact that $1/\log^2 x$ is monotone decreasing for $x > 1$. It is clear that (4) and (5) show that (2) and (3) are equivalent to one another. The advantage of the version (3) is that the function

$$\mathrm{Li}(x) = \int_2^x \frac{dy}{\log x},$$

called the *logarithmic integral*, provides a much closer numerical approximation to $\pi(x)$ than does $x/\log x$. This is a rather deep fact and we shall return to it.

In this article, I should like to explore the history of the ideas which led up to the prime number theorem and to its proof, which was not supplied until some one hundred years after the first conjecture was made. The history of the prime number theorem provides a beautiful example of the way in which great ideas develop and interrelate, feeding upon one another ultimately to yield a coherent theory which rather completely explains observed phenomena.

The very conception of a prime number goes back to antiquity, although it is not possible to say precisely when the concept first was clearly formulated. However, a number of elementary facts concerning the primes were known to the Greeks. Let us cite three examples (the first two appearing in Euclid):

(i) (*Fundamental Theorem of Arithmetic*): Every positive integer n can be written as a product of primes. Moreover, this expression of n is unique up to a rearrangement of the factors.

(ii) There exist infinitely many primes.

(iii) The primes may be effectively listed using the so-called *sieve of Eratosthenes*.

We will not comment on (i) or (iii) any further, since they are part of the curriculum of most undergraduate courses in number theory, and hence are probably familiar to most of you. However, there is a proof of (ii) which is quite different from Euclid's well-known proof and which is very significant to the history of the prime number theorem. This proof is due to the Swiss mathematician Leonhard Euler and dates from the middle of the eighteenth century. It runs as follows:

Assume that $p_1, \ldots, p_N$ is a complete list of all primes, and consider the product

$$\prod_{i=1}^{N}\left(1-\frac{1}{p_i}\right)^{-1} = \prod_{i=1}^{N}\left(1+\frac{1}{p_i}+\frac{1}{p_i^2}+\cdots\right). \quad (6)$$

Since every positive integer n can be written uniquely as a product of prime powers, every unit fraction $1/n$ appears in the formal expansion of the product (6). For example, if $n = p_1^{a_1} \cdots p_N^{a_N}$, then $1/n$ occurs from multiplying the terms

$$1/p_1^{a_1}, \ 1/p_2^{a_2}, \ \cdots, \ 1/p_N^{a_N}.$$

Therefore, if R is any positive integer,

$$\prod_{i=1}^{N}\left(1-\frac{1}{p_i}\right)^{-1} \geq \sum_{n=1}^{R} 1/n, \quad (7)$$

However, as $R \to \infty$, the sum on the right hand side of (7) tends to infinity, which contradicts (7). Thus, $p_1, \ldots, p_N$ cannot be a complete list of all primes.

We should make two comments about Euler's proof: First, it links the Fundamental Theorem of Arithmetic with the infinitude of primes. Second, it uses an analytic fact, namely the divergence of the harmonic series, to conclude an arithmetic result. It is this latter feature which became the cornerstone upon which much of nineteenth-century number theory was erected.

The first published statement which came close to the prime number theorem was due to Legendre in 1798 [7]. He asserted that $\pi(x)$ is of the form

$$x/(A \log x + B)$$

for constants A and B. On the basis of numerical work, Legendre refined his conjecture in 1808 [8] by asserting that

$$\pi(x) = \frac{x}{\log x + A(x)},$$

where $A(x)$ is "approximately $1.08366 \cdots$". Presumably, by this latter statement, Legendre meant that

$$\lim_{x \to \infty} A(x) = 1.08366.$$

It is precisely in regard to $A(x)$ where Legendre was in error, as we shall see below. In his memoir [9] of 1808, Legendre formulated another famous conjecture. Let k and l be integers which are relatively prime to one another. Then Legendre asserted that there exist infinitely many primes of the form $l + kn$ ($n = 0, 1, 2, 3, \ldots$). In other words, if $\pi_{k,l}(x)$ denotes the number of primes p of the form $l + kn$ for which $p \leq x$, then Legendre conjectured that

$$\pi_{k,l}(x) \to \infty \text{ as } x \to \infty. \quad (8)$$

Actually, the proof of (8) by Dirichlet in 1837 [1] provided several crucial ideas on how to approach the prime number theorem.

Although Legendre was the first person to publish a conjectural form of the prime number theorem, Gauss had already done extensive work on the theory of primes in 1792–1793. Evidently Gauss considered the tabulation of primes as some sort of pastime and amused himself by compiling extensive tables on how the primes distribute themselves in various intervals of length 1000. We have included some of Gauss's tabulations as appendices. The first table, excerpted from ([2], p.436), covers the primes from 1 to 50,000. Each entry in the table represents an interval of length 1000. Thus, for example, there are 168 primes from 1 to 1000; 135 from 1001 to 2000; 127 from 3001 to 4000; and so forth. Gauss suspected that the density with which primes occurred in the neighborhood of the integer n was $1/\log n$, so that the number of primes in the interval $[a, b)$ should be approximately equal to

$$\int_a^b \frac{dx}{\log x}.$$

In the second set of tables, samples from ([3], pp.442–443]), Gauss investigates the distribution of primes up to 3,000,000 and compares the number of primes found with the above integral. The agreement is striking. For example, between 2,600,000 and 2,700,000, Gauss found 6762 primes, whereas

$$\int_{2,600,000}^{2,700,000} \frac{dx}{\log x} = 6761.332.$$

Gauss never published his investigations on the distribution of primes. Nevertheless, there is little reason to doubt Gauss's claim that he first undertook his work in 1792–1793, well before the memoir of Legendre was written. Indeed, there are several other known examples of results of the first rank which Gauss proved, but never communicated to anyone until years after the original work had been done. This was the case, for example, with the elliptic functions, where Gauss preceded Jacobi, and with Riemannian geometry, where Gauss anticipated Riemann. The only information beyond Gauss's tables concerning Gauss's work in the distribution of primes is contained in an 1849 letter to the astronomer Encke. We have included a translation of Gauss's letter.

In his letter Gauss describes his numerical experiments and his conjecture concerning $\pi(x)$. There are a number of remarkable features of Gauss's letter. On the second page of the letter, Gauss compares his approximation to $\pi(x)$, namely $\text{Li}(x)$, with Legendre's formula. The results are tabulated at the top of the second page and Gauss's formula yields a much larger numerical error. In a very prescient statement, Gauss defends his formula by noting that although Legendre's formula yields a smaller error, the rate of increase of Legendre's error term is much greater than his own. We shall see below that Gauss anticipated what is known as the *Riemann hypothesis*. Another feature of Gauss's letter is that he casts doubt on Legendre's assertion about $A(x)$. He asserts that the numerical evidence does not support any conjecture about the limiting value of $A(x)$.

Gauss's calculations are awesome to contemplate, since they were done long before the days of high-speed computers. Gauss's persistence is most impressive. However, Gauss's tables are not error-free. My student, Edward Korn, has checked Gauss's tables using an electronic computer and has found a number of errors. We include the corrected entries in an appendix. In spite of these (remarkably few) errors, Gauss's calculations still provide overwhelming evidence in favor of the prime number theorem. Modern students of mathematics should take note of the great care with which data was compiled by such giants as Gauss. Conjectures in those days were rarely idle guesses. They were usually supported by piles of laboriously gathered evidence.

The next step toward a proof of the prime number theorem was a step in a completely different direction, and was taken by Dirichlet in 1837 [1]. In a beautiful memoir, Dirichlet proved Legendre's conjecture (8) concerning the infinitude of primes in an arithmetic progression. Dirichlet's work contained two radically new ideas, which we should discuss in some detail.

Let $\mathbb{Z}_n$ denote the ring of residue classes modulo n, and let $\mathbb{Z}_n^\times$ denote the group of units of $\mathbb{Z}_n$. Then $\mathbb{Z}_n^\times$ is the so-called *group of reduced residue classes modulo n* and consists of those residue classes containing an element relatively prime to n. If k is an integer, let us denote by $\bar{k}$ its residue class modulo n. Dirichlet's first brilliant idea was to introduce the *characters* of the group $\mathbb{Z}_n^\times$; that is, the homomorphisms of $\mathbb{Z}_n^\times$ into the multiplicative group $\mathbb{C}^\times$ of non-zero complex numbers. If χ is such a character, then we may associate with χ a function (also denoted by χ) from the semi-group $\mathbb{Z}^*$ of non-zero integers as follows. Set

$$\chi(a) = \begin{cases} \chi(\bar{a}) & \text{if } (a,n) = 1 \\ 0 & \text{otherwise.} \end{cases}$$

Then it is clear that $\chi : \mathbb{Z}^* \to \mathbb{C}$ and has the following properties:

(i) $\chi(a+n) = \chi(a)$,

(ii) $\chi(aa') = \chi(a)\chi(a')$,

(iii) $\chi(a) = 0$ if $(a,n) \neq 1$,

(iv) $\chi(1) = 1$.

A function $\chi : \mathbb{Z}^* \to \mathbb{C}$ satisfying (i)–(iv) is called a *numerical character* modulo n. Dirichlet's main result about such numerical characters was the so-called *orthogonality relations*, which assert the following:

$$\sum_a \chi(a) = \begin{cases} \phi(n) & \text{if } \chi \text{ is identically } 1 \\ 0 & \text{otherwise,} \end{cases} \quad (A)$$

where a runs over a complete system of residues modulo n;

$$\sum_\chi \chi(a) = \begin{cases} \phi(n) & \text{if } a \equiv 1 \pmod{n} \\ 0 & \text{otherwise,} \end{cases} \quad (B)$$

where χ runs over all numerical characters modulo n. Dirichlet's ideas gave birth to the modern theory of duality on locally compact abelian groups.

Dirichlet's second great idea was to associate to each numerical character modulo n and each real number $s > 1$, the following infinite series

$$L(s, \chi) = \sum_{n=1}^{\infty} \frac{\chi(n)}{n^s}. \quad (9)$$

It is clear that the series converges absolutely and represents a continuous function for $s > 1$. However, a more delicate analysis shows that the series (9) converges (although not absolutely) for $s > 0$ and represents a continuous function of s in this semi-infinite interval *provided that χ is not identically* 1. The function $L(s, \chi)$ has come to be called a *Dirichlet L-function*.

Note the following facts about $L(s, \chi)$: First $L(s, \chi)$ has a product formula of the form

$$L(s, \chi) = \prod_p \left(1 - \frac{\chi(p)}{p^s}\right)^{-1} \quad (s > 1), \quad (10)$$

where the product is taken over all primes p. The proof of (10) is very similar to the argument given above in Euler's proof of the infinity of prime numbers. Therefore, by (10),

$$\log L(s, \chi) = -\sum_p \log\left(1 - \frac{\chi(p)}{p^s}\right)$$

$$= -\sum_p \sum_{m=1}^{\infty} \frac{\chi(p^m)}{mp^{ms}}. \quad (11)$$

Dirichlet's idea in proving the infinitude of primes in the arithmetic progression $a, a+n, a+2n, \cdots$, $(a, n) = 1$, was to imitate, somehow, Euler's proof of the infinitude of primes, by studying the function $L(s, \chi)$ for s near 1. The basic quantity to consider is

$$\sum_\chi \chi(a)^{-1} \log L(s, \chi)$$

$$= -\sum_p \sum_{m=1}^{\infty} \sum_\chi \frac{\chi(a)^{-1}\chi(p^m)}{mp^{ms}}$$

$$= -\sum_p \sum_{m=1}^{\infty} \frac{1}{mp^{ms}} \sum_\chi \chi(a)^{-1}\chi(p^m), \quad (12)$$

where we have used (11). Let a^* be an integer such that $aa^* \equiv 1 \pmod{n}$. Then $\chi(a^*) = \chi(a)^{-1}$ by (i)–(iv). Moreover,

$$\sum_\chi \chi(a)^{-1}\chi(p^m)$$

$$= \sum_\chi \chi(a^* p^m)$$

$$= \begin{cases} \phi(n) & \text{if } a^* p^m \equiv 1 \pmod{n} \\ 0 & \text{otherwise}. \end{cases} \quad (13)$$

However, $a^* p^m \equiv 1 \pmod{n}$ is equivalent to $p^m \equiv a \pmod{n}$. Therefore, by (12) and (13), we have

$$\sum_\chi \chi(a)^{-1} \log L(s, \chi)$$

$$= -\phi(n) \sum_{\substack{p \\ p^m \equiv a \pmod{n}}} \sum_{m=1}^{\infty} \frac{1}{mp^{ms}}. \quad (14)$$

Thus, finally, we have

$$-\frac{1}{\phi(n)} \sum_\chi \chi(a) \log L(s, \chi)$$

$$-\sum_{\substack{p \\ p^m \equiv a \pmod{n}}} \sum_{m=2}^{\infty} \frac{1}{mp^{mp}} \quad (15)$$

$$= \sum_{\substack{p \\ p \equiv a \pmod{n}}} \frac{1}{p^s} \quad (s > 1).$$

From (15), we immediately see that in order to prove that there are infinitely many primes $p \equiv a \pmod{n}$, it is enough to show that the function

$$\sum_{\substack{p, \\ p \equiv a \pmod{n}}} \frac{1}{p^s}$$

tends to $+\infty$ as s approaches 1 from the right. But it is fairly easy to see that as $s \to 1+$, the sum

$$\sum_{\substack{p, \\ p^m \equiv a \pmod{n}}} \sum_{m=2}^{\infty} \frac{1}{mp^{ms}}$$

remains bounded. Thus, it suffices to show that

$$-\frac{1}{\phi(n)} \sum_\chi \chi(a)^{-1} \log L(s, \chi) \to +\infty \quad (s \to 1+).$$

However, if χ_0 denotes the character which is identically 1, then it is easy to see that

$$-\frac{1}{\phi(n)} \chi_0(a)^{-1} L(s, \chi_0) \to +\infty \text{ as } s \to 1+.$$

Therefore, it is enough to show that if $\chi \neq \chi_0$, then $\log L(s, \chi)$ remains bounded as $s \to 1+$. We have already mentioned that $L(s, \chi)$ is continuous for $s > 0$ if $\chi \neq \chi_0$. Therefore, it suffices to show that $L(1, \chi) \neq 0$. And this is precisely what Dirichlet showed.

Dirichlet's theorem on primes in arithmetic progressions was one of the major achievements of nineteenth-century mathematics, because it introduced a fertile new idea into number theory—that analytic methods (in this case the study of the Dirichlet L-series) could be fruitfully applied to arithmetic problems (in this case the problem of primes in arithmetic progressions). To the novice, such an application of analysis to number theory would seem to be a waste of time. After all, number theory is the study of the discrete, whereas analysis is the study of the continuous; and what should one have to do with the other? However, Dirichlet's 1837 paper was the beginning of a revolution in number-theoretic thought, the substance of which was to apply analysis to number theory. At first, undoubtedly, mathematicians were very uncomfortable with Dirichlet's ideas. They regarded them as very clever devices, which would eventually be supplanted by completely arithmetic ideas. For although analysis might be useful in proving results about the integers, surely the analytic tools were not intrinsic. Rather, they entered the theory of the integers in an inessential way and could be eliminated by the use of suitably sophisticated arithmetic. However, the history of number theory in the nineteenth century shows that this idea was eventually repudiated and the rightful connection between analysis and number theory came to be recognized.

The first major progress toward a proof of the prime number theorem after Dirichlet was due to the Russian mathematician Tchebycheff in two memoirs ([11], [12]) written in 1851 and 1852. Tchebycheff introduced the following two functions of a real variable x:

$$\theta(x) = \sum_{p \leq x} \log p,$$

$$\psi(x) = \sum_{p^m \leq x} \log p,$$

where p runs over primes and m over positive integers. Tchebycheff proved that the prime number theorem (1) is equivalent to either of the two statements

$$\lim_{x \to \infty} \frac{\theta(x)}{x} = 1, \qquad (16)$$

$$\lim_{x \to \infty} \frac{\psi(x)}{x} = 1. \qquad (17)$$

Moreover, Tchebycheff proved that if $\lim_{x \to \infty} (\theta(x)/x)$ exists, then its value must be 1. Furthermore, Tchebycheff proved that

$$0.92129 \leq \liminf_{x \to \infty} \frac{\pi(x)}{x/\log x} \leq 1$$

$$\leq \limsup_{x \to \infty} \frac{\pi(x)}{x/\log x} \leq 1.10555. \quad (18)$$

Tchebycheff's methods were of an elementary combinatorial nature, and as such were not powerful enough to prove the prime number theorem.

The first giant strides toward a proof of the prime number theorem were taken by B. Riemann in a memoir [9] written in 1860. Riemann followed Dirichlet in connecting problems of an arithmetic nature with the properties of a function of a continuous variable. However, where Dirichlet considered the functions $L(s, \chi)$ as functions of a real variable s, Riemann took the decisive step in connecting arithmetic with the theory of functions of a complex variable. Riemann introduced the following function:

$$\zeta(s) = \sum_{n=1}^{\infty} \frac{1}{n^s}, \qquad (19)$$

which has come to be known as the *Riemann zeta function*. It is reasonably easy to see that the series (19) converges absolutely and uniformly for s in a compact subset of the half-plane $\Re(s) > 1$. Thus, $\zeta(s)$ is analytic for $\Re(s) > 1$. Moreover, by using the same sort of argument used in Euler's proof of the infinitude of primes, it is easy to prove that

$$\zeta(s) = \prod_{p} \left(1 - \frac{1}{p^s}\right)^{-1} \quad (\Re(s) > 1), \qquad (20)$$

where the product is extended over all primes p. Euler's proof of the infinitude of primes suggests that the behavior of $\zeta(s)$ for $s = 1$ is somehow connected with the distribution of primes. And, indeed, this is the case.

Riemann proved that $\zeta(s)$ can be analytically continued to a function which is meromorphic in the whole s-plane. The only singularity of $\zeta(s)$ occurs at $s = 1$ and the Laurent series about $s = 1$ looks like

$$\zeta(s) = \frac{1}{s-1} + a_0 + a_1(s-1) + \cdots. \qquad (21)$$

Moreover, if we set

$$R(s) = s(s-1)\pi^{-s/2}\Gamma(s/2)\zeta(s), \qquad (22)$$

then $R(s)$ is an entire function of s and satisfies the functional equation

$$R(s) = R(1-s). \qquad (23)$$

To see the immediate connection between the Riemann zeta function and the distribution of primes, let us return to Euler's proof of the infinitude of primes. A variation on the idea of Euler's proof is as follows. Suppose that there were only finitely many primes $p_1, \ldots, p_N$. Then by (20), $\zeta(s)$ would be bounded as s tends to 1, which contradicts equation (21). Thus, the presence of a pole of $\zeta(s)$ at $s = 1$ immediately implies that there are infinitely many primes. But the connection between the zeta function and the distribution of primes runs even deeper.

Let us consider the following heuristic argument. From equation (20), it is easy to deduce that

$$\frac{\zeta'(s)}{\zeta(s)} = \sum_p \sum_{m=1}^{\infty} (\log p) p^{-ms} \quad (\Re(s) > 1). \quad (24)$$

Moreover, by residue calculus, it is easy to verify that

$$\lim_{T \to \infty} \frac{1}{2pi-1} \int_{2-iT}^{2+iT} \frac{a^s}{s} ds = \begin{cases} 1, & x < 1. \\ 0, & x > 1. \end{cases} \quad (25)$$

Therefore, assuming that interchange of limit and summation is justified, we see that, for x not equal to an integer, we have

$$\lim_{T \to \infty} \frac{1}{2pi-1} \int_{2-iT}^{2+iT} \frac{x^s}{s} \frac{\zeta'(s)}{\zeta(s)} ds$$

$$= \sum_p \sum_{m=1}^{\infty} (\log p) \lim_{T \to \infty} \frac{1}{2\pi i} \int_{2-iT}^{2+iT} \left(\frac{x}{p^m}\right)^s \frac{1}{s} ds$$

$$= \sum_{p^m \leq x} \log p \quad \text{(by equation (25))} \quad (26)$$

$$= \psi(x).$$

Thus, we see that there is an intimate connection between the function $\psi(x)$ and $\zeta(s)$. This connection was first exploited by Riemann, in his 1860 paper.

Note that the function

$$\frac{x^s}{s} \frac{\zeta'(s)}{\zeta(s)} \quad (27)$$

has poles at $s = 0$ and at all zeros ρ of $\zeta(s)$. Moreover, note that by equation (20), we see that $\zeta(s) \neq 0$ for $\Re(s) > 1$. Therefore, all zeros of $\zeta(s)$ lie in the half-plane $\Re(s) \leq 1$. Further, since $R(s)$ is entire and $\zeta(s) \neq 0$ for $\Re(s) > 1$, the functional equation (23) implies that the only zeros of $\zeta(s)$ for which $\Re(s) < 0$ are at $s = -2, -4, -6, -8, \cdots$, and these are all simple zeros and are called the *trivial zeros* of $\zeta(s)$. Thus, we have shown that all non-trivial zeros of $\zeta(s)$ lie in the strip $0 \leq \Re(s) \leq 1$. This strip is called the *critical strip*. The residue of (27) at a non-trivial zero ρ is

$$\frac{x^\rho}{\rho}.$$

Thus, if σ is a large negative number, and if $C_{\sigma,T}$ denotes the rectangle with vertices $\sigma \pm iT, 2 \pm iT$, then Cauchy's theorem implies that

$$\frac{1}{2\pi i} \int_{2-iT}^{2+iT} \frac{x^s}{s} \frac{\zeta'(s)}{\zeta(s)} ds$$

$$= \frac{1}{2\pi i} \left[\int_{\sigma-iT}^{\sigma+iT} + \int_{\sigma+iT}^{2+iT} + \int_{2-iT}^{\sigma-iT} \right] \frac{x^s}{s} \frac{\zeta'(s)}{\zeta(s)} ds$$

$$+ R(\sigma, T), \quad (28)$$

where $R(\sigma, T)$ denotes the sum of the residues of the function (27) at the poles inside $C_{\sigma,T}$. By letting σ and T tend to infinity and by applying equations (26) and (28), Riemann arrived at the following remarkable formula, known as *Riemann's explicit formula*

$$\psi(x) = x - \sum_\rho \frac{x^\rho}{\rho} - \frac{\zeta'(0)}{\zeta(0)}$$

$$- \frac{1}{2} \log\left(1 - x^{-2}\right), \quad (29)$$

where ρ runs over all non-trivial zeros of the Riemann zeta function. Riemann's formula is surprising for at least two reasons. First, it connects the function $\psi(x)$, which is connected with the distribution of primes, with the distribution of the zeros of the Riemann zeta function. That there should be any connection at all is amazing. But, secondly, the formula (29) explicitly puts in evidence a form of the prime number theorem by equating $\psi(x)$ with x plus an error term which depends on the zeros of the zeta function. If we denote this error term by $E(x)$, then we see that the prime number theorem is equivalent to the assertion

$$\lim_{x \to \infty} \frac{E(x)}{x} = 0, \quad (30)$$

which, in turn, is equivalent to the assertion

$$\lim_{x \to \infty} \frac{1}{x} \sum_\rho \frac{x^\rho}{\rho} = 0. \quad (31)$$

Riemann was unable to prove (31), but he made a number of conjectures concerning the distributions of the zeros ρ from which the statement (31) follows immediately. The most famous of Riemann's conjectures is the so-called *Riemann hypothesis*, which asserts that all non-trivial zeros of $\zeta(s)$ lie on the line $\Re(s) = \frac{1}{2}$, which is the line of symmetry of the functional equation (23). This conjecture has resisted all attempts to prove it for more than a century and is one of the most celebrated open problems in all of mathematics. However, if the Riemann hypothesis is true, then

$$\left|\frac{x^\rho}{\rho}\right| = x^{1/2}\frac{1}{|\rho|},$$

and from this fact and equation (29), it is possible to prove that

$$\psi(x) = x + O(x^{\frac{1}{2}+\varepsilon}) \qquad (32)$$

for every $\varepsilon > 0$, where $O(x^{\frac{1}{2}+\varepsilon})$ denotes a function $f(x)$ such that $f(x)/x^{\frac{1}{2}+\varepsilon}$ is bounded for all large x. Thus, the Riemann hypothesis implies (31) in a trivial way, and hence the prime number theorem follows from the Riemann hypothesis. What is perhaps more striking is the fact that *if* (32) *holds then the Riemann hypothesis is true*. Thus, the prime number theorem in the sharp form (32) is equivalent to the Riemann hypothesis. We see, therefore, that the connection between the zeta function and the distribution of primes is no accidental affair, but somehow is woven into the fabric of nature.

In his memoir, Riemann made many other conjectures. For example, if $N(T)$ denotes the number of non-trivial zeros ρ of $\zeta(s)$ such that $-T \leq \text{Im}(\rho) \leq T$, then Riemann conjectured that

$$N(T) = \frac{1}{2\pi}T\log T$$
$$-\frac{1+\log(2\pi)}{2\pi}T + O(\log T). \quad (33)$$

The formula (33) was first proven by von Mangoldt in 1895 [14]. An interesting line of research has been involved in obtaining estimates for the number of non-trivial zeros ρ on the line $\Re(s) = \frac{1}{2}$. Let $M(T)$ denote the number of ρ such that $\Re(s) = \frac{1}{2}$ and $-T \leq \text{Im}(s) \leq T$. Then Hardy [5] in 1912 proved that $M(T)$ tends to infinity as T tends to infinity. Later, Hardy [6] improved his argument to prove that $M(T) > AT\log T$, where A is a positive constant, not depending on T. The ultimate result of this sort was obtained by Atle Selberg in 1943 [10]. He proved that $M(T) > AT\log T$ for some positive constant A. In view of equation (33), Selberg's result shows that a positive percentage of the zeros of $\zeta(s)$ actually lie on the line $\Re(s) = \frac{1}{2}$. This result represents the best progress made to date in attempting to prove the Riemann hypothesis.

Fortunately, it is not necessary to prove the Riemann hypothesis in order to prove the prime number theorem in the form (17). However, it is necessary to obtain some information about the distribution of the zeros of $\zeta(s)$. Such information was obtained independently by Hadamard [4] and de la Vallée Poussin [13] in 1896, thereby providing the first complete proofs of the prime number theorem. Although their proofs differ in detail, they both establish the existence of a zero-free region for $\zeta(s)$, the existence of which serves as a substitute for the Riemann hypothesis in the reasoning presented above. More specifically, they proved that there exist constants a, t_0 such that

$$\zeta(\sigma + it) \neq 0 \text{ if } \sigma \geq 1 - 1/a\log|t|, |t| \geq t_0.$$

This zero-free region allows one to prove the prime number theorem in the form

$$\psi(x) = x + O\left(xe^{c(\log x)^{1/14}}\right). \qquad (34)$$

Please note, however, that the error term in (34) is much larger than the error term predicted by the Riemann hypothesis.

Thus, the prime number theorem was finally proved after a century of hard work by many of the world's best mathematicians. It is grossly unfair to attribute proof of such a theorem to the genius of a single individual. For, as we have seen, each step in the direction of a proof was conditioned historically by the work of preceding generations. On the other hand, to deny that there is genius in the work which led up to the ultimate proof would be equally unfair. For at each step in the chain of discovery, brilliant and fertile ideas were discovered, and provided the material out of which to fashion the next link.

Appendices

APPENDIX A: Samples from Gauss's Tables

Table 1. (*Werke*, II, p.436)

1	168	14	105	27	101	40	96
2	135	15	102	28	94	41	88
3	127	16	108	29	98	42	101
4	120	17	98	30	92	43	102
5	119	18	104	31	95	44	85
6	114	19	94	32	92	45	96
7	117	20	102	33	106	46	86
8	107	21	98	34	100	47	90
9	110	22	104	35	94	48	95
10	112	23	100	36	92	49	89
11	106	24	104	37	99	50	98
12	103	25	94	38	94		
13	109	26	98	39	90		

The frequency of primes

Table 2. (*Werke*, II, p.443) 2000000 − 3000000

	210	220	230	240	250	260	270	280	290	300	
0							1				1
1	3	2	2	4	1	3	4	2	2	2	25
2	10	9	9	11	9	5	10	7	15	13	98
3	32	27	29	32	37	35	28	43	30	44	337
4	69	69	73	86	78	88	71	95	85	64	778
5	119	146	138	136	147	136	158	135	140	153	1408
6	197	183	179	176	193	194	195	195	179	187	1878
7	204	201	205	194	189	180	201	188	222	214	1998
8	157	168	168	158	151	170	142	145	132	134	1525
9	115	109	113	112	102	88	96	87	109	103	1034
10	63	52	44	55	58	58	53	67	53	58	561
11	21	18	30	28	23	24	22	24	18	15	223
12	8	9	10	7	7	13	17	9	8	11	99
13	2	4		1	5	6	1	2	5	1	27
14		3					1		2		6
15										1	1
16											
17								1			1
	6874	6857	6849	6787	6766	6804	6762	6714	6744	6705	67862

APPENDIX B: Gauss's Letter to Encke

My distinguished friend:

Your remarks concerning the frequency of primes were of interest to me in more ways than one. You have reminded me of my own endeavors in this field which began in the very distant past, in 1792 or 1793, after I had acquired the Lambert supplements to the logarithmic tables. Even before I had begun my more detailed investigations into higher arithmetic, one of my first projects was to turn my attention to the decreasing frequency of primes, to which end I counted the primes in several chiliads [intervals of length 1000], and recorded the results on the attached white pages. I soon recognized that behind all of its fluctuations, this frequency is on the average inversely proportional to the logarithm, so that the number of primes below a given bound n is approximately equal to

$$\int \frac{dn}{\log n},$$

where the logarithm is understood to be hyperbolic. Later on, when I became acquainted with the list in Vega's tables (1796) going up to 400031, I extended my computation further, confirming that estimate. In 1811, the appearance of Chernau's cribrum gave me much pleasure and I have frequently (since I lack the patience for a continuous count) spent an idle quarter of an hour to count another chiliad here and there; although I eventually gave it up without quite getting through a million. Only some time later did I make use of the diligence of Goldschmidt to fill some of the remaining gaps in the first million and to continue the computation according to Burkhardt's tables. Thus (for many years now) the first three million have been counted and checked against the integral. A small excerpt follows: (See Table A)

I was not aware that Legendre had also worked on this subject; your letter caused me to look in his *Théorie des Nombres*, and in the second edition I found a few pages on the subject which I must have previously overlooked (or, by now, forgotten). Legendre used the formula

$$\frac{n}{\log n - A},$$

where A is a constant which he sets equal to 1.08366. After a hasty computation, I find in the above cases the deviations

$$-23.3 \;\; +42.2 \;\; +68.1 \;\; +92.8 \;\; +159.1 \;\; +167.6$$

These differences are even smaller than those from the integral, but they seem to grow faster with n so that it is quite possible they may surpass them. To make the count and the formula agree, one would have to use, respectively, instead of $A = 1.08366$, the following numbers:

1.09040 1.07682 1.07582 1.07529 1.07179 1.07297

Table A

below	here are prime	integral + error	your formula + error
500000	41556	41606.4+50.4	41596.9+40.9
1000000	78501	79627.5+126.5	78672.7+171.7
1500000	114112	114263.1+151.1	114374.0+264.0
2000000	148883	149054.8+171.8	149233.0+350.0
2500000	183016	183245.0+229.0	183495.1+479.1
3000000	216745	216970.6+225.6	217308.5+563.5

It appears that, with increasing n, the (average) value of A decreases; however, I dare not conjecture whether the limit as n approaches infinity is 1 or a number different from 1. I cannot say that there is any justification for expecting a very simple limiting value; on the other hand, the excess of A over 1 might well be a quantity of the order of $1/\log n$. I would be inclined to believe that the differential of the function must be simpler than the function itself.

If $dn/\log n$ is postulated for the function, Legendre's formula would suggest that the differential function might be something of the form $dn/(\log n - (A-1))$. By the way, for large n, your formula could be considered to coincide with

$$\frac{n}{\log n - (1/2k)},$$

where k is the modulus of Briggs's logarithms; that is, with Legendre's formula, if we put $A = 1/2k = 1.1513$.

Finally, I want to remark that I noticed a couple of disagreements between your counts and mine.

Between 59000 and 60000, you have 95, while I have 94; between 101000 and 102000, you have 94, while I have 93.

The first difference possibly results from the fact that, in Lambert's *Supplement*, the prime 59023 occurs twice. The chiliad from 101000–102000 in Lambert's *Supplement* is virtually crawling with errors; in my copy, I have indicated seven numbers which are not primes at all, and supplied two missing ones. Would it not be possible to induce young Mr. Dase to count the primes in the following (few) millions, using the tables at the Academy which, I am afraid, are not intended for public distribution? In this case, let me remark that in the 2nd and 3rd million, the count is, according to my instructions, based on a special scheme which I myself have employed in counting the first million. The counts for each 100000 are indicated on a single page in 10 columns, each column belonging to one myriad [an interval of length 1000]; an additional column in front (left) and another column following it on the right; for example here is a vertical column and the two additional columns for the interval 10000000 to 11000000 —

As an illustration, take the first vertical column. In the myriad 1000000 to 1010000 there are 100 Hecatontades [intervals of length 100]; among them one containing a single prime, none containing two or three primes; two containing four each; eleven containing 5 each, etc., yielding altogether

$$752 = 1 \cdot 1 + 4 \cdot 2 + 5 \cdot 11 + 6 \cdot 14 + \cdots$$

primes. The last column contains the totals from the other ten. The numbers 14, 15, 16 in the first vertical column are superfluous since no hecatontades occur containing that many primes; but on the following pages they are needed. Finally the 10 pages are again combined into one and thus comprise the entire second million.

It is high time to quit—. With most cordial wishes for your good health

Yours, as ever,

C. F. Gauss

Göttingen, 24 December 1849.

APPENDIX C: Corrections to Gauss's Tables

thousands	Gauss	actual	Δ
20	102	104	-2
159	87	77	$+10$
199	96	86	$+10$
206	85	83	$+2$
245	78	88	-10
289	85	77	$+8$
290	84	85	-1
334	80	81	-1
352	80	81	-1
354	79	76	$+3$
500	up to here		$+18$

	Totals		Δ
500000	41556	41538	$+18$
1000000	78501	78498	$+3$
1500000	114112	114156	-44
2000000	148883	148934	-51
2500000	183016	183073	-57
3000000	216745	216817	-72

(The entries in the third column are from *List of Prime Numbers from 1 to 10,006,771*, by D. N. Lehmer.)

References

1. L. Dirichlet, Über den Satz: das jede unbegrenzte arithmetische Progression, deren erstes Glied und Differenz keinen gemeinschaftlichen Factor sind, unendlichen viele Primzahlen enthalt, 1837, *Mathematische Abhandlungen* **I** (1889), 313–342.

2. C. F. Gauss, Tafel der Frequenz der Primzahlen, *Werke* II, 1872, pp. 436–442.

3. C. F. Gauss, Gauss an Encke, *Werke* II, 1872, pp. 444–447.

4. J. Hadamard, Sur la distribution des zéros de la fonction $\zeta(s)$ et ses conséquences arithmétiques, *Bull. Soc. Math. de France* **24** (1896), 199–220.

5. G. H. Hardy, Sur les zéros de la fonction $\zeta(s)$ de Riemann, *C. R. Acad. Sci. Paris* **158** (1914), 1012–1014.

6. G. H. Hardy and J. E. Littlewood, The zeros of Riemann's zeta function on the critical line, *Math. Zeitschrift* **10** (1921), 283–317.

7. A. M. Legendre, *Essai sur la Théorie de Nombres*, 1st ed., 1798, p. 19.

8. A. M. Legendre, *Essai sur la Théorie de Nombres*, 2nd ed., 1808, p. 394.

9. B. Riemann, Über die Anzahl der Primzahlen unter einer gegebenen Grösse, *Gesammelte Mathematische Werke*, 2nd ed., 1892, pp. 145–155.

10. A. Selberg, On the zeros of Riemann's zeta function, *Skr. Norske Vid. Akad., Oslo* **10** (1942).

11. P. Tchebycheff, Sur la fonction qui détermine la totalité de nombres premiers inférieurs à une limite donnée, *Oeuvres* I, 1899, pp. 27–48.

12. P. Tchebycheff, Mémoire sur les nombres premiers, *Oeuvres* I, 1899, pp. 49–70.

13. Ch. de la Vallée Poussin, Recherches analytiques sur la théorie des nombres premiers. Première partie: La fonction $\zeta(s)$ de Riemann et les nombres premiers en général. Deuxième partie: Les fonctions de Dirichlet et les nombres premiers de la forme linéaire $Mx + N$. Troisième partie: Les formes quadratiques de déterminant négatif, *Ann. Soc. Sci. Bruxelles* **20** (1896), 183–256, 281–397.

14. H. von Mangoldt, Auszug aus einer Arbeit unter dem Titel: Zu Riemann's Abhandlung über die Anzahl der Primzahlen unter einer gegebenen Grösse, *Sits. König. Preus. Akad. Wiss. zu Berlin* (1894), 337–350, 883–895.

A Hundred Years of Prime Numbers

PAUL T. BATEMAN AND HAROLD G. DIAMOND

American Mathematical Monthly **103** (1996), 729–741

1 Early work on primes

This year marks the hundredth anniversary of the proof of the Prime Number Theorem (PNT), one of the most celebrated results in mathematics. The theorem is an asymptotic formula for the counting function of primes $\pi(x) := \#\{p \leq x : p \text{ prime}\}$ asserting that

$$\pi(x) \sim \frac{x}{\log x}. \qquad \text{(PNT)}$$

The twiddle notation is shorthand for the statement

$$\lim_{x \to \infty} \frac{\pi(x)}{(x/\log x)} = 1.$$

Here we shall survey early work on the distribution of primes, the proof of the PNT, and some later developments.

Since the time of Euclid, the primes, $2, 3, 5, 7, 11, 13, \ldots$, have been known to be infinite in number. They appear to be distributed quite irregularly, and early attempts to find a closed formula for the nth prime were unsuccessful. By the end of the eighteenth century many mathematical tables had been computed, and examination of tables of prime numbers led C. F. Gauss and A. M. Legendre to change the question under investigation. Instead of seeking an exact formula for the nth prime, they considered the counting function $\pi(x)$ and asked for approximations to this function, evidently a new kind of question in number theory. Each of the two men conjectured the PNT, though neither did so in the form we have given. In 1808 Legendre published the formula

$$\pi(x) = x/(\log x + A(x)),$$

where $A(x)$ tends to a constant as $x \to \infty$.

Gauss recorded his conjecture in one of his favorite books of tables around 1792 or 1793 but first disclosed it, in a mathematical letter, over fifty years later. He actually found a better approximation for $\pi(x)$ in terms of the logarithmic integral function, defined for $x > 0$ by

$$\text{li}(x) := \lim_{\varepsilon \to 0+} \left\{ \int_0^{1-\varepsilon} + \int_{1+\varepsilon}^x \right\} \frac{1}{\log t} dt.$$

It is easy to show that $\text{li}(x) \sim x/\log x$, so either expression can be used in the asymptotic formula for $\pi(x)$. It has been shown that $\text{li}(x)$ is a more accurate estimate of $\pi(x)$ than either $x/\log x$ or Legendre's proposed formula, so today $\text{li}(x)$ is used in PNT error estimates. For more details about Gauss's meditations on the PNT, see [10], included in this volume.

The function that we now call the Riemann zeta function, which was to play a decisive role in the proof of the PNT, was introduced by L. Euler in the eighteenth century. For s real and $s > 1$, define

$$\zeta(s) = \sum_{n=1}^{\infty} n^{-s}.$$

Using the unique factorization of positive integers, Euler proved that

$$\zeta(s) = \prod_p \{1 + p^{-s} + p^{-2s} + \cdots\}$$
$$= \prod_p \{1 - p^{-s}\}^{-1},$$

where the product extends over all primes p. Further, he gave another proof of the infinitude of the primes by observing that if the number of primes were finite, then the product for $\zeta(1)$ would converge, while in fact the sum for ζ at $s = 1$ is the harmonic series, which diverges. Euler's proof shows further that the primes are sufficiently numerous that the sum of their reciprocals diverges.

Legendre conjectured and incorrectly believed he had proved that there are an infinite number of primes in each arithmetic progression for which the first term and common difference are relatively prime. This theorem was established by P. L. Dirichlet in 1837 by greatly extending the method of Euler described above. In two papers, Dirichlet introduced characters (periodic completely multiplicative arithmetic functions) to select the elements of an arithmetic progression; he generalized the ζ function by multiplying terms of the series for ζ by characters to make what we today call Dirichlet L functions; he related the value of an L function $L(1,\chi)$ with the class number of quadratic forms of a given discriminant, and from the positivity of the class number he deduced his key lemma that each of the L functions is non-zero at the point $s = 1$. The subject of analytic number theory is generally considered to have begun with Dirichlet.

The first person to establish the true order of $\pi(x)$ was P. L. Chebyshev. In the middle of the nineteenth century he found an ingenious elementary method to estimate $\pi(x)$ and established the bounds

$$\frac{0.921x}{\log x} < \pi(x) < \frac{1.106x}{\log x}$$

for all sufficiently large values of x. Chebyshev's work was based on the use of the arithmetic identity

$$\sum_{d|n} \Lambda(d) = \log n,$$

where von Mangoldt's function Λ is a weighted prime and prime-power counting function defined by $\Lambda(d) = \log p$ if $d = p^\alpha$ for some prime p and positive integer α and $\Lambda(d) = 0$ otherwise. Chebyshev's formula is the arithmetic equivalent of the zeta function identity $\{-\zeta'(s)/\zeta(s)\} \cdot \zeta(s) = -\zeta'(s)$. Chebyshev showed also that if $\pi(x)/(x/\log x)$ had a limit as $x \to \infty$, then its value would be 1. Attempts at improving Chebyshev's methods led to slightly sharper estimates and much more elaborate calculations, but the PNT was not to be established by an elementary method for another hundred years.

A few years after the appearance of Chebyshev's paper, a path to the proof of the PNT was laid out by G. F. B. Riemann ([9], [14]) in his only published paper on number theory. Riemann's revolutionary idea was to consider ζ as a function of a complex variable and express $\pi(x)$ in terms of a complex integral

P. L. Chebyshev (1821–1894)

involving ζ. By formally deforming the integration contour, Riemann achieved an explicit formula for $\pi(x)$ as an infinite series whose leading term was $\text{li}(x)$ and that involved the zeros of $\zeta(s)$. However, there was not enough analysis available at that time to rigorously deduce the PNT following Riemann's program. It was not until the end of the nineteenth century that the missing essential ingredient was supplied: this was the theory of entire functions of finite order, which was developed by J. Hadamard for the purpose of proving the PNT.

Riemann proved that the ζ function has an analytic continuation to $\mathbb{C}$ with just one singularity, a simple pole with residue 1 at the point $s = 1$, and that ζ satisfies a functional equation connecting its values at complex arguments s and $1 - s$. Incidentally, we owe to Riemann the unusual notation for a complex number $s = \sigma + it$ that has become standard in analytic number theory. Riemann recognized the key role that zeros of the ζ function play in prime number theory. He conjectured several properties of these zeros, all

B. Riemann (1826–1866)

but one of which were proved around the end of the nineteenth century by Hadamard and H. von Mangoldt. The one conjecture that remains to this day, and is generally considered to be the most famous unsolved problem in mathematics, is the so-called Riemann hypothesis:

All non-real zeros of the ζ function have real part $\frac{1}{2}$.
(RH)

Riemann evidently perceived the greater difficulty of the RH, for while he stated his other conjectures with no qualification, he prefaced the statement of the RH with the phrase "it is very likely that [es ist sehr wahrscheinlich dass]...."

Activity in prime number theory increased toward the end of the nineteenth century. The term *Prime Number Theorem* appears to have originated at this time in the Göttingen dissertation of H. von Schaper, *Über die Theorie der Hadamardschen Funktionen und ihre Anwendung auf das Problem der Primzahlen*, 1898. There were several false starts before correct proofs of the PNT were given. For example, in 1885 Stieltjes [18] claimed to have proved the RH. With this result one could establish the PNT with an essentially optimal error term

$$\pi(x) - \operatorname{li}(x) = O(x^{\frac{1}{2}+\varepsilon}). \qquad (1)$$

Here we have used the notation $f(x) = O(g(x))$, where g is a positive function for all x from some point onward, if $|f(x)|/g(x) < B$ holds for some positive constant B and all sufficiently large positive values of x. The deduction of (1) under the assumption of the RH was later carried out by von Koch. Stieltjes died in 1894 without having either substantiated or retracted his claim of having proved the RH.

First proofs of the prime number theorem

The PNT was established in 1896 by Jacques Hadamard and by Charles-Jean de la Vallée Poussin. It was the first major achievement for each at the start of long and distinguished careers. Hadamard was born at Versailles, France, in 1865. After studies at the École Normale Supérieure he obtained his doctorate in 1892. He spent most of his career in Paris, working principally in complex function theory, partial differential equations, and differential geometry. He died in 1963, within two months of his 98th birthday. De la Vallée Poussin was born in 1866 in Louvain, Belgium, where his father was a professor of mineralogy and geology at the University. After studying at Louvain, he too joined the faculty of the University, at the age of 26, as Professor of Mathematics. His elegant and lucid *Cours d'Analyse* has educated generations of mathematicians in the methods of Borel and Lebesgue. De la Vallée Poussin died in 1962, in his 96th year.

The arguments of both Hadamard and de la Vallée Poussin followed the scheme laid out by Riemann. Both papers made essential use of Riemann's functional equation for the zeta function, several other properties of ζ conjectured by Riemann and established by Hadamard, and Hadamard's new theory of entire functions.

Hadamard's paper on the PNT [11] consists of two parts. Here are the opening paragraphs of Part I, *On the distribution of zeros of the zeta function* (in our translation). It is interesting to see how he treats Stieltjes's claim.

J. Hadamard (1865–1963)

The Riemann zeta function is defined, when the real part of s is greater than 1, by the formula

$$\log \zeta(s) = -\sum_p \log\left(1 - 1/p^s\right), \qquad (2)$$

where p runs over the prime numbers [Translators' remark: Use the principal branch for the logarithms on the right side of (2).] It is holomorphic in the entire plane, except at the point $s = 1$, which is a simple pole. It does not vanish for any value of s with real part greater than 1, since the right-hand side of (2) is finite. But it admits an infinity of complex zeros with real part between 0 and 1. Stieltjes proved, in accordance with Riemann's expectations, that these zeros are all of the form $\frac{1}{2} + it$ (where t is real); but his proof has never been published, and it has not even been established that the function ζ has no zeros on the line $\Re s = 1$.

It is this last assertion that I propose to prove here.

Hadamard's proof that $\zeta \neq 0$ on the line

$$L\{\sigma = 1\} := \{s \in \mathbb{C} : \Re s = 1\}$$

used formula (2) for $\log \zeta(s)$, where $s = \sigma + it$ with $\sigma > 1$ and t real, and the representation

$$-\Re \log\left(1 - p^{-s}\right) = \Re \sum_{m=1}^{\infty} \frac{1}{mp^{ms}}$$

$$= \sum_{m=1}^{\infty} \frac{\cos(mt \log p)}{mp^{m\sigma}}.$$

Thus

$$\log |\zeta(s)| = \sum_p \sum_{m=1}^{\infty} \frac{\cos(mt \log p)}{mp^{m\sigma}}. \qquad (3)$$

In the analysis of ζ, one can ignore the contribution of the higher prime powers, because that part of the series is uniformly bounded for $\sigma > 1$, while the sum over just the primes in (3) is not. Hadamard observed first that, because of the simple pole of ζ at $s = 1$,

$$\sum_p p^{-\sigma} \sim \log \zeta(\sigma) \sim \log \frac{1}{\sigma - 1} \quad (\sigma \to 1+). \quad (4)$$

He next noted that if $1 + it_0$ were a zero of ζ, necessarily simple, then it would follow that

$$\sum_p p^{-\sigma} \cos(t_0 \log p) \sim -\log \frac{1}{\sigma - 1} \quad (\sigma \to 1+).$$

$$(5)$$

Comparing (4) and (5), he concluded in succession that

(a) $\cos(t_0 \log p) \approx -1$ for most primes p;

(b) hence $\cos(2t_0 \log p) \approx +1$ for most primes p;

(c) finally $1 + 2it_0$ would be a pole of ζ, contradicting the fact that ζ has no singularities in $\mathbb{C}$ other than at $s = 1$.

We have omitted some details that Hadamard gave to make this argument complete; they can be found also in Chapter 3 of [19]. Today it is customary to use a cleaner method, due to F. Mertens, that combines formula (3) with a trigonometric inequality to get an inequality that expresses Hadamard's idea. For example, the choice

$$3 + 4\cos\theta + \cos 2\theta \geq 0$$

yields

$$\zeta(s)^3 |\zeta(\sigma + it)|^4 |\zeta(\sigma + 2it)| > 1 \quad (\sigma > 1).$$

Part II of Hadamard's 1896 paper, "Arithmetic Consequences", contains his deduction of the PNT. It begins with the following modest words:

> As one can see, we are quite far from having proved the assertion of Riemann-Stieltjes; we have not even been able to exclude the hypothesis of an infinity of zeros of $\zeta(s)$ approaching arbitrarily close to the limiting line $\Re s = 1$. However, the result which we have obtained suffices by itself to prove the principal arithmetic consequences which people have, up to now, sought to deduce from the properties of $\zeta(s)$.

Ch. J. de la Vallée Poussin (1866–1962)

Here are the main ingredients in Hadamard's deduction of the PNT. He first established the following "smoothed" form of the Mellin inversion formula,

$$\sum_{n<x} a_n \log(x/n) = \frac{1}{2\pi i} \int_{2-i\infty}^{2+i\infty} \frac{x^s}{s^2} \sum_{n=1}^{\infty} \frac{a_n}{n^s} ds,$$

valid for x positive and $\sum a_n n^{-s}$ a Dirichlet series that is absolutely convergent for $\Re s > 1$. The arithmetic function to which he applied the formula was the von Mangoldt function $\Lambda(n)$ that appeared in Chebyshev's work. The associated Dirichlet series satisfies the zeta function formula

$$\sum_{n=1}^{\infty} \Lambda(n) n^{-s} = -\zeta'(s)/\zeta(s),$$

which is shown by differentiating formula (2) for $\log \zeta(s)$. Using the Weierstrass–Hadamard product representation for $(s-1)\zeta(s)$, the convergence of $\sum |\rho|^{-2}$ (where ρ runs over the non-real zeros of ζ), and a contour deformation and estimation of the above Mellin integral, Hadamard deduced that

$$\sum_{n \leq x} \Lambda(n) \log(x/n) \sim x.$$

From this relation the PNT follows quite easily.

Like Hadamard, de la Vallée Poussin [6] began his proof by establishing that ζ has no zeros with real part 1 (by a rather more complicated argument than that of Hadamard). He also used a smoothed form of the Mellin inversion formula, but with an expression $x^s/(s-\mu)(s-\nu)$ in place of x^s/s^2. In 1899, de la Vallée Poussin published another article in which he obtained the PNT with an error estimate

$$\pi(x) - \mathrm{li}(x) = O\left(x \exp\left\{-c \log^\alpha x\right\}\right), \quad (6)$$

where $\alpha = \frac{1}{2}$ and c is some positive constant. In the last paper he made use of Mertens' trigonometric inequality. A quarter of a century went by before de la Vallée Poussin's error bound was improved.

We note that the estimate (6) with a fixed positive value of α is superior to any estimate of the form

$$\pi(x) - \mathrm{li}(x) = O(x/\log^k x)$$

with fixed $k > 0$. The RH implies that (6) holds with $\alpha = 1, c = \frac{1}{2} - \varepsilon$, as stated in (1).

Later developments

In just over a decade after the proof of the PNT, prime number theory moved from obscurity to mainstream. So little was known on the subject in England at the turn of the century that J. E. Littlewood was assigned the task of proving the RH by E. W. Barnes, his Cambridge research supervisor, and at one point, according to G. H. Hardy, it was believed that the RH had been proved. The publication of E. Landau's *Handbuch der Lehre von der Verteilung der Primzahlen* [14] in 1909 quickly changed the status of the subject. Landau's book presented in accessible form nearly everything that was then known about the distribution of primes. Incidentally, the O notation we use was popularized by Landau.

In addition to writing about prime number theory, Landau made significant contributions to the subject, including the simplification of some of the main arguments and extension of the results. For example, he was the first to prove the PNT without making use of the functional equation of ζ. His idea was to combine an analytic continuation of the zeta function a bit to the left of $L\{\sigma = 1\}$, e.g., via

$$\zeta(s) - \frac{s}{s-1} = s\int_1^\infty \frac{[x]-x}{x^{s+1}}dx, \quad \Re s > 0,$$

with an upper bound for the logarithmic derivative of the zeta function in a suitable zero-free region. With the aid of his new methods, Landau was able to treat some related problems, such as estimating the number of prime ideals of norm at most x in the ring of integers of an arbitrary algebraic number field (Section 242 of [14]). This result solved part of the eighth problem posed in Hilbert's famous 1900 address to the International Congress of Mathematicians.

E. Landau (1877–1938)

It had long been noted, possibly already by Gauss, that

$$\pi(x) - \mathrm{li}(x) < 0 \tag{7}$$

for $x = 2, 3, \ldots$ to whatever point it was checked. In addition to this empirical evidence, theoretical support for the conjecture that (7) holds for all $x \geq 2$ was provided by Riemann, who observed that his formula for $\pi(x)$ begins with the terms $\mathrm{li}(x) - \mathrm{li}\left(\sqrt{x}\right)/2$. However, this conjecture was disproved by Littlewood [12], who used almost periodic functions and diophantine ideas to show that in fact the difference changes sign infinitely often. Littlewood's proof did not provide an estimate of where the first change of sign might be, and this question attracted further attention. The suggestion was raised that the question might be undecidable. However, it was proved by S. Skewes that there is a number

$$x < \exp\exp\exp\exp 7.705$$

for which (7) does not hold. Skewes's number, which is among the largest that have occurred in mathematics, has subsequently been replaced by a more modest number with fewer than 400 decimal digits. There is a moral here: vast amounts of empirical evidence, together with a "philosophical" explanation for a mathematical phenomenon, are not the same as a proof.

What is the relation between the PNT and the nonvanishing of the Riemann zeta function on $L\{\sigma = 1\}$? It is quite easy to see that the PNT implies that ζ has no zeros on the line. Proofs of the PNT were given first by Landau ([14], Section 241) and then by Hardy and Littlewood that used, besides the non-vanishing of ζ on $L\{\sigma = 1\}$, only very weak growth conditions for $\zeta(\sigma + it)$ for $\sigma > 1$ and $|t| \to \infty$. The question arose whether the PNT could be proved using just the fact that ζ has no zeros on $L\{\sigma = 1\}$. This was answered affirmatively around 1930 by work of N. Wiener using Fourier analysis. Wiener created an approximate integral formula for $\pi(x)$ involving a compactly supported smoothing function. The following tauberian theorem [4] provides one of the most direct proofs now known for the PNT.

Wiener–Ikehara Theorem. *Suppose f is a nondecreasing real-valued function on $[1, \infty)$ such that*

$$\int_1^\infty f(u)u^{-\sigma-1}du < \infty$$

J. E. Littlewood (1885–1977)

weaker hypothesis on the behavior of the function g near $L\{\sigma = 1\}$ can be used in the proof of Beurling's theorem.

Generalized prime number theory has several applications, and it has raised interesting new problems. For example, Landau's prime ideal theorem is easily deduced from Beurling's result. Also, there are generalized prime models for which the counting function of generalized integers is quite close to that of the usual integers, but for which the analogue of the RH is false. This means that a successful proof of the RH will require more than just the facts that the positive integers are a multiplicative semigroup and that the counting function of positive integers $[x]$ is close to x; presumably, the additive structure of the integers must be taken into account. More on this topic can be found in the authors' survey article [2].

Many different proofs have been given for the PNT. A very concise argument that uses only the analyticity and non-vanishing of $(s-1)\zeta(s)$ on the closed half plane $\{s : \Re s \geq 1\}$ was found by D. J. Newman [15]. In place of the Wiener–Ikehara theorem or an application of the Mellin inversion integral, Newman's method uses basic complex function theory to

for each real $\sigma > 1$. Suppose further that

$$\int_1^\infty f(u)u^{-s-1}du = \frac{\alpha}{s-1} + g(s), \quad \Re s > 1,$$

where $\alpha \in \mathbb{R}$ and g is the restriction to $\{s : \Re s > 1\}$ of a continuous function on the closed half plane $\{s : \Re s \geq 1\}$. Then

$$\lim_{u \to \infty} u^{-1} f(u) = \alpha.$$

In 1937 A. Beurling introduced an abstraction of prime number theory in which multiplicative structure was preserved but the additive structure of the integers was dropped. A sequence of real numbers $p_1 \leq p_2 \leq p_3 \leq \cdots$, called *generalized primes*, was introduced, and the free abelian semigroup generated from them under multiplication was called the associated sequence of *generalized integers*. From the assumption that the counting function of the generalized integers satisfies the condition

$$I(x) = Ax + O\left(x \log^{-\gamma} x\right), \quad \gamma > \frac{3}{2},$$

an analogue of the PNT was established. Moreover, the condition that $\gamma > \frac{3}{2}$ was shown to be best possible. A form of the Wiener-Ikehara theorem with a

N. Wiener (1894–1964)

estimate the integral

$$\int \frac{x^2}{s}\left\{1+\frac{s^2}{R^2}\right\}\sum_{n=1}^{\infty}\frac{a_n}{n^s}ds$$

over a finite contour for large values of R. Some other interesting proofs of the PNT include that of H. Daboussi, which uses elements of sieve theory, and a method of A. Hildebrand based on the large sieve.

De la Vallée Poussin's PNT error term was improved by Littlewood, who used exponential sum methods to find bounds for Dirichlet series. These estimates led to enlarged regions on which the zeta function is guaranteed to be non-zero and consequently to better PNT estimates. The method was developed and improved by the school of I. M. Vinogradov, leading to the bound in which (6) holds with $\alpha = \frac{3}{5} - \varepsilon$.

The failure of the Chebyshev methods and the success of Riemann's program in proving the PNT led to the opinion, voiced by Hardy and others, that the PNT could be proved only with the use of the Riemann zeta function. This belief was strengthened by Wiener's proof of the equivalence of the PNT and the non-vanishing of ζ on $L\{\sigma = 1\}$. Inspired by work in sieve theory, A. Selberg developed a kind of weighted analogue of Chebyshev's identity. With this formula and an argument of P. Erdős he succeeded in giving an "elementary" proof of the PNT. Subsequently, Selberg and Erdős each discovered an independent proof. Their arguments are considered elementary in the sense that they do not involve the zeta function, complex analysis, or Fourier methods; however, the methods are quite intricate. Subsequently, elementary estimates were sought for the PNT error term, and by use of higher-order analogues of Selberg's formula and more elaborate tauberian arguments, error terms of type (6) with $\alpha = \frac{1}{6}\varepsilon$ were achieved. For a survey of the use of elementary methods in prime number theory, see [7].

We conclude with a summary of what is now known about the truth of the RH. If the RH is false and has even a single non-real zero off the critical line $\{s \in \mathbb{C} : \Re s = \frac{1}{2}\}$, there would be consequences for prime number theory, such as in the quality of the PNT error term. The numerical evidence in support of the RH is very great—by comparing the sign changes of a real-valued equivalent of $\zeta(\frac{1}{2} + it)$ with the zeros predicted by use of the argument principle, van de Lune, te Riele, and Winter showed that the first one and a half billion (!) non-real zeros of zeta lie on the critical line and are simple. In the 1920s, Littlewood showed that almost all the non-real zeros lie in any given strip of positive width that contains the critical line. Hardy proved that there were infinitely many zeros of zeta on the critical line, and later Selberg showed that a positive proportion of the non-real zeros were on the line. Near the end of his life, N. Levinson introduced an efficient zero counting method, which B. Conrey has developed to show that more than $\frac{2}{5}$ of the non-real zeta zeros are simple and lie on the critical line.

Sources

The theory of the distribution of prime numbers is a rich and fascinating topic. In this survey we have had to treat fleetingly or omit entirely many interesting topics. Also, it was not feasible to list the sources for all the facts cited. The following books and articles discuss further topics and provide references to original sources.

Landau's *Handbuch* [14] remains an excellent introduction to prime number theory and is a reference for virtually all early results in the area. The second edition of the *Handbuch*, edited by the first author, contains information on work up to about 1950 on the distribution of primes. The books of Chandrasekharan [4], [5], Ingham [12], and Ellison and Mendès-France [8] provide very readable introductions to the subject. Titchmarsh and Heath-Brown [19] and Ivić [13] are standard references on the Riemann zeta function, and Edwards [9] provides a historical view of this subject. There are detailed and authoritative encyclopedia articles on prime number theory by Hadamard [1] and by Bohr and Cramér [3]. The recent survey article of W. Schwarz [17] describes the development of prime number theory in the twentieth century, including several topics that we have not treated. Finally, Ribenboim [16] provides a kind of Guinness record book about primes and includes extensive references.

References

1. P. Bachmann, J. Hadamard and E. Maillet, Propositions transcendantes de la théorie des nombres, *Encyclopédie des Sciences Mathématiques I* **17** (1910), 215–387.

2. P. T. Bateman and H. G. Diamond, Asymptotic distribution of Beurling's generalized prime numbers, *Studies in Number Theory* (ed. W. J. LeVeque), Math. Assoc. America, 1969, pp. 152–210.

3. H. Bohr and H. Cramér, Die neuere Entwicklung des analytischen Zahientheorie, *Encyklopädie der Mathematischen Wissenschaften* II C 8 (1923), 722–849. (Also in Vol. 3 of Harald Bohr's *Collected Mathematical Works*, Copenhagen, 1952).

4. K. Chandrasekharan, *Introduction to Analytic Number Theory*, Springer-Verlag, 1968.

5. K. Chandrasekharan, *Arithmetical Functions*, Springer-Verlag, 1970.

6. C.-J. de la Vallée Poussin, Recherches analytiques sur la théorie des nombres premiers, *Ann. de la Soc. Scientifique de Bruxelles* **20** (1896), 183–256, 281–397.

7. H. G. Diamond, Elementary methods in the study of the distribution of prime numbers, *Bull. Amer. Math. Soc.* **7** (1982), 553–589.

8. W. J. Ellison and M. Mendès-France, *Les Nombres Premiers*, Hermann, 1975.

9. H. M. Edwards, *Riemann's Zeta Function*, Academic Press, 1974.

10. L. J. Goldstein, A history of the prime number theorem, *Amer. Math. Monthly* **80** (1973), 599–615; elsewhere in this volume.

11. J. Hadamard, Sur la distribution des zéros de la fonction $\zeta(s)$ et ses consequences arithmétiques, *Bull. de la Soc. Math. de France* **24** (1896), 199–220.

12. A. E. Ingham, *The Distribution of Prime Numbers*, Cambridge Univ. Press, 1932.

13. A. Ivić, *The Riemann Zeta-function*, Wiley-Interscience, 1985.

14. E. Landau, *Handbuch der Lehre von der Verteilung der Primzahlen*, 2nd ed., with an appendix by P. T. Bateman, Chelsea, 1953.

15. D. J. Newman, Simple analytic proof of the prime number theorem, *Amer. Math. Monthly* **87** (1980), 693–696.

16. P. Ribenboim, *The New Book of Prime Number Records*, 3rd ed., Springer-Verlag, 1996.

17. W. Schwarz, Some remarks on the history of the prime number theorem from 1896 to 1960, *Development of Mathematics 1900–1950* (ed. J.-P. Pier), Birkhäuser, 1994.

18. T. J. Stieltjes, Sur une fonction uniforme, *Comptes Rendus Acad. Sci. Paris* **101** (1885), 153–154.

19. E. C. Titchmarsh and D. R. Heath-Brown, *The Theory of the Riemann Zeta-function*, 2nd ed., Oxford Univ. Press, 1986.

The Indian Mathematician Ramanujan

G. H. HARDY

American Mathematical Monthly **44** (1937), 137–155

I have set myself a task in these lectures which is genuinely difficult and which, if I were determined to begin by making every excuse for failure, I might represent as almost impossible. I have to form myself, as I have never really formed before, and to try to help you to form, some sort of reasoned estimate of the most romantic figure in the recent history of mathematics; a man whose career seems full of paradoxes and contradictions, who defies almost all the canons by which we are accustomed to judge one another, and about whom all of us will probably agree in one judgment only, that he was in some sense a very great mathematician.

The difficulties in judging Ramanujan are obvious and formidable enough. Ramanujan was an Indian, and I suppose that it is always a little difficult for an Englishman and an Indian to understand one another properly. He was, at the best, a half-educated Indian; he never had the advantages, such as they are, of an orthodox Indian training; he never was able to pass the First Arts Examination of an Indian university, and never could rise even to be a "Failed B.A." He worked, for most of his life, in practically complete ignorance of modern European mathematics, and died when he was a little over 30 and when his mathematical education had in some ways hardly begun. He published abundantly—his published papers make a volume of nearly 400 pages—but he also left a mass of unpublished work which had never been analysed properly until the last few years. This work includes a great deal that is new, but much more that is rediscovery, and often imperfect rediscovery; and it is sometimes still impossible to distinguish between what he must have rediscovered and what he may somehow have learnt. I cannot imagine anybody saying with any confidence, even now, just how great a mathematician he was and still less how great a mathematician he might have been.

These are genuine difficulties, but I think that we shall find some of them less formidable than they look, and the difficulty which is the greatest for me has nothing to do with the obvious paradoxes of Ramanujan's career. The real difficulty for me is that Ramanujan was, in a way, my discovery. I did not invent him—like other great men, he invented himself—but I was the first really competent person who had the chance to see some of his work, and I can still remember with satisfaction that I could recognise at once what a treasure I had found. And I suppose that I still know more of Ramanujan than any one else, and am still the first authority on this particular subject. There are other people in England, Professor Watson in particular, and Professor Mordell, who know parts of his work very much better than I do, but neither Watson nor Mordell knew Ramanujan himself as I did. I saw him and talked with him almost every day for several years, and above all I actually collaborated with him. I owe more to him than to any one else in the world with one exception, and my association with him is the one romantic incident in my life. The difficulty for me then is not that I do not know enough about him, but that I know and feel too much and that I simply cannot be impartial.

I rely, for the facts of Ramanujan's life, on Seshu Aiyar and Ramachaundra Rao, whose memoir of Ramanujan is printed, along with my own, in his *Collected Papers* [3]. He was born in 1887 in a Brahmin family at Erode near Kumbakonam, a fair-sized town in the Tanjore district of the Presidency of Madras. His father was a clerk in a cloth-merchant's office in Kumbakonam, and all his relatives, though of high caste, were very poor.

He was sent at 7 to the High School of Kumbakonam, and remained there nine years. His exceptional abilities had begun to show themselves before he was 10, and by the time that he was 12 or 13 he was recognised as a quite abnormal boy. His biographers tell some curious stories of his early years. They say for example that, soon after he had begun the study of trigonometry, he discovered for himself "Euler's theorems for the sine and cosine" (by which I understand the relations between the circular and exponential functions), and was very disappointed when he found later, apparently from the second volume of Loney's *Trigonometry*, that they were known already. Until he was 16 he had never seen a mathematical book of any higher class. Whittaker's *Modern Analysis* had not yet spread so far, and Bromwich's *Infinite Series* did not exist. There can be no doubt that either of these books would have made a tremendous difference to him if they could have come his way. It was a book of a very different kind, Carr's *Synopsis*, which first aroused Ramanujan's full powers.

George Shoobridge Carr was formerly Scholar of Gonville and Caius College, Cambridge; his book (*A Synopsis of Elementary Results in Pure and Applied Mathematics* [2]) is almost unprocurable now. There is a copy in the Cambridge University Library, and there happened to be one in the library of the Government College of Kumbakonam, which was borrowed for Ramanujan by a friend. The book is not in any sense a great one, but Ramanujan has made it famous, and there is no doubt that it influenced him profoundly and that his acquaintance with it marked the real starting point of his career. Such a book must have had its qualities, and Carr's, if not a book of any high distinction, is no mere third-rate textbook, but a book written with some real scholarship and enthusiasm and with a style and individuality of its own. Carr himself was a private coach in London, who came to Cambridge as an undergraduate when he was nearly 40, and was 12th Senior Optime in the Mathematical Tripos of 1880 (the same year in which he published the first volume of his book). He is now completely forgotten, even in his own college, except in so far as Ramanujan has kept his name alive; but he must have been in some ways rather a remarkable man.

I suppose that the book is substantially a summary of Carr's coaching notes. If you were a pupil of Carr, you worked through the appropriate sections of the *Synopsis*. It covers roughly the subjects of Schedule A of the present Tripos (as these subjects were understood in Cambridge in 1880), and is effectively the "synopsis" it professes to be. It contains the enunciations of 6165 theorems, systematically and quite scientifically arranged, with proofs which are often little more than cross-references and are decidedly the least interesting part of the book. All this is exaggerated in Ramanujan's famous note-books (which contain practically no proofs at all), and any student of the note-books can see that Ramanujan's ideal of presentation had been copied from Carr's.

Carr has sections on the obvious subjects, algebra, trigonometry, calculus and analytical geometry, but some sections are developed disproportionately, and particularly the formal side of the integral calculus. This seems to have been Carr's pet subject, and the treatment of it is very full and in its way definitely good. There is no theory of functions; and I very much doubt whether Ramanujan, to the end of his life, ever understood at all clearly what an analytic function is. What is more surprising, in view of Carr's own tastes and Ramanujan's later work, is that there are no elliptic functions. However Ramanujan may have acquired his very peculiar knowledge of this theory, it was not from Carr.

On the whole, considered as an inspiration for a boy of such abnormal gifts, Carr was not too bad, and Ramanujan responded amazingly. According to his Indian biographers [3]:

Through the new world thus opened to him, Ramanujan went ranging with delight. It was this book which awakened his genius. He set himself to establish the formulae given therein. As he was without the aid of other books, each solution was a piece of research so far as he was concerned... Ramanujan used to say that the goddess of Namakkal inspired him with the formulae in dreams. It is a remarkable fact that frequently, on rising from bed, he would note down results and rapidly verify them, though he was not always able to supply a rigorous proof...

I have quoted the last sentences deliberately, not because I attach any importance to them—I am no more interested in the goddess of Namakkal than you are—but because we are now approaching the difficult and tragic part of Ramanujan's career, and we must try to understand what we can of his psychology and of the atmosphere surrounding him in his early years.

I am sure that Ramanujan was no mystic and that religion, except in a strictly material sense, played no important part in his life. He was an orthodox high-caste Hindu, and always adhered (indeed with a severity most unusual in Indian residents in England) to all the observances of his caste. He had promised his parents to do so, and he kept his promises to the letter. He was a vegetarian in the strictest sense—this proved a terrible difficulty later when he fell ill—and all the time he was in Cambridge he cooked all his food himself, and never cooked it without first changing into pyjamas.

Now the two memoirs of Ramanujan printed in the *Papers* (and both written by men who, in their different ways, knew him very well) contradict one another flatly about his religion. Seshu Aiyar and Ramachaundra Rao say

> Ramanujan had definite religious views. He had a special veneration for the Namakkal goddess... He believed in the existence of a Supreme Being and in the attainment of Godhead by men... He had settled convictions about the problem of life and after...

while I say

> his religion was a matter of observance and not of intellectual conviction, and I remember well his telling me (much to my surprise) that all religions seemed to him more or less equally true...

Which of us is right? For my part I have no doubt at all; I am quite certain that I am.

Classical scholars have, I believe, a general principle, *difficilior lectio potior*—the more difficult reading is to be preferred—in textual criticism. If the Archbishop of Canterbury tells one man that he believes in God, and another that he does not, then it is probably the second assertion which is true, since otherwise it is very difficult to understand why he should have made it, while there are many excellent reasons for his making the first whether it be true or false. Similarly, if a strict Brahmin like Ramanujan told me, as he certainly did, that he had no definite beliefs, then it is 100 to 1 that he meant what he said.

This was no sufficient reason why Ramanujan should outrage the feelings of his parents or his Indian friends. He was not a reasoned infidel, but an "agnostic" in its strict sense, who saw no particular good, and no particular harm, in Hinduism or in any other religion. Hinduism is, far more for example than Christianity, a religion of observance, in which belief counts for extremely little in any case, and, if Ramanujan's friends assumed that he accepted the conventional doctrines of such a religion, and he did not disillusion them, he was practising a quite harmless, and probably necessary, economy of truth.

This question of Ramanujan's religion is not itself important, but it is not altogether irrelevant, because there is one thing which I am really anxious to insist upon as strongly as I can. There is quite enough about Ramanujan that is difficult to understand, and we have no need to go out of our way to manufacture mystery. For myself, I liked and admired him enough to wish to be a rationalist about him; and I want to make it quite clear to you that Ramanujan, when he was living in Cambridge in good health and comfortable surroundings, was, in spite of his oddities, as reasonable, as sane, and in his way as shrewd a person as anyone here. The last thing which I want you to do is to throw up your hands and exclaim "here is something unintelligible, some mysterious manifestation of the immemorial wisdom of the East!" I do not believe in the immemorial wisdom of the East, and the picture I want to present to you is that of a man who had his peculiarities like other distinguished men, but a man in whose society one could take pleasure, with whom one could take tea and discuss politics or mathematics; the picture in short, not of a wonder from the East, or an inspired idiot, or a psychological freak, but of a rational human being who happened to be a great mathematician.

Until he was about 17, all went well with Ramanujan.

> In December 1903 he passed the Matriculation Examination of the University of Madras, and in the January of the succeeding year he joined the Junior First in Arts class of the Government College, Kumbakonam, and won the Subrahmanyam scholarship, which is generally awarded for proficiency in English and Mathematics...,

but after this there came a series of tragic checks.

> By this time, he was so absorbed in the study of mathematics that in all lecture hours—whether devoted to English, History, or Physiology—he used to engage himself in some mathematical investigation, unmindful of what was happening in the class. This excessive devotion to mathematics and his consequent neglect of the other subjects resulted in his failure to secure promotion to the senior class and in the consequent discontinuance of the scholarship. Partly owing to disappointment and partly owing to the influence of a friend, he ran away northward into the Telugu country, but returned to Kumbakonam after some wandering and rejoined the college. As owing to his absence he failed to make sufficient attendances to obtain his term certificate in 1905, he entered Pachaiyappa's College, Madras, in

1906, but falling ill returned to Kumbakonam. He appeared as a private student for the F. A. examination of December 1907 and failed...

Ramanujan does not seem to have had any definite occupation, except mathematics, until 1912. In 1909 he married, and it became necessary for him to have some regular employment, but he had great difficulty in finding any because of his unfortunate college career. About 1910 he began to find more influential Indian friends, Ramaswami Aiyar and his two biographers, but all their efforts to find a tolerable position for him failed, and in 1912 he became a clerk in the office of the Port Trust of Madras, at a salary of about £30 a year. He was then nearly 25. The years between 18 and 25 are the critical years in a mathematician's career, and the damage had been done. Ramanujan's genius never had again its chance of full development.

There is not much to say about the rest of Ramanujan's life. His first substantial paper had been published in 1911, and in 1912 his exceptional powers began to be understood. It is significant that, though Indians could befriend him, it was only the English who could get anything effective done. Sir Francis Spring and Sir Gilbert Walker obtained a special scholarship for him, £60 a year, sufficient for a married Indian to live in tolerable comfort. At the beginning of 1913 he wrote to me, and Professor Neville and I, after many difficulties, got him to England in 1914. Here he had three years of uninterrupted activity, the results of which you can read in his *Papers*. He fell ill in the summer of 1917, and never really recovered, though he continued to work, rather spasmodically, but with no real sign of degeneration, until his death in 1920. He became a Fellow of the Royal Society early in 1918, and a Fellow of Trinity College, Cambridge, later in the same year (and was the first Indian elected to either society). His last mathematical letter on "Mock-Theta functions", the subject of Professor Watson's presidential address to the London Mathematical Society last year, was written about two months before he died.

The real tragedy about Ramanujan was not his early death. It is of course a disaster that any great man should die young, but a mathematician is often comparatively old at 30, and his death may be less of a catastrophe than it seems. Abel died at 26 and, although he would no doubt have added a great deal more to mathematics, he could hardly have become a greater man. The tragedy of Ramanujan was not that he died young, but that, during his five unfortunate years, his genius was misdirected, side-tracked, and to a certain extent distorted.

I have been looking again through what I wrote about Ramanujan 16 years ago, and, although I know his work a good deal better now than I did then, and can think about him more dispassionately, I do not find a great deal which I should particularly want to alter. But there is just one sentence which now seems to me indefensible. I wrote

> Opinions may differ about the importance of Ramanujan's work, the kind of standard by which it should be judged, and the influence which it is likely to have on the mathematics of the future. It has not the simplicity and the inevitableness of the very greatest work; it would be greater if it were less strange. One gift it shows which no one can deny, profound and invincible originality. He would probably have been a greater mathematician if he could have been caught and tamed a little in his youth; he would have discovered more that was new, and that, no doubt, of greater importance. On the other hand he would have been less of a Ramanujan, and more of a European professor, and the loss might have been greater than the gain...

and I stand by that except for the last sentence, which is quite ridiculous sentimentalism. There was no gain at all when the College at Kumbakonam rejected the one great man they had ever possessed, and the loss was irreparable; it is the worst instance that I know of the damage that can be done by an inefficient and inelastic educational system. So little was wanted, £60 a year for five years, occasional contact with almost anyone who had real knowledge and a little imagination, for the world to have gained another of its greatest mathematicians.

Ramanujan's letters to me, which are reprinted in full in the *Papers*, contain the bare statements of about 120 theorems, mostly formal identities extracted from his note-books. I quote fifteen which are fairly representative. They include two theorems, (14) and (15), which are as interesting as any but of which one is false and the other, as stated, misleading. The rest have all been verified since by somebody; in particular Rogers and Watson found the proofs of the extremely difficult theorems (10)–(12).

I should like you to begin by trying to reconstruct the immediate reactions of an ordinary professional mathematician who receives a letter like this from an unknown Hindu clerk.

(1) $1 - \dfrac{3!}{(1!2!)^3} x^2 + \dfrac{6!}{(2!4!)^3} x^4 - \cdots = \left(1 + \dfrac{x}{(1!)^3} + \dfrac{x^2}{(2!)^3} + \cdots\right)\left(1 - \dfrac{x}{(1!)^3} + \dfrac{x^2}{(2!)^3} - \cdots\right)$.

(2) $1 - 5\left(\dfrac{1}{2}\right)^3 + 9\left(\dfrac{1 \cdot 3}{2 \cdot 4}\right)^3 - 13\left(\dfrac{1 \cdot 3 \cdot 5}{2 \cdot 4 \cdot 6}\right)^3 + \cdots = \dfrac{2}{\pi}$.

(3) $1 + 9\left(\dfrac{1}{4}\right)^4 + 17\left(\dfrac{1 \cdot 5}{4 \cdot 8}\right)^4 + 25\left(\dfrac{1 \cdot 5 \cdot 9}{4 \cdot 8 \cdot 12}\right)^4 + \cdots = \dfrac{2^{3/2}}{\pi^{1/2}\left\{\Gamma\left(\frac{3}{4}\right)\right\}^2}$.

(4) $1 - 5\left(\dfrac{1}{2}\right)^5 + 9\left(\dfrac{1 \cdot 3}{2 \cdot 4}\right)^5 - 13\left(\dfrac{1 \cdot 3 \cdot 5}{2 \cdot 4 \cdot 6}\right)^5 + \cdots = \dfrac{2}{\left\{\Gamma\left(\frac{3}{4}\right)\right\}^4}$.

(5) $\displaystyle\int_0^\infty \dfrac{1 + \left(\frac{x}{b+1}\right)^2}{1 + \left(\frac{x}{a}\right)^2} \cdot \dfrac{1 + \left(\frac{x}{b+2}\right)^3}{1 + \left(\frac{x}{a+1}\right)^3} \cdots dx = \dfrac{1}{2}\pi^{1/2}\dfrac{\Gamma\left(a + \frac{1}{2}\right)\Gamma(b+1)\Gamma\left(b - a + \frac{1}{2}\right)}{\Gamma(a)\Gamma\left(b + \frac{1}{2}\right)\Gamma(b - a + 1)}$.

(6) $\displaystyle\int_0^\infty \dfrac{dx}{(1 + x^2)(1 + r^2 x^2)(1 + r^4 x^2)\cdots} = \dfrac{\pi}{2(1 + r + r^3 + r^6 + r^{10} + \cdots)}$.

(7) If $\alpha\beta = \pi^2$, then $\alpha^{-1/4}\left(1 + 4\alpha \displaystyle\int_0^\infty \dfrac{xe^{-\alpha x^2}}{e^{2\pi x} - 1} dx\right) = \beta^{-1/4}\left(1 + 4\beta \displaystyle\int_0^\infty \dfrac{xe^{-\beta x^2}}{e^{2\pi x} - 1} dx\right)$.

(8) $\displaystyle\int_0^a e^{-x^2} dx = \dfrac{1}{2}\pi^{1/2} - \dfrac{e^{-a^2}}{2a+} \dfrac{1}{a+} \dfrac{2}{2a+} \dfrac{3}{a+} \dfrac{4}{2a + \cdots}$.

(9) $4\displaystyle\int_0^\infty \dfrac{xe^{-x\sqrt{5}}}{\cosh x} dx = \dfrac{1}{1+} \dfrac{1^2}{1+} \dfrac{1^2}{1+} \dfrac{2^2}{1+} \dfrac{2^2}{1+} \dfrac{3^2}{1+} \dfrac{3^2}{1 + \cdots}$.

(10) If $u = \dfrac{x}{1+} \dfrac{x^5}{1+} \dfrac{x^{10}}{1+} \dfrac{x^{15}}{1 + \cdots}$ and $v = \dfrac{x^{1/5}}{1+} \dfrac{x}{1+} \dfrac{x^2}{1+} \dfrac{x^3}{1 + \cdots}$, then $v^5 = u \dfrac{1 - 2u + 4u^2 - 3u^3 + u^4}{1 + 3u + 4u^2 + 2u^3 + u^4}$.

(11) $\dfrac{1}{1+} \dfrac{e^{-2\pi}}{1+} \dfrac{e^{-4\pi}}{1 + \cdots} = \left\{\sqrt{\dfrac{5 + \sqrt{5}}{2}} - \dfrac{\sqrt{5} + 1}{2}\right\} e^{2\pi/5}$.

(12) $\dfrac{1}{1+} \dfrac{e^{-2\pi\sqrt{5}}}{1+} \dfrac{e^{-4\pi\sqrt{5}}}{1+} \cdots = \left[\dfrac{\sqrt{5}}{1 + \sqrt[5]{5^{3/4}\left(\frac{\sqrt{5}-1}{2}\right)^{5/2} - 1}} - \dfrac{\sqrt{5} + 1}{2}\right] e^{2\pi/\sqrt{5}}$.

(13) If $F(k) = 1 + \left(\dfrac{1}{2}\right)^2 k + \left(\dfrac{1 \cdot 3}{2 \cdot 4}\right)^2 k^2 + \cdots$ and $F(1 - k) = \sqrt{210} F(k)$, then
$k = (\sqrt{2} - 1)^4 (2 - \sqrt{3})^2 (\sqrt{7} - \sqrt{6})^4 (8 - 3\sqrt{7})^2 \cdot (\sqrt{10} - 3)^4 (4 - \sqrt{15})^4 (\sqrt{15} - \sqrt{14})^2 (6 - \sqrt{35})^2$.

(14) The coefficient of x^n in $\left(1 - 2x + 2x^4 - 2x^9 + \cdots\right)^{-1}$ is the integer nearest to
$\dfrac{1}{4n}\left(\cosh(\pi\sqrt{n}) - \dfrac{\sinh(\pi\sqrt{n})}{\pi\sqrt{n}}\right)$.

(15) The number of numbers between A and x which are either squares or sums of two squares is
$K \displaystyle\int_A^x \dfrac{dt}{\sqrt{\log t}} + \theta(x)$, where $K = 0.764\ldots$ and $\theta(x)$ is very small compared to the previous integral.

The first question was whether I could recognize anything. I had proved things rather like (7) myself, and seemed vaguely familiar with (8). Actually (8) is classical; it is a formula of Laplace first proved properly by Jacobi; and (9) occurs in a paper published by Rogers in 1907. I thought that, as an expert in definite integrals, I could probably prove (5) and (6), and did so, though with a good deal more trouble than I had expected. On the whole the integral formulas seemed the least impressive.

The series formulas (1)–(4) I found much more intriguing, and it soon became obvious that Ramanujan must possess much more general theorems and was keeping a great deal up his sleeve. The second is a formula of Bauer well known in the theory of Legendre series, but the others are much harder than they look. The theorems required in proving them can all be found now in Bailey's *Cambridge Tract* on hypergeometric functions.

The formulas (10)–(13) are on a different level and obviously both difficult and deep. An expert in elliptic functions can see at once that (13) is derived somehow from the theory of "complex multiplication", but (10)–(12) defeated me completely; I had never seen anything in the least like them before. A single look at them is enough to show that they could only be written down by a mathematician of the highest class. They must be true because, if they were not true, no one would have had the imagination to invent them. Finally (you must remember that I knew nothing whatever about Ramanujan, and had to think of every possibility), the writer must be completely honest, because great mathematicians are commoner than thieves or humbugs of such incredible skill.

The last two formulas stand apart because they are not right and show Ramanujan's limitations, but that does not prevent them from being additional evidence of his extraordinary powers. The function in (14) is a genuine approximation to the coefficient, though not at all so close as Ramanujan imagined, and Ramanujan's false statement was one of the most fruitful he ever made, since it ended by leading us to all our joint work on partitions. Finally (15), though literally "true", is definitely misleading (and Ramanujan was under a real misapprehension). The integral has no advantage, as an approximation, over the simpler function

$$\frac{Kx}{\sqrt{\log x}} \qquad (16)$$

found in 1908 by Landau. Ramanujan was deceived by a false analogy with the problem of the distribution of primes. I must postpone till later what I have to say about Ramanujan's work on this side of the theory of numbers.

It was inevitable that a very large part of Ramanujan's work should prove on examination to have been anticipated. He had been carrying an impossible handicap, a poor and solitary Hindu pitting his brains against the accumulated wisdom of Europe. He had had no real teaching at all; there was no one in India from whom he had anything to learn. He can have seen at the outside three or four books of good quality, all of them English. There had been periods in his life when he had access to the library in Madras, but it was not a very good one; it contained very few French or German books; and in any case Ramanujan did not know a word of either language. I should estimate that about two-thirds of Ramanujan's best Indian work was rediscovery, and comparatively little of it was published in his life-time, though Watson, who has worked systematically through his notebooks, has since disinterred a good deal more.

The great bulk of Ramanujan's published work was done in England. His mind had hardened to some extent, and he never became at all an "orthodox" mathematician, but he could still learn to do new things, and do them extremely well. It was impossible to teach him systematically, but he gradually absorbed new points of view. In particular he learnt what was meant by proof, and his later papers, while in some ways as odd and individual as ever, read like the works of a well-informed mathematician. His methods and his weapons, however, remained essentially the same. One would have thought that such a formalist as Ramanujan would have revelled in Cauchy's Theorem, but he practically never used it, and the most astonishing testimony to his formal genius is that he never seemed to feel the want of it in the least.

It is easy to compile an imposing list of theorems which Ramanujan re-discovered. Such a list naturally cannot be quite sharp, since sometimes he found a part only of a theorem, and sometimes, though he found the whole theorem, he was without the proof which is essential if the theorem is to be properly understood. For example, in the analytic theory of numbers he had, in a sense, discovered a great deal, but he was a very long way from understanding the real difficulties of the subject. And there is some of his work, mostly in the theory of elliptic functions, about which some mystery still remains; it is not possible, after all the work of Watson and Mordell, to draw the line between what he may have picked up somehow and what he must have found for himself.

I will take only cases in which the evidence seems to me tolerably clear.

Here I must admit that I am to blame, since there is a good deal which we should like to know now and which I could have discovered quite easily. I saw Ramanujan almost every day, and could have cleared up most of the obscurity by a little cross-examination. Ramanujan was quite able and willing to give a straight answer to a question, and not in the least disposed to make a mystery of his achievements. I hardly asked him a single question of this kind; I never even asked him whether (as I think he must have done) he had seen Cayley's or Greenhill's *Elliptic Functions*.

I am sorry about this now, but it does not really matter very much, and it was entirely natural. In the first place, I did not know that Ramanujan was going to die. He was not particularly interested in his own history or psychology; he was a mathematician anxious to get on with the job. And after all I too was a mathematician, and a mathematician meeting Ramanujan had more interesting things to think about than historical research. It seemed ridiculous to worry him about how he had found this or that known theorem, when he was showing me half a dozen new ones almost every day.

I do not think that Ramanujan discovered much in the classical theory of numbers, or indeed that he ever knew a great deal. He had no knowledge at all, at any time, of the general theory of arithmetical forms. I doubt whether he knew the law of quadratic reciprocity before he came here. Diophantine equations should have suited him, but he did comparatively little with them, and what he did do was not his best. Thus he gave solutions of Euler's equation

$$x^3 + y^3 + z^3 = w^3, \qquad (17)$$

such as

$$x = 3a^2 + 5ab - 5b^2, \quad y = 4a^2 - 4ab + 6b^2,$$
$$z = 5a^2 - 5ab - 3b^2, \quad w = 6a^2 - 4ab + 4b^2; \qquad (18)$$

and

$$x = m^7 - 3m^4(1+p) + m(2 + 6p + 3p^2),$$
$$y = 2m^6 - 3m^3(1+2p) + 1 + 3p + 3p^2,$$
$$z = m^6 - 1 - 3p - 3p^2,$$
$$w = m^7 - 3m^4 p + m(3p^2 - 1); \qquad (19)$$

but neither of these is the general solution.

He re-discovered the famous theorem of von Staudt about the Bernoullian numbers:

$$(-1)^n B_n = G_n + \frac{1}{2} + \frac{1}{p} + \frac{1}{q} + \cdots + \frac{1}{r}, \qquad (20)$$

where $p, q, \ldots$ are those odd primes such that $p-1, q-1, \ldots$ are divisors of $2n$, and G_n is an integer. In what sense he had proved it it is difficult to say, since he found it at a time of his life when he had hardly formed any definite concept of proof. As Littlewood says:

the clear-cut idea of what is meant by a proof, nowadays so familiar as to be taken for granted, he perhaps did not possess at all; if a significant piece of reasoning occurred somewhere, and the total mixture of evidence and intuition gave him certainty, he looked no further.

I shall have something to say later about this question of proof, but I postpone it to another context in which it is much more important. In this case there is nothing in the proof that was not obviously within Ramanujan's powers.

There is a considerable chapter of the theory of numbers, in particular the theory of the representation of integers by sums of squares, which is closely bound up with the theory of elliptic functions. Thus the number of representations of n by two squares is

$$r(n) = 4\{d_1(n) - d_3(n)\}, \qquad (21)$$

where $d_1(n)$ is the number of divisors of n of the form $4k+1$ and $d_3(n)$ the number of divisors of the form $4k+3$. Jacobi gave similar formulas for 4, 6 and 8 squares. Ramanujan found all these, and much more of the same kind.

He also found Gauss's theorem that n is the sum of 3 squares except when it is of the form

$$4^a(8k+7), \qquad (22)$$

but I do not attach much importance to this. The theorem is quite easy to guess and difficult to prove. All known proofs depend upon the general theory of ternary forms, of which Ramanujan knew nothing, and I agree with Professor Dickson in thinking it very unlikely that he possessed one. In any case he knew nothing about the number of representations.

Ramanujan, then, before he came to England, had added comparatively little to the theory of numbers; but no one can understand him who does not understand his passion for numbers in themselves. I wrote before:

He could remember the idiosyncrasies of numbers in an almost uncanny way. It was Littlewood who said that every positive integer was one of Ramanujan's personal friends. I remember going to see him once when he was lying ill in Putney. I had ridden in taxi-cab No. 1729, and remarked that the number seemed to me rather a dull one, and that I hoped that it was not an unfavorable omen. "No," he replied, "it is a very interesting number; it is the smallest number expressible as a sum of two cubes in two different ways" ($1729 = 12^3 + 1^3 = 10^3 + 9^3$). I asked him, naturally, whether he could tell me the solution of the corresponding problem for fourth powers; and he replied, after a moment's thought, that he knew no obvious example, and supposed that the first such number must be very large. (The smallest known is Euler's example

$$635318657 = 158^4 + 59^4 = 134^4 + 133^4.)$$

In algebra, Ramanujan's main work was concerned with hypergeometric series and continued fractions (I use the word algebra, of course, in its old-fashioned sense). These subjects suited him exactly, and here he was unquestionably one of the great masters. There are three now famous identities, the *Dougall-Ramanujan identity*

$$\sum_{n=0}^{\infty}(-1)^n (s+2n) \frac{s^{(n)}}{1^{(n)}} \frac{(x+y+z+u+2s+1)^{(n)}}{(x+y+z+u+s)_{(n)}}$$
$$\times \prod_{x,y,z,u} \frac{x_{(n)}}{(x+s+1)^{(n)}}$$
$$= \frac{s}{\Gamma(s+1)\Gamma(x+y+z+u+s+1)} \qquad (23)$$
$$\times \prod_{x,y,z,u} \frac{\Gamma(x+s+1)\Gamma(y+z+u+s+1)}{\Gamma(z+u+s+1)},$$

where

$$a^{(n)} = a(a+1)\cdots(a+n-1),$$
$$a_{(n)} = a(a-1)\cdots(a-n+1),$$

and the *Rogers-Ramanujan identities*

$$1 + \frac{q}{1-q} + \frac{q^4}{(1-q)(1-q^2)}$$
$$+ \frac{q^9}{(1-q)(1-q^2)(1-q^3)} + \cdots$$
$$= \frac{1}{(1-q)(1-q^6)\cdots(1-q^4)(1-q^9)\cdots},$$

$$1 + \frac{q^2}{1-q} + \frac{q^6}{(1-q)(1-q^2)}$$
$$+ \frac{q^{12}}{(1-q)(1-q^2)(1-q^3)} + \cdots \qquad (24)$$
$$= \frac{1}{(1-q^2)(1-q^7)\cdots(1-q^3)(1-q^8)\cdots},$$

in which he had been anticipated by British mathematicians, and about which I shall speak in other lectures. As regards hypergeometric series one may say, roughly, that he re-discovered the formal theory, set out in Bailey's tract, as it was known up to 1920. There is something about it in Carr, and more in Chrystal's *Algebra*, and no doubt he got his start from that. The four formulas (1)–(4) are highly specialized examples of this work.

His masterpiece in continued fractions was his work on

$$\frac{1}{1+} \frac{x}{1+} \frac{x^2}{1+\cdots}, \qquad (25)$$

which includes the theorems (10)–(12). The theory of this fraction depends upon the Rogers-Ramanujan identities, in which he had been anticipated by Rogers, but he had gone beyond Rogers in other ways and the theorems which I have quoted are his own. He had many other very general and very beautiful formulas, of which formulas like Laguerre's

$$\frac{(x+1)^n - (x-1)^n}{(x+1)^n + (x-1)^n} = \frac{n}{x+} \frac{n^2-1}{3x+} \frac{n^2-2^2}{5x+\cdots} \qquad (26)$$

are extremely special cases. Watson [6] has recently published a proof of the most imposing of them.

It is perhaps in his work in these fields that Ramanujan shows at his very best. I wrote before:

It was his insight into algebraical formulae, transformation of infinite series, and so forth, that was most amazing. On this side most certainly I have never met his equal, and I can compare him only with Euler or Jacobi. He worked, far more than the majority of modern mathematicians, by induction from numerical examples; all his congruence properties of partitions, for example, were discovered in this way. But with his memory, his patience, and his power of calculation he combined a power of generalization, a feeling for form, and a capacity for rapid modification of his hypotheses, that were often really startling, and made him, in his own peculiar field, without a rival in his day.

I do not think now that this extremely strong language is extravagant. It is possible that the great days of formulas are finished, and that Ramanujan ought to have been born 100 years ago; but he was by far the greatest formalist of his time. There have been a good many more important, and I suppose one must say greater, mathematicians than Ramanujan during the last 50 years, but not one who could stand up to him on his own ground. Playing the game of which he knew the rules, he could give any mathematician in the world fifteen.

In analysis proper Ramanujan's work is inevitably less impressive, since he knew no theory of functions, and you cannot do real analysis without it, and since the formal side of the integral calculus, which was all that he could learn from Carr or any other book, has been worked over so repeatedly and so intensively. Still, Ramanujan rediscovered an astonishing number of the most beautiful analytic identities. Thus the functional equation for the Riemann zeta-function

$$\zeta(s) = \sum_{n=1}^{\infty} \frac{1}{n^s},$$

namely

$$\zeta(1-s) = 2(2\pi)^{-s} \cos\left(\frac{1}{2}s\pi\right) \Gamma(s)\zeta(s), \quad (27)$$

stands (in an almost unrecognizable notation) in the notebooks. So does Poisson's summation formula

$$\alpha^{1/2} \left\{ \frac{1}{2}\phi(0) + \phi(\alpha) + \phi(2\alpha) + \cdots \right\}$$
$$= \beta^{1/2} \left\{ \frac{1}{2}\psi(0) + \psi(\beta) + \psi(2\beta) + \cdots \right\}, \quad (28)$$

where

$$\alpha\beta = 2\pi \text{ and } \psi(x) = \sqrt{\frac{2}{\pi}} \int_0^\infty \phi(t) \cos(xt) dt.$$

So also does Abel's functional equation ([1], [4]):

$$L(x) + L(y) + L(xy) + L\left(\frac{x(1-y)}{1-xy}\right)$$
$$+ L\left(\frac{y(1-x)}{1-xy}\right) = 3L(1) \quad (29)$$

for

$$L(x) = \frac{x}{1^2} + \frac{x^2}{2^2} + \frac{x^3}{3^2} + \cdots.$$

He had most of the formal ideas which underlie the recent work of Watson and of Titchmarsh and myself on *Fourier kernels* and *reciprocal functions*; and he could of course evaluate any evaluable definite integral. There is one particularly interesting formula, viz.

$$\int_0^\infty x^{s-1} \left\{ \phi(0) - x\phi(1) + x^2\phi(2) - \cdots \right\} dx$$
$$= \frac{\pi\phi(-s)}{\sin s\pi}, \quad (30)$$

of which he was especially fond and made continual use. This is really an *interpolation formula*, which enables us to say, for example, that, under certain conditions, a function which vanishes for all positive integral values of its argument must vanish identically. I have never seen this formula stated explicitly by any one else, though it is closely connected with the work of Mellin and others.

I have left till last the two most intriguing sides of Ramanujan's early work, his work on elliptic functions and in the analytic theory of numbers. The first is probably too specialized and intricate for anyone but an expert to understand, and I shall say nothing about it now. The second subject is still more difficult (as anyone who has read Landau's book on primes or Ingham's tract will know), but anyone can understand roughly what the problems of the subject are, and any decent mathematician can understand roughly why they defeated Ramanujan. For this was Ramanujan's one real failure; he showed, as always, astonishing imaginative power, but he proved next to nothing, and a great deal even of what he imagined was false.

Here I am obliged to interpolate some remarks on a very difficult subject, *proof* and its importance in mathematics. All physicists, and a good many quite respectable mathematicians, are contemptuous about proof. I have heard Professor Eddington, for example, maintain that proof, as pure mathematicians understand it, is really quite uninteresting and unimportant, and that no one who is really certain that he has found something good should waste his time looking for a proof. It is true that Eddington is inconsistent, and has sometimes even descended to proof himself. It is not enough for him to have direct knowledge that there are exactly $136 \cdot 2^{256}$ protons in the universe; he cannot resist the temptation of proving it; and I cannot help thinking that the proof, whatever it may be worth, gives him a certain amount of intellectual satisfaction. His apology would no doubt be that *proof* means something quite different for him from what it means for a pure mathematician, and in any case we need not take him too literally. But the opinion which I have attributed to him, and with which I am sure that almost all physicists agree at the bottom of their hearts, is one to which a mathematician ought to have some reply.

I am not going to get entangled in the analysis of a particularly prickly concept, but I think that there are a few points about proof where nearly all mathematicians are agreed. In the first place, even if we do not understand exactly what proof is, we can, in ordinary analysis at any rate, recognise a proof when we see one. Secondly, there are two different motives in any presentation of a proof. The first motive is simply to secure conviction. The second is to exhibit the conclusion as the climax of a conventional pattern of propositions, a sequence of propositions whose truth is admitted and which are arranged in accordance with rules. These are the two ideals, and experience shows that, except in the simplest mathematics, we can hardly ever satisfy the first ideal without also satisfying the second. We may be able to recognise directly that 5, or even 17, is prime, but nobody can convince himself that

$$2^{127} - 1$$

is prime except by studying a proof. No one has ever had an imagination so vivid and comprehensive as that.

A mathematician usually discovers a theorem by an effort of intuition; the conclusion strikes him as plausible, and he sets to work to manufacture a proof. Sometimes this is a matter of routine, and any well-trained professional could supply what is wanted, but more often imagination is a very unreliable guide. In particular this is so in the analytic theory of numbers, where even Ramanujan's imagination led him very seriously astray.

There is a striking example, which I have very often quoted, of a false conjecture which seems to have been endorsed even by Gauss and which took about 100 years to refute. The central problem of the analytic theory of numbers is that of the distribution of the primes. The number $\pi(x)$ of primes less than a large number x is approximately

$$\frac{x}{\log x}; \qquad (31)$$

this is the *Prime Number Theorem*, which had been conjectured for a very long time, but was never established properly until Hadamard and de la Vallée Poussin proved it in 1896. The approximation errs by defect, and a much better one is

$$\mathrm{Li}(x) = \int_2^x \frac{dt}{\log t} \qquad (32)$$

In some ways a still better one is

$$\mathrm{Li}(x) - \frac{1}{2}\mathrm{Li}(x^{1/2}) - \frac{1}{3}\mathrm{Li}(x^{1/3}) - \frac{1}{5}\mathrm{Li}(x^{1/5})$$
$$- \frac{1}{6}\mathrm{Li}(x^{1/6}) - \frac{1}{7}\mathrm{Li}(x^{1/7}) + \cdots \qquad (33)$$

(we need not trouble now about the law of formation of the series). It is extremely natural to infer that

$$\pi(x) < \mathrm{Li}(x), \qquad (34)$$

at any rate for large x, and Gauss and other mathematicians commented on the high probability of this conjecture. The conjecture is not only plausible but is supported by *all* the evidence of the facts. The primes are known up to $10,000,000$, and their number at intervals up to $1,000,000,000$, and (34) is true for every value of x for which data exist.

In 1912 Littlewood proved that the conjecture is false, and that there are an infinity of values of x for which the sign of inequality in (34) must be reversed. In particular, there is a number X such that (34) is false for some x less than X. Littlewood proved the existence of X, but his method did not give any particular value, and it is only very recently that an admissible value, viz.

$$X = 10^{10^{10^{34}}}$$

was found by Skewes [5]. I think that this is the largest number which has ever served any definite purpose in mathematics.

The number of protons in the universe is about 10^{80}. The number of possible games of chess is much bigger, perhaps

$$10^{10^{50}}$$

(in any case a second-order exponential). If the universe were the chessboard, the protons the chessmen, and any interchange in the position of two protons a move, then the number of possible games would be something like the Skewes number. However much the number may be reduced by refinements on Skewes's argument, it does not seem at all likely that we shall ever know a single instance of the truth of Littlewood's theorem.

This is an example in which the truth has defeated not only all the evidence of the facts and of common sense but even a mathematical imagination so powerful and profound as that of Gauss; but of course it is taken from the most difficult parts of the theory. No part of the theory of primes is really easy, but up to a point simple arguments, although they will prove very little, do not actually mislead us. For example, there are simple arguments which might lead any good mathematician to the conclusion

$$\pi(x) \sim \frac{x}{\log x} \tag{35}$$

of the *Prime Number Theorem*, or, what is the same thing, to the conclusion that

$$p_n \sim n \log n, \tag{36}$$

where p_n is the nth prime number (here $f \sim g$ means that the ratio f/g tends to unity).

In the first place, we may start from Euler's identity

$$\prod_p \frac{1}{1-p^{-s}} = \frac{1}{(1-2^{-s})(1-3^{-s})(1-5^{-s})\cdots}$$
$$= \frac{1}{1^s} + \frac{1}{2^s} + \frac{1}{3^s} + \cdots$$
$$= \sum_n \frac{1}{n^s}. \tag{37}$$

This is true for $s > 1$, but both series and product become infinite for $s = 1$. It is natural to argue that, when $s = 1$, the series and the product should diverge in the same sort of way. Also,

$$\log \prod \frac{1}{1-p^{-s}} = \sum \log \frac{1}{1-p^{-s}}$$
$$= \sum \frac{1}{p^s} + \sum \left(\frac{1}{2p^{2s}} + \frac{1}{3p^{3a}} + \cdots \right), \tag{38}$$

and the last series remains finite for $s = 1$. It is natural to infer that $\sum \frac{1}{p}$ diverges like

$$\log \left(\sum \frac{1}{n} \right),$$

or, more precisely, that

$$\sum_{p \leq x} \frac{1}{p} \sim \log \left(\sum_{n \leq x} \frac{1}{n} \right) \sim \log \log x \tag{39}$$

for large x. Since also

$$\sum_{n \leq x} \frac{1}{n \log n} \sim \log \log x,$$

formula (39) indicates that p_n is about $n \log n$.

There is a slightly more sophisticated argument which is really simpler. It is easy to see that the highest power of a prime p which divides $x!$ is

$$\left[\frac{x}{p} \right] + \left[\frac{x}{p^2} \right] + \left[\frac{x}{p^3} \right] + \cdots,$$

where $[y]$ denotes the integral part of y. Hence

$$x! = \prod_{p \leq x} p^{[x/p]+[x/p^2]+\cdots} \quad \text{and}$$

$$\log x! = \sum_{p \leq x} \left(\left[\frac{x}{p} \right] + \left[\frac{x}{p^2} \right] + \cdots \right) \log p. \tag{40}$$

The left-hand side of (40) is practically $x \log x$, by Stirling's Theorem. As regards the right-hand, one may argue: squares, cubes, ... of primes are comparatively rare, and the terms involving them should be unimportant, and it should also make comparatively little difference if we replace $[x/p]$ by x/p. We thus infer that

$$x \sum_{p \leq x} \frac{\log p}{p} \sim x \log x, \quad \sum_{p \leq x} \frac{\log p}{p} \sim \log x,$$

and this again just fits the view that p_n is approximately $n \log n$.

This is broadly the argument used, naturally in a less naive form, by Tchebychef, who was the first to make substantial progress in the theory of primes, and I imagine that Ramanujan began by arguing in the same sort of way, though there is nothing in the note-books to show. All that is plain is that Ramanujan found the form of the Prime Number Theorem for himself. This was a considerable achievement; for the men who had found the form of the theorem before him, like Legendre, Gauss, and Dirichlet, had all been very great mathematicians; and Ramanujan found other formulas which lie still further below the surface. Perhaps the best instance is (15). The integral is better replaced by the simpler function (16), but what Ramanujan says is correct as it stands and was proved by Landau in 1908; and there is nothing obvious to suggest its truth.

The fact remains that hardly any of Ramanujan's work in this field had any permanent value. The analytic theory of numbers is one of those exceptional branches of mathematics in which proof really is everything and nothing short of absolute rigour counts. The achievement of the mathematicians who found the Prime Number Theorem was quite a small thing compared with that of those who found the proof. It is not merely that in this theory (as Littlewood's theorem shows) you can never be sure of the facts without the proof, though this is important enough. The whole history of the Prime Number Theorem, and the other big theorems of the subject, shows that you cannot reach any real understanding of the structure and meaning of the theory, or have any sound instincts to guide you in further research, until you have mastered the proofs. It is comparatively easy to make clever guesses; indeed there are theorems, like *Goldbach's Theorem* (that any even number greater than 2 is a sum of two primes), which have never been proved and which any fool could have guessed.

The theory of primes depends upon the properties of Riemann's function $\zeta(s)$, considered as an analytic function of the complex variable s, and in particular on the distribution of its zeros; and Ramanujan knew nothing at all about the theory of analytic functions. I wrote before:

> Ramanujan's theory of primes was vitiated by his ignorance of the theory of functions of a complex variable. It was (so to say) what the theory might be if the zeta-function had no complex zeros. His method depended upon a wholesale use of divergent series... That his proofs should have been invalid was only to be expected. But the mistakes went deeper than that, and many of the actual results were false. He had obtained the dominant terms of the classical formulae, although by invalid methods; but none of them are such close approximations as he supposed. This may be said to have been Ramanujan's one great failure...

and if I had stopped there I should have had nothing to add, but I allowed myself again to be led away by sentimentalism. I went on to argue that "his failure was more wonderful than any of his triumphs", and that is an absurd exaggeration. It is no use trying to pretend that failure is something else. This much perhaps we may say, that his failure is one which, on the balance, should increase and not diminish our admiration for his gifts, since it gives us additional, and surprising, evidence of his imagination and versatility.

But the reputation of a mathematician cannot be made by failures or by re-discoveries; it must rest primarily, and rightly, on actual and original achievement. I have to justify Ramanujan on this ground, and that I hope to do in my later lectures.

Note: This lecture was originally given at the Harvard Tercentenary Conference of Arts and Sciences, on August 31, 1936.

References

1. N. H. Abel, *Oeuvres*, Vol. 2, 1935, p. 7.
2. G. S. Carr, *A Synopsis of Elementary Results in Pure and Applied Mathematics*, 2 vols., C. F. Hodgson and Sons, 1880, 1886.
3. S. Ramanujan, *Collected Papers* (ed. G. H. Hardy, P. V. Seshu Aiyar and B. M. Wilson), Cambridge Univ. Press, 1927.
4. L. J. Rogers, *Papers*, p. 337.
5. S. Skewes, On the difference $\pi(x) - \text{Li}(x)$, *J. London Math. Soc.* **8** (1933), 277–283.
6. G. N. Watson, Ramanujan's continued fraction, *Proc. Cambridge Phil. Soc.* **31** (1935), 7–17.

Emmy Noether

CLARK H. KIMBERLING

American Mathematical Monthly **79** (1972), 136–149

The past two years have seen a surge of interest in Emmy Noether and her mathematics. Along with Auguste Dick's biography of her, listed below, Constance Reid's biography, *Hilbert*, frequently mentions Emmy Noether. New mathematics books, such as *Introduction to the Calculus of Variations* by Hans Sagan, and *Commutative Rings* by Irving Kaplansky, are spreading anew her methods, and the adjective "noetherian" abounds in titles to papers in mathematics research journals. The State University of New York at Buffalo has just set up a George William Hill–Emmy Noether Fellowship. A high school textbook, *Modern Introductory Analysis*, by Dolciani, Donnelly, Jurgensen, and Wooten, devotes a page to Emmy Noether. And one finds such remarks in periodical literature as "The woman mathematician today is better off than Emmy Noether, who taught without pay. But . . . ".

Despite all this recent interest, it is difficult to find much about Emmy Noether in mathematics history books. Although she was dubbed "der Noether" by P. S. Alexandroff, and that name with its masculine German article has stuck, she is given only a footnote in E. T. Bell's *Men of Mathematics* and hardly more in comparable books. In fact, little else can be found about her than three obituary addresses and the biography published just last year:

1. Emmy Noether, by Hermann Weyl (memorial address at Bryn Mawr College, April 26, 1935), *Scripta Mathematica* **3** (1935), 201–220.

2. Nachruf auf Emmy Noether, by B. L. van der Waerden (in German), *Mathematische Annalen* **111** (1935), 469–476.

3. Emmy Noether, by P. S. Alexandroff, address to the Moscow Mathematical Society, Sept. 5, 1935.

4. *Emmy Noether*, by Auguste Dick (in German), Birkhäuser Verlag, 1970.

Since (3) was not published I shall draw from it more than from the others.

1 Her passing

A note in the files of the *Bryn Mawr Alumnae Bulletin* reads, "The above was inspired, if not written, by Dr. Hermann Weyl, eminent German mathematician. Mr. Einstein had never met Miss Noether." The "above" is the following, as it appeared in *The New York Times*, May 3, 1935:

> The efforts of most human beings are consumed in the struggle for their daily bread, but most of those who are, either through fortune or some special gift, relieved of this struggle are largely absorbed in further improving their worldly lot. Beneath the effort directed toward the accumulation of worldly goods lies all too frequently the illusion that this is the most substantial and desirable end to be achieved; but there is, fortunately, a minority composed of those who recognize early in their lives that the most beautiful and satisfying experiences open to humankind are not derived from the outside, but are bound up with the development of the individual's own feeling, thinking and acting. The genuine artists, investigators and thinkers have always been persons of this kind. However inconspicuously the life of these individuals runs its course, none the less the fruits of their endeavors are the most valuable contributions which one generation can make to its successors.
>
> Within the past few days a distinguished mathematician, Professor Emmy Noether, formerly connected with the University of Göttingen and for the past two years at Bryn Mawr College, died in her fifty-third year. In the judgment of the

most competent living mathematicians, Fräulein Noether was the most significant creative mathematical genius thus far produced since the higher education of women began. In the realm of algebra, in which the most gifted mathematicians have been busy for centuries, she discovered methods which have proved of enormous importance in the development of the present-day younger generation of mathematicians. Pure mathematics is, in its way, the poetry of logical ideas. One seeks the most general ideas of operation which will bring together in simple, logical and unified form the largest possible circle of formal relationships. In this effort toward logical beauty spiritual formulae are discovered necessary for the deeper penetration into the laws of nature.

Born in a Jewish family distinguished for the love of learning, Emmy Noether, who, in spite of the efforts of the great Göttingen mathematician, Hilbert, never reached the academic standing due her in her own country, none the less surrounded herself with a group of students and investigators at Göttingen, who have already become distinguished as teachers and investigators. Her unselfish, significant work over a period of many years was rewarded by new rulers of Germany with a dismissal, which cost her the means of maintaining her simple life and the opportunity to carry on her mathematical studies. Farsighted friends of science in this country were fortunately able to make such arrangements at Bryn Mawr College and at Princeton that she found in America up to the day of her death not only colleagues who esteemed her friendship but grateful pupils whose enthusiasm made her last years the happiest and perhaps the most fruitful of her entire career.

ALBERT EINSTEIN
Princeton University, May 1, 1935.

Freeman J. Dyson of The Institute for Advanced Study has written me concerning the statement above that the letter written by Einstein to *The New York Times* was "inspired, if not written, by Dr. Hermann Weyl." Professor Dyson discussed this statement with Miss Dukas, Einstein's former secretary, who is presently in charge of the Einstein archive at The Institute for Advanced Study.

I quote from Professor Dyson's letter:

> Miss Dukas has the original German draft of the letter. She confirms that this was written by Einstein himself at the request of Weyl. She does not remember whether Weyl or somebody else afterwards translated it into English.
>
> Miss Dukes also has a letter from Einstein to Hilbert dated May 24, 1918, including the following passage: "Gestern erhielt ich von Fr. Noether eine sehr interessante Arbeit ueber Invariantenbildung. Es imponiert mir, dass man diese Dinge von so allgemeinem Standpunkt übersehen kann. Es hätte den Goettinger Felgrauen nichts geschadet, wenn sie zu Frl. Noether in die Schule geschickt worden waeren. Sie scheint ihr Handwerk zu verstehen!"

Here "Feldgrauen" is slang for "Warriors". From the letter you can see that, while it may be true that Einstein and Emmy Noether never met (Miss Dukes is not sure about this), Einstein certainly knew her work well and understood its importance early and at first hand.

2 Early years

We are indebted to Dr. Auguste Dick of Vienna for much of what we know today about Emmy Noether's early life and her forebears. Most of the information in this present section may be found in Dr. Dick's biography.

Among those affected by an 1809 Tolerance Edict in the German state of Baden was one Elias Samuel, who as the head of a Jewish household was required to change his name and the names of his nine children. He chose the surname *Nöther*, and one of his sons, Hertz, he renamed Hermann. At the age of eighteen, Hermann left his birthplace, Bruchsal, and went to Mannheim to study theology. However, in 1837, he and his older brother Joseph founded an iron-wholesaling firm. The firm lasted for nearly a century, when it fell to anti-Jewish forces.

Born to Hermann and Amalia Nöther were five children, and the third, in 1844, was named Max. During his fourteenth year, Max suffered from infantile paralysis and was somewhat handicapped for the rest of his life. Nevertheless, he became a mathematician of great stature, arriving at the University of Erlangen as a professor in 1875, where he remained until his death in 1921. In 1880, Max married Ida Amalia Kaufmann. Although their marriage certificate bears the name *Nöther*, Max and all his children used the name *Noether* instead.

Amalie Emmy Noether was born on March 23, 1882 in the South German town of Erlangen. She was the first child of Max and Ida Noether and soon had brothers, Albert, born in 1883, and Fritz, in 1884. Still another brother was born in 1889. The family rented a large flat in the first story of an apartment house at Nürnberger Strasse 30–32. Another tenant there for many years was Professor Eilhard Wiedemann, remembered as an Islamist as well as a physicist. The Noether family occupied their flat for about forty-five years.

As a child, Emmy was acutely near-sighted, not outwardly attractive, and not exceptional in any way. Her teachers and classmates remember that she favored the study of language and that little she did reflected teachings of the Jewish religion. Like many other girls, she took clavier lessons and dancing lessons, but apparently with little fervor.

Three years after leaving her "high school", the Städtischen Höheren Töchterschule in Erlangen, Emmy took tests for prospective schoolteachers of French and English. These tests were given in Ansbach in April, 1900. No sooner had she passed these and thus qualified as a language teacher than she became interested in university studies.

Among nearly a thousand students at the University of Erlangen in the winter of 1900, Emmy Noether was one of two women. As a rule, female students could not be registered in the usual sense, and they could take an examination for course credit only upon consent of the professor teaching the course. This consent was often withheld. Nevertheless, whether passing through the prerequisite courses in the usual manner or not, a woman could eventually take an examination for a university certificate.

Among Emmy's early professors at Erlangen, one was a historian and another, Julius Pirson, a professor of romance languages. Between 1900 and 1902, Emmy must have chosen to pursue mathematics rather than languages, since during that time she must have been preparing for the final university examination, which she passed in July, 1903. This examination was given in Nürnberg at the royal Realgymnasium, now the Willstätter-Gymnasium. Quite possibly it was administered by the mathematician Aurel Voß, from whom Emmy's brother Fritz later received his doctorate.

In the winter of 1903 Emmy attended classes at the University of Göttingen. There she heard such eminent mathematicians as Hermann Minkowski, Otto Blumenthal, Felix Klein, and David Hilbert. After just one semester, however, she returned to Erlangen, for it had become possible for women to be matriculated and tested in the manner formerly reserved for men.

In October of 1904, Emmy Noether was officially registered as a student at the University of Erlangen. As a member of Section II of the Philosophical Faculty, she studied only mathematics. On December 13, 1907, she passed her doctoral oral examination, and in July of 1908 her dissertation was registered with the Erlanger Universitätsschriften as Number 202.

3 Excerpts from Weyl's address

Concerning the dissertation and the professor, Paul Gordan, under whom Emmy wrote it, Weyl spoke as follows in his memorial address:

> Side by side with [Max] Noether acted in Erlangen as a mathematician the closely befriended Gordan, an offspring of Clebsch's school like Noether himself. Gordan had come to Erlangen shortly before, in 1874, and he, too, remained associated with that university until his death in 1912. Emmy wrote her doctor's thesis under him in 1907: *On complete systems of invariants for ternary biquadratic forms*; it is entirely in line with the Gordan spirit and his problems. The *Mathematische Annalen* contains a detailed obituary of Gordan and an analysis of his work, written by Max Noether with Emmy's collaboration. Besides her father, Gordan must have been wellnigh one of the most familiar figures in Emmy's early life, first as a friend of the house, later as a mathematician also; she kept a profound reverence for him though her own mathematical taste soon developed in quite a different direction. I remember that his picture decorated the wall of her study in Göttingen. These two men, the father and Gordan, determined the atmosphere in which she grew up. Therefore I shall venture to describe them with a few strokes.
>
> Riemann had developed the theory of algebraic functions of one variable and their integrals, the so-called Abelian integrals, by a function-theoretic transcendental method resting on the minimum principle of potential theory which he named after Dirichlet, and had uncovered the purely topological foundations of the manifold function-theoretic relations governing this domain. (Stringent proof of Dirichlet's principle which seemed so evident from the physicist's standpoint was only given about fifty years later by Hilbert.) There remained the task of replacing and securing his transcendental existential proofs by the explicit algebraic construction starting with the equation of the algebraic curve. Weierstrass solved this problem (in his

lectures published in detail only later) in his own half function-theoretic, half algebraic way, but Clebsch had introduced Riemann's ideas into the geometric theory of algebraic curves and Noether became, after Clebsch had passed away young, his executor in this matter: he succeeded in erecting the whole structure of the algebraic geometry of curves on the basis of the so-called Noether residual theorem. This line of research was taken up later on, mainly in Italy; the vein Noether struck is still a profusely gushing spring of investigations; among us, men like Lefschetz and Zariski bear witness thereto. Later on there arose, beside Riemann's transcendental and Noether's algebraic-geometric method, an arithmetical theory of algebraic functions due to Dedekind and Weber on the one side, to Hensel and Landsberg on the other. Emmy Noether stood closer to this trend of thought. A brief report on the arithmetical theory of algebraic functions that parallels the corresponding notions in the competing theories was published by her in 1920 in the *Jahresberichte der Deutschen Mathematikervereinigung*. She thus supplemented the well-known report by Brill and her father on the algebraic-geometric theory that had appeared in 1894 in one of the first volumes of the *Jahresberichte*. Noether's residual theorem was later fitted by Emmy into her general theory of ideals in arbitrary rings. This scientific kinship of father and daughter—who became in a certain sense his successor in algebra, but stands beside him independent in her fundamental attitude and in her problems—is something extremely beautiful and gratifying. The father was—such is the impression I gather from his papers and even more from the many obituary biographies he wrote for the *Mathematische Annalen*—a very intelligent, warm-hearted harmonious man of many-sided interests and sterling education.

Gordan was of a different stamp. A queer fellow, impulsive and one-sided. A great walker and talker—he liked that kind of walk to which frequent stops at a beer-garden or a cafe belong. Either with friends, and then accompanying his discussions with violent gesticulations, completely irrespective of his surroundings; or alone, and then murmuring to himself and pondering over mathematical problems; or if in an idler mood, carrying out long numerical calculations by heart. There always remained something of the eternal "Bursche" of the 1848 type about him—an air of dressing gown, beer and tobacco, relieved however by a keen sense of humor and a strong dash of wit. When he had to listen to others, in classrooms or at meetings, he was always half asleep. As a mathematician not of Noether's rank, and of an essentially different kind, Noether himself concludes his characterization of him with the short sentence: "Er war ein Algorithmiker." His strength rested on the invention and calculative execution of formal processes. There exist papers of his where twenty pages of formulas are not interrupted by a single text word; it is told that in all his papers he himself wrote the formulas only, the text being added by his friends. Noether says of him: "The formula always and everywhere was the indispensable support for the formation of his thoughts, his conclusions and his mode of expression... In his lectures he carefully avoided any fundamental definition of conceptual kind, even that of the limit."

He, too, had belonged to Clebsch's most intimate collaborators, had written with Clebsch their book on Abelian integrals; he later shifted over to the theory of invariants following his formal talent; here he added considerably to the development of the so-called *symbolic method*, and he finally succeeded in proving by means of this computative method of explicit construction the finiteness of a rational integral basis for binary invariants. Years later Hilbert demonstrated the theorem much more generally for an arbitrary number of variables—by an entirely new approach, the characteristic Hilbertian species of methods, putting aside the whole apparatus of symbolic treatment and attacking the thing itself as directly as possible. *Ex ungue leonem*— the young lion Hilbert showed his claws. It was, however, at first only an existential proof providing for no actual, finite algebraic construction. Hence Gordan's characteristic exclamation: "This is not mathematics, but theology!" What then would he have said about his former pupil Emmy Noether's later "theology", that abhorred all calculation and operated in a much thinner air of abstraction than Hilbert ever dared!

It is queer enough that a formalist like Gordan was the mathematician from whom her mathematical orbit set out; a greater contrast is hardly imaginable than between her first paper, the dissertation, and her works of maturity; for the former is an extreme example of formal computations and the latter constitute an extreme and grandiose example of conceptual axiomatic thinking in mathematics. Her thesis ends with a table of the complete system of covariant forms for a given ternary quartic consisting of not less than 331 forms in symbolic representation. It is an awe-inspiring piece of work; but today I am afraid we should be inclined to rank it among those achievements with regard to which Gordan himself once said when asked about the use of the theory of invariants: "Oh, it is very useful indeed; one can write many theses about it."

In 1910 Gordan retired, soon to be replaced by Ernst Fischer. In Weyl's judgment Fischer had a more penetrating influence on Emmy Noether's work than Gordan did. Weyl wrote as follows:

> Under his direction the transition from Gordan's formal standpoint to the Hilbert method of approach was accomplished. She refers in her papers at this time again and again to conversations with Fischer. This epoch extends until about 1919. The main interest is concentrated on finite rational and integral bases; the proof of finiteness is given by her for the invariants of a finite group (without using Hilbert's general basis theorem for ideals), for invariants with restriction to integral coefficients, and finally she attacks the same question along with the question of a minimum basis consisting of independent elements for fields of rational functions.

4 Her contribution to physics

In 1916, Emmy Noether left Erlangen and went to the University of Göttingen. At that time Hilbert was working on the general theory of relativity and Emmy was especially welcome because of her knowledge of the theory of invariants.

Weyl described her major contribution to two important aspects of relativity as

> the genuine and universal mathematical formulation: first, the reduction of the problem of differential invariants to a purely algebraic one by use of 'normal coordinates'; second, the identities between the left sides of Euler's equations of a problem of variation which occur when the (multiple) integral is invariant with respect to a group of transformations involving arbitrary functions (identities that contain the conservation theorem of energy and momentum in the case of invariance with respect to arbitrary transformations of the four world coordinates).

During my own inquiries about Emmy Noether, it was once hinted that "young physicists are using her theories," and I was eventually referred to Professor Eugene Wigner (1963 Noble Prize in Physics), who wrote,

> We physicists pay lip service to the great accomplishments of Emmy Noether, but we do not really use her work. Her contribution to physics that is most often quoted arose from a suggestion of Felix Klein. It concerns the conservation laws of physics, which she derived in a way which was at that time novel and should have excited physicists more than it did. However, most physicists know little else about her, even though many of us who have a marginal interest in mathematics have read much else by and about her.

Professor Peter G. Bergmann of Syracuse University gave the following account of Emmy Noether's influence in physics:

> Noether's Theorem, so-called, forms one of the corner stones of work in general relativity as well as in certain aspects of elementary particle physics. The idea is, briefly, that to every invariance or symmetry property of the laws of nature (or of a proposed theory) there corresponds a conservation law, and *vice versa*. Accordingly, if a physical quantity is known to satisfy a conservation law (known as a "good quantum number" in quantum physics), the theorist attempts to construct a theory with appropriate symmetry properties. Conversely, if a theory is known to possess certain symmetries, then this fact alone entails the existence of certain integrals of the dynamical equations.
>
> General relativity is characterized by the principle of general covariance, according to which the laws of nature are invariant with respect to arbitrary curvilinear coordinate transformations that satisfy minimal conditions of continuity and differentiability. A discussion of the consequences in terms of Noether's theorem (whether explicitly quoted as such or not) would have to include all of the work on ponderomotive laws, *inter alia*.
>
> Goldstein's text, *Classical Mechanics*, contains a treatment of Noether's theorem on

pp. 47 ff., without, however, calling it by that (or any other) name. J. L. Anderson's book, *Principles of Relativity Physics* (Academic Press, 1967), explicitly refers to Noether's Theorem on p. 92. These references, picked at random from my book shelves at home, will indicate to you that a list of papers involving Noether's theorem in one way or other would probably amount to hundreds of items.

5 World War I years

At Göttingen it was still difficult, as it had been in Erlangen, for anyone to push through any provision for remuneration for Dr. Noether. The philologists and historians of the Göttingen Philosophical Faculty opposed Hilbert's efforts in her behalf, and Hilbert once declared during a University Senate meeting, "I do not see that the sex of the candidate is an argument against her admission as *Privatdozent*. After all, we are a university and not a bathing establishment." Finally, in 1919, her habilitation as *Privatdozent* was made possible, and three years later she became a "nicht-beamteter ausserordentlicher Professor", under which title she received no salary. A small salary was soon afforded her, however, as a lecturer in algebra.

Weyl's description of Emmy Noether's political life is interesting as a commentary on pre-World War II Germany:

> During the wild times after the Revolution of 1918, she did not keep aloof from the political excitement, she sided more or less with the Social Democrats; without being actually in party life she participated intensely in the discussion of the political and social problems of the day. One of her first pupils, Grete Hermann, belonged to Nelson's philosophic-political circle in Göttingen. It is hardly imaginable nowadays how willing the young generation in Germany was at that time for a fresh start, to try to build up Germany, Europe, society in general, on the foundations of reason, humaneness, and justice. But alas! the mood among the academic youth soon enough veered around; in the struggles that shook Germany during the following years and which took on the form of civil war here and there, we find them mostly on the side of the reactionary and nationalistic forces. Responsible for this above all was the breaking by the Allies of the promise of Wilson's Fourteen Points, and the fact that Republican Germany came to feel the victors' fist not less hard than the Imperial Reich could have; in particular, the youth were embittered by the national defamation added to the enforcement of a grim peace treaty. It was then that the great opportunity for the pacification of Europe was lost, and the seed sown for the disastrous development we are the witnesses of. In later years Emmy Noether took no part in matters political. She always remained, however, a convinced pacifist, a stand which she held very important and serious.

6 Excerpts from Alexandroff's address

Emmy Noether's mathematical activities from 1919 to 1923 and her influence on the mathematical community are covered by Alexandroff in his 1935 address to the Moscow Mathematical Society:

> Emmy Noether entered upon her wholly individual path of mathematical work in 1919–1920. She herself dated the beginning of this principal period of activity with the well-known collaborative work with V. Schmeidler (*Mathematische Zeitschrift* **8** (1920)). This work serves as a prologue to her general theory of ideals, opening with the classical memoir of 1921, *Idealtheorie in Ringbereiche*. I think that of all that Emmy Noether did, the bases of the general theory of ideals and all the work related to them have exerted, and will continue to exert, the greatest influence on mathematics as a whole... If the development of today's mathematics undoubtedly proceeds under the aegis of algebra, and algebraic concepts and algebraic methods have penetrated into the various mathematical theories themselves, then all that has become possible only after the works of Emmy Noether. She taught us just to think in simple, and thus general, terms: homomorphic representation, the group or ring with operators, the ideal—and not in complicated algebraic calculations, and she therefore opened a path to the discovery of algebraic regularities where before these regularities had been obscured by complicated specific conditions.
>
> It is enough to glance at the work of Pontryagin in the theory of continuous groups, at the just completed work of Kolmogoroff in the combinatorial topology of locally-bicompact spaces,

at the works of Hopf in the theory of continuous representations, not to mention the works of van der Waerden in algebraic geometry, to feel the influence of Emmy Noether's ideas. This influence is vividly clear also in the book by H. Weyl, *Gruppentheorie und Quantenmechanik*.

For all the concreteness and constructiveness of Emmy Noether's various findings, as related to the various working periods of her life, there is no doubt that her greatest energy and the major thrust of her talent were directed toward general mathematical conceptions which had to be axiomatically tinctured to a considerable degree. It is quite appropriate to analyze this aspect of her work in more detail—especially because now the question of general and specific, abstract and concrete, axiomatic and constructive, appears as one of the most acute questions of mathematical practice. Interest in the problem as a whole is sharpened by the fact that, on the one hand, mathematical journals are, without doubt unnecessarily, burdened with an abundance of all sorts of generalizing, axiomatic, and similar articles, often devoid of concrete mathematical content; while on the other hand, here and there declarations are heard that only that which is "classical" comprises the true mathematics. Under this latter slogan, important mathematical problems are rejected only because they oppose one or another habit of thought, or because they employ concepts that were not current several decades ago... H. Weyl, in the obituary that I have already cited, also raises this general question. What he says in this regard penetrates so far into the heart of the matter that I cannot but quote him in full.

In a conference on topology and abstract algebra as two ways of mathematical understanding, in 1931, I said this:

> Nevertheless I should not pass over in silence the fact that today the feeling among mathematicians is beginning to spread that the fertility of these abstracting methods is approaching exhaustion. The case is this: that all these nice general notions do not fall into our laps by themselves. But definite concrete problems were first conquered in their undivided complexity, single-handed by brute force, so to speak. Only afterwards the axiomaticians came along and stated: instead of breaking in the door with all your might and bruising your hands, you should have constructed such and such a key of skill, and by it you would have been able to open the door quite smoothly. But they can construct the key only because they are able, after the breaking in was successful, to study the lock from within and without. Before you can generalize, formalize and axiomatize, there must be a mathematical substance. I think that the mathematical substance in the formalizing of which we have trained ourselves during the last decades, becomes gradually exhausted. And so I foresee that the generation now rising will have a hard time in mathematics.

Emmy Noether protested against that: and indeed, she could point to the fact that the axiomatic method in her hands had opened new, concrete, profound problems and pointed the way to their solution.

In this quotation there is much that deserves attention: First of all, of course, the indisputable point of view that a concrete, I would say, naive, seizure of mathematical material must precede any axiomatic treatment of it; that, further, the axiomatic treatment is only of interest when it touches upon real mathematical knowledge (the "mathematical substance", of which H. Weyl speaks), and does not appear, to speak crudely, as a milling of the wind. All this is indisputable, and it is not against this that Emmy Noether protested. But she did protest against that pessimism which is seen in the last words cited by Weyl himself from his speech of 1931; the substance of human knowledge, including mathematical knowledge, is inexhaustible, at least for many long years to come—in this Emmy Noether firmly believed. The "substance of the *last decades*" is exhausting itself, but not mathematical substance in general, which by a thousand complicated threads is connected with the reality of the world's and mankind's existence. Emmy Noether intensely felt this connection of every great mathematical system, even the most abstract, with real existence, and even if she did not think this connection out philosophically, she felt it with the whole being of a learned, lively person, who was by no means shackled within abstract schemes. For Emmy Noether mathematics was always knowledge of the world and not a game of symbols, and she avidly protested when representatives of those areas of mathematics which are immediately concerned with applications wanted to secure privilege for practical knowledge.

In 1924–1925 the school of Emmy Noether made one of its most brilliant acquisitions: a graduating Amsterdam student, B. L. van der Waerden, became her pupil. He was then 22 years old and one of the brightest young mathematical talents in Europe. Van der Waerden quickly mastered the theories of Emmy Noether, enlarged them with important new findings, and like no one else, promoted her ideas. A course in the general theory of ideals, given by van der Waerden in 1927 in Göttingen, was enormously successful. The ideas of Emmy Noether in the brilliant exposition of van der Waerden subdued public mathematical opinion, first at Göttingen, then in the other leading mathematical centers of Europe. It was no accident that Emmy Noether required a popularizer of her ideas: her lectures were intended for a small group of students, working in the direction of her own investigations and listening constantly to her. From external appearances, Emmy Noether's delivery was poor, hurried, and inconsistent; but in her lectures there was immense strength of mathematical thought and extraordinary animation and fervor. Of such a kind, too, were her reports to mathematical societies and at meetings. For the mathematician who had already been captured by her ideas and become interested in her work, her reports provided much; but the mathematician who stood far from her work often could understand her exposition only with great difficulty.

From 1927 the influence of the ideas of Emmy Noether on contemporary mathematics continually grew, and along with it grew scientific praise for the author of those ideas. The direction of her work at this time moved more and more into the region of non-commutative algebra, the theory of representation and of the general arithmetic of hypercomplex quantities. Two fundamental works of the last period of her activity are *Hyperkomplexe Grossen und Darstellungstheorie* (1929) and *Nichtcommutative Algebra* (1933), both published in *Mathematische Zeitschrift* (vols. 30 and 37). These and related works evoked considerable response from spokesmen for the algebraic theory of numbers, especially from Helmut Hasse. Among her pupils during this period of her activities, the most outstanding was M. Deuring; in addition there was a whole row of young, beginning mathematicians (Witt, Fitting, and others).

Emmy Noether at last received recognition for her ideas. If in the years 1923–1925 she had to demonstrate the importance of the theories that she had developed, in 1932, at the International Mathematical Congress in Zürich, she was crowned with the laurel of her success. A summary of her work read by her at this gathering was the real triumph of the direction she represented, and she could look, not only with inner satisfaction, but now also with consciousness of full recognition, upon the mathematical path that she had traveled. The Zürich congress was the high point of her international scientific reputation. In a few months there would burst over German culture, and in particular over her home, which the University of Göttingen had become, the catastrophe of the Fascist revolution, which in a few weeks scattered to the wind all that had been built up over a long period of decades. One of the greatest tragedies that human culture has undergone since the time of the Renaissance took place, a tragedy which a few years ago appeared improbable and impossible in the Europe of the twentieth century. One of its numerous victims was the Göttingen School of Algebra, which had been founded by Emmy Noether. Its directress was banished from the walls of the University; and having lost the right to teach, Emmy Noether had to emigrate from Germany. She accepted the invitation from the women's college at Bryn Mawr, where she lived out the last year and a half of her life.

If what I have just quoted is the main strand of the material of Alexandroff's address, another is his description of Emmy Noether's influence on Soviet mathematics and her regard for Soviet ideals:

Emmy Noether was closely connected with Moscow. This connection began in 1923, when the late Pavel Samuelovitch Urysohn and I first arrived in Göttingen and immediately found ourselves in a mathematical circle whose leader was Emmy Noether. The basic traits of the Noether school struck us right away: the scientific enthusiasm of the directress of the school which was passed along to all her students, her deep belief in the importance and mathematical fruitfulness of her ideas (a belief that was not at all shared then by everyone, even in Göttingen), and the extraordinary simplicity and sincerity of relations between the head of the school and its members. In those days this school was almost entirely made up of young Göttingen students; the time was still in the future when it would become, for its membership and for its acknowledged worldwide influence, an outstanding international center of algebraic thought.

The mathematical interests of Emmy Noether (centered at that time in the full swing of her work on the general theory of ideals) and the mathematical interests of Urysohn and myself (centered around the problems of so-called abstract topology) had many points in common and quickly led to continual, almost daily, mathematical discussions. Emmy Noether was interested, however, not only in our topological work, but also in what had been taking place in the whole area of mathematics (and not only in the area of mathematics) in Soviet Russia; she did not hide her sympathies with our country and our social and governmental system, in spite of the fact that the manifestation of these sympathies seemed outrageous and unseemly to the majority of representatives of Western European academic circles. The matter had reached the point where Emmy Noether was literally banished from one of the Göttingen boarding houses (where she had settled and lived) at the demand of the student corporation, resident in the same house, who did not want to live under the same roof with a "Marxist-inclined Jewess."

And Emmy Noether was truly gladdened by the scientific, and particularly the mathematical successes of the Soviet country, since she saw in this the final refutation of all the old wives' tales to the effect that "the Bolsheviks are destroying culture." A spokesman of the most abstract areas of mathematical science, she distinguished herself at the same time by a surprising sensitivity in understanding the great historical movements of our epoch; always vitally interested in politics, hating war with her whole being, and hating chauvinism in all its manifestations, she never in this area knew any vacillation: her sympathies always and unchangingly belonged to the Soviet Union, in which she saw the beginning of a new era in the history of mankind and firm support for everything progressive for which human thought has lived and lives still.

The scientific and personal friendship which sprang up between Emmy Noether and me in 1923 did not come to an end even with her death. Recalling this friendship in his obituary speech, Weyl advances the supposition that the general system of thought of Emmy Noether did not remain without influence on my own topological research. I am happy now to affirm the truth of Weyl's supposition: Emmy Noether's influence on my own, and on other topological research in Moscow, was very great, and it affected the whole essence of our work. In particular, my theory of the continuous breakdown of topological spaces arose to a significant degree under the influence of conversations with her in December and January of 1925–1926, when we were in Holland together.

Emmy Noether spent the winter of 1928–1929 in Moscow. She taught a course in abstract algebra at the University of Moscow and conducted a seminar in algebraic geometry at the Communist Academy. She quickly established contact with a majority of Moscow's mathematicians, in particular and especially, with L. S. Pontryagin and O. U. Schmidt. It is not difficult to trace the influence of Emmy Noether on the mathematical talent of L. S. Pontryagin; a strong algebraic note in his work was undoubtedly benefited in its development by contact with Emmy Noether. In Moscow, Emmy Noether very easily fit herself in with our life, both in her scientific and her nonprofessional relationships. She lived in a modest room in the KSU hostel near the Crimean Bridge, and most of the time she walked to the University. She was very much interested in the life of our country, especially in the life of Soviet young people and students.

In the winter of 1928–1929 I was as usual on a visit to Smolensk and was giving lectures on algebra at the Pedagogical Institute there. Inspired by my continual conversations with Emmy Noether, I gave my lectures along the lines established by her. Among my students there, A. G. Kurosh immediately stood out, and the theories that I was expounding, wholly steeped as they were in Emmy Noether's ideas, appealed very much to him. In this way, through my teaching, Emmy Noether acquired a disciple who has since grown into an independent and learned man, as is well known, and whose works through the present day have proceeded in the principal circle of ideas created by her.

In the spring of 1929, she left Moscow for Göttingen with the firm intention of coming to visit us again in the near future. Several times she was close to carrying out that intention, and closest to doing so in the last year of her life. After her exile from Germany, she seriously considered a final trip to Moscow, and I exchanged letters with her in this regard. She clearly understood that nowhere could she find the means to create a new brilliant mathematical school in exchange for the one that had been taken from her in Göttingen. I had already conducted talks with the Narkompros [The People's Commissariat for Education] about assigning her a chair at the University of Moscow. However, at the Commissariat, as usual, they were slow in making a decision, and they did not give me a final answer. Meanwhile, time passed, and Emmy Noether, deprived even of that modest work which she had in Göttingen, could wait no longer and had to accept the invitation of the women's college...

Such was Emmy Noether, the greatest of women mathematicians, a great scientist, an amazing teacher, and an unforgettable person... True, Weyl has said that "the Graces did not stand at her cradle," and he is right, if one has in mind the generally known heaviness of her appearance. But here Weyl is speaking of her not only as a great scholar, but also as a great woman. And she was that—her femininity appeared in that gentle and subtle lyricism which lay at the heart of the far-flung but never superficial concerns which she maintained for people, for her profession, and for the interests of all mankind. She loved people, science, life, with all the warmth, all the cheerfulness, all the unselfishness, and all the tenderness of which a deeply sensitive—and feminine—soul is capable.

7 In America

Among the scientists who left Germany during the early thirties and sooner or later took refuge in the United States were E. Artin, R. Courant, P. Debye, M. Dehn, A. Einstein, P. Ewald, W. Feller, J. Franck, K. Friedrichs, K. Gödel, E. Hellinger, O. Neugebauer, J. von Neumann, Emmy Noether, L. Nordheim, O. Ore, G. Pólya, G. Szegö, A. Tarski, Olga Taussky (Todd), H. Weyl, and E. Wigner.

Arrangements were made for Emmy Noether to teach at Bryn Mawr College, just outside Philadelphia, beginning in the autumn of 1933. Conveniently close was the Institute for Advanced Study at Princeton, where, beginning in February, 1934, she gave weekly lectures. At the Institute were Einstein, Weyl, Oswald Veblen, and Abraham Flexner.

Emmy Noether returned to Germany to visit during the summer of 1934 and then resumed her work at Bryn Mawr and Princeton in the early fall. Richard Brauer had joined the Institute, and after her lectures, she usually visited with Brauer, Weyl, and Veblen before returning to Bryn Mawr.

One of Emmy Noether's associates at Bryn Mawr was Grace Shover (Quinn), now a professor at the American University. Awarded the Emmy Noether Fellowship for post-doctoral study, she became acquainted with Emmy Noether in September, 1934. There were three other graduate students in mathematics. Marie Weiss of Newcomb College held the Emmy Noether Scholarship. Olga Taussky held the foreign fellowship. Ruth Stauffer (McKee) was a doctoral candidate, Emmy Noether's only American Ph.D. student.

Professor Quinn recalls that Emmy Noether

was around 5'4" tall and was slightly rotund in build. Her complexion was swarthy. Her dark hair, flecked with grey, was cropped short. She wore thick glasses to cover her near-sighted eyes, and she had a way of turning her head aside and looking into the distance when trying to think while talking. Her looks and dress were most unconventional, such as to attract attention, but such a result was far from her thoughts. She was sincere, straightforward, kindly, thoughtful, and considerate.

Her lectures were delivered in broken English. She often lapsed into her native German when she was bothered by some idea in lecturing. She loved to walk. She would take her students off for a jaunt on a Saturday afternoon. On these trips she would become so absorbed in her conversation on mathematics that she would forget about the traffic and her students would need to protect her.

The chairman of the Bryn Mawr mathematics department was Anna Pell Wheeler, now deceased. Having studied at Göttingen a few years before receiving her doctorate from the University of Chicago in 1910, Professor Wheeler became a very close friend of Emmy Noether. In this connection, Mrs. McKee has written:

Probably the greatest difference in her life in America was her close friendship with the head of the mathematics department. In Germany at that time women were neither expected nor encouraged to study. It was rather assumed that their role in life was that of a homemaker. Therefore, to have as a friend a woman who was a nationally recognized mathematician who had earlier studied at Göttingen and who thoroughly understood the problems of a woman scholar in Germany, was a unique experience for Miss Noether... Many of Miss Noether's former students and colleagues stopped to see [her] in Bryn Mawr and she always took them to see her 'good friend.'

8 Missing letters recovered

Together with Jean Cavaillés, Emmy Noether edited the correspondence between Richard Dedekind and Georg Cantor. Although completed in March, 1933, the book, *Briefwechsels—G. Cantor und R. Dedekind*, did not appear until 1937, when it was published by Hermann of Paris.

The Cantor-Dedekind letters were still in Emmy Noether's possession when she died, along with correspondence from G. Frobenius and H. Weber (Hilbert's predecessor at Göttingen). A representative of the law firm which settled Emmy Noether's estate wrote to her brother Fritz, exiled in Tomsk, asking what should be done with the letters. Professor Noether's reply was that they be returned to their (unspecified) owner. This directive was not carried out. Instead, the letters lay lost in the files of a Philadelphia law office, until, after some 33 years, a member of the firm wrote me, "Inasmuch as you are researching her life, a rather valuable bit of information was unearthed by me in going through the Estate file. Under separate cover you will shortly receive them via parcel post. I suppose you will be agreeable to the modest charge of $25.00..."

I had no idea what the "valuable bit of information" was, but as promised, the letter collection arrived a few days later. Included with the famous Cantor-Dedekind letters, a few of whose paragraphs may be found in English in Sherman K. Stein's popular *Mathematics: The Man-made Universe* (Freeman), are 47 letters written by Weber in Königsberg and Heidelberg to Dedekind in Braunschweig. Together with 20 post cards, telegrams, and printed circulars, these letters span the years 1876–1879.

Three letters each by Frobenius and Dedekind are dated 1882–1883. Their remaining 38 letters and 7 postcards are dated from 1895 to 1901. Most were written in 1886. Their content is more mathematical than that of the Weber letters. A few reach a length of 20 pages. I counted a total of 178 pages from Frobenius to Dedekind and 113 from Dedekind to Frobenius.

At present, the letters are kept in the Clifford Memorial Library at the University of Evansville.

"A Marvelous Proof"

FERNANDO Q. GOUVÊA

American Mathematical Monthly **101** (1994), 203–222

No one really knows when it was that the story of what came to be known as "Fermat's Last Theorem" really started. Presumably it was sometime in the late 1630s that Pierre de Fermat made that famous inscription in the margin of Diophantus's *Arithmetica* claiming to have found "a marvelous proof". It seems now, however, that the story may be coming close to an end. In June 1993, Andrew Wiles announced that he could prove Fermat's assertion. Since then, difficulties seem to have arisen, but Wiles's strategy is fundamentally sound and may yet succeed.

The argument sketched by Wiles is an artful blend of various topics that have been, for years now, the focus of intensive research in number theory: elliptic curves, modular forms, and Galois representations. The goal of this article is to give mathematicians who are not specialists in the subject access to a general outline of the strategy proposed by Wiles. Of necessity, we concentrate largely on background material giving first a brief description of the relevant topics, and only afterwards describe how they come together and relate to Fermat's assertion. Readers who are mainly interested in the structure of the argument and who do not need or want too many details about the background concepts may want to skim through Section 2, then concentrate on Section 3. Our discussion includes a few historical remarks, but history is not our main intention, and therefore we only touch on a few highlights that are relevant to our goal of describing the main ideas in Wiles's attack on the problem.

1 Preliminaries

We all know the basic statement that Fermat wrote in his margin. The claim is that for any exponent $n \geq 3$ there are no non-trivial integer solutions of the equation $x^n + y^n = z^n$. (Here, "non-trivial" will just mean that none of the integers x, y, and z will be equal to zero.) Fermat claims, in his marginal note, to have found "a marvelous proof" of this fact, which unfortunately would not fit in the margin.

This statement became known as "Fermat's Last Theorem", not, apparently, due to any belief that the "theorem" was the last one found by Fermat, but rather due to the fact that by the 1800s all of the other assertions made by Fermat had been either proved or refuted. This one was the last one left open, whence the name. In what follows, we will adopt the abbreviation FLT for Fermat's statement, and we will refer to "$x^n + y^n = z^n$" as the *Fermat equation*.

The first important results relating to FLT were theorems that showed that Fermat's claim was true for specific values of n. The first of these is due to Fermat himself: very few of his proofs were ever made public, but in one that was he shows that the equation

$$x^4 + y^4 = z^2$$

has no non-trivial integer solutions. Since any solution of the Fermat equation with exponent 4 gives a solution of this equation also, it follows that Fermat's claim is true for $n = 4$.

Once that is done, it is easy to see that we can restrict our attention to the case in which n is a prime number. To see this, notice that any number greater than 2 is either divisible by 4 or by an odd prime, and then notice that we can rewrite an equation

$$x^{mk} + y^{mk} = z^{mk}$$

as

$$(x^m)^k + (y^m)^k = (z^m)^k,$$

so that any solution for $n = mk$ yields at once a solution for $n = k$. If n is not prime, we can always choose k to be either 4 or an odd prime, so that the problem reduces to these two cases.

In the 1750s, Euler became interested in Fermat's work on number theory, and began a systematic investigation of the subject. In particular, he considered the Fermat equation for $n = 3$ and $n = 4$, and once again proved that there were no solutions. (Euler's proof for $n = 3$ depends on studying the "numbers" one gets by adjoining $\sqrt{-3}$ to the rationals, one of the first instances where one meets "algebraic numbers".) A good historical account of Euler's work is to be found in [35]. In the following years, several other mathematicians extended this step by step to $n = 5, 7, \ldots$ A general account of the fortunes of FLT during this time can be found in [24].

Since then, ways for testing Fermat's assertion for any specific value of n have been developed, and the range of exponents for which the result was known to be true kept getting pushed up. As of 1992, one knew that FLT was true for exponents up to four million (by work of S. Buhler).

It is clear, however, that to get general results one needs a general method, i.e., a way to connect the Fermat equation (for any n) with some mathematical context which would allow for its analysis. Over the centuries, there have been many attempts at doing this; we mention only the two biggest successes (omitting quite a lot of very good work, for which see, for example, [24]).

The first of these is the work of E. Kummer, who, in the mid-nineteenth century established a link between FLT and the theory of cyclotomic fields. This link allowed Kummer to prove Fermat's assertion when the exponent was a prime that had a particularly nice property (Kummer named such primes *regular*). The proof is an impressive bit of work, and was the first general result about the Fermat equation. Unfortunately, while in numerical tests a good percentage of primes seem to turn out to be regular, no one has yet managed to prove even that there are infinitely many regular primes. (And, ironically, we do have a proof that there are infinitely many primes that are *not* regular.) A discussion of Kummer's approach can be found in [24]; for more detailed information on the cyclotomic theory, one could start with [33].

The second accomplishment we should mention is that of G. Faltings, who, in the early 1980s, proved Mordell's conjecture about rational solutions to certain kinds of polynomial equations. Applying this to the Fermat equations, one sees that for any $n > 4$ one can have only a *finite* number of non-trivial solutions. Once again, this is an impressive result, but its impact on FLT itself turns out to be minor because we have not yet found a way to actually determine how many solutions should exist. For an introduction to Faltings's work, check [7], which contains an English translation of the original paper.

Wiles's attack on the problem turns on another such linkage, also developed in the early 1980s by G. Frey, J.-P. Serre, and K. A. Ribet. This one connects FLT with the theory of elliptic curves, which has been much studied throughout this century, and thereby to all the machinery of modular forms and Galois representations that is the central theme of Wiles's work. The main goal of this paper is to describe this connection and then to explain how Wiles attempts to use it to prove FLT.

Notation We will use the usual symbols $\mathbb{Q}$ for the rational numbers and $\mathbb{Z}$ for the integers. The integers modulo m will be written as $\mathbb{Z}/m\mathbb{Z}$; we will most often need them when m is a power of a prime number p. If p is prime, then $\mathbb{Z}/p\mathbb{Z}$ is a field, and we commemorate that fact by using an alternative notation: $\mathbb{F}_p = \mathbb{Z}/p\mathbb{Z}$.

2 The actors

We begin by introducing the main actors in the drama. First, we briefly (and very informally) introduce the p-adic numbers. These are not so much actors in the play as they are part of the stage set: tools to allow the actors to do their job. Then we give brief and impressionistic outlines of the theories of elliptic curves, modular forms, and Galois representations.

p-adic numbers

The p-adic numbers are an extension of the field of rational numbers which are, in many ways, analogous to the real numbers. Like the real numbers, they can be obtained by defining a notion of distance between rational numbers, and then passing to the completion with respect to that distance. For our purposes, we do not really need to know much about them. The crucial facts are:

1. For each prime number p there exists a field $\mathbb{Q}_p$, which is complete with respect to a certain notion of distance and contains the rational numbers as a dense subfield.

2. Proximity in the p-adic metric is closely related to congruence properties modulo powers of p. For example, two integers whose difference is divisible by p^n are "close" in the p-adic world (the bigger the n the closer they are).

3. As a consequence, one can think of the p-adics as encoding congruence information: whenever one knows something modulo p^n for every n, one can translate this into p-adic information, and vice-versa.

4. The field $\mathbb{Q}_p$ contains a subring $\mathbb{Z}_p$, which is called the *ring of p-adic integers*. (In fact, $\mathbb{Z}_p$ is the closure of $\mathbb{Z}$ in $\mathbb{Q}_p$.)

There is, of course, a lot more to say, and the reader will find it said in many references, such as [15], [3], [1], and even [12]. The p-adic numbers were introduced by K. Hensel (a student of Kummer), and many of the basic ideas seem to appear, in veiled form, in Kummer's work; since then, they have become a fundamental tool in number theory.

Elliptic curves

Elliptic curves are a special kind of algebraic curves which have a very rich arithmetical structure. (Perhaps it's best to dispel the obvious confusion right up front: ellipses are not elliptic curves. In fact, the connection between elliptic curves and ellipses is a rather subtle one. What happens is that elliptic curves over the complex numbers are the "natural habitat" of the elliptic integrals which arise, among other places, when one attempts to compute the arc length of an ellipse. For us, this connection will be of very little importance.)

There are several fancy ways of defining elliptic curves, but for our purposes we can just define them as the set of points satisfying a polynomial equation of a certain form.

To be specific, consider an equation of the form

$$y^2 + a_1 xy + a_3 y = x^3 + a_2 x^2 + a_4 x + a_6,$$

where the a_i are integers (there is a reason for the strange choice of indices on the a_i but we won't go into it here). We want to consider the set of points (x, y) which satisfy this equation. Since we are doing number theory, we don't want to tie ourselves down too seriously as to what sort of numbers x and y are: it makes sense to take them in the real numbers, in the complex numbers, in the rational numbers, and even, for any prime number p, in $\mathbb{F}_p$ (in which case we think of the equation as a congruence modulo p). We will describe the situation by saying that there is an underlying object which we call *the curve E* and, for each one of the possible fields of definition for points (x, y), we call the set of possible solutions the "points of E" over that field. So, if we consider all possible complex solutions, we get the set $E(\mathbb{C})$ of the complex points of E. Similarly, we can consider the real points $E(\mathbb{R})$, the rational points $E(\mathbb{Q})$, and even the $\mathbb{F}_p$-points $E(\mathbb{F}_p)$.

We haven't yet said when it is that such equations define elliptic curves. The condition is simply that the curve be *smooth*. If we consider the real or complex points, this means exactly what one would expect — the set of points contains no "singular" points, that is, at every point there is a well-defined tangent line. We know, from elementary analysis, that an equation $f(x, y) = 0$ defines a smooth curve exactly when there are no points on the curve at which both partial derivatives of f vanish. In other words, the curve will be smooth if there are no common solutions of the equations

$$f(x,y) = 0 \qquad \frac{\partial f}{\partial x}(x,y) = 0 \qquad \frac{\partial f}{\partial y}(x,y) = 0.$$

Notice, though, that this condition is really algebraic (the derivatives are derivatives of polynomials, and hence can be taken formally). In fact, we can boil it down to a (complicated) polynomial condition in the a_i. There is a polynomial $\Delta(E) = \Delta(a_1, a_2, a_3, a_4, a_6)$ in the a_i such that E is smooth if and only if $\Delta(E) \neq 0$. This gives us the means to give a completely formal definition (which makes sense even over $\mathbb{F}_p$). The number $\Delta(E)$ is called the *discriminant* of the curve E.

Definition 1. *Let K be a field. An elliptic curve over K is an algebraic curve determined by an equation of the form*

$$y^2 + a_1 xy + a_3 y = x^3 + a_2 x^2 + a_4 x + a_6,$$

where each of the a_i belongs to K and such that $\Delta(a_1, a_2, a_3, a_4, a_6) \neq 0$.

Specialists would want to rephrase that definition to allow other equations, provided that a well-chosen change of variables could transform them into equations of this form.

It's about time to give some examples. To make things easier, let us focus on the special case in which the equation is of the form $y^2 = g(x)$, with $g(x)$ a cubic polynomial (in other words, we're assuming $a_1 = a_3 = 0$). In this case, it's very easy to determine when there can be singular points, and even what sort of singular points they will be. If we put $f(x, y) = y^2 - g(x)$, then we have

$$\frac{\partial f}{\partial x}(x,y) = -g'(x) \quad \text{and} \quad \frac{\partial f}{\partial y}(x,y) = 2y,$$

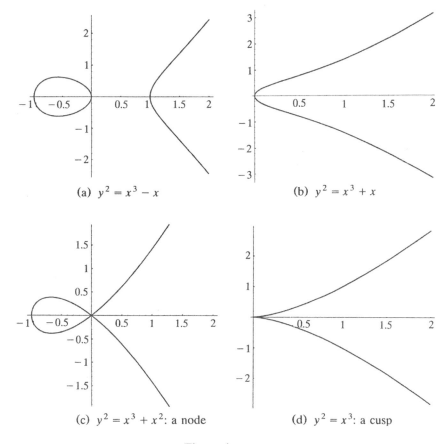

Figure 1

and the condition for a point to be "bad" becomes

$$y^2 = g(x) \qquad -g'(x) = 0 \qquad 2y = 0,$$

which boils down to $y^2 = g(x) = g'(x) = 0$. In other words, a point will be bad exactly when its y-coordinate is zero and its x-coordinate is a *double root* of the polynomial $g(x)$. Since $g(x)$ is of degree 3, this gives us only three possibilities:

- $g(x)$ has no multiple roots, and the equation defines an elliptic curve;
- $g(x)$ has a double root;
- $g(x)$ has a triple root.

Let's look at one example of each case, and graph the real points of the corresponding curve.

For the first case, consider the curve given by $y^2 = x^3 - x$. Its graph is in Figure 1a (to be precise, this is the graph of its real points). A different example of the same case is given by $y^2 = x^3 + x$; see Figure 1b. (The reason these look so different is that we are only looking at the real points of the curve; in fact, over the complex numbers these two curves are isomorphic.)

When there are "bad" points, what has happened is that either two roots of $g(x)$ have "come together" or all three roots have done so. In the first case, we get a loop. At the crossing point, which is usually called a *node*, the curve has two different tangent lines; see Figure 1c, where we have the graph of the equation $y^2 = x^3 + x^2$ (double root at zero).

In the final case, not only have all three roots of $g(x)$ come together, but also the two tangents in the node have come together to form a sort of "double tangent" (this can be made precise with some easy algebra of polynomials, but it's more fun to think of it geometrically). The graph now looks like Figure 1d, and we call this kind of singular point a *cusp*.

How does all this relate to the discriminant Δ we mentioned above? Well, if r_1, r_2, and r_3 are the roots of the polynomial $g(x)$, the discriminant for the equation $y^2 = g(x)$ turns out to be

$$\Delta = K \left(r_1 - r_2\right)^2 \left(r_1 - r_3\right)^2 \left(r_2 - r_3\right)^2,$$

where K is a constant. This does just what we want: if two of the roots are equal, it is zero, and if not, not. Furthermore, it is not too hard to see that Δ is actually a polynomial in the coefficients of $g(x)$, which is what we claimed. In other words, all that the discriminant is doing for us is giving a direct algebraic procedure for determining whether there are singular points.

While this analysis applies specifically to curves of the form $y^2 = g(x)$, it actually extends to all equations of the sort we are considering: there is at most one singular point, and it is either a node or a cusp.

One final geometric point: as one can see from the graphs, these curves are not closed. It is often convenient to "close them up." This is done by adding one more point to the curve, usually referred to as the *point at infinity*. This can be done in a precise way by embedding the curve into the projective plane, and then taking the closure. For us, however, the only important thing is to remember that we actually have one extra point on our curves. (One should imagine it to be "infinitely far up the y-axis," but keep in mind that there is only one "point at infinity" on the y-axis, so that it is also "infinitely far down.")

With some examples in hand, we can proceed to deeper waters. In order to understand the connection we are going to establish between elliptic curves and FLT, we need to review quite a large portion of what is known about the rich arithmetic structure of these curves.

The first thing to note is that one can define an operation on the set of points of an elliptic curve that makes it, in a natural way, an abelian group. This operation is usually referred to as "addition". The identity element of this group turns out to be the point at infinity (it would be more honest to say that we *choose* the point at infinity for this role).

We won't enter into the details of how one adds points on an elliptic curve. In fact, there are several equivalent definitions, each of which has its advantages! The reader should see the references for more details of how it is done (and the proof that one does get a group). The main thing to know about the definition, for now, is that it preserves the field of definition of the points: adding two rational points gives a rational point, and so on.

What this means is that for every choice of a base field, we can get a group of points on the curve with coordinates in that field, so that in fact an elliptic curve gives us a whole bunch of groups, which are, of course all related (though sometimes related in a mysterious way). So, given an E, we can look at the complex points $E(\mathbb{C})$, which form a complex Lie group which is topologically a torus, or we can look at the real Lie group $E(\mathbb{R})$, which turns out to be either isomorphic to the circle S^1 or to the direct product

$\mathbb{Z}/2\mathbb{Z} \times S^1$. (Look back at the examples above; can you see which is which?)

From an arithmetical point of view, however, the most interesting of these groups is the group of rational points, $E(\mathbb{Q})$. A point $P \in E(\mathbb{Q})$ gives a solution in the rational numbers of our cubic equation, and looking for such solutions is, of course, an example of solving a Diophantine equation, a sort of problem that is quite important in number theory. What is especially nice about $E(\mathbb{Q})$ is the fact, proved by L. Mordell (and extended by A. Weil) in the 1920s, that it is a *finitely generated* abelian group. What this means is just the following: there is a finite list of rational points on the curve (or, if one prefers, of rational solutions to the equation) such that any other rational solution is obtained by combining (using the addition law) these points with one another. These points are called the *generators* of the group $E(\mathbb{Q})$, which is usually called the *Mordell-Weil group* of E.

The curves we considered earlier have very simple Mordell-Weil groups. For the curve given by $y^2 = x^3 - x$ (Figure 1a), it has four points, and for $y^2 = x^3 + x$ (Figure 1b) it has two. It is easy, though, to give more interesting examples. Here is one, chosen at random from [6]: if E is the curve defined by $y^2 + y = x^3 - x^2 - 2x + 2$, the Mordell-Weil group $E(\mathbb{Q})$ is an infinite cyclic group, generated by the point $(2, 1)$.

Of course, knowing that we have a finitely generated group raises the obvious question of estimating or computing the number of generators needed and of how one might go about actually finding these generating points. Both of these questions are still open, even though there are rather precise conjectures about what their answers should be. For many specific curves, both the number and the generators themselves have been completely worked out (see, for example, the tables in [6]), but the general problem still seems quite difficult.

A fundamental component of the conjectural plan for determining the generators is considering, for each prime number p, the reduction of our curve modulo p. The basic idea is quite simple: since our equation has integer coefficients, we can reduce it modulo p and look for solutions in the field $\mathbb{F}_p$ of integers modulo p. This should give a finite group $E(\mathbb{F}_p)$, whose structure should be easier to analyze than that of the big group $E(\mathbb{Q})$. It's a rather simple idea, but several complications arise.

The main thing that can go wrong is that the reduction modulo p may fail to be an elliptic curve. That is actually very easy to see. To tell whether the curve is elliptic (that is, if it has no singular points), one needs to look at Δ. It is perfectly possible for Δ to be non-zero (so that the curve over $\mathbb{Q}$ is elliptic) while being at the same time congruent to zero modulo p (so that the curve over $\mathbb{F}_p$ is singular). This phenomenon is called *bad reduction*, and it is easy to come up with examples. One might take $p = 5$, and look at the curve $y^2 = x^3 - 5$. This turns out to be an elliptic curve over $\mathbb{Q}$, but its reduction modulo 5 is going to have a cusp. One says, then, that the curve has bad reduction at 5. In fact, the discriminant turns out to be $\Delta = -10,800$, which is clearly divisible by 2, 3 and 5, so that the curve has bad reduction at each of these. (In each case, it's easy to verify that the reduced curve has a cusp.)

We want to clarify the possible types of reduction, but there is one further glitch that we have to deal with before we can do so. To see what it is, consider the curve $y^2 = x^3 - 625x$. At first glance, it seems even worse than the first, and the discriminant, which turns out to be $\Delta = -15{,}625{,}000{,}000$, looks *very* divisible by 5. But look what we can do: let's change variables by setting $x = 25u = 5^2 u$ and $y = 125v = 5^3 v$. Then our equation becomes

$$\left(5^3 v\right)^2 = \left(5^2 u\right)^3 - 625(5^2 u),$$

which simplifies to

$$5^6 v^2 = 5^6 u^3 - 5^6 u,$$

and hence to

$$v^2 = u^3 - u,$$

which is not only a nice elliptic curve, but has good reduction at 5. In other words, this example shows that *curves which are isomorphic over $\mathbb{Q}$ can have very different reductions modulo p.*

It turns out that among all the possible equations for our curve, one can choose an equation that is minimal, in the sense that its discriminant will be divisible by fewer primes than the discriminant for other equations. Since the primes that divide the discriminant are the primes of bad reduction, a minimal equation will have reduction properties that are as good as possible. When studying the reduction properties of the curve, then, one must also pass to such a minimal equation (and there are algorithms to do this).

Well, then, suppose we have done so, and have an elliptic curve E given by a minimal equation. Then we can classify all prime numbers into three groups:

- *Primes of good reduction*: those which do not divide the discriminant of the minimal equation. The curve modulo p is an elliptic curve, and we have a group $E(\mathbb{F}_p)$.

- *Primes of multiplicative reduction*: those for which the curve modulo p has a node. If the singular point is (x_0, y_0), it turns out that the set $E(\mathbb{F}_p)\setminus\{(x_0, y_0)\}$ has a group structure, and is isomorphic to the multiplicative group $\mathbb{F}_p\setminus\{0\}$.

- *Primes of additive reduction*: those for which the curve modulo p has a cusp. If the singular point is (x_0, y_0), the set $E(\mathbb{F}_p)\setminus\{(x_0, y_0)\}$ once again has a group structure, and is isomorphic to the additive group $\mathbb{F}_p$.

No curve can have good reduction everywhere, so there will always be some bad primes, but the feeling one should get is that multiplicative reduction is somehow not as bad as additive reduction. There are various technical reasons for this, which we don't really need to go into. Instead, we codify the information about the reduction types of the curve into a number, called the *conductor* of the curve. We define the conductor to be a product $N = \prod p^{n(p)}$, where

$$n(p) = \begin{cases} 0 & \text{if } E \text{ has good reduction at } p \\ 1 & \text{if } E \text{ has multiplicative reduction at } p \\ \geq 2 & \text{if } E \text{ has additive reduction at } p. \end{cases}$$

(The exact value for $n(p)$ for the case of additive reduction depends on some rather subtle properties of the reduction modulo such primes; most of the time, the exponent is 2.) The result is that one can tell, by looking at the conductor, exactly what the reduction type of E at each prime is.

The elliptic curves we will want to consider are those whose reduction properties are as good as possible. Since good reduction at all primes is not possible, we opt for the next best thing: good reduction at almost all primes, multiplicative reduction at the others. Such curves are called *semistable*:

Definition 2. *An elliptic curve is called semistable if all of its reductions are either good or multiplicative. Equivalently, a curve is semistable if its conductor is square-free.*

A crucial step in the application of Wiles's theorem to FLT will be verifying that a certain curve is semistable. Just to give us some reference points, let's look at a few examples.

1. Let E_1 be the curve $y^2 = x^3 - 5$, which we considered above. One checks that this equation is minimal, and that the curve has additive reduction

at 2, 3, and 5, so that it is not semistable. The conductor turns out to be equal to 10,800 (essentially, the same as the discriminant!).

2. Let E_2 be the curve $y^2 + y = x^3 + x$. This has multiplicative reduction at 7 and 13 (checking this makes a nice exercise) and good reduction at all other primes. Hence, E_2 is semistable and its conductor is 91.

3. Let E_3 be the curve $y^2 = x^3 + x^2 + 2x + 2$ (which is minimal). This has discriminant $\Delta = -1152 = -2^7 \cdot 3^2$, so that the bad primes are 2 and 3. It turns out that the reduction is multiplicative at 3 and additive at 2, and the conductor is 384: the curve is not semistable.

4. *The main example for the purpose at hand*: Let a, b and c be relatively prime integers such that $a + b + c = 0$. Consider the curve E_{abc} whose equation is $y^2 = x(x-a)(x+b)$. Depending on what a, b and c are, this equation may or may not be minimal, so let's make the additional assumption that $a \equiv -1 \pmod{4}$ and that $b \equiv 0 \pmod{32}$. In this case, the equation is *not* minimal. A minimal equation for this curve turns out to be

$$y^2 + xy = x^3 + \frac{b-a-1}{4}x^2 - \frac{ab}{16},$$

which we get by the change in variables $x \to 4x, y \to 8y + 4x$. One can then compute that the discriminant is $\Delta = a^2 b^2 c^2 / 256$ (not surprising: a constant times the product of the squares of the differences of the roots of the original cubic), and that the curve is semistable. The primes of bad reduction are those that divide abc (this would be easy to see directly from the equation, by checking when there is a multiple root modulo p), and therefore the conductor is equal to the product of the primes that divide abc:

$$N = \prod_{p | abc} p$$

(this number is sometimes called the *radical* of abc). We will be using curves of the form E_{abc} (for very special a, b and c) when we make the link with FLT.

We need a final bit of elliptic curve theory. It is interesting to look at the number of points in the groups $E(\mathbb{F}_p)$ as p ranges through the primes of good reduction for E. Part of the motivation for this is the reasoning that if the group $E(\mathbb{Q})$ is large (i.e., there are many rational solutions), one would expect that for many choices of the prime p, many of the points in $E(\mathbb{Q})$ would survive reduction modulo p, so that the group $E(\mathbb{F}_p)$ would be large. Therefore, one would like to make some sort of conjecture that said that if the $E(\mathbb{F}_p)$ are very large for many primes p, then the group $E(\mathbb{Q})$ will be large.

Elaborating and refining this idea leads to the conjecture of Birch and Swinnerton-Dyer, which we won't get into here. But even this coarse version suggests that variation of the size of $E(\mathbb{F}_p)$ as p runs through the primes should tell us something about the arithmetic on the curve. To "encode" this variation, we start by observing that the (projective) line over F_p has exactly $p + 1$ points (the p elements of $\mathbb{F}_p$ plus the point at infinity). We take this as the "standard" number of points for a curve over $\mathbb{F}_p$ and, when we look at $E(\mathbb{F}_p)$, record how far from the standard we are. To be precise, given an elliptic curve E and a prime number p at which E has good reduction, we define a number a_p by the equation

$$\#E(\mathbb{F}_p) = p + 1 - a_p.$$

For primes of bad reduction, we extend the definition in a convenient way; it turns out that we get $a_p = \pm 1$ when the reduction is multiplicative (with a precise rule to decide which) and $a_p = 0$ when it is additive.

The usual way to "record" the sequence of the a_p is to use them to build a complex analytic function called the *L-function* of the curve E. It then is natural to conjecture that this L-function has properties similar to those of other L-functions that arise in number theory, and that one can read off properties of E from properties of its L-function. This is a huge story which we cannot tell in this article, but which is really very close to some of the issues which we do discuss later on. Suffice it to say, for now, that we get a function

$$L(E, s) = \sum_{n=1}^{\infty} \frac{a_n}{n^s},$$

where the a_p are exactly the same as the ones just introduced, the a_n are determined from the a_p by "Euler product" expansion for the L-function, and the series can be shown to converge when $\Re(s) > \frac{3}{2}$. The L-function is conjectured to have an analytic continuation to the whole complex plane and to satisfy a certain functional equation.

It is time to introduce the other actors in the play and to explain how they relate to elliptic curves. The reader who would like to delve further into this theory has a lot to choose from. As an informal introduction, one could look at J. Silverman's article [30], which relates elliptic curves to "sums of two cubes" and Ramanujan's taxicab number. Various introductory texts are available, including [4], [13], [14], [29], and [32]. Each of these has particular strengths; the last is intended as an undergraduate text. In addition to these and other texts, the interested reader might enjoy looking at symbolic manipulation software that will handle elliptic curves well. Such capabilities are built into GP-PARI and SIMATH, and can be added to *Mathematica* by using Silverman's *EllipticCurveCalc* package (which is what we used for most of the computations in this paper), and to *Maple* by using Connell's *Apecs* package. See [2], [36], [32], [5].

Modular forms

Modular forms start their lives as analytic objects (or, to be more honest, as objects of group representation theory), but end up playing a very intriguing role in number theory. In this section, we will very briefly sketch out their definition and explain their relation to elliptic curves.

Let $\mathfrak{h} = \{x + iy : y > 0\}$ be the complex upper half-plane. As is well known (and, in any case, easy to check), matrices in $\mathrm{SL}_2(\mathbb{Z})$ act on $\mathfrak{h}$ in the following way. If $\gamma \in \mathrm{SL}_2(\mathbb{Z})$ is the matrix

$$\gamma = \begin{pmatrix} a & b \\ c & d \end{pmatrix}$$

(so that a, b, c and d are integers and $ad - bc = 1$), and $z \in \mathfrak{h}$, we define

$$\gamma \cdot z = \frac{az + b}{cz + d}.$$

It is easy to check that if $z \in \mathfrak{h}$ then $\gamma \cdot z \in \mathfrak{h}$, and that $\gamma_1 \cdot (\gamma_2 \cdot z) = (\gamma_1 \gamma_2) \cdot z$.

We want to consider functions on the upper half-plane which are "as invariant as possible" under this action, perhaps when restricted to a smaller group. The subgroups we will need to consider are the "congruence subgroups" which we get by adding a congruence condition to the entries of the matrix. Thus, for any positive integer N, we want to look at the group

$$\Gamma_0(N) = \left\{ \gamma = \begin{pmatrix} a & b \\ c & d \end{pmatrix} \in \mathrm{SL}_2(\mathbb{Z}) : c \equiv 0 (\bmod N) \right\}.$$

We are now ready to begin defining modular forms. They will be functions $f : \mathfrak{h} \to \mathbb{C}$, holomorphic, which "transform well" under one of the subgroups $\Gamma_0(N)$. To be specific, we require that there exist an integer k such that

$$f\left(\frac{az + b}{cz + d}\right) = (cz + d)^k f(z).$$

Applying this formula to the special case in which the matrix is

$$\begin{pmatrix} 1 & 1 \\ 0 & 1 \end{pmatrix}$$

shows that any such function must satisfy $f(z + 1) = f(z)$, and hence must have a Fourier expansion

$$f(z) = \sum_{n=-\infty}^{\infty} a_n q^n, \quad \text{where } q = e^{2\pi i z}.$$

We require that this expression in fact only involve non-negative powers of q (and in fact we extend that requirement to a finite number of other, similar, expansions, which the experts call the "Fourier expansions at the other cusps"). A function satisfying all these constraints is called *a modular form of weight k on $\Gamma_0(N)$*.

We will need to consider one special subspace of the space of modular forms of a given weight and level. Rather than having a Fourier expansion with non-negative powers only, we might require *positive* powers only (in the main expansion and in the ones "at the other cusps"). We call such modular forms *cusp forms*; they turn out to be the more interesting part of the space of modular forms.

Finally, one must make a remark on the relation between the theory at various levels: if N divides M, then every form of level N (and weight k) gives rise to (a number of) forms of level M (and the same weight). The subspace generated by all forms of level M and weight k which arise in this manner (from the various divisors of M) is called the space of *old forms* of level M. With respect to a natural inner product structure on the space of modular forms, one can then take the orthogonal complement of the space of old forms. This complement is called the space of *new forms*, which are the ones we will be most interested in.

What really makes the theory of modular forms interesting for arithmetic is the existence of a family of commuting operators on each space of modular forms, called the *Hecke operators*. We will not go into the definition of these operators (they are quite natural from the point of view of representation theory); for us the crucial things will be:

- For each positive integer n relatively prime to the level N, there is a Hecke operator T_n acting on the space of modular forms of fixed weight and level N.

- The Hecke operators commute with each other.
- If m and n are relatively prime, then $T_{nm} = T_n T_m$.

We will be especially interested in modular forms which are eigenvectors for the action of all the Hecke operators, i.e., forms for which there exist numbers λ_n such that $T_n(f) = \lambda_n f$ for each n which is relatively prime to the level. We will call such forms *eigenforms*.

This is all quite strange and complicated, so let's immediately point out one connection between modular forms and elliptic curves. Suppose one has a modular form which is

- of weight 2 and level N,
- a cusp form,
- new,
- an eigenform.

If that is the case, one can normalize the form so that its Fourier expansion looks like

$$f(z) = \sum_{n=1}^{\infty} a_n q^n \text{ with } a_1 = 1.$$

Suppose that, once we have done the normalization,

- all of the Fourier coefficients a_n are integers.

Then there exists an elliptic curve whose equation has integer coefficients, whose conductor is N, and whose a_n are exactly the ones that appear in the Fourier expansion of f. In particular, the L-function of E can be expressed in terms of f (as a Mellin transform), and the nice analytic properties of f then allow us to prove that the L-function does have an analytic continuation and does satisfy a functional equation.

This connection between forms and elliptic curves is so powerful that it led people to investigate the matter further. The first one to suggest that *every* elliptic curve should come about in this manner was Y. Taniyama, in the mid-fifties. The suggestion only penetrated the mathematical culture much later, largely due to the work of G. Shimura, and it was made more precise by A. Weil's work pinning down the role of the conductor. We now call this the *Shimura-Taniyama-Weil Conjecture*. Here it is:

Conjecture 1 (Shimura-Taniyama-Weil). *Let E be an elliptic curve whose equation has integer coefficients. Let N be the conductor of E, and for each n let a_n be the number appearing in the L-function of E. Then there exists a modular form of weight 2, now of level N, an eigenform under the Hecke operators, and (when normalized) with Fourier expansion equal to $\sum a_n q^n$.*

For any specific curve, it is not too hard to check that this is true. One takes E, determines the conductor and the a_n for a range of n. Since the space of modular forms of weight 2 and level N is finite-dimensional, knowing enough of the a_n must determine the form, and we can go and look if it is there. (In general, given a list of a_n it is not at all easy to determine whether $\sum a_n q^n$ is the Fourier expansion of a modular form, so we need to do it the other way: we generate a basis of the space of modular forms, then try to find our putative form as a linear combination of the basis.) If we find a form with the right (initial chunk of) Fourier expansion, this gives *prima facie* evidence that the curve satisfies the STW conjecture. To clinch the matter, one can use a form of the Čebotarev density theorem to show that if *enough* (in an explicit sense) of the a_n are right, then they all are.

This method has been used to verify the STW conjecture for any number of specific curves (see, for example, [6]). The conjecture has a really crucial role in the theory of elliptic curves; in fact, curves that satisfy the conjecture are known as "modular elliptic curves", and many of the fundamental new results in the theory have only been proved for curves that have this property.

As our final remark on modular forms, we point out that it is possible, for any given N, to determine (essentially using the Riemann-Roch theorem) the exact dimension of the space of cusp forms of weight 2 and level N. This gives us a very good handle on what curves of that conductor should exist (if the STW conjecture is true). For more information on modular forms, one might look at [17], [23], or [28]. There is an intriguing account of the Shimura-Taniyama-Weil conjecture, in a very different spirit, in Mazur's article [21], and a useful survey in [18].

Galois representations

The final actors in our play are Galois representations. One starts with the Galois group of an extension of the field of rational numbers. To understand this Galois group, one can try to "represent" the elements of the group as matrices. In other words, one can try to find a vector space on which our Galois group acts, which gives a way to associate a matrix to each element of the group. This in fact gives a group homomorphism from the Galois group to a group of matrices (this need not be injective; when it is, one calls the representation *faithful*).

Rather than work with specific finite extensions of $\mathbb{Q}$, we work with the Galois group $G = \text{Gal}(\bar{\mathbb{Q}}/\mathbb{Q})$ of the algebraic closure of $\mathbb{Q}$. This is a huge group (which one makes more manageable by giving it a topology) that hides within itself an enormous amount of arithmetic information. The representations we will be considering will be into 2×2 matrices over various fields and rings, and they will (for the most part) be obtained from elliptic curves and from modular forms.

To see how to get Galois representations from an elliptic curve, let's start with an elliptic curve E, whose equation has coefficients in $\mathbb{Z}$. Choose a prime p. Since the (complex, say) points of E form a group, one can look in this group for points which are of order p (that is, for points (x, y) such that adding them to themselves p times gives the identity). It turns out that (over $\mathbb{C}$) there are p^2 such points, and they form a subgroup which we denote by $E[p]$. In fact, this group is isomorphic to the product of two copies of $\mathbb{F}_p$:

$$E[p] \cong \mathbb{F}_p \times \mathbb{F}_p.$$

Now, the points in $E[p]$ are *a priori* complex, but on closer look one sees that in fact they are all defined over some extension of $\mathbb{Q}$, and in particular that transforming the coefficients of a point of order p by the Galois group G yields another point of order p. In fact, it's even better than that: since the rule for adding points is defined in rational terms, the whole group structure is preserved. Since $E[p]$ looks like a vector space of dimension 2 over $\mathbb{F}_p$, this means that each element of G acts as a linear transformation on this space, and hence that we get a representation

$$\bar{\rho}_{E,p} : G \to \text{GL}_2(\mathbb{F}_p).$$

(We use a bar to remind ourselves that this is a representation "modulo p".)

Now, $\text{GL}_2(\mathbb{F}_p)$ is a finite group, and G is very infinite, so this representation, while it tells us a lot, can't be the whole story. It turns out, however, that we can use p-adic numbers to get a whole lot more. Instead of considering only the points of order p, we can consider points of order p^n for each n. This gives a whole bunch of subgroups

$$E[p] \subset E[p^2] \subset E[p^3] \subset \cdots$$

and a whole bunch of representations, into $\text{GL}_2(\mathbb{F}_p)$, then into $\text{GL}_2(\mathbb{Z}/p^2\mathbb{Z})$, then into $\text{GL}_2(\mathbb{Z}/p^3\mathbb{Z})$, ... Putting all of these together ends up by giving us a p-adic representation

$$\rho_{E,p} : G \to \text{GL}_2(\mathbb{Q}_p),$$

which hides within itself all of the others. The representations $\rho_{E,p}$ contain a lot of arithmetic information about the curve E.

And how does it look on the modular forms side? Well, it follows from the work of several mathematicians (M. Eichler, G. Shimura, P. Deligne, and J.-P. Serre) that, whenever we have a modular form f (of any weight) which is an eigenform for the action of the Hecke operators and whose Fourier coefficients (after normalization) are integers, we can construct a representation

$$\rho_{f,p} : G \to \text{GL}_2(\mathbb{Q}_p)$$

which is attached to f in a precise sense which is too technical to explain here. (The construction of the representation is quite difficult, and in fact no satisfactory expository account is yet available.)

The crucial thing to know, for our purposes, is that *when an elliptic curve E arises from a modular form f, then the representations $\rho_{E,p}$ and $\rho_{f,p}$ are the same*. In fact, a converse is also true: given a curve E, if one can find a modular form f such that $\rho_{E,p}$ is the same as $\rho_{f,p}$, then E will be modular.

3 The play

We are now ready to take the plunge and try to see how all of the theory relates to Fermat's Last Theorem. The idea is to assume that FLT is false, and then, using this assumption, to construct an elliptic curve that contradicts just about every conjecture under the sun.

Linking FLT to elliptic curves

So let's start by assuming FLT is false, i.e., that there exist three non-zero integers u, v and w such that $u^p + v^p + w^p = 0$ (as we know, we only need to consider the case of prime exponent p, which is therefore odd, so that we can recast a solution in Fermat's form to be in the form above). Since we know that the theorem is true for $p = 3$, we might as well assume that $p \geq 5$. We may clearly assume that u, v and w are relatively prime, which means that precisely one of them must be even. Let's say v is even. Since p is bigger than 2, we can see, by looking at the equation modulo 4, that one of u and w must be congruent to -1 modulo

4, and the other must be congruent to 1. Let's say $u \equiv -1 \pmod 4$.

Let's use this data to build an elliptic curve, following an idea due to G. Frey (see [9], [10], [11]). We consider the curve

$$y^2 = x(x - u^p)(x + v^p).$$

This is usually known as the *Frey curve*. Following our discussion, above, of the curve E_{abc}, we already know quite a bit about the Frey curve. Here's a summary:

1. Since v is even and $p \geq 5$, we know that we have $v^p \equiv 0 \pmod{32}$. We also know that $u^p \equiv -1 \pmod 4$. This puts us in the right position to use what we know about the curves E_{abc}.

2. The minimal discriminant of the Frey curve is
$$\Delta = \frac{(uvw)^{2p}}{256}.$$

3. The conductor of the Frey curve is the product of all the primes dividing $u^p v^p w^p$, which is, of course, the same as the product of all the primes dividing uvw.

4. The Frey curve is semistable.

Now, as Frey observed in the mid-1980s, this curve seems much too strange to exist. For one thing, its conductor is extremely small when compared to its discriminant (because of that exponent of $2p$). For another, its Galois representations are pretty weird. Very soon, people were pointing out that there were several conjectures that would rule out the existence of Frey's curve, and therefore would prove that Fermat was correct in saying that his equation had no solutions.

FLT follows from the Shimura-Taniyama-Weil conjecture

It was already clear to Frey that it was likely that the existence of his curve would contradict the Shimura-Taniyama-Weil conjecture, but he was unable to give a solid proof of this. A few months after Frey's work, Serre pinpointed, in a letter to J.-F. Mestre, exactly what one would need to prove to establish the link. In this letter (published as [26]), Serre describes the situation with the phrase "STW $+ \varepsilon$ implies Fermat." Because of this, the missing theorem became known, for a while, as "conjecture epsilon". This conjecture was proved by K. A. Ribet in [25] about a year later, and this established the link. A survey of these results can be found in [19].

What Serre noticed was that the representation modulo p

$$\bar{\rho}_{E,p} : G \to \mathrm{GL}_2(\mathbb{F}_p)$$

obtained from the Frey curve was rather strange. It looked like the sort of representation one would get from a modular form of weight 2, but if one applied the "usual recipe" for guessing the level of that modular form, the answer came out to be $N = 2$. He also showed that the modular form must be a cusp form. The problem is that *there are no cusp forms of weight 2 and level 2*!

So suppose there is a solution of the Fermat equation for some prime p, and use this solution to build a Frey curve E. Let N be the conductor of E (which we determined above). Suppose, also, that STW holds for E, so that there exists a modular form of weight 2 and level N whose Galois representation is the same as the one for E. Then we have the following curious situation: we have a representation $\bar{\rho}$ which we know comes from a modular form of weight 2 and level N, but which *looks* as if it should come from a modular form of smaller level.

Here is where Ribet's theorem comes in: he proves that (under certain hypotheses which will hold in our case) whenever this happens the modular form of smaller level must actually exist! Notice that this doesn't mean that the original modular form came from lower level; what it means is that there is a form of lower level whose representation reduces modulo p to the same representation.

The upshot of Ribet's theorem is the following:

Theorem 1 (Ribet). *Suppose STW holds for all semistable elliptic curves. Then FLT is true.*

This is true because if FLT were false, one could choose a solution of the Fermat equation and use it to construct a Frey curve, which would be a semistable elliptic curve. By STW, this curve would be attached to a modular form, so that its Galois representation is attached to a modular form. By Ribet's theorem, there must exist a modular form of weight 2 and level 2 which gives the same representation modulo p. Just a little more work allows one to check that this modular form must be a cusp form. But this is a contradiction, because there are no cusp forms of weight 2 and level 2.

Deforming Galois representations

It is now that we come to Wiles's work. His idea was that one can attack the problem of proving STW by using the Galois representations, and in particular by thinking of "deformations" of Galois representations. The idea is to consider not only a representation modulo p, but also all the possible p-adic representations attached to it (one speaks of "all the possible lifts" of the representation modulo p). These can be thought of as "deformations" because, from the p-adic point of view, they are "close" to the original representation.

This sort of idea had been introduced by B. Mazur in [20]. Mazur showed that one could often obtain a "universal lift", i.e., a representation into GL_2 of a big ring such that all possible lifts were "hidden" in this representation. If one knew that the representations modulo p were modular, then one could make another big ring "containing" all the lifts which are attached to modular forms. The abstract deformation theory then provides us with a homomorphism between these two rings: and one can try to prove that this is an isomorphism. If so, it follows that all lifts are modular.

What Wiles proposes to do is very much in this spirit, except that he restricts himself to lifts that have especially nice properties. He starts with a representation modulo p, and supposes it is modular and that it satisfies certain technical assumptions. Then he considers all possible deformations which "look like they could be attached to forms at weight 2," and gets a deformation ring. Considering all deformations which are attached to modular forms of weight 2 gives a second ring (which is closely related to the algebra generated by the Hecke operators, in fact). Wiles then attempts to prove, using a vast array of recent results, including ideas of Mazur, Ribet, Faltings, V. Kolyvagin, and M. Flach, that these two rings are the same.

It is not hard to see that the homomorphism between the two rings we want to consider is surjective. The difficulty is to prove it is also injective. Wiles reduces this question to bounding the size of a certain cohomology group. It is here that the brilliant ideas of Kolyvagin and of Flach come in. About five years ago, Kolyvagin came up with a very powerful method for controlling the size of certain cohomology groups using what he calls "Euler systems" (see [16] and the survey of the method in [22]). This method seems to be adaptable to any number of situations, and has been used to prove several important recent results. The initial breakthrough showing how one could begin to use Kolyvagin's method in our context is due to Flach (see [8]), who found a way to construct something that can be thought of as the beginning of an Euler system applicable to our situation. Wiles called on all these ideas to construct a "geometric Euler system" which plays a central role in the argument. (It is at this point that the current difficulty lies.)

From the bound on the cohomology group one will get a proof that the two rings are in fact isomorphic. Translated back to the language of representations, this means that if one starts with a representation modulo p which satisfies Wiles's technical assumptions (and is modular), then any lift of the kind Wiles considers is also modular.

Put it all together . . .

Assume, then, that one can prove that all lifts of a modular representation are still modular. Now suppose we have an elliptic curve E whose representation modulo p we can prove (by some means) to be modular. Suppose also that this representation satisfies Wiles's technical assumptions. Then any lift of this representation is modular. But the p-adic representation $\rho_{E,p}$ attached to E is one such lift! It follows that this representation is modular, and hence that E is modular.

All we need, now, is to prime the pump; we must find a way to decide that the representation modulo p is modular, and then use that to clinch the issue. What Wiles does is quite beautiful. First of all, he takes a semistable elliptic curve and looks at the Galois representation modulo 3 attached to this curve. At this point, there are two possibilities. The representation, as we pointed out above, amounts to an action of the Galois group on the vector space $\mathbb{F}_3 \times \mathbb{F}_3$. Now, it may happen that there is a subspace of that vector space which is invariant under every element of the Galois group. In that case, one says that the representation is *reducible*. If not, it is *irreducible*.

One has to be just a little more careful. Just as it sometimes happens that a real matrix has complex eigenvalues, it can happen that the invariant subspace only exists after we enlarge the base field. We will say a representation is *absolutely irreducible* when this does not happen: even over bigger fields, there is no invariant subspace.

Well, look at $\bar{\rho}_{E,3}$. It may or may not be absolutely irreducible. If it is, Wiles calls upon a famous theorem of J. Tunnell, based on work of R. P. Langlands (see [33], [19]) to show that it is modular, and hence, using the deformation theory, that the curve is modular.

If $\bar{\rho}_{E,3}$ is not absolutely irreducible, Wiles shows that there is another elliptic curve which has the same representation modulo 5 as our initial curve, but whose representation modulo 3 is absolutely irreducible. By the first case, it is modular. Hence, its representation modulo 5 is modular. But since this is the same as the representation modulo 5 attached to our original curve, we can apply the deformation theory for $p = 5$ to conclude that our original curve is modular.

If Wiles's strategy is successful, we get:

Theorem 2. *The Shimura-Taniyama-Weil conjecture holds for any semistable elliptic curve.*

And, since the Frey curve is semistable,

Corollary 1. *For any $n \geq 3$, there are no non-zero integer solutions to the equation $x^n + y^n = z^n$.*

And of course, this is just *one* corollary of the proof of the STW conjecture for semistable curves, and it is certain that there will he many others still. For example, as Serre pointed out in [27], one can apply Frey's ideas to many other Diophantine equations that are just as hard to handle as Fermat's. These are equations that are closely related to the Fermat equation, of the form

$$x^p + y^p = Mz^p,$$

where p is a prime number and M is some integer. From Serre's argument and Wiles's result, one gets something like this:

Corollary 2. *Let p be a prime number, and M be a power of one of the following primes:*

$$3, 5, 7, 11, 13, 17, 19, 23, 29, 53, 59.$$

Suppose that $p \geq 11$ and that p does not divide m. Then there are no non-zero integer solutions of the equation $x^p + y^p = Mz^p$.

The proof is precisely parallel to what we have done before: given a solution, construct a Frey curve, and consider the resulting modular form. Apply Ribet's theorem to lower its level, and then study the space of modular forms of that level to see if the form predicted by Ribet is there. If there is no such form, there can be no solution.

In fact, one can even get a general result, as Mazur pointed out:

Corollary 3. *Let M be a power of a prime number ℓ, and assume that ℓ is not of the form $2^n \pm 1$. Then there exists a constant C_ℓ such that the equation $x^p + y^p = Mz^p$ has no non-zero solutions for any $p \geq C_\ell$.*

However successful they may be in the end at proving the Shimura-Taniyama-Weil conjecture, Wiles's ideas are certain to have enormous impact.

References

1. Y. Amice, *Les Nombres p-adiques*, Univ. Press of France, 1975.

2. Henri Cohen et al., *GP-PARI*, a number-theoretic calculator and C library. Available by anonymous ftp from *math.ucla.edu*.

3. J. W. S. Cassels, *Local Fields*, Cambridge Univ. Press, 1986.

4. J. W. S. Cassels, *Lectures on Elliptic Curves*, Cambridge Univ. Press, 1991.

5. I. Connell, *APECS: arithmetic of plane elliptic curves*, an add-on to *Maple*. Available by anonymous ftp from *math.mcgill.edu*.

6. J. E. Cremona, *Algorithms for Modular Elliptic Curves*, Cambridge Univ. Press, 1992.

7. G. Cornell and J. H. Silverman (eds.), *Arithmetic Geometry*, Springer-Verlag, 1986.

8. M. Flach, A finiteness theorem for the symmetric square of an elliptic curve, *Invent. Math.* **109** (1992), 307–327.

9. G. Frey, Links between stable elliptic curves and certain Diophantine equations, *Annales Universitatis Saraviensis* **1** (1986), 1–40.

10. G. Frey, Links between elliptic curves and solutions of $A - B = C$, *Indian Math. Soc.* **51** (1987), 117–145.

11. G. Frey, Links between solutions of $A - B = C$ and elliptic curves, in *Number Theory, Ulm 1987* (eds. H. P. Shlickewei and E. Wirsing), Lecture Notes in Mathematics **1380**, Springer-Verlag, 1987.

12. F. Q. Gouvêa, *p-adic Numbers: an Introduction*, Springer-Verlag, 1993.

13. D. Husemöller, *Elliptic Curves*, Springer-Verlag, 1987.

14. A. W. Knapp, *Elliptic Curves*, Princeton Univ. Press, 1992.

15. N. Koblitz, *p-adic Numbers, p-adic Analysis, and Zeta Functions*, Springer-Verlag, 2nd ed., 1984.

16. V. Kolyvagin, Euler systems, in *The Grothendieck Festshrift*, Vol. 2, Birkhäuser, 1991, pp. 435–483.

17. S. Lang, *Introduction to Modular Forms*, Springer-Verlag, 1974.

18. R. P. Langlands, *Base change for $GL(2)$*, Ann. Math. Stud. **96**, Princeton Univ. Press, 1980.

19. S. Lang, Number Theory III, in *Encyclopedia of Mathematical Sciences*, Vol. 3, Springer-Verlag, 1991.

20. B. Mazur, Deforming Galois representations, in *Galois Groups over $\mathbb{Q}$* (eds. Y. Ihara, K. A. Ribet and J.-P. Serre) Springer-Verlag, 1989.

21. B. Mazur, Number theory as gadfly, *American Math. Monthly* **98** (1991), 593–614.

22. B. Mazur, On the passage from local to global in number theory, *Bull. Amer. Math. Soc.* **29** (1993), 14–50.

23. T. Miyake, *Modular Forms*, Springer-Verlag, 1989.

24. P. Ribenboim, *13 Lectures on Fermat's Theorem*, Springer-Verlag, 1979.

25. K. A. Ribet, On modular representations of $\mathrm{Gal}(\bar{\mathbb{Q}}/\mathbb{Q})$ arising from modular forms, *Invent. Math.* **100** (1900), 431–476.

26. J.-P. Serre, Lettre à J-F Mestre, in *Current Trends in Arithmetical Algebraic Geometry* (ed. K. A. Ribet), Contemporary Mathematics **67**, American Mathematical Society, 1987.

27. J.-P. Serre, Sur les représentations modulaires de degré 2 de $\mathrm{Gal}(\bar{\mathbb{Q}}/\mathbb{Q})$, *Duke Math. J.* **54** (1987), 179–230.

28. G. Shimura, *Introduction to the Arithmetic Theory of Automorphic Forms*, Princeton Univ. Press, 1971.

29. J. H. Silverman, *The Arithmetic of Elliptic Curves*, Springer-Verlag, 1986.

30. J. H. Silverman, Taxicabs and sums of two cubes, *Amer. Math. Monthly* **100** (1993), 331–340.

31. J. H. Silverman and J. Tate, *Rational Points on Elliptic Curves*, Springer-Verlag, 1992.

32. J. H. Silverman and P. van Mulbregt, *EllipticCurveCalc*, a *Mathematica* package. Available by anonymous ftp from *gauss.math.brown.edu*; contact *jhs@gauss.math.brown.edu* for information.

33. J. Tunnell, Artin's conjecture for representations of octahedral type, *Bull. Amer. Math. Soc.* **5** (1981), 173–175.

34. L. C. Washington, *Introduction to Cyclotomic Fields*, Springer-Verlag, 1982.

35. A. Weil, *Number Theory: an Approach through History, from Hammurapi to Legendre*, Birkhäuser, 1983.

36. H. G. Zimmer *et al.*, *SIMATH*, a computer algebra system with main focus on algebraic number theory. Contact *simath@math.uni-sb.de* for more information.

Afterword

In chapter 10 of his *A History of Algebra* [26], B. L. van der Waerden repeats much of what he wrote about Hamilton's discovery of quaternions. Interestingly, there he mentions Caspar Wessel as one of the originators of the geometric interpretation of complex numbers, while in the current article he ignores Wessel. But he also goes on to discuss Cayley's own use of quaternions to describe rotations in three-space, meanwhile pointing out the earlier results of Rodrigues. In addition, he deals with some applications of quaternions to the question of representing integers as sums of four squares. He concludes by discussing Hermann Hankel's 1867 book that presents many of Grassmann's results, but in a form that was easier to understand.

Simon Altmann writes in his article that we know "next to nothing" about Olinde Rodrigues, but in the next fifteen years he remedied this situation, publishing the results in his recent biography, *Mathematics and Social Utopias in France: Olinde Rodrigues and His Times* [1]. Similarly, Karen Parshall went on to do further research on the life and work of Sylvester. Her results appear in her edition of Sylvester's letters [18] as well as in her magnificent biography of the English mathematician [19].

Israel Kleiner has expanded his paper on group theory and some of his other work on the history of algebra into a new book, *A History of Abstract Algebra* [15]. Leo Corry's *Modern Algebra and the Rise of Mathematical Structures* [5] is another recent work that concentrates specifically on the development of abstraction in the nineteenth and twentieth centuries, but claims that true abstraction did not come into being until the work of Emmy Noether in the 1920s. Corry's conclusion is based on his detailed study of algebra texts of the period, and he reports on this in an article [6] on the change in the conception of algebra between 1895 and 1930. Sujoy Chakraborty and Munibur Rahman Chowdhury [4] have studied the immediate motivations for Arthur Cayley's first paper on group theory and argue that, despite Cayley's seemingly faulty definition of an abstract group, he really understood the central ideas and did in fact include all the necessary conditions. There is also a new study of the idea of quotient groups by Julia Nicholson [17], who detailed the reasons for the development of this idea throughout the nineteenth century.

Joseph Gallian wrote his report on finite simple groups in 1976, when it was not yet known that no more sporadic groups would be found. But in February, 1981, Daniel Gorenstein, one of the leading group theorists of the late twentieth century, confirmed that the "Monster" was the last of the sporadic groups and the classification of the finite simple groups was complete. Two recent articles by Ron Solomon [25] and Michael Aschbacher [2] discuss the current status of the enormous proof of the classification. Two recent popular books discuss the long search for all of the sporadic simple groups, Mark Ronan's *Symmetry and the Monster* [23], and Marcus du Sautoy's *Symmetry: A Journey into the Patterns of Nature* [10].

Both versions of Hermann Grassmann's *Ausdehnungslehre* have been translated by Lloyd Kannenberg ([12], [13]). In addition, in 1994 there was a major conference in Germany to celebrate the 150th anniversary of the publication of the first edition. The proceedings have been published as *Hermann Günther Grassmann (1809–1877): Visionary Mathematician, Scientist, and Neohumanist Scholar* [24] and contain numerous essays on Grassmann's work and influence.

In Kleiner's discussion of Dedekind's development of ideals, he notes that Leopold Kronecker developed an alternative approach to the subject of algebraic number theory around the same time. Although Dedekind's approach prevailed, Harold Edwards has written a detailed study of Kronecker's theory in *Divisor Theory* [11]. There is also a more general study of divisibility theory in a recent paper by Olaf Neumann [16].

Edward Waring's book, in which Waring's problem was first stated, has been translated into English by Dennis Weeks [27]. There continue to be new developments in the mathematics of this problem, which are frequently reported in the mathematics press. There have also been several recent books dealing with the Riemann hypothesis and its relationship to the prime number theorem, including John Derbyshire's *Prime Obsession* [7], Marcus du Sautoy's *The Music of the Primes* [9] and Don Rockmore's *Stalking the Riemann Hypothesis* [22].

As to Fermat's Last Theorem, it turned out that Wiles's announced proof of 1993 had a flaw, but one that he and a former student, Richard Taylor, were finally able to correct about a year later. Nevertheless, Gouvêa's outline of the proof is still the best study of the proof aimed at non-specialist mathematicians. More details are in Griffith's article in Part 4 of this book as well as in the reference mentioned in the afterword to that chapter. Recently, David Pengelley and Reinhard Laubenbacher have shown that Sophie Germain in the early nineteenth century had developed her own strategy for proving the theorem, although, as it turned out, the strategy could not be successful. See [20] for details.

Robert Kanigel has written a wonderful biography of Ramanujan: *The Man Who Knew Infinity* [14]. Bruce Berndt and George Andrews have spent many years working through Ramanujan's notebooks. They have written a series of volumes in recent years detailing the mathematics in these notebooks, all published by Springer. Finally, Auguste Dick's biography of Emmy Noether was translated into English and published in 1981 [8]. At the same time J. W. Brewer and M. K. Smith edited a volume of essays paying tribute to her: *Emmy Noether: A Tribute to her Life and Work* [3].

References

1. Simon Altmann, *Mathematics and Social Utopias in France: Olinde Rodrigues and His Times*, American Mathematical Society and London Mathematical Society, 2005.

2. Michael Aschbacher, The Status of the Classification of the Finite Simple Groups, *Notices Amer. Math. Soc.* **51** (2004), 736–740.

3. J. W. Brewer and M. K. Smith (eds.), *Emmy Noether: A Tribute to her Life and Work*, Marcel Dekker, 1981.

4. Sujoy Chakraborty and Munibur Rahman Chowdhury, Arthur Cayley and the abstract group concept, *Math. Magazine* **78** (2005), 269–282.

5. Leo Corry, *Modern Algebra and the Rise of Mathematical Structures*, Birkhäuser, 1996.

6. Leo Corry, From *Algebra* (1895) to *Moderne Algebra* (1930): changing conceptions of a discipline — a guided tour using the *Jahrbuch über die Fortschritte der Mathematik*, *Episodes in the History of Modern Algebra (1800–1950)* (ed. Jeremy Gray and Karen Parshall), American Mathematical Society and London Mathematical Society, 2007, pp. 221–243.

7. John Derbyshire, *Prime Obsession: Bernhard Riemann and the Greatest Unsolved Problem in Mathematics*, Joseph Henry Press, 2003.

8. Auguste Dick, *Emmy Noether, 1882–1935*, Birkhäuser, 1981.
9. Marcus du Sautoy, *The Music of the Primes: Searching to Solve the Greatest Mystery in Mathematics*, Harper, 2004.
10. Marcus du Sautoy, *Symmetry: A Journey into the Patterns of Nature*, Harper, 2008.
11. Harold Edwards, *Divisor Theory*, Birkhäuser, 1990.
12. Hermann Grassmann, *A New Branch of Mathematics: The Ausdehnungslehre of 1844 and Other Works* (transl. Lloyd C. Kannenberg), Open Court, 1995.
13. Hermann Grassmann, *Extension Theory* (transl. Lloyd C. Kannenberg), American Mathematical Society, 2000.
14. Robert Kanigel, *The Man Who Knew Infinity: A Life of the Genius Ramanujan*, Charles Scribner, 1991.
15. Israel Kleiner, *A History of Abstract Algebra*, Birkhäuser, 2007.
16. Olaf Neumann, Divisibility Theories in the Early History of Commutative Algebra and the Foundations of Algebraic Geometry, *Episodes in the History of Modern Algebra (1800–1950)* (ed. Jeremy Gray and Karen Parshall), American Mathematical Society and London Mathematical Society, 2007, pp. 73–105.
17. Julia Nicholson, The development and understanding of the concept of quotient group, *Historia Mathematica* **20** (1993), 68–88.
18. Karen Parshall, *James Joseph Sylvester: Life and Work in Letters*, Oxford Univ. Press, 1998.
19. Karen Parshall, *James Joseph Sylvester: Jewish Mathematician in a Victorian World*, Johns Hopkins Univ. Press, 2006.
20. David Pengelley and Reinhard Laubenbacher, "Voici ce que j'ai trouvé": Sophie Germain's grand plan to prove Fermat's Last Theorem, available at *www.math.nmsu.edu/~davidp/germain.pdf*.
21. Laura Toti Rigatelli, *Evariste Galois: 1811–1832* (transl. John Denton), Birkhäuser, 1996.
22. Don Rockmore, *Stalking the Riemann Hypothesis: The Quest to Find the Hidden Law of Prime Numbers*, Pantheon, 2005.
23. Mark Ronan, *Symmetry and the Monster: The Story of One of the Greatest Quests of Mathematics*, Oxford Univ. Press, 2006.
24. Gert Schubring (ed.), *Hermann Günther Grassmann (1809–1877): Visionary Mathematician, Scientist and Neohumanist Scholar*, Springer-Verlag, 1996.
25. Ron Solomon, On finite simple groups and their classification, *Notices Amer. Math. Soc.* **42** (1995), 231–239.
26. B. L. van der Waerden, *A History of Algebra from al-Khwarizmi to Emmy Noether*, Springer-Verlag, 1985.
27. Edward Waring, *Meditationes Algebraicae* (transl. Dennis Weeks), American Mathematical Society, 1991.

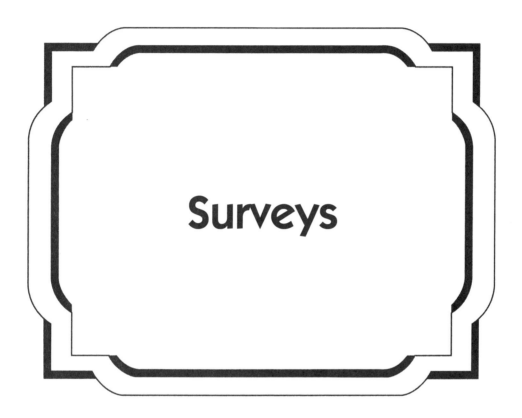

Surveys

Foreword

This final chapter contains three survey articles on mathematics, dating from 1900, 1951, and 2000, as well as a brief and subjective account of the Second International Congress of Mathematicians, held in Paris in August 1900.

George Bruce Halsted was one of the American delegates to the Congress and wrote a report for the *Monthly* shortly after he returned. The major part of the paper deals with his reactions to Hilbert's famous address on the problems of mathematics, an address that set the agenda for twentieth-century work in mathematics. But the other talk that particularly interested Halsted was one on Japanese mathematics, by Rikitaro Fujisawa (1861–1933). Fujisawa's conclusions as to the Japanese independent discovery of both zero and the square root of -1 are not accepted today.

In a report written for the beginning of the twentieth century, G. A. Miller discusses some "new fields" of mathematics, fields that seemed to him to be particularly fertile. Among the important areas currently under active investigation, Miller picked the arithmetization of analysis, the development of set theory, and the study of groups as particularly worthy of further attention. He also noted that practical applications of mathematics were important; in particular, he was impressed with the discovery of a linkage that would construct a straight line.

A half-century later, Hermann Weyl discussed the mathematics of the first half of the twentieth century. In the areas of algebra and number theory, Weyl emphasizes the advances in the abstract theory of rings, ideals, and fields, pointing out how van der Waerden's text *Moderne Algebra* explaining these subjects was written from a structural and abstract point of view, far different from the textbooks of the turn of the century. Weyl particularly stresses how the development of p-adic analysis enabled mathematicians to develop class field theory, "that most fascinating branch of mathematics." He also deals with the theory of group representations, especially its application to continuous groups. In analysis, Weyl considers spectral theory in Hilbert spaces as one of the most important new developments (see Steen's article earlier in this volume). He also notes the importance of Lebesgue's theory of integrals and its generalizations, as well as homology theory and algebraic topology in general. Turning next to geometry, Weyl discusses differential geometry and its relation to general relativity, calling it "the most important development of geometry in the twentieth century." He concludes with a few words about the foundations of mathematics, in particular the effect of the work of Gödel on the various philosophies of mathematics.

Passing on to the beginning of the twenty-first century, Phillip Griffiths deals with the major achievements of the late twentieth century and the challenges for the future. In particular, he discusses Andrew Wiles's proof of Fermat's Last Theorem, Thomas Hales's settling of Kepler's Sphere Packing Conjecture, and Kenneth Appel and Wolfgang Haken's solution to the four-color problem. Griffiths then shows how late twentieth-century mathematics defied the earlier trends toward fragmentation of the subject by developing the connections among many seemingly disparate fields, and also continued to develop new interactions with other sciences. Among the most important of the interactions is with the life sciences, as mathematical biology now seems to be ever more important for the understanding of our world. Griffiths concludes his articles with a list of some of the major mathematical challenges for mathematics in the twenty-first century, including the Riemann hypothesis, the Poincaré conjecture, and the question whether $P = NP$, as well as the more general challenges of mathematics education and outreach to the general public.

The International Congress of Mathematicians

GEORGE BRUCE HALSTED

American Mathematical Monthly **7** (1900), 188–189

On the sixth of August at the Palais des Congrès in the Paris Exposition, was held the opening session of the second International Congress of Mathematicians. The president, Poincaré, is regarded as the greatest of living mathematicians. Among the vice presidents in attendance were Gordan, Lindeloef, Lindemann, Mittag-Leffler.

Representing Japan was Fujisawa; Spain sent Zoel de Galdeano; the United States, Miss Scott. The president of the section of Arithmetic and Algebra was Hilbert; of Geometry was Darboux, of Bibliography and History was Prince Roland Bonaparte. Among the most interesting personalities present may be mentioned Dickstein of Warsaw, Gutzmer of Jena, Hagen of Washington, Laisant of Paris, Langel of Golfe Juan, Lemoine of Paris, Delury of Toronto, Padoa of Rome, Shroeder of Carlsruhe, Sintsof of Yekaterinoslav, Stringham of Berkeley, Tannery of Paris, Vasiliev of Kazan, Whitehead of Cambridge.

Of the many important papers presented two may be selected for their general interest and the enthusiasm with which they were received. These are: *The Mathematics of the Old Japanese School* by Fujisawa, and *The Problems of Mathematics* by Hilbert.

Among other matters of extraordinary importance, Fujisawa showed his astonished audience that the Japanese had independently discovered the zero and by a mysterious coincidence used for it a circular symbol as did the Hindus and as do we. He showed that the Japanese had rectified the circle with an accuracy far exceeding Archimedes and only paralleled in our modern developments of pure mathematics. He showed that the Japanese had recognized $\sqrt{-1}$ (the square root of minus one) as a number, as a new unit, a neomon, and thus had reached the basis for the theory of the complex numbers. This paper is epoch-making for the history of mathematics.

Hilbert's beautiful paper on *The Problems of Mathematics* shows that when a science progresses continuously we may from the problems which actually occupy it judge of its ulterior development. The existence of precise problems has a capital importance both for the progress of mathematics and for the work of each investigator.

Whence come the problems of mathematics? It is experience that in each domain puts before us the primary problems (e.g., duplication of the cube, quadrature of the circle, etc.). In the later development of the science it is the mind which by logical reasonings (combination, generalization, specialization) creates itself problems new and fertile (e.g., problems of prime numbers). We say that a problem is solved when starting from a finite number of assumptions furnished by the problem itself we demonstrate the justness of the solution by a finite number of deductions. This mathematical rigor which we require does not necessitate complicated demonstrations; the most rigorous method is often the simplest and the easiest to comprehend.

The conceptions of arithmetic or those of analysis are not the only ones susceptible of rigorous treatment. Those of geometry and the physical sciences are equally so, provided that by means of a complete system of assumptions they are as well fixed as the conceptions of arithmetic.

When a problem presents serious difficulties, by what methods can we attack it?

First by *generalization*, in attacking the problem considered to a group of questions of the same order (e.g., introduction of ideal numbers into the theory of algebraic numbers; employment of complex paths in the theory of definite integrals).

Or else by *specialization*, in deepening the study of more simple analogous problems already solved.

The failure of attempts at the solution of a problem comes often from the problem being impossible to solve under the form given. Then we require a rigorous *demonstration of the impossibility* (parallel postulate, quadrature of the circle, algebraic solution of the equation of the fifth degree).

We say that a conception exists from the mathematical point of view when the assumptions which define it are compatible, that is to say when a finite chain or system of logical deductions starting from these assumptions can never lead to a contradiction.

Mathematics in developing, far from losing its character of unique science, manifests it from day to day more clearly. Each real progress brings necessarily the discovery of methods more incisive and more simple, permitting to each geometer an access relatively facile to all the parts of our science.

The magnificent reception given by the President of France M. Loubet and his wife Madame Emilie Loubet in which the members of the Congress participated, was only surpassed in charm by the delightful entertainment given in our honor by Prince Roland Bonaparte.

A Popular Account of Some New Fields of Thought in Mathematics

G. A. MILLER

American Mathematical Monthly 7 (1900), 91–99

At the beginning of the nineteenth century, elementary arithmetic was a Freshman subject in our best colleges. In 1802 the standard of admission to Harvard College was raised so as to include a knowledge of arithmetic to the 'Rule of Three'. A boy could enter the oldest college in America prior to 1803 without a knowledge of a multiplication table ([3], p. 60). From that time on the entrance requirements in mathematics were rapidly increased, but it was not until after the founding of Johns Hopkins University that the spirit of mathematical investigation took deep root in this country.

The lectures of Sylvester and Cayley at Johns Hopkins University, the founding of the *American Journal of Mathematics* and the young men who received their training abroad cooperated to spread the spirit of mathematical investigation throughout our land. This has led to the formation of the American Mathematical Society eight years ago, as well as to the starting of a new research journal, *The Transactions of the American Mathematical Society*, at the beginning of this year. While these were some of the results of mathematical activity, they, in a still stronger sense, tend to augment this activity.

In Europe such men as Descartes, Newton, Leibniz, Lagrange and Euler laid the foundation for the development of mathematics in many directions. These men, as well as a few of the most prominent names in the early part of the nineteenth century, were not specialists in mathematics. They were familiar with all the fields of mathematical activity in their day and some of them were well known for their contributions in other fields of knowledge. The last three-quarters of a century, and especially the last two or three decades, have witnessed a marvelous change in the mathematical activity of Europe. Mathematical periodicals have sprung up on all sides. A number of mathematical societies have been organized and many of the leading mathematicians have confined their investigations to comparatively small fields of mathematics.

The rapid increase of the mathematical literature created an imperative need of bibliographical reviews. This need was met in part by the establishment at Berlin, in 1869, of a year-book devoted exclusively to the reviews of mathematical articles, *Jahrbuch über die Fortschritte der Mathematik*. The 28th volume of this work reached our library a short time ago. It contains over 900 pages, and gives a review of over 2000 memoirs and books. With a view towards further increasing the facilities to keep in touch with the growing literature, the Amsterdam Mathematical Society commenced the publication of a semi-annual review, *Revue Semestrielle*, in 1893. In the last number of this, 236 periodicals are quoted, each of which contains, at times, mathematical articles that are of sufficient merit to be noted. Each of the four countries, France, Germany, Italy, and America publishes over thirty such periodicals.

One of the characteristic features of our times is the prominence of the spirit of cooperation. The mathematical periodicals and the mathematical societies are evidences of this spirit. In quite recent years international mathematical congresses have given further expression of the widespread desire to cooperate with even the most remote workers in the same fields. The first of these congresses was held in Zurich in 1897, and the second is to be held during the coming summer in connection with the Paris Exposition.

The same spirit led in 1894 to the starting of a periodical, *L'Intermédiaire des Mathématiciens*, which is devoted exclusively to the publishing of queries and answers in regard to different mathematical subjects.

This desire for extensive cooperation is tending towards unifying mathematics and towards laying especial stress on those subjects which have the widest application in the different mathematical disciplines. This explains why the theory of functions of a complex variable and the theory of groups are occupying such prominent places in recent mathematical thought ([7], p. 134).

Before entering upon a description of some of the fields included in these subjects and the interesting problems which they present, it may be well to state explicitly that our remarks on mathematics will have very little reference to its application to other sciences. To the pure mathematician a result that has extensive application in mathematics is just as important and useful as one which applies to the other sciences. Mathematics is a science which deserves to be developed for its own sake. The thought that some of its results may find application in other sciences is, however, a continual inspiration, and those who investigate such applications sometimes add materially to the development of mathematics.

The curve representing a function of a single variable was the principal object of investigation during the eighteenth and a great part of the nineteenth century [9]. The investigations of Abel and Cauchy on power series during the early part of the nineteenth century furnished the foundation for the modern theory of analytic functions—a theory which has been adorned by the labors of some of the most brilliant mathematicians of the preceding generation and which is claiming the attention of some of the foremost thinkers of the present time. Quite recently this theory has been made more accessible to English readers by the treatises of Forsyth and Harkness and Morley.

The critical spirit of our age is, in a large measure, due to the study of the theory of analytic functions. "Newton assumes without hesitation the existence, in every case, of a velocity in a moving point, without troubling himself with the inquiry whether there might not be continuous functions having no derivative" ([8], p. 41). When it was discovered that such functions exist and that the works of some of the greatest mathematicians of the preceding centuries had to be modified in some instances, mathematicians naturally became much more exacting in regard to rigor, and thus ushered in an age which may be compared with the times of Euclid with respect to its demands for rigor. Whether our critical age will produce a work which, like Euclid, will serve as a model for millennia cannot be foretold, but it seems certain that works which can stand the critical tests of this age will stand the tests of all ages.

The critical spirit of our times is the foundation of what has been styled the *arithmetization of mathematics*. This movement which the late Weierstrass knew so well to lead is pervading more and more the whole mathematical world. We are rapidly banishing from our treatises the term *quantity* and replacing it by the word *number*. Our geometric intuitions are forced into the background and logical deductions from intuitions are taking their places. Who can conceive of curves which have no tangent at any of their rational points in a given interval? Nevertheless it is well known that such curves exist. An account of such functions was first published by Hankel (see [11], p. 398) in 1870.

Mathematicians find themselves in a great dilemma at this point. Geometric intuition has been such a strong instrument of research and has given so much life and beauty to mathematical investigation that mathematicians cling to it as to their own lives. It is an enormous price when rigor can be purchased only with geometric intuition. Yet, in the present stage of mathematical thought, this seems to be the only thing that will be accepted, and mathematicians stand helpless before this decree.

A few examples may throw some light on this subject. What do we understand by the length of a continuous curve? The intuitionalist says, if we connect different points of the continuous curve by straight lines and find the sum of the lengths of these straight lines, then the length of the curve will be the limit of this sum as the number of the points is indefinitely increased. Jordan was the first to call attention to the fact that this sum need not have a limit. Hence there are continuous curves which do not have any length according to the ordinary definition of length. In fact, a number of area-filling curves have recently been studied, and Cantor has shown that a multiplicity of any number of dimensions can be put in a one-to-one correspondence with a multiplicity of one dimension.

These are some of the facts which have compelled mathematicians to construct their own worlds—the number worlds. Conclusions drawn in one number world do not necessarily apply to another. When a problem is under consideration, the number world is so chosen as to meet the demands of the problem. For instance, the constructions and demonstrations of Euclid's geometry seem to require only a space composed of quadratic numbers (see [14], p. 443). Hence it appears that we do not need to assume that space is continuous in order to demonstrate the theorems of elementary geometry. Similarly in algebra, we are laying more and more stress upon a distinct statement of the number world (the domain of rationality) in which we are operating. Such specifications add a clearness and rigor to our work which would otherwise scarcely be possible.

This refinement which the mathematical thought of today is so actively cultivating is not restricted to the finite region. Mathematical infinity is receiving its share of attention. It is well known that it is sometimes desirable to regard the infinite region as a single point. This is, for instance, the case in the transformation known as *inversion*. Again, in ordinary projective geometry, it is generally convenient to regard the infinite region as of one lower dimension than the finite, so that the infinite region of a plane is merely a line and the infinite region of space is a plane. The student of differential calculus is, moreover, familiar with the infinite variable and the many simplifications which its use makes possible.

The most fruitful investigations along this line are those on multiplicities (*Mengenlehre, ensembles*). Any total of definite and clearly defined elements is said to form a *multiplicity*. If two multiplicities are simply isomorphic, i.e., if there is a one-to-one correspondence between the elements of the multiplicities, they are said to be *equivalent*, or to have the same power. For example, it is easy to prove that all the positive rational numbers are equivalent to the natural numbers. To do this we may associate all the rational numbers p/q for which the sum $p+q=n$, any positive integer. We thus have the $n-1$ numbers.

$$\frac{n-1}{1}, \frac{n-2}{2}, \frac{n-3}{3}, \ldots, \frac{2}{n-2}, \frac{1}{n-1}.$$

We may let 1 correspond to 1; the numbers for which $n=3$ correspond to 2 and 3; the numbers for which $n=4$ correspond to 4, 5 and 6, etc. We thus obtain the following one-to-one correspondence between all the rational numbers and the positive integers

$\frac{1}{1}$	$\frac{2}{1}$	$\frac{1}{2}$	$\frac{3}{1}$	$\frac{2}{2}$	$\frac{1}{3}$	$\frac{4}{1}$	$\frac{3}{2}$
1	2	3	4	5	6	7	8
$\frac{2}{3}$	$\frac{1}{4}$	$\frac{5}{1}$	$\frac{4}{2}$	$\frac{3}{3}$	$\frac{2}{4}$	$\frac{1}{5}$	$\ldots$
9	10	11	12	13	14	15	$\ldots$

It may be observed that we do not need to reduce the rational fractions to their lowest terms to effect this correspondence. This method of proof is due to Cantor, who has also proved that all algebraic numbers are equivalent to the natural numbers ([4], [5]). How important and far-reaching the investigations along this line are may be inferred from the fact that Jordan has employed them to serve as a foundation of the elementary part of the second edition of his magistral *Cours d'Analyse*.

A large number of mathematical problems may be reduced to equations involving a single unknown. The solution of such equations has occupied a prominent place in the mathematical literature for centuries. The problem is so difficult that it has been attacked from a number of different points and by means of a large variety of instruments. The instrument which has proved to be the most powerful and far-reaching is substitution groups. By means of it Abel succeeded in 1826 to prove that an equation whose degree exceeds four cannot generally be solved by the successive extraction of roots [11] and Galois a few years later sketched a far-reaching theory of equations which rests upon the theory of these groups.

In recent years it has been recognized (especially through the labors of Sophus Lie) that the theory of groups has very extensive and fundamental application in a large number of the other domains of mathematics. About a year ago the great French mathematician H. Poincaré showed in an article, published in the Chicago *Monist* [12], how this concept may be employed in laying the foundations of elementary geometry. It should be observed that the theory of groups is intrinsically based upon the fundamental concepts of mathematics. It is not a superstructure. It stands on its own foundation and supports more or less a number of other mathematical structures.

As this theory is less known than most of the other extensive branches of mathematics, it may be desirable to enter into some details. It is evident that there are rational functions of n independent variables

$$(x_1, x_2, x_3, \ldots, x_n)$$

which are not changed when these variables are permuted in every possible manner. Such functions are said to be *symmetric* in regard to these variables. The sum of any given power of these variables

$$x_1^a + x_2^a + x_3^a + \cdots + x_n^a$$

is an instance of a symmetric function. These functions occupy one extreme. The other extreme is furnished by functions such as

$$x_1 + 2x_2 + 3x_3 + \cdots nx_n$$

which change their value for every possible interchange of the variables. Most of the functions are of neither of these extreme types.

Suppose that a function $\phi(x_1, x_2, \ldots x_n)$ is not changed by either of two interchanges of its independent variables. Such interchanges are called *substitutions* and they may be represented by S_1 and S_2. Since ϕ is not changed by either of the substitutions S_1, S_2, it cannot be changed by the substitutions which are equivalent to the succession of these substitutions taken in any order. All such substitutions may be represented by

$$S_1^\alpha S_2^\beta S_1^\gamma S_2^\delta \cdots$$

(where the exponents indicate the number of times the substitution is employed in succession). Since only a finite number of permutations are possible with n letters it follows that $S_1^\alpha S_2^\beta S_1^\gamma S_2^\delta \cdots$ can represent only a finite number of distinct substitutions. The totality of these substitutions is said to be a *substitution group*. Hence we observe that every rational function belongs to some substitution group.

It was soon observed that an infinite number of functions belong to the same substitution group and that all of these functions can be expressed rationally in terms of one of them. The researches of Abel, Galois, and Jordan were based upon these facts and they show that the most important problems in the theory of equations involve the theory of substitution groups. The theory of groups was thus founded with a view to its application to a subject of paramount importance. A broad mathematical subject can, however, not grow vigorously and harmoniously as long as it is studied with a view to its direct applications to other mathematical subjects. It must be free to expand in all directions. That freedom for which the human race has ever been struggling must be vouchsafed to such fundamental subjects before they will exhibit their great fertility and far-reaching connections. Less than three years ago the first work on the theory of groups that does not consider its application [2] was given to the public, but the mathematical journals have been publishing a large number of memoirs along the same line for a number of years.

In defining a substitution group we implied only two conditions; *viz.*, no two substitutions of the group are identical, and if we combine the substitutions in any way we obtain only substitutions which are already in the group. Substitutions obey *per se* some other conditions; i.e., when they are combined (multiplied together) they obey the associative law, and if we multiply a substitution by (or into) two different substitutions the products will be different. In general we say that any finite number of operations which obey these four conditions constitute a group; e.g., all the rotations around the center of a regular solid which make the solid coincide with itself constitute a group, the n nth roots of unity constitute a group with respect to multiplication, but not with respect to addition, etc.

While the bulk of the mathematicians are revelling in the new fields of thought which are opening up on all sides, without any thought in reference to any immediate practical application of their results, there is fortunately a goodly number whose main efforts are devoted towards making some of these new results useful to the investigators in some of the other sciences. As an instance of fairly recent work of the latter kind, we may mention the study of linkages with a view towards describing a straight line. Although the straight line is of such fundamental importance both in pure and applied mathematics, yet it seems it was not until the latter half of the nineteenth century that a mechanical device has been discovered by means of which such a line can be described.

In 1864 M. Peaucellier, an officer of engineers in the French army, discovered the well-known device to describe a straight line by means of an instrument composed of seven links.

> His discovery was not at first estimated at its true value, fell almost into oblivion, and was rediscovered by a Russian named Lipkin, who got a substantial reward from the Russian government for his supposed originality. However, M. Peaucellier's merit was finally recognized and he has been awarded the great mechanical prize of the Institute of France, the Prix Montyon ([6], p. 12).

Although the straight line and the circle occupy such a prominent place in elementary geometry and have been before the eyes of the mathematicians for thousands of years, yet less than half a century has passed since the invention of a mechanical device by means of which the straight line can be drawn. Such discoveries go far towards emphasizing the need of investigations, even in the most elementary subjects.

Such investigations should, however, be preceded by a thorough knowledge of what has been done along the same lines.

If elementary mathematics is to continue to furnish the best possible preparation for the study of advanced mathematics, it is evident that it has to adapt itself to the rapid changes which are going on in the different branches of mathematics. A need is thus created for elementary textbooks which meet the new requirements, and we are happy to be able to state that such books are being produced in our midst. How radical such changes may become cannot be foretold. In his address before the New York Mathematical Society, Simon Newcomb said, "The mathematics of the twenty-first century may be very different from our own; perhaps the schoolboy will begin algebra with the theory of substitution groups, as he might now but for inherited habits" [10]. It is to be hoped that our inherited habits will not furnish an insurmountable barrier to progress in this direction.

In modern times the continent of Europe has always been the most progressive, and most of the new theories were first developed in these countries. The theory of invariants seems to be an exception to this rule. The two great English mathematicians, Cayley and Sylvester, developed this theory with great vigor; when their important results became generally known on the continent (largely through the work of Clebsch), they aroused a great deal of interest and they furnished a starting point for many important investigations.

One of the fundamental processes of mathematics is transformation—the deducing of truths from given facts and relations. The expressions which remain invariant when given transformations are performed are naturally objects of great interest and of fundamental importance. Imbued with this thought, Lie once said, "What do the natural phenomena present to us if it is not a succession of infinitesimal transformations of which the laws of the universe are the invariants?"

It need scarcely be added that all mathematical thought, even on the same subject, is not running in the same channel. Klein has divided mathematicians into three main categories [8], *viz.*, the logicians, the formalists, and the intuitionists. The term logician is "intended to indicate that the main strength of the men belonging to this class lies in their logical and critical powers, in their ability to give strict definitions and to derive rigid deductions therefrom. The great and wholesome influence exerted in Germany by Weierstrass in this direction is well known. The formalists among the mathematicians excel mainly in the skillful formal treatment of a given question, in devising for it an algorithm. Gordan, or let us say Cayley or Sylvester, must be ranged in this group. To the intuitionists, finally, belong those who lay particular stress on geometrical intuition, not in pure geometry only, but in all branches of mathematics. What Benjamin Peirce has called "geometrizing a mathematical question" seems to express the same idea. Lord Kelvin and von Staudt may be mentioned as types of this category."

In his address before the Zurich International Congress Poincaré says [13],

Mathematics has a triple end. It should furnish an instrument for the study of nature. Furthermore, it has a philosophical end, and, I venture to say, an aesthetic end. It ought to incite the philosopher to search into the notions of number, space, and time; and above all, adepts find in mathematics delights analogous to those that painting and music give. They admire the delicate harmony of numbers and of forms; they are amazed when a new discovery discloses for them an unlooked-for perspective; and the joy they experience has it not the aesthetic character, although the senses take no part in it? Only the privileged few are called to enjoy it fully, it is true, but is it not the same with all the noblest arts? Hence I do not hesitate to say that mathematics deserves to be cultivated for its own sake and that the theories not admitting of application to physics deserve to be studied as well as others. Moreover, a science produced with a view single to its applications is impossible; truths are fruitful only if they are concatenated, if we cleave to those only of which we expect immediate results the connecting links will be lacking, and there will be no longer a chain.

In closing we may remark that no effort has been made to mention all the new fields of mathematical thought. Mathematics, like the other sciences, seems to offer inexhaustible fields of investigation. As it expands its perimeter increases and hence there is a continually increasing demand for investigators. The fields that have been examined present many difficulties which cannot at present be surmounted. Some of the old difficulties are being removed by the light of the new discoveries. Still we know only a few things even about the fields which have been investigated. It is the exception that something can be done by known methods, the rule is that it cannot yet be done.

When we study the literature of some of the older subjects we are sometimes surprised by the large number of known facts, but when we come to study the subjects themselves and ask independent questions we are generally surprised to learn that so few properties are known. Hence it seems very desirable that the advanced student, at least, should study subjects, rather than the known facts in regard to these subjects. In this way a more accurate idea of the true state of knowledge can be obtained. Besides, the knowledge of having discovered facts and relations which will enter into the structure of a growing science is the greatest source of pleasure that the student can obtain.

References

1. N. Abel, Beweis der Unmöglichkeit algebraischer Gleichungen von höhren Graden als dem vierten allgemein aufzulösen, *J. Reine Angew. Math.* **1** (1826), 65–84.

2. W. Burnside, *Theory of Groups of Finite Order*, Cambridge Univ. Press, 1897.

3. F. Cajori, *The History and Teaching of Mathematics*, U.S. Bureau of Education, 1890.

4. G. Cantor, Über eine Eigenschaft des Inbegriffes aller reellen algebraischen Zahlen, *J. Reine Angew. Math.* **77** (1874), 258–262; **84** (1881), 250.

5. G. Cantor, Ein Beitrag zur Mannigfältigkeitslehre, *J. Reine Angew. Math.* **84** (1878), 242–258.

6. A. B. Kempe, *How to Draw a Straight Line*, Macmillan, 1886, p. 12.

7. F. Klein, The present state of mathematics, *Mathematical Papers read at the International Mathematical Congress* (ed. E. H. Moore), Macmillan, 1893.

8. F. Klein, *Evanston Colloquium*, Macmillan, 1894.

9. S. Lie, *Leipziger Berichte* **47** (1895).

10. S. Newcomb, Modern mathematical thought, *Bull. Amer. Math. Soc.* **3** (1894), 95–107.

11. J. Pierpont, On the arithmetization of mathematics, *Bull. Amer. Math. Soc.* **5** (1899), 394–406.

12. H. Poincaré, On the foundations of geometry, *Monist* **9** (1898), 1–43.

13. H. Poincaré, Sur les rapports de l'analyse pur et de la physique mathématique, *Report Internat. Cong. Math. Zurich*, 1897, p. 82.

14. W. Strong, Is continuity of space necessary to Euclid's geometry?, *Bull. Amer. Math. Soc.* **4** (1898), 443–448.

A Half-Century of Mathematics

HERMANN WEYL

American Mathematical Monthly **58** (1951), 523–553

Introduction. Axiomatics

Mathematics, beside astronomy, is the oldest of all sciences. Without the concepts, methods and results found and developed by previous generations right down to Greek antiquity, one cannot understand either the aims or the achievements of mathematics in the last fifty years. Mathematics has been called the science of the infinite; indeed, the mathematician invents finite constructions by which questions are decided that by their very nature refer to the infinite. That is his glory. Kierkegaard once said religion deals with what concerns man unconditionally. In contrast (but with equal exaggeration) one may say that mathematics talks about the things which are of no concern at all to man. Mathematics has the inhuman quality of starlight, brilliant and sharp, but cold.

But it seems an irony of creation that man's mind knows how to handle things the better the farther removed they are from the center of his existence. Thus we are cleverest where knowledge matters least: in mathematics, especially in number theory. There is nothing in any other science that, in subtlety and complexity, could compare even remotely with such mathematical theories as for instance that of algebraic class fields. Whereas physics in its development since the turn of the century resembles a mighty stream rushing on in one direction, mathematics is more like the Nile delta, its waters fanning out in all directions. In view of all this—dependence on a long past, otherworldliness, intricacy, and diversity—it seems an almost hopeless task to give a non-esoteric account of what mathematicians have done during the last fifty years. What I shall try to do here is, first to describe in somewhat vague terms general trends of development, and then in more precise language explain the most outstanding mathematical notions devised, and list some of the more important problems solved, in this period.

One very conspicuous aspect of twentieth-century mathematics is the enormously increased role which the axiomatic approach plays. Whereas the axiomatic method was formerly used merely for the purpose of elucidating the foundations on which we build, it has now become a tool for concrete mathematical research. It is perhaps in algebra that it has scored its greatest successes. Take for instance the system of real numbers. It is like a Janus head facing in two directions: on the one side it is the field of the algebraic operations of addition and multiplication; on the other hand it is a continuous manifold, the parts of which are so connected as to defy exact isolation from each other. The one is the algebraic, the other the topological face of numbers. Modern axiomatics, simple-minded as it is (in contrast to modern politics), does not like such ambiguous mixtures of peace and war, and therefore cleanly separated both aspects from each other.

In order to understand a complex mathematical situation it is often convenient to separate in a natural manner the various sides of the subject in question, make each side accessible by a relatively narrow and easily surveyable group of notions and of facts formulated in terms of these notions, and finally return to the whole by uniting the partial results in their proper specialization. The last synthetic act is purely mechanical. The art lies in the first, the analytic act of suitable separation and generalization. Our mathematics of the last decades has wallowed in generalizations and formalizations. But one misunderstands this tendency if one thinks that generality was sought merely for generality's sake. The real aim is simplicity: every natural generalization simplifies since it reduces the assumptions that have to be taken into account.

It is not easy to say what constitutes a natural separation and generalization. For this there is ultimately no other criterion but fruitfulness: the success decides. In following this procedure the individual investigator is guided by more-or-less obvious analogies and by an instinctive discernment of the essential acquired through accumulated previous research experience. When systematized, the procedure leads straight to axiomatics. Then the basic notions and facts of which we spoke are changed into undefined terms and into axioms involving them. The body of statements deduced from these hypothetical axioms is at our disposal now, not only for the instance from which the notions and axioms were abstracted, but wherever we come across an interpretation of the basic terms which turns the axioms into true statements.

It is a common occurrence that there are several such interpretations with widely different subject matter. The axiomatic approach has often revealed inner relations between, and has made for unification of methods within, domains that apparently lie far apart. This tendency of several branches of mathematics to coalesce is another conspicuous feature in the modern development of our science, and one that goes side by side with the apparently opposite tendency of axiomatization. It is as if you took a man out of a milieu in which he had lived, not because it fitted him but from ingrained habits and prejudices, and then allowed him, after thus setting him free, to form associations in better accordance with his true inner nature.

In stressing the importance of the axiomatic method I do not wish to exaggerate. Without inventing new constructive processes no mathematician will get very far. It is perhaps proper to say that the strength of modern mathematics lies in the interaction between axiomatics and construction. Take algebra as a representative example. It is only in this century that algebra has come into its own by breaking away from the one universal system Ω of numbers which used to form the basis of all mathematical operations, as well as all physical measurements. In its newly-acquired freedom algebra envisages an infinite variety of "number fields", each of which may serve as an operational basis; no attempt is made to embed them into the one system Ω. Axioms limit the possibilities for the number concept; constructive processes yield number fields that satisfy the axioms.

In this way algebra has made itself independent of its former master analysis, and in some branches has even assumed the dominant role. This development in mathematics is paralleled in physics to a certain degree by the transition from classical to quantum physics, inasmuch as the latter ascribes to each physical structure its own system of observables or quantities. These quantities are subject to the algebraic operations of addition and multiplication; but as their multiplication is non-commutative, they are certainly not reducible to ordinary numbers.

At the International Mathematical Congress in Paris in 1900 David Hilbert, convinced that problems are the life-blood of science, formulated twenty-three unsolved problems which he expected to play an important role in the development of mathematics during the next era. How much better he predicted the future of mathematics than any politician foresaw the gifts of war and terror that the new century was about to lavish upon mankind! We mathematicians have often measured our progress by checking which of Hilbert's questions had been settled in the meantime. It would be tempting to use his list as a guide for a survey like the one attempted here. I have not done so because it would necessitate explanation of too many details. I shall have to tax the reader's patience enough anyhow.

Part I. Algebra, Number Theory, Groups

Rings, Fields, Ideals

Indeed, at this point it seems impossible for me to go on without illustrating the axiomatic approach by some of the simplest algebraic notions. Some of them are as old as Methuselah. For what is older than the sequence of *natural numbers* $1, 2, 3, \ldots$, by which we count? Two such numbers a, b may be added and multiplied ($a + b$ and $a \cdot b$). The next step in the genesis of numbers adds to these positive *integers* the negative ones and zero; in the wider system thus created the operation of addition permits of a unique inversion, subtraction. One does not stop here: the integers in their turn get absorbed into the still wider range of *rational numbers* (fractions). Thereby division, the operation inverse to multiplication, also becomes possible, with one notable exception however: division by zero. (Since $b \cdot 0 = 0$ for every rational number b, there is no inverse b of 0 such that $b \cdot 0 = 1$.) I now formulate the fundamental facts about the operations "plus" and "times" in the form of a table of axioms:

Table T

1. The commutative and associative laws for addition,
$$a + b = b + a. \quad a + (b + c) = (a + b) + c.$$

2. The corresponding laws for multiplication.

3. The distributive law connecting addition with multiplication
$$c \cdot (a + b) = (c \cdot a) + (c \cdot b).$$

4. The axioms of subtraction:
   4.1 There is an element o (0, "zero") such that $a + o = o + a = a$ for every a.
   4.2 To every a there is a number $-a$ such that $a + (-a) = (-a) + a = o$.

5. The axioms of division:
   5.1 There is an element e (1, "unity") such that $a \cdot e = e \cdot a = a$ for every a.
   5.2 To every $a \neq o$ there is an a^{-1} such that $a \cdot a^{-1} = a^{-1} a = e$.

By means of (4.2) and (5.2) one may introduce the difference $b - a$ and the quotient b/a as $b + (-a)$ and ba^{-1}, respectively.

When the Greeks discovered that the ratio $\sqrt{2}$ between the diagonal and side of a square is not measurable by a rational number, a further extension of the number concept was called for. However, all measurements of continuous quantities are possible only approximately, and always have a certain range of inaccuracy. Hence rational numbers, or even finite decimal fractions, can and do serve the ends of mensuration provided they are interpreted as approximations, and a calculus with approximate numbers seems the adequate numerical instrument for all measuring sciences.

But mathematics ought to be prepared for any subsequent refinement of measurements. Hence dealing, say, with electric phenomena, one would be glad if one could consider the approximate values of the charge e of the electron which the experimentalist determines with ever greater accuracy as approximations of one definite exact value e.

And thus, during more than two millennia from Plato's time until the end of the nineteenth century, the mathematicians worked out an exact number concept, that of *real numbers*, that underlies all our theories in natural science. Not even to this day are the logical issues involved in that concept completely clarified and settled. The rational numbers are but a small part of the real numbers. The latter satisfy our axioms no less than the rational ones, but their system possesses a certain completeness not enjoyed by the rational numbers, and it is this, their "topological" feature, on which the operations with infinite sums and the like, as well as all continuity arguments, rest. We shall come back to this later.

Finally, during the Renaissance *complex numbers* were introduced. They are essentially pairs $z = (x, y)$ of real numbers x, y, pairs for which addition and multiplication are defined in such a way that all axioms hold. On the ground of these definitions $e = (1, 0)$ turns out to be the unity, while $i = (0, 1)$ satisfies the equation $i \cdot i = -e$. The two members x, y of the pair z are called its real and imaginary parts, and z is usually written in the form $xe + yi$, or simply $x + yi$. The usefulness of the complex numbers rests on the fact that every algebraic equation (with real or even complex coefficients) is solvable in the field of complex numbers. The analytic functions of a complex variable are the subject of a particularly rich and harmonious theory, which is the show-piece of classical nineteenth-century analysis.

A set of elements for which the operations $a + b$ and $a \cdot b$ are so defined as to satisfy the axioms (1)–(4) is called a *ring*; it is called a *field* if also the axioms (5) hold. Thus the common integers form a ring I, the rational numbers form a field ω; so do the real numbers (field Ω) and the complex numbers (field Ω^*). But these are by no means the only rings or fields. The polynomials of all possible degrees h,

$$f = f(x) = a_0 + a_1 x + a_2 x^2 + \cdots + a_h x^h, \quad (1)$$

with coefficients a_i taken from a given ring R (e.g., the ring I of integers, or the field ω), called "polynomials over R", form a ring $R[x]$. Here the variable or indeterminate x is to be looked upon as an empty symbol; the polynomial is really nothing but the sequence of its coefficients $a_0, a_1, a_2, \cdots$. But writing it in the customary form (1) suggests the rules for the addition and multiplication of polynomials which I will not repeat here. By substituting for the variable x a definite element ("number") γ of R, or of a ring P containing R as a subring, one projects the elements f of $R[x]$ into elements α of P, $f \to \alpha$: the polynomial $f = f(x)$ goes over into the number $\alpha = f(\gamma)$. This mapping $f \to \alpha$ is *homomorphic*, i.e., it preserves addition and multiplication. Indeed, if the substitution of γ for x carries the polynomial f into α and the polynomial g into β, then it carries $f + g, f \cdot g$ into $\alpha + \beta, \alpha \cdot \beta$, respectively.

If the product of two elements of a ring is never zero unless one of the factors is, one says that the ring is without null-divisor. This is the case for the rings discussed so far. A field is always a ring without null-divisor. The construction by which one rises from the integers to the fractions can be used to show that any ring R with unity and without null-divisor may be imbedded in a field k, the quotient field, such that every element of k is the quotient a/b of two elements a and b of R the second of which (the denominator) is not zero.

Writing $1a, 2a, 3a, \ldots$, for $a, a+a, a+a+a$, etc., we use the natural numbers $n = 1, 2, 3, \ldots$, as multipliers for the elements a of a ring or a field. Suppose the ring contains the unity e. It may happen that a certain multiple ne of e equals zero; then one readily sees that $na = 0$ for every element a of the ring. If the ring is without null divisors, in particular if it is a field and p is the least natural number for which $pe = 0$, then p is necessarily a prime number like 2 or 3 or 5 or 7 or 11.... One thus distinguishes fields of prime characteristic p from those of characteristic 0 in which no multiple of e is zero.

Plot the integers $\ldots, -2, -1, 0, 1, 2, \ldots$ as equidistant marks on a line. Let n be a natural number ≥ 2 and roll this line upon a wheel of circumference n. Then any two marks a, a' coincide, the difference $a - a'$ of which is divisible by n. (The mathematicians write $a \equiv a' \pmod{n}$; they say: a is congruent to a' modulo n.) By this identification the ring of integers I goes over into a ring I_n consisting of n elements only (the marks on the wheel), as which one may take the "residues" $0, 1, \ldots, n - 1$. Indeed, congruent numbers give congruent results under both addition and multiplication: $a \equiv a', b \equiv b' \pmod{n}$ imply $a + b \equiv a' + b', a \cdot b \equiv a' \cdot b' \pmod{n}$. For instance, modulo 12 we have $7 + 8 = 3$ and $5 \cdot 8 = 4$ because 15 leaves the residue 3 and 40 the residue 4 if divided by 12. The ring I_{12} is not without null divisors since $3 \cdot 4$ is divisible by 12, but neither 3 nor 4 is. However, if p is a natural prime number, then I_p has no null divisor and is even a field; for as the ancient Greeks proved by an ingenious procedure (Euclid's algorism), for every integer a not divisible by p there is one, a', such that $a \cdot a' \equiv 1 \pmod{p}$. This Euclidean theorem is at the basis of the whole of number theory. The example shows that there are fields of any given prime characteristic p.

In any ring R one may introduce the notions of unit and prime element as follows. The ring element a is a unit if it has a reciprocal a' in the ring, such that $a' \cdot a = e$. The element a is composite if it may be decomposed into two factors $a_1 \cdot a_2$, neither of which is a unit. A prime number is one that is neither a unit nor composite. The units of I are the numbers $+1$ and -1. The units of the ring $k[x]$ of polynomials over a field k are the non-vanishing elements of k (polynomials of degree 0). According to the Greek discovery of the irrationality of $\sqrt{2}$, the polynomial $x^2 - 2$ is prime in the ring $\omega[x]$; but, of course, not in $\Omega[x]$, for there it splits into the two linear factors $(x - \sqrt{2}) \cdot (x + \sqrt{2})$.

Euclid's algorism is also applicable to polynomials $f(x)$ of one variable x over any field k. Hence they satisfy Euclid's theorem: given a prime element $P = P(x)$ in this ring $k[x]$ and an element $f(x)$ of $k[x]$ not divisible by $P(x)$, there exists another polynomial $\bar{f}(x)$ over k such that $\{f(x) \cdot \bar{f}(x)\} - 1$ is divisible by $P(x)$. Identification of any elements f and g of $k[x]$, the difference of which is divisible by P, therefore changes the ring $k[x]$ into a field, the "residue field κ of $k[x]$ modulo P." Example: $\omega[x]$ (mod $x^2 - 2$). (Incidentally, the complex numbers may be described as the elements of the residue field of $\Omega[x]$ (mod $x^2 + 1$).)

Strangely enough, the fundamental Euclidean theorem does not hold for polynomials of two variables x, y. For instance, $P(x, y) = x - y$ is a prime element of $\omega[x, y]$, and $f(x, y) = x$ an element not divisible by $P(x, y)$. But a congruence

$$x \cdot \bar{f}(x, y) \equiv 1 \pmod{x - y}$$

is impossible. Indeed, it would imply $-1 + x \cdot \bar{f}(x, x) = 0$, contrary to the fact that the polynomial of one indeterminate x,

$$-1 + x \cdot \bar{f}(x, x) = -1 + c_1 x + c_2 x^2 + \cdots,$$

is not zero. Thus the ring $\omega[x, y]$ does not obey the simple laws prevailing in I and in $\omega[x]$.

Consider κ, the residue field of $\omega[x]$ (mod $x^2 - 2$). Since for any two polynomials $f(x), \bar{f}(x)$ which are congruent mod $x^2 - 2$ the numbers $f(\sqrt{2}), \bar{f}(\sqrt{2})$ coincide, the transition $f(x) \to f(\sqrt{2})$ maps κ into a subfield $\omega[\sqrt{2}]$ of Ω consisting of the numbers $a + b\sqrt{2}$ with rational a, b. Another such projection would be $f(x) \to f(-\sqrt{2})$. In former times one looked upon κ as the part $\omega[\sqrt{2}]$ of the continuum Ω or Ω^* of all real or all complex numbers; one wished to embed everything into this universe Ω or Ω^* in which analysis and physics operate. But as we have introduced it here, κ is an algebraic entity whose elements are not numbers in the ordinary sense. It requires for its construction no other numbers but the rational ones. It has nothing to do with Ω, and ought not to be confused with the one or the other of its two projections into Ω. More generally, if $P = P(x)$ is any prime element in $\omega[x]$, we can form the residue field κ_P of $\omega[x]$ modulo P. To be sure, if δ is any of the real or complex roots of the equation $P(x) = 0$ in Ω^*, then $f(x) \to f(\delta)$ defines a homomorphic projection of κ_P into Ω^*. But the projection is not κ_P itself.

Let us return to the ordinary integers $\cdots$, $-2, -1, 0, 1, 2, \cdots$, which form the ring I. The multiples of 5, i.e., the integers divisible by 5, clearly form a ring. It is a ring without unity, but it has another important peculiarity: not only does the product of any two of its elements lie in it, but all the integral multiples of an element do. The queer term *ideal* has been introduced for such a set: Given a ring R, an R-ideal $(\mathfrak{a})$ is a set of elements of R such that

1. the sum and difference of any two elements of $(\mathfrak{a})$ are in $(\mathfrak{a})$,
2. the product of an element in $(\mathfrak{a})$ by any element of R is in $(\mathfrak{a})$.

We may try to describe a divisor $\mathfrak{a}$ by the set of all elements divisible by $\mathfrak{a}$. One would certainly expect this set to be an ideal $(\mathfrak{a})$ in the sense just defined. Given an ideal $(\mathfrak{a})$, there may not exist an actual element $\mathfrak{a}$ of R such that $(\mathfrak{a})$ consists of all multiples $j = m \cdot \mathfrak{a}$ of $\mathfrak{a}$ (m any element in R). But then we would say that $(\mathfrak{a})$ stands for an "ideal divisor" $\mathfrak{a}$: the words "the element j of R is divisible by $\mathfrak{a}$" would simply mean: "j belongs to $(\mathfrak{a})$." In the ring I of common integers all divisors are actual.

But this is not so in every ring. An algebraic surface in the three-dimensional Euclidean space with the Cartesian coordinates x, y, z is defined by an equation $F(x, y, z) = 0$, where F is an element of $^3\Omega = \Omega[x, y, z]$, i.e., a polynomial of the variables x, y, z with real coefficients. F is zero in all the points of the surface; but the same is true for every multiple $L \cdot F$ of F (L being any element of $^3\Omega$), in other words, for every polynomial of the ideal (F) in $^3\Omega$. Two simultaneous polynomial equations

$$F_1(x, y, z) = 0, \qquad F_2(x, y, z) = 0$$

will in general define a curve, the intersection of the surface $F_1 = 0$ and the surface $F_2 = 0$. The polynomials $(L_1 \cdot F_1) + (L_2 \cdot F_2)$ formed by arbitrary elements L_1, L_2 of $^3\Omega$ form an ideal (F_1, F_2), and all these polynomials vanish on the curve. This ideal will in general not correspond to an actual divisor F, for a curve is not a surface. Examples like this should convince the reader that the study of algebraic manifolds (curves, surfaces, etc., in 2, 3, or any number of dimensions) amounts essentially to a study of polynomial ideals. The field of coefficients is not necessarily Ω or Ω^*, but may be a field of a more general nature.

Some achievements of algebra and number theory

I have finally reached a point where I can hint, I hope, with something less than complete obscurity, at some of the accomplishments of algebra and number theory in our century. The most important is probably the freedom with which we have learned to manage these abstract axiomatic concepts, like field, ring, ideal, etc. The atmosphere in a book like van der Waerden's *Moderne Algebra*, published about 1930, is completely different from that prevailing, e.g., in the articles on algebra written for the *Mathematical Encyclopaedia* around 1900. More specifically, a general theory of ideals, and in particular of polynomial ideals, was developed. (However, it should be said that the great pioneer of abstract algebra, Richard Dedekind, who first introduced the ideals into number theory, still belonged to the nineteenth century.)

Algebraic geometry, before and around 1900 flourishing chiefly in Italy, was at that time a discipline of a type uncommon in the sisterhood of mathematical disciplines: it had powerful methods, plenty of general results, but they were of somewhat doubtful validity. By the abstract algebraic methods of the twentieth century all this was put on a safe basis, and the whole subject received a new impetus. Admission of fields other than Ω^*, as the field of coefficients, opened up a new horizon.

A new technique, the "primadic numbers", was introduced into algebra and number theory by K. Hensel shortly after the turn of the century, and since then has become of ever increasing importance. Hensel shaped this instrument in analogy to the power series which played such an important part in Riemann's and Weierstrass's theory of algebraic functions of one variable and their integrals (Abelian integrals). In this theory, one of the most impressive accomplishments of the previous century, the coefficients were supposed to vary over the field Ω^* of all complex numbers. Without pursuing the analogy, I may illustrate the idea of p-adic numbers by one typical example, that of quadratic norms. Let p be a prime number, and let us first agree that a congruence $a \equiv b$ modulo a power p^h of p for rational numbers a, b has

this meaning, that $(a-b)/p^h$ equals a fraction whose denominator is not divisible by p; for example,
$$\frac{39}{4} - \frac{12}{5} \equiv 0 \pmod{7^2} \text{ because } \frac{39}{4} - \frac{12}{5} = 7^2 \cdot \frac{3}{20}.$$
Let now a, b be rational numbers, $a \neq 0$, and b not the square of a rational number. In the quadratic field $\omega[\sqrt{b}]$ the number a is a *norm* if there are rational numbers x, y such that
$$a = \left(x + \sqrt{b}y\right)\left(x - \sqrt{b}y\right), \text{ or } a = x^2 - by^2.$$
Necessary for the solvability of this equation is that for every prime p and every power p^h of p the congruence $a \equiv x^2 - by^2 \pmod{p^h}$ has a solution. This is what we mean by saying the equation has a p-adic solution. Moreover, there must exist rational numbers x and y such that $x^2 - by^2$ differs as little as one wants from a. This is what we mean by saying that the equation has an ∞-adic solution. The latter condition is clearly satisfied for every a, provided b is positive; however, if b is negative it is satisfied only for positive a. In the first case every a is a ∞-adic norm; in the second case only half of the a's are, namely, the positive ones. A similar situation prevails with respect to p-adic norms. One proves that these necessary conditions are also sufficient: if a is a norm locally everywhere, i.e., if $a = x^2 - by^2$ has a p-adic solution for every "finite prime spot p" and also for the "infinite prime spot ∞", then it has a "global" solution, namely an exact solution in rational numbers x, y.

This example, the simplest I could think of, is closely connected with the theory of genera of quadratic forms, a subject that goes back to Gauss's *Disquisitiones Arithmeticae*, but in which the twentieth century has made some decisive progress by means of the p-adic technique, and it is also typical for that most fascinating branch of mathematics mentioned in the introduction: class field theory. Around 1900 David Hilbert had formulated a number of interlaced theorems concerning class fields, proved some of them at least in special cases, and left the rest to his twentieth-century successors, among whom I name Takagi, Artin and Chevalley. His norm residue symbol paved the way for Artin's general reciprocity law. Hilbert had used the analogy with the Riemann-Weierstrass theory of algebraic functions over Ω^* for his orientation, but the ingenious, partly transcendental, methods which he applied had nothing to do with the much simpler ones that had proved effective for the functions. By the primadic technique a rapprochement of methods has occurred, although there is still a considerable gap separating the theory of algebraic functions and the much subtler algebraic numbers.

Hensel and his successors have expressed the p-adic technique in terms of the non-algebraic "topological" notion of ("valuation" or) *convergence*. An infinite sequence of rational numbers $a_1, a_2, \cdots$ is convergent if the difference $a_i - a_j$ tends to zero, $a_i - a_j \to 0$, provided i and j independently of each other tend to infinity; more explicitly, if for every positive rational number ε there exists a positive integer N such that $-\varepsilon < a_i - a_j < \varepsilon$ for all i and $j > N$. The completeness of the real number system is expressed by Cauchy's convergence theorem: To every convergent sequence $a_1, a_2, \cdots$ of rational numbers there exists a *real* number α to which it converges: $a_i - \alpha \to 0$ for $i \to \infty$. With the ∞-adic concept of convergence we have now confronted the p-adic one induced by a prime number p. Here the sequence is considered convergent if for every exponent $h = 1, 2, 3, \cdots$, there is a positive integer N such that $a_i - a_j$ is divisible by p^h as soon as i and $j > N$. By the introduction of p-adic numbers one can make the system of rational numbers complete in the p-adic sense, just as the introduction of real numbers makes them complete in the ∞-adic sense. The rational numbers are embedded in the continuum of all real numbers, but they may be embedded as well in that of all p-adic numbers. Each of these embedments corresponding to a finite or the infinite prime spot p is equally interesting from the arithmetical viewpoint. Now it is more evident than ever how wrong it was to identify an algebraic number field with one of its homomorphic projections into the field Ω of real numbers; along with the (real) infinite prime spots one must pay attention to the finite prime spots which correspond to the various prime ideals of the field. This is a golden rule abstracted from earlier, and then made fruitful for later, arithmetical research; and here is one bridge (others will be pointed out later) joining the two most fascinating branches of modern mathematics: abstract algebra and topology.

Besides the introduction of the primadic treatment and the progress made in the theory of class fields, the most important advances of number theory during the last fifty years seem to lie in those regions where the powerful tool of analytic functions can be brought to bear upon its problems. I mention two such fields of investigation: I. distribution of primes and the zeta function, II. additive number theory.

I. The notion of prime number is of course as old and as primitive as that of the multiplication of natural numbers. Hence it is most surprising to find that the distribution of primes among all natural numbers is of such a highly irregular and almost mysterious character. While on the whole the prime numbers thin out the further one gets in the sequence of numbers, wide gaps are always followed again by clusters. An old conjecture maintains that there even come along again and again pairs of primes of the smallest possible difference 2, like 71 and 73. However, the distribution of primes obeys at least a fairly simple asymptotic law: the number $\pi(n)$ of primes among all numbers from 1 to n is asymptotically equal to $n/\log n$. [Here $\log n$ is not the Briggs logarithm which our logarithmic tables give, but the natural logarithm as defined by the integral $\int_1^n dx/x$.] By "asymptotic" is meant that the quotient between $\pi(n)$ and the approximating function $n/\log n$ tends to 1 as n tends to infinity.

In antiquity Eratosthenes had devised a method to sift out the prime numbers. By this sieve method the Russian mathematician Tchebycheff had obtained, during the nineteenth century, the first non-trivial results about the distribution of primes. Riemann used a different approach: his tool is the so-called *zeta-function* defined by the infinite series

$$\zeta(s) = 1^{-s} + 2^{-s} + 3^{-s} + \cdots. \qquad (2)$$

Here s is a complex variable, and the series converges for all values of s, the real part of which is greater than 1, $\Re s > 1$. Already in the eighteenth century the fact that every positive integer can be uniquely factorized into primes had been translated by Euler into the equation

$$1/\zeta(s) = \left(1 - 2^{-s}\right)\left(1 - 3^{-s}\right)\left(1 - 5^{-s}\right)\cdots,$$

where the (infinite) product extends over all primes $2, 3, 5, \ldots$. Riemann showed that the zeta-function has a unique "analytic continuation" to all values of s other than 1 and that it satisfies a certain functional equation connecting its values for s and $1 - s$. Decisive for the prime number problem are the zeros of the zeta-function, i.e., the values s for which $\zeta(s) = 0$. Riemann's equation showed that, except for the "trivial" zeros at $s = -2, -4, -6, \ldots$, all zeros have real parts between 0 and 1. Riemann conjectured that their real parts actually equal $\frac{1}{2}$. His conjecture has remained a challenge to mathematics now for almost a hundred years. However, enough had been learned about these zeros at the close of the nineteenth century to enable mathematicians, by means of some profound and newly-discovered theorems concerning analytic functions, to prove the above-mentioned asymptotic law. This was generally considered a great triumph of mathematics. Since the turn of the century Riemann's functional equation with the attending consequences has been carried over from the "classical" zeta-function of the field of rational numbers to that of an arbitrary algebraic number field (E. Hecke). For certain fields of prime characteristic, one succeeded in confirming Riemann's conjecture, but this provides hardly a clue for the classical case. About the classical zeta-function we know now that it has infinitely many zeros on the critical line $\Re s = \frac{1}{2}$, and even that at least a fixed percentage, say 10 per cent, of them lie on it. (What this means is the following: Some percentage of those zeros whose imaginary part lies between arbitrary fixed limits $-T$ and $+T$ will have a real part equal to $\frac{1}{2}$, and this percentage will not sink below a certain positive limit, like 10 per cent, when T tends to infinity.) Finally, about two years ago, Atle Selberg succeeded, to the astonishment of the mathematical world, in giving an "elementary" proof of the prime number law by an ingenious refinement of old Eratosthenes's sieve method.

II. It has been known for a long time that every natural number may be written as the sum of at most four square numbers, e.g.,

$$7 = 2^2 + 1^2 + 1^2 + 1^2,$$
$$87 = 9^2 + 2^2 + 1^2 + 1^2 = 7^2 + 5^2 + 3^2 + 2^2.$$

The same question arises for cubes, and generally for any kth powers ($k = 2, 3, 4, 5, \ldots$). In the eighteenth century Waring had conjectured that every non-negative integer n may be expressed as the sum of a limited number M of kth powers,

$$n = n_1^k + n_2^k + \cdots + n_M^k, \qquad (3)$$

where the n_i are also non-negative integers and M is independent of it. The first decade of the twentieth century brought two events: first one found that every n is expressible as the sum of at most 9 cubes (and that, excepting a few comparatively small n, even 8 cubes will do); and shortly afterwards Hilbert proved Waring's general theorem. His method was soon replaced by a different approach, the Hardy-Littlewood circle method, which rests on the use of a certain analytic function of a complex variable and yields asymptotic formulas for the number of different representations of it in the form (3). With some precautions demanded by the nature of the problem, and by overcoming some quite serious obstacles, the result was later carried over to arbitrary algebraic number fields; and by a further refinement of the circle method in a different direction Vinogradoff proved that every sufficiently large n is the sum of at most 3 primes. Is it even true that every even n is the sum of 2 primes? To show this seems to transcend our present mathematical powers as much as the twin-prime conjecture. The prime numbers remain very elusive fellows.

III. Finally, a word ought to be said about investigations concerning the arithmetical nature of numbers originating in analysis. One of the most elementary such constants is π, the area of the circle of radius 1. By proving that π is a transcendental number (not satisfying an algebraic equation with rational coefficients), the age-old problem of "squaring the circle" was settled in 1882 in the negative sense; that is, one cannot square the circle by constructions with ruler and compass. In general it is much harder to establish the transcendency of numbers than of functions. Whereas it is easy to see that the exponential function

$$e^x = 1 + \frac{x}{1} + \frac{x^2}{1 \cdot 2} + \frac{x^3}{1 \cdot 2 \cdot 3} + \cdots$$

is not algebraic, it is quite difficult to prove that its basis e is a transcendental number. C. L. Siegel was the first who succeeded, around 1930, in developing a sort of general method for testing the transcendency of numbers. But the results in this field remain sporadic.

Groups, vector spaces and algebras

This ends our report on number theory, but not on algebra. For now we have to introduce the *group* concept, which, since the young genius Evariste Galois blazed the trail in 1830, has penetrated the entire body of mathematics. Without it an understanding of modern mathematics is impossible.

Groups first occurred as *groups of transformations*. Transformations may operate in any set of elements, whether it is finite like the integers from 1 to 10, or infinite like the points in space. Set is a pre-mathematical concept: whenever we deal with a realm of objects, a set is defined by giving a criterion which decides for any object of the realm whether it belongs to the set or not. Thus we speak of the set of prime numbers, or of the set of all points on a circle, or of all points with rational coordinates in a given coordinate system, or of all people living at this moment in the State of New Jersey. Two sets are considered equal if every element of the one belongs to the other and *vice versa*. A *mapping* S of a set σ into a set σ' is defined if, with every element a of σ, there is associated an element a' of σ', $a \to a'$. Here a rule is required which allows one to find the "image" a' for any given element a of σ. This general notion of mapping we may also call of a pre-mathematical nature. Examples: a real-valued function of a real variable is a mapping of the continuum Ω into itself. Perpendicular projection of the space points upon a given plane is a mapping of the space into the plane. Representing every space-point by its three coordinates x, y, z with respect to a given coordinate system is a mapping of space into the continuum of real number triples (x, y, z). If a mapping S, $a \to a'$ of σ into σ', is followed by a mapping S', $a' \to \sigma''$ of σ' into a third set σ'', the result is a mapping SS', $a \to a''$ of σ into σ''. A *one-to-one mapping* between two sets σ, σ' is a pair of mappings, $S : a \to a'$ of σ into σ', and $S' : a' \to a$ of σ' into σ, which are inverse to each other. This means that the mapping SS' of σ into σ is the identical mapping E of σ which sends every element a of σ into itself, and that $S'S$ is the identical mapping of σ'. In particular, we are interested in one-to-one mappings of a set σ into itself. For them we shall use the word *transformation*. Permutations are nothing but transformations of a finite set.

The inverse S' of a transformation S, $a \to a'$ of a given set σ, is again a transformation and is usually denoted by S^{-1}. The result ST of any two transformations S and T of σ is again a transformation, and its inverse is $T^{-1}S^{-1}$ (according to the rule of dressing and undressing: if in dressing one begins with the shirt and ends with the jacket, one must in undressing begin with the jacket and end with the shirt: the order of the two "factors" S, T is essential.) A *group of transformations* is a set of transformations of a given manifold which

1. contains the identity E,

2. contains with every transformation S its inverse S^{-1}, and

3. contains with any two transformations S, T their "product" ST.

Example: One could define congruent configurations in space as point sets of which one goes into the other by a congruent transformation of space. The congruent transformations, or "motions", of space form a group, a statement which, according to the above definition of group, is equivalent to the three-fold statement that

1. every figure is congruent to itself,

2. if a figure F is congruent to F', then F' is congruent to F, and

3. if F is congruent to F' and F' is congruent to F'', then F is congruent to F''.

This example at once illuminates the inner significance of the group concept. Symmetry of a configuration F in space is described by the group of motions that carry F into itself.

Often manifolds have a structure. For instance, the elements of a field are connected by the two operations of plus and times; or in Euclidean space we have the relationship of congruence between figures. Hence we have the idea of structure-preserving mappings; they are called *homomorphisms*. Thus a homomorphic mapping of a field k into a field k' is a mapping $a \to a'$ of the "numbers" a of k into the numbers a' of k' such that $(a+b)' = a' + b'$ and $(a \cdot b)' = a' \cdot b'$. A homomorphic mapping of space into itself would be one that carries any two congruent figures into two mutually congruent figures. The following terminology (suggestive to anyone who knows a little Greek) has been agreed upon: homomorphisms which are one-to-one mappings are called *isomorphisms*; when a homomorphism maps a manifold σ into itself, it is called an *endomorphism*, and an *automorphism* when it is both: a one-to-one mapping of σ into itself. Isomorphic systems, i.e., any two systems mapped isomorphically upon each other, have the same structure; indeed nothing can be said about the structure of the one system that is not equally true for the other.

The *automorphisms* of a manifold with a well-defined structure form a group. Two sub-sets of the manifold that go over into each other by an automorphism deserve the name of *equivalent*. This is the precise idea at which Leibniz hints when he says that two such sub-sets are "indiscernible when each is considered in itself"; he recognized this general idea as lying behind the specific geometric notion of similitude. The general problem of relativity consists in nothing else but to find the group of automorphisms. Here then is an important lesson the mathematicians learned in the twentieth century: whenever you are concerned with a structured manifold, study its group of automorphisms. Also the inverse problem, which Felix Klein stressed in his famous Erlangen program (1872), deserves attention: Given a group of transformations of a manifold σ, determine such relations or operations as are invariant with respect to the group.

If in studying a group of transformations we ignore the fact that it consists of transformations and look merely at the way in which any two of its transformations S, T give rise to a composite ST, we obtain the abstract composition schema of the group. Hence an *abstract group* is a set of elements (of unknown or irrelevant nature) for which an operation of composition is defined generating an element st from any two elements s, t, such that the following axioms hold:

1. There is a unit element e such that $es = se = s$ for every s.

2. Every element s has an inverse s^{-1} such that $ss^{-1} = s^{-1}s = e$.

3. The associative law $(st)u = s(tu)$ holds.

It is perhaps the most astonishing experience of modern mathematics how rich in consequences these three simple axioms are.

A realization of an abstract group by transformations of a given manifold σ is obtained by associating with every element s of the group a transformation S of σ, $s \to S$ such that $s \to S, t \to T$ imply $st \to ST$. In general, the commutative law $st = ts$ will not hold. If it does, the group is called commutative or *Abelian* (after the Norwegian mathematician Niels Henrik Abel). Because composition of group elements in general does not satisfy the commutative law, it has proved convenient to use the term "ring" in the wider sense in which it does not imply the commutative law for multiplication. (However, in speaking of a field one usually assumes this law.)

The simplest mappings are the linear ones. They operate in a vector space. The vectors in our ordinary three-dimensional space are directed segments AB leading from a point A to a point B. The vector AB is considered equal to $A'B'$ if a parallel displacement (translation) carries AB into $A'B'$. In consequence of this convention one can add vectors and one can also multiply a vector by a number (integral, rational or even real). Addition satisfies the same axioms as enumerated for numbers in the table T above, and it is also easy to formulate the axioms for the second operation. These axioms constitute the general axiomatic notion of *vector space*, which is therefore an algebraic and not a geometric concept. The numbers which serve as multipliers of the vectors may be the elements of any ring; this generality is actually required in the application of the axiomatic vector concept to topology. However, here we shall assume that they form a field. Then one sees at once that one can ascribe to the vector space a natural number n as its dimensionality in this sense: there exist n vectors $e_1, \cdots, e_n$ such that every vector may be expressed in one and only one way as a linear combination $x_1 e_1 + \cdots + x_n e_n$, where the "coordinates" x_i are definite numbers of the field. In our three-dimensional space n equals 3, but mechanics and physics give ample occasion to use the general algebraic notion of an n-dimensional vector space for higher n.

The endomorphisms of a vector space are called its *linear mappings*; as such they allow composition ST (perform first the mapping S, then T), but they also allow addition and multiplication by numbers γ: if S sends the arbitrary vector x into xS, T into xT, then $S + T, \gamma S$ are those linear mappings which send x into $(xS) + (xT)$ and $\gamma \cdot xS$, respectively. We must forgo to describe how in terms of a vector basis $e_1, \ldots, e_n$, a linear mapping is represented by a square matrix of numbers.

Often rings occur—they are then called *algebras*—which are at the same time vector spaces, i.e., for which three operations, addition of two elements, multiplication of two elements and multiplication of an element by a number, are defined in such manner as to satisfy the characteristic axioms. The linear mappings of an n-dimensional vector space themselves form such an algebra, called the complete matrix algebra (in n dimensions). According to quantum mechanics the observables of a physical system form an algebra of special type with a non-commutative multiplication. In the hands of the physicists abstract algebra has thus become a key that unlocked to them the secrets of the atom. A realization of an abstract group by linear transformations of a vector space is called a *representation*. One may also speak of representations of a ring or an algebra: in each case the representation can be described as a homomorphic mapping of the group or ring or algebra into the complete matrix algebra (which indeed is a group and a ring and an algebra, all in one).

Finale

After spending so much time on the explanation of the notions I can be brief in my enumeration of some of the essential achievements for which they provided the tools. If g is a subgroup of the group G, one may identify elements s, t of G that are congruent mod g, i.e., for which st^{-1} is in g; g is a "self-conjugate" subgroup if this process of identification carries G again into a group, the "factor group" G/g. The group-theoretic core of Galois's theory is a theorem due to C. Jordan and O. Hölder which deals with the several ways in which one may break down a given finite group G in steps $G = G_0, G_1, G_2, \ldots$, each G_i being a self-conjugate subgroup of the preceding group G_{i-1}. Under the assumption that this is done in

as small steps as possible, the theorem states, the steps (factor groups) G_{i-1}/G_i ($i = 1, 2, \cdots$) in one such "composition series" are isomorphic to the steps, suitably rearranged, in a second such series. The theorem is very remarkable in itself, but perhaps the more so as its proof rests on the same argument by which one proves what I consider the most fundamental proposition in all mathematics, namely the fact that if you count a finite set of elements in two ways, you end up with the same number n both times.

The Jordan-Hölder theorem in recent times received a much more natural and general formulation by

1. abandoning the assumption that the breaking down is done in the smallest possible steps, and

2. by admitting only such subgroups as are invariant with respect to a given set of endomorphic mappings of G.

It thus has become applicable to infinite as well as finite groups, and has provided a common denominator for quite a number of important algebraic facts.

The theory of representations of finite groups, the most systematic and substantial part of group theory developed shortly before the turn of the century by G. Frobenius, taught us that there are only a few irreducible representations, of which all others are composed. This theory was greatly simplified after 1900 and later carried over, first to continuous groups that have the topological property of compactness, but then also to all infinite groups, with a restrictive imposition (called almost-periodicity) on the representations. With these generalizations one trespasses the limits of algebra, and a few more words will have to be said about it under the title *analysis*.

New phenomena occur if representations of finite groups in fields of prime characteristic are taken into account, and from their investigation profound number-theoretic consequences have been derived. It is easy to embed a finite group into an algebra, and hence facts about representations of a group are best deduced from those of the embedding algebra. At the beginning of the century algebras seemed to be ferocious beasts of unpredictable behavior, but after fifty years of investigation they, or at least the variety called semi-simple, have become remarkably tame; indeed the wild things do not happen in these superstructures, but in the underlying commutative "number" fields.

In the nineteenth century geometry seemed to have been reduced to a study of invariants of groups; Felix Klein formulated this standpoint explicitly in his

Erlangen program. But the full linear group was practically the only group whose invariants were studied. We have now outgrown this limitation and no longer ignore all the other continuous groups one encounters in algebra, analysis, geometry and physics.

Above all we have come to realize that the theory of invariants has to be subsumed under that of representations. Certain infinite discontinuous groups, like the unimodular and the modular groups, which are of special importance to number theory, witness Gauss's class theory of quadratic forms, have been studied with remarkable success and profound results. The macroscopic and microscopic symmetries of crystals are described by discontinuous groups of motions, and it has been proved for n dimensions, what had long been known for 3 dimensions, that in a certain sense there is but a finite number of possibilities for these crystallographic groups.

In the nineteenth century Sophus Lie had reduced a continuous group to its "germ" of infinitesimal elements. These elements form a sort of algebra in which the associative law is replaced by a different type of law. A *Lie algebra* is a purely algebraic structure, especially if the numbers which act as multipliers are taken from an algebraically defined field rather than from the continuum of real numbers Ω. These Lie groups have provided a new playground for our algebraists.

The constructions of the mathematical mind are at the same time free and necessary. The individual mathematician feels free to define his notions and to set up his axioms as he pleases. But the question is, will he get his fellow mathematicians interested in the constructs of his imagination? We can not help feeling that certain mathematical structures which have evolved through the combined efforts of the mathematical community bear the stamp of a necessity not affected by the accidents of their historical birth. Everybody who looks at the spectacle of modern algebra will be struck by this complementarity of freedom and necessity.

Part II. Analysis, Topology, Geometry, Foundations

Linear operators and their spectral decomposition. Hilbert space

A mechanical system of n degrees of freedom in stable equilibrium is capable of oscillations deviating "infinitely little" from the state of equilibrium. It is a fact of fundamental significance not only for physics but also for music that all these oscillations are superpositions of n "harmonic" oscillations with definite frequencies.

Mathematically the problem of determining the harmonic oscillations amounts to constructing the principal axes of an ellipsoid in an n-dimensional Euclidean space. Representing the vectors x in this space by their coordinates $(x_1, x_2, \ldots, x_n)$ one has to solve an equation

$$x\lambda \cdot Kx = 0,$$

where K denotes a given linear operator (= linear mapping); λ is the square of the unknown frequency ν of the harmonic oscillation, whereas the "eigenvector" x characterizes its amplitude. Define the scalar product (x, y) of two vectors x and y by the sum $x_1 y_1 + \cdots + x_n y_n$. Our "affine" vector space is made into a metric one by assigning to any vector x the length $||x||$ given by $||x||^2 = (x, x)$, and this metric is the Euclidean one so familiar to us from the 3-dimensional case and epitomized by the Pythagorean theorem. The linear operator K is symmetric, in the sense that $(x, Ky) = (Kx, y)$. The field of numbers in which we operate here is, of course, the continuum of all real numbers. Determination of the n frequencies ν, or rather of the corresponding eigenvalues $\lambda = \nu^2$, requires the solution of an algebraic equation of degree n (often known as the *secular equation*, because it first appeared in the theory of the secular perturbations of the planetary system).

More important in physics than the oscillations of a mechanical system of a finite number of degrees of freedom are the oscillations of continuous media, as the mechanical-acoustical oscillations of a string, a membrane or a 3-dimensional elastic body, and the electromagnetic-optical oscillations of the "ether". Here the vectors with which one has to operate are continuous functions $x(s)$ of points with one or several coordinates that vary over a given domain, and consequently K is a linear *integral* operator.

Take for instance a straight string of length 1, the points of which are distinguished by a parameter varying from 0 to 1. Here (x, x) is the integral $\int_0^1 x^2(s)ds$, and the problem of harmonic oscillations (which first suggested to the early Greeks the idea of a universe ruled by harmonious mathematical laws) takes the form of the integral equation

$$x(s) - \lambda \int_0^1 K(s,t)x(t)dt = 0, (0 \leq s \leq 1), \quad (4)$$

where

$$K(s,t) = \left(\frac{a}{\pi}\right) \cdot \begin{cases} s(1-t) & \text{for } s \leq t \\ (1-s)t & \text{for } s \geq t, \end{cases} \quad (5)$$

and a is a constant determined by the physical conditions of the string. The solutions are

$$\lambda = (na)^2, \qquad x(s) = \sin n\pi s,$$

where n is capable of all positive integral values $1, 2, 3, \ldots$. This fact that the frequencies of a string are integral multiples na of a ground frequency a is the basic law of musical harmony. If one prefers an optical to an acoustic language, one speaks of the *spectrum* of eigenvalues λ.

After Fredholm at the very close of the nineteenth century had developed the theory of linear integral equations, it was Hilbert who in the next decade established the general *spectral theory of symmetric linear operators* K. Only twenty years earlier it had required the greatest mathematical efforts to prove the existence of the ground frequency for a membrane, and now constructive proofs for the existence of the whole series of harmonic oscillations and their characteristic frequencies were given under very general assumptions concerning the oscillating medium. This was an event of great consequence both in mathematics and theoretical physics. Soon afterwards, Hilbert's approach made it possible to establish those asymptotic laws for the distribution of eigenvalues the physicists had postulated in their statistical treatment of the thermodynamics of radiation and elastic bodies.

Hilbert observed that an arbitrary continuous function $x(s)$ defined in the interval $0 \leq s \leq 1$ may be replaced by the sequence

$$x_n = \sqrt{2} \int_0^1 x(s) \sin n\pi s \, ds, \quad n = 1, 2, 3, \ldots,$$

of its Fourier coefficients. Thus there is no inner difference between a vector space whose elements are functions $x(s)$ of a continuous variable and one whose elements are infinite sequences of numbers $(x_1, x_2, x_3, \cdots)$.

The square of the "length", $\int_0^1 x^2(s)ds$, equals $x_1^2 + x_2^2 + x_3^2 + \cdots$. Between the two forms in which one may pass from a finite sum to a limit, the infinite sum $a_1 + a_2 + a_3 + \cdots$ and the integral $\int_0^1 x^2(s)ds$, there is therefore here no essential difference. Thus an axiomatic formulation is called for. To the axioms for an (affine) vector space one adds the postulate of the existence of a scalar product (x, y) of any two vectors x, y with the properties characteristic for the Euclidean metric: (x, y) is a number depending linearly on either of the two argument vectors x and y; it is symmetric, $(x, y) = (y, x)$; and $(x, x) = ||x||^2$ is positive, except for $x = 0$. The axiom of finite dimensionality is replaced by a denumerability axiom of more general character.

All operations in such a space are greatly facilitated if it is assumed to be complete in the same sense that the system of real numbers is complete; i.e., if the following is true: Given a "convergent" sequence $x', x'', \ldots$ of vectors, namely, one for which $||x^{(m)} - x^{(n)}||$ tends to zero with m and n tending to infinity, there exists a vector a toward which this sequence converges,

$$||x^{(n)} - a|| \to 0 \text{ for } n \to \infty.$$

A non-complete vector space can be made complete by the same construction by which the system of rational numbers is completed to form that of real numbers. Later authors have coined the name "Hilbert space" for a vector space satisfying these axioms.

Hilbert himself first tackled only integral operators in the strict sense as exemplified by (4). But soon he extended his spectral theory to a far wider class, that of bounded (symmetric) linear operators in Hilbert space. Boundedness of the linear operator requires the existence of a constant M such that $||Kx||^2 \leq M||x||^2$, for all vectors x of finite length $||x||$. Indeed, the restriction to integral operators would be unnatural since the simplest operator, the identity $x \to x$, is not of this type. And now one of those events happened, unforeseeable by the wildest imagination, the like of which could tempt one to believe in a pre-established harmony between physical nature and mathematical mind. Twenty years after Hilbert's investigations, *quantum mechanics* found that the observables of a physical system are represented by the linear symmetric operators in a Hilbert space and that the eigenvalues and eigenvectors of that operator which represents *energy* are the energy levels and corresponding stationary quantum states of the system. Of course this quantum-physical interpretation added greatly to the interest in the theory and led to a more scrupulous investigation of it, resulting in various simplifications and extensions.

Oscillations of continua, the boundary value problems of classical physics and the problem of energy levels in quantum physics, are not the only titles for applications of the theory of integral equations and their spectra. One other somewhat isolated application is the solution of *Riemann's monodromy problem* concerning analytic functions of a complex variable z. It concerns the determination of n analytic functions of z which remain regular under analytic continuation along arbitrary paths in the z-plane, provided these avoid a finite number of singular points, whereas the functions undergo a given linear transformation with constant coefficients when the path circles one of these points.

Another surprising application is to the establishment of the fundamental facts, in particular of the completeness relation, in the theory of *representations of continuous compact groups*. The simplest such group consists of the rotations of a circle, and in that case the theory of representations is nothing but the theory of the so-called Fourier series, which expresses an arbitrary periodic function $f(s)$ of period 2π in terms of the harmonic oscillations

$$\cos ns, \quad \sin ns, \quad n = 0, 1, 2, \cdots.$$

In Nature functions often occur with hidden non-commensurable periodicities. The mathematician Harald Bohr, the brother of the physicist Niels Bohr, prompted by certain of his investigations concerning the Riemann zeta function, developed the general mathematical theory of such *almost periodic functions*. One may describe his theory as that of almost periodic representations of the simplest continuous group one can imagine, namely, the group of all translations of a straight line.

His main results could be carried over to arbitrary groups. No restriction is imposed on the group, but the representations one studies are supposed to be almost-periodic. For a function $x(s)$, the argument s of which runs over the group elements, while its values are real or complex numbers, almost-periodicity amounts to the requirement that the group be compact in a certain topology induced by the function. This relative compactness instead of absolute compactness is sufficient. Even so the restriction is severe. Indeed the most important representations of the classical continuous groups are not almost-periodic. Hence the theory is in need of further extension, which has busied a number of American and Russian mathematicians during the last decade.

Lebesgue's integral, Measure theory, Ergodic hypothesis

Before turning to other applications of operators in Hilbert space I must mention the, in all probability final, form given to the idea of integration by Lebesgue at the beginning of our century. Instead of speaking of the area of a piece of the 2-dimensional plane referred to coordinates x, y, or the volume of a piece of the 3-dimensional Euclidean space, we use the neutral term *measure* for all dimensions. The notions of measure and integral are interconnected. Any piece of space, any set of space points, can be described by its characteristic function $\chi(P)$, which equals 1 or 0 according to whether the point P belongs or does not belong to the set. The measure of the point set is the integral of this characteristic function.

Before Lebesgue one first defined the integral for continuous functions; the notion of measure was secondary; it required transition from continuous to such discontinuous functions as $\chi(P)$. Lebesgue goes the opposite and perhaps more natural way: for him measure comes first and the integral second.

The one-dimensional space is sufficient for an illustration. Consider a real-valued function $y = f(x)$ of a real variable x which maps the interval $0 \leq x \leq 1$ into a finite interval $a \leq y \leq b$. Instead of subdividing the interval of the argument x, Lebesgue subdivides the interval (a, b) of the dependent variable y into a finite number of small subintervals $a_i \leq y < a_{i+1}$ say, of lengths $< \varepsilon$, and then determines the measure m_i of the set S_i on the x-axis, the points of which satisfy the inequality $a_i \leq f(x) < a_{i+1}$. The integral lies between the two sums $\sum_i a_i m_i$ and $\sum_i a_{i+1} m_i$ which differ by less than ε, and thus can be computed with any degree of accuracy.

In determining the measure of a point set—and this is the more essential modification—Lebesgue covers the set with infinite sequences, rather than finite ensembles, of intervals. Thus, to the set of rational x in the interval $0 \leq x \leq 1$ no measure could be ascribed before Lebesgue. But these rational numbers can be arranged in a denumerable sequence $a_1, a_2, a_3, \ldots$, and, after choosing a positive number ε as small as one likes, one can surround the point a_n by an interval of length $\varepsilon/2^n$ with the center a_n. Thus the whole set of rational points is enclosed in a sequence of intervals of total length

$$\varepsilon \left(\frac{1}{2} + \frac{1}{2^2} + \frac{1}{2^3} + \cdots \right) = \varepsilon;$$

and according to Lebesgue's definition its measure is therefore less than (the arbitrary positive) ε and hence zero. The notion of *probability* is tied to that of measure, and for this reason mathematical statisticians are deeply interested in measure theory.

Lebesgue's idea has been generalized in several directions. The two fundamental operations one can perform with sets are: forming the intersection and the union of given sets, and thus sets may be considered as elements of a "Boolean algebra" with these two operations, the properties of which may be laid down in a number of axioms resembling the arithmetical axioms for addition and multiplication. Hence one of the questions which has occupied the more axiomatically minded among the mathematicians and statisticians is concerned with the introduction of measure in abstract Boolean algebras.

Lebesgue's integral is important in our present context, because those real-valued functions $f(x)$ of a real variable x ranging over the interval $0 \leq x \leq 1$, the squares of which are Lebesgue-integrable, form a complete Hilbert space—provided two functions $f(x), g(x)$ are considered equal if those values x for which $f(x) \neq g(x)$ form a set of measure zero (Riesz-Fischer theorem).

The mechanical equations for a system of n degrees of freedom in Hamilton's form uniquely determine the state tP at the moment t if the state P at the moment $t = 0$ is given. Such is the precise formulation of the law of causality in mechanics. The possible states P form the points of a $(2n)$-dimensional phase space, and for a fixed t and an arbitrary P the transition $P \to tP$ is a measure-preserving mapping (t). These transformations form a group: $(t_1)(t_2) = (t_1 + t_2)$. For a given P and a variable t the point tP describes the consecutive states which this system assumes if at the moment $t = 0$ it is in the state P. Considering P as a particle of a $(2n)$-dimensional fluid which fills the phase-space and ascribing to the particle P the position tP at the time t, one obtains the picture of an incompressible fluid in stationary flow. The statistical derivation of the laws of thermodynamics makes use of the so-called *ergodic hypothesis*, according to which the path of an arbitrary individual particle P (excepting initial states P which form a set of measure zero) covers the phase-space (or at least that $(2n - 1)$-dimensional sub-space of it where the energy has a given value) everywhere densely, so that in the course of its history the probability of finding it in this or that part of the space is the same for any parts of equal measure.

Nineteenth-century mathematics seemed to be a long way off from proving this hypothesis with any degree of generality. Strangely enough it was proved shortly after the transition from classical to quantum mechanics had rendered the hypothesis almost valueless, and it was proved by making use of the mathematical apparatus of quantum physics. Under the influence of the mapping (t), $P \to tP$, any function $f(P)$ in phase-space is transformed into the function $\bar{f} = U_i f$, defined by the equation $\bar{f}(tP) = f(P)$. The U_i form a group of operators in the Hilbert space of arbitrary functions $f(P)$, $U_{i_1} U_{i_2} = U_{i_1 + i_2}$ and, application of spectral decomposition to this group enabled J. von Neumann to deduce the ergodic hypothesis with two provisos:

1. Convergence of a sequence of functions $f_n(P)$ toward a function $f(P)$, $f_n \to f$, is understood (as it would be in quantum mechanics), namely as convergence in Hilbert space where it means that the total integral of $(f_n f)^2$ tends to zero with $n \to \infty$.

2. One assumes that there are no subspaces of the phase-space which are invariant under the group of transformations (t), except those spaces that are in Lebesgue's sense equal either to the empty or the total space.

Shortly afterwards proofs were also given for other interpretations of the notion of convergence.

The laws of nature can either be formulated as differential equations or as "principles of variation," according to which certain quantities assume extremal values under given conditions. For instance, in an optically homogeneous or non-homogeneous medium, the light travels along that road from a given point A to a given point B for which the time of travel assumes minimal value. In potential theory the quantity which assumes a minimum is the so-called *Dirichlet integral*. Attempts to establish directly the existence of a minimum had been discouraged by Weierstrass's criticism in the nineteenth century. Our century, however, restored the direct methods of the Calculus of Variation to a position of honor after Hilbert in 1900 gave a direct proof of the Dirichlet principle and later showed how it can be applied not only in establishing the fundamental facts about functions and integrals ("algebraic" functions and "abelian" integrals) on a compact Riemann surface (as Riemann had suggested 50 years earlier), but also for deriving the basic propositions of the *theory of uniformization*. That theory occupies a central position in the theory of functions of one complex variable, and the first decade of the twentieth century witnessed the first proofs by P. Koebe and H. Poincaré of these propositions conjectured about 25 years before by Poincaré himself and by Felix Klein. As in an Euclidean vector space of finite dimensionality, so in the Hilbert space of infinitely many dimensions, this fact is true: Given a linear (complete) subspace E, any vector may be split in a uniquely determined manner into a component lying in E (orthogonal projection) and one perpendicular to E. Dirichlet's principle is nothing but a special case of this fact. But since the function-theoretic applications of orthogonal projection in Hilbert space which we alluded to are closely tied up with topology, we had better turn first to a discussion of this important branch of modern mathematics: topology.

Topology and harmonic integrals

Essential features of the modern approach to *topology* can be brought to light in its connection with the, only recently developed, theory of *harmonic integrals*. Consider a stationary magnetic field h in a domain G which is free from electric currents. At every point of G it satisfies two differential conditions which, in the usual notations of vector analysis, are written in the form div $h = 0$, rot $h = 0$. A field of this type is called *harmonic*. The second of these conditions states that the line integral of h along a closed curve (cycle) C, $\int_C h$, vanishes provided C lies in a sufficiently small neighborhood of an arbitrary point of G. This implies $\int_C h = 0$ for any cycle C in G that is the boundary of a surface in G. However, for an arbitrary cycle C in G the integral is equal to the electric current surrounded by C.

Let the phrase "C homologous to zero", $C \sim 0$, indicate that the cycle C in G bounds a surface in G. One can travel over a cycle C in the opposite sense, thus obtaining $-C$, or travel over it $2, 3, \ldots$ times, thus obtaining $2C, 3C, \ldots$; and cycles may be added and subtracted from each other (if one does not insist that cycles are of one piece). Two cycles C, C' are called *homologous*, $C \sim C'$, if $C - C' \sim 0$. Note that $C \sim 0$, $C' \sim 0$ imply $-C \sim 0$, $C + C' \sim 0$. Hence the cycles form a commutative group under addition, the "Betti group", if homologous cycles are considered as one and the same group element.

These notions of *cycles and their homologies* may be carried over from a three-dimensional domain in Euclidean space to any n-dimensional manifold, in particular to closed (compact) manifolds like the two-dimensional surfaces of the sphere or the torus; and on an n-dimensional manifold we can speak not only of 1-dimensional, but also of 2-, 3-, ..., n-dimensional cycles. The notion of a harmonic vector field permits a similar generalization, harmonic tensor field (harmonic form) of rank r ($r = 1, 2, \ldots, n$), provided the manifold bears a Riemannian metric, an assumption the meaning of which will be discussed later in the section on geometry. Any tensor field (linear differential form) of rank r may be integrated over an r-dimensional cycle.

The fundamental problem of *homology theory* consists in determining the structure of the Betti group, not only for 1-, but also for 2-, ..., n-dimensional cycles, in particular in determining the number of linearly independent cycles (the Betti number). [ν cycles $C_1, \ldots, C_\nu$, are linearly independent if there exists no homology $k_1 C_1 + \cdots + k_\nu C_\nu \sim 0$ with integral coefficients k, except $k_1 = \cdots = k_\nu = 0$.] The fundamental theorem for harmonic forms on compact manifolds states that, given ν linearly independent cycles $C_1, \ldots, C_\nu$, there exists a harmonic form h with pre-assigned periods

$$\int_{C_1} h = \pi_1, \cdots, \int_{C_\nu} h = \pi_\nu.$$

H. Poincaré developed the algebraic apparatus necessary to formulate exactly the notions of cycle and homology. In the course of the twentieth century it turned out that in most problems cohomologies are easier to handle than homologies. I illustrate this for 1-dimensional cycles.

A line C_1 leading from a point p_1 to p_2, when followed by a line C_2 leading from p_2 to a third point p_3, gives rise to a line $C_1 + C_2$ leading from p_1 to p_3. The line integral $\int_C h$ of a given vector field h along an arbitrary (closed or open) line C is an additive function $\phi(C)$ of C, $\phi(C_1 + C_2) = \phi(C_1) + \phi(C_2)$. If moreover rot h vanishes everywhere, then $\phi(C) = 0$ for any closed line C that lies in a sufficiently small neighborhood of a point, whatever this point may be. Any real-valued function ϕ satisfying these two conditions may be called an *abstract integral*. The cohomology $\phi \sim 0$ means that $\phi(C) = 0$ for any closed line C, and thus it is clear what the cohomology $k_1 \phi_1 + \cdots + k_\nu \phi_\nu \sim 0$ with arbitrary real coefficients $k_1, \ldots, k_\nu$, means. The homology $C \sim 0$ could now be defined, not by the condition that the cycle C bounds, but by the requirement that $\phi(C) = 0$ for every abstract integral ϕ. With the convention that any two abstract integrals ϕ, ϕ' are identified if $\phi - \phi' \sim 0$, these integrals form a vector space, and the dimensionality of this vector space is now introduced as the *Betti number*. And the fundamental theorem for harmonic integrals on a compact manifold now asserts that for any given abstract integral ϕ there exists one and only one harmonic vector field h whose integral is cohomologous to ϕ, $\int_C h = \phi(C)$ for every cycle C (realization of the abstract integral *in concreto* by a harmonic integral).

J. W. Alexander discovered an important result connecting the Betti numbers of a manifold M that is embedded in the n-dimensional Euclidean space R_n with the Betti numbers of the complement $R_n \setminus M$ (*Alexander's duality theorem*).

The difficulties of topology spring from the double aspect under which one can consider continuous manifolds. Euclid looked upon a figure as an assemblage of a finite number of geometric elements, like points, straight lines, circles, planes, spheres. But after replacing each line or surface by the set of points lying on it, one may also adopt the set-theoretic view that there is only one sort of element, points, and that any (in general infinite) set of points can serve as a figure. This modern standpoint obviously gives geometry far greater generality and freedom. In topology, however, it is not necessary to descend to the points as the ultimate atoms, but one can construct the manifold like a building from "blocks" or cells, and a finite number of such cells serving as units will do, provided the manifold is compact. Thus it is possible here to revert to a treatment in Euclid's "finitistic" style (combinatorial topology).

On the first standpoint, a manifold as a point set, the task is to formulate that continuity by which a point p approaching a given point p_0 becomes gradually indistinguishable from p_0. This is done by associating with p_0 the *neighborhoods* of p_0, an infinite shrinking sequence of sub-sets $U_1 \supset U_2 \supset U_3 \supset \cdots$, all containing p_0. For example, in a plane referred to Cartesian coordinates x, y we may choose as the nth neighborhood U of a point $p_0 = (x_0, y_0)$ the interior of the circle of radius $1/2^n$ around p_0. The notion of convergence, basic for all continuity considerations, is defined in terms of the sequence of neighborhoods as follows: A sequence of points $p_1, p_2, \ldots$ converges to p_0 if for every natural number n there is an N so that all points p_ν with $\nu > N$ lie in the nth neighborhood U_n of p_0. Of course, the choice of the neighborhoods U_n is arbitrary to a certain extent. For instance, one could also have chosen as the nth neighborhood V_n of (x_0, y_0) the square of side $2/n$ around (x_0, y_0), to which a point (x, y) belongs if

$$-1/n < x - x_0 < 1/n, \quad -1/n < y - y_0 < 1/n.$$

However, the sequence V_n is equivalent to the sequence U_n in the sense that for every n there is an n' such that $U_{n'} \subset V_n$ (and thus $U_\nu \subset V_n$ for $\nu \geq n'$), and also for every m there is an m' such that $V_{m'} \subset U_m$; and consequently the notion of convergence for points is the same, whether based on the one or the other sequence of neighborhoods.

It is clear how to define continuity of a mapping of one manifold into another. A one-to-one mapping of two manifolds upon each other is called *topological* if it is continuous in both directions, and two manifolds that can be mapped topologically upon each other are topologically equivalent. Topology investigates such properties of manifolds as are invariant with respect to topological mappings (in particular, with respect to continuous deformations).

A continuous function $y = f(x)$ may be approximated by piecewise linear functions. The corresponding device in higher dimensions, the method of *simplicial approximations* of a given continuous mapping of one manifold into another, is of great importance in set-theoretic topology. It has served to develop a general theory of dimensions, to prove the topological invariance of the Betti groups, to define the decisive notion of the degree of mapping ("Abbildungsgrad", L. E. J. Brouwer), and to prove a number of interesting fixed-point theorems. For instance, a continuous mapping of a square into itself has necessarily a fixed point, i.e., a point carried by the mapping into itself. Given two continuous mappings of a (compact) manifold M into another M', one can ask more generally for which points p on M both images on M' coincide. A famous formula by S. Lefschetz relates the "total index" of such points with the homology theory of cycles on M and M'.

Application of fixed-point theorems to functional spaces of infinitely many dimensions has proved a powerful method to establish the existence of solutions for non-linear differential equations. This is particularly valuable, because the hydrodynamical and aerodynamical problems are almost all of this type.

Poincaré found that a satisfactory formulation of the homology theory of cycles was possible only from the second standpoint where the n-dimensional manifold is considered as a conglomerate of n-dimensional cells. The boundary of an n-dimensional cell (n-cell) consists of a finite number of $(n-1)$-cells, the boundary of an $(n-1)$-cell consists of a finite number of $(n-2)$-cells, etc. The combinatorial skeleton of the manifold is obtained by assigning symbols to these cells and then stating in terms of their symbols which $(i-1)$-cells belong to the boundary of any of the occurring i-cells ($i = 1, 2, \ldots, n$). From the cells one descends to the points of the manifold by a repeated process of subdivision which catches the points in an ever finer net. Since this subdivision proceeds according to a fixed combinatorial scheme, the manifold is in topological regard completely fixed by its combinatorial skeleton. And at once the question arises under what circumstances two given combinatorial skeletons represent the same manifold, i.e., lead by iterated subdivision to topologically equivalent manifolds. We are far from being able to solve this fundamental problem. Algebraic topology, which operates with the combinatorial skeletons, is in itself a rich and beautiful theory, linked in various ways with the basic notions and theorems of algebra and group theory.

The connection between algebraic and set-theoretic topology is fraught with serious difficulties which are not yet overcome in a quite satisfactory manner. So much, however, seems clear that one had better start, not with a division into cells, but with a covering by patches which are allowed to overlap. From such a pattern the fundamental topologically invariant concepts are to be developed. The above notion of an abstract integral, which relates homology and cohomology, is an indication; it can indeed be used for a direct proof of the invariance of the first Betti number without the tool of simplicial approximation.

Conformal mapping, meromorphic functions, Calculus of variations in the large

Homology theory, in combination with the Dirichlet principle or the method of orthogonal projection in Hilbert space, leads to the theory of harmonic integrals, in particular for the lowest dimension $n = 2$ to the theory of abelian integrals on Riemann surfaces. But for Riemann surfaces the Dirichlet principle also yields the fundamental facts concerning uniformization of analytic functions of one variable if one combines it with the *homotopy* (not homology) theory of closed curves. Whereas a cycle is homologous to zero if it bounds, it is homotopic to zero if it can be contracted into a point by continuous deformation. The homotopy theory of 1- and more-dimensional cycles has recently come to the fore as an important branch of topology, and the group-theoretic aspect of homotopy has led to some surprising discoveries in abstract group theory.

Homotopy of 1-dimensional cycles is closely related with the idea of the *universal covering manifold* of a given manifold. Given a continuous mapping $p \to p'$ of one manifold M into another M', the point p' may be considered as the trace or projection in M' of the arbitrary point p on M, and thus M becomes a manifold covering M'. There may be no point or several points p on M which lie over a given point p' of M' (which are mapped into p'). The mapping is without ramifications if, for any point p_0 of M, it is one-to-one (and continuous both ways) in a sufficiently small neighborhood of p_0. Let p_0 be a point on M, p_0' its trace on M', and C' a curve on M' beginning at p_0'. If M covers M' without ramifications we can follow this curve on M by starting at p_0, at least up to a certain point where we would run against a "boundary of M relative to M'".

Of chief interest are those covering manifolds M over a given M' for which this never happens and which therefore cover M' without ramifications and relative boundaries. The best way of defining the central topological notion "simply connected" is by describing a simply connected manifold as one having no other unramified unbounded covering but itself. There is a strongest of all unramified unbounded covering manifolds, the universal covering manifold, which can be described by the statement that on it a curve C is closed only if its trace C' is (closed and) homotopic to zero. The proof of the fundamental theorem on uniformization consists of two parts:

1. constructing the universal covering manifold of the given Riemann surface;

2. constructing by means of the Dirichlet principle a one-to-one conformal mapping of the covering manifold upon the interior of a circle of finite or infinite radius.

All we have discussed so far in our account of analysis is in some way tied up with operators and projections in Hilbert space, the analogue in infinitely many dimensions of Euclidean space. In H. Minkowski's *Geometry of Numbers*, distances $|AB|$, which are different from the Euclidean distance but satisfy the axioms that $|BA| = |AB|$ and that in a triangle ABC the inequality $|AC| \leq |AB| + |BC|$ holds, were used to great advantage for obtaining numerous results concerning the solvability of inequalities by integers. We do not find time here to report on the progress of this attractive branch of number theory during the last fifty years. In infinitely many dimensions spaces endowed with a metric of this sort, of a more general nature than the Euclid-Hilbert metric, have been introduced by Banach, not however for number-theoretic but for purely analytic purposes. Whether the importance of the subject justifies the large number of papers written on *Banach spaces* is perhaps questionable.

The Dirichlet principle is but the simplest example of the direct methods of the Calculus of Variations as they came into use with the turn of the century. It was by these methods that the theory of *minimal surfaces*, so closely related to that of analytic functions, was put on a new footing. What we know about non-linear differential equations has been obtained either by the topological fixed-point method (see above), or by the so-called continuity method, or by constructing their solutions as extremals of a suitable functional.

A continuous function on an n-dimensional compact manifold assumes somewhere a minimum and somewhere else a maximum value. Interpret the function as altitude. Besides summit (local maximum) and bottom (local minimum) one has the further possibility of a saddle point (pass) as a point of "stationary" altitude. In n dimensions the several possibilities are indicated by an inertial index k which is capable of the values $k = 0, 1, 2, \ldots, n$, the value $k = 0$ characterizing a minimum and $k = n$ a maximum. Marston Morse discovered the inequality $M_k \geq B_k$ between the number M_k of stationary points of index k and the Betti number B_k of linearly independent homology classes of k-dimensional cycles. In their generalization to functional spaces these relations have opened a line of study adequately described as *Calculus of Variation in the large*.

Development of the theory of uniformization for analytic functions led to a closer investigation of *conformal mapping* of 2-dimensional manifolds in the large, which resulted in a number of theorems of surprising simplicity and beauty. In the same field there is to register an enormous extension of our knowledge of the behavior of *meromorphic functions*, i.e., single-valued analytic functions of the complex variable z which are regular everywhere with the exception of isolated "poles" (points of infinity). Towards the end of the previous century Riemann's zeta function had provided the stimulus for a deeper study of "entire functions" (functions without poles).

The greatest stride forward, both in methods and results, was marked by a paper on meromorphic functions published in 1925 by the Finnish mathematician Rolf Nevanlinna. Besides meromorphic functions in the z-plane one can study such functions on a given Riemann surface; and in the way in which the theory of algebraic functions (equal to meromorphic functions on a compact Riemann surface) as a theory of algebraic curves in two complex dimensions may be generalized to any number of dimensions, so one can pass from meromorphic functions to meromorphic curves.

The theory of *analytic functions of several complex variables*, in spite of a number of deep results, is still in its infancy.

Geometry

After having dealt at some length with the problems of analysis and topology I must be brief about geometry. Of subjects mentioned before, minimal surfaces, conformal mapping, algebraic manifolds and the whole of topology could be subsumed under the title of geometry.

In the domain of *elementary axiomatic geometry* one strange discovery, that of von Neumann's pointless "continuous geometries" stands out, because it is intimately interrelated with quantum mechanics, logic and the general algebraic theory of "lattices". The 1-, 2-, ..., n-dimensional linear manifolds of an n-dimensional vector space form the 0-, 1-, ..., $(n-1)$-dimensional linear manifolds in an $(n-1)$-dimensional projective point space. The usual axiomatic foundation of projective geometry uses the points as the primitive elements or atoms of which the higher-than-zero-dimensional manifolds are composed.

However, there is possible a treatment where the linear manifolds of all dimensions figure as elements, and the axioms deal with the relation "B contains A" ($A \subset B$) between these elements and the operation of intersection, $A \cap B$, and of union, $A \cup B$, performed on them; the union $A \cup B$ consists of all sums $x + y$ of a vector x in A and a vector y in B. In quantum logic this relation and these operations correspond to the relation of implication ("The statement A implies B") and the operations 'and', 'or' in classical logic. But whereas in classical logic the distributive law

$$A \cap (B \cup C) = (A \cap B) \cup (A \cap C)$$

holds, this is not so in quantum logic; it must be replaced by the weaker axiom:

If $C \subset A$, then $A \cap (B \cup C) = (A \cap B) \cup C$.

On formulating the axioms without the implication of finite dimensionality one will come across several possibilities; one leads to the Hilbert space in which quantum mechanics operates, another to von Neumann's continuous geometry with its continuous scale of dimensions, in which elements of arbitrarily low dimensions exist but none of dimension zero.

The most important development of geometry in the twentieth century took place in differential geometry and was stimulated by general relativity, which showed that the world is a 4-dimensional manifold endowed with a Riemannian metric. A piece of an n-dimensional manifold can be mapped in one-to-one continuous fashion upon a piece of the n-dimensional "arithmetical space" which consists of all n-tuples $(x_1, x_2, \ldots, x_n)$ of real numbers x_i. A *Riemann metric* assigns to a line element which leads from the point $P = (x_1, x_2, \ldots, x_n)$ to the infinitely near point $P' = (x_1 + dx_1, x_2 + dx_2, \ldots, x_n + dx_n)$ a distance ds, the square of which is a quadratic form of the relative coordinates dx_i,

$$ds^2 = \sum g_{ij} dx_i dx_j, \qquad (i, j = 1, \ldots, n)$$

with coefficients g_{ij} depending on the point P but not on the line element. This means that, in the infinitely small, Pythagoras's theorem and hence Euclidean geometry are valid, but in general not in a region of finite extension. The line elements at a point may be considered as the infinitesimal vectors of an n-dimensional vector space in P, the tangent space or the compass at P; indeed an arbitrary (differentiable) transformation of the coordinates x_i induces a linear transformation of the components dx_i of any line element at a given point P.

As Levi-Civita found in 1915, the development of Riemannian geometry hinges on the fact that a Riemannian metric uniquely determines an infinitesimal parallel displacement of the vector compass at P to any infinitely near point P'. From this a general scheme for differential geometry arose in which each point P of the manifold is associated with a homogeneous space Σ_p described by a definite group of "authomorphisms", this space now taking over the role of the tangent space (whose group of authomorphisms consists of all non-singular linear transformations). One assumes that one knows how this associated space Σ_p is transferred by infinitesimal displacement to the space $\Sigma_{p'}$ associated with any infinitely near point P'. The most fundamental notion of Riemannian geometry, that of curvature, which figures so prominently in Einstein's equations of the gravitational field, can be carried over to this general scheme. Thus one has erected general differential affine, projective, conformal, geometries, etc.

One has also tried by their structures to account for the other physical fields existing in nature beside the gravitational one, namely the electromagnetic field, the electronic wave-field and further fields corresponding to the several kinds of elementary particles. But it seems to the author that so far all such speculative attempts of building up a unified field theory have failed. There are very good reasons for interpreting gravitation in terms of the basic concepts of differential geometry. But it is probably unsound to try to "geometrize" all physical entities.

Differential geometry in the large is an interesting field of investigation which relates the differential properties of a manifold with its topological structure. The schema of differential geometry explained above with its associated spaces Σ_p and their displacements has a purely topological kernel which has recently developed under the name of *fibre spaces* into an important topological technique.

Our account of progress made during the last fifty years in analysis, geometry and topology had to touch on many special subjects. It would have failed completely had it not imparted to the reader some feeling of the close relationship connecting all these mathematical endeavors. As the last example of fibre spaces (beside many others) shows, this unity in diversity even makes a clear-cut division into analysis, geometry, topology (and algebra) practically impossible.

Foundations

Finally, a few words about the foundations of mathematics. The nineteenth century had witnessed the critical analysis of all mathematical notions including that of natural numbers to the point where they got reduced to pure logic and the ideas "set" and "mapping". At the end of the century it became clear that the unrestricted formation of sets, subsets of sets, sets of sets etc., together with an unimpeded application to them as to the original elements of the logical quantifiers "there exists" and "all" [cf. the sentences: the (natural) number n is even if there exists a number x such that $n = 2x$; it is odd if n is different from $2x$ for all x] inexorably leads to antinomies. The three most characteristic contributions of the twentieth century to the solution of this Gordian knot are connected with the names of L. E. J. Brouwer, David Hilbert and Kurt Gödel. Brouwer's critique of "mathematical existentialism" not only dissolved the antinomies completely, but also destroyed a good part of classical mathematics that had heretofore been universally accepted.

If only the historical event that somebody has succeeded in constructing a (natural) number n with the given property P can give a right to the assertion that "there exists a number with that property," then the alternative that there either exists such a number or that all numbers have the opposite property non-P is without foundation. The principle of excluded middle for such sentences may be valid for God who surveys the infinite sequence of all natural numbers, as it were, with one glance, but not for human logic. Since the quantifiers "there is" and "all" are piled upon each other in the most manifold way in the formation of mathematical propositions, Brouwer's critique makes almost all of them meaningless, and therefore Brouwer set out to build up a new mathematics which makes no use of that logical principle.

I think that everybody has to accept Brouwer's critique who wants to hold on to the belief that mathematical propositions tell the sheer truth, truth based on evidence. At least Brouwer's opponent, Hilbert, accepted it tacitly. He tried to save classical mathematics by converting it from a system of meaningful propositions into a game of meaningless formulas, and by showing that this game never leads to two formulas, F and non-F, which are inconsistent. Consistency, not truth, is his aim. His attempts at proving consistency revealed the astonishingly complex logical structure of mathematics. The first steps were promising indeed. But then Gödel's discovery cast a deep shadow over Hilbert's enterprise. Consistency itself may be expressed by a formula. What Gödel showed was this: If the game of mathematics is actually consistent, then the formula of consistency cannot be proved within this game. How can we then hope to prove it at all?

This is where we stand now. It is pretty clear that our theory of the physical world is not a description of the phenomena as we perceive them, but is a bold symbolic construction. However, one may be surprised to learn that even mathematics shares this character. The success of the anti-phenomenological constructive method is undeniable. And yet the ultimate foundations on which it rests remain a mystery, even in mathematics.

Mathematics at the Turn of the Millennium

PHILLIP A. GRIFFITHS

American Mathematical Monthly **107** (2000), 1–14

1 Introduction

The last century has been a golden age for mathematics. Many important, long-standing problems have been resolved, in large part because of the growing understanding of the complex interactions among the subfields of mathematics. As those relationships continue to expand and deepen, mathematics is beginning to reach out to explore interactions with other areas of science. These interactions, both within diverse areas of mathematics and between mathematics and other fields of science, have led to some great insights and to the broadening and deepening of the field of mathematics. I discuss some of these interactions and insights, describe a few mathematical achievements of the twentieth century, and pose some challenges and opportunities that await us in the twenty-first century.

2 The world of mathematics

In discussing our subject, we mathematicians face a dilemma. The most effective way to explain mathematics to general readers is to use metaphors, which entails a loss of precision and carries the risk of misunderstanding. On the other hand, advanced mathematical terms are incomprehensible to most people—including other scientists. As my colleague David Mumford, former president of the International Mathematics Union, has said, "I am accustomed, as a professional mathematician, to living in a sort of vacuum, surrounded by people who... declare with an odd sort of pride that they are mathematically illiterate."

Within the mathematical community, however, the use of a precise language is a distinct advantage. Because of its abstract nature and universality, mathematics knows neither linguistic nor political boundaries. It is one reason that mathematics has always carried a distinctly international flavor. A mathematician in Japan can usually read the paper of a colleague in Germany without translation.

The number of highly active research mathematical scientists worldwide is small—probably well under 10,000—so that a given subfield may be populated by a small number of highly specialized individuals. By necessity, these colleagues know one another well, regardless of country of residence, and collaborate over long distances. During the present century, a growing number of papers have been co-authored by mathematicians from different nations (the number rose by about 50% between 1981 and 1993). And so mathematicians are well adapted to the current trend toward a world of vanishing borders.

But what is it that these mathematicians do? In general, mathematics can be described as the search for structures and patterns that bring order and simplicity to our universe. It may be said that the object or beginning point of a mathematical study is not as important as the patterns and coherence that emerge. And it is these patterns and coherence that give mathematics its power, because they often bring clarity to a completely different object or process—to another branch of mathematics, another science, or to society at large.

When mathematicians speak of their work, two words carry great importance. Mathematics is a field where a "problem" is not a bad thing. In fact, a good problem is what mathematicians yearn for; it signifies interesting work. The second word is "proof", which strongly suggests the rigor of the discipline. Sir Arthur Eddington once said, "Proof is an idol before which the mathematician tortures himself." A mathematical proof is a formal and logical line of reasoning that begins with a set of axioms and moves through logical steps to a conclusion. A proof, once given, is permanent; some have existed since the times of the Greeks. A proof confirms truth for the mathematician the way experiment or observation does for the natural scientist.

The twentieth century has been a fertile time for the resolution of long-standing problems, and for a wealth of accomplishments that would require at least an encyclopedia to describe. Let us look at just two of the more interesting achievements—solutions to problems that were over 300 years old. Both occurred toward the end of the century and could have succeeded only because of the mathematics that preceded them.

Fermat's Last Theorem

The first is the proof of Fermat's Last Theorem by Andrew Wiles, which made news around the globe in 1993. This example is interesting because of Fermat, an eccentric jurist and amateur mathematician who published no papers; because of Wiles, who toiled on the problem alone for seven years; and because of the problem itself, whose solution depended on fundamental advances in number theory by many mathematicians over a period of 350 years, especially during the last half-century.

The theorem was written in 1637, when Pierre de Fermat was studying an ancient text on number theory called *Arithmetica*, by Diophantus. Interest in number theory had waned since the time of the ancient Greeks, but Fermat loved numbers. He came across the famous Pythagorean equation most of us learn in school: $x^2 + y^2 = z^2$. Even today, countless schoolchildren learn to say, "The square of the hypotenuse equals the sum of the squares of the other two sides."

Of particular interest are solutions to the Pythagorean equation in whole numbers, such as the beautiful 3–4–5 right triangle. When Fermat saw this, he noted that for any exponent greater than two, the equation could not have solutions in whole numbers. He also wrote, in Latin, that he had discovered his own wonderful proof, but that the margin was too small to contain it. No such proof has ever been found. Fermat made many such marginal comments (some regarded as taunts to his fellow mathematicians), and over the centuries they were all resolved except this one, Fermat's Last Theorem.

Andrew Wiles first came across Fermat at age 10, in a library in Cambridge, England, where he grew up. He decided that some day he would prove it. By the time he was a young mathematician, however, he had learned that pursuing Fermat by itself was not advisable, and decided to work instead, in a complex area of algebraic number theory known as *Iwasawa theory*. But he never forgot about Fermat.

In 1986 he learned of a breakthrough: a colleague, Ken Ribet at the University of California at Berkeley, had linked Fermat's Last Theorem to another unsolved problem, the Taniyama-Shimura conjecture, a surprising and brilliant formulation in algebraic geometry posed in 1955. To summarize a very complex sequence of reasoning, this linkage showed that proving the Taniyama-Shimura conjecture would also prove Fermat's Last Theorem as well. It constructed a logical bridge between the intricate worlds of elliptic curves and modular forms, a kind of dictionary that allows questions and insights to be translated between the two worlds. It also meant that Wiles's earlier work in algebraic number theory would be helpful, and that he would probably generate some interesting problems—whether or not he found a proof.

He did find a proof—after a series of baffling obstacles and sudden insights. Even after he had presented his results, a crucial error was found during the refereeing process, which led to a further year's work. Again, there seemed to be no solution—and again, there was one. Wiles called this last insight "the most important moment of my working life. It was so indescribably beautiful, it was so simple and elegant and I just stared in disbelief for twenty minutes."

> Perhaps I can best describe my experience of doing mathematics in terms of a journey through a dark unexplored mansion. You enter the first room of the mansion and it's completely dark. You stumble around bumping into furniture, but generally you learn where each piece of furniture is, Finally, after six months or so, you find the light switch, you turn it on, and suddenly it's all illuminated. You can see exactly where you were. Then you move into the next room and spend another six months in the dark. So each of these breakthroughs, while sometimes they're momentary, sometimes over a period of a day or two, they are the culmination of—and couldn't exist without—the many months of stumbling around in the dark that precede them.
>
> *Andrew Wiles, who proved Fermat's Last Theorem in 1993*

Did Fermat really complete his own proof in the seventeenth century? Undoubtedly, some will continue to look for evidence that he did, but it is highly unlikely. Wiles's work made use of whole subfields of nineteenth and twentieth century mathematics that didn't exist in Fermat's time. Beneath Fermat's equation now lies an enormous and elaborate formal structure—the kind of structure that mathematicians strive for. The solution to Fermat arises from the implications of understanding that structure.

Kepler's Sphere Packing Conjecture

The second problem is Kepler's Sphere Packing Conjecture. Like the Fermat problem, sphere packing could have been solved in the way it was only in the last few decades. Even so, it took Thomas Hales, professor of mathematics at the University of Michigan, ten years to do so. Like Fermat, sphere packing sounds simple, but it defeated mathematicians for nearly four centuries. Moreover, both problems had subtle difficulties that led countless mathematicians to believe that they had found solutions—but those "solutions" turned out to be false.

The question was posed in the latter half of the sixteenth century, when Sir Walter Raleigh asked the English mathematician Thomas Harriot for a quick way to estimate the number of cannonballs that could be stacked on the deck of a ship. In turn, Harriot wrote to Johannes Kepler, the German astronomer, who was already interested in stacking: how could spheres be arranged to minimize the gaps among them? Kepler could find no system more efficient than the way sailors naturally stack cannonballs, or grocers stack oranges, known as face-centered cubic packing. Kepler declared that by this technique "the packing will be the tightest possible, so that in no other arrangement could more pellets be stuffed into the same container." This assertion became known as the *Kepler conjecture*.

Major progress was made in the nineteenth century, when the legendary German mathematician Carl Friedrich Gauss proved that the orange-pile arrangement was the most efficient among all "lattice-packings", but this didn't rule out the possibility of a more efficient non-lattice arrangement. By the twentieth century the Kepler conjecture was deemed sufficiently important for David Hilbert to include it in his list of 23 great unsolved problems.

The problem is difficult because of the immense number of possibilities that must be eliminated. By the mid-twentieth century, mathematicians knew how to reduce it to a finite problem, but the problem was too complex to compute. A major advance came in 1953 when the Hungarian mathematician László Fejes Tóth reduced the problem to a huge calculation involving many specific cases and also suggested how it might be solved by computer.

Even for Hales the challenge was immense. His equation has 150 variables, each of which must be changed to describe every conceivable stacking arrangement. The proof that Kepler was right, explained in a 250-page argument that contains 3 gigabytes of computer files, relies extensively on methods from the theory of global optimization, linear programming, and interval arithmetic. Hales acknowledged that for a proof so long and complex, it would be some time before anyone could confirm all its details.

It is worth noting, however, that this exercise was far from frivolous. The topic of sphere packing belongs to a crucial part of the mathematics that lies behind the error-detecting and error-correcting codes that are widely used to store information on compact disks and to compress information for transmission around the world. In today's information society, it is difficult to think of a more significant application.

The four-color problem

As an addendum to sphere-packing, a related problem is worth mentioning—the so-called *four-color problem* of map-making. This is the assertion that only four colors are needed to color any map so that no neighboring countries have the same color. This problem is similar to sphere packing: it is an elementary problem that probably seemed straightforward enough when first proposed by the English mathematician Francis Guthrie in 1852. It is also similar in that the existing proof reduces it to a finite problem, which then required heavy amounts of computing capability.

The proof, accomplished in 1976 by Wolfgang Haken and Kenneth Appel, involved showing that if every one of a list of x maps is four-colorable, then all maps are four-colorable. Although the number of conceivable maps is infinite, Haken and Appel showed that the colorability of all of them depends only on the colorability of a large but finite set of fundamental maps. This was the first significant problem to succumb to the raw power of the computer. At the same time, it has caused some people to suggest that "brute-force" computer proofs lack the clarity of traditional proof: that is, they prove that the conjecture is true, but do not explain why. We may expect further debate on this point.

The dual nature of mathematics

We have spoken of mathematics' reputation for smugness; in fact, it has been called the Queen of Sciences: somewhat superior, unto herself. There is a feeling of blue sky, of airy exercises performed for their own sake. Indeed, the mathematician G. H. Hardy once said that the practice of mathematics can be justified only as an art form.

In fact, there is a parallel with the arts here. Mathematicians, like artists, rely heavily on aesthetics as well as intuition, and it is not uncommon to solve problems while taking a shower or a walk. But with respect to utility, the argument in mathematicians' favor is a strong one. To mention just a few examples, the modern computer would not exist without Leibniz's binary number code; Einstein could not have formulated his Theory of Relativity without the development of Riemannian geometry, and the edifices of quantum mechanics, crystallography, and communications technology all rest firmly on the platform of group theory.

Moreover, the "reach" of mathematicians seems large in relation to their number. That is, the mental constructs of mathematicians (a tiny sub-population of a single species on a single planet) seem to reflect principles that apply throughout the universe. Earlier in the century, the physicist Eugene Wigner called this phenomenon "the unreasonable effectiveness of [abstract] mathematics in the natural sciences"; nowadays, one would add effectiveness in drug design, finance, and many other fields. There are many opinions about the ultimate origin of mathematical constructs. Arthur Jaffe suggests: "Mathematical ideas do not spring full grown from the minds of researchers. Mathematics often takes its inspirations from patterns in nature. Lessons distilled from one encounter with nature continue to serve as well when we explore other natural phenomena."

Mathematicians have always carried their discoveries into adjacent fields where they have produced new insights and whole new subfields. Francis Bacon, at the dawn of the Enlightenment in 1605, prefigured this principle of integrative science with an apt image: "No perfect discovery can be made upon a flat or a level: neither is it possible to discover the more remote or deeper parts of any science, if you stand but upon the level of the same science and ascend not to a higher science."

Repeatedly in the twentieth century, mathematics has ascended to that higher place. For example, the development of x-ray tomography (the CAT and MRI scanning technologies) was built upon integral geometry; the generation of codes for secure transmission of data depends on the arithmetic of prime numbers; and the design of large, efficient networks in telecommunications uses infinite-dimensional representations of groups.

Thus mathematics has a dual nature: it is both an independent discipline valued for precision and intrinsic beauty, and it is a rich source of tools for the world of applications. And the two parts of this duality are intimately connected. As we see in the following section, it is the strengthening of these connections during the twentieth century that has allowed the field steadily to gain effectiveness—both within mathematics and in the world beyond.

3 Trends of the twentieth century

A principal reason that mathematics is healthy today is the breakdown of barriers within the field. At first glance, the full span of mathematics—an enormous body of concepts, conjectures, hypotheses, and theorems amassed over more than 2000 years—seems to defy the possibility of unity. Gone are the days when a single giant—an Euler or a Gauss—could command its entirety. With the rapid development of subfields after World War II, mathematics became so specialized that practitioners had difficulty communicating with anyone outside their own specialty. And today these specialists are commonly scattered among Bonn, Princeton, Berkeley, and Tokyo.

But this trend toward fragmentation is complemented by the growing tendency to address interesting problems in an overarching manner. Subfields, once viewed as quite disjoint, are now seen as parts of a whole, as new connections emerge between them. For example, algebraic geometry, the field I am most familiar with, combines algebra, geometry, topology, and analysis. As we near the end of this century, synergies in this strongly interconnected area have played a major role in some of the crowning achievements of pure mathematics. One, of course, is the solution of Fermat's Last Theorem. Another is the solution of the Mordell Conjecture, which states that a polynomial equation of degree 4 or more with rational coefficients can have at most a finite number of rational solutions (the Fermat equation has no such solutions). A third is the solution of the Weil conjectures, which are the analogs for finite fields of the Riemann hypothesis. All these accomplishments reflect the ability of mathematicians simultaneously to draw on multiple subfields and to perceive their subject as a whole.

Solitons

One of the most remarkable achievements of mathematics of the latter half of the twentieth century is the theory of *solitons*, which illustrates the underlying unity of the field. Solitons are non-linear waves that exhibit extremely unexpected and interesting behavior.

Let me give a bit of background. Traditionally, we talk about two different kinds of waves. The first, *linear waves*, are familiar in everyday life; examples are light waves or sound waves. Linear waves move uniformly through space without changing. That is, they have a constant velocity, no matter what their shape; C-sharp travels at the same speed as F-flat. And they have a constant wavelength; a C-sharp remains C-sharp if you hear it a block away. Linear waves also obey the property of superposition: if you play multiple notes on the piano simultaneously, you always hear the sum of all notes at once, which brings us harmony. Even a very complicated sound can be resolved into its constituent harmonics.

The second kind of waves, *non-linear waves*, are less familiar and quite different. A simple example can be seen as an ocean wave approaches the shore. The amplitude, wavelength, and velocity, which are constant in linear waves, all can be seen to change. The distance between wave-tops decreases, the height increases as the waves "feel" the bottom, and the velocity changes; the upper part overtakes the bottom part and falls over it as the wave breaks. In an even more intricate event, two waves may come together, interact in a complicated, non-linear way, and give rise to three waves instead of two.

Now we come to solitons. The story begins in 1834 when a Scottish engineer named John Scott Russell was trying to determine the most efficient design for a canal boat. One day, he observed that waves in a shallow canal sometimes behaved in a very peculiar fashion. Waves might travel long distances at a constant velocity without changing shape, but those with large amplitude went faster than those with small amplitude. A large wave might overtake a smaller one, resulting in a complex interaction, whereupon the large wave would emerge traveling faster than the smaller one. After this non-linear interaction, they would again act like linear waves.

In the middle of the twentieth century, a group of mathematicians were studying a non-linear wave equation. Because it described non-linear waves, they expected that its solution would develop singularities, or breaks, at some point, just as intersecting waves interact and break in a non-linear fashion. They wrote a computer program to solve the equation numerically and found that the wave did not break as expected. This led them to look at the Korteweg-de Vries equation, which was written down a century ago to describe the behavior of waves in shallow water. It was found that the phenomena observed by Russell were provable mathematically for the Korteweg-de Vries equation; in other words, the solution to that equation exhibited soliton behavior. These are extremely unusual equations, because solitons are in some ways like linear waves and in other ways like non-linear waves.

This discovery provoked a rush of activity, which exhibited in the most beautiful way the unity of mathematics. It involved developments in computation, and in mathematical analysis, which is the traditional way to study differential equations. It turns out that one can understand the solutions to these differential equations through certain very elegant constructions in algebraic geometry. The solutions are also intimately related to representation theory, in that these equations turn out to have an infinite number of hidden symmetries.

Finally, they relate back to problems in elementary geometry. For example, an interesting problem is to find the surface of a cone with a fixed volume but least area among all surfaces with a given boundary. It's not at all evident that this has anything to do with a shallow water wave, but in fact it does. The differential equations that describe the solution turn out to have soliton behavior like that of the equations describing shallow water waves, So we have started with two mathematical problems—one in mathematical physics and one in differential geometry—and found that each exhibits the same extremely rare and interesting soliton behavior.

Mathematics and other sciences

Beyond the breakdown of internal barriers, mathematics has become much more interactive with other sciences and with business, finance, security, management, decision-making, and the modeling of complex systems. And some of these disciplines, in turn, are challenging mathematicians with interesting new types of problems, which then lead to new applications.

Mathematics and theoretical physics

This is nowhere better illustrated than in theoretical physics. Beyond the purely mathematical contributions, algebraic geometry has been used by theoretical physicists in their search for a unified field theory—or more precisely, for a theory that unifies gravity with the three fundamental forces of physics: the strong nuclear force, the weak nuclear force, and electromagnetism.

One exciting candidate for a new unifying theory is *string theory*. The name comes from the idea that the most elementary building blocks of matter are tiny vibrating loops or segments that are string-like in shape and vibrate in many different modes, like violin strings. The effort to understand this extremely complex theory has led a group of theoretical physicists deep into mathematics, where they have made a series of spectacular predictions about mathematics; these predictions are beginning to be verified. These results have stimulated a flurry of work that continues to add to the plausibility of the theory. It has also spawned a new branch of four-dimensional mathematics called *quantum geometry*, which, in turn, is opening new insights to physics.

Another indication of the close relationship between mathematics and physics was seen in the 1998 awards of the Fields Medals, the highest honor in mathematics. Of the four medalists, three worked in areas with strong physical influences, and a special award was given for work in quantum computing, which has its roots in quantum mechanics.

Mathematics and the life sciences

One of the fastest-growing new partnerships is the collaboration between mathematics and biology. The partnership began in the field of ecology in the 1920s, when the Italian mathematician Vito Volterra developed the first models of *predator-prey relationships*. He found that the waxing and waning of predator and prey populations of fish could be described mathematically. After World War II, the modeling methods that Volterra developed for populations were extended to epidemiology, the study of diseases in large populations.

Most recently, the insights of molecular genetics have inspired scientists to adapt these same methods to infectious diseases, where the objects of study are not populations of organisms or people, but populations of cells. In a cellular system, the predator is a population of viruses, for example, and the prey is a population of human cells. These two populations rise and fall in a complex Darwinian struggle for survival that lends itself to mathematical descriptions. In the last decade, the ability to use mathematical models that describe infectious agents as predators and host cells as prey has redefined many aspects of immunology, genetics, epidemiology, neurology, and drug design. The reason this partnership is successful is that mathematical models offer powerful tools to describe the immensity of numbers and relationships found in biological systems.

For example, mathematical biologists have been able to make quantitative predictions about how viruses and other microbes grow in their hosts, how they change the genetic structure of hosts, and how they interact with the host's immune system. Some of the most surprising results have emerged in the study of the AIDS epidemic, reversing our understanding of HIV viruses in infected patients. The prevailing view had been that HIV viruses lie dormant for a period of 10 or so years before beginning to infect host cells and cause disease. Mathematical modeling has shown that the HIV viruses that cause the most disease are not dormant; they grow steadily and rapidly, with a half-life of only about two days.

Why, then, does it take an average of ten years for infection to begin? Again, mathematical modeling has shown that disease progression may be caused by viral evolution. The immune system is capable of suppressing the virus for a long time, but eventually new forms of viruses mutate and become abundant and overwhelm the immune defense. This happens because viruses, like other infectious agents, can reproduce faster than their hosts, and the reproduction of their genetic material is less accurate. Virtually every HIV infection is seen as an evolutionary process in which the virus population constantly changes and new virus mutants continuously emerge. Natural selection favors variants able to escape the immune response, or to infect more kinds of cells in the human body, or to reproduce faster. The models show that all evolutionary changes increase virus abundance in the patient and thereby accelerate disease.

These same mathematical models have brought an understanding of why anti-HIV drugs should be given in combination, and given as early as possible during infection. These drugs are most effective in combination because viruses seldom produce multiple mutations at once. And they should be given early before viral evolution can progress very far.

A major threat to human health in the next century will be microbial resistance to drug therapy, and this will be another area where mathematical models can contribute. They can present clear guidelines for collection and analysis of data, which can make drugs more effective. Good models of the complex interactions between infectious agents and the immune system can eventually create a new discipline of quantitative immunology.

There are many more new partnerships between mathematics and the other sciences; much of the most innovative and productive work is being done at the frontiers between fields and disciplines. An excellent example is the study of *fluid dynamics*. Describing the complex movement of fluids—hurricanes, blood flow through the heart, oil in porous ground—was virtually impossible before the discovery that a purely mathematical construct called the Navier-Stokes equation can do just that. Another example is *control theory*, a branch of the theory of dynamical systems. In just one application, much testing of high-performance aircraft can now be done by computer simulations, greatly reducing the expense and danger of wind tunnels and test flights.

It is important to emphasize that while modeling and simulation are modern and important topics, we are still not very good at addressing the uncertainties that are present in these complex simulations. Learning to grapple with uncertainty is high on the list of priorities for mathematicians, who must develop fundamentally new approaches if they are to understand how uncertainties arise in models and how they propagate through systems. Our models will be only as accurate as our ability to smooth out their uncertainties.

4 Research challenges for the twenty-first century

Despite the tremendous achievements of the twentieth century, dozens of outstanding problems still await solution. Most of us would probably agree that the following three problems are among the most challenging and interesting.

The Riemann hypothesis

The first is the *Riemann hypothesis*, which has tantalized mathematicians for 150 years. It has to do with the concept of prime numbers, which are the basic building blocks of arithmetic. A *prime number* is a positive whole number greater than 1 that cannot be divided by any positive number except itself and 1. The primes begin with 2, 3, 5, 7, 11, 13, and continue without bound. As long ago as the third century B.C., Euclid proved that no one could ever find the "largest" prime number; in other words, the primes are infinite in number.

But is there a pattern to these primes? For someone studying them by pencil and paper, they at first appear to occur randomly. But in the nineteenth century, the German mathematician Bernhard Riemann extended Euclid's observation to assert that primes are not only infinite in number, but they occur in a very subtle and precise pattern. Proving that this is so (or not so) is perhaps the deepest existing problem in pure mathematics.

The Poincaré conjecture

This problem is baffling both because it is so fundamental and it seems so simple. In Poincaré's day, a century ago, it was even regarded as trivial, as was the whole field of topology—a field that he essentially invented. But today topology is a vital and significant subfield of mathematics.

In rough terms, topology is concerned with the fundamental properties of structures and spaces. A sphere, for example, can be stretched, compressed, or warped in any number of ways (in a topologist's eyes) and still remain a sphere, as long as it isn't torn or punctured. A topologist sees a donut and a coffee cup as identical, because either one of them can be massaged into the same basic shape as the other—a ring with a hole in it, or torus. Of special interest to topologists are manifolds, which means "having multiple features or forms". A soccer ball, for example, is a two-dimensional manifold, or two-sphere; we can manipulate it any way we want as long as we don't rip it, and it will still be a soccer ball.

Topologists seek to identify all possible manifolds, including the shape of the universe—which is the subject of Poincaré's conjecture. This is relatively easy in two dimensions, and was done by the end of the nineteenth century. The test of whether a manifold is really a two-sphere is also straightforward. Imagine placing a rubber band on the surface of a soccer ball. If the rubber band can be shrunk to a point without leaving the surface, and if this can be done anywhere on the surface, the ball is a two-sphere, and we say that it is simply connected.

In 1904, Poincaré conjectured that what is true for two dimensions is also true for three—that any simply-connected, three-dimensional manifold (such as the universe) has to be a three-sphere. This sounds intuitively obvious, but nobody has ever shown that there aren't some false three-spheres, so the conjecture still hasn't been proved. Surprisingly, proofs are known for the equivalent of Poincaré's conjecture for all dimensions strictly greater than 3, but not yet for 3.

Does P = NP?

A third problem is closely related to the philosophical question of what is knowable and what is unknowable. In 1931, the Austrian-born logician Kurt Gödel established that complete certainty could not be found in arithmetic—assuming that arithmetic is founded on certain "self-evident" properties, or axioms, of whole numbers. In the theory of computing, Alan Turing set down rules in the 1930s to decide what is computable and what is not. A more refined question is to ask what is computable in polynomial time, or P time. In the familiar traveling salesman problem, for example, polynomial time means that you can write a computer program to compute the best route for him to visit n cities in a reasonable amount of computer time, which would be roughly n-squared or n-cubed time.

As the problem becomes more complex, computer time may increase exponentially, until the problem becomes computationally intractable, or NP. For example, most cryptographic codes today are based on the assumption that factoring large integers is a computationally intractable problem.

In fact there are some very interesting current developments on the "P vs. NP" question, which may be related to the Gödel incompleteness theorem. It seems possible that certain mathematical statements that eventually include lower bounds on computation, such as "P does not equal NP", cannot be proved within the framework of Peano arithmetic, the standard, or most natural, version of arithmetic. This thesis is not yet proved; but its resolution appears feasible in the foreseeable future. What is known is that all techniques used so far to prove lower bounds on computational models reside in a specific low fragment of Peano arithmetic. Moreover, proved techniques in these fragments cannot separate P from NP—unless the integer factorization has much faster algorithms than we currently know or suspect. In other words, whether a problem is P or NP will depend on whether or not we can factor integers much faster than we have thought possible.

Theoretical computer science

The field we are discussing, theoretical computer science, is one of the most important and active areas of scientific study today. It was actually founded half a century ago, before computers existed, when Alan Turing and his contemporaries set out to define mathematically the concept of "computation" and to study its powers and limits. These questions led to the practical construction by von Neumann of the first computer, followed by the computer revolution we are witnessing today.

The practical use of computers, and the unexpected depth of the concept of "computation", have significantly expanded theoretical computer science, or TCS. In the last quarter-century TCS has grown into a rich and beautiful field, making connections to other sciences and attracting first-rate scientists. Here are just a few aspects of this evolution:

First, the focus of the field has changed from the notion of "computation" to the much more elusive concept of "efficient computation". The fundamental notion of NP-completeness was formulated, its near universal impact was gradually understood, and long-term goals, such as resolving the P vs. NP question, were established. The theory of algorithms and a variety of computational models were developed. Randomness became a key tool and resource, revolutionizing the theory of algorithms.

Significantly, the emergence of the complexity-based notion of one-way function, together with the use of randomness, led to the development of modern cryptography. What many people at first thought were just mental games, such as trying to play poker without cards, has turned into a powerful theory and practical systems of major economic importance. Complexity theory, which attempts to classify problems according to their computational difficulty, has integrated many of these ideas and has given rise to the field of proof complexity, where the goal is to quantify what constitutes a difficult proof.

Beyond these activities, which are internal to TCS, is important cross-fertilization between TCS and other subfields, such as combinatorics, algebra, topology, and analysis. Moreover, the fundamental problems of TCS, notably "P vs. NP", have gained prominence as central problems of mathematics. More and more mathematicians are considering the computational aspects of their areas. In other words, they start with the fact "An object exists" and follow it with the problem "How fast can this object be found?"

A final aspect of TCS, which is to some people the most interesting, is that the field now overlaps with a whole new set of algorithmic problems from the sciences. In these problems the required output is not well defined in advance, and it may begin with almost any kind of data: a picture, a sonogram, readings from the Hubble Space Telescope, stock-market share values, DNA sequences, neuron recordings of animals reacting to stimuli. Mathematical models are used to try to make sense of the data or predict their future values.

In general, the very notion of "computation", and the major problems surrounding it, have taken on deep philosophical meaning and consequences. In addition to the "P vs. NP" question, the field is focused on a few clear and deep questions: For example: Does randomization help computation? What makes a theorem difficult to prove? and, Can quantum mechanics be effectively simulated by classical means? The time is ripe for exciting growth and fundamental new understanding throughout this new field of theoretical computer science.

Quantum computing

Another new and exciting area of research is the investigation of quantum computing. This topic is closely related to the "P vs. NP" question because of a surprising demonstration in 1994. It was discovered that if a quantum computer could be built, it would be capable of breaking any computer code now used or thought to be secure.

The need for a fundamentally new form of computing is very real, especially for running complex simulation models. Although modern computers are extremely fast, they still use the classical binary calculating system of 0s and 1s that dates back 150 years to George Boole's adding machine. For many years this has seemed sufficient, especially in the presence of "Moore's Law": the observation that the capacity of computer chips doubles every two years or so while chip prices drop by half. This has been accomplished through better engineering and the production of smaller and smaller chips. But now, we are approaching quantum mechanical limits on how small a chip can be.

These limits were foreseen as early as 1982, when Richard Feynman predicted that efforts to simulate quantum mechanical systems on digital computers would carry an inherent "overhead". But in a lengthy side remark, he proposed that this difficulty might be circumvented by some form of quantum computer. In 1985, David Deutsch advanced the dialog by suggesting that if quantum computers could be fast enough to solve quantum mechanical problems, they might also solve classical problems more quickly.

It appears that this is the case. In 1994, Peter W. Shor showed that a quantum computer could factor large numbers in a time that is polynomial in the length of the numbers, a nearly exponential speed-up over classical algorithms. This was surprising for two reasons. First, modern cryptographers use long numbers as security codes because they are so difficult to factor—a job that a quantum computer could do rather quickly. Second, theoretical computer scientists had believed that no type of computing could be much faster than conventional digital computers.

On the other hand, experimentalists are not at all sure they can build a quantum computer. As they pursue that possibility, there are numerous parallel efforts to design other kinds of computers based on principles other than Boolean arithmetic—all with the same goal of greatly expanding computer capacity. We can expect exciting and intense work in this field for years to come.

5 Maintaining the strength of mathematics in the twenty-first century

The bulk of this essay has been devoted to trends and problems of research. However, it is irresponsible to discuss research without mentioning the context in which it occurs. The success of research depends on the quality of the people doing it and the degree to which it receives sustained support from society; in other words, it requires "patient capital". The next millennium will bring a set of contextual questions every bit as challenging as the research problems we want to solve.

Education

First, how can we attract the best young talent into mathematics? Here we have seen a significant change in the last half-century. During World War II, the system and techniques of science and technology generated much excitement, attracting post-war students to careers in research. This trend received a powerful stimulus in 1957, when the Soviet satellite Sputnik was launched and science was recognized for the political and economic power it could generate. Research became as important to society as it was fascinating to practitioners.

Toward the end of the century, however, society's interest in many areas of research appears to have diminished in most countries around the world. Mathematics and science, with some exception in biomedical areas, appear not to be felt as important to society as they once were, nor to provide desirable career opportunities. In both developed and developing countries, many bright students who once would have chosen careers in mathematics are not doing so; they are choosing applied information science, business, or other areas where the future looks more interesting. There appears to be a fundamental lack of appreciation for the richness and relevance of the subject itself.

It is ironic that student interest is low at a time when career opportunities for professional mathematicians have never been greater or more diverse. This is true both for the traditional disciplinary areas, which are rich with new developments and challenging problems, and in the applied fields and other areas of science, where demand for mathematicians with proper training will continue to grow rapidly in the foreseeable future.

An obvious reason for student disinterest is that we are not communicating a full picture of mathematics as a field where one may choose among so many intellectually rewarding and challenging careers. The people best positioned to effect this communication are high school teachers, college professors, and fellow students. However, these groups can describe current opportunities and fast-growing fields only if they, in turn, are informed by those in the profession. Thus the mathematics community as a whole faces a critical challenge: to foster more interaction at every level of teaching and practice, and to widen the channels of communication with the students who will eventually replace us and extend our work into the next century.

Outreach

Closely allied with our educational needs is the opportunity to better communicate with and educate the public about mathematical issues. Mathematicians clearly understand the purpose and value of their work, but many people in government, business, and even education do not share this understanding. If mathematicians expect their research at universities to be supported by public funds, we must present to the public a more vivid picture of that research and its power to deepen our understanding and improve our well-being. It is no longer possible to remain aloof from the passing needs of the world or to work in an ivory tower.

The culture of the next millennium will be highly interactive and collaborative. Mathematicians must seize the opportunity not only to collaborate with other mathematicians and scientists, but also to reach out to the community at large. They are uniquely qualified to articulate the value of mathematics in catalyzing major advances in science and health, in promoting powerful economic tools and efficiencies, and to probing the patterns and truths of the universe in which we live.

Interactivity

Finally, the trend toward interactivity merits a final mention. We have seen that within mathematics and throughout the sciences much of the most productive work is being done at frontiers between subfields, fields, and disciplines. Mathematics loses something when it is isolated or fragmented according to disciplinary paradigms. However, many institutions have been slow to adapt to this reality. Universities around the world, and many industries and government agencies, stand to gain much by removing barriers to collaborations.

In particular, much can be done to enhance interactions between academic and industrial mathematicians. The primary missions of academia and industry are indeed different; nonetheless, the two cultures have much to learn from one another and can gain much from collaboration. In general, the scientific enterprise can function at full potential only when there is a fast flow of knowledge between the creators and users of mathematics.

The next millennium

The globalization, interactivity, and "opening out" of the mathematical enterprise is a new and powerful trend. As a harbinger of things to come, note the mode in which Thomas Hales chose to announce his proof of the Kepler sphere-packing problem. Rather than publish it in a journal, where his results would be seen by a small number of specialists, he opened it to an unlimited audience via the Internet. In addition, he frankly invited scrutiny of his proof and further contributions to its accuracy—a significant step in the competitive world of top-level mathematics.

In general, then, we mathematicians have two objectives as we enter the next millennium. The first is to maintain traditional strengths in basic research, which is the seedbed of new thinking and new applications. The second is to broaden our exploration of the terrain outside the traditional boundaries of our field—to the other sciences and to the world beyond science. With each passing year, mathematicians achieve more effectiveness in their work to the degree that they offer it to others and include others in the world of mathematics.

Afterword

G. B. Halsted names several mathematicians who participated in the Second International Congress, most of whom are not household names today. These include, first, the representatives of the U.S., Charlotte Angas Scott (1858–1931), from Bryn Mawr College [14]; of Japan, Rikitaro Fujisawa (1861–1933); and of Spain, Zoel Garcia de Galdeano y Yanguas (1846–1924). The last of these was the academic advisor of Julio Rey Pastor, who later became the central figure in the development of mathematics in Argentina in the twentieth century. Then there were Halsted's "interesting personalities". One of these was Samuel Dickstein (1851–1939), a Jew from Russian Poland who at the time of the Congress was the principal of a science-oriented secondary school in Warsaw that had introduced Hebrew into its curriculum and who later was one of the first professors of mathematics at the University of Warsaw. There were also Karl Gutzmer (1860–1924), a German mathematician who worked on differential equations; Emile Lemoine (1840–1912), a French civil engineer who did some work in geometry; Alessandro Padoa (1868–1937), an Italian logician who lectured on a new system of definitions for Euclidean geometry at the Congress; and Dmitrii Sintsov (1867–1946), who created a school of geometry at Kharkov University in Russia. One wonders what Halsted's criteria for "interesting personalities" were. (Biographies of most of the mathematicians Halsted mentions are available online at Wikipedia or at the St. Andrews MacTutor website.)

There are several available books dealing with Hilbert's Congress lecture on problems. These include the 1976 AMS publication *Mathematical Developments Arising from Hilbert Problems* [2], which contains a copy of Hilbert's lecture, and the more recent books by Jeremy Gray, *The Hilbert Challenge* [12], and Ben Yandell, *The Honors Class: Hilbert's Problems and Their Solvers* [24].

G. A. Miller mentions that the last quarter of the nineteenth century saw the beginnings of a research mathematics tradition in the United States. This theme is explored in great detail in Parshall and Rowe's *The Emergence of the American Mathematical Research Community* [17]. More details on the arithmetization of analysis can be found in Chapter 7 of Umberto Bottazzini's *The Higher Calculus* [1] and on the nineteenth-century developments in set theory in the chapter "Developments in the Foundations of Mathematics", in I. Grattan-Guinness's *From the Calculus to Set Theory* [3]. The history of the theory of groups is well covered in Hans Wussing's *The Genesis of the Abstract Group Theory* [23] and in earlier articles in this volume.

H. Weyl's article covers a good bit of ground, and there are many new works that detail various aspects of his story. The development of structure in algebra is dealt with by Leo Corry in *Modern Algebra and the Rise of Mathematical Structures* [6]. A detailed introduction to the theory of p-adic numbers is found in Fernando Gouvêa's *p-adic Numbers: An Introduction* [11], while one of the better treatments of the history of the Riemann Hypothesis is found in *Prime Obsession* [7], by John Derbyshire. Lynn Steen's article in Chapter 1 of this volume gives more details on spectral theory and Hilbert spaces, while Thomas Hawkins has written the definitive treatment of the history of Lebesgue integration [13] and Jean Dieudonné's *History of Algebraic and Differential Topology* [9] covers the developments in homology theory and other areas of algebraic topology. There is a brief treatment by S. S. Chern on differential geometry in the twentieth century [5], which expands on Weyl's comments. Similarly, the logician Stephen Kleene has written an analysis of the work of Gödel [14].

Phillip Griffiths refers to several long-standing problems that were resolved late in the twentieth century. There are book-length treatments of many of these. Simon Singh has written a solid history of Fermat's Last Theorem; *Fermat's Enigma: The Epic Quest to Solve the World's Greatest Mathematical Problem* [18]. The solution of Kepler's Conjecture is well-covered in George Szpiro's *Kepler's Conjecture* [21]. Similarly, Robin Wilson has written the definitive book on the four-color problem and its solution: *Four Colors Suffice* [22]. As to Griffith's trends of the twentieth century, there are numerous books and articles dealing with solitons. But string theory has been in the news lately, mostly because it has not reached its goals. A book-length study debunking the theory has been written by Lee Smolin: *The Trouble with Physics: The Rise of String Theory, The Fall of a Science, and What Comes Next* [19]. For a good treatment of the relationship between mathematics and biology, and its implications in education, consult the MAA's *Math and Bio 2010: Linking Undergraduate Disciplines* [20].

Finally, although the Riemann conjecture has still not been solved, the Poincaré conjecture is now a theorem. A history of this conjecture and of its recent proof by Grigory Perelman is available as *The Poincaré Conjecture: In Search of the Shape of the Universe* [16], by Donal O'Shea. We also note that, like Hales, Perelman announced his proof by placing it on the Internet in late 2002 and early 2003, whereupon it took three years for experts to agree that his proof was in fact valid. The Poincaré conjecture is the first of the seven Clay Institute Millennium Prize Problems to be solved. A survey of all seven problems is found in the AMS publication, *The Millennium Prize Problems* [4], and in a more accessible form in Keith Devlin's *The Millennium Problems: The Seven Greatest Unsolved Mathematical Puzzles of Our Time* [8]. Besides the Riemann Hypothesis, the P versus NP problem is also on this list, and the reader should consult Devlin's book or the book by Garey and Johnson [10] for a more complete description than Griffiths gives.

References

1. Umberto Bottazzini, *The Higher Calculus: A History of Real and Complex Analysis from Euler to Weierstrass* (transl. Warren Vann Egmond), Springer-Verlag, 1986.
2. Felix Browder (ed.), *Mathematical Developments Arising from Hilbert Problems*, American Mathematical Society, 1976.
3. R. Bunn, Developments in the Foundations of Mathematics, *From the Calculus to Set Theory, 1630–1910* (I. Grattan-Guinness, ed.), Duckworth, 1980, pp. 220–255.
4. J. Carlson, A. Jaffe, and A. Wiles (eds.), *The Millennium Prize Problems*, American Mathematical Society, 2006.
5. S.-S. Chern, Differential Geometry: Its Past and Its Future, *Actes du Congrès International des Mathématiciens, Paris, 1970*, Vol. 1, Gauthier-Villars, 1971, pp. 41–53.
6. Leo Corry, *Modern Algebra and the Rise of Mathematical Structures*, Birkhäuser, 1996.
7. John Derbyshire, *Prime Obsession: Bernhard Riemann and the Greatest Unsolved Problem in Mathematics*, Joseph Henry Press, 2003.
8. Keith Devlin, *The Millennium Problems: The Seven Greatest Mathematical Puzzles of Our Time*, Basic Books, 2002.
9. Jean Dieudonné, *A History of Algebraic and Differential Topology. 1900-1960*, Birkhäuser, 1989.

10. M. R. Garey and D. S. Johnson, *Computers and Intractibility: A Guide to the Theory of NP-completeness*, W. H. Freeman, 1979.

11. Fernando Gouvêa, *p-adic Numbers: An Introduction*, Springer-Verlag, 1997.

12. Jeremy Gray, *The Hilbert Challenge*, Oxford Univ. Press, 2001.

13. Thomas Hawkins, *Lebesgue's Theory of Integration: Its Origins and Development*, Chelsea, 1975.

14. Patricia Kenschaft, Charlotte Angas Scott, *College Math. Journal* **18** (1987), 98–110.

15. Stephen C. Kleene, The Work of Kurt Gödel, *J. Symbolic Logic* **41** (1976), 761–778.

16. Donal O'Shea, *The Poincaré Conjecture: In Search of the Shape of the Universe*, Walker and Company, 2007.

17. Karen H. Parshall and David E. Rowe, *The Emergence of the American Mathematical Research Community, 1876–1900: J. J. Sylvester, Felix Klein, and E. H. Moore*, American Mathematical Society, 1994.

18. Simon Singh, *Fermat's Enigma: The Epic Quest to Solve the World's Greatest Mathematical Problem*, Walker and Company, 1997.

19. Lee Smolin, *The Trouble with Physics: The Rise of String Theory, The Fall of a Science, and What Comes Next*, Mariner Books, 2007.

20. Lynn Steen (ed.), *Math and Bio 2010: Linking Undergraduate Disciplines*, Mathematical Association of America, 2005.

21. George Szpiro, *Kepler's Conjecture: How Some of the Greatest Minds in History Helped Solve One of the Oldest Math Problems in the World*, Wiley, 2003.

22. Robin Wilson, *Four Colors Suffice: How the Map Problem Was Solved*, Princeton Univ. Press, 2003.

23. Hans Wussing, *The Genesis of the Abstract Group Theory*, MIT Press, 1984.

24. Ben H. Yandell, *The Honors Class: Hilbert's Problems and Their Solvers*, A. K. Peters, 2002.

Index

Abel, N.H., 231, 241
Abelian groups, 244–246
Abstraction in group theory, 247–250
Alexandroff, P.S., 149–150, 355–358
Algebra, 392–401
Algebraic number theory, 299–307
Algebraic topology, 148–156
Axiomatics, 157–159, 391

Baire, 22–23
Banach algebra, 45
Berkeley, G., 8
Bienaymé, I.-J., 222–224
Blackwell, D.H., 98–107
Bletchley Park, 58–59
Bolyai, J., 115–116
Bolzano, B., 143
Burnside, W., 256–258, 260

Calculus, 5–13
Calculus of variation, 408
Cantor, G., 137–140, 143–144, 163–164
Cantor set, 137–140
Carr, G.S., 338
Cartan, E., 85, 153
Cartwright, M.L., 88–96
Category theory, 25–26
Cauchy, A.-L., 5, 10–12, 32–33, 81, 231–236, 242–243, 275, 293
Cayley, A., 220, 247–248
Central limit theorem, 53–59
Chebyshev, P.L., 329
Chevalier, A., 288–289
Chevalley, C., 260
Cole, F.N., 255
Commutative algebra, 299–307
Complex numbers, 200
Computability, 176
Conformal mapping, 407
Connected set, 142–147
Continued fractions, 186

D'Alembert, 17–18
D'Herbinville, P., 271, 278, 285, 288
De la Vallée-Poussin, C., 330, 332

De Séguier, J.A., 250, 252
Dedekind, R., 139, 185, 303–305
Desargues, G., 126
Dickson, L.E., 258–259, 314, 316
Dirichlet, L., 19–21, 320–322
Dirichlet function, 20
Distribution, 25
Divergence theorems, 62–63, 78–79
Dougall-Ramanujan identity, 343
Duality, 134
Dumotel, S., 286–287

Einstein, A., 349–350
Eisenstein, G., 309–312
Electricity and magnetism, 61–67
Elliptic curves, 362–367
Encke, J.F., 325
Epsilon, 5
Erdős, P., 335
Euclid, 120, 328
Euler, L., 6, 15–18, 319, 361

Fermat, P., 299, 314, 360
Fermat's last theorem, 360–372, 412
Finite simple groups, 254–267
Formalism, 174–175
Four-color problem, 413–414
Fourier, J., 18–19, 275–276
Fourier series, 18–19
Franklin, F., 227
Frey curve, 370
Frobenius, G., 245–246, 250
Fujisawa, R., 383
Function, 14–26

Galois, E., 231, 241–242, 254, 271–290
Galois representations, 368–369
Gauss, C.F., 78–79, 115–119, 122, 238–239, 301, 309, 311–312, 319–320, 324–326, 328
Gaussian distribution, 55–57
Gelfand–Naimark theorem, 45–46
Gergonne, J., 135
Gödel, K., 175
Goldbach's conjecture, 347
Grassmann, H., 291–297

427

Green, G., 61–67, 69–77, 80–81
Group theory, 230–252, 398–400

Hadamard, J., 330–331
Hamilton, W.R., 201–204, 206–213, 216–217
Hardy, G.H., 88–89, 315, 337–348
Hausdorff, F., 142, 146
Helmholtz, H. von, 72–74
Hermite, C., 222
Hilbert, D., 38–39, 42, 123, 174–175, 178, 315, 383
Hilbert–Schmidt spectral theory, 39–40
Hilbert space, 24, 43, 402
Hölder, O., 255–256
Homogeneous coordinates, 133
Homology and homotopy, 148–156
Hopf, H., 149, 151
Hurewicz, W., 154

Imprimitive groups, 234
Inequalities, 7
Infinite series, 6–7
Infinite sets, 161–170
Integral equations, 38
International Congress of Mathematicians, 383–384
Intransitive groups, 233–234
Intuitionism, 177

Jordan, C., 144, 243–244

Kelvin, Lord, 71, 74
Kepler's sphere packing problem, 413
Klein, F., 239–241
Knots, 75
Kovalevskaya, S., 27–35
Kronecker, L., 244–245, 304–306
Kummer, E., 300, 302–303, 307, 361

La Hire, P., 126
Lagrange, J.–L., 8–9, 11, 230, 238, 313
Lamé, G., 300
Landau, E., 332–333
Lebesgue, H., 23
Lebesgue integral, 41, 403–404
Legendre, A.–M., 319, 328
Leibniz, G.W., 6, 291
Lennes, N.L., 142, 145–147
Lie, S., 240, 260
Limit, 5
Linear algebra, 291–298
Linear equations, 37
Linear operator, 43, 401–402
Liouville, J., 300
Littlewood, J.E., 88–96, 315, 333–334, 346
Lobachevski, N. 115–116
Logic, 172–181

Mathieu, E., 259
Maxwell, J.C., 69
Measure theory, 403–404
Mittag-Leffler, 30–31
Möbius, A.F., 133–134, 297
Modular forms, 367–368

Newton, I., 6
Noether, Emmy, 349–359
Non-Euclidean geometry, 115–124
Number theory, 392–401

Octonians, 204–205
Ostrogradsky, M., 78–80

P = NP?, 418
p-adic numbers, 361–362
Parallel postulate, 120–124
Pascal, B., 126
Pasch, M., 157
Peano, G., 158–159, 292
Peirce, C.S., 161–170
Permutation groups, 241–244
Playfair, J., 121
Plücker, J., 134
Poincaré, H., 148, 241
Poincaré conjecture, 417–418
Poisson, S.–D., 11, 62
Polish logic, 179–180
Poncelet, J.–V., 127, 135
Prime numbers, 318, 325–326, 328–335
Prime number theorem, 318–324, 328–335
Principal axes theorem, 36
Proclus, 120
Projective geometry, 125–136

Quadratic reciprocity, 309–312
Quantum computing, 419
Quantum mechanics, 41
Quaternions, 200–219

Ramanujan, S., 337–348
Real numbers, 185–190
Recursion, 175–176
Ribet, K.A., 370
Riemann, B., 21, 72, 322–323, 329–330, 397
Riemann hypothesis, 323, 417
Riesz, F., 145
Rigor, 7–9
Rodrigues, O., 206, 213–218
Rogers-Ramanujan identities, 344
Rotations, 211–213
Ruffini, P., 230, 241
Russell, B., 172–174

Saccheri, G.G., 121–122
Schmidt, E., 39
Schoenflies, A., 144
Schweikart, F.K., 117–118
Selberg, A., 335
Shimura–Taniyama–Weil conjecture, 368
Simple groups, 254–267
Skewes, S., 346
Smith, H.J.S., 137–138
Smoke-rings, 74–75
Solitons, 415
Spectral theory, 36–47
Spectrum, 39–40
Sphere packing, 413
Sporadic groups, 261–262
Steiner, J., 128
Stokes, G.G., 81–82
Stokes's theorem, 78–86
Surface integrals, 63–64
Sylvester, J.J., 220–227

Tait, P.G., 74–77
Theoretical computer science, 418–419
Thompson, J.G., 262–263

Thomson, W. (Lord Kelvin), 71, 74
Topology, 405
Transformation groups, 246–247
Turing, A., 52–59

Van der Pol's equation, 92–93
Vibrating string, 16–18
Von Neumann, J., 43
Von Neumann–Stone spectral theory, 44–45
Von Staudt, J.K.C., 128–131
Vortices, 73

Waring's problem, 313–317
Weber, H., 248–249
Weierstrass, K., 21, 28–30
Weyl, Hermann, 349–353
Whitehead, A.N., 172–174
Whitney, H., 152
Wiener, N., 334
Wiener–Ikehara theorem, 333–334
Wiles, Andrew, 360

Young, W.H. & G.C., 145–147

Zeta function, 323, 397

About the Editors

Marlow Anderson is a professor of mathematics at The Colorado College, in Colorado Springs. He has been a member of the mathematics department there since 1982. He was born in Seattle, and received his undergraduate degree from Whitman College. He studied partially ordered algebra at the University of Kansas and received his PhD in 1978. He has written over 20 research papers, and co-authored a monograph on lattice-ordered groups. He has also co-written an undergraduate textbook on abstract algebra. In addition to algebra, he is interested the history of mathematics. When not teaching, reading or researching mathematics, he may be found with his wife Audrey scuba-diving in far-flung parts of the world.

Victor J. Katz received his PhD in mathematics from Brandeis University in 1968 and was for many years Professor of Mathematics at the University of the District of Columbia. He has long been interested in the history of mathematics and, in particular, in its use in teaching. His well-regarded textbook, *A History of Mathematics: An Introduction*, is now in its third edition. Its first edition received the Watson Davis Prize of the History of Science Society, a prize awarded annually by the Society for a book in any field of the history of science suitable for undergraduates. Professor Katz is also the editor of *The Mathematics of Egypt, Mesopotamia, China, India and Islam: A Sourcebook*, which was published in July, 2007 by Princeton University Press. Currently, he is the PI on an NSF grant to the MAA supporting *Loci: Convergence*, the online magazine in the history of mathematics and its use in teaching. k

Robin Wilson is Professor of Pure Mathematics at the Open University (UK), a Fellow in Mathematics at Keble College, Oxford University, and Emeritus Gresham Professor of Geometry, London (the oldest mathematical Chair in England). He has written and edited about thirty books, mainly on graph theory and the history of mathematics. His research interests focus mainly on British mathematics, especially in the 19th and early 20th centuries, and on the history of graph theory and combinatorics. He is an enthusiastic popularizer of mathematics, having produced books on mathematics and music, mathematical philately, and sudoku, and gives about forty public lectures per year. He has an Erdős number of 1 and has won two MAA awards (a Lester Ford Award (1975) and a George Pólya award (2005).

LIBRARY
Lyndon State College
Lyndonville, VT 05851